David Scalcione

Geometry

SECOND EDITION

Geometry

SECOND EDITION

DAVID A. BRANNAN
MATTHEW F. ESPLEN
JEREMY J. GRAY

The Open University

CAMBRIDGE UNIVERSITY PRESS
Cambridge, New York, Melbourne, Madrid, Cape Town,
Singapore, São Paulo, Delhi, Tokyo, Mexico City

Cambridge University Press
The Edinburgh Building, Cambridge CB2 8RU, UK

Published in the United States of America by Cambridge University Press, New York

www.cambridge.org
Information on this title: www.cambridge.org/9781107647831

First published 1999
Second edition 2012

Printed in the United Kingdom at the University Press, Cambridge

A catalogue record for this publication is available from the British Library

Library of Congress Cataloguing in Publication data

Brannan, D. A.
Geometry / David A. Brannan, Matthew F. Esplen, Jeremy J. Gray. – 2nd ed.
p. cm.
ISBN 978-1-107-64783-1 (Paperback)
1. Geometry. I. Esplen, Matthew F. II. Gray, Jeremy, 1947– III. Title.

QA445.B688 2011
516–dc23

2011030683

ISBN 978-1-107-64783-1 Paperback

Additional resources for this publication at www.cambridge.org/9781107647831

2.1

In memory of Wilson Stothers

Contents

Preface

Geometry! For over two thousand years it was one of the criteria for recognition as an educated person to be acquainted with the subject of geometry. Euclidean geometry, of course.

In the golden era of Greek civilization around 400 BC, geometry was studied rigorously and put on a firm theoretical basis – for intellectual satisfaction, the intrinsic beauty of many geometrical results, and the utility of the subject. For example, it was written above the door of Plato's Academy 'Let no-one ignorant of Geometry enter here!' Indeed, Archimedes is said to have used the reflection properties of a parabola to focus sunlight on the sails of the Roman fleet besieging Syracuse and set them on flame.

Plato (c. 427–347 BC) was an Athenian philosopher who established a school of theoretical research (with a mathematical bias), legislation and government.

Archimedes (c. 287–212 BC) was a Greek geometer and physicist who used many of the basic limiting ideas of differential and integral calculus.

For two millennia the children of those families sufficiently well-off to be educated were compelled to have their minds trained in the noble art of rigorous mathematical thinking by the careful study of translations of the work of Euclid. This involved grasping the notions of axioms and postulates, the drawing of suitable construction lines, and the careful deduction of the necessary results from the given facts and the Euclidean axioms – generally in two-dimensional or three-dimensional Euclidean space (which we shall denote by $\mathbb{R}^2$ and $\mathbb{R}^3$, respectively). Indeed, in the 1700s and 1800s popular publications such as *The Lady's and Gentleman's Diary* published geometric problems for the consideration of gentlefolk at their leisure. And as late as the 1950s translations of Euclid's *Elements* were being used as standard school geometry textbooks in many countries.

Euclid (c. 325–265 BC) was a mathematician in Hellenistic Alexandria during the reign of Ptolemy I (323–283 BC), famous for his book *The Elements*.

We give a careful algebraic definition of $\mathbb{R}^2$ and $\mathbb{R}^3$ in Appendix 2.

Just as nowadays, not everyone enjoyed Mathematics! For instance, the German poet and philosopher Goethe wrote that 'Mathematicians are like Frenchmen: whatever you say to them, they translate into their own language, and forthwith it is something entirely different!'

Johann Wolfgang von Goethe (1749–1832) is said to have studied all areas of science of his day except mathematics – for which he had no aptitude.

The Golden Era of geometry came to an end rather abruptly. When the USSR launched the Sputnik satellite in 1957, the Western World suddenly decided for political and military reasons to give increased priority to its research and educational efforts in science and mathematics, and redeveloped the curricula in these subjects. In order to make space for subjects newly developed or perceived as more 'relevant in the modern age', the amount of geometry taught in schools and universities plummeted. Interest in geometry languished: it was thought 'old-fashioned' by the fashionable majority.

Nowadays it is being realized that geometry is still a subject of abiding beauty that provides tremendous intellectual satisfaction in return for effort put into its study, and plays a key underlying role in the understanding, development and applications of many other branches of mathematics. More and more universities are reintroducing courses in geometry, to give students a 'feel' for the reasons for studying various areas of mathematics (such as Topology), to service the needs of Computer Graphics courses, and so on. Geometry is having a revival!

Topics in computer graphics such as 'hidden' surfaces and the shading of curved surfaces involve much mathematics.

Since 1971, the Open University in the United Kingdom has taught mathematics to students via specially written correspondence texts, and has traditionally given geometry a central position in its courses. This book arises from those correspondence texts.

We adopt the Klein approach to geometry. That is, we regard the various geometries as each consisting of an underlying set together with a group of transformations acting on that set. Those properties of the set that are not altered by any of the transformations are called *the properties of that geometry.*

Following a historical review of the development of the various geometries, we look at conics (and at the related quadric surfaces) in Euclidean geometry. Then we address a whole series of different geometries in turn. First, affine geometry (that provides simple proofs of some results in Euclidean geometry). Then projective geometry, which can be regarded as the most basic of all geometries; we divide this material into a chapter on projective lines and a chapter on projective conics. We then return to study inversive geometry, which provides beautiful proofs of many results involving lines and circles in Euclidean geometry. This leads naturally to the study of hyperbolic geometry in the unit disc, in which there are two lines through any given point that are parallel to a given line. Via the link of stereographic projection, this leads on to spherical geometry: a natural enough concept for a human race that lives on the surface of a sphere! Finally we tie things together, explaining how the various geometries are inter-related.

Chapter 0
Chapter 1
Chapter 2
Chapters 3 and 4
Chapter 5
Chapter 6
Chapter 7
Chapter 8

Study Guide

The book assumes a basic knowledge of Group Theory and of Linear Algebra, as these are used throughout. However, for completeness and students' convenience we give a very rapid review of both topics in the appendices.

Appendices 1 and 2.

The book follows many of the standard teaching styles of The Open University. Thus, most chapters are divided into five sections (each often further divided into subsections); sections are numbered using two digits (such as 'Section 3.2') and subsections using three digits (such as 'Subsection 3.2.4'). Generally a section is considered to be about one evening's hard work for an average student.

We number in order the theorems, examples, problems and equations within each section.

We use wide pages with margins in which we place various historical notes, cross-references, teaching comments and diagrams; the cross-references need not be consulted by students unless they wish to remind themselves of some point on that topic, but the other margin notes should be read carefully. We use boxes in the main text to highlight definitions, strategies, and the statements of theorems and other key results. The end of the proof of a theorem is indicated by a solid symbol '■', and the end of the solution of a worked example by a hollow symbol '□'. Occasionally the text includes a set of 'Remarks'; these are comments of the type that an instructor would give orally to a class, to clarify a definition, result, or whatever, and should be read carefully. There are many worked examples within the text to explain the concepts being taught, and it is important that students read these carefully as they contain many key teaching points; in addition, there is a good stock of in-text problems to reinforce the teaching, and solutions to these are given in Appendix 3. At the end of each chapter there are exercises covering the material of that chapter, some of which are fairly straight-forward and some are more challenging; solutions are not given to the exercises.

Our philosophy is to provide clear and complete explanations of all geometric facts, and to teach these in such a way that students can understand them without much external help. As a result, students should be able to learn (and, we hope, to enjoy) the key concepts of the subject in an uncluttered way.

Most students will have met many parts of Chapter 1 already, and so can proceed fairly quickly through it. Thereafter it is possible to tackle Chapters 2 to 4 or Chapters 5 and 6, in either order. It is possible to omit Chapters 7 or 8, if the time in a course runs short.

Notation for Functions as Mappings

Suppose that a function f maps some set A into some set B, and that it maps a typical point x of A onto some *image point* y of B. Then we say that A is the *domain* (or *domain of definition*) of f, B the *codomain* of f, and denote the function f as a *mapping* (or *map*) as follows:

Note that we use two different arrows here, to distinguish between the mapping of a set and the mapping of an element.

$$\begin{aligned} f : A &\to B \\ x &\mapsto y \end{aligned}$$

We often denote y by the expression $f(x)$ to indicate its dependence on f and x.

Acknowledgements

This material has been critically read by, or contributed to in some way, by many colleagues in The Open University and the BBC/OU TV Production Centre in Milton Keynes, including Andrew Adamyk, Alison Cadle, Anne-Marie Gallen, Ian Harrison, John Hodgson, Roy Knight, Alan Pears, Alan Slomson, Wilson Stothers and Robin Wilson. Its appearance in book form owes a great deal to the work of Toni Cokayne, Pat Jeal and the OU Mathematics and Computing Faculty's Course Materials Production Unit.

Without the assistance and the forbearance of our families, the writing of the original OU course and its later rewriting in this form would have been impossible. It was Michael Brannan's idea to produce it as a book.

Changes in the Second Edition

In addition to correcting typos and errors, the authors have changed the term 'gradient' to 'slope', and avoided the use of 'reversed square brackets' — so that, for instance, the interval $\{x : 0 < x \leq 1\}$ is now written as (0,1] rather than]0,1]. Also, they have clarified the difference between a geometry and models of that geometry; in particular, the term 'non-Euclidean' geometry has now been largely replaced by 'hyperbolic' geometry, and the term 'elliptic' geometry has been introduced where appropriate. The problems and exercises have been revised somewhat, and more exercises included. Each chapter now includes a summary of the material in that chapter, and before the appendices there are now lists of symbols and suggestions for further reading.

Solutions to the exercises appear in an Instructors' Manual available from the publisher.

The authors have taken the opportunity to add some new material to enrich the reader's diet: a treatment of conics as envelopes of tangent families, barycentric coordinates, Poncelet's Porism and Ptolemy's Theorem, and planar maps. Also, the treatment of a number of existing topics has been significantly changed: the geometric interpretation of projective transformations, the analysis of the formula for hyperbolic distance, and the treatment of asymptotic d-triangles.

The authors appreciate the warm reception of the first edition, and have tried to take on board as many as possible of the helpful comments received. Special thanks are due to John Snygg and Jonathan I. Hall for invaluable comments and advice.

Instructors' Manual

Complete solutions to all of the end-of-chapter exercises are available in an Instructors' Manual, which can be downloaded from www.cambridge.org/9781107647831.

0 Introduction: Geometry and Geometries

Geometry is the study of shape. It takes its name from the Greek belief that geometry began with Egyptian surveyors of two or three millennia ago measuring the Earth, or at least the fertile expanse of it that was annually flooded by the Nile.

The word comes from the Greek words *geo* (Earth) and *metria* (measuring).

It rapidly became more ambitious. Classical Greek geometry, called *Euclidean geometry* after Euclid, who organized an extensive collection of theorems into his definitive text *The Elements*, was regarded by all in the early modern world as the true geometry of space. Isaac Newton used it to formulate his *Principia Mathematica* (1687), the book that first set out the theory of gravity. Until the mid-19th Century, Euclidean geometry was regarded as one of the highest points of rational thought, as a foundation for practical mathematics as well as advanced science, and as a logical system splendidly adapted for the training of the mind. We shall see in this book that by the 1850s geometry had evolved considerably – indeed, whole new geometries had been discovered.

Isaac Newton (1643–1727) was an English astronomer, physicist and mathematician. He was Professor of Mathematics at Cambridge, Master of the Royal Mint, and successor of Samuel Pepys as President of the Royal Society.

The idea of using coordinates in geometry can be traced back to Apollonius's treatment of conic sections, written a generation after Euclid. But their use in a systematic way with a view to simplifying the treatment of geometry is really due to Fermat and Descartes. Fermat showed how to obtain an equation in two variables to describe a conic or a straight line in 1636, but his work was only published posthumously in 1679. Meanwhile in 1637 Descartes published his book *Discourse on Method*, with an extensive appendix entitled *La Géometrie*, in which he showed how to introduce coordinates to solve a wide variety of geometrical problems; this idea has become so central a part of mathematics that whole sections of *La Géometrie* read like a modern textbook.

Apollonius of Perga (c. 255–170 BC) was a Greek geometer, whose only surviving work is a text on conics.

Pierre de Fermat (1601–1665) was a French lawyer and amateur mathematician, who claimed to have a proof of the recently proved Fermat's Last Theorem in Number Theory.

A contemporary of Descartes, Girard Desargues, was interested in the ideas of perspective that had been developed over many centuries by artists (anxious to portray three-dimensional scenes in a realistic way on two-dimensional walls or canvases). For instance, how do you draw a picture of a building, or a staircase, which your client can understand and commission, and from which artisans can deduce the correct dimensions of each stone? Desargues also realized that since any two conics can always be obtained as sections of the same cone in $\mathbb{R}^3$, it is possible to present the theory of conics in a unified

René Descartes (1596–1650) was a French scientist, philosopher and mathematician. He is also known for the phrase 'Cogito, ergo sum' (I think, therefore I am).

way, using concepts which later mathematicians distilled into the notion of the cross-ratio of four points. Desargues' discoveries came to be known as *projective geometry*.

Girard Desargues (1591–1661) was a French architect.
We deal with these ideas in Chapters 4 and 5.

Blaise Pascal was the son of a mathematician, Étienne, who attended a group of scholars frequented by Desargues. He heard of Desargues's work from his father, and quickly came up with one of the most famous results in the geometry of conics, Pascal's Theorem, which we discuss in Chapter 4. By the late 19th century projective geometry came to be seen as the most basic geometry, with Euclidean geometry as a significant but special case.

Blaise Pascal (1623–1662) was a French geometer, probabilist, physicist and philosopher.

At the start of the 19th century the world of mathematics began to change. The French Revolution saw the creation of the École Polytechnique in Paris in 1794, an entirely new kind of institution for the training of military engineers. It was staffed by mathematicians of the highest calibre, and run for many years by Gaspard Monge, an enthusiastic geometer who had invented a simple system of *descriptive geometry* for the design of forts and other military sites. Monge was one of those rare teachers who get students to see what is going on, and he inspired a generation of French geometers. The École Polytechnique, moreover, was the sole entry-point for any one seeking a career in engineering in France, and the stranglehold of the mathematicians ensured that all students received a good, rigorous education in mathematics before entering the specialist engineering schools. Thus prepared they then assisted Napoleon's armies everywhere across Europe and into Egypt.

Gaspard Monge (1746–1818) was a French analyst and geometer. A strong republican and supporter of the Revolution, he was French Minister of the Navy in 1792–93, but deprived of all his honours on the restoration of the French monarchy.

One of the École's former students, Jean Victor Poncelet, was taken prisoner in 1812 in Napoleon's retreat from Moscow. He kept his spirits up during a terrible winter by reviewing what his old teacher, Monge, had taught him about descriptive geometry. This is a system of projections of a solid onto a plane – or rather two projections, one vertically and one horizontally (giving what are called to this day the *plan* and *elevation* of the solid). Poncelet realized that instead of projecting 'from infinity' so to speak, one could adapt Monge's ideas to the study of projection from a point. In this way he re-discovered Desargues' ideas of projective geometry. During his imprisonment he wrote his famous book *Traité des propriétés projectives des figures* outlining the foundations of projective geometry, which he extensively rewrote after his release in 1814 and published in 1822.

Jean Victor Poncelet (1788–1867) followed a career as a military engineer by becoming Professor of Mechanics at Metz, where he worked on the efficiency of turbines.

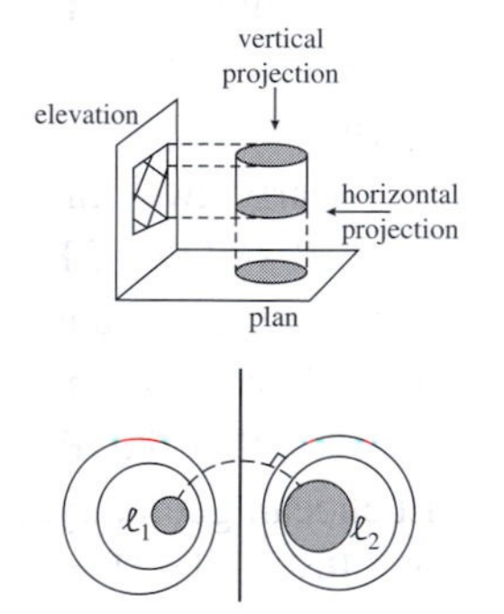

Around the same time that projective geometry was emerging, mathematicians began to realize that there was more to be said about circles than they had previously thought. For instance, in the study of electrostatics let ℓ_1 and ℓ_2 be two infinitely long parallel cylinders of opposite charge. Then the intersection of the surfaces of equipotential with a vertical plane is two families of circles (and a single line), and a point charge placed in the electrostatic field moves along a circular path through a specific point inside each cylinder, at right angles to circles in the families. The study of properties of such families of circles gave rise to a new geometry, called *inversive geometry*, which was able to provide particularly striking proofs of previously known results in Euclidean geometry as well as new results.

In inversive geometry mathematicians had to add a 'point at infinity' to the plane, and had to regard circles and straight lines as equivalent figures under the natural mappings, inversions, as these can turn circles into lines, and vice-versa. Analogously, in projective geometry mathematicians had to add a whole 'line at infinity' in order to simplify the geometry, and found that there were projective transformations that turned hyperbolas into ellipses, and so on. So mathematicians began to move towards thinking of geometry as the study of shapes and the transformations that preserve (at least specified properties of) those shapes.

For example, there are very few theorems in Euclidean geometry that depend on the size of the figure. The ability to make scale copies without altering 'anything important' is basic to mathematical modelling and a familiar fact of everyday life. If we wish to restrict our attention to the transformations that preserve length, we deal with Euclidean geometry, whereas if we allow arbitrary changes of scale we deal with *similarity geometry*.

Another interesting geometry was discovered by Möbius in the 1820s, in which transformations of the plane map lines to lines, parallel lines to parallel lines, and preserve ratios of lengths along lines. He called this geometry *affine geometry* because any two figures related by such a transformation have a likeness or affinity to one another. This is the geometry appropriate, in a sense, to Monge's descriptive geometry, and the geometry that describes the shadows of figures in sunlight.

August Ferdinand Möbius (1790–1868) was a German geometer, topologist, number theorist and astronomer; he discovered the famous Möbius Strip (or Band).

Since the days of Greek mathematics, with a stimulus provided by the needs of commercial navigation, mathematicians had studied *spherical geometry* too; that is, the geometry of figures on the surface of a sphere. Here geometry is rather different from plane Euclidean geometry; for instance the area of a triangle is proportional to the amount by which its angle sum exceeds π, and there is a nice generalization of Pythagoras' Theorem, which says that in a right-angled triangle with sides a, b and the hypotenuse c, then $\cos c = \cos a \cdot \cos b$. It turns out that there is a close connection between spherical geometry and inversive geometry.

For the surface of the Earth is very nearly spherical.

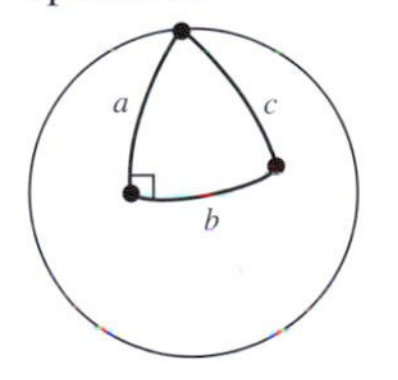

For nearly two millennia mathematicians had accepted as obvious the *Parallel Postulate* of Euclid: namely, that given any line ℓ and any point P not on ℓ, there is a unique line m in the same plane as P and ℓ which passes through P and does not meet ℓ. Indeed much effort had been put into determining whether this Postulate could be deduced from the other assumptions of Euclidean geometry. In the 1820s two young and little-known mathematicians, Bolyai in Hungary and Lobachevskii in Russia, showed that there were perfectly good so-called '*non-Euclidean geometries*', namely *hyperbolic geometry* and *elliptic geometry*, that share all the initial assumptions of Euclidean geometry except the parallel postulate.

Janos Bolyai (1802–1860) was an officer in the Hungarian Army.

Nicolai Ivanovich Lobachevskii (1792–1856) was a Russian geometer who became Rector of the University of Kazan.

In hyperbolic geometry given any line ℓ and any point P not on ℓ, there are infinitely many lines in the same plane as P and ℓ which pass through P and do not meet ℓ; in elliptic geometry all lines intersect each other. However, it still makes sense in both hyperbolic and elliptic geometries to talk about the length of line segments, the distance between points, the angles between lines, and so forth. Around 1900 Poincaré did a great deal to popularise these geometries by demonstrating their applications in many surprising areas of mathematics, such as Analysis.

Jules Henri Poincaré (1854–1912) was a prolific French mathematician, physicist, astronomer and philosopher at the University of Paris.

By 1870, the situation was that there were many geometries: Euclidean, affine, projective, inversive, hyperbolic and elliptic geometries. One way mathematicians have of coping with the growth of their subject is to re-define it so that different branches of it become branches of the same subject. This was done for geometry by Klein, who developed a programme (the *Erlangen Programme*) for classifying geometries. His elegant idea was to regard a *geometry* as a space together with a group of transformations of that space; the properties of figures that are not altered by any transformation in the group are their geometrical properties.

Christian Felix Klein (1849–1925) was a German algebraist, geometer, topologist and physicist; he became a professor at the University of Erlangen at the remarkable age of 22.

For example, in two-dimensional Euclidean geometry the space is the plane and the group is the group of all length-preserving transformations of the plane (or *isometries*). In projective geometry the space is the plane enlarged (in a way we make precise in Chapter 6) by a line of extra points, and the group is the group of all continuous transformations of the space that preserve *cross-ratio*.

Klein's approach to a geometry involves three components: a set of points (the space), a set of transformations (that specify the invariant properties – for example, congruence in Euclidean geometry), and a group (that specifies how the transformations may be composed). The transformations and their group are the fundamental components of the geometry that may be applied to different spaces. A *model of a geometry* is a set which possesses all the properties of the geometry; two different models of any geometry will be isomorphic. There may be several different models of a given geometry, which have different advantages and disadvantages. Therefore, we shall use the terms 'geometry' and 'model (of a geometry)' interchangeably whenever we think that there is no risk of confusion.

For example, you will meet two models of hyperbolic geometry.

In fact as Klein was keen to stress, most geometries are examples of projective geometry with some extra conditions. For example, affine geometry emerges as the geometry obtained from projective geometry by selecting a line and considering only those transformations that map that line to itself; the line can then be thought of as lying 'at infinity' and safely ignored. The result was that Klein not only had a real insight into the nature of geometry, he could even show that projective geometry was almost the most basic geometry.

This philosophy of geometry, called the *Kleinian view of geometry*, is the one we have adopted in this book. We hope that you will enjoy this introduction to the various geometries that it contains, and go on to further study of one of the oldest, and yet most fertile, branches of mathematics.

1 Conics

The study of conics is well over 2000 years old, and has given rise to some of the most beautiful and striking results in the whole of geometry.

In Section 1.1 we outline the Greek idea of a *conic section* – that is, a conic as defined by the curve in which a double cone is intersected by a plane. We then look at some properties of circles, the simplest of the non-degenerate conics, such as the condition for two circles to be *orthogonal* and the equations of the family of all circles through two given points.

That is, they intersect at right angles.

We explain the focus–directrix definition of the parabola, ellipse and hyperbola, and study the focal-distance properties of the ellipse and hyperbola. Finally, we use the so-called *Dandelin spheres* to show that the Greek conic sections are just the same as the conics defined in terms of a focus and a directrix.

In Section 1.2 we look at tangents to conics, and the reflection properties of the parabola, ellipse and hyperbola. It turns out that these are useful in practical situations as diverse as anti-aircraft searchlights and astronomical optical telescopes! We also see how we can construct each non-degenerate conic as the 'envelope' of lines in a suitably-chosen family of lines.

The equations of conics are all second degree equations in x and y. In Section 1.3 we show that the converse result holds – that is, that every second degree equation in x and y represents a conic. We also find an algorithm for determining from its equation in x and y which type of non-degenerate conic a given second degree equation represents, and for finding its principal features.

The analogue in $\mathbb{R}^3$ of a plane conic in $\mathbb{R}^2$ is a *quadric surface*, specified by a suitable second degree equation in x, y and z. A well-known example of a quadric surface is the cooling tower of an electricity generating station. In Section 1.4 we find an algorithm for identifying from its equation which type of non-degenerate quadric a given second degree equation in x, y and z represents. We also discover that two of the non-degenerate quadric surfaces can be generated by two different families of straight lines, and that this feature is of practical importance.

We use the notation $\mathbb{R}^2$ and $\mathbb{R}^3$ to denote 2-dimensional and 3-dimensional Euclidean space, respectively.

1.1 Conic Sections and Conics

1.1.1 Conic Sections

Conic Section is the name given to the shapes that we obtain by taking different plane slices through a double cone. The shapes that we obtain from these cross-sections are as drawn below.

It is thought that the Greek mathematician Menaechmus discovered the conic sections around 350 BC.

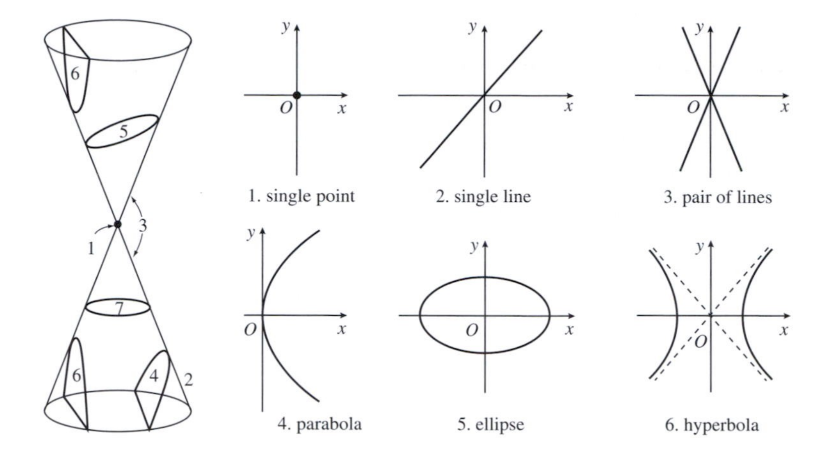

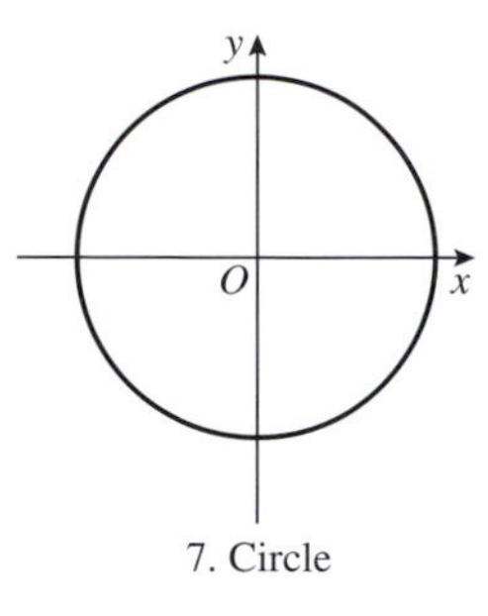

7. Circle

Notice that the circle shown in slice 7 can be regarded as a special case of an ellipse.

Notice, also, that the ellipse and the hyperbola both have a *centre;* that is, there is a point C such that rotation about C through an angle π is a symmetry of the conic. For example, for the ellipse and hyperbola illustrated above, the centre is in fact just the origin. On the other hand, the parabola does not have a centre.

In Subsection 1.1.5 we shall verify that the curves, the 'conic sections', obtained by slicing through a double cone are exactly the same curves, the 'conics', obtained as the locus of points in the plane whose distance from a fixed point is a constant multiple of its distance from a fixed line. As a result, we often choose not to distinguish between the terms 'conic section' and 'conic'!

We use the term *non-degenerate conics* to describe those *conics* that are parabolas, ellipses or hyperbolas; and the term *degenerate conics* to describe the single point, single line and pair of lines.

In this chapter we study conics for their own interest, and we will meet them frequently throughout our study of geometry as the book progresses.

1.1.2 Circles

The first conic that we investigate is the circle. Recall that a *circle* in $\mathbb{R}^2$ is the set of points (x, y) that lie at a fixed distance, called the *radius,* from a fixed point, called the *centre* of the circle. We can use the techniques of coordinate geometry to find the equation of a circle with given centre and radius.

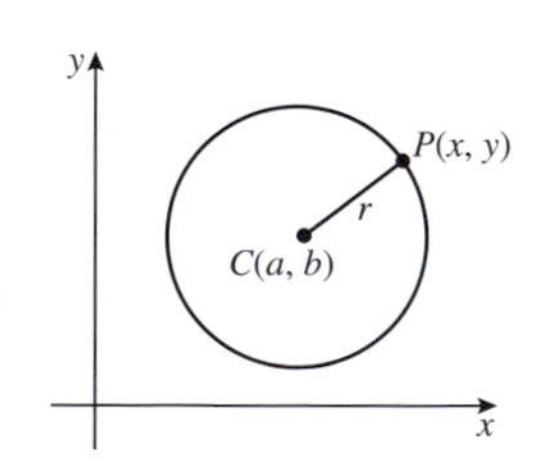

Let the circle have centre $C(a, b)$ and radius r. Then, if $P(x, y)$ is an arbitrary point on the circumference of the circle, the distance CP equals r. It follows from the formula for the distance between two points in the plane that

$$r^2 = (x - a)^2 + (y - b)^2. \quad (1)$$

Here we use the *Distance Formula* for the distance d between two points (x_1, y_1), (x_2, y_2) in $\mathbb{R}^2$:

$$d^2 = (x_1 - x_2)^2 + (y_1 - y_2)^2.$$

If we now expand the brackets in equation (1) and collect the corresponding terms, we can rewrite equation (1) in the form

$$x^2 + y^2 - 2ax - 2by + (a^2 + b^2 - r^2) = 0.$$

Then, if we write f for $-2a$, g for $-2b$ and h for $a^2 + b^2 - r^2$, this equation takes the form

$$x^2 + y^2 + fx + gy + h = 0. \quad (2)$$

Note here that the coefficients of x^2 and y^2 are equal.

It turns out that in many situations, however, equation (1) is more useful than equation (2) for determining the equation of a particular circle.

Theorem 1 The equation of a circle in $\mathbb{R}^2$ with centre (a, b) and radius r is

$$(x - a)^2 + (y - b)^2 = r^2.$$

For example, it follows from this formula that the circle with centre $(-1, 2)$ and radius $\sqrt{3}$ has equation

$$(x + 1)^2 + (y - 2)^2 = \left(\sqrt{3}\right)^2;$$

this can be simplified to give

$$x^2 + 2x + 1 + y^2 - 4y + 4 = 3,$$

or

$$x^2 + y^2 + 2x - 4y + 2 = 0.$$

Problem 1 Determine the equation of each of the circles with the following centre and radius:

(a) centre the origin, radius 1;
(b) centre the origin, radius 4;
(c) centre (3, 4), radius 2;
(d) centre (3, 4), radius 3.

Problem 2 Determine the condition on the numbers f, g and h in the equation

$$x^2 + y^2 + fx + gy + h = 0$$

for the circle with this equation to pass through the origin.

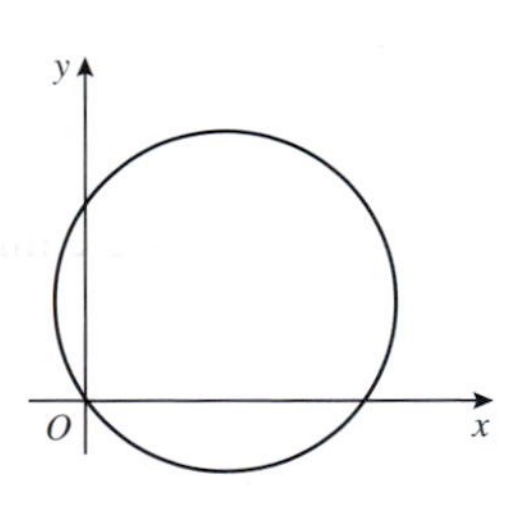

We have seen that the equation of a circle can be written in the form

$$x^2 + y^2 + fx + gy + h = 0. \quad (2)$$

In the opposite direction, given an equation of the form (2), can we determine whether it represents a circle? If it does represent a circle, can we determine its centre and radius?

For example, consider the set of points (x, y) in the plane that satisfy the equation:

$$x^2 + y^2 - 4x + 6y + 9 = 0. \qquad (3)$$

Note that in equation (3) the coefficients of x^2 and y^2 are both 1.

In order to transform equation (3) into an equation of the form (1), we use the technique called 'completing the square' – we rewrite the terms that involve only xs and the terms that involve only ys as follows:

$$x^2 - 4x = (x - 2)^2 - 4,$$
$$y^2 + 6y = (y + 3)^2 - 9.$$

Note that -2 is half the coefficient of x, and $+3$ is half the coefficient of y, in equation (3).

Substituting these expressions into equation (3), we obtain

$$(x - 2)^2 + (y + 3)^2 = 4.$$

We can 'read off' the centre and radius of the circle from this equation.

It follows that the equation represents a circle whose centre is $(2, -3)$ and whose radius is 2.

In general, we can use the same method of 'completing the square' to rewrite the equation

$$x^2 + y^2 + fx + gy + h = 0$$

in the form

$$\left(x + \tfrac{1}{2}f\right)^2 + \left(y + \tfrac{1}{2}g\right)^2 = \tfrac{1}{4}f^2 + \tfrac{1}{4}g^2 - h, \qquad (4)$$

from which we can 'read off' the centre and radius.

Here we start with the coefficients of x^2 and y^2 both equal (to 1). Otherwise the equation cannot be reformulated in the form (1).

Theorem 2 An equation of the form

$$x^2 + y^2 + fx + gy + h = 0$$

represents a circle with

$$\text{centre } \left(-\tfrac{1}{2}f, -\tfrac{1}{2}g\right) \text{ and radius } \sqrt{\tfrac{1}{4}f^2 + \tfrac{1}{4}g^2 - h},$$

provided that $\frac{1}{4}f^2 + \frac{1}{4}g^2 - h > 0$.

Remark

It follows from equation (4) above that if $\frac{1}{4}f^2 + \frac{1}{4}g^2 - h < 0$, then there are no points (x, y) that satisfy the equation $x^2 + y^2 + fx + gy + h = 0$; and if $\frac{1}{4}f^2 + \frac{1}{4}g^2 - h = 0$, then the given equation simply represents the single point $\left(-\frac{1}{2}f, -\frac{1}{2}g\right)$.

Problem 3 Determine the centre and radius of each of the circles given by the following equations:

(a) $x^2 + y^2 - 2x - 6y + 1 = 0$; (b) $3x^2 + 3y^2 - 12x - 48y = 0$.

Problem 4 Determine the set of points (x, y) in $\mathbb{R}^2$ that satisfies each of the following equations:

(a) $x^2 + y^2 + x + y + 1 = 0$;
(b) $x^2 + y^2 - 2x + 4y + 5 = 0$;
(c) $2x^2 + 2y^2 + x - 3y - 5 = 0$.

Orthogonal Circles

We shall sometimes be interested in whether two intersecting circles are *orthogonal*: that is, whether they meet at right angles. The following result answers this question if we know the equations of the two circles.

For example, in Chapters 5 and 6.

Theorem 3 Orthogonality Test
Two intersecting circles C_1 and C_2 with equations

$$x^2 + y^2 + f_1x + g_1y + h_1 = 0 \quad \text{and}$$
$$x^2 + y^2 + f_2x + g_2y + h_2 = 0,$$

respectively, are orthogonal if and only if

$$f_1f_2 + g_1g_2 = 2(h_1 + h_2).$$

Proof The circle C_1 has centre $A = \left(-\frac{1}{2}f_1, -\frac{1}{2}g_1\right)$ and radius $r_1 = \sqrt{\frac{1}{4}f_1^2 + \frac{1}{4}g_1^2 - h_1}$; the circle C_2 has centre $B = \left(-\frac{1}{2}f_2, -\frac{1}{2}g_2\right)$ and radius $r_2 = \sqrt{\frac{1}{4}f_2^2 + \frac{1}{2}g_2^2 - h_2}$.

You met these formulas in Theorem 2.

Let P be one of their points of intersection, and look at the triangle ΔABP. If the circles meet at right angles, then the line AP is tangential to the circle C_2, and is therefore at right angles to the line BP. So the triangle ΔABP is right-angled, and we may apply Pythagoras' Theorem to it to obtain

We use the symbol Δ to indicate a triangle.

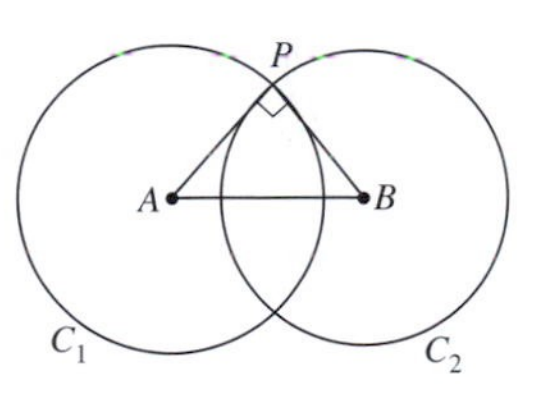

$$AP^2 + BP^2 = AB^2. \tag{5}$$

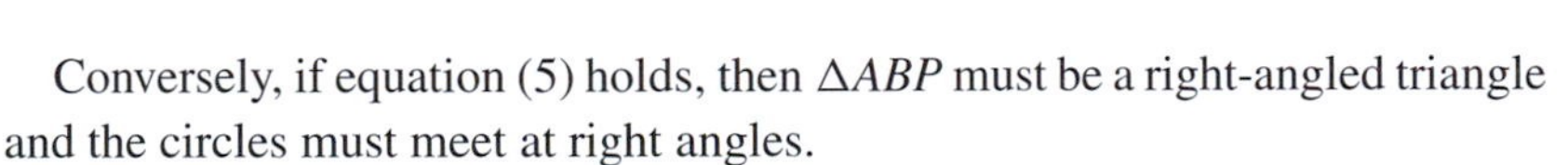

Conversely, if equation (5) holds, then ΔABP must be a right-angled triangle and the circles must meet at right angles.

Now

$$AP^2 = r_1^2 = \tfrac{1}{4}f_1^2 + \tfrac{1}{4}g_1^2 - h_1 \quad \text{and}$$
$$BP^2 = r_2^2 = \tfrac{1}{4}f_2^2 + \tfrac{1}{4}g_2^2 - h_2.$$

Also

$$AB^2 = \left(\tfrac{1}{2}f_1 - \tfrac{1}{2}f_2\right)^2 + \left(\tfrac{1}{2}g_1 - \tfrac{1}{2}g_2\right)^2$$
$$= \left(\tfrac{1}{4}f_1^2 - \tfrac{1}{2}f_1f_2 + \tfrac{1}{4}f_2^2\right) + \left(\tfrac{1}{4}g_1^2 - \tfrac{1}{2}g_1g_2 + \tfrac{1}{4}g_2^2\right).$$

Substituting for AP^2, BP^2 and AB^2 into equation (5), and cancelling common terms, we deduce that equation (5) is equivalent to

$$-h_1 - h_2 = -\tfrac{1}{2} f_1 f_2 - \tfrac{1}{2} g_1 g_2,$$

that is,

$$f_1 f_2 + g_1 g_2 = 2(h_1 + h_2).$$

This is the required result. ■

Problem 5 Determine which, if any, of the following pairs of intersecting circles are mutually orthogonal.

(a) $C_1 = \{(x, y) : x^2 + y^2 - 4x - 4y + 7 = 0\}$ and
$C_2 = \{(x, y) : x^2 + y^2 + 2x - 8y + 5 = 0\}$

(b) $C_1 = \{(x, y) : x^2 + y^2 + 3x - 6y + 5 = 0\}$ and
$C_2 = \{(x, y) : 3x^2 + 3y^2 + 4x + y - 15 = 0\}$.

Circles through Two Points

We shall also be interested later in the family of circles through two given points. So, let two circles C_1 and C_2 with equations

Section 5.5

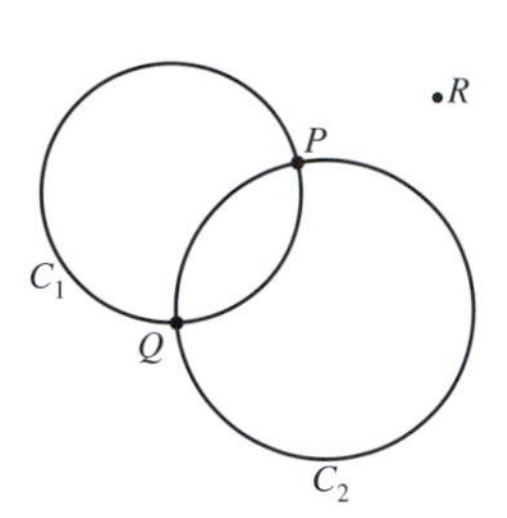

$$x^2 + y^2 + f_1 x + g_1 y + h_1 = 0 \quad \text{and}$$

$$x^2 + y^2 + f_2 x + g_2 y + h_2 = 0 \qquad (6)$$

intersect at the distinct points P and Q, say. Then, if $k \neq -1$, the equation

$$x^2 + y^2 + f_1 x + g_1 y + h_1 + k(x^2 + y^2 + f_2 x + g_2 y + h_2) = 0 \qquad (7)$$

represents a circle since it is a second degree equation in x and y with equal (non-zero) coefficients of x^2 and y^2 and with no terms in xy. This circle passes through both P and Q; for the coordinates of P and Q both satisfy the equations in (6) and so must satisfy equation (7).

If $k = -1$, equation (7) is linear in x and y, and so represents a line; since P and Q both lie on it, it must be the line through P and Q.

Conversely, given any point R in the plane that does not lie on the circle C_2 we can substitute the coordinates of R into equation (7) to find the unique value of k such that the circle with equation (7) passes through R. We can think of the circle C_2 as corresponding to the case '$k = \infty$' of equation (7). For, if we rewrite equation (7) in the form

This is possible because, since R does not lie on C_2, the term in the bracket in (7) does not vanish at R.

$$\frac{1}{k}(x^2 + y^2 + f_1 x + g_1 y + h_1) + x^2 + y^2 + f_2 x + g_2 y + h_2 = 0 \qquad (8)$$

and let $k \to \infty$, then $1/k \to 0$ and equation (8) becomes the equation of C_2.

Theorem 4 Let C_1 and C_2 be circles with equations

$$x^2 + y^2 + f_1 x + g_1 y + h_1 = 0 \quad \text{and}$$

$$x^2 + y^2 + f_2 x + g_2 y + h_2 = 0$$

that intersect at distinct points P and Q. Then the line and all circles (other than C_2) through P and Q have an equation of the form

$$x^2 + y^2 + f_1x + g_1y + h_1 + k(x^2 + y^2 + f_2x + g_2y + h_2) = 0$$

for some number k.

If $k \neq -1$, this equation is one of the circles; if $k = -1$, this is the equation of the line.

Example 1 Find the equation of the circle that passes through (1, 2) and the points of intersection of the circles

$$x^2 + y^2 - 3x + 4y - 1 = 0 \quad \text{and} \quad x^2 + y^2 + \tfrac{5}{2}x - 3y + \tfrac{3}{2} = 0.$$

Solution By Theorem 4, the required equation is of the form

$$x^2 + y^2 - 3x + 4y - 1 + k\left(x^2 + y^2 + \tfrac{5}{2}x - 3y + \tfrac{3}{2}\right) = 0 \qquad (9)$$

for some number k. Since (1, 2) must satisfy this equation, it follows that

$$1 + 4 - 3 + 8 - 1 + k\left(1 + 4 + \tfrac{5}{2} - 6 + \tfrac{3}{2}\right) = 0,$$

so that $k = -3$. Substituting $k = -3$ back into equation (9), we deduce that the equation of the required circle is

$$x^2 + y^2 - 3x + 4y - 1 - 3\left(x^2 + y^2 + \tfrac{5}{2}x - 3y + \tfrac{3}{2}\right) = 0,$$

which we can simplify to the form

$$4x^2 + 4y^2 + 21x - 26y + 11 = 0. \qquad \square$$

Problem 6 Find the equation of the line through the points of intersection of the circles

$$x^2 + y^2 - 3x + 4y - 1 = 0 \quad \text{and}$$
$$2x^2 + 2y^2 + 5x - 6y + 3 = 0.$$

1.1.3 Focus-Directrix Definition of the Non-Degenerate Conics

Earlier we defined the conic sections as the curves of intersection of a double cone with a plane. We have seen that the circle can be defined in a different way: as the set of points at a fixed distance from a fixed point.

Subsection 1.1.1
Subsection 1.1.2

Here we give a method for constructing the other non-degenerate conics, the parabola, ellipse and hyperbola, as sets of points that satisfy a *somewhat* similar condition involving distances. Later we shall give a careful proof

Subsection 1.1.5

that each non-degenerate conic section is a non-degenerate (plane) conic, and vice-versa.

Theorem 4 of Subsection 4.1.4

The three *non-degenerate conics* (the parabola, ellipse and hyperbola) can be defined as the set of points P in the plane that satisfy the following condition: The distance of P from a fixed point (called the *focus* of the conic) is a constant multiple (called its *eccentricity*, e) of the distance of P from a fixed line (called its *directrix*).

The different conics arise according to the value of the eccentricity:

Eccentricity A non-degenerate conic is an ellipse if $0 \leq e < 1$, a parabola if $e = 1$, or a hyperbola if $e > 1$.

When $e = 0$, the ellipse is actually a circle; the focus is the centre of the circle, and the directrix is 'at infinity'.

Parabola ($e = 1$)

A *parabola* is defined to be the set of points P in the plane whose distance from a fixed point F is equal to their distance from a fixed line d. We obtain a parabola *in standard form* if we choose

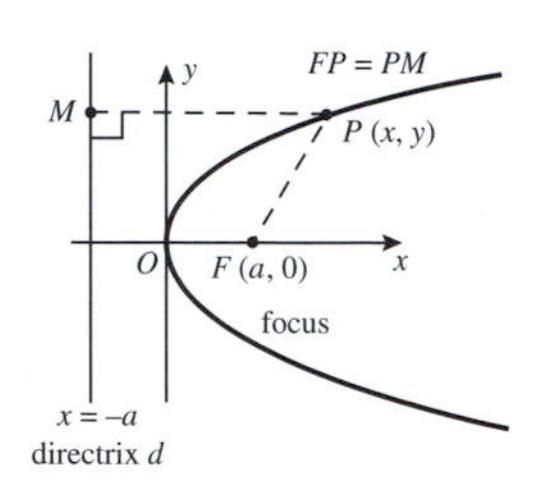

1. the focus F to lie on the x-axis, and to have coordinates $(a, 0)$, $a > 0$;
2. the directrix d to be the line with equation $x = -a$.

Notice in particular that the origin $O(0, 0)$ lies on the parabola since it is equidistant from F and d.

Let $P(x, y)$ be an arbitrary point on the parabola, and let M be the foot of the perpendicular from P to the directrix. Since $FP = PM$, by the definition of the parabola, it follows that $FP^2 = PM^2$; we may rewrite this equation in terms of coordinates as

$$(x - a)^2 + y^2 = (x + a)^2.$$

Multiplying out the brackets we get

$$x^2 - 2ax + a^2 + y^2 = x^2 + 2ax + a^2,$$

which simplifies to the equation $y^2 = 4ax$.

Notice that each point with coordinates $\left(at^2, 2at\right)$, where $t \in \mathbb{R}$, lies on the parabola, since $(2at)^2 = 4a \cdot at^2$. Conversely, we can write the coordinates of each point on the parabola in the form $\left(at^2, 2at\right)$. For if we choose $t = y/(2a)$, then $y = 2at$ and

We use the notation $\mathbb{R}$ to denote the 'real numbers' or the 'real line'.

$$\begin{aligned} x &= \frac{y^2}{4a} \quad \text{(from the equation } y^2 = 4ax\text{)} \\ &= \frac{(2at)^2}{4a} = at^2, \end{aligned}$$

as required. It follows that there is a one–one correspondence between the real numbers t and the points of the parabola.

We summarize the above facts as follows.

Parabola in Standard Form A parabola in standard form has equation

$$y^2 = 4ax, \quad \text{where } a > 0.$$

It has focus $(a, 0)$ and directrix $x = -a$; and it can be described by the parametric equations

$$x = at^2, \quad y = 2at \quad (t \in \mathbb{R}).$$

We call the x-axis the *axis* of the parabola in standard form, since the parabola is symmetric with respect to this line, and we call the origin the *vertex* of a parabola in standard form, since it is the point of intersection of the axis of the parabola with the parabola. A parabola has no centre.

'Centre' was defined in Subsection 1.1.1.

Example 2 This question concerns the parabola E with equation $y^2 = 2x$ and parametric equations $x = \frac{1}{2}t^2$, $y = t$ $(t \in \mathbb{R})$.

We generally use the letter E to denote a conic.

(a) Write down the focus, vertex, axis and directrix of E.

(b) Determine the equation of the chord that joins distinct points P and Q on E with parameters t_1 and t_2, respectively. Determine the condition on t_1 and t_2 such that the chord PQ passes through the focus of E.

Such a chord is called *focal chord*.

Solution

(a) The parabola E is the parabola in standard form where $4a = 2$, or $a = \frac{1}{2}$. It follows that the focus of E is $\left(\frac{1}{2}, 0\right)$, its vertex is (0, 0), its axis is the x-axis, and the equation of its directrix is $x = -\frac{1}{2}$.

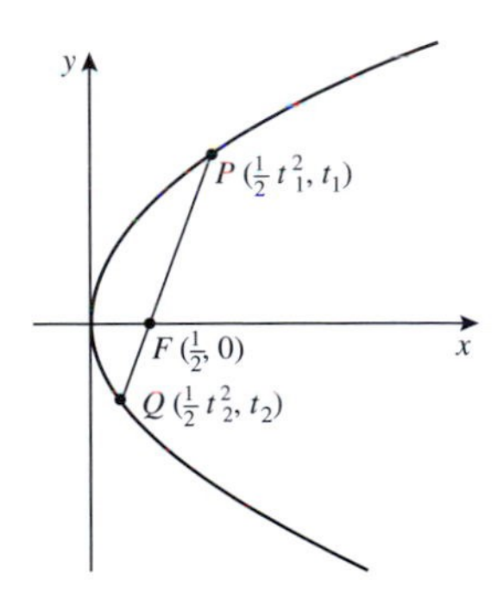

(b) The coordinates of P and Q are $\left(\frac{1}{2}t_1^2, t_1\right)$ and $\left(\frac{1}{2}t_2^2, t_2\right)$, respectively. So, if $t_1^2 \neq t_2^2$, the *slope* (or *gradient*, as it is sometimes called) of PQ is given by

$$m = \frac{t_1 - t_2}{\frac{1}{2}t_1^2 - \frac{1}{2}t_2^2} = \frac{t_1 - t_2}{\frac{1}{2}\left(t_1^2 - t_2^2\right)} = \frac{2}{t_1 + t_2}.$$

Since $\left(\frac{1}{2}t_1^2, t_1\right)$ lies on the line PQ, it follows that the equation of PQ is

$$y - t_1 = \frac{2}{t_1 + t_2}\left(x - \frac{1}{2}t_1^2\right).$$

Multiplying both sides by $t_1 + t_2$, we get

$$(t_1 + t_2)(y - t_1) = 2x - t_1^2,$$

so that

$$(t_1 + t_2)y - t_1^2 - t_1t_2 = 2x - t_1^2,$$

or

$$(t_1 + t_2)y = 2x + t_1t_2. \qquad (10)$$

If, however, $t_1^2 = t_2^2$, then since $t_1 \neq t_2$ we have $t_1 = -t_2$. Thus PQ is parallel to the y-axis, and so has equation $x = \frac{1}{2}t_1^2$; so in this case too, PQ has equation given by (10).

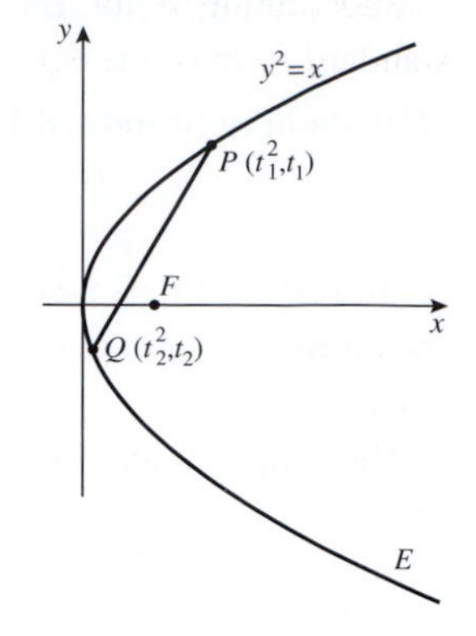

The chord PQ with equation (10) passes through the focus $\left(\frac{1}{2}, 0\right)$ if $(t_1 + t_2)0 = 1 + t_1t_2$; in other words, if $t_1t_2 = -1$. □

Problem 7 This question concerns the parabola E with equation $y^2 = x$ and parametric equations $x = t^2$, $y = t$ ($t \in \mathbb{R}$).

(a) Write down the focus, vertex, axis and directrix of E.
(b) Determine the equation of the chord that joins distinct points P and Q on E with parameters t_1 and t_2, respectively.
(c) Determine the condition on t_1 and t_2 (and so on P and Q) that the focus of E is the midpoint of the chord PQ.

Ellipse ($0 \leq e < 1$)

We define an ellipse *with eccentricity zero* to be a circle. We have already discussed circles.

Subsection 1.1.2

We define an *ellipse* with eccentricity e (where $0 < e < 1$) to be the set of points P in the plane whose distance from a fixed point F is e times their distance from a fixed line d. We obtain such an ellipse *in standard form* if we choose

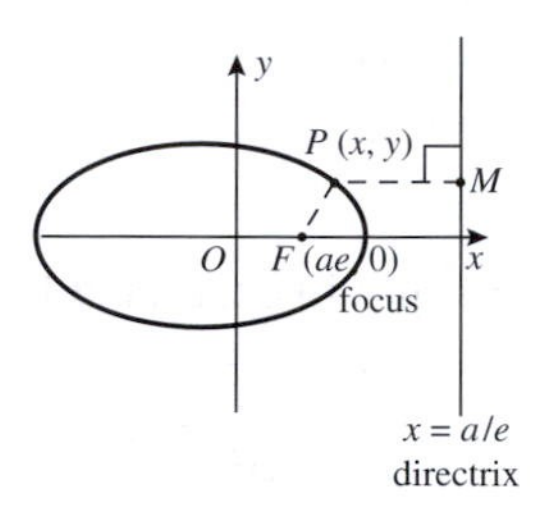

1. the focus F to lie on the x-axis, and to have coordinates $(ae, 0)$, $a > 0$;
2. the directrix d to be the line with equation $x = a/e$.

Let $P(x, y)$ be an arbitrary point on the ellipse, and let M be the foot of the perpendicular from P to the directrix. Since $FP = e \cdot PM$, by the definition of the ellipse, it follows that $FP^2 = e^2 \cdot PM^2$; we may rewrite this equation in terms of coordinates as

$$(x - ae)^2 + y^2 = e^2\left(x - \frac{a}{e}\right)^2 = (ex - a)^2.$$

Multiplying out the brackets we get

$$x^2 - 2aex + a^2e^2 + y^2 = e^2x^2 - 2aex + a^2,$$

which simplifies to the equation

$$x^2\left(1 - e^2\right) + y^2 = a^2\left(1 - e^2\right)$$

or

$$\frac{x^2}{a^2} + \frac{y^2}{a^2\left(1 - e^2\right)} = 1.$$

Substituting b for $a\sqrt{1-e^2}$, so that $b^2 = a^2(1-e^2)$, we obtain the standard form of the equation of the ellipse

Since $0 < e < 1$, we have that $0 < b < a$.

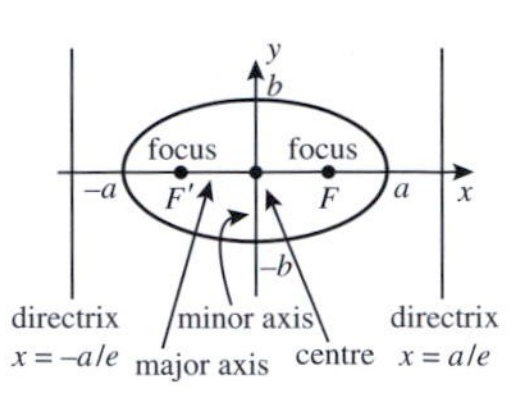

$$\frac{x^2}{a^2} + \frac{y^2}{b^2} = 1.$$

Notice that this equation is symmetrical in x and symmetrical in y, so that the ellipse also has a second focus $F'(-ae, 0)$ and a second directrix d' with equation $x = -a/e$.

The ellipse intersects the axes at the points $(\pm a, 0)$ and $(0, \pm b)$. We call the segment joining the points $(\pm a, 0)$ the *major axis* of the ellipse, and the segment joining the points $(0, \pm b)$ the *minor axis* of the ellipse. Since $b < a$, the minor axis is shorter than the major axis. The origin is the centre of this ellipse.

Notice that each point with coordinates $(a\cos t, b\sin t)$ lies on the ellipse, since

$$\frac{(a\cos t)^2}{a^2} + \frac{(b\sin t)^2}{b^2} = \cos^2 t + \sin^2 t = 1.$$

Then, just as for the parabola, we can check that

$$x = a\cos t, \quad y = b\sin t \quad (t \in (-\pi, \pi])$$

gives a parametric representation of the ellipse.

Sometimes it is convenient to assume that $t \in [0, 2\pi)$, for instance, instead of $(-\pi, \pi]$. Notice our notation for intervals:

$$(p, q), [p, q], [p, q), (p, q]$$

denote those real numbers x for which $p < x < q$, $p \le x \le q$, $p \le x < q$, $p < x \le q$, respectively.

We now summarize the above facts about ellipses (including circles) as follows.

Ellipse in Standard Form An ellipse in standard form has equation

$$\frac{x^2}{a^2} + \frac{y^2}{b^2} = 1, \quad \text{where } a \ge b > 0,\ b^2 = a^2\left(1-e^2\right), 0 \le e < 1.$$

It can be described by the parametric equations

$$x = a\cos t, \quad y = b\sin t \quad (t \in (-\pi, \pi]).$$

If $e > 0$, it has foci $(\pm ae, 0)$ and directrices $x = \pm a/e$.

Another parametric representation of this ellipse is

$$x = a\frac{1-t^2}{1+t^2},$$

$$y = b\frac{2t}{1+t^2},$$

$t \in \mathbb{R}$.

Example 3 Let PQ be an arbitrary chord of the ellipse with equation

$$\frac{x^2}{a^2} + \frac{y^2}{b^2} = 1.$$

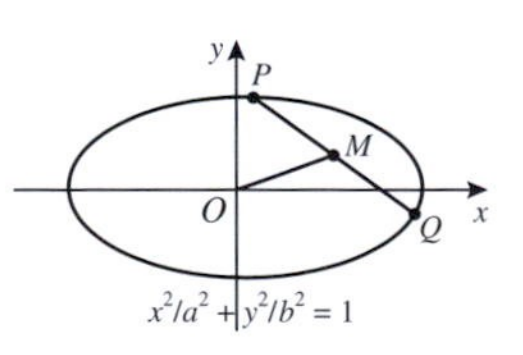

Let M be the midpoint of PQ. Prove that the following expression is independent of the choice of P and Q:

$$\text{slope of } OM \times \text{slope of } PQ.$$

Solution Let P and Q have the parametric coordinates $(a\cos t_1, b\sin t_1)$ and $(a\cos t_2, b\sin t_2)$, respectively. It follows that M has coordinates $(\frac{a}{2}(\cos t_1 + \cos t_2), \frac{b}{2}(\sin t_1 + \sin t_2))$.

Now,

$$\text{the slope of } OM = \frac{b(\sin t_1 + \sin t_2)}{a(\cos t_1 + \cos t_2)}$$

and

$$\text{the slope of } PQ = \frac{b(\sin t_1 - \sin t_2)}{a(\cos t_1 - \cos t_2)},$$

so

$$\begin{aligned}
&\text{slope of } OM \times \text{slope of } PQ \\
&= \frac{b(\sin t_1 + \sin t_2)}{a(\cos t_1 + \cos t_2)} \cdot \frac{b(\sin t_1 - \sin t_2)}{a(\cos t_1 - \cos t_2)} \\
&= \frac{b^2}{a^2} \cdot \frac{\sin^2 t_1 - \sin^2 t_2}{\cos^2 t_1 - \cos^2 t_2} \\
&= \frac{b^2}{a^2} \cdot \frac{\sin^2 t_1 - \sin^2 t_2}{(1 - \sin^2 t_1) - (1 - \sin^2 t_2)} \\
&= -\frac{b^2}{a^2},
\end{aligned}$$

which is independent of the values of t_1 and t_2. □

In general,

$$\cos^2\theta = 1 - \sin^2\theta.$$

Problem 8 Let P be an arbitrary point on the ellipse with equation $\frac{x^2}{a^2} + \frac{y^2}{b^2} = 1$ and focus $F(ae, 0)$. Let M be the midpoint of FP. Prove that M lies on an ellipse whose centre is midway between the origin and F.

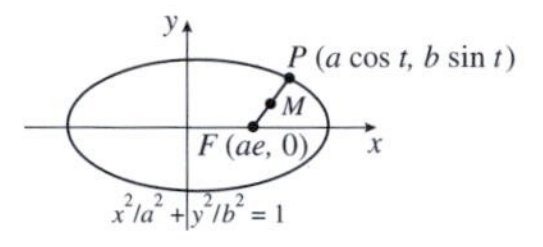

Hyperbola ($e > 1$)

A *hyperbola* is the set of points P in the plane whose distance from a fixed point F is e times their distance from a fixed line d, where $e > 1$. We obtain a hyperbola *in standard form* if we choose

1. the focus F to lie on the x-axis, and to have coordinates $(ae, 0)$, $a > 0$;
2. the directrix d to be the line with equation $x = a/e$.

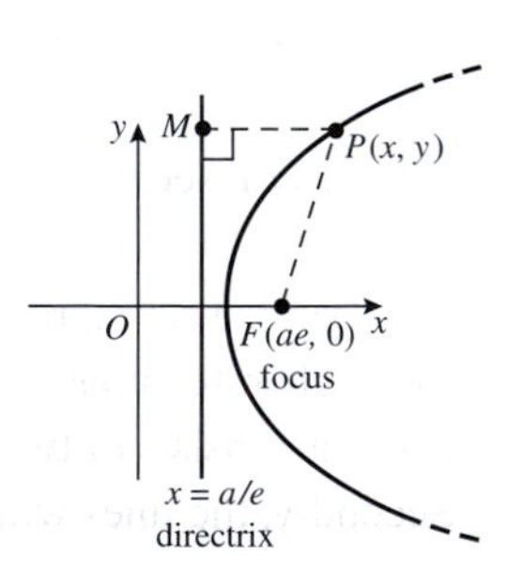

Let $P(x, y)$ be an arbitrary point on the hyperbola, and let M be the foot of the perpendicular from P to the directrix. Since $FP = e \cdot PM$, by the definition of the hyperbola, it follows that $FP^2 = e^2 \cdot PM^2$; we may rewrite this equation in terms of coordinates as

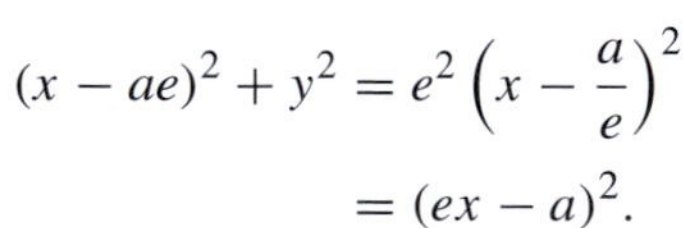

$$\begin{aligned}
(x - ae)^2 + y^2 &= e^2\left(x - \frac{a}{e}\right)^2 \\
&= (ex - a)^2.
\end{aligned}$$

Multiplying out the brackets we get

$$x^2 - 2aex + a^2e^2 + y^2 = e^2x^2 - 2aex + a^2,$$

which simplifies to

$$x^2\left(e^2-1\right)-y^2=a^2\left(e^2-1\right),$$

or

$$\frac{x^2}{a^2}-\frac{y^2}{a^2\left(e^2-1\right)}=1.$$

Substituting b for $a\sqrt{e^2-1}$, so that $b^2=a^2\left(e^2-1\right)$, we obtain the standard form of the equation of the hyperbola

$$\frac{x^2}{a^2}-\frac{y^2}{b^2}=1.$$

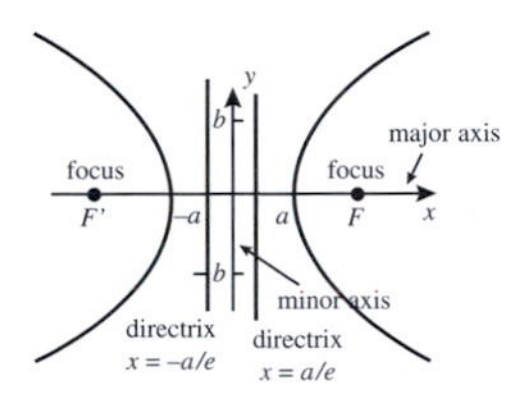

Notice that this equation is symmetrical in x and symmetrical in y, so that the hyperbola also has a second focus $F'(-ae,0)$ and a second directrix d' with equation $x=-a/e$.

The hyperbola intersects the x-axis at the points $(\pm a,0)$. We call the segment joining the points $(\pm a,0)$ the *major axis* or *transverse axis* of the hyperbola, and the segment joining the points $(0,\pm b)$ the *minor axis* or *conjugate axis* of the hyperbola (notice that this is NOT a chord of the hyperbola). The origin is the centre of this hyperbola.

Notice also that each point with coordinates $(a\sec t, b\tan t)$, where t is not an odd multiple of $\pi/2$, lies on the hyperbola, since

$$\frac{a^2\sec^2 t}{a^2}-\frac{b^2\tan^2 t}{b^2}=1.$$

In general,

$$\sec^2\theta=1+\tan^2\theta.$$

Then, just as for the parabola, we can check that

$$x=a\sec t,\quad y=b\tan t\quad (t\in(-\pi/2,\pi/2)\cup(\pi/2,3\pi/2))$$

gives a parametric representation of the hyperbola.

Points for which $t\in(-\pi/2,\pi/2)$ lie on the right branch of the hyperbola, and points for which $t\in(\pi/2,3\pi/2)$ lie on the left branch of the hyperbola.

Two other features of the shape of the hyperbola stand out. Firstly, the hyperbola consists of two separate curves or *branches.*

Secondly, the lines with equations

$$\frac{x^2}{a^2}-\frac{y^2}{b^2}=0,\quad\text{or}\quad y=\pm\frac{b}{a}x,$$

divide the plane into two pairs of opposite sectors; the branches of the hyperbola lie in one pair. As $x\to\pm\infty$ the branches of the hyperbola get closer and closer to these two lines. We call the lines $y=\pm(b/a)x$ the *asymptotes* of the hyperbola.

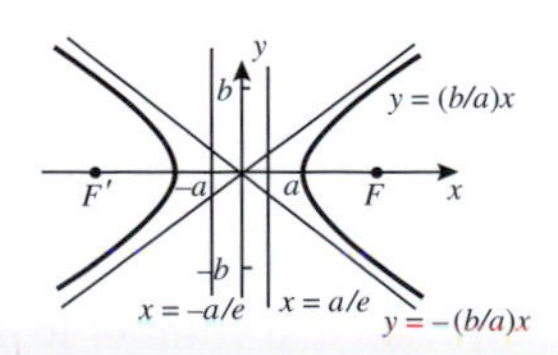

We summarize the above facts as follows.

Hyperbola in Standard Form A hyperbola in standard form has equation

$$\frac{x^2}{a^2} - \frac{y^2}{b^2} = 1, \quad \text{where } b^2 = a^2\left(e^2 - 1\right), \quad a > 0, e > 1.$$

It has foci $(\pm ae, 0)$ and directrices $x = \pm a/e$; and it can be described by the parametric equations

$$x = a \sec t, \quad y = b \tan t \quad (t \in (-\pi/2, \pi/2) \cup (\pi/2, 3\pi/2)).$$

Another parametric representation of this hyperbola is

$$x = a\frac{1+t^2}{1-t^2}, y = b\frac{2t}{1-t^2}$$

$t \in \mathbb{R} - \{\pm 1\}$.

Problem 9 Let P be a point $(\sec t, \frac{1}{\sqrt{2}} \tan t)$, where $(t \in (-\pi/2, \pi/2) \cup (\pi/2, 3\pi/2))$, on the hyperbola E with equation $x^2 - 2y^2 = 1$.

In this problem you will find the identity

$$\sec^2 \theta = 1 + \tan^2 \theta$$

useful.

(a) Determine the foci F and F' of E.
(b) Determine the slopes of FP and $F'P$, when these lines are not parallel to the y-axis.
(c) Determine the point P in the first quadrant on E for which FP is perpendicular to $F'P$.

Rectangular Hyperbola ($e = \sqrt{2}$)

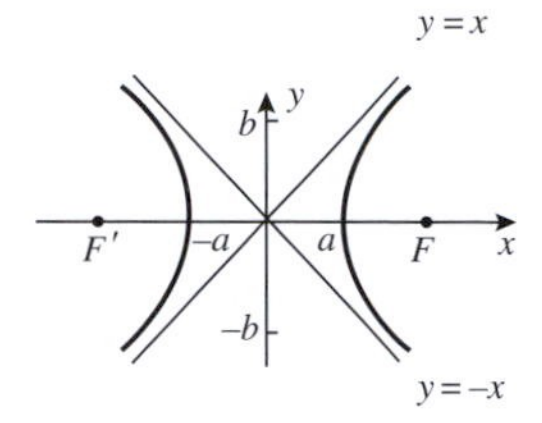

When the eccentricity e of a hyperbola takes the value $\sqrt{2}$, then $e^2 = 2$ and $b = a$. Then the asymptotes of the hyperbola have equations $y = \pm x$, so that in particular they are at right angles. A hyperbola whose asymptotes are at right angles is called a *rectangular hyperbola.*

Then, if we use the asymptotes as new x- and y-axes (instead of the original x- and y-axes), it turns out that the equation of the hyperbola can be written in the form $xy = c^2$, for some positive number c.

We omit the details.

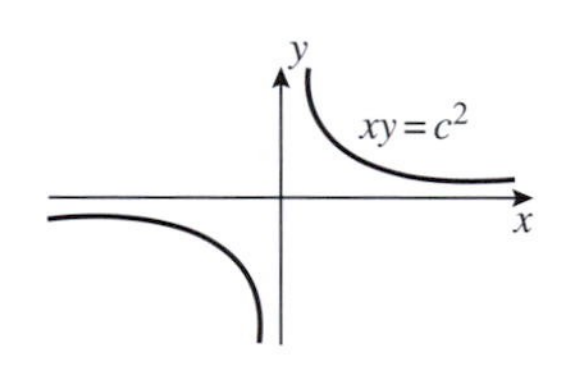

The rectangular hyperbola with equation $xy = c^2$ has the origin as its centre, and the x- and y-axes as its asymptotes. Also, each point on it can be uniquely represented by the parametric representation

$$x = ct, \quad y = \frac{c}{t} \quad \text{where } t \neq 0.$$

We shall use rectangular hyperbolas later on.

Section 2.5

Polar Equation of a Conic

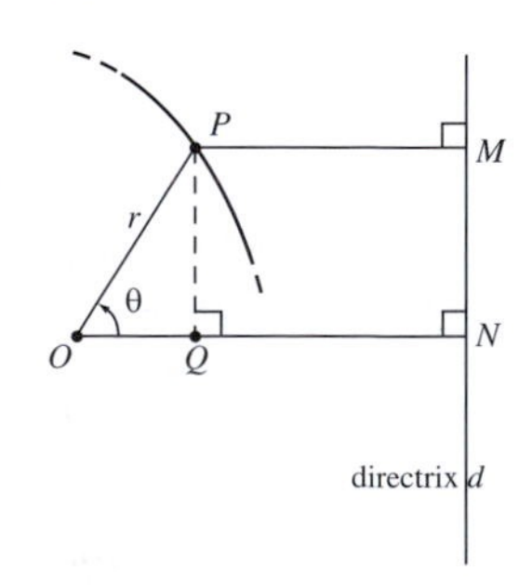

For many applications it is useful to describe the equation of a non-degenerate conic in terms of *polar coordinates* r and θ. A point $P(x, y)$ in the plane has polar coordinates (r, θ) if r is the distance OP (where O is the origin) and θ is the anticlockwise angle between OP and the positive direction of the x-axis.

Take the origin O to be the focus of the conic, d the directrix, M the foot of the perpendicular from a point P on the conic to d, N the foot of the perpendicular from O to d, and Q the foot of the perpendicular from P to ON.

Then by the definition of the conic, we have $OP = e \cdot PM$. We can rewrite this as

$$\begin{aligned} r &= e(ON - OQ) \\ &= e \cdot ON - er\cos\theta, \end{aligned}$$

or

$$\begin{aligned} r(1 + e\cos\theta) &= e \cdot ON \\ &= l, \quad \text{a constant.} \end{aligned}$$

It follows that the equation of the conic can be expressed in the form

$$r = \frac{l}{1 + e\cos\theta}.$$

The polar form of the equation of a conic is often used in problems in Dynamics: for example, in determining the motion of a planet or of a comet round the Sun.

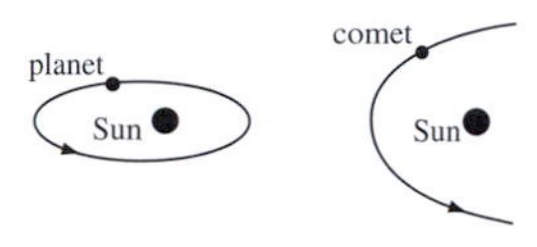

1.1.4 Focal Distance Properties of Ellipse and Hyperbola

We now prove two simple but surprising results. We deal with the ellipse first.

Theorem 5 Sum of Focal Distances of Ellipse
Let E be an ellipse with major axis $(-a, a)$ and foci F and F'. Then, if P is a point on the ellipse, $FP + PF' = 2a$. In particular, $FP + PF'$ is constant for all points P on the ellipse.

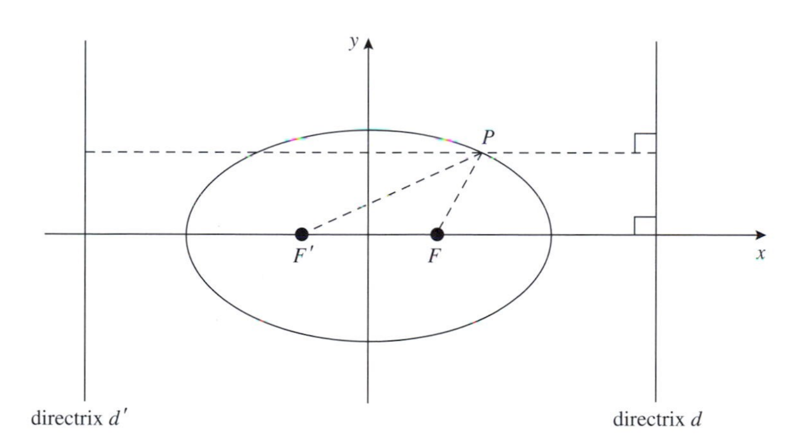

Proof Let d and d' be the directrices of the ellipse that correspond to the foci F and F', respectively. Then, since

$$PF = e \times (\text{distance from } P \text{ to } d)$$

and

$$PF' = e \times (\text{distance from } P \text{ to } d'),$$

it follows that

$$PF + PF' = e \times (\text{distance between } d \text{ and } d').$$
$$= 2a,$$

which is a constant. ■

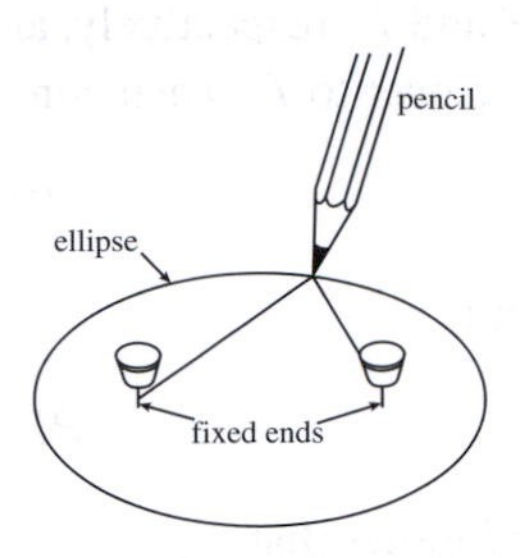

The result of Theorem 5 can be used to draw an ellipse, using a piece of string fixed at both ends. A pencil is used to pull the string taut; then, as we move the pencil round, the shape that it traces out is an ellipse whose foci are the two ends of the string.

Notice that, if we are given any three points F, F' and P (not on the line segment $F'F$) in the plane, then there is only one ellipse through P with F and F' as its foci. Its centre is the midpoint, O, of the segment $F'F$, its axes are the line along $F'F$ and the line through O perpendicular to $F'F$, and its major axis has length $PF + PF'$.

For, the location of the foci and the length of the major axis specify an ellipse uniquely.

Also, if we are given any two points F and F' in the plane, the locus of points P (not on the line segment $F'F$) in the plane for which $PF + PF'$ is a constant is necessarily an ellipse. Thus the converse of Theorem 5 holds.

As a result, some books take the 'Sum of Focal Distances Property' as the definition of the ellipse.

There is an analogous result for the hyperbola.

Theorem 6 Difference of Focal Distances of Hyperbola
Let H be a hyperbola with major axis $(-a, a)$ and foci F and F'. Then, if P is a point on the branch of the hyperbola that is closer to F,

$$PF' - PF = 2a;$$

and, if P is a point on the branch of the hyperbola closer to F',

$$PF' - PF = -2a.$$

In particular, $|PF' - PF|$ is constant for all points P on the hyperbola.

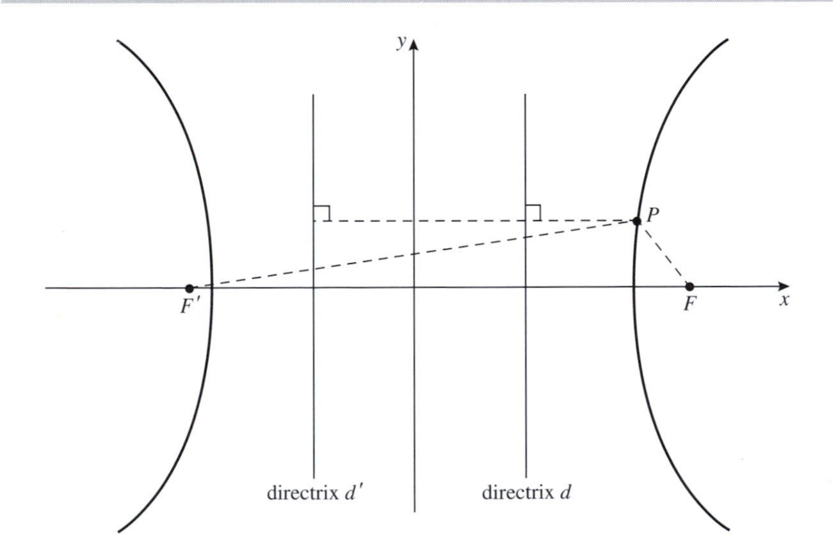

Proof We shall prove only the first formula; the proof of the second is similar.

Let d and d' be the directrices of the hyperbola that correspond to the foci F and F' respectively, and let P be a point on the branch of the hyperbola that is closer to F. Then, since

$$PF = e \times (\text{distance from } P \text{ to } d)$$

and

$$PF' = e \times (\text{distance from } P \text{ to } d'),$$

it follows that

$$\begin{aligned} PF' - PF &= e \times (\text{distance between } d \text{ and } d') \\ &= 2a, \end{aligned}$$

which is a constant. ■

The result of Theorem 6 can be used to draw a hyperbola, this time using a piece of string and a stick. Choose two points F and F' on the x-axis, equidistant from and on opposite sides of the origin. Hinge one end of a movable stick $F'X$ at the focus F'; attach one end of a string of length ℓ (where ℓ is less than the length of $F'X$) to the end X of the stick and the other end of the string to F, and keep the string taut by holding a pencil tight against the stick, as shown.

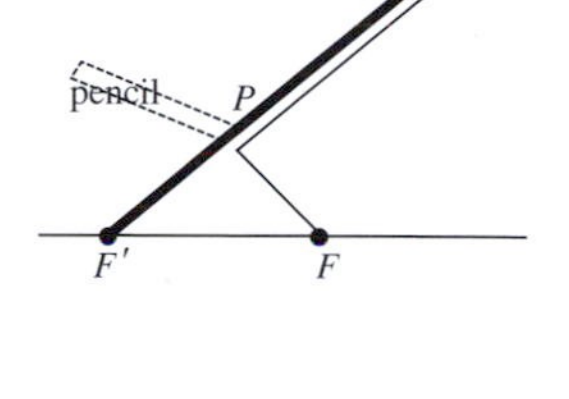

Then, as we move the pencil along the stick, the shape that it traces out is part of one branch of a hyperbola with foci F and F'. For,

$$\begin{aligned} PF' - PF &= XF' - (XP + PF) \\ &= XF' - \ell \\ &= \text{a constant independent of } P. \end{aligned}$$

We obtain the other branch of the hyperbola by interchanging the roles of F and F' in the construction.

Notice that, if we are given any three points F, F' and P (not on the line through $F'F$ or its perpendicular bisector) in the plane, then there is only one hyperbola through P with F and F' as its foci. Its centre is the midpoint, O, of the segment $F'F$, its axes are the line along $F'F$ and the line through O perpendicular to $F'F$, and its major axis has length $|PF' - PF|$.

For, the location of the foci and the length of the major axis specify a hyperbola uniquely.

Also, if we are given any two points F and F' in the plane, the locus of points P (not on the line segment $F'F$) in the plane for which $PF' - PF$ is a non-zero constant is necessarily one branch of a hyperbola. Thus the converse of Theorem 6 holds, in the following sense: Given any three points F, F' and P (where P must lie strictly between F and F' if it lies on the line through $F'F$) in the plane for which $PF' - PF \neq 0$, the locus of points Q in the plane for which $QF' - QF = \pm|PF' - PF|$ is a hyperbola.

As a result, some books take the 'Difference of Focal Distances Property' as the definition of the hyperbola.

1.1.5 Dandelin Spheres

We now give a beautiful proof due to Dandelin of the fact that a slant plane π that cuts one portion of a right circular cone in a 'complete' curve E is an ellipse, just as it appears to be!

Germinal Pierre Dandelin (1794–1847) was a French–Belgian Professor of Mechanics at Liège University; he made his discovery in 1822.

To do this, first fit a sphere inside the cone so that touches the plane π (at a point F) and the cone (in a circle C with centre O), as shown in the figure below. The circle C lies in a horizontal plane which intersects π in a line d. Take an arbitrary point P on the curve E, and extend the line from the vertex V of the cone through P to meet C at the point L, and let D be the point on d such that PD is perpendicular to d.

The line PD lies in a vertical plane which intersects the vertical plane VLO in the line PM, so that $\triangle PMD$ and $\triangle PML$ are both right-angled triangles.

Denote by α the angle between the slant plane π and the horizontal plane through C, and by β the angle $\angle PLM$ (the base angle of the cone).

We use the symbol $\angle$ to indicate an angle.

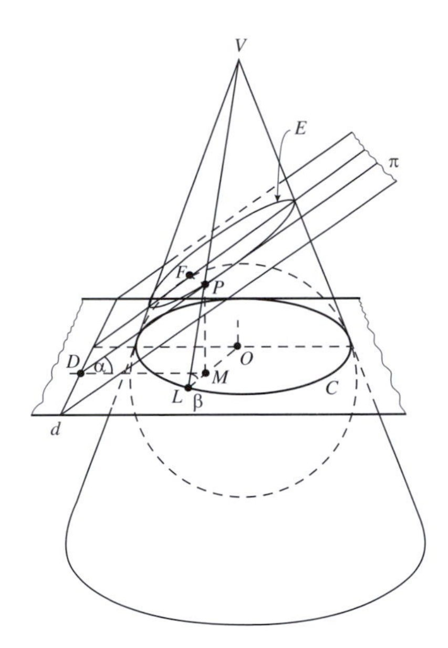

From the right-angled triangles $\triangle PMD$ and $\triangle PML$, we see that

$$PL = \frac{PM}{\sin\beta} \quad \text{and} \quad PD = \frac{PM}{\sin\alpha},$$

so that

$$\frac{PL}{PD} = \frac{\sin\alpha}{\sin\beta}.$$

Now $PF = PL$ since they are both tangents from a given point to a given sphere; it follows that

$$\frac{PF}{PD} = \frac{\sin\alpha}{\sin\beta}. \tag{11}$$

Now $0 < \alpha < \beta < \frac{\pi}{2}$, since the plane π is less steep than the base angle of the cone. So if we let $e = \sin\alpha / \sin\beta$, it follows that $0 < e < 1$.

It then follows from equation (11) that for any point P on the curve E, its distance PF from the fixed point F is e times its distance PD from the fixed line d. Since $0 < e < 1$, it follows from the focus–directrix definition of an ellipse that the curve E must be an ellipse, with focus F and directrix d.

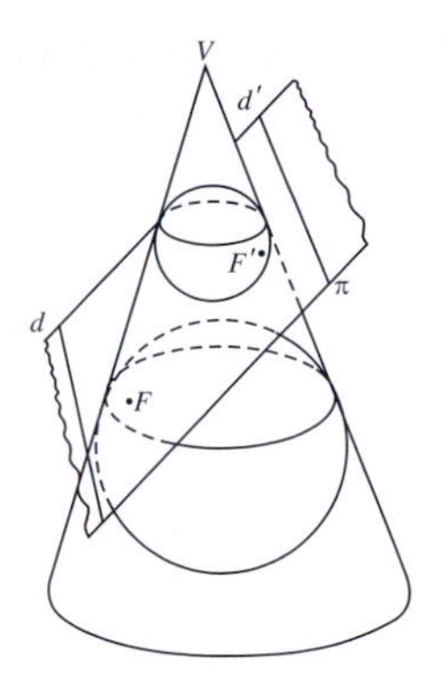

If we then construct the other sphere that touches both π and the cone, a similar argument shows that the point of contact of the sphere with π is the other focus F' of the ellipse; and the other directrix of the ellipse is the line of intersection of π with the horizontal plane through the circle in which the sphere touches the cone.

A similar construction involving spheres proves that in the cases 4 and 6 illustrated in the sketch in Subsection 1.1.1 the curve of intersection is a parabola and a hyperbola, respectively.

We ask you to look at the two cases 4 and 6 in Exercise 7 of Subsection 1.5.

This completes the proof of our claim in Subsection 1.1.1 that the curves of intersection of certain planes with a double cone are an ellipse, a parabola or a hyperbola. We shall investigate the converse in Theorem 4 of Subsection 4.1.4.

1.2 Properties of Conics

1.2.1 Tangents

In the previous section you met the parametric equations of the parabola, ellipse and hyperbola in standard form.

Subsection 1.1.3

We now tackle a rather natural question: given parametric equations $x = x(t)$, $y = y(t)$ describing a curve, what is the slope of the tangent to the curve at the point with parameter t? This information will enable us to determine the equation of the tangent to the curve at that point.

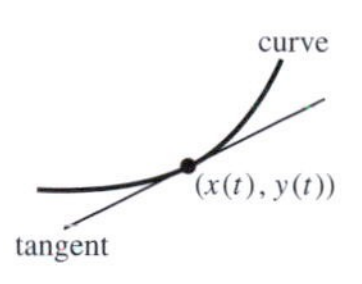

Theorem 1 The slope of the tangent to a curve in $\mathbb{R}^2$ with parametric equations $x = x(t)$, $y = y(t)$ at the point with parameter t is

$$\frac{y'(t)}{x'(t)},$$

provided that $x'(t) \neq 0$.

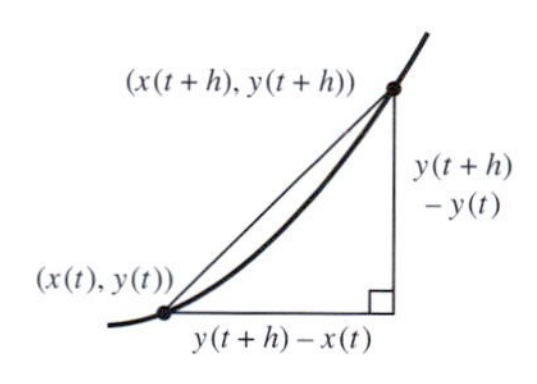

Proof The points on the curve with parameters t and $t + h$ have coordinates $(x(t), y(t))$ and $(x(t + h), y(t + h))$, respectively. Then, if $h \neq 0$, the slope of the chord joining these two points is

$$\frac{y(t+h) - y(t)}{x(t+h) - x(t)},$$

which we can write in the form

$$\frac{(y(t+h) - y(t))/h}{(x(t+h) - x(t))/h}.$$

We then take the limit of this ratio as $h \to 0$. The slope of the chord tends to the slope of the tangent, namely $y'(t)/x'(t)$. ■

Example 1

(a) Determine the equation of the tangent at the point with parameter t to the ellipse with parametric equations

$$x = a\cos t, \quad y = b\sin t,$$

where $t \in (-\pi, \pi]$, $t \neq 0, \pi$.

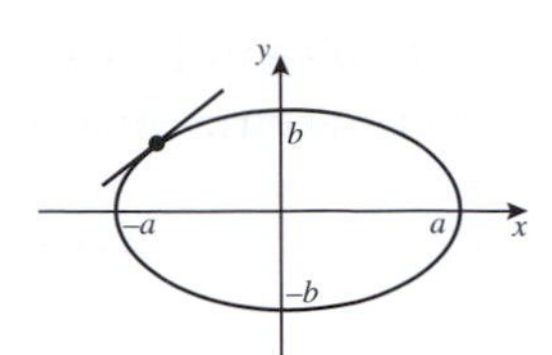

(b) Hence determine the equation of the tangent to the ellipse with parametric equations $x = 3\cos t$, $y = \sin t$ at the point with parameter $t = \pi/4$. Deduce the coordinates of the point of intersection of this tangent with the x-axis.

Solution

(a) Now, $y'(t) = b\cos t$ and $x'(t) = -a\sin t$ for $t \in (-\pi, \pi]$; it follows that, for $t \neq 0$ or π, the slope of the tangent at the point with parameter t is

$$\frac{y'(t)}{x'(t)} = \frac{b\cos t}{-a\sin t}.$$

Hence the equation of the tangent at the point $(a\cos t, b\sin t)$, $t \neq 0$, π, is

$$y - b\sin t = -\frac{b\cos t}{a\sin t}(x - a\cos t).$$

Multiplying both sides and rearranging terms, we get

$$xb\cos t + ya\sin t = ab\cos^2 t + ab\sin^2 t = ab,$$

and dividing both sides by ab gives the equation

$$\frac{x}{a}\cos t + \frac{y}{b}\sin t = 1. \quad (1)$$

We shall use this equation in Subsection 1.2.2.

The point on the ellipse where $t = 0$ is $(a, 0)$, at which the tangent has equation $x = a$. Similarly, the point on the ellipse where $t = \pi$ is $(-a, 0)$, at which the tangent has equation $x = -a$. It follows that equation (1) covers these cases also.

(b) Here the curve is the ellipse in part (a) in the particular case that $a = 3$, $b = 1$. When $t = \pi/4$, it follows from equation (1) that the equation of the tangent at the point with parameter $t = \pi/4$ is

$$\tfrac{x}{3}\cdot\tfrac{1}{\sqrt{2}} + y\cdot\tfrac{1}{\sqrt{2}} = 1,$$

or

$$\tfrac{1}{3}x + y = \sqrt{2}.$$

Hence, at the point T where the tangent crosses the x-axis, $y = 0$ and so $x = 3\sqrt{2}$. Thus, T is the point $\left(3\sqrt{2}, 0\right)$. □

Problem 1 Determine the slope of the tangent to the curve in $\mathbb{R}^2$ with parametric equations

$$x = 2\cos t + \cos 2t + 1, \qquad y = 2\sin t + \sin 2t$$

at the point with parameter t, where t is not a multiple of π. Hence determine the equation of the tangent to this curve at the points with parameters $t = \pi/3$ and $t = \pi/2$.

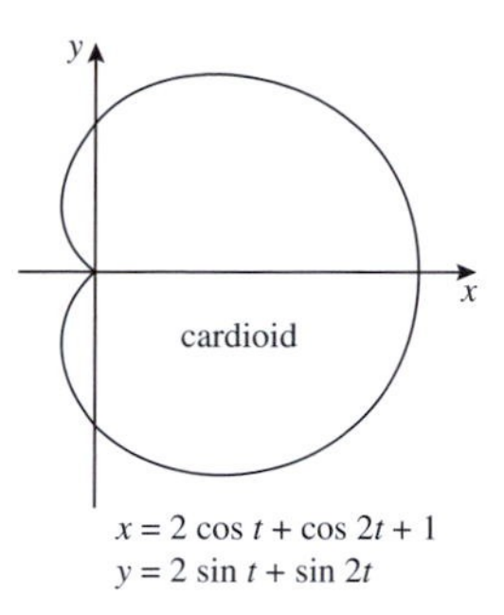

$x = 2\cos t + \cos 2t + 1$
$y = 2\sin t + \sin 2t$

Problem 2

(a) Determine the equation of the tangent at a point P with parameter t on the rectangular hyperbola with parametric equations $x = t$, $y = 1/t$.
(b) Hence determine the equations of the two tangents to the rectangular hyperbola from the point $(1, -1)$.

We can modify the result of Example 1(a) to find the equation of the tangent at the point (x_1, y_1) on the ellipse with equation $\frac{x^2}{a^2} + \frac{y^2}{b^2} = 1$. We take $x = a\cos t$, $y = b\sin t$ as parametric equations for the ellipse, and let $x_1 = a\cos t_1$ and $y_1 = b\sin t_1$. Then it follows from equation (1) above that the equation of the tangent is

$$\frac{x}{a}\cos t_1 + \frac{y}{b}\sin t_1 = 1,$$

which we can rewrite in the form $\frac{xx_1}{a^2} + \frac{yy_1}{b^2} = 1$.

We can determine the equations of tangents to the hyperbola and the parabola in a similar way; the results are given in the following theorem.

Theorem 2 The equation of the tangent at the point (x_1, y_1) to a conic in standard form is as follows.

Conic	*Tangent*
Ellipse $\frac{x^2}{a^2} + \frac{y^2}{b^2} = 1$	$\frac{xx_1}{a^2} + \frac{yy_1}{b^2} = 1$
Hyperbola $\frac{x^2}{a^2} - \frac{y^2}{b^2} = 1$	$\frac{xx_1}{a^2} - \frac{yy_1}{b^2} = 1$
Parabola $y^2 = 4ax$	$yy_1 = 2a(x + x_1)$

Problem 3 Prove that the equation of the tangent at the point (x_1, y_1) to the rectangular hyperbola $xy = 1$ is $\frac{1}{2}(xy_1 + x_1y) = 1$.

Problem 4 For each of the following conics, determine the equation of the tangent to the conic at the indicated point.

(a) The unit circle $x^2 + y^2 = 1$ at $\left(-\frac{1}{2}, \frac{1}{2}\sqrt{3}\right)$.

(b) The hyperbola $xy = 1$ at $\left(-4, -\frac{1}{4}\right)$.

(c) The parabola $y^2 = x$ at $(1, -1)$.

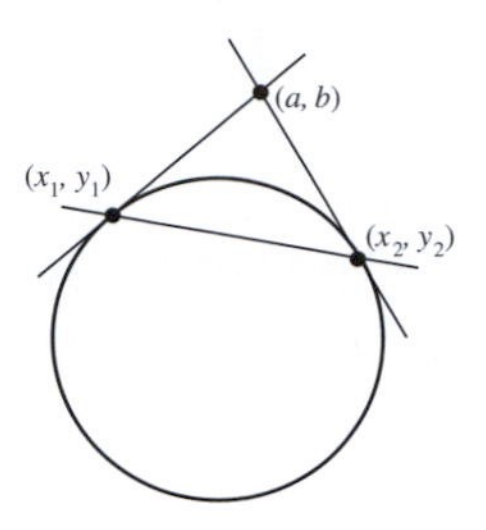

We can deduce a useful fact from the equation $xx_1 + yy_1 = 1$ for the tangent at the point (x_1, y_1) to the unit circle $x^2 + y^2 = 1$. Let (a, b) be some point on this tangent, so that

$$ax_1 + by_1 = 1. \quad (2)$$

Next, let the other tangent to the unit circle through the point (a, b) touch the circle at the point (x_2, y_2); it follows that

$$ax_2 + by_2 = 1. \quad (3)$$

This is because (a, b) lies on the tangent at (x_2, y_2), whose equation is $xx_2 + yy_2 = 1$.

From equations (2) and (3) we deduce that the points (x_1, y_1) and (x_2, y_2) both satisfy the equation $ax + by = 1$. Since this is the equation of a line, it must be the equation of the line through the points (x_1, y_1) and (x_2, y_2). For historical reasons, this line is called the *polar of* (a, b) *with respect to the unit circle*.

We shall meet polars of other conics in Subsection 4.2.1.

Theorem 3 Let (a, b) be a point outside the unit circle, and let the tangents to the circle from (a, b) touch the circle at P_1 and P_2. Then the equation of the line through P_1 and P_2 is

$$ax + by = 1.$$

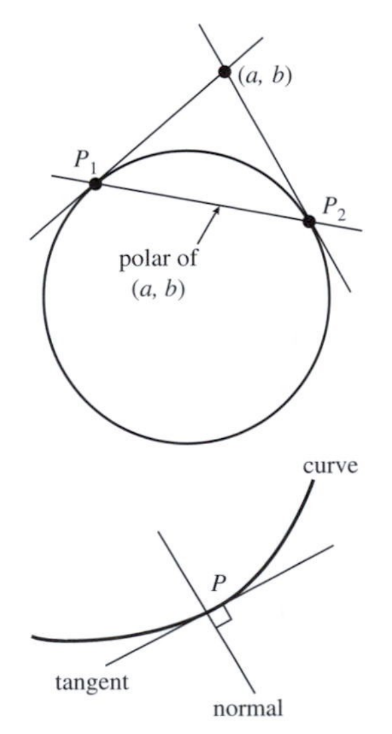

For example, the polar of $(2, 0)$ with respect to the unit circle is the line $2x = 1$.

Problem 5 Determine the equation of the polar of the point $(2, 3)$ with respect to the unit circle.

In the next example we meet the idea of the normal to a curve.

Definition The **normal** to a curve C at a point P on C is the line through P that is perpendicular to the tangent to C at P.

Example 2

(a) Determine the equation of the tangent at the point with parameter t to the parabola with parametric equations

$$x = at^2, \quad y = 2at \quad (t \in \mathbb{R}).$$

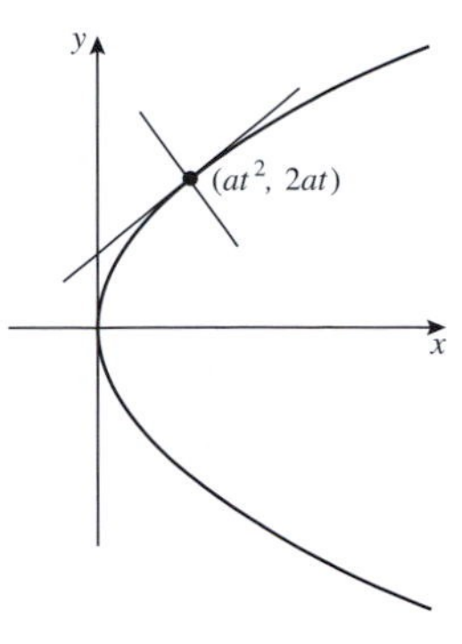

(b) Hence determine the equations of the tangent and the normal to the parabola with parametric equations $x = 2t^2$, $y = 4t$ at the point with parameter $t = 3$.

Solution

(a) Since $y'(t) = 2a$ and $x'(t) = 2at$, it follows that, for $t \neq 0$, the slope of the tangent at this point is

$$\frac{y'(t)}{x'(t)} = \frac{2a}{2at} = \frac{1}{t}.$$

Hence the equation of the tangent at the point $(at^2, 2at)$, $t \neq 0$, is

$$y - 2at = \frac{1}{t}(x - at^2),$$

which can be rearranged in the form

$$ty = x + at^2. \qquad (4)$$

We shall use this equation in Subsection 1.2.2.

The point on the parabola at which $t = 0$ is $(0, 0)$; there the tangent to the parabola is the y-axis, with equation $x = 0$. It follows that equation (4) covers this case also.

(b) Here the curve is the parabola in part (a) in the particular case that $a = 2$. When $t = 3$, it follows from equation (4) that the equation of the tangent is $3y = x + 2 \cdot 3^2$, or $3y = x + 18$.

To find the equation of the normal, we must find its slope and the coordinates of the point on the parabola at which $t = 3$.

Recall that lines of (non-zero) slope m_1 and m_2 are perpendicular if and only if $m_1 \cdot m_2 = -1$.

When $t = 3$, it follows from the equation of the tangent that the slope of the tangent is $\frac{1}{3}$. Since the tangent and normal are perpendicular to each other, it follows that the slope of the normal must be -3. Also, when $t = 3$, we have that $x = 2 \cdot 3^2 = 18$ and $y = 4 \cdot 3 = 12$; so the corresponding point on the parabola has coordinates $(18, 12)$.

It follows that the equation of the normal to the parabola at the point $(18, 12)$ is

$$\begin{aligned} y - 12 &= -3(x - 18) \\ &= -3x + 54, \end{aligned}$$

or

$$y = -3x + 66. \qquad \square$$

Problem 6 The normal to the parabola with parametric equations $x = t^2$, $y = 2t$ ($t \in \mathbb{R}$) at the point P with parameter t, $t \neq 0$, meets the parabola at a second point Q with parameter T.

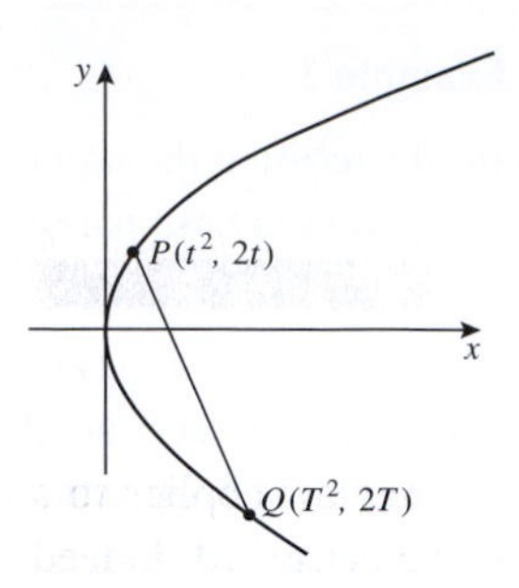

(a) Prove that the slope of the normal to the parabola at P is $-t$.
(b) Find the equation of the normal to the parabola at P.
(c) By substituting the coordinates of Q into your equation from part (b), prove that $T = -\frac{2}{t} - t$.

Problem 7 This question concerns the parabola with parametric equations $x = at^2$, $y = 2at$ ($t \in \mathbb{R}$).

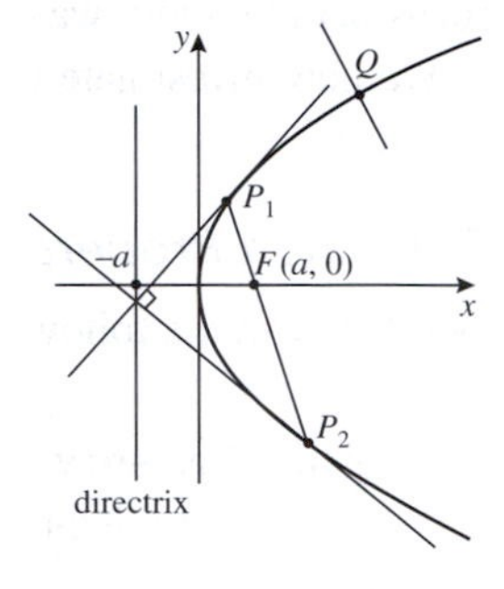

(a) Determine the equation of the chord joining the points P_1 and P_2 on the parabola with parameters t_1 and t_2, respectively, where t_1 and t_2 are unequal and non-zero.

Now assume that the chord P_1P_2 passes through the focus $(a, 0)$ of the parabola.

(b) Prove that $t_1t_2 = -1$.
(c) Use the result of Example 2(a) to write down the equations of the tangents to the parabola at P_1 and P_2, and to prove that these tangents are perpendicular.
(d) Find the point of intersection P of the two tangents in part (c), and verify that it lies on the directrix $x = -a$ of the parabola.
(e) Find the equation of the normal at the point $Q(at^2, 2at)$ to the parabola. Hence prove that if the normal at Q passes through the focus $F(a, 0)$, then Q is the vertex of the parabola.

1.2.2 Reflections

We use the reflection properties of mirrors all the time. For example, we look in plane mirrors while shaving or combing our hair, and we use electric fires with reflecting rear surfaces to throw radiant heat out into a room.

A line is *normal* to a surface in $\mathbb{R}^3$ if it is perpendicular to the tangent plane to the surface at its point of intersection with the surface.

All reflecting surfaces – mirrors, for example – obey the same Reflection Law. The Reflection Law is often expressed in terms of the angles made with the normal to the surface rather than the surface itself. However in this section we shall state and use it in the following form.

The Reflection Law The angle that incoming light makes with the tangent to a surface is the same as the angle that the reflected light makes with the tangent.

Radio waves or radiant heat, etc. obey the same Reflection Law.

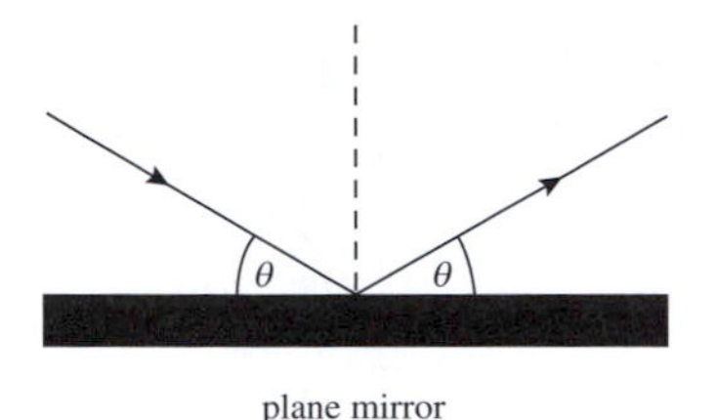

plane mirror

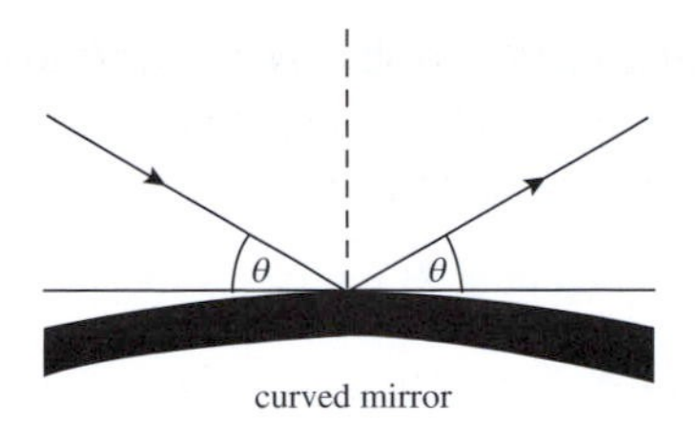

curved mirror

This law applies to all mirrors, no matter whether the reflecting surface is plane or curved. Indeed, in many practical applications the mirror is designed to have a cross-section that is a conic curve – for example, the Lovell radio-telescope at Jodrell Bank in Cheshire, England uses a parabolic reflector to focus parallel radio waves from space onto a receiver.

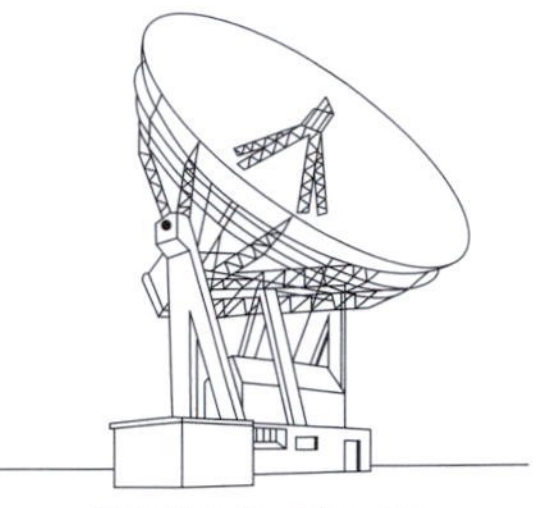
Lovell radio-telescope.

We now investigate the reflection properties of mirrors in the shape of the non-degenerate conics.

Reflection Property of the Ellipse

We start with the following interesting property of the ellipse.

Reflection Property of the Ellipse Light which comes from one focus of an elliptical mirror is reflected at the ellipse to pass through the second focus.

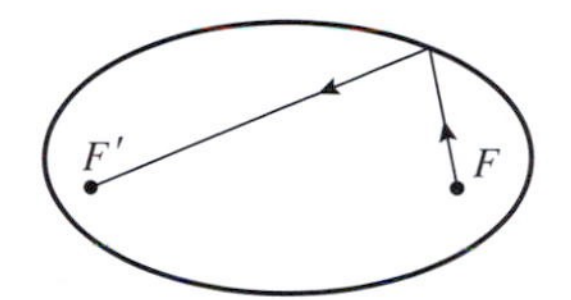

In our proof we use the following trigonometric result for triangles.

Sine Formula In a triangle $\triangle ABC$ with sides *a, b, c* opposite the vertices *A, B, C,* respectively,

$$\frac{a}{\sin \angle BAC} = \frac{b}{\sin \angle ABC} = \frac{c}{\sin \angle ACB}.$$

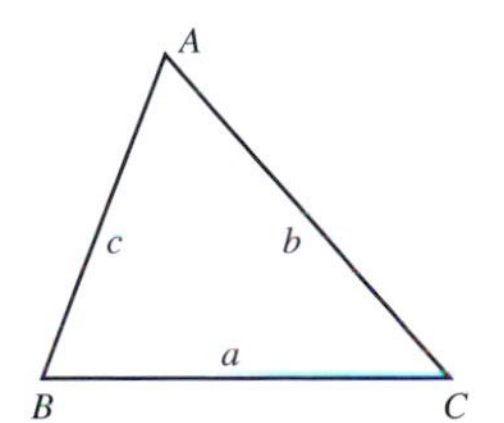

Proof of Reflection Property Let E be the ellipse in standard form, and $P(a\cos t, b\sin t)$ an arbitrary point on E; for simplicity, we shall assume that P lies in the first quadrant.

Then, as we saw earlier,

Subsection 1.1.3

$$\begin{aligned} PF &= e \times (\text{distance from } P \text{ to corresponding directrix } d) \\ &= e \times \left(\frac{a}{e} - a\cos t\right) = a - ae\cos t, \end{aligned}$$

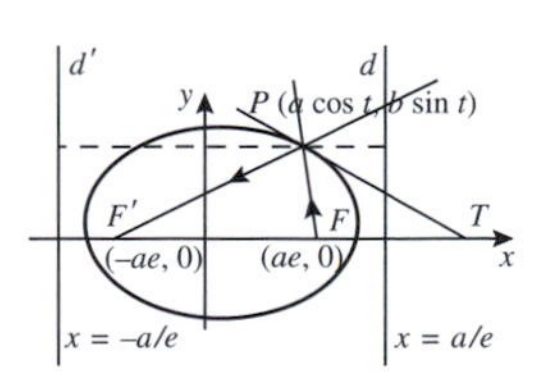

and

$$\begin{aligned} PF' &= e \times (\text{distance from } P \text{ to } d') \\ &= e \times \left(\frac{a}{e} + a\cos t\right) = a + ae\cos t. \end{aligned}$$

Hence,

$$\frac{PF}{PF'} = \frac{a - ae\cos t}{a + ae\cos t} = \frac{1 - e\cos t}{1 + e\cos t},$$

Next, we saw earlier that the equation of the tangent at P to the ellipse is

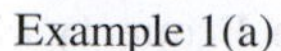

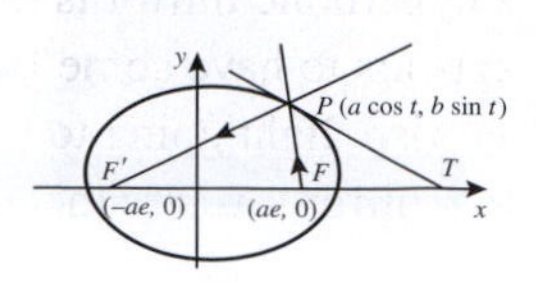

$$\frac{x}{a}\cos t + \frac{y}{b}\sin t = 1;$$

hence at the point T where the tangent at P intersects the x-axis, we have

$$\frac{x}{a}\cos t = 1, \quad \text{or} \quad x = a/\cos t.$$

It follows that

$$\frac{TF}{TF'} = \frac{(a/\cos t) - ae}{(a/\cos t) + ae} = \frac{1 - e\cos t}{1 + e\cos t}.$$

We deduce that

$$\frac{PF}{PF'} = \frac{TF}{TF'}, \quad \text{or} \quad \frac{PF}{TF} = \frac{PF'}{TF'}.$$

By applying the Sine Formula to the triangles $\triangle PFT$ and $\triangle PF'T$, we obtain that

$$\frac{PF}{TF} = \frac{\sin\angle PTF}{\sin\angle TPF} \quad \text{and} \quad \frac{PF'}{TF'} = \frac{\sin\angle PTF'}{\sin\angle TPF'},$$

so that

$$\frac{\sin\angle PTF}{\sin\angle TPF} = \frac{\sin\angle PTF'}{\sin\angle TPF'}.$$

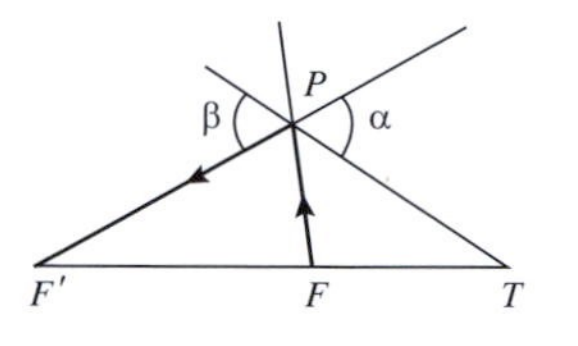

Since $\angle PTF = \angle PTF'$ it follows that $\sin\angle TPF = \sin\angle TPF'$, so that $\angle TPF = \pi - \angle TPF'$ since $\angle TPF \neq \angle TPF'$. Hence $\angle TPF$ equals the angle denoted by the symbol α in the diagram, and this is equal to the angle β (as α and β are vertically opposite).

This completes the proof of the Reflection Property. ■

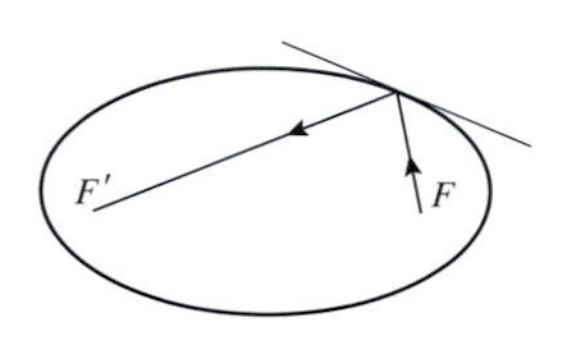

An amusing illustration of the property is as follows. A poor snooker player could appear to be a 'crack shot' if he used a snooker table in the shape of an ellipse: for if he places his snooker ball on the table at one focus and a target ball at the other focus, then *no matter what* direction he hits his ball, he is certain to reach his target!

Reflection Property of the Hyperbola

The hyperbola has a reflection property similar to that of the ellipse, with an appropriate modification.

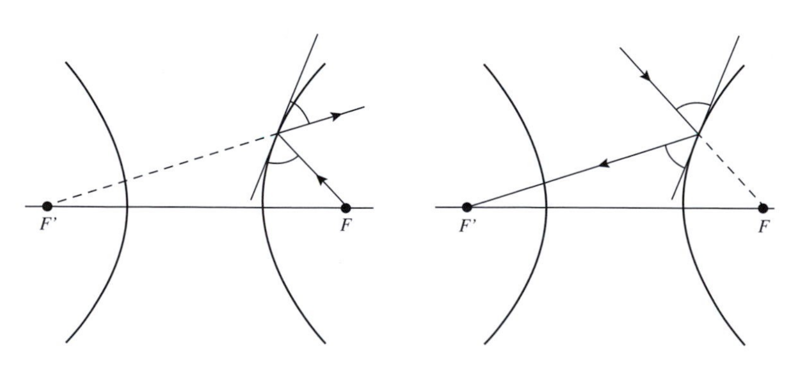

Reflection Property of the Hyperbola Light coming from one focus of a hyperbolic mirror is reflected at the hyperbola in such a way that the light appears to have come from the other focus.

Also, light going towards one focus of a hyperbolic mirror is reflected at the mirror towards the other focus.

This is called the *Internal Reflection Property*.

This is called the *External Reflection Property*.

We omit a proof of this result, as it is similar to the proof of the Reflection Property of the ellipse.

Reflection Property of the Parabola

The Reflection Property of the parabola is also similar to the reflection property of the ellipse.

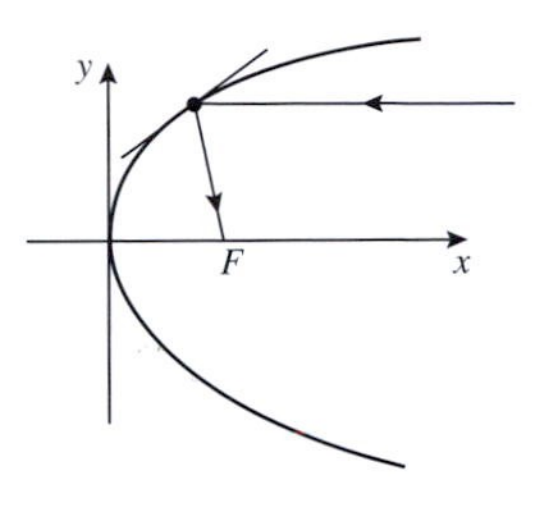

Reflection Property of the Parabola Incoming light parallel to the axis of a parabolic mirror is reflected at the parabola to pass through the focus.

Conversely, light coming from the focus of a parabolic mirror is reflected at the parabola to give a beam of light parallel to the axis of the parabola.

Proof Let E be the parabola in standard form, and let $P(at^2, 2at)$ be an arbitrary point on E.

We have seen that the equation of the tangent at P to the parabola has equation $ty = x + at^2$. If T is the point where this tangent meets the x-axis, then at T we have $y = 0$ and $t \cdot 0 = x + at^2$, so that $x = -at^2$.

Example 2

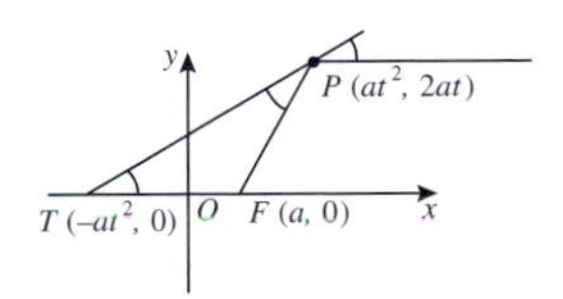

In the triangle $\triangle PTF$ we have

$$TF = TO + OF = at^2 + a$$

and, by the Distance Formula,

$$FP = \sqrt{(a - at^2)^2 + (2at)^2} = \sqrt{a^2 + 2a^2t^2 + a^2t^4} = a + at^2.$$

Then, since $TF = FP$, the triangle $\triangle PTF$ is isosceles, and so $\angle TPF = \angle FTP$.

Now since the horizontal line through P is parallel to the x-axis, the angle between the tangent at P and the horizontal line through P is equal to $\angle FTP$ (as they are corresponding angles), and so also to $\angle TPF$. This completes the proof of the reflection property. ∎

The reflection property of the parabola is also the principle behind the design of searchlights as well as radio-telescopes. The reflector of a searchlight is a parabolic mirror, with the bulb at its focus. Light from the bulb hits the mirror and is reflected outwards as a parallel beam (see p. 32).

The design of optical telescopes sometimes uses the Reflection Properties of other conics too. For example, the 4.2 metre William Herschel telescope at the Roque de los Muchachos Observatory on the island of La Palma in the Canary Islands, has an arrangement of mirrors known as a *Cassegrain focus:*

a primary parabolic mirror reflects light towards a secondary hyperbolic mirror, which reflects it again to a focus behind the primary mirror.

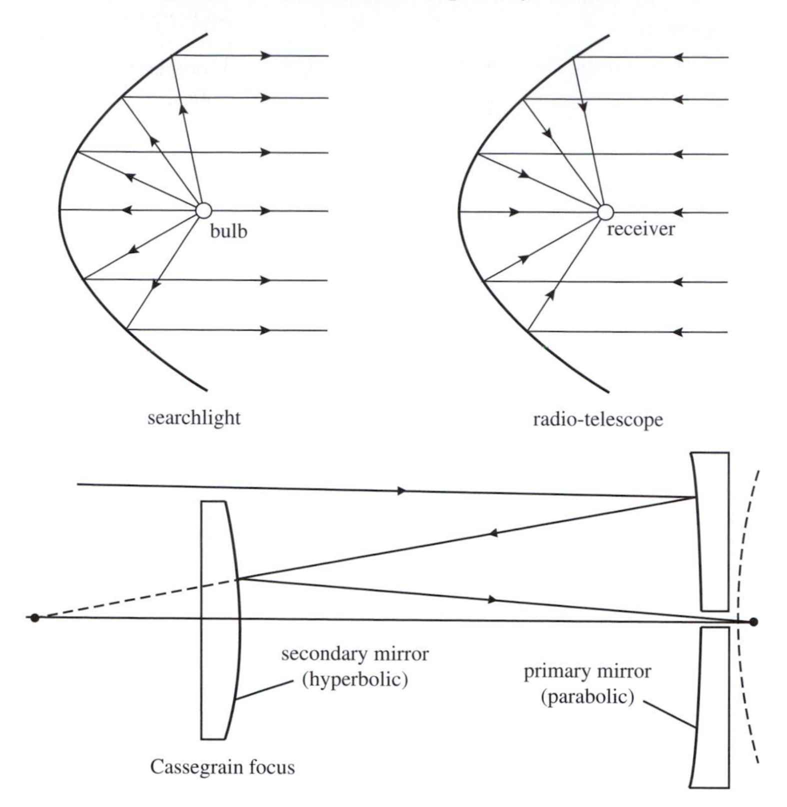

The secondary mirror is used to focus the light to a much more convenient place than the focus of the primary mirror, and to increase the effective focal length of the telescope (and so its resolution).

We can summarize the above three Reflection Properties concisely as follows. *All mirrors in the shape of a non-degenerate conic reflect light coming from or going to one focus towards the other focus.*

In the case of the parabola, we regard the second focus as 'lying at infinity'.

> **Problem 8** Let E and H be an ellipse and a hyperbola, both having the same points F and F' as their foci. Use the reflection properties of the ellipse and hyperbola to prove that at each point of intersection, E and H meet at right angles.

1.2.3 Conics as envelopes of tangent families

We now show how we can construct the non-degenerate conics as the *envelope* of a family of lines that are tangents to the conics. In other words, the conic being constructed is the curve in the plane that has each of the lines in the family as a tangent.

Such methods are often used in exhibitions to display the shapes of the conics in a visually appealing way, using coloured threads or string.

The method depends on the use of a circle associated with each non-degenerate conic, called its *auxiliary circle*. The auxiliary circle of an ellipse or hyperbola is the circle whose diameter is its major axis; analogously we shall define the tangent to a parabola at its vertex to be the auxiliary circle of the parabola.

This definition for the parabola enables us to give succinct statements of properties for all conics.

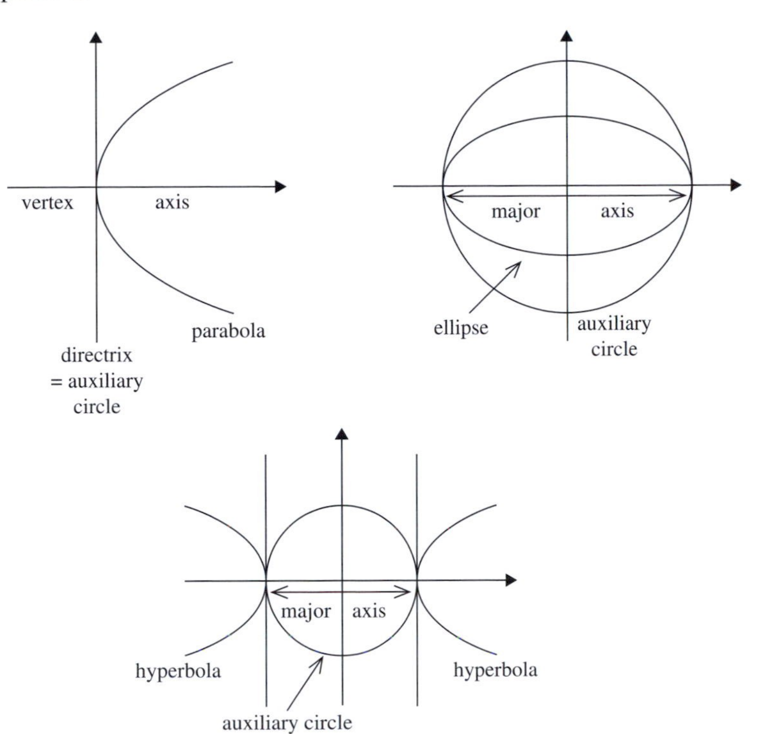

The mathematical tool that we use in our construction is the following result.

Theorem 4 A perpendicular from a focus of a non-degenerate conic to a tangent meets the tangent on the auxiliary circle of the conic.

Here by 'non-degenerate conic' we mean a parabola, a (non-circular) ellipse or a hyperbola.

Proof (for a parabola) Let the point $P(at^2, 2at)$ lie on the parabola in standard form with equation $y^2 = 4ax$, and let the perpendicular from the focus $F(a, 0)$ to the tangent at P meet it at T.

Analytic proofs for an ellipse or hyperbola are similar; however for these conics there are two perpendiculars, both of which meet the tangent on the auxiliary circle.

By Theorem 2 of Subsection 1.2.1, the tangent at P has equation

$$y \cdot 2at = 2a(x + at^2),$$

which we may rewrite in the form

$$y = \frac{1}{t}x + at. \tag{5}$$

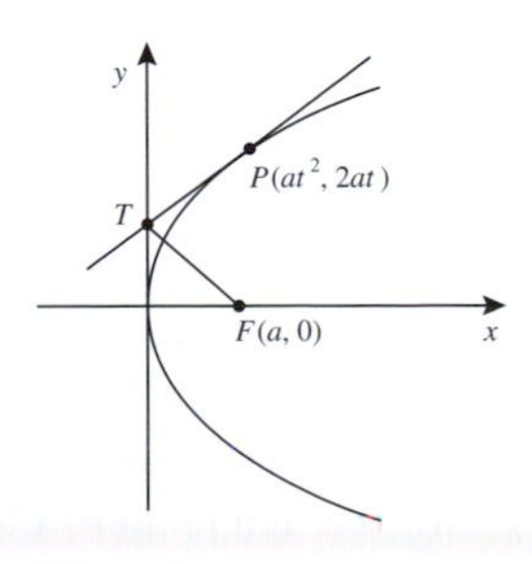

From this we see that the slope of the tangent PT is $1/t$, so that the slope of the perpendicular FT must be $-t$. Since FT also passes through $F(a, 0)$, FT must have equation

$$y + tx = 0 + t \cdot a$$

which we may rewrite in the form

$$y = -tx + at. \tag{6}$$

The equations (5) for PT and (6) for FT clearly have the solution $x = 0$, $y = at$. This means that the point of intersection T of the lines PT and FT has coordinates $(0, at)$. Hence T lies on the directrix of the parabola, as required. ■

Remark

Given a parabola and its axis, we can use Theorem 4 to identify the focus of the parabola. We draw the tangent at any point P on the parabola, and then the perpendicular to the tangent at the point T where the tangent meets the directrix. This perpendicular crosses the parabola's axis at its focus.

The directrix is simply the tangent where the axis cuts the parabola.

Problem 9 Prove Theorem 4 for an ellipse.

To construct the envelopes of the conics, you will need a sheet of paper, a pair of compasses, a set square and a pin.

Parabola

Draw a line d for the directrix of the parabola and a point F (not on d) for its focus. Place a set square so that its right-angled vertex lies at a point of d and one of its adjacent sides passes through F; draw the line ℓ along the other adjacent side of the set square. By Theorem 4, ℓ is a tangent to the parabola with focus F and directrix d.

Repeating the process with the vertex of the set square at different points of d gives a family of lines ℓ that is the envelope of tangents to the parabola, as shown below.

The family of lines forming the envelope is $\{\ell$: vertex of set square $\in d\}$.

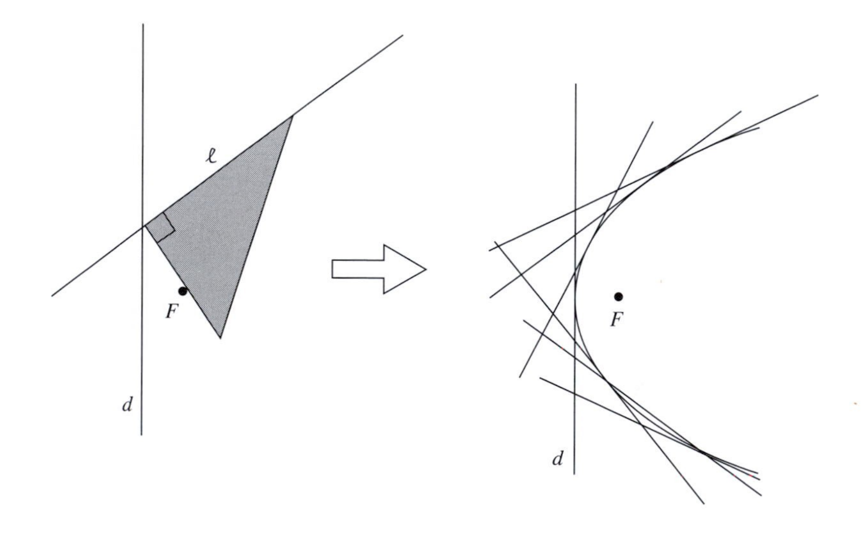

Ellipse

Draw a circle C for the auxiliary circle of the ellipse and a point F inside C (but not at its centre) for a focus. Place a set square so that its right-angled vertex lies at a point of C and one of its adjacent sides passes through F; draw the line ℓ along the other adjacent side of the set square. By Theorem 4, ℓ is a tangent to the ellipse with focus F and auxiliary circle C.

Repeating the process with the vertex of the set square at different points of C gives a family of lines that is the envelope of tangents to the ellipse, as shown below.

The family of lines forming the envelope is $\{\ell$: vertex of set square $\in C\}$.

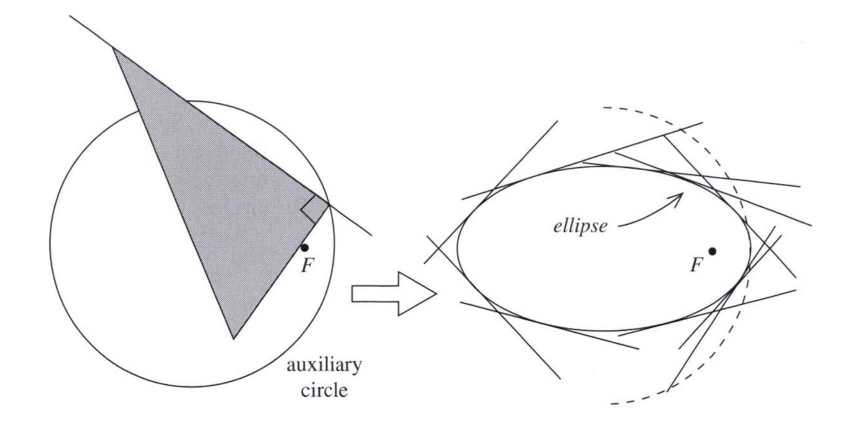

Hyperbola

Draw a circle C for the auxiliary circle of the hyperbola and a point F outside C for a focus. Place a set square so that its right-angled vertex lies at a point of C and one of its adjacent sides passes through F; draw the line ℓ along the other adjacent side of the set square. By Theorem 4, ℓ is a tangent to the hyperbola with focus F and auxiliary circle C.

Repeating the process with the vertex of the set square at different points of C gives a family of lines that is the envelope of tangents to one branch of the hyperbola, as shown below.

The family of lines forming the envelope is $\{\ell$: vertex of set square $\in C\}$.

Repeating the construction with the other focus F' (diametrically opposite F with respect to C) gives the other branch of the hyperbola.

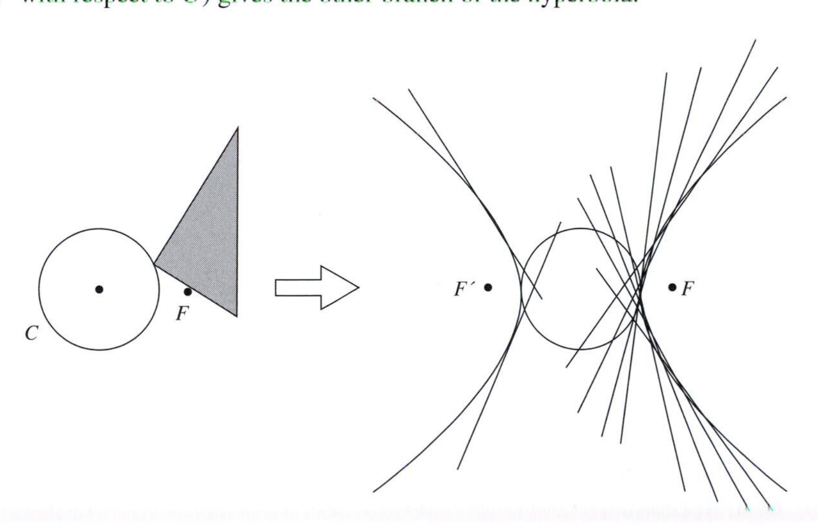

1.3 Recognizing Conics

So far, we have considered the equation of a conic largely when it is in 'standard form'; that is, when the centre of the conic (if it has a centre) is at the origin, and the axes of the conic are parallel to the x- and y-axes. However, most of the conics which arise in calculations are not in standard form; thus we need some way of determining from the equation of a conic which type of conic it describes.

The equations we have met were:

$(x-a)^2 + (y-b)^2 = r^2$,

$y^2 = 4ax$,

$\frac{x^2}{a^2} + \frac{y^2}{b^2} = 1$,

$\frac{x^2}{a^2} - \frac{y^2}{b^2} = 1$.

First we observe that all the equations of all (non-degenerate) conics in standard form can be expressed in the form

$$Ax^2 + Bxy + Cy^2 + Fx + Gy + H = 0, \qquad (1)$$

where not all of A, B and C are zero. For example, the equation of the circle

$$x^2 + y^2 + 4x + 6y - 23 = 0 \qquad (2)$$

is of the form (1), with $A = C = 1$, $B = 0$, $F = 4$, $G = 6$ and $H = -23$.

Now we can obtain any non-degenerate conic from a conic in standard form by a suitable rotation

This rotates the axes through an anticlockwise angle θ to align them with the axes of the conic.

$$(x, y) \mapsto (x\cos\theta - y\sin\theta, x\sin\theta + y\cos\theta)$$

followed by a suitable translation

This moves the centre or vertex of the conic to the origin.

$$(x, y) \mapsto (x - a, y - b).$$

Both of these transformations are linear, so that the equation of the conic at each stage is a second degree equation of the type (1); in other words, any non-degenerate conic has an equation of type (1).

The equations of degenerate conics can also be expressed in the form (1). For example,

$x^2 + y^2 = 0$ represents the single point (0, 0);

$y^2 - 2xy + x^2 = 0$ represents the single line $y = x$, since
$$y^2 - 2xy + x^2 = (y - x)^2;$$

$y^2 - x^2 = 0$ represents the pair of lines $y = \pm x$, since
$$y^2 - x^2 = (y + x)(y - x).$$

However, an equation of the form (1) can also describe the empty set; an example of this is the equation $x^2 + y^2 + 1 = 0$, as there are no points (x, y) in $\mathbb{R}^2$ for which $x^2 + y^2 = -1$. For simplicity in the statement of the theorem below, therefore, we add the *empty set* to our existing list of degenerate conics.

This is an unexpected possibility!

In the above discussion, we proved one part of the following result.

Theorem 1 Any conic has an equation of the form

$$Ax^2 + Bxy + Cy^2 + Fx + Gy + H = 0, \tag{3}$$

where A, B, C, F, G and H are real numbers, and not all of A, B and C are zero. Conversely, any set of points in $\mathbb{R}^2$ whose coordinates (x, y) satisfy equation (3) is a conic.

We omit a proof of the converse part. It would simply be a reworking of the classification methods in the rest of the section.

In this section we investigate the classification of conics in terms of equation (3). In particular, if we are given the equation of a non-degenerate conic in the form (3) how can we determine whether it is a parabola, an ellipse or a hyperbola? And how can we identify its vertex or centre? And its axis, or its major and minor axes? A key tool in this work is the matrix representation of the equation of a conic.

Introducing Matrices

We can express a general second degree equation in x and y

$$Ax^2 + Bxy + Cy^2 + Fx + Gy + H = 0, \tag{4}$$

where A, B and C are not all zero, in terms of matrices as follows.

This will be useful, since we can then use the whole armoury of Linear Algebra to study such equations.

Let $\mathbf{A} = \begin{pmatrix} A & \frac{1}{2}B \\ \frac{1}{2}B & C \end{pmatrix}$, $\mathbf{J} = \begin{pmatrix} F \\ G \end{pmatrix}$ and $\mathbf{x} = \begin{pmatrix} x \\ y \end{pmatrix}$. Then

$$\begin{aligned} \mathbf{x}^T\mathbf{A}\mathbf{x} &= (x \quad y)\begin{pmatrix} A & \frac{1}{2}B \\ \frac{1}{2}B & C \end{pmatrix}\begin{pmatrix} x \\ y \end{pmatrix} \\ &= \left(Ax + \frac{1}{2}By \quad \frac{1}{2}Bx + Cy\right)\begin{pmatrix} x \\ y \end{pmatrix} \\ &= Ax^2 + Bxy + Cy^2 \end{aligned}$$

Here we choose to regard 1×1 matrices and real numbers as equivalent; this will cause no problems.

and

$$\begin{aligned} \mathbf{J}^T\mathbf{x} &= (F \quad G)\begin{pmatrix} x \\ y \end{pmatrix} \\ &= Fx + Gy. \end{aligned}$$

We may therefore write the equation (4) in the form

$$\mathbf{x}^T\mathbf{A}\mathbf{x} + \mathbf{J}^T\mathbf{x} + H = 0. \tag{5}$$

This is called the *matrix form* of the equation (4).

For example, let E be the conic with equation

$$3x^2 - 10xy + 3y^2 + 14x - 2y + 3 = 0.$$

The equation of E is of the form (4) with $A = 3$, $B = -10$, $C = 3$, $F = 14$, $G = -2$ and $H = 3$. It follows from the above discussion that we can express the equation of E in matrix form as $\mathbf{x}^\mathbf{T}\mathbf{A}\mathbf{x} + \mathbf{J}^\mathbf{T}\mathbf{x} + H = 0$, where

$$\mathbf{A} = \begin{pmatrix} 3 & -5 \\ -5 & 3 \end{pmatrix}, \quad \mathbf{J} = \begin{pmatrix} 14 \\ -2 \end{pmatrix}, \quad H = 3 \quad \text{and} \quad \mathbf{x} = \begin{pmatrix} x \\ y \end{pmatrix}.$$

Problem 1 Write the equation of each of the following conics in matrix form.

(a) $11x^2 + 4xy + 14y^2 - 4x - 28y - 16 = 0$
(b) $x^2 - 4xy + 4y^2 - 6x - 8y + 5 = 0$

A key tool in our use of matrices will be the following result.

Theorem 2 A 2×2 matrix $\mathbf{P}$ represents a rotation of $\mathbb{R}^2$ about the origin if and only if it satisfies the following two conditions:

(a) $\mathbf{P}$ is orthogonal;
(b) $\det \mathbf{P} = 1$.

If $\mathbf{P}$ is orthogonal, then $\det \mathbf{P} = \pm 1$; when $\det \mathbf{P} = -1$, $\mathbf{P}$ represents reflection in the x-axis followed by a rotation.

Proof A matrix $\mathbf{P}$ represents a rotation about the origin (anticlockwise through an angle θ) if and only it is of the form

$$\begin{pmatrix} \cos\theta & -\sin\theta \\ \sin\theta & \cos\theta \end{pmatrix}. \tag{6}$$

It is easy to verify that $\mathbf{P}$ satisfies conditions (a) and (b).

Next, let $\mathbf{P} = \begin{pmatrix} a & b \\ c & d \end{pmatrix}$ be a matrix that satisfies conditions (a) and (b).

Then, since $\mathbf{P}$ is orthogonal, the vector $\begin{pmatrix} a \\ c \end{pmatrix}$ has length 1; that is, $a^2 + c^2 = 1$. Thus there is a number θ for which

$$a = \cos\theta \quad \text{and} \quad c = \sin\theta.$$

Also, since $\mathbf{P}$ is orthogonal, the vectors $\begin{pmatrix} a \\ c \end{pmatrix} = \begin{pmatrix} \cos\theta \\ \sin\theta \end{pmatrix}$ and $\begin{pmatrix} b \\ d \end{pmatrix}$ are orthogonal; that is, $(\cos\theta \quad \sin\theta)\begin{pmatrix} b \\ d \end{pmatrix} = 0$ or

$$\cos\theta \cdot b + \sin\theta \cdot d = 0.$$

So there exists some number λ, say, such that

$$b = -\lambda \sin\theta \quad \text{and} \quad d = \lambda \cos\theta.$$

Then since $\det \mathbf{P} = 1$, we have

$$1 = ad - bc = \lambda \cos^2\theta + \lambda \sin^2\theta,$$

so that $\lambda = 1$. It follows that $\mathbf{P}$ must be of the form (6), and so represent a rotation of $\mathbb{R}^2$ about the origin. ■

Using Matrices

We now use the methods of Linear Algebra to recognize conics specified by their equations.

Example 1 Prove that the conic E with equation

$$3x^2 - 10xy + 3y^2 + 14x - 2y + 3 = 0$$

is a hyperbola. Determine its centre, and its major and minor axes.

Solution We saw above that the equation of E can be written in matrix form as $\mathbf{x}^T\mathbf{A}\mathbf{x} + \mathbf{J}^T\mathbf{x} + H = 0$, where

$$\mathbf{A} = \begin{pmatrix} 3 & -5 \\ -5 & 3 \end{pmatrix}, \quad \mathbf{J} = \begin{pmatrix} 14 \\ -2 \end{pmatrix}, \quad H = 3 \text{ and } \quad \mathbf{x} = \begin{pmatrix} x \\ y \end{pmatrix};$$

that is, as

$$(x \quad y)\begin{pmatrix} 3 & -5 \\ -5 & 3 \end{pmatrix}\begin{pmatrix} x \\ y \end{pmatrix} + (14 \quad -2)\begin{pmatrix} x \\ y \end{pmatrix} + 3 = 0.$$

We start by diagonalizing the matrix $\mathbf{A}$. Its characteristic equation is

$$\begin{aligned} 0 = \det(\mathbf{A} - \lambda\mathbf{I}) &= \begin{vmatrix} 3-\lambda & -5 \\ -5 & 3-\lambda \end{vmatrix} \\ &= \lambda^2 - 6\lambda - 16 \\ &= (\lambda - 8)(\lambda + 2), \end{aligned}$$

so that the eigenvalues of $\mathbf{A}$ are $\lambda = 8$ and $\lambda = -2$. The eigenvector equations of $\mathbf{A}$ are

$$\begin{aligned} (3-\lambda)x - 5y &= 0, \\ -5x + (3-\lambda)y &= 0. \end{aligned}$$

When $\lambda = 8$, these equations both become

$$-5x - 5y = 0,$$

so that we may take as a corresponding eigenvector $\begin{pmatrix} 1 \\ -1 \end{pmatrix}$, which we normalize to have unit length as $\begin{pmatrix} 1/\sqrt{2} \\ -1/\sqrt{2} \end{pmatrix}$.

When $\lambda = -2$, the eigenvector equations of $\mathbf{A}$ become

$$\begin{aligned} 5x - 5y &= 0, \\ -5x + 5y &= 0, \end{aligned}$$

so that we may take as a corresponding eigenvector $\begin{pmatrix} 1 \\ 1 \end{pmatrix}$, which we normalize to have unit length as $\begin{pmatrix} 1/\sqrt{2} \\ 1/\sqrt{2} \end{pmatrix}$.

Now

$$\begin{vmatrix} 1/\sqrt{2} & 1/\sqrt{2} \\ -1/\sqrt{2} & 1/\sqrt{2} \end{vmatrix} = \frac{1}{2} + \frac{1}{2} = 1,$$

so we take as our rotation of the plane the transformation $\mathbf{x} = \mathbf{P}\mathbf{x}'$ where $\mathbf{P} = \begin{pmatrix} 1/\sqrt{2} & 1/\sqrt{2} \\ -1/\sqrt{2} & 1/\sqrt{2} \end{pmatrix}$. The transformation $\mathbf{x} = \mathbf{P}\mathbf{x}$ changes the equation of the conic to the form

Note that we need to check the order in which the normalised eigenvectors appear in $\mathbf{P}$ to ensure that $\mathbf{P}$ represents a rotation.

$$(\mathbf{P}\mathbf{x}')^T \mathbf{A}(\mathbf{P}\mathbf{x}') + \mathbf{J}^T(\mathbf{P}\mathbf{x}') + H = 0$$

or

$$(\mathbf{x}')^T(\mathbf{P}^T\mathbf{A}\mathbf{P})\mathbf{x}' + (\mathbf{J}^T\mathbf{P})\mathbf{x}' + H = 0.$$

Since $\mathbf{P}$ is orthogonal and $\det \mathbf{P} = 1$, $\mathbf{P}$ represents a rotation – in fact, an anticlockwise rotation through $-\pi/4$; that is, a clockwise rotation through $\pi/4$.

Since $\mathbf{P}^T\mathbf{A}\mathbf{P} = \begin{bmatrix} 8 & 0 \\ 0 & -2 \end{bmatrix}$, the equation of the conic is now

$$\begin{pmatrix} x' & y' \end{pmatrix} \begin{pmatrix} 8 & 0 \\ 0 & -2 \end{pmatrix} \begin{pmatrix} x' \\ y' \end{pmatrix} + \begin{pmatrix} 14 & -2 \end{pmatrix} \begin{pmatrix} 1/\sqrt{2} & 1/\sqrt{2} \\ -1/\sqrt{2} & 1/\sqrt{2} \end{pmatrix} \begin{pmatrix} x' \\ y' \end{pmatrix} + 3 = 0,$$

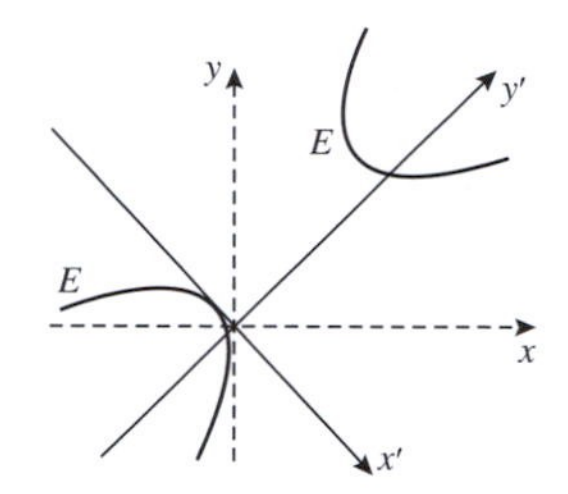

which we can rewrite in the form

$$8x'^2 - 2y'^2 + 8\sqrt{2}x' + 6\sqrt{2}y' + 3 = 0.$$

We may rewrite this equation in the form

$$8\left(x'^2 + \sqrt{2}x'\right) - 2\left(y'^2 - 3\sqrt{2}y'\right) + 3 = 0$$

so that, on completing the square, we have

$$8\left(x' + 1/\sqrt{2}\right)^2 - 4 - 2\left(y' - 3/\sqrt{2}\right)^2 + 9 + 3 = 0,$$

For

$$x'^2 + \sqrt{2}x' = \left(x' + \tfrac{1}{\sqrt{2}}\right)^2 - \tfrac{1}{2}$$

and

$$y'^2 - 3\sqrt{2}y' = \left(y' - \tfrac{3}{\sqrt{2}}\right)^2 - \tfrac{9}{2}.$$

which we can rewrite in the form

$$8\left(x' + 1/\sqrt{2}\right)^2 - 2\left(y' - 3/\sqrt{2}\right)^2 = -8,$$

or

$$\frac{\left(y' - 3/\sqrt{2}\right)^2}{4} - \frac{\left(x' + 1/\sqrt{2}\right)^2}{1} = 1. \qquad (7)$$

This is the equation of a hyperbola.

We can write this equation in (nearly) standard form as

$$\frac{(y'')^2}{4} - \frac{(x'')^2}{1} = 1,$$

where $x'' = x' + \frac{1}{\sqrt{2}}$ and $y'' = y' - \frac{3}{\sqrt{2}}$.

From equation (7) it follows that the centre of the hyperbola E is the point where $x' = -1/\sqrt{2}$ and $y' = 3/\sqrt{2}$. From the equation $\mathbf{x} = \mathbf{P}\mathbf{x}'$, it follows that in terms of the original coordinate system this is the point

$$\begin{pmatrix} x \\ y \end{pmatrix} = \begin{pmatrix} 1/\sqrt{2} & 1/\sqrt{2} \\ -1/\sqrt{2} & 1/\sqrt{2} \end{pmatrix} \begin{pmatrix} -1/\sqrt{2} \\ 3/\sqrt{2} \end{pmatrix} = \begin{pmatrix} 1 \\ 2 \end{pmatrix},$$

that is, the point (1, 2).

It also follows from equation (7) that the major axis of E has equation $x' + 1/\sqrt{2} = 0$, or $x' = -1/\sqrt{2}$; and the minor axis of E has equation $y' - 3/\sqrt{2} = 0$, or $y' = 3/\sqrt{2}$.

Recall that the equation $\frac{x^2}{a^2} - \frac{y^2}{b^2} = 1$ represents a hyperbola with major axis $y = 0$ and minor axis $x = 0$.

Finally, since the matrix $\mathbf{P}$ is orthogonal we can rewrite the equation $\mathbf{x} = \mathbf{P}\mathbf{x}'$ in the form $\mathbf{x}' = \mathbf{P}^{-1}\mathbf{x} = \mathbf{P}^T\mathbf{x}$, so that

$$\begin{pmatrix} x' \\ y' \end{pmatrix} = \begin{pmatrix} \frac{1}{\sqrt{2}} & \frac{-1}{\sqrt{2}} \\ \frac{1}{\sqrt{2}} & \frac{1}{\sqrt{2}} \end{pmatrix} \begin{pmatrix} x \\ y \end{pmatrix}$$

or as a pair of equations

$$x' = \frac{1}{\sqrt{2}}x - \frac{1}{\sqrt{2}}y,$$
$$y' = \frac{1}{\sqrt{2}}x + \frac{1}{\sqrt{2}}y.$$

It follows that the equation, $x' = -1/\sqrt{2}$, of the major axis of the hyperbola E can be expressed in terms of the original coordinate system as

Recall that $x'' = x' + \frac{1}{\sqrt{2}}$ and $y'' = y' - \frac{3}{\sqrt{2}}$.

$$\frac{1}{\sqrt{2}}x - \frac{1}{\sqrt{2}}y = -\frac{1}{\sqrt{2}}, \quad \text{or} \quad x - y = -1.$$

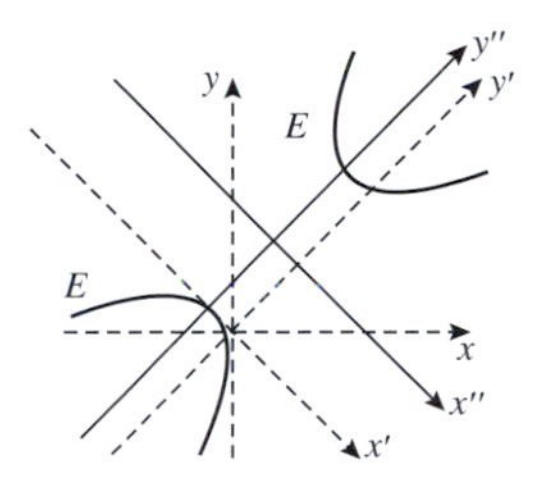

Similarly, the equation, $y' = 3/\sqrt{2}$, of the minor axis of the hyperbola can be expressed in terms of the original coordinate system as

$$\frac{1}{\sqrt{2}}x + \frac{1}{\sqrt{2}}y = \frac{3}{\sqrt{2}}, \quad \text{or} \quad x + y = 3. \qquad \square$$

The above problem illustrates a general strategy for identifying conics from their second degree equations.

Strategy To classify a conic E with equation

$$Ax^2 + Bxy + Cy^2 + Fx + Gy + H = 0:$$

1. Write the equation of E in matrix form $\mathbf{x}^T\mathbf{A}\mathbf{x} + \mathbf{J}^T\mathbf{x} + H = 0$.
2. Determine an orthogonal matrix $\mathbf{P}$, with determinant 1, that diagonalizes $\mathbf{A}$.
3. Make the change of coordinate system $\mathbf{x} = \mathbf{P}\mathbf{x}'$. The equation of E then becomes of the form

$$\lambda_1 x'^2 + \lambda_2 y'^2 + f x' + g y' + h = 0,$$

where λ_1 and λ_2 are the eigenvalues of $\mathbf{A}$.
4. 'Complete the squares', if necessary, to rewrite the equation of E in terms of an (x'', y'')-coordinate system as the equation of a conic in standard form.
5. Use the equation $\mathbf{x}' = \mathbf{P}^T\mathbf{x}$ to determine the centre and axes of E in terms of the original coordinate system.

Reorder the columns of $\mathbf{P}$ if necessary to ensure that $\det \mathbf{P} = 1$ rather than -1.

This is a rotation of $\mathbb{R}^2$.

Here λ_1 corresponds to the first column in $\mathbf{P}$, and λ_2 to the second column in $\mathbf{P}$.

This is a translation of $\mathbb{R}^2$.

Problem 2 Classify the conics in $\mathbb{R}^2$ with the following equations. Determine the centre of those that have a centre.

(a) $11x^2 + 4xy + 14y^2 - 4x - 28y - 16 = 0$
(b) $x^2 - 4xy + 4y^2 - 6x - 8y + 5 = 0$

In fact, using the above strategy we can prove the following result.

Theorem 3 A non-degenerate conic with equation

$$Ax^2 + Bxy + Cy^2 + Fx + Gy + H = 0$$

and matrix $\mathbf{A} = \begin{pmatrix} A & \frac{1}{2}B \\ \frac{1}{2}B & C \end{pmatrix}$ can be classified as follows:

(a) If $\det \mathbf{A} < 0$, E is a hyperbola.
(b) If $\det \mathbf{A} = 0$, E is a parabola.
(c) If $\det \mathbf{A} > 0$, E is an ellipse.

We omit a proof of this result.

Since $\det \mathbf{A} = AC - \frac{1}{4}B^2 = -\frac{1}{4}(B^2 - 4AC)$. Theorem 3 is often referred to as 'the $B^2 - 4AC$ test' for conics.

Note that $\det \mathbf{A}$ and $B^2 - 4AC$ have *opposite* signs.

Problem 3 Use Theorem 3 to classify the non-degenerate conics in $\mathbb{R}^2$ with the following equations.

(a) $3x^2 - 8xy + 2y^2 - 2x + 4y - 16 = 0$
(b) $x^2 + 8xy + 16y^2 - x + 8y - 12 = 0$
(c) $52x^2 - 72xy + 73y^2 - 32x - 74y + 28 = 0$

You may assume that these conics are non-degenerate.

1.4 Quadric Surfaces

1.4.1 Quadric Surfaces in $\mathbb{R}^3$

Quadric surfaces (or *quadrics*) are surfaces in $\mathbb{R}^3$ that are the natural analogues of those curves in $\mathbb{R}^2$ that we call conics.

Definition A **quadric surface** in $\mathbb{R}^3$ is a set given by an equation of the form

$$Ax^2 + By^2 + Cz^2 + Fxy + Gyz + Hxz + Jx + Ky + Lz + M = 0, \quad (1)$$

where $A, B, C, F, G, H, J, K, L$ and M are real numbers, and not all of A, B, C, F, G and H are zero.

We use the term *degenerate quadrics* to describe those quadrics that are the empty set, a single point, a single line, a single plane, a pair of planes and a cylinder.

By a *cylinder* we mean a surface that consists of an ellipse, parabola or hyperbola in some plane π, together with all the lines in $\mathbb{R}^3$ through that conic that are normal to π.

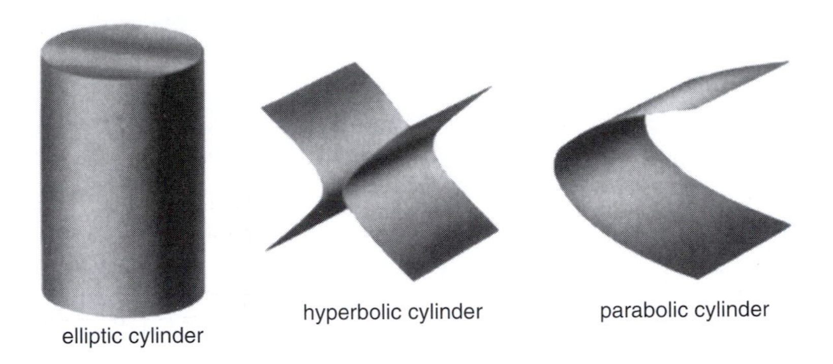

This leaves six different types of quadric surface, the so-called *non-degenerate quadric surfaces*. We illustrate these below, with a typical equation for each. In each case, we state also the curve of intersection of a plane parallel to a coordinate plane that meets the surface in a non-trivial intersection.

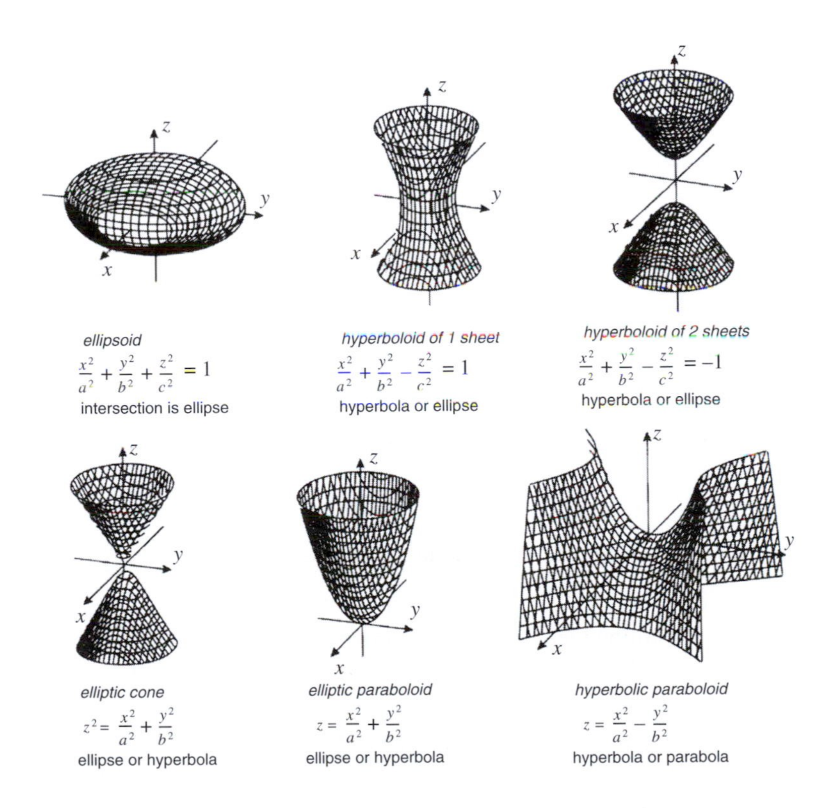

As well as being attractive visually, quadrics arise naturally in various areas of applied mathematics.

1.4.2 Recognizing Quadric Surfaces

We use the notion of orthogonal diagonalization of 3×3 matrices to classify non-degenerate quadrics, just as we used 2×2 matrices to classify conics.

Introducing Matrices

Consider a quadric surface with equation

$$Ax^2 + By^2 + Cz^2 + Fxy + Gyz + Hxz + Jx + Ky + Lz + M = 0,$$

and let

$$\mathbf{A} = \begin{pmatrix} A & \frac{1}{2}F & \frac{1}{2}H \\ \frac{1}{2}F & B & \frac{1}{2}G \\ \frac{1}{2}H & \frac{1}{2}G & C \end{pmatrix}, \quad \mathbf{J} = \begin{pmatrix} J \\ K \\ L \end{pmatrix} \quad \text{and} \quad \mathbf{x} = \begin{pmatrix} x \\ y \\ z \end{pmatrix}.$$

Then a calculation similar to that for conics shows that

Section 1.3

$$\mathbf{x}^T\mathbf{A}\mathbf{x} = Ax^2 + By^2 + Cz^2 + Fxy + Gyz + Hxz$$

and

$$\mathbf{J}^T\mathbf{x} = Jx + Ky + Lz.$$

We may therefore write the equation of the quadric surface in the form

$$\mathbf{x}^T\mathbf{A}\mathbf{x} + \mathbf{J}^T\mathbf{x} + M = 0. \qquad (2)$$

This is called the *matrix form* of the equation of the quadric surface.

For example, the equation of the quadric surface given by

$$5x^2 + 3y^2 + 3z^2 - 2xy + 2yz - 2xz - 10x + 6y - 2z - 10 = 0$$

may be written in matrix form $\mathbf{x}^T\mathbf{A}\mathbf{x} + \mathbf{J}^T\mathbf{x} + M = 0$ where

$$\mathbf{A} = \begin{pmatrix} 5 & -1 & -1 \\ -1 & 3 & 1 \\ -1 & 1 & 3 \end{pmatrix}, \quad \mathbf{J} = \begin{pmatrix} -10 \\ 6 \\ -2 \end{pmatrix}, \quad \mathbf{x} = \begin{pmatrix} x \\ y \\ z \end{pmatrix} \quad \text{and}$$

$$M = -10.$$

Problem 1 Write the equation of the following quadrics in matrix form.

$$2x^2 + 5y^2 - z^2 + xy - 3yz - 2xz - 2x - 6y + 10z - 12 = 0$$

$$y - yz = xz$$

A key tool for classifying quadrics is the following result about matrices.

We omit a proof, as it is similar to that of Theorem 2 in Section 1.3.

Theorem 1 A 3×3 matrix $\mathbf{P}$ represents a rotation of $\mathbb{R}^3$ about the origin if and only if it satisfies the following two conditions:

(a) $\mathbf{P}$ is orthogonal;
(b) $\det \mathbf{P} = 1$.

If $\mathbf{P}$ is orthogonal and $\det \mathbf{P} = -1$, then $\mathbf{P}$ represents a rotation about the origin composed with a reflection in a plane through the origin.

Using Matrices

Our approach to classifying quadrics in $\mathbb{R}^3$ using matrices is broadly similar to that for classifying conics in $\mathbb{R}^2$.

Strategy To classify a quadric E with equation

$$Ax^2 + By^2 + Cz^2 + Fxy + Gyz + Hxz + Jx + Ky + Lz + M = 0:$$

1. Write the equation of E in matrix form $\mathbf{x}^T\mathbf{A}\mathbf{x} + \mathbf{J}^T\mathbf{x} + M = 0$.
2. Determine an orthogonal matrix $\mathbf{P}$, with determinant 1, that diagonalizes $\mathbf{A}$.
3. Make the change of coordinate system $\mathbf{x} = \mathbf{P}\mathbf{x}'$. The equation of E then becomes of the form

$$\lambda_1 x'^2 + \lambda_2 y'^2 + \lambda_3 z'^2 + jx' + ky' + lz' + m = 0,$$

where λ_1, λ_2 and λ_3 are the eigenvalues of $\mathbf{A}$.

4. 'Complete the squares', if necessary, to rewrite the equation of E in terms of an (x'', y'', z'')-coordinate system as the equation of a quadric in standard form.
5. Use the equation $\mathbf{x}' = \mathbf{P}^T\mathbf{x}$ to determine the centre and planes of symmetry of E in terms of the original coordinate system.

Reorder the columns of $\mathbf{P}$ if necessary to ensure that $\det \mathbf{P} = 1$ rather than -1.

This is a rotation of $\mathbb{R}^3$.

Here λ_i corresponds to the ith column in $\mathbf{P}$.

This is a translation of $\mathbb{R}^3$.

Example 1 Prove that the quadric E with equation

$$5x^2 + 3y^2 + 3z^2 - 2xy + 2yz - 2xz - 10x + 6y - 2z - 10 = 0$$

is an ellipsoid. Determine its centre.

Solution We saw above that the equation of E can be written in matrix form as $\mathbf{x}^T\mathbf{A}\mathbf{x} + \mathbf{J}^T\mathbf{x} + M = 0$, where

Just before Problem 1

$$\mathbf{A} = \begin{pmatrix} 5 & -1 & -1 \\ -1 & 3 & 1 \\ -1 & 1 & 3 \end{pmatrix}, \quad \mathbf{J} = \begin{pmatrix} -10 \\ 6 \\ -2 \end{pmatrix}, \quad \mathbf{x} = \begin{pmatrix} x \\ y \\ z \end{pmatrix} \quad \text{and}$$

$$M = -10;$$

that is , as

$$(x \quad y \quad z)\begin{pmatrix} 5 & -1 & -1 \\ -1 & 3 & 1 \\ -1 & 1 & 3 \end{pmatrix}\begin{pmatrix} x \\ y \\ z \end{pmatrix} + (-10 \quad 6 \quad -2)\begin{pmatrix} x \\ y \\ z \end{pmatrix} - 10 = 0.$$

We start by diagonalizing the matrix $\mathbf{A}$. Its characteristic equation is

$$\begin{aligned} 0 = \det(\mathbf{A} - \lambda\mathbf{I}) &= \begin{vmatrix} 5-\lambda & -1 & -1 \\ -1 & 3-\lambda & 1 \\ -1 & 1 & 3-\lambda \end{vmatrix} \\ &= -\lambda^3 + 11\lambda^2 - 36\lambda + 36 \\ &= -(\lambda - 2)(\lambda - 3)(\lambda - 6), \end{aligned}$$

so that the eigenvalues of $\mathbf{A}$ are $\lambda = 2$, 3 and 6. The eigenvector equations of $\mathbf{A}$ are

$$\begin{aligned}(5-\lambda)x - y - z &= 0,\\ -x + (3-\lambda)y + z &= 0,\\ -x + y + (3-\lambda)z &= 0.\end{aligned}$$

When $\lambda = 2$, these equations become

$$\begin{aligned}3x - y - z &= 0,\\ -x + y + z &= 0,\\ -x + y + z &= 0.\end{aligned}$$

Adding the first two equations we get $x = 0$; it then follows from all the equations that $y + z = 0$. So we may take as a corresponding eigenvector $\begin{pmatrix}0\\1\\-1\end{pmatrix}$, which we normalize to have unit length as $\begin{pmatrix}0\\1/\sqrt{2}\\-1/\sqrt{2}\end{pmatrix}$.

Similarly, when $\lambda = 3$, we may take as a corresponding eigenvector $\begin{pmatrix}1\\1\\1\end{pmatrix}$, which we normalize to have unit length as $\begin{pmatrix}1/\sqrt{3}\\1/\sqrt{3}\\1/\sqrt{3}\end{pmatrix}$; and when $\lambda = 6$, we may take as a corresponding eigenvector $\begin{pmatrix}2\\-1\\-1\end{pmatrix}$, which we normalize to have unit length as $\begin{pmatrix}2/\sqrt{6}\\-1/\sqrt{6}\\-1/\sqrt{6}\end{pmatrix}$.

We omit the details.

Now

$$\begin{vmatrix}0 & \frac{1}{\sqrt{3}} & \frac{2}{\sqrt{6}}\\ \frac{1}{\sqrt{2}} & \frac{1}{\sqrt{3}} & \frac{-1}{\sqrt{6}}\\ \frac{-1}{\sqrt{2}} & \frac{1}{\sqrt{3}} & \frac{-1}{\sqrt{6}}\end{vmatrix} = \begin{vmatrix}0 & \frac{1}{\sqrt{3}} & \frac{2}{\sqrt{6}}\\ \frac{1}{\sqrt{2}} & \frac{1}{\sqrt{3}} & \frac{-1}{\sqrt{6}}\\ 0 & \frac{2}{\sqrt{3}} & \frac{-2}{\sqrt{6}}\end{vmatrix}$$

Adding row 2 to row 3

$$= -\frac{1}{\sqrt{2}}\begin{vmatrix}\frac{1}{\sqrt{3}} & \frac{2}{\sqrt{6}}\\ \frac{2}{\sqrt{3}} & \frac{-2}{\sqrt{6}}\end{vmatrix}$$

Expanding in terms of the first column of the determinant

$$= -\frac{1}{\sqrt{2}}\left(-\frac{6}{\sqrt{18}}\right) = +1,$$

so we take as a convenient rotation of $\mathbb{R}^3$ the transformation $\mathbf{x} = \mathbf{P}\mathbf{x}'$, where $\mathbf{P} = \begin{pmatrix}0 & 1/\sqrt{3} & 2/\sqrt{6}\\ 1/\sqrt{2} & 1/\sqrt{3} & -1/\sqrt{6}\\ -1/\sqrt{2} & 1/\sqrt{3} & -1/\sqrt{6}\end{pmatrix}$. This transformation changes the equation of the quadric to the form

$\mathbf{P}$ represents a rotation of $\mathbb{R}^3$ since it is orthogonal and $\det \mathbf{P} = 1$.

$$(\mathbf{P}\mathbf{x}')^T\mathbf{A}(\mathbf{P}\mathbf{x}') + \mathbf{J}^T(\mathbf{P}\mathbf{x}') + M = 0$$

or

$$(\mathbf{x}')^T(\mathbf{P}^T\mathbf{A}\mathbf{P})\mathbf{x}' + (\mathbf{J}^T\mathbf{P})\mathbf{x}' + M = 0.$$

Since $\mathbf{P}^T\mathbf{A}\mathbf{P} = \begin{pmatrix} 2 & 0 & 0 \\ 0 & 3 & 0 \\ 0 & 0 & 6 \end{pmatrix}$, this is the equation

$$\begin{pmatrix} x' & y' & z' \end{pmatrix} \begin{pmatrix} 2 & 0 & 0 \\ 0 & 3 & 0 \\ 0 & 0 & 6 \end{pmatrix} \begin{pmatrix} x' \\ y' \\ z' \end{pmatrix}$$

$$+ \begin{pmatrix} -10 & 6 & -2 \end{pmatrix} \begin{pmatrix} 0 & \frac{1}{\sqrt{3}} & \frac{2}{\sqrt{6}} \\ \frac{1}{\sqrt{2}} & \frac{1}{\sqrt{3}} & \frac{-1}{\sqrt{6}} \\ \frac{-1}{\sqrt{2}} & \frac{1}{\sqrt{3}} & \frac{-1}{\sqrt{6}} \end{pmatrix} \begin{pmatrix} x' \\ y' \\ z' \end{pmatrix} - 10 = 0,$$

which we can rewrite in the form

$$2x'^2 + 3y'^2 + 6z'^2 + 4\sqrt{2}x' - 2\sqrt{3}y' - 4\sqrt{6}z' - 10 = 0.$$

'Completing the square' in this equation, we get

$$2\left(x'^2 + 2\sqrt{2}x'\right) + 3\left(y'^2 - \tfrac{2}{\sqrt{3}}y'\right) + 6\left(z'^2 - \tfrac{4}{\sqrt{6}}z'\right) - 10 = 0$$

so that

$$2\left(x' + \sqrt{2}\right)^2 - 4 + 3\left(y' - \tfrac{1}{\sqrt{3}}\right)^2 - 1 + 6\left(z' - \tfrac{2}{\sqrt{6}}\right)^2 - 4 - 10 = 0.$$

We now make the transformation

This is a translation of $\mathbb{R}^3$.

$$\mathbf{x}'' = \mathbf{x}' + \begin{pmatrix} \sqrt{2} \\ \frac{-1}{\sqrt{3}} \\ \frac{-2}{\sqrt{6}} \end{pmatrix}, \qquad (3)$$

so that we can rewrite the equation of E in the form

$$2(x'')^2 + 3(y'')^2 + 6(z'')^2 = 19,$$

or

$$\frac{(x'')^2}{19/2} + \frac{(y'')^2}{19/3} + \frac{(z'')^2}{19/6} = 1. \qquad (4)$$

In the general form of the equation of an ellipsoid in Subsection 1.4.1, E has $a^2 = 19/2$, $b^2 = 19/3$ and $c^2 = 19/6$.

It follows from equation (4) that E must be an ellipsoid.

From equations (3) and (4) it follows that the centre of the ellipsoid E is the point where $x' = -\sqrt{2}$, $y' = \frac{1}{\sqrt{3}}$ and $z' = \frac{2}{\sqrt{6}}$. From the equation $\mathbf{x} = \mathbf{P}\mathbf{x}'$, it follows that in terms of the original coordinate system this is the point

$$\begin{pmatrix} x \\ y \\ z \end{pmatrix} = \begin{pmatrix} 0 & \frac{1}{\sqrt{3}} & \frac{2}{\sqrt{6}} \\ \frac{1}{\sqrt{2}} & \frac{1}{\sqrt{3}} & \frac{-1}{\sqrt{6}} \\ \frac{-1}{\sqrt{2}} & \frac{1}{\sqrt{3}} & \frac{-1}{\sqrt{6}} \end{pmatrix} \begin{pmatrix} -\sqrt{2} \\ \frac{1}{\sqrt{3}} \\ \frac{2}{\sqrt{6}} \end{pmatrix}$$

$$= \begin{pmatrix} 1 \\ -1 \\ 1 \end{pmatrix},$$

that is, the point $(1, -1, 1)$. □

Problem 2 Prove that the quadric surface E with equation

$$y - yz = xz$$

is a hyperbolic paraboloid. Determine its centre.

1.4.3 Rulings of Quadric Surfaces

We now turn our attention to two of the quadric surfaces, the hyperboloid of one sheet and the hyperbolic paraboloid. Each of these can be very beautifully constructed entirely from a family of straight lines.

Definition A **ruled surface** in $\mathbb{R}^3$ is a surface that can be made up from a family of straight lines.

The Hyperboloid of One Sheet

Firstly, we look at the hyperboloid of one sheet E with equation

$$x^2 + y^2 - z^2 = 1,$$

illustrated below.

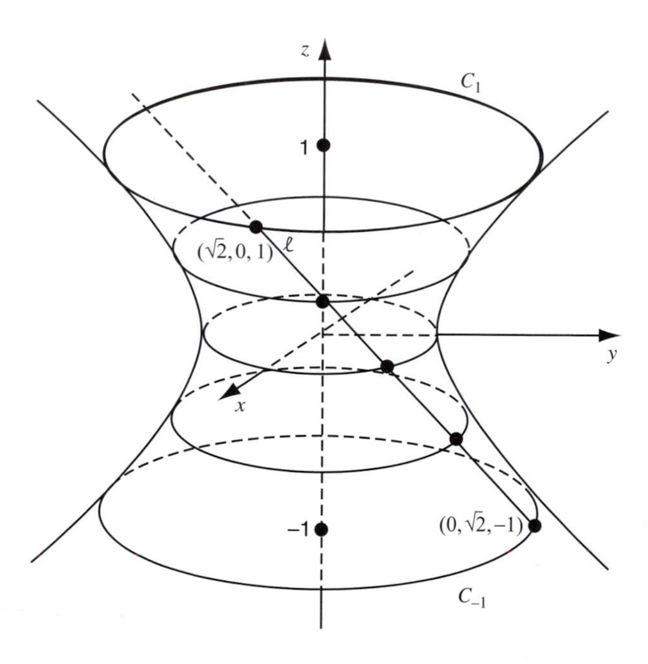

The surface meets each horizontal plane in a circle whose centre lies on the z-axis; for example, the circles C_1 and C_{-1} drawn in the figure, where the surface meets the planes $z = 1$ and $z = -1$, respectively; both of these circles have radius $\sqrt{2}$. The surface meets each plane containing the z-axis in a rectangular hyperbola. The surface appears rather like a cooling tower at a power station.

That is, each plane parallel to the (x, y)-plane.

Let ℓ be the line through the points $\left(\sqrt{2}, 0, 1\right)$ on C_1 and $\left(0, \sqrt{2}, -1\right)$ on C_{-1}. Any point on ℓ has coordinates

$$\lambda\left(\sqrt{2}, 0, 1\right) + (1-\lambda)\left(0, \sqrt{2}, -1\right) = \left(\lambda\sqrt{2}, (1-\lambda)\sqrt{2}, 2\lambda - 1\right) \quad (\lambda \in \mathbb{R}). \tag{5}$$

Clearly each point with coordinates given by equation (5) lies on the surface E, since

$$\begin{aligned}
&\left(\lambda\sqrt{2}\right)^2 + \left((1-\lambda)\sqrt{2}\right)^2 - (2\lambda - 1)^2 \\
&= 2\lambda^2 + 2(1 - 2\lambda + \lambda^2) - (4\lambda^2 - 4\lambda + 1) \\
&= 1.
\end{aligned}$$

In other words, the point lies on E for any choice of the parameter λ; so the whole of the line ℓ lies in the surface E.

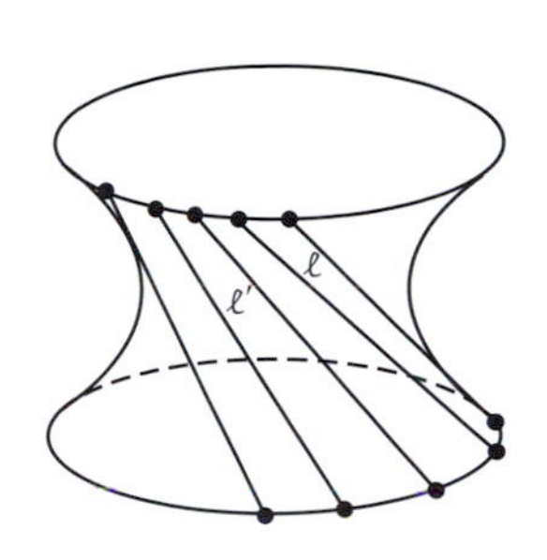

We now use the fact that the surface is symmetric about the z-axis; in other words, a rotation about the z-axis carries the surface to itself. Our line ℓ meets each horizontal circle in E in a single point; and, if we rotate the surface about the z-axis, ℓ is moved to a new line, ℓ' say, which also lies in the surface, and which does not meet ℓ.

We can see that ℓ' does not meet ℓ because ℓ' meets each horizontal circle in a point different to that in which ℓ meets that horizontal circle – namely, in the point obtained by the rotation of the intersection point with ℓ.

So we say that the hyperboloid of one sheet E is *generated* by the straight line ℓ and the rotations of ℓ described above. These straight lines are called a *family of generators* (or *generating lines*), $\mathscr{L}$ say, of E, and E is called a *ruled surface.*

In fact, E possesses another family of generators too.

> **Problem 3** Verify that the line m through the points $(\sqrt{2}, 0, 1)$ and $(0, -\sqrt{2}, -1)$ lies entirely in the quadric surface E with equation $x^2 + y^2 - z^2 = 1$.

There is thus a second family, $\mathscr{M}$, say, of lines that are also generators of the surface E, and this is obtained by rotating the line m about the z-axis (as shown dotted in the diagram in the margin).

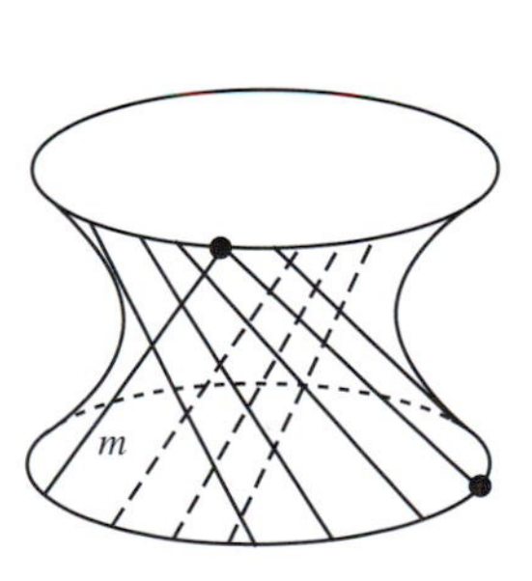

From the construction of the families $\mathscr{L}$ and $\mathscr{M}$, it is clear that any two distinct lines in a given family do not meet. However, each line in $\mathscr{L}$ meets each line in $\mathscr{M}$ (– with one exception, as we shall explain below).

To prove this claim, it is sufficient (in view of the rotational symmetry of the surface E) to verify that the given line ℓ in $\mathscr{L}$ meets each line in $\mathscr{M}$. Now, recall (from equation (5) above) that each point of ℓ has coordinates

$$\left(\lambda\sqrt{2}, (1-\lambda)\sqrt{2}, 2\lambda - 1\right), \tag{6}$$

for some $\lambda \in \mathbb{R}$. In a similar way as you saw in your solution to Problem 3, a typical point of the line m in $\mathcal{M}$ has coordinates

$$\mu\left(\sqrt{2},0,1\right)+(1-\mu)\left(0,-\sqrt{2},-1\right)$$
$$=\left(\mu\sqrt{2},(\mu-1)\sqrt{2},2\mu-1\right), \qquad (7)$$

for some $\mu \in \mathbb{R}$. We can then find the coordinates of points on any other line m' in the family $\mathcal{M}$ by rotating m about the z-axis through a suitable angle θ; that is, by the transformation

$$(x,y,z)\mapsto(x\cos\theta-y\sin\theta,x\sin\theta+y\cos\theta,z),$$

Here we have

$x=\mu\sqrt{2}$,

$y=(\mu-1)\sqrt{2}$, and

$z=2\mu-1$.

which sends the point on m with coordinates (7) to some point

$$\left(\sqrt{2}(\mu\cos\theta-\mu\sin\theta+\sin\theta),\right.$$
$$\left.\sqrt{2}(\mu\sin\theta+\mu\cos\theta-\cos\theta),2\mu-1\right) \qquad (8)$$

on a line m' in $\mathcal{M}$.

To find the point(s) of intersection of the lines ℓ and m', we have to find the values of λ and μ for which the points (6) and (8) are equal.

By comparing the third coordinates in (6) and (8), we see that we must have $\lambda=\mu$. Then, by comparing the first coordinates, we find

Here we have used the fact that $\lambda=\mu$.

$$\lambda\sqrt{2}=\sqrt{2}(\lambda\cos\theta-\lambda\sin\theta+\sin\theta),$$

so that

$$\lambda=\frac{\sin\theta}{1+\sin\theta-\cos\theta}. \qquad (9)$$

Finally, by comparing the second coordinates in (6) and (8), we find

Here we have again used the fact that $\lambda=\mu$.

$$(1-\lambda)\sqrt{2}=\sqrt{2}(\lambda\sin\theta+\lambda\cos\theta-\cos\theta),$$

so that

$$\lambda=\frac{1+\cos\theta}{1+\sin\theta+\cos\theta}. \qquad (10)$$

The expressions for λ in (9) and (10) are equal, since

$$\frac{\sin\theta}{1+\sin\theta-\cos\theta}\Bigg/\frac{1+\cos\theta}{1+\sin\theta+\cos\theta}$$
$$=\frac{\sin\theta(1+\sin\theta+\cos\theta)}{(1+\sin\theta-\cos\theta)(1+\cos\theta)}$$
$$=\frac{\sin\theta+\sin^2\theta+\sin\theta\cos\theta}{1+\sin\theta+\sin\theta\cos\theta-\cos^2\theta}$$
$$=1.$$

Here we use the fact that $1-\cos^2\theta=\sin^2\theta$.

It follows that the value of λ (and so μ) given by (6) gives us the (unique) point of intersection of ℓ and m'.

There is one exceptional case, however, when the expression (9) for λ makes no sense because its denominator is zero: that is, when $\theta=3\pi/2$, in which case $\cos\theta = 0$ and $\sin\theta = -1$, and the expression for λ has denominator zero. Geometrically this corresponds to the situation in which m has been rotated around to the back of the surface E, when m' is actually parallel to ℓ.

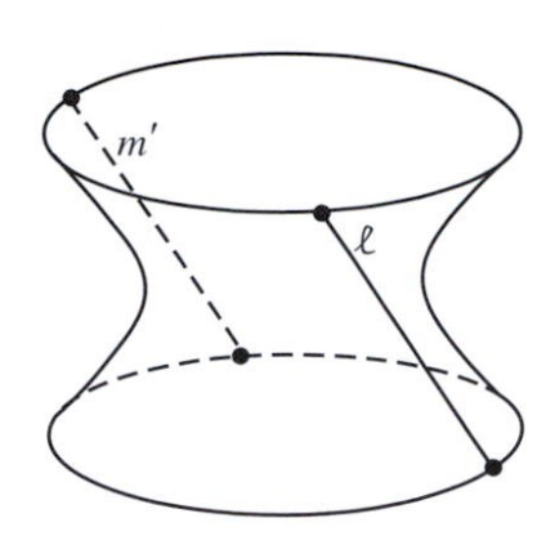

We have thus proved a special case of the following general result.

Theorem 2 A hyperboloid of one sheet contains two families of generating lines. The members of each family are disjoint, and each member of either family intersects each member of the other – with exactly one exception.

It is precisely the existence of these two families of generators that give power station cooling towers great intrinsic structural strength.

The Hyperbolic Paraboloid

It turns out that the result of Theorem 2 is also valid for hyperbolic paraboloids. In the following problem we ask you to verify this in a particular instance.

Problem 4 Let E be the hyperbolic paraboloid with equation $x^2 - y^2 + z = 0$.

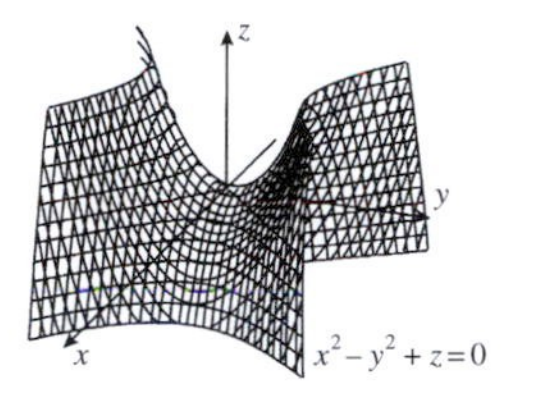

(a) Verify that the lines $A = \{(x, y, z) : x + y = 0, z = 0\}$ and $B = \{(x, y, z) : x - y = 0, z = 0\}$ lie in the surface E.

(b) Verify that the point $(\lambda - \mu, \lambda + \mu, 4\lambda\mu)$ lies in E, for each value of λ and μ.

(c) For each value of λ, let ℓ_λ denote the set $\ell_\lambda = \{(\lambda - \mu, \lambda + \mu, 4\lambda\mu) : \mu \in \mathbb{R}\}$ lying in the surface E. Prove that ℓ_λ is a line which meets B at $(\lambda, \lambda, 0)$, and which passes through the point $(\lambda - 1, \lambda + 1, 4\lambda)$. Identify the line ℓ_λ when $\lambda = 0$.

(d) For each value of μ, let m_μ denote the set $m_\mu = \{(\lambda - \mu, \lambda + \mu, 4\lambda\mu) : \lambda \in \mathbb{R}\}$ lying in the surface E. Prove that m_μ is a line which meets A at $(-\mu, \mu, 0)$, and which passes through the point $(1 - \mu, 1 + \mu, 4\mu)$. Identify the line m_μ when $\mu = 0$.

(e) Let $\mathscr{L}$ denote the set of all lines ℓ_λ and $\mathscr{M}$ the set of all lines m_μ. Prove that the members of each family are disjoint, and that each member of either family intersects each member of the other in exactly one point.

A particularly astonishing result is the following. Let E be any surface in $\mathbb{R}^3$ that has two families of generating lines with the following property: the members of each family are disjoint, and each member of either family intersects each member of the other (– with at most one exception). Then E is necessarily a quadric surface, and is in fact either a hyperboloid of one sheet or a hyperbolic paraboloid.

We omit a proof.

In the 1950s in the UK, a Lincoln artist and architect, Sam Scorer, saw the beauty of the hyperbolic paraboloid shape, and realized that the shape had intrinsic structural strength that could be utilized for the building of large roofs.

The bracing effect of the two families of generators meant that roofs in this shape could be built of very thin reinforced concrete; these require relatively few vertical supports, leaving wide unobstructed space underneath. Among the remaining examples of his work are a library, a church and a former garage that is now a restaurant. The roof of the Brisbane Exhibition and Conference Centre in Australia also uses the same shape.

The *Little Chef* at Markham Moor on the A1 road in England.

1.5 Exercises

Section 1.1

1. Determine the equation of the circle with centre (2, 1) and radius 3.
2. Determine the points of intersection of the line with equation $y = x + 2$ and the circle in Exercise 1.
3. Determine whether the circles with equations

$$2x^2 + 2y^2 - 3x - 4y + 2 = 0 \quad \text{and} \quad x^2 + y^2 - 4x + 2y = 0$$

intersect orthogonally. Find the equation of the line through their points of intersection.

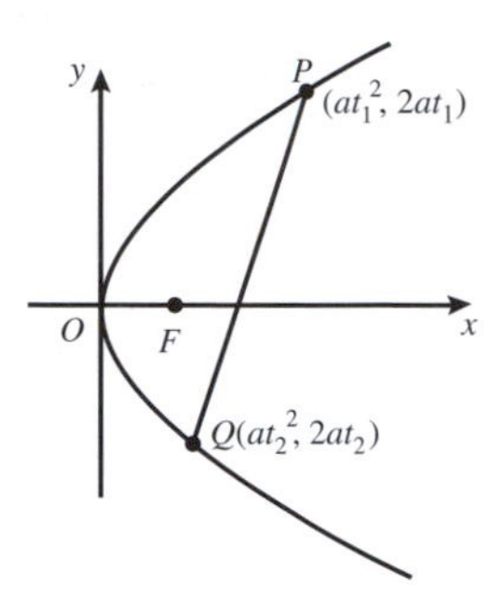

4. This question concerns the parabola $y^2 = 4ax$ $(a > 0)$ with parametric equations $x = at^2$, $y = 2at$ and focus F. Let P and Q be points on the parabola with parameters t_1 and t_2, respectively.
   (a) If PQ subtends a right angle at the vertex O of the parabola, prove that $t_1 \cdot t_2 = -4$.
   (b) If $t_1 = 2$ and PQ is perpendicular to OP, prove that $t_2 = -4$.

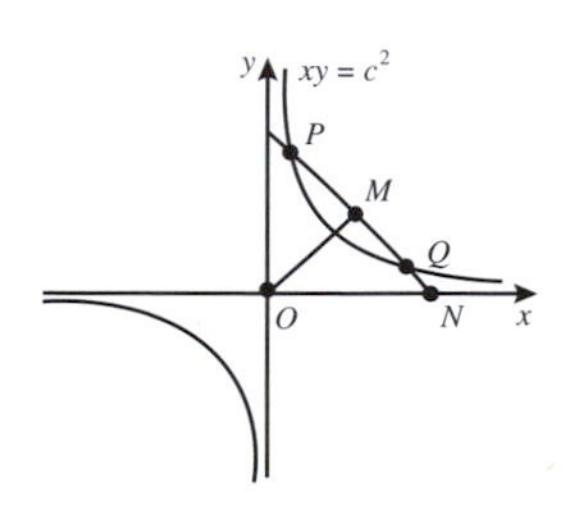

5. This question concerns the rectangular hyperbola $xy = c^2$ $(c > 0)$ with parametric equations $x = ct$, $y = c/t$. Let P and Q be points on the hyperbola with parameters t_1 and t_2, respectively.
   (a) Determine the equation of the chord PQ.
   (b) Determine the coordinates of the point N where PQ meets the x-axis.
   (c) Determine the midpoint M of PQ.
   (d) Prove that $OM = MN$, where O is the origin.
6. Let P be a point in the plane and C a circle with centre O and radius r. Then we define the *power of P with respect to C* as

You will use the results of this exercise in Chapter 5.

$$\text{power of } P \text{ with respect to } C = OP^2 - r^2.$$

   (a) Determine the sign of the power of P with respect to C when
      (i) P lies inside C;
      (ii) P lies on C;
      (iii) P lies outside C.

   In parts (b) and (c) we regard distances as *directed distances*; that is, distances along a line in one direction have a positive sign associated with their length and distances in the opposite direction have a negative sign associated with their length.

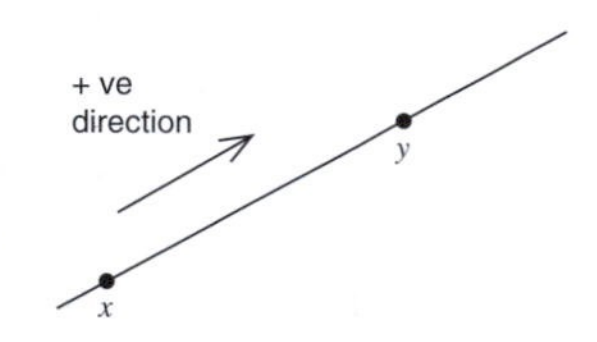

(b) If P lies inside C and a line through P meets C at two distinct points A and B, prove that

$$\text{power of } P \text{ with respect to } C = PA \cdot PB.$$

(c) If P lies outside C, a line through P meets C at two distinct points A and B, and PT is one of the tangents from P to C, prove that

$$\begin{aligned}\text{power of } P \text{ with respect to } C &= PA \cdot PB \\ &= PT^2.\end{aligned}$$

(d) If C has equation $x^2 + y^2 + fx + gy + h = 0$ and P has coordinates (x, y), find the power of P with respect to C in terms of x, y, f, g and h.

7. (a) Let a plane π in $\mathbb{R}^3$ meet both portions of a right circular cone, in two separate portions of a curve E. Let the two spheres inside the cone (on the same side of π as the vertex) that each touch both the cone in a horizontal circle (C_1 and C_2, respectively) and π touch π at F and F', respectively. Let P be any point of E, and the generator of the cone through P meet C_1 and C_2 at A and B, respectively. Prove that $PF' - PF = AB$. Deduce that E is a hyperbola.

We promised to tackle these two situations earlier, in Subsection 1.1.5.

(b) Let a plane π in $\mathbb{R}^3$ that is parallel to a generator of a right circular cone meet the cone in a curve E. Let the sphere inside the cone (on the same side of π as the vertex) that touches both the cone in a horizontal circle C and π meet π at F. Let P be any point of E, and the generator of the cone through P meet C at A. Let N be the foot of the perpendicular from P to the line of intersection of the horizontal plane and π, and let NA meet C again at M. Prove that $PF = PN$. Deduce that E is a parabola.

Section 1.2

1. Determine the slope of the tangent to the cycloid in $\mathbb{R}^2$ with parametric equations

$$x = t - \sin t, \quad y = 1 - \cos t$$

at the point with parameter t, where t is not a multiple of 2π.

2. Determine the equation of the tangent to the curve in $\mathbb{R}^2$ with parametric equations

$$x = 1 + 4t + t^2, \quad y = 1 - t$$

at the point where $t = 1$.

3. Let P be a point on the ellipse with equation $\frac{x^2}{a^2} + \frac{y^2}{b^2} = 1$, where $a > b > 0$, $b^2 = a^2(1 - e^2)$, and $0 < e < 1$.

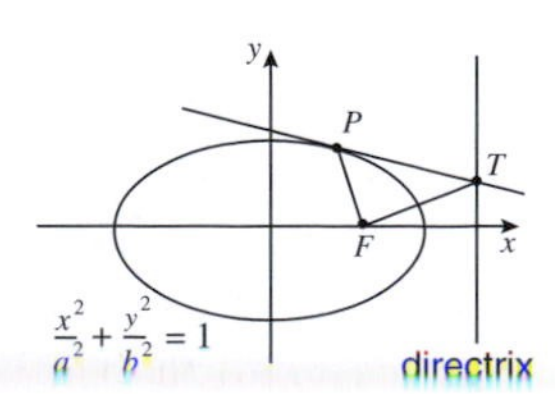

(a) If P has coordinates $(a \cos t, b \sin t)$, determine the equation of the tangent at P to the ellipse.

(b) Determine the coordinates of the point T where the tangent in part (a) meets the directrix $x = a/e$.

(c) Let F be the focus with coordinates $(ae, 0)$. Prove that PF is perpendicular to TF.

4. The perpendicular from a point P on the hyperbola H with parametric equations $x = 2\sec t$, $y = 3\tan t$, to the x-axis meets the x-axis at the point N. The tangent at P to H meets the x-axis at the point T.

   (a) Write down the coordinates of N.

   (b) Find the coordinates of T.

   (c) Prove that $ON \cdot OT = 4$, where O is the origin.

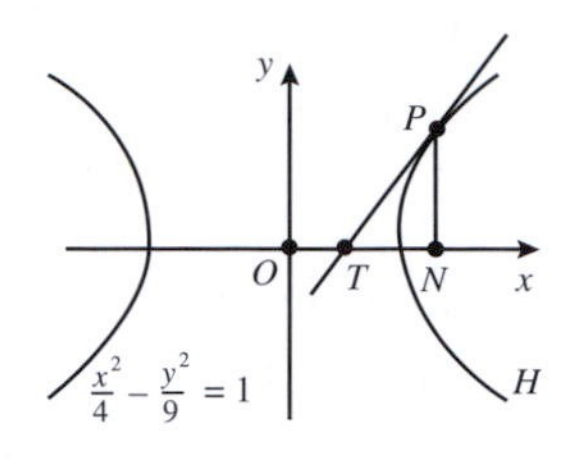

5. Let P be a point on the ellipse with equation $\frac{x^2}{a^2} + \frac{y^2}{b^2} = 1$, where $a > b > 0$, $b^2 = a^2(1 - e^2)$, and $0 < e < 1$.

   (a) If P has coordinates $(a\cos t, b\sin t)$, determine the equation of the normal at P to the ellipse.

   (b) Determine the coordinates of the point Q where the normal in part (a) meets the axis $y = 0$.

   (c) Let F be the focus with coordinates $(ae, 0)$. Prove that $QF = e \cdot PF$.

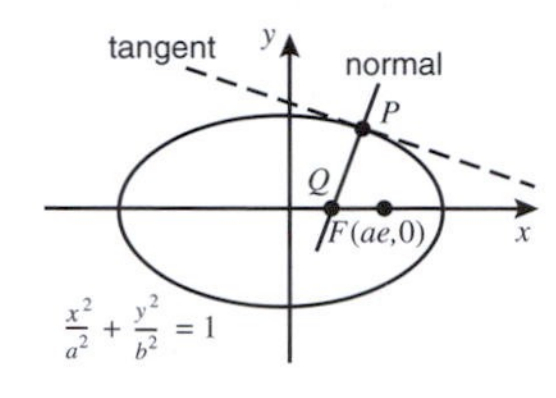

6. Let $\mathscr{F}$ denote the family of parabolas $\{(x, y) : y^2 = 4a(x + a)\}$ as a takes all positive values, and $\mathscr{G}$ denote the family of parabolas $\{(x, y) : y^2 = 4a(-x + a)\}$ as a takes all positive values. Use the reflection property of the parabola to prove that, if $F \in \mathscr{F}$ and $G \in \mathscr{G}$, then, at each point of intersection, F and G cross at right angles.

7. Prove that a perpendicular from the focus nearer to a point P on an ellipse meets the tangent at P on the auxiliary circle of the ellipse, in the following geometric way.

   This is part of Theorem 4 of Subsection 1.2.3. A similar argument works for the other focus, but we do not look at that here.

   It is sufficient to prove the result for the ellipse $E : \frac{x^2}{a^2} + \frac{y^2}{b^2} = 1$, $a > b > 0$, and points P of E in the first quadrant. Let T be the foot of the perpendicular from $F(ae, 0)$ to the tangent at P, let T' be the foot of the perpendicular from $F'(-ae, 0)$ to the tangent at P, and let FT meet $F'P$ at X.

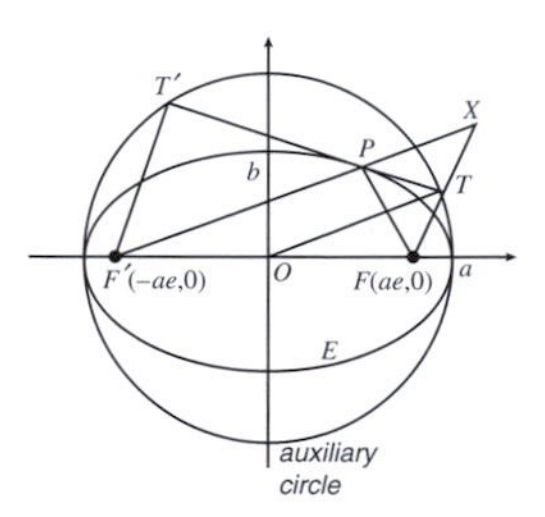

   (a) Prove that the triangles ΔFPT and ΔXPT are congruent.

   (b) Using the sum of focal distances property for E, prove that $F'X = 2a$.

   (c) Prove that OT is parallel to $F'X$, where O is the centre of E.

   (d) Prove that $OT = a$, so that T lies on the auxiliary circle of E.

   Remark: A similar argument to that in parts (a)–(d) shows that $OT' = a$, so that T' also lies on the auxiliary circle of E.

8. (a) Let E be an ellipse with major axis AB and minor axis CD, and let the tangents to E at A and B meet the tangent at D at the points T and T', respectively. Prove that the circle with diameter TT' cuts the major axis of E at its foci.

   This gives a method of locating the foci of an ellipse, given its major and minor axes.

   (b) Let H be a hyperbola with major axis AB, whose midpoint is O, and let the perpendicular at A to the major axis meet an asymptote at a point T. Prove that the circle with centre O and radius OT cuts the major axis of H at its foci.

   This gives a method of locating the foci of a hyperbola, given its major and minor axes.

Section 1.3

1. Classify the conics in $\mathbb{R}^2$ with the following equations. Determine the centre/vertex and axis of each.
   (a) $x^2 - 4xy - 2y^2 + 6x + 12y + 21 = 0$
   (b) $5x^2 + 4xy + 5y^2 + 20x + 8y - 1 = 0$
   (c) $x^2 - 4xy + 4y^2 - 6x - 8y + 5 = 0$
   (d) $21x^2 - 24xy + 31y^2 + 6x + 4y - 25 = 0$
   (e) $3x^2 - 10xy + 3y^2 + 14x - 2y + 3 = 0$
2. Determine the eccentricities of the conics in parts (a), (b) and (c) of Exercise 1.

Section 1.4

1. Classify the quadrics in $\mathbb{R}^3$ with the following equations. Determine the centre of each.
   (a) $4x^2 + 3y^2 - 2z^2 + 4xy + 4yz + 12x + 12z + 18 = 0$
   (b) $xy - y + yz = xz$
   (c) $5x^2 + 5y^2 + 6z^2 + 2\sqrt{2}yz + 2\sqrt{2}xz + 2xy + 2\sqrt{3}x - 4\sqrt{6}y - 1 = 0$
   (d) $-3x^2 + 7y^2 + 72x + 126y + z + 95 = 0$
2. Determine the equations of the generators of the hyperboloid of one sheet E with equation $2x^2 - 3y^2 + 4z^2 = 3$ through the point $(1, 1, 1)$.

Summary of Chapter 1

Section 1.1: Conic Sections and Conics

1. **Conics** (or **conic sections**) are the shapes that we obtain by taking different plane slices through a double cone.
   The **non-degenerate conic** sections are parabolas, ellipses and hyperbolas; the **degenerate conic** sections are the single point, single line and pair of lines.
2. The ellipse and the hyperbola both have a **centre**; that is, there is a point C such that rotation about C through an angle π is a symmetry of the conic.
3. The equation of a circle in $\mathbb{R}^2$ with centre (a, b) and radius r is $(x - a)^2 + (y - b)^2 = r^2$.
   An equation of the form $x^2 + y^2 + fx + gy + h = 0$ represents a circle with centre $\left(-\frac{1}{2}f, -\frac{1}{2}g\right)$ and radius $\sqrt{\frac{1}{4}f^2 + \frac{1}{4}g^2 - h}$, provided that $\frac{1}{4}f^2 + \frac{1}{4}g^2 - h > 0$.
4. Two intersecting circles C_1 and C_2 with equations $x^2 + y^2 + f_1x + g_1y + h_1 = 0$ and $x^2 + y^2 + f_2x + g_2y + h_2 = 0$, respectively, are orthogonal if and only if $f_1f_2 + g_1g_2 = 2(h_1 + h_2)$.
5. Let C_1 and C_2 be circles with equations $x^2 + y^2 + f_1x + g_1y + h_1 = 0$ and $x^2 + y^2 + f_2x + g_2y + h_2 = 0$ that intersect at distinct points P and Q. Then the line and all circles (other than C_2) through P and Q have an equation

of the form $x^2+y^2+f_1x+g_1y+h_1+k(x^2+y^2+f_2x+g_2y+h_2)=0$ for some number k.

If $k \neq -1$, this equation is one of the circles; if $k=-1$, this is the equation of the line.

6. The non-degenerate conics can be defined as the set of points P in the plane that satisfy the following condition: The distance of P from a fixed point (the **focus**) is a constant multiple e (the **eccentricity**) of the distance of P from a fixed line (the **directrix**).

   A non-degenerate conic is an ellipse if $0 \leq e < 1$, a parabola if $e=1$, or a hyperbola if $e>1$.
7. A **parabola in standard form** has equation $y^2=4ax$, where $a>0$.

   It has focus $(a,0)$ and directrix $x=-a$; and it can be described by the parametric equations $x=at^2$, $y=2at$ $(t \in \mathbb{R})$.
8. An **ellipse in standard form** has equation $\frac{x^2}{a^2}+\frac{y^2}{b^2}=1$, where $a \geq b > 0$, $b^2=a^2(1-e^2)$, $0 \leq e < 1$.

   It can be described by the parametric equations $x=a\cos t$, $y=b\sin t$ $(t \in (-\pi,\pi])$; or by $x=a\frac{1-t^2}{1+t^2}$, $y=b\frac{2t}{1+t^2}$ $(t \in \mathbb{R})$.

   If $e>0$, it has foci $(\pm ae,0)$ and directrices $x=\pm a/e$.
9. A **hyperbola in standard form** has equation $\frac{x^2}{a^2}-\frac{y^2}{b^2}=1$, where $b^2=a^2(e^2-1)$, $e>1$.

   It has foci $(\pm ae,0)$ and directrices $x=\pm a/e$; and it can be described by the parametric equations $x=a\sec t$, $y=b\tan t$ $(t \in (-\pi/2,\pi/2) \cup (\pi/2,3\pi/2))$; or by $x=a\frac{1+t^2}{1-t^2}$, $y=b\frac{2t}{1-t^2}$ $(t \in \mathbb{R}-\{\pm 1\})$.
10. A **rectangular hyperbola** has its asymptotes at right angles, and has eccentricity $e=\sqrt{2}$. In standard form it has equation $xy=c^2$, $c>0$.

    It can be described by the parametric equations $x=ct$, $y=\frac{c}{t}$ $(t \neq 0)$.
11. The polar equation of a conic with focus O can be expressed in the form $r=l/(1+e\cos\theta)$ $(\theta \in \mathbb{R})$.
12. **Sum of Focal Distances of Ellipse** Let E be an ellipse with major axis $(-a,a)$ and foci F and F'. Then, if P is a point on the ellipse, $FP+PF'=2a$. In particular, $FP+PF'$ is constant for all points P on the ellipse.

    Given any two points F and F' in the plane, the locus of points P in the plane for which $PF+PF'$ is a constant is an ellipse.
13. **Difference of Focal Distances of Hyperbola** Let H be a hyperbola with major axis $(-2a,2a)$ and foci F and F'. Then, if P is a point on the branch of the hyperbola that is closer to F, $PF'-PF=2a$; and, if P is a point on the branch of the hyperbola closer to F', $PF'-PF=-2a$. In particular, $|PF'-PF|$ is constant for all points P on the hyperbola.

    Given any three points F, F' and P in the plane for which $PF'-PF \neq 0$, the locus of points Q in the plane for which $QF'-QF=\pm|PF'-PF|$ is a hyperbola.
14. **Dandelin spheres** Let a plane π cut one portion of a right circular cone in a curve E. Let F and F' be the points of contact with π of two spheres that touch that portion of the cone (in circles C and C', respectively)

and π. Then E is an ellipse, with foci F and F', and with directrices the lines of intersection of π with the planes through C and C', respectively).

Section 1.2: Properties of Conics

1. The **slope** (or **gradient**) of the tangent to a curve in $\mathbb{R}^2$ with parametric equations $x = x(t)$, $y = y(t)$ at the point with parameter t is $\frac{y'(t)}{x'(t)}$, provided that $x'(t) \neq 0$.
2. The equation of the tangent at the point (x_1, y_1) to a conic in standard form is as follows.

Conic		Tangent
Ellipse	$\frac{x^2}{a^2} + \frac{y^2}{b^2} = 1$	$\frac{xx_1}{a^2} + \frac{yy_1}{b^2} = 1$
Hyperbola	$\frac{x^2}{a^2} - \frac{y^2}{b^2} = 1$	$\frac{xx_1}{a^2} - \frac{yy_1}{b^2} = 1$
Parabola	$y^2 = 4ax$	$yy_1 = 2a(x + x_1)$

3. Let (a, b) be a point outside the unit circle, and let the tangents to the circle from (a, b) touch the circle at P_1 and P_2. The **polar of** (a, b) **with respect to the unit circle** is the line through P_1 and P_2; this has equation $ax + by = 1$.
4. The **normal** to a curve C at a point P of C is the line through P that is perpendicular to the tangent to C at P.
5. **The Reflection Law** The angle that incoming light makes with the tangent to a surface is the same as the angle that the reflected light makes with the tangent.
6. **Reflection Property of the Ellipse** Light which comes from one focus of an elliptical mirror is reflected at the ellipse to pass through the second focus.
7. **Sine Formula** In a triangle ΔABC with sides a, b, c opposite the vertices A, B, C, respectively,
$$\frac{a}{\sin \angle BAC} = \frac{b}{\sin \angle ABC} = \frac{c}{\sin \angle ACB}.$$
8. **Reflection Property of the Hyperbola** Light coming from one focus of a hyperbolic mirror is reflected at the hyperbola in such a way that the light appears to have come from the other focus.

   Also, light going towards one focus of a hyperbolic mirror is reflected at the mirror towards the other focus.
9. **Reflection Property of the Parabola** Incoming light parallel to the axis of a parabolic mirror is reflected at the parabola to pass through the focus.

   Conversely, light coming from the focus of a parabolic mirror is reflected at the parabola to give a beam of light parallel to the axis of the parabola.

10. **Reflection Property of Conics** All mirrors in the shape of a non-degenerate conic reflect light coming from or going to one focus towards the other focus.
11. The **auxiliary circle** of an ellipse or hyperbola is the circle whose diameter is its major axis. The auxiliary circle of a parabola is the tangent to the parabola at its vertex.
12. A perpendicular from a focus of a non-degenerate conic to a tangent meets the tangent on the auxiliary circle of the conic.
    This property can be used to construct a parabola, ellipse or hyperbola as the **envelope** of a family of lines that are tangents to the conics.

Section 1.3: Recognising Conics

1. Any conic has an equation of the form $Ax^2+Bxy+Cy^2+Fx+Gy+H=0$, where A, B, C, F, G and H are real numbers, and not all of A, B and C are zero. Conversely, any set of points in $\mathbb{R}^2$ whose coordinates (x, y) satisfy this equation is a conic.
   This equation can be expressed in matrix form as
   $$\mathbf{x}^T\mathbf{A}\mathbf{x}+\mathbf{J}^T\mathbf{x}+H=0, \text{ where } \mathbf{A}=\begin{pmatrix} A & \frac{1}{2}B \\ \frac{1}{2}B & C \end{pmatrix},$$
   $$\mathbf{J}=\begin{pmatrix} F \\ G \end{pmatrix} \text{ and } \mathbf{x}=\begin{pmatrix} x \\ y \end{pmatrix}.$$
2. A 2×2 matrix $\mathbf{P}$ represents a rotation of $\mathbb{R}^2$ about the origin if and only if it satisfies the following two conditions:
   (a) $\mathbf{P}$ is orthogonal;
   (b) $\det \mathbf{P}=1$.
3. **Strategy** To classify a conic E with equation
   $$Ax^2+Bxy+Cy^2+Fx+Gy+H=0:$$
   1. Write the equation of E in matrix form $\mathbf{x}^T\mathbf{A}\mathbf{x}+\mathbf{J}^T\mathbf{x}+H=0$.
   2. Determine an orthogonal matrix $\mathbf{P}$, with determinant 1, that diagonalizes $\mathbf{A}$.
   3. Make the change of coordinate system $\mathbf{x}=\mathbf{P}\mathbf{x}'$. The equation of E then becomes of the form $\lambda_1 x'^2+\lambda_2 y'^2+fx'+gy'+h=0$, where λ_1 and λ_2 are the eigenvalues of $\mathbf{A}$.
   4. 'Complete the squares', if necessary, to rewrite the equation of E in terms of an (x'', y'')-coordinate system as the equation of a conic in standard form.
   5. Use the equation $\mathbf{x}'=\mathbf{P}^T\mathbf{x}$ to determine the centre and axes of E in terms of the original coordinate system.
4. A non-degenerate conic with equation $Ax^2+Bxy+Cy^2+Fx+Gy+H=0$ and matrix $\mathbf{A}$ can be classified as follows:
   (a) If $\det\mathbf{A}<0$, E is a hyperbola.
   (b) If $\det\mathbf{A}=0$, E is a parabola.
   (c) If $\det\mathbf{A}>0$, E is an ellipse.

Section 1.3: Quadric surfaces

1. A **quadric** (or **quadric surface**) in $\mathbb{R}^3$ is a set given by an equation of the form $Ax^2 + By^2 + Cz^2 + Fxy + Gyz + Hxz + Jx + Ky + Lz + M = 0$, where $A, B, C, F, G, H, J, K, L$ and M are real numbers, and not all of A, B, C, F, G and H are zero.

   This equation can be expressed in matrix form as

$$\mathbf{x}^T\mathbf{A}\mathbf{x} + \mathbf{J}^T\mathbf{x} + M = 0, \text{ where } \mathbf{A} = \begin{pmatrix} A & \frac{1}{2}F & \frac{1}{2}H \\ \frac{1}{2}F & B & \frac{1}{2}G \\ \frac{1}{2}H & \frac{1}{2}G & C \end{pmatrix},$$

$$\mathbf{J} = \begin{pmatrix} J \\ K \\ L \end{pmatrix} \text{ and } \mathbf{x} = \begin{pmatrix} x \\ y \\ z \end{pmatrix}.$$

2. The **degenerate quadrics** are the empty set, a single point, a single line, a single plane, a pair of planes and a cylinder. A **cylinder** is a surface that consists of an ellipse, parabola or hyperbola in some plane π, together with all the lines in $\mathbb{R}^3$ through that conic that are normal to π.
3. There are six **non-degenerate quadrics**:

Quadric	Typical equation
Ellipsoid	$\frac{x^2}{a^2} + \frac{y^2}{b^2} + \frac{z^2}{c^2} = 1$
Hyperboloid of one sheet	$\frac{x^2}{a^2} + \frac{y^2}{b^2} - \frac{z^2}{c^2} = 1$
Hyperboloid of two sheets	$\frac{x^2}{a^2} + \frac{y^2}{b^2} - \frac{z^2}{c^2} = -1$
Elliptic cone	$z^2 = \frac{x^2}{a^2} + \frac{y^2}{b^2}$
Elliptic paraboloid	$z = \frac{x^2}{a^2} + \frac{y^2}{b^2}$
Hyperbolic paraboloid	$z = \frac{x^2}{a^2} - \frac{y^2}{b^2}$

4. A 3×3 matrix $\mathbf{P}$ represents a rotation of $\mathbb{R}^3$ about the origin if and only if it satisfies the following two conditions:
   (a) $\mathbf{P}$ is orthogonal;
   (b) $\det \mathbf{P} = 1$.
5. **Strategy** To classify a quadric E with equation $Ax^2 + By^2 + Cz^2 + Fxy + Gyz + Hxz + Jx + Ky + Lz + M = 0$:
   1. Write the equation of E in matrix form $\mathbf{x}^T\mathbf{A}\mathbf{x} + \mathbf{J}^T\mathbf{x} + M = 0$.

2. Determine an orthogonal matrix $\mathbf{P}$, with determinant 1, that diagonalizes $\mathbf{A}$.
3. Make the change of coordinate system $\mathbf{x} = \mathbf{P}\mathbf{x}'$. The equation of E then becomes of the form $\lambda_1 x'^2 + \lambda_2 y'^2 + \lambda_3 z'^2 + jx' + ky' + lz' + m = 0$, where λ_1, λ_2 and λ_3 are the eigenvalues of $\mathbf{A}$.
4. 'Complete the squares', if necessary, to rewrite the equation of E in terms of an (x'', y'', z'')-coordinate system as the equation of a quadric in standard form.
5. Use the equation $\mathbf{x}' = \mathbf{P}^T\mathbf{x}$ to determine the centre and planes of symmetry of E in terms of the original coordinate system.

6. A **ruled surface** in $\mathbb{R}^3$ is a surface that can be made up from a family of straight lines, called its **generating lines** or **generators**.
7. A hyperboloid of one sheet and a hyperbolic paraboloid contain two families of generating lines. The members of each family are disjoint, and each member of either family intersects each member of the other – with exactly one exception.

   These are the only two quadrics that are ruled surfaces.

2 Affine Geometry

In Chapter 1 we studied conics in Euclidean geometry. In the rest of the book we prove a whole range of results about figures such as lines and conics, in geometries other than Euclidean geometry. In the process of doing this, we meet two particular features of our approach to geometry which may be new to you.

The first feature is the use of transformations in geometry to simplify problems and bring out their essential character. You may have met some of these transformations previously in courses on Group Theory or on Linear Algebra.

In this book, we shall often use matrices to simplify our work.

The second feature arises from the fact that the transformations we introduce form groups. Generally, we restrict our attention to geometry in the plane, $\mathbb{R}^2$, but even in this familiar setting there may be more than one group of transformations at our disposal. This leads to the exciting new idea that there are many different geometries!

Each geometry consists of a space, some properties possessed by figures in that space, and a group of transformations of the space that preserve these properties. For example, Euclidean plane geometry uses the space $\mathbb{R}^2$, and is concerned with those properties of figures that depend on the notion of distance. The group associated with Euclidean geometry is the group of isometries of the plane.

This is because isometries of the plane preserve distances.

This idea, that geometry can be thought of in terms of a space and a group acting on it, is called the *Kleinian view of geometry*, after the 19th-century German mathematician Felix Klein who proposed it first. It has the virtue of enabling us to generate many geometries, while seeing how they are related.

For instance, we can take $\mathbb{R}^2$ as our space and use the group of all transformations of the form $t(\mathbf{x}) = \mathbf{Ax} + \mathbf{a}$, where $\mathbf{a} \in \mathbb{R}^2$ and $\mathbf{A}$ is a 2×2 invertible matrix. These are the so-called *affine transformations* of $\mathbb{R}^2$. But what properties of figures in $\mathbb{R}^2$ are preserved by such transformations, and what is the corresponding geometry? The geometry is called *affine geometry*, and it is the subject of this chapter. As you will see, it has some features in common with Euclidean geometry, but also some very different features.

In Section 2.1 we examine Euclidean geometry from the Kleinian point of view, and explain why geometries other than Euclidean geometry exist.

In Sections 2.2 and 2.3 we introduce affine geometry and consider its properties. In particular, we show that affine transformations map straight lines to straight lines, map parallel lines to parallel lines, and preserve ratios of lengths along a given line. We also discover that in affine geometry all triangles are congruent, in the sense that any triangle can be mapped onto any other triangle by an affine transformation. This result is known as the *Fundamental Theorem of Affine Geometry*.

In Section 2.4 we establish two important theorems, due to Ceva and Menelaus, which involve ratios of lengths along the sides of a triangle.

Finally, in Section 2.5 we investigate the effect of affine transformations on conics, and discover that we can use the methods of affine geometry to obtain very simple proofs of certain types of theorems about conics.

2.1 Geometry and Transformations

Before embarking on a study of various other geometries, it is useful first to look back at our familiar Euclidean geometry.

2.1.1 What is Euclidean Geometry?

To help us answer this question, we begin by considering the following well-known result.

Example 1 Let ΔABC be a triangle in which $\angle ABC = \angle ACB$. Prove that $AB = AC$.

Solution First, reflect the triangle in the perpendicular bisector of BC, so that the points B and C change places and the point A moves to some point A', say. Since reflection preserves angles, it follows that $\angle A'BC = \angle ACB$.

Also, we are given that $\angle ACB = \angle ABC$, so

$$\angle A'BC = \angle ABC.$$

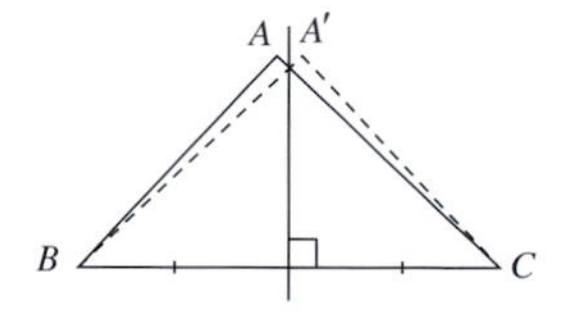

But this can happen only if A' lies on the line through A and B. Similarly,

$$\angle A'CB = \angle ABC = \angle ACB,$$

so A' must also lie on the line through A and C. This means that A' and A must coincide. Hence the line segment AB reflects to the line segment AC, and vice versa. Since reflection preserves lengths, it follows that $AB = AC$. □

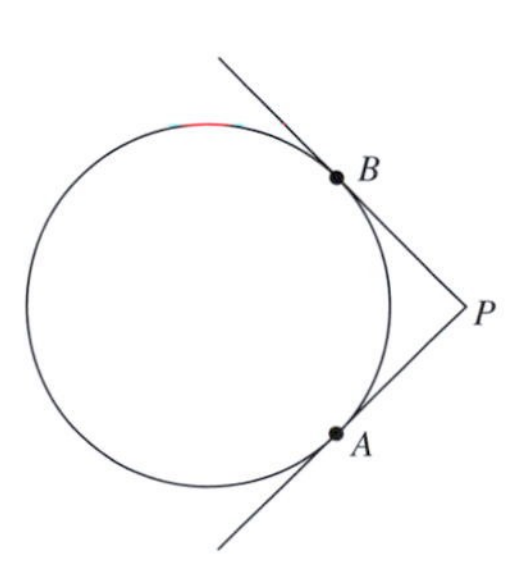

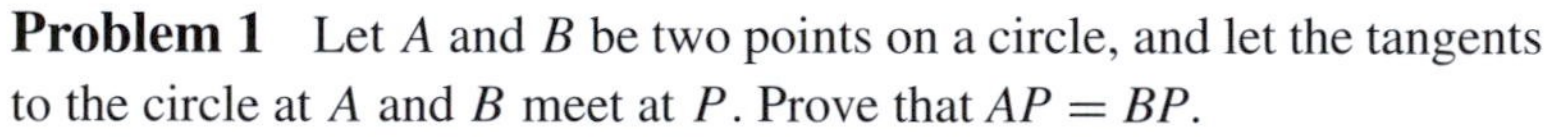

Problem 1 Let A and B be two points on a circle, and let the tangents to the circle at A and B meet at P. Prove that $AP = BP$.

Hint: Consider a reflection in the line which passes through P and the centre of the circle.

The result in Example 1 is concerned with the properties of length and angle associated with the triangle ΔABC. To investigate these properties, we introduced a reflection that enabled us to compare various lengths and angles. We were able to do this because reflections leave lengths and angles unchanged.

Of course, reflections are not the only transformations that preserve lengths and angles: other examples include rotations and translations. In general, any transformation that preserves lengths and angles can be used to tackle problems which involve these properties. In fact, we need worry only about leaving distances unchanged, since any transformation from $\mathbb{R}^2$ onto $\mathbb{R}^2$ that changes angles must also change lengths. Transformations that leave distances unchanged are called *isometries*.

This is because, once we know the lengths of the sides of a triangle, the angles are uniquely determined.

Definition An **isometry** of $\mathbb{R}^2$ is a function which maps $\mathbb{R}^2$ onto $\mathbb{R}^2$ and preserves distances.

In fact, every isometry has one of the following forms:

a translation along a line in $\mathbb{R}^2$;

a reflection in a line in $\mathbb{R}^2$;

a rotation about a point in $\mathbb{R}^2$;

a composite of translations, reflections and rotations in $\mathbb{R}^2$.

The identity isometry can be regarded as a rotation through an angle that is a multiple of 2π.

The composite of any two isometries is also an isometry, and so it is easy to verify that the set $S(\mathbb{R}^2)$ of isometries of $\mathbb{R}^2$ forms a group under composition of functions. These observations can be used to build up the transformations we need in order to prove Euclidean results.

Example 2 Prove that if ΔABC and ΔDEF are two triangles such that

$$AB = DE, \quad AC = DF \quad \text{and} \quad \angle BAC = \angle EDF,$$

then $BC = EF$, $\angle ABC = \angle DEF$ and $\angle ACB = \angle DFE$.

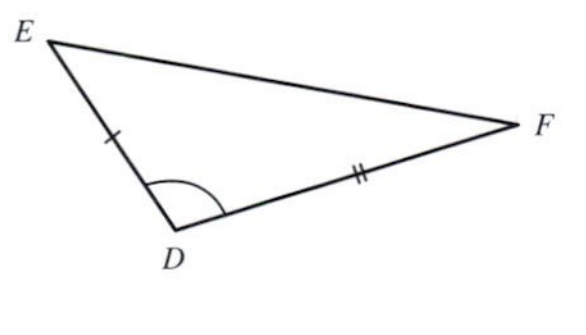

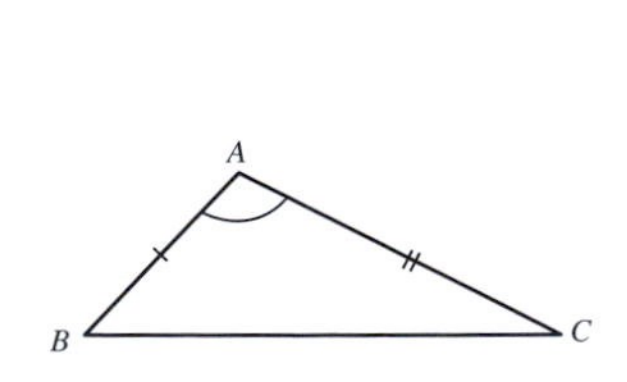

Solution It is sufficient to show that there is an isometry which maps ΔABC onto ΔDEF. We construct this isometry in stages, starting with the translation which maps A to D. This translation maps ΔABC onto $\Delta DB'C'$, where B' and C' are the images of B and C under the translation.

Of course, A and D may already coincide, in which case we omit the translation stage.

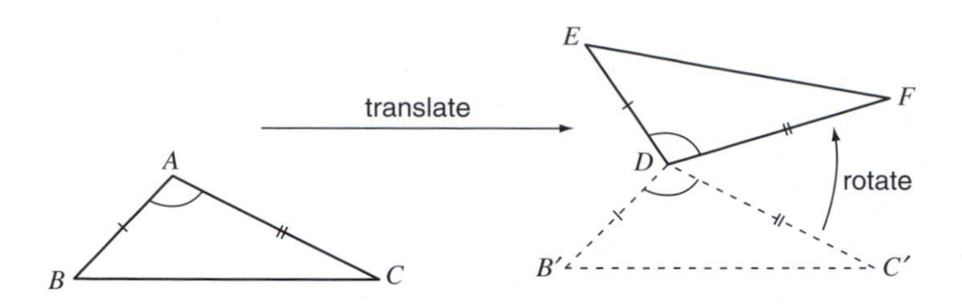

Since we are given that $DF = AC$, and since the translation maps AC onto DC', it follows that $DF = DC'$. We can therefore rotate the point C', about D, until it coincides with the point F. This rotation maps $\Delta DB'C'$ onto $\Delta DB''F$, as shown in the margin, where B'' is the image of B' under the rotation.

If C' already coincides with F, then we omit the rotation stage.

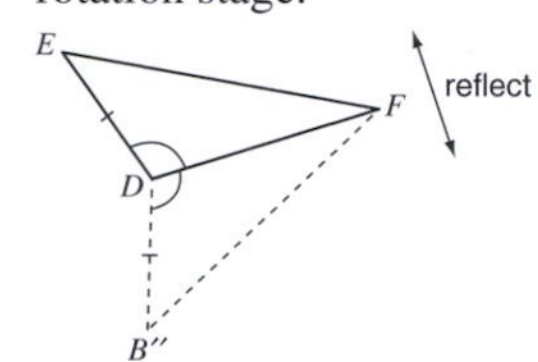

Finally, notice that

$$\begin{aligned} \angle FDE &= \angle CAB && \text{(given)} \\ &= \angle C'DB' && \text{(translation)} \\ &= \angle FDB'' && \text{(rotation)}, \end{aligned}$$

so either B'' lies on DE or the reflection of B'' in the line FD lies on DE. Also

$$\begin{aligned} DE &= AB && \text{(given)} \\ &= DB' && \text{(translation)} \\ &= DB'' && \text{(rotation)}. \end{aligned}$$

It follows that either B'' coincides with E or the reflection of B'' in the line FD coincides with E.

So, composing the translation, the rotation, and (if necessary) a reflection, we obtain the required isometry that maps ΔABC onto ΔDEF. Since isometries preserve length and angle, it follows that $BC = EF$, $\angle ABC = \angle DEF$ and $\angle ACB = \angle DFE$. □

Problem 2 Prove that if ΔABC and ΔDEF are two triangles such that

$$AC = DF, \quad \angle BAC = \angle EDF \quad \text{and} \quad \angle ACB = \angle DFE,$$

then $BC = EF$, $AB = DE$ and $\angle ABC = \angle DEF$.

We can now answer the question 'What is Euclidean geometry?'. **Euclidean geometry** is the study of those properties of figures that are unchanged by the group of isometries. We call these properties **Euclidean properties**. Roughly speaking, a Euclidean property is one that is preserved by a rigid figure as it moves around the plane. Of course, these properties include distance and angle, but they also include other properties such as collinearity of points and concurrence of lines.

We consider only Euclidean geometry in the plane $\mathbb{R}^2$.

This idea, that geometry can be thought of in terms of a group of transformations acting on a space, is known as the *Kleinian view of geometry*. It enables us to generate many geometries, without losing sight of the relationship between them.

Here a space is simply a collection of points; for example, the plane is $\mathbb{R}^2 = \{(x, y) : x, y \text{ are real}\}$.

When we consider geometries in this way, it is often convenient to have an algebraic representation for the transformations involved. This not only enables us to solve problems in the geometry algebraically, but also provides us with formulas that can be used to compare different geometries.

In the case of Euclidean geometry, perhaps the easiest way to represent isometries algebraically is to use matrices. For example, the function defined by

$$t : \begin{pmatrix} x \\ y \end{pmatrix} \mapsto \begin{pmatrix} \cos\theta & -\sin\theta \\ \sin\theta & \cos\theta \end{pmatrix} \begin{pmatrix} x \\ y \end{pmatrix} + \begin{pmatrix} e \\ f \end{pmatrix} \quad \left((x, y) \in \mathbb{R}^2\right) \tag{1}$$

is an isometry because it is the composite of an anticlockwise rotation through an angle θ about the origin, followed by a translation through the vector (e, f).

Similarly, the function

$$t : \begin{pmatrix} x \\ y \end{pmatrix} \mapsto \begin{pmatrix} \cos\theta & \sin\theta \\ \sin\theta & -\cos\theta \end{pmatrix} \begin{pmatrix} x \\ y \end{pmatrix} + \begin{pmatrix} e \\ f \end{pmatrix} \quad \left((x, y) \in \mathbb{R}^2\right) \tag{2}$$

is an isometry because it is the composite of a reflection in the line through the origin that makes an angle $\theta/2$ with the x-axis, followed by a translation through the vector (e, f).

Recall that the effect of the matrix multiplication in (1) and (2) can be interpreted geometrically by examining what happens to the vectors (1, 0) and (0, 1). For example, in (2) the matrix multiplication sends (1, 0) and (0, 1) to $(\cos\theta, \sin\theta)$ and $(\sin\theta, -\cos\theta)$, respectively, so it corresponds to the reflection shown in the figure below.

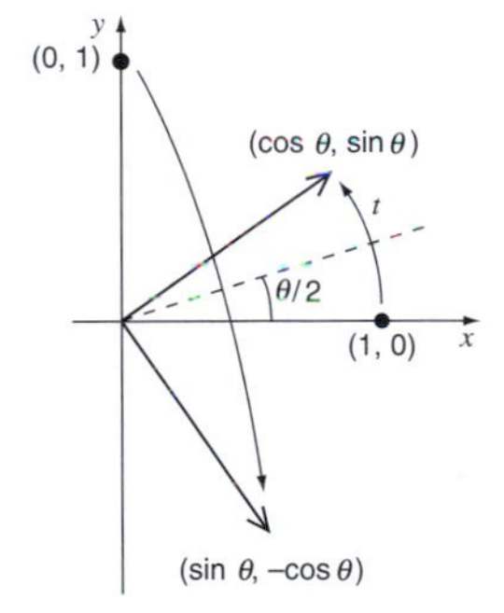

Remarkably, we can represent *any* isometry by one or other of the forms given in (1) and (2). To see this, notice that any isometry t can be written in the form

$$t(\mathbf{x}) = t_0(\mathbf{x}) + (e, f) \quad \left(\mathbf{x} \in \mathbb{R}^2\right), \tag{3}$$

where t_0 is an isometry which fixes the origin. Indeed, if we let $(e, f) = t(\mathbf{0})$, then we can let t_0 be the transformation defined by $t_0(\mathbf{x}) = t(\mathbf{x}) - (e, f)$. This is an isometry because it is the composite of the isometry t and the translation through the vector $-(e, f)$. It fixes the origin since $t_0(\mathbf{0}) = t(\mathbf{0}) - (e, f) = \mathbf{0}$.

Now an isometry that fixes the origin must be either a rotation about the origin, or a reflection in a line through the origin. If t_0 is a rotation about the origin, then (3) can be written in the matrix form given in (1), whereas if t_0 is a reflection in a line through the origin, then (3) can be written in the matrix form given in (2).

So together, equations (1) and (2) provide us with an algebraic representation of all possible isometries of the plane. The next problem indicates how we can obtain a more concise description of this algebraic representation by using orthogonal matrices to combine equations (1) and (2).

Recall that a matrix $\mathbf{U}$ is *orthogonal* if $\mathbf{U}^{-1} = \mathbf{U}^T$, that is, if $\mathbf{U}^T\mathbf{U} = \mathbf{I}$. This is equivalent to saying that the columns of $\mathbf{U}$ are *orthonormal*.

Problem 3 Show that both the matrices

$$\begin{pmatrix} \cos\theta & -\sin\theta \\ \sin\theta & \cos\theta \end{pmatrix} \quad \text{and} \quad \begin{pmatrix} \cos\theta & \sin\theta \\ \sin\theta & -\cos\theta \end{pmatrix},$$

which appear in (1) and (2), are orthogonal for each real number θ.

By applying the solution of Problem 3 to equations (1) and (2), we see that every isometry t has an algebraic representation of the form

$$t(\mathbf{x}) = \mathbf{Ux} + \mathbf{a},$$

where $\mathbf{U}$ is an orthogonal 2×2 matrix, and $\mathbf{a}$ is a vector in $\mathbb{R}^2$.

This equation shows the matrix $\mathbf{U}$ acting on the vector $\mathbf{x} = (x, y)$. Strictly speaking, $\mathbf{U}$ acts on the coordinates $\begin{pmatrix} x \\ y \end{pmatrix}$ of $\mathbf{x}$ with respect to the standard basis of the vector space $\mathbb{R}^2$ as in equations (1) and (2). However, since the numbers x and y are the same for the vector and its standard coordinates, no confusion should arise.

Definition A **Euclidean transformation** of $\mathbb{R}^2$ is a function $t : \mathbb{R}^2 \to \mathbb{R}^2$ of the form

$$t(\mathbf{x}) = \mathbf{Ux} + \mathbf{a},$$

where $\mathbf{U}$ is an orthogonal 2×2 matrix and $\mathbf{a} \in \mathbb{R}^2$. The set of all Euclidean transformations of $\mathbb{R}^2$ is denoted by $E(2)$.

We may summarize the discussion above by saying that every isometry of the plane is a Euclidean transformation of $\mathbb{R}^2$.

In fact, the converse is also true, for if $\mathbf{U}$ is any orthogonal matrix, then its columns are orthonormal. In particular, its first and second columns have unit length and can therefore be written in the form $\begin{pmatrix} \cos\theta \\ \sin\theta \end{pmatrix}$ and $\begin{pmatrix} \cos\phi \\ \sin\phi \end{pmatrix}$, respectively, for some real θ, ϕ. For these to be orthonormal, we must have $\cos\theta \cdot \cos\phi + \sin\theta \cdot \sin\phi = 0$, so that $\tan\theta \cdot \tan\phi = -1$ and hence $\phi = \theta \pm \frac{\pi}{2}$. It follows that the second column must be

$$\begin{pmatrix} \cos(\theta + \pi/2) \\ \sin(\theta + \pi/2) \end{pmatrix} = \begin{pmatrix} -\sin\theta \\ \cos\theta \end{pmatrix} \quad \text{or}$$

$$\begin{pmatrix} \cos(\theta - \pi/2) \\ \sin(\theta - \pi/2) \end{pmatrix} = \begin{pmatrix} \sin\theta \\ -\cos\theta \end{pmatrix}.$$

So

$$\mathbf{U} = \begin{pmatrix} \cos\theta & -\sin\theta \\ \sin\theta & \cos\theta \end{pmatrix} \quad \text{or} \quad \mathbf{U} = \begin{pmatrix} \cos\theta & \sin\theta \\ \sin\theta & -\cos\theta \end{pmatrix}.$$

It follows that every Euclidean transformation $t(\mathbf{x}) = \mathbf{Ux} + \mathbf{a}$ of $\mathbb{R}^2$ has one of the forms given in equations (1) and (2). Since both of these forms represent isometries of the plane,we have the following theorem.

Theorem 1 Every isometry of $\mathbb{R}^2$ is a Euclidean transformation of $\mathbb{R}^2$, and vice versa.

Now the set of all isometries of $\mathbb{R}^2$ forms a group under composition of functions, so it follows from Theorem 1 that the same must be true of the set of all Euclidean transformations of $\mathbb{R}^2$. We therefore have the following theorem.

It is not difficult to prove that this is a group, so we omit the proof.

Theorem 2 The set of Euclidean transformations of $\mathbb{R}^2$ forms a group under the operation of composition of functions.

It is instructive to check the group axioms algebraically, for in the process of doing so we obtain formulas for the composites and inverses of Euclidean transformations.

We start by considering closure. Suppose that t_1 and t_2 are two Euclidean transformations given by

$$t_1(\mathbf{x}) = \mathbf{U}_1\mathbf{x} + \mathbf{a}_1 \quad \text{and} \quad t_2(\mathbf{x}) = \mathbf{U}_2\mathbf{x} + \mathbf{a}_2,$$

where $\mathbf{U}_1$ and $\mathbf{U}_2$ are orthogonal 2×2 matrices. Then the composite $t_1 \circ t_2$ is given by

$$\begin{aligned} t_1 \circ t_2(\mathbf{x}) &= t_1(\mathbf{U}_2\mathbf{x} + \mathbf{a}_2) \\ &= \mathbf{U}_1(\mathbf{U}_2\mathbf{x} + \mathbf{a}_2) + \mathbf{a}_1 \\ &= \mathbf{U}_1\mathbf{U}_2\mathbf{x} + (\mathbf{U}_1\mathbf{a}_2 + \mathbf{a}_1). \end{aligned}$$

This is a Euclidean transformation since $\mathbf{U}_1\mathbf{U}_2$ is orthogonal. Indeed,

$$(\mathbf{U}_1\mathbf{U}_2)^T = \mathbf{U}_2^T\mathbf{U}_1^T = \mathbf{U}_2^{-1}\mathbf{U}_1^{-1} = (\mathbf{U}_1\mathbf{U}_2)^{-1}.$$

Here we are using the result that $(\mathbf{AB})^T = \mathbf{B}^T\mathbf{A}^T$.

So the set of Euclidean transformations is closed under composition of functions.

Problem 4 Let the Euclidean transformations t_1 and t_2 of $\mathbb{R}^2$ be given by

$$t_1(\mathbf{x}) = \begin{pmatrix} \frac{3}{5} & -\frac{4}{5} \\ \frac{4}{5} & \frac{3}{5} \end{pmatrix}\mathbf{x} + \begin{pmatrix} 1 \\ -2 \end{pmatrix}$$

and

$$t_2(\mathbf{x}) = \begin{pmatrix} -\frac{4}{5} & \frac{3}{5} \\ \frac{3}{5} & \frac{4}{5} \end{pmatrix}\mathbf{x} + \begin{pmatrix} -2 \\ 1 \end{pmatrix}.$$

Determine $t_1 \circ t_2$ and $t_2 \circ t_1$.

Next recall that under composition of functions the identity is the transformation given by $i(\mathbf{x}) = \mathbf{x}$. This is a Euclidean transformation since it can be written in the form

$$i(\mathbf{x}) = \mathbf{Ix} + \mathbf{0},$$

where $\mathbf{I}$ is the 2×2 identity matrix, which is orthogonal.

The next problem asks you to show that inverses exist.

Problem 5 Prove that if t_1 is a Euclidean transformation of $\mathbb{R}^2$ given by

$$t_1(\mathbf{x}) = \mathbf{Ux} + \mathbf{a} \qquad (\mathbf{x} \in \mathbb{R}^2),$$

then:

(a) the transformation of $\mathbb{R}^2$ given by

$$t_2(\mathbf{x}) = \mathbf{U}^{-1}\mathbf{x} - \mathbf{U}^{-1}\mathbf{a} \qquad (\mathbf{x} \in \mathbb{R}^2)$$

is also a Euclidean transformation;

(b) the transformation t_2 is the inverse of t_1.

The solution of Problem 5 shows that we can calculate the inverse of a Euclidean transformation by using the following result.

The inverse of the Euclidean transformation $t(\mathbf{x}) = \mathbf{Ux} + \mathbf{a}$ is given by

$$t^{-1}(\mathbf{x}) = \mathbf{U}^{-1}\mathbf{x} - \mathbf{U}^{-1}\mathbf{a}.$$

Problem 6 Determine the inverse of the Euclidean transformation given by

$$t(\mathbf{x}) = \begin{pmatrix} \frac{3}{5} & -\frac{4}{5} \\ \frac{4}{5} & \frac{3}{5} \end{pmatrix} \mathbf{x} + \begin{pmatrix} 1 \\ -2 \end{pmatrix}.$$

Finally, composition of functions is always associative. So all four group properties hold, as we expected.

Earlier, we described Euclidean geometry as the study of those properties of figures that are preserved by isometries. Having identified these isometries with the group of Euclidean transformations, we can now give the equivalent algebraic description of Euclidean geometry. **Euclidean geometry** is the study of those properties of figures that are preserved by Euclidean transformations of $\mathbb{R}^2$.

2.1.2 Euclidean-Congruence

In the solution to Example 2 we showed that if two triangles ΔABC and ΔDEF are such that $AB = DE$, $AC = DF$ and $\angle BAC = \angle EDF$, then there is a Euclidean transformation which maps ΔABC onto ΔDEF.

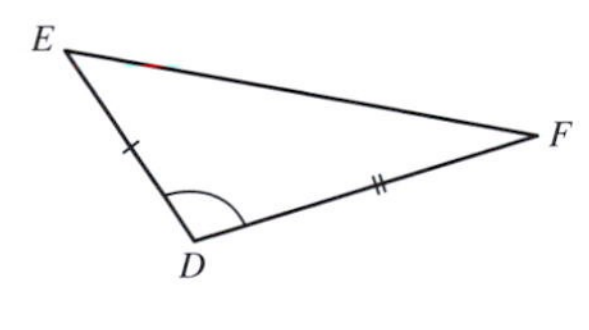

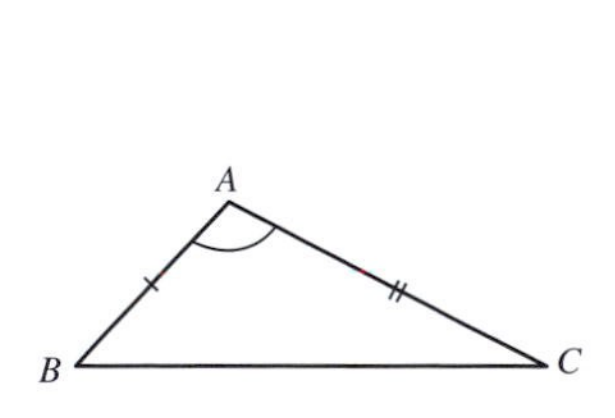

The existence of this transformation enabled us to deduce that both triangles have the same Euclidean properties. In particular, we were able to deduce that $BC = EF$, $\angle ABC = \angle DEF$ and $\angle ACB = \angle DFE$.

In order to formalize this way of relating two figures, we say that two figures are *congruent* if one can be moved to fill exactly the position of the other by means of a Euclidean transformation. Loosely speaking, two figures are congruent if they have the same size and shape.

Later we consider congruence with respect to other groups of transformations (that is, congruence in other geometries), so if there is any danger of confusion we sometimes say that two figures are *Euclidean-congruent*.

Definition A figure F_1 is **Euclidean-congruent** to a figure F_2 if there is a Euclidean transformation which maps F_1 onto F_2.

For example, any two circles of unit radius are Euclidean-congruent to each other because we can map one of the circles onto the other by means of a translation that makes their centres coincide.

Of course, there are many other Euclidean transformations which map one of the circles onto the other.

Problem 7 Which of the following sets consist of figures that are Euclidean-congruent to each other?

(a) The set of all ellipses
(b) The set of all line segments of length 1
(c) The set of all triangles
(d) The set of all squares that have sides of length 2

Earlier, we emphasized that the Euclidean transformations form a group. This is important because it ensures that Euclidean-congruence has the kind of properties that we should expect. For example, we should expect every figure to be congruent to itself. Also, if a figure F_1 is congruent to a figure F_2, then we should expect F_2 to be congruent to F_1. We can, in fact, establish the following result.

Theorem 3 Euclidean-congruence is an equivalence relation.

Proof We show that the three equivalence relation axioms E1, E2 and E3 hold.

E1 REFLEXIVE For all figures F in $\mathbb{R}^2$, the identity transformation maps F onto itself; so Euclidean-congruence is reflexive.

This uses the existence of an *identity* transformation.

E2 SYMMETRIC Let a figure F_1 in $\mathbb{R}^2$ be congruent to a figure F_2, and let t be a Euclidean transformation which maps F_1 onto F_2. Then the inverse Euclidean transformation t^{-1} maps F_2 onto F_1, so that F_2 is congruent to F_1. Thus Euclidean-congruence is symmetric.

This uses the existence of *inverse* transformations.

E3 TRANSITIVE Let a figure F_1 in $\mathbb{R}^2$ be congruent to a figure F_2, and let F_2 be congruent to a figure F_3. Then there exist Euclidean transformations t_1 mapping F_1 onto F_2 and t_2 mapping F_2 onto F_3. Thus the Euclidean transformation $t_2 \circ t_1$ maps F_1 onto F_3, so that F_1 is congruent to F_3. Hence Euclidean-congruence is transitive.

This uses the closure axiom for the group of Euclidean transformations.

It follows that Euclidean-congruence is an equivalence relation, because it satisfies the axioms El, E2 and E3. ■

Problem 8 Prove that if two figures in $\mathbb{R}^2$ are each Euclidean-congruent to a third figure, then they are Euclidean-congruent to each other.

Since Euclidean-congruence is an equivalence relation, it partitions the set of all figures into disjoint equivalence classes. Each class consists of figures which are Euclidean-congruent to each other, and hence share the same Euclidean properties (for example, one class consists of all circles of unit radius, another class consists of all equilateral triangles with sides of length 3, and so on). If we wish to show that two figures have the same Euclidean properties, then it is sufficient to show that they are Euclidean-congruent.

For example, to show that two triangles $\triangle ABC$ and $\triangle DEF$ have the same Euclidean properties, it is sufficient to show that $AB = DE$, $AC = DF$ and $\angle BAC = \angle EDF$, as you saw in Example 2. This congruence condition is frequently used in Euclidean geometry. It is known as the 'side angle side' (SAS) condition for congruence.

Now Euclidean geometry is just one of several different geometries. Each geometry is defined by a group G of transformations that act on a space. In general, we say that two figures are ***G*-congruent** if there is a transformation in G which maps one of the figures onto the other. Since the only properties used in the proof of Theorem 3 are the group properties of Euclidean transformations, the theorem holds also with 'G-congruent' in place of 'Euclidean-congruent'. Thus, like Euclidean-congruence, G-congruence is an equivalence relation that partitions the set of all figures into disjoint equivalence classes.

This idea of partitioning figures into equivalence classes is central to geometry. It enables us to distinguish between figures in different equivalence classes, without having to worry about the differences between figures in the same equivalence class. For example, if we are interested in whether a conic is an ellipse rather than a hyperbola or a parabola, but do not care about its shape (that is, the ratio of the lengths of its axes), we might choose to work with some geometry whose group of transformations makes all ellipses congruent to each other – but not congruent to any hyperbola or parabola. We describe a group of transformations which defines such a geometry in Section 2.2.

2.2 Affine Transformations and Parallel Projections

2.2.1 Affine Transformations

In Section 2.1 you met a new approach to Euclidean geometry in $\mathbb{R}^2$ – namely, the idea that Euclidean geometry of $\mathbb{R}^2$ can be interpreted as a space, $\mathbb{R}^2$, together with the group of Euclidean transformations which act on that space. Recall that a Euclidean transformation is a function $t : \mathbb{R}^2 \to \mathbb{R}^2$ of the form

$$t(\mathbf{x}) = \mathbf{Ux} + \mathbf{a} \quad (\mathbf{x} \in \mathbb{R}^2),$$

where $\mathbf{U}$ is an orthogonal 2×2 matrix. Euclidean properties of figures are those, like distance and angle, that are preserved by these transformations.

In this section we meet the first of our new geometries in $\mathbb{R}^2$ – *affine geometry*. This geometry consists of the space $\mathbb{R}^2$ together with a group of transformations, the *affine transformations*, acting on $\mathbb{R}^2$.

Affine geometry can be defined in $\mathbb{R}^n$, for any $n \geq 2$; we restrict our attention here to the case when $n = 2$.

Definition An **affine transformation** of $\mathbb{R}^2$ is a function $t : \mathbb{R}^2 \to \mathbb{R}^2$ of the form

$$t(\mathbf{x}) = \mathbf{Ax} + \mathbf{b},$$

where $\mathbf{A}$ is an invertible 2×2 matrix and $\mathbf{b} \in \mathbb{R}^2$. The set of all affine transformations of $\mathbb{R}^2$ is denoted by $A(2)$.

Remark

Note that every Euclidean transformation of $\mathbb{R}^2$ is an affine transformation of $\mathbb{R}^2$ since every orthogonal matrix is invertible. (In terms of groups, the group of Euclidean transformations of $\mathbb{R}^2$ is a proper subgroup of the group of affine transformations of $\mathbb{R}^2$.) This means that all properties of figures that are preserved by affine transformations must be preserved also by Euclidean transformations.

Problem 1 Determine whether or not each of the following transformations of $\mathbb{R}^2$ is an affine transformation.

(a) $t_1(\mathbf{x}) = \begin{pmatrix} 1 & 3 \\ 1 & 2 \end{pmatrix}\mathbf{x} + \begin{pmatrix} 4 \\ -2 \end{pmatrix}$ (b) $t_2(\mathbf{x}) = \begin{pmatrix} -6 & 5 \\ 3 & 2 \end{pmatrix}\mathbf{x} + \begin{pmatrix} 2 \\ 1 \end{pmatrix}$

(c) $t_3(\mathbf{x}) = \begin{pmatrix} -2 & -1 \\ 8 & 4 \end{pmatrix}\mathbf{x} + \begin{pmatrix} 1 \\ 3 \end{pmatrix}$ (d) $t_4(\mathbf{x}) = \begin{pmatrix} 5 & -3 \\ -2 & 2 \end{pmatrix}\mathbf{x}$

The algebra required to compose affine transformations is similar to the algebra that we used to compose Euclidean transformations.

Problem 2 For the transformations of $\mathbb{R}^2$ given in Problem 1, determine formulas for the following composites. In each case, state whether or not the composite is an affine transformation.

(a) $t_1 \circ t_2$ (b) $t_2 \circ t_4$

We now verify our assertion above that the set of affine transformations forms a group.

Theorem 1 The set of affine transformations $A(2)$ forms a group under the operation of composition of functions.

Proof We check that the four group axioms hold.

G1 CLOSURE Let t_1 and t_2 be affine transformations given by

$$t_1(\mathbf{x}) = \mathbf{A}_1\mathbf{x} + \mathbf{b}_1 \quad \text{and} \quad t_2(\mathbf{x}) = \mathbf{A}_2\mathbf{x} + \mathbf{b}_2,$$

where $\mathbf{A}_1$ and $\mathbf{A}_2$ are invertible 2×2 matrices. Then, for each $\mathbf{x} \in \mathbb{R}^2$,

$$\begin{aligned}(t_1 \circ t_2)(\mathbf{x}) &= t_1(\mathbf{A}_2\mathbf{x} + \mathbf{b}_2)\\ &= \mathbf{A}_1(\mathbf{A}_2\mathbf{x} + \mathbf{b}_2) + \mathbf{b}_1\\ &= (\mathbf{A}_1\mathbf{A}_2)\mathbf{x} + (\mathbf{A}_1\mathbf{b}_2 + \mathbf{b}_1).\end{aligned}$$

Since $\mathbf{A}_1$ and $\mathbf{A}_2$ are invertible, it follows that $\mathbf{A}_1\mathbf{A}_2$ is also invertible. So by definition $t_1 \circ t_2$ is an affine transformation.

G2 IDENTITY Let i be the affine transformation given by

$$i(\mathbf{x}) = \mathbf{I}\mathbf{x} + \mathbf{0} \quad (\mathbf{x} \in \mathbb{R}^2),$$

where $\mathbf{I}$ is the 2×2 identity matrix. If t is an affine transformation given by

$$t(\mathbf{x}) = \mathbf{A}\mathbf{x} + \mathbf{b} \quad (\mathbf{x} \in \mathbb{R}^2),$$

then, for each $\mathbf{x} \in \mathbb{R}^2$,

$$(t \circ i)(\mathbf{x}) = \mathbf{A}(\mathbf{I}\mathbf{x} + \mathbf{0}) + \mathbf{b} = \mathbf{A}\mathbf{x} + \mathbf{b} = t(\mathbf{x})$$

and

$$(i \circ t)(\mathbf{x}) = \mathbf{I}(\mathbf{A}\mathbf{x} + \mathbf{b}) + \mathbf{0} = \mathbf{A}\mathbf{x} + \mathbf{b} = t(\mathbf{x}).$$

Thus $t \circ i = i \circ t = t$. Hence i is the identity transformation.

G3 INVERSES If t is an arbitrary affine transformation given by

$$t(\mathbf{x}) = \mathbf{A}\mathbf{x} + \mathbf{b} \quad (\mathbf{x} \in \mathbb{R}^2),$$

then we can define another affine transformation t' by

$$t'(\mathbf{x}) = \mathbf{A}^{-1}\mathbf{x} - \mathbf{A}^{-1}\mathbf{b}.$$

Now for each $\mathbf{x} \in \mathbb{R}^2$, we have

$$\begin{aligned}(t \circ t')(\mathbf{x}) &= t(\mathbf{A}^{-1}\mathbf{x} - \mathbf{A}^{-1}\mathbf{b})\\ &= \mathbf{A}(\mathbf{A}^{-1}\mathbf{x} - \mathbf{A}^{-1}\mathbf{b}) + \mathbf{b}\\ &= (\mathbf{A}\mathbf{A}^{-1}\mathbf{x} - \mathbf{A}\mathbf{A}^{-1}\mathbf{b}) + \mathbf{b}\\ &= (\mathbf{x} - \mathbf{b}) + \mathbf{b}\\ &= \mathbf{x}.\end{aligned}$$

Also,

$$\begin{aligned}(t' \circ t)(\mathbf{x}) &= t'(\mathbf{Ax} + \mathbf{b}) \\ &= \mathbf{A}^{-1}(\mathbf{Ax} + \mathbf{b}) - \mathbf{A}^{-1}\mathbf{b} \\ &= (\mathbf{A}^{-1}\mathbf{Ax} + \mathbf{A}^{-1}\mathbf{b}) - \mathbf{A}^{-1}\mathbf{b} \\ &= (\mathbf{x} + \mathbf{A}^{-1}\mathbf{b}) - \mathbf{A}^{-1}\mathbf{b} \\ &= \mathbf{x}.\end{aligned}$$

Thus $t \circ t' = t' \circ t = i$. Hence t' is an inverse for t.

G4 ASSOCIATIVITY Composition of functions is always associative.

It follows that the set of affine transformations $A(2)$ forms a group under composition of functions. ■

The above proof shows that we can calculate the inverse of an affine transformation by using the following result.

The inverse of the affine transformation $t(\mathbf{x}) = \mathbf{Ax} + \mathbf{b}$ is given by

$$t^{-1}(\mathbf{x}) = \mathbf{A}^{-1}\mathbf{x} - \mathbf{A}^{-1}\mathbf{b}.$$

Problem 3 Find the inverse of the affine transformation

$$t(\mathbf{x}) = \begin{pmatrix} 1 & 3 \\ 1 & 2 \end{pmatrix}\mathbf{x} + \begin{pmatrix} 4 \\ -2 \end{pmatrix}.$$

Having shown that the set of affine transformations forms a group under composition of functions, we now define **affine geometry** to be the study of those properties of figures in the plane $\mathbb{R}^2$ that are preserved by affine transformations. These are the so-called **affine properties** of figures. We begin our investigation of affine geometry by considering the three affine properties listed below.

Basic Properties of Affine Transformations

Affine transformations:

1. map straight lines to straight lines;
2. map parallel straight lines to parallel straight lines;
3. preserve ratios of lengths along a given straight line.

There are two approaches that we shall use to investigate these properties. One approach is to use the definition of an affine transformation to investigate the properties algebraically; we do this in Section 2.3. First, however, we investigate the properties geometrically. We begin to do this in the next subsection by introducing a special type of affine transformation for which there is a simple geometric interpretation.

2.2.2 Parallel Projections

A *parallel projection* is a one–one mapping from $\mathbb{R}^2$ onto itself, defined in the following way. First, we think of its domain and codomain as two separate copies of $\mathbb{R}^2$.

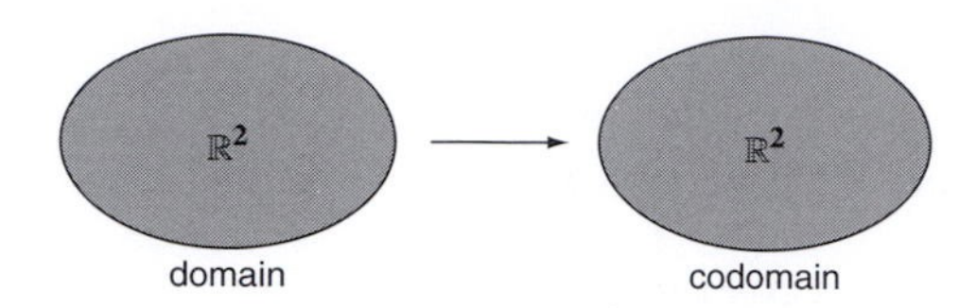

Geometrically, we can represent these copies of $\mathbb{R}^2$ by two separate planes, each equipped with a pair of rectangular axes.

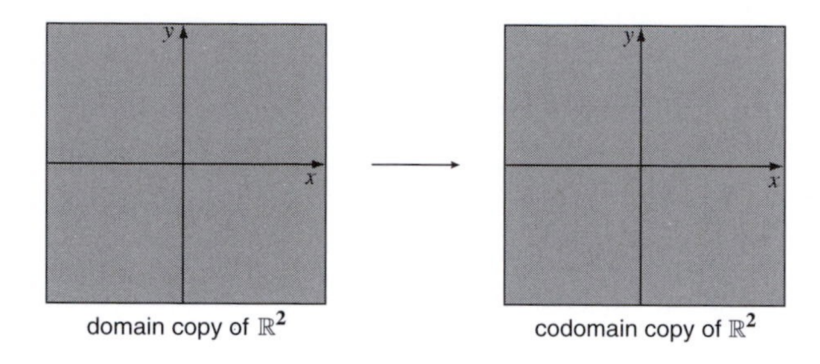

Next we place these planes into three-dimensional space; we denote the domain plane by π_1 and the codomain plane by π_2.

Now imagine parallel rays of light shining through π_1 and π_2. Each point P in the plane π_1 has a (unique) ray passing through it, that also passes through a point P', say, in the plane π_2. This provides us with a one–one correspondence between points in the two planes π_1 and π_2. We call the function p which maps each point P in π_1 to the corresponding point P' in π_2 a **parallel projection from π_1 onto π_2.**

Of course, since π_1 and π_2 represent copies of $\mathbb{R}^2$, a parallel projection is really a function from $\mathbb{R}^2$ onto itself. In Subsection 2.2.3 we show that parallel projections are affine transformations.

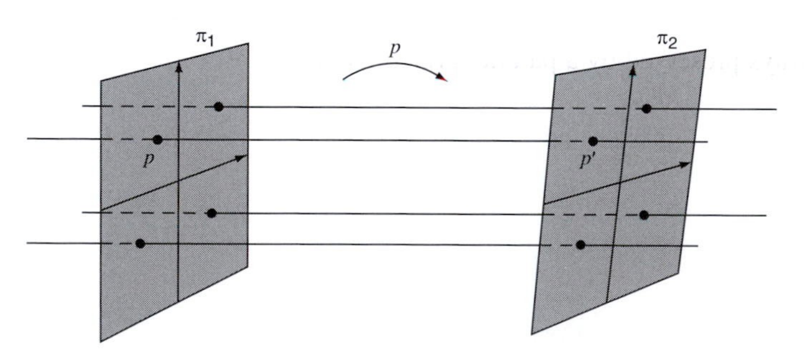

If the roles of the planes π_1 and π_2 are reversed, so that π_2 becomes the domain plane and π_1 becomes the codomain plane, then we obtain the inverse function p^{-1} which maps points P' in π_2 back to the corresponding points P in π_1. Clearly, p^{-1} is a parallel projection of π_2 onto π_1.

Each choice of location for the domain plane π_1, and the codomain plane π_2, and each choice of direction for the rays of light, yields a parallel projection. The only constraint is that the rays of light must not be parallel to either plane.

If the planes π_1 and π_2 are parallel to each other, then any parallel projection p from π_1 onto π_2 is an isometry, since the distance between any two points is unaltered.

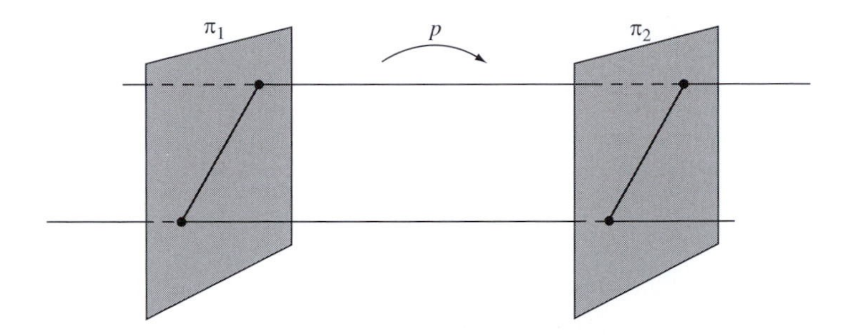

You can envisage the mapping p from π_1 onto π_2 as 'sliding π_1 parallel to itself along the family of rays'.

On the other hand, if the planes are not parallel to each other, then some distances are changed under the projection, and so the parallel projection is not an isometry; notice, however, that distances along the line of intersection of the planes π_1 and π_2 do remain unchanged by the parallel projection.

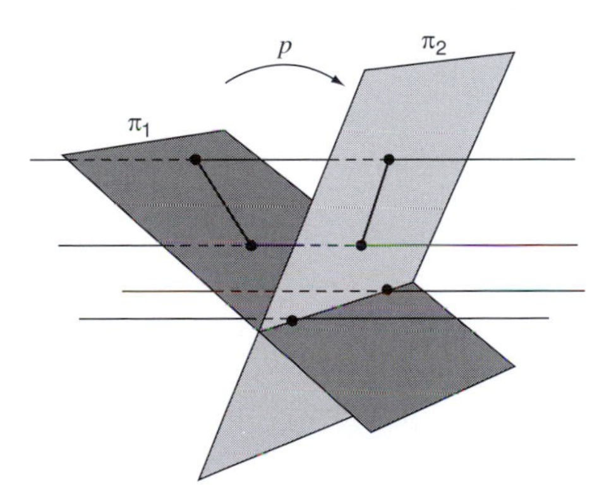

Although distances are not always preserved by a parallel projection, there are some basic properties that are preserved; three of these are listed below. As you will see, these are the same as the basic affine properties that we mentioned at the end of Subsection 2.2.1.

Basic Properties of Parallel Projections

Parallel projections:

1. map straight lines to straight lines;
2. map parallel straight lines to parallel straight lines;
3. preserve ratios of lengths along a given straight line.

Later, we will show that each basic affine property follows directly from the corresponding property for parallel projections. In anticipation of this, we first show that the properties hold for parallel projections.

Property 1 A parallel projection maps straight lines to straight lines.

Proof Let ℓ be a line in the plane π_1, and let p be a parallel projection mapping π_1 onto the plane π_2. Now consider all the rays associated with p that pass through ℓ. Since these rays are parallel, they must fill a plane. Call this plane π.

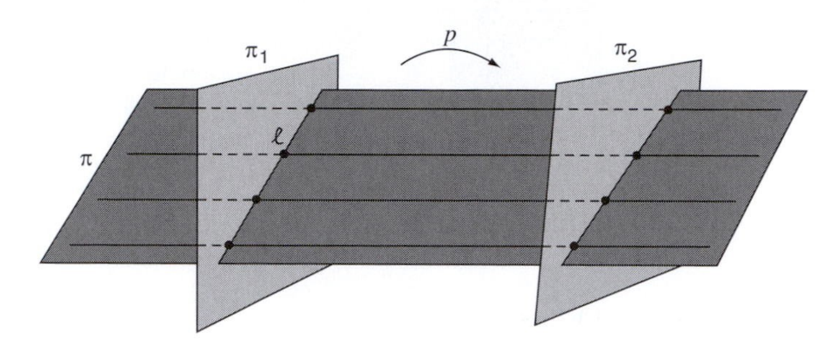

The image of ℓ under p consists of those points where the rays that pass through ℓ meet π_2. But these points are simply the points of intersection of π with π_2. Since any two intersecting planes in $\mathbb{R}^3$ meet in a line, it follows that the image of ℓ under p is a straight line. ■

Property 2 A parallel projection maps parallel straight lines to parallel straight lines.

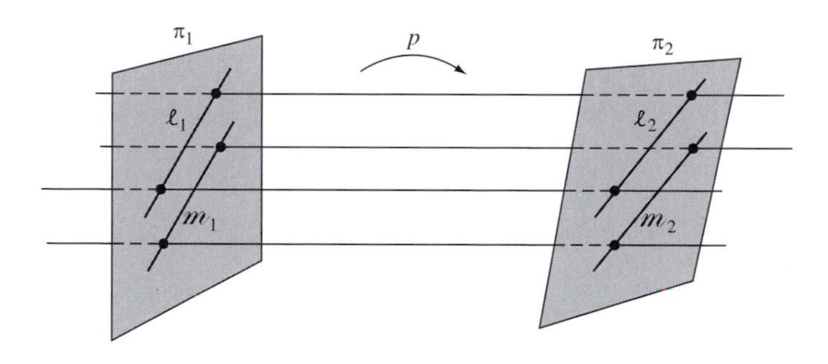

Proof Let ℓ_1 and m_1 be parallel lines in the plane π_1, and let p be a parallel projection mapping π_1 onto the plane π_2. Let ℓ_2 and m_2 be the lines in π_2 that are the images under p of ℓ_1 and m_1.

If ℓ_2 and m_2 are not parallel, they meet at some point, P_2 say. Let P_1 be the point of π_1 which maps to P_2. Then P_1 must lie on both ℓ_1 and m_1. Since ℓ_1 and m_1 are parallel, no such point of intersection can exist, which is a contradiction. It follows that ℓ_2 and m_2 must indeed be parallel. ■

P_1 is the point $p^{-1}(P_2)$.

Property 3 A parallel projection preserves ratios of lengths along a given straight line.

We shall give a slightly more general form of this property in Theorem 4 of Subsection 2.3.3.

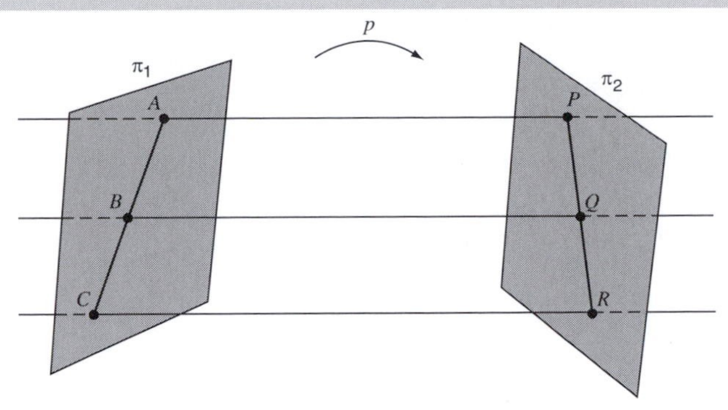

Proof Let A, B, C be three points on a line in the plane π_1, and let p be a parallel projection mapping π_1 onto the plane π_2. Let P, Q, R be the points in π_2 that are the images under p of A, B, C. We know from Property 1 that P, Q, R lie on a line; we have to show that the ratio $AB : AC$ is equal to the ratio $PQ : PR$.

If the planes π_1 and π_2 are parallel, then the parallel projection p is an isometry, and so the ratios $AB : AC$ and $PQ : PR$ are equal, as required. On the other hand, if π_1 and π_2 are not parallel, then we can construct a plane π through the point P which is parallel to π_1, as shown in the margin. This plane intersects the ray through B and Q at some point B', and the ray through C and R at some point C'. So in this case the ratios $AB : AC$ and $PB' : PC'$ are equal.

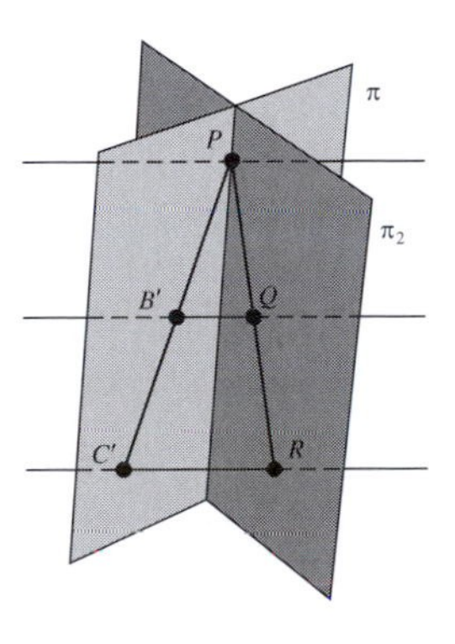

Now consider $\triangle PC'R$. The lines $B'Q$ and $C'R$ are parallel, since they are rays from the parallel projection. Hence $B'Q$ meets the sides PR and PC' in equal ratios. Thus $PQ : PR = PB' : PC'$. It follows that $PQ : PR = AB : AC$, as required. ■

Notice, in particular, that if a point is the midpoint of a line segment, then under a parallel projection the image of the point is the midpoint of the image of the line segment.

We make use of this fact in Subsection 2.2.3.

In Subsection 2.2.3 you will see why the basic properties of affine transformations and of parallel projections are the same, and you will meet some further properties of each.

2.2.3 Affine Geometry

In this subsection we explore further the ideas of affine geometry and of parallel projection in order to prove two attractive and unexpected results about ellipses. Also, we examine the relationship between affine transformations and parallel projections.

Two Results about Ellipses

First, starting with any chord ℓ of an ellipse, draw all the chords parallel to ℓ and construct their midpoints. We claim that these midpoints lie on a chord through the centre of the ellipse – that is, on a *diameter* of the ellipse.

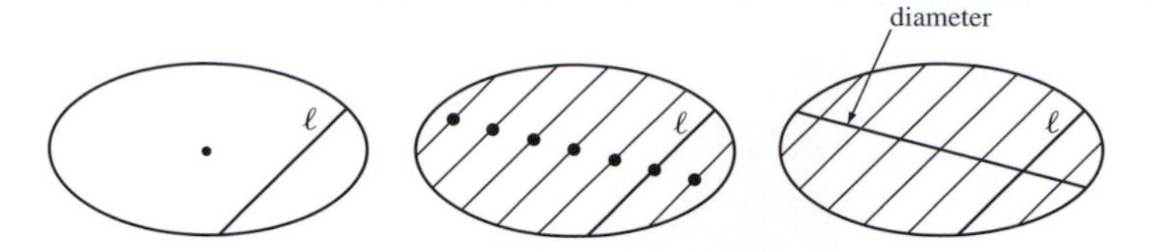

Theorem 2 Midpoint Theorem
Let ℓ be a chord of an ellipse. Then the midpoints of the chords parallel to ℓ lie on a diameter of the ellipse.

Next, start with any diameter ℓ of an ellipse and construct a second diameter m by following the construction used in Theorem 2, as shown below. Then repeat the construction starting this time with the diameter m; this might reasonably be expected to give us a third diameter of the ellipse – but, surprisingly, it gives us the diameter ℓ with which we started.

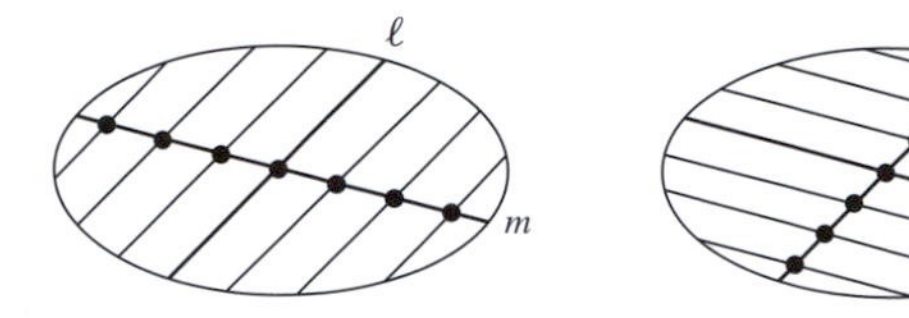

Theorem 3 Conjugate Diameters Theorem
Let ℓ be a diameter of an ellipse. Then there is another diameter m of the ellipse such that:

(a) the midpoints of all chords parallel to ℓ lie on m;
(b) the midpoints of all chords parallel to m lie on ℓ.

The directions of these two diameters are called *conjugate directions*, and the diameters are called *conjugate diameters*.

Proofs for the Special Case of a Circle

We now investigate these theorems for the special case when the ellipse is a circle. To prove the Midpoint Theorem in this case, start with a chord ℓ. If necessary, rotate the circle to ensure that ℓ is horizontal. It is then sufficient to prove that every horizontal chord is bisected by the vertical diameter, m.

Recall that a circle is an ellipse with eccentricity zero.

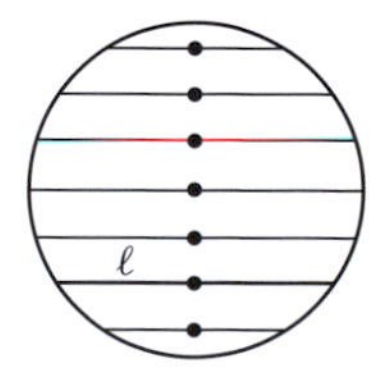

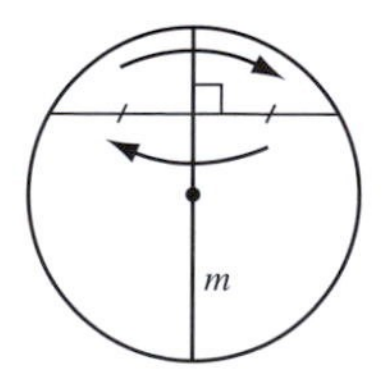

To do this note that the circle is symmetrical about m; so, reflection in m maps that part of every horizontal chord to the left of m exactly onto the part to the right of m. Since reflection preserves length, these two parts must be the same length; in other words, m bisects each horizontal chord, as required.

What about the Conjugate Diameters Theorem for the special case of the circle?

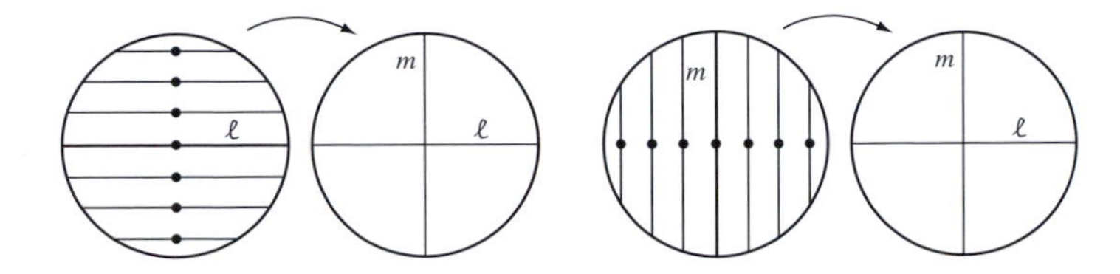

Start with the horizontal diameter ℓ, and carry out the construction of another diameter as in Theorem 2; this yields the vertical diameter m. If we then start with the vertical diameter m and repeat the construction, we obtain ℓ, the horizontal diameter of the circle. So Theorem 3 certainly holds when the ellipse is a circle.

Generalizing the Proof

We now investigate how the proofs of Theorems 2 and 3 for the circle can be turned into proofs for any kind of ellipse. The crucial fact is as follows.

Theorem 4 Given any ellipse, there is a parallel projection which maps the ellipse onto a circle.

An algebraic proof of a related theorem is given in Theorem 1 of Subsection 2.5.1.

A suitable parallel projection is illustrated below. Here the plane π_1 (initially parallel to π_2) has been tilted about the minor axis of the ellipse. Under the projection distances which are parallel to the minor axis remain unchanged, but distances parallel to the major axis are scaled by a factor which depends on the 'angle of tilt'. By choosing just the right amount of tilt we can ensure that the image of the major axis is equal in length to the image of the minor axis, thereby ensuring that the image of the ellipse is a circle.

Algebraically, in terms of a suitable coordinate system, the mapping

$$x \mapsto \frac{b}{a}x, \quad y \mapsto y,$$

maps the ellipse

$$\frac{x^2}{a^2} + \frac{y^2}{b^2} = 1,$$

to the circle $x^2 + y^2 = b^2$.

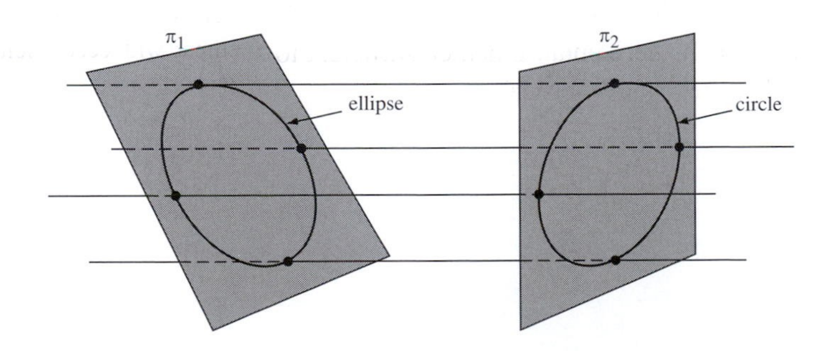

Both Theorems 2 and 3 may now be proved using the following technique. First, map the given ellipse onto a circle, using a suitable parallel projection p.

Since we have seen that the theorems hold in the case of the circle, we then map the circle back to the ellipse, using the inverse parallel projection p^{-1}. Now collinearity and parallelism are preserved under a parallel projection, as is the property of being the midpoint of a line segment, so the above two theorems, which hold for a circle, must hold also for the ellipse.

Here we are using Properties 1, 2 and 3 of parallel projections, in a crucial way.

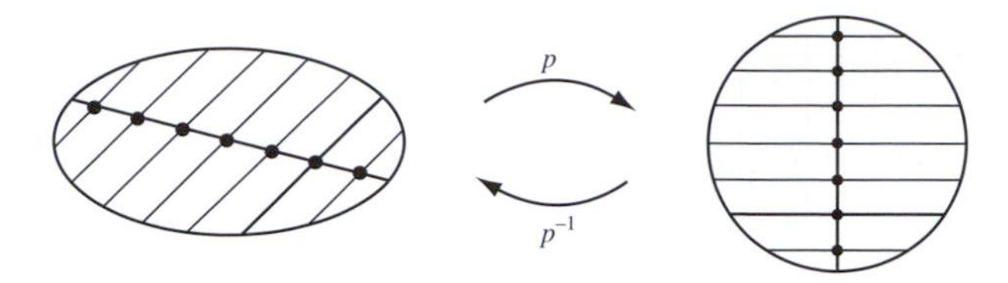

Notice that certain properties of figures, such as length and angle, are not preserved under a parallel projection. This is one difference between Euclidean geometry and affine geometry. The difference arises because the group of affine transformations is larger than the group of Euclidean transformations. In general, the larger the group that is used to define a geometry, the fewer properties the geometry has.

Affine Transformations and Parallel Projections

Earlier we mentioned that a parallel projection is a special type of affine transformation. We now show why this is indeed the case.

First, consider a parallel projection p of a plane π_1 onto a plane π_2. For the moment, suppose that the planes are aligned so that the origin in π_1 is mapped to the origin in π_2. Since ratios of lengths are preserved along a straight line, we must have, for any vector $\mathbf{v} \in \mathbb{R}^2$ and any $\lambda \in \mathbb{R}$,

Property 3, Subsection 2.2.2

$$p(\lambda \mathbf{v}) = \lambda p(\mathbf{v}). \tag{1}$$

Next, let $\mathbf{v}$ and $\mathbf{w}$ be two position vectors in π_1. Their sum, $\mathbf{v} + \mathbf{w}$, is found from the Parallelogram Law for addition of vectors, as shown in the diagram below. The images under p in π_2 are $p(\mathbf{v})$ and $p(\mathbf{w})$, and the sum of these two vectors is $p(\mathbf{v}) + p(\mathbf{w})$.

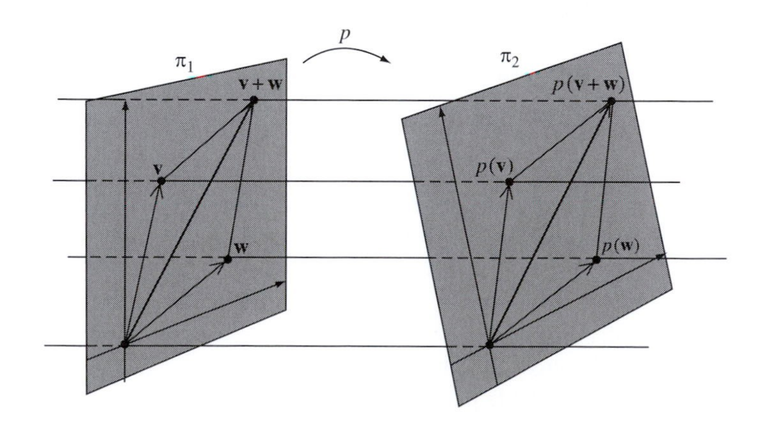

But a parallel projection maps parallel lines onto parallel lines, so it must map parallelograms onto parallelograms. Hence it must map the parallelogram in π_1 onto the parallelogram in π_2, and, in particular, it must map $\mathbf{v}+\mathbf{w}$ to $p(\mathbf{v})+p(\mathbf{w})$. We may write this as

$$p(\mathbf{v}+\mathbf{w}) = p(\mathbf{v}) + p(\mathbf{w}). \tag{2}$$

It follows from equations (1) and (2) that p must be a *linear* transformation of $\mathbb{R}^2$ onto itself.

Remember that π_1 and π_2 represent copies of $\mathbb{R}^2$.

Hence there exists some matrix $\mathbf{A}$ such that for each $\mathbf{v} \in \mathbb{R}^2$,

$$p(\mathbf{v}) = \mathbf{A}\mathbf{v}. \tag{3}$$

Since the linear transformation p is invertible, it follows that $\mathbf{A}$ is invertible.

Now suppose that the parallel projection maps the origin in π_1 to some point B with position vector $\mathbf{b}$ in π_2, as shown below. If we temporarily construct a new set of axes in π_2 that are parallel to the original axes, but which intersect at the point B, then with respect to these new axes $p(\mathbf{v}) = \mathbf{A}\mathbf{v}$ for some invertible matrix $\mathbf{A}$, as before. To express $p(\mathbf{v})$ with respect to the original axes, we simply add on the vector $\mathbf{b}$ to obtain

$$p(\mathbf{v}) = \mathbf{A}\mathbf{v} + \mathbf{b} \tag{4}$$

for some invertible 2×2 matrix $\mathbf{A}$.

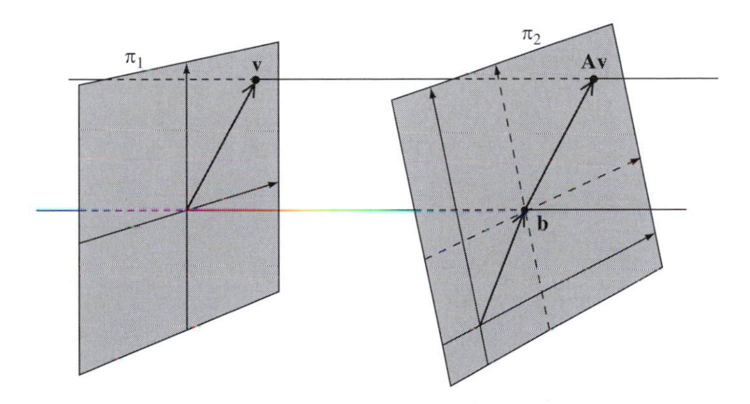

It follows from equation (4) that p must be an affine transformation.

Theorem 5 Each parallel projection is an affine transformation.

The converse is false, for it is *not* true that every affine transformation can be represented as a parallel projection.

For example, consider the so-called 'doubling map' of $\mathbb{R}^2$ to itself given by

$$t(\mathbf{v}) = 2\mathbf{v} \quad (\mathbf{v} \in \mathbb{R}^2). \tag{5}$$

This is an affine transformation, since it can be written in the form $t(\mathbf{x}) = \mathbf{Ax}+\mathbf{b}$ with $\mathbf{A} = 2\mathbf{I}$ and $\mathbf{b} = \mathbf{0}$. However, a parallel projection is *either* between two parallel planes, in which case all lengths are unchanged, *or* between two intersecting planes, in which case distances along the line of intersection are unchanged. The doubling map has neither of these properties and so is not a parallel projection.

Observation An affine transformation is not necessarily a parallel projection.

Although the doubling map is not a parallel projection, it *is* possible to double lengths in $\mathbb{R}^2$ by following one parallel projection by another: the first doubles all horizontal lengths, and the second doubles all vertical lengths. Thus the doubling map (5) can be represented as the *composition* of two parallel projections.

We end this subsection by showing that *every* affine transformation can be expressed as a composition of two parallel projections.

Recall that any affine transformation $t : \mathbb{R}^2 \to \mathbb{R}^2$ has the form

$$t(\mathbf{x}) = \mathbf{Ax} + \mathbf{b} \quad (\mathbf{x} \in \mathbb{R}^2), \tag{6}$$

where $\mathbf{A}$ is an invertible 2×2 matrix. Now, t is not a linear transformation unless $\mathbf{b} = \mathbf{0}$, but we can use methods similar to those for linear transformations to determine $\mathbf{A}$ and $\mathbf{b}$.

First, it follows from equation (6) that $t(\mathbf{0}) = \mathbf{b}$; so $\mathbf{b}$ is the image of the origin under t. If we let e and f be the coordinates of $t(\mathbf{0})$, then we can write

$$\mathbf{A} = \begin{pmatrix} a & b \\ c & d \end{pmatrix} \quad \text{and} \quad \mathbf{b} = \begin{pmatrix} e \\ f \end{pmatrix},$$

where a, b, c, d are real numbers that have yet to be found. It follows from equation (6) that the images under t of the points (1, 0) and (0, 1) are given by

$$\begin{pmatrix} a & b \\ c & d \end{pmatrix}\begin{pmatrix} 1 \\ 0 \end{pmatrix} + \begin{pmatrix} e \\ f \end{pmatrix} = \begin{pmatrix} a \\ c \end{pmatrix} + \begin{pmatrix} e \\ f \end{pmatrix}$$

and

$$\begin{pmatrix} a & b \\ c & d \end{pmatrix}\begin{pmatrix} 0 \\ 1 \end{pmatrix} + \begin{pmatrix} e \\ f \end{pmatrix} = \begin{pmatrix} b \\ d \end{pmatrix} + \begin{pmatrix} e \\ f \end{pmatrix}.$$

So if, in addition to $t(\mathbf{0}) = (e, f)$, we know the points onto which (1, 0) and (0, 1) are mapped by t, then we can determine the values of a, b, c and d. Indeed, we have

$$(a, c) = t(1, 0) - (e, f) \quad \text{and} \quad (b, d) = t(0, 1) - (e, f).$$

Notice that for an affine transformation t, the images $t(1, 0)$, $t(0, 1)$ and $t(0, 0) = (e, f)$ cannot be collinear, for if they were, then (a, c) and (b, d) would be linearly dependent, and $\mathbf{A}$ would not be invertible.

It follows that an affine transformation is *uniquely* determined by its effect on the three non-collinear points (0, 0), (1, 0) and (0, 1). We shall return to this method of determining affine transformations in Section 2.3.

So suppose that a given affine transformation t maps the points (0, 0), (1, 0) and (0, 1) to three non-collinear points P, Q and R, respectively. In order to express t as the composition of two parallel projections p_1 and p_2, we need to define p_1 and p_2 in such a way that $p_2 \circ p_1$ has the same effect as t on (0, 0), (1, 0) and (0, 1). To do this, we first define p_1 so that it maps (0, 0) to P, (1, 0) to Q, and (0, 1) to some point X, say, and then define p_2 so that it maps X to R while leaving P and Q fixed.

Uniqueness then guarantees that $t = p_2 \circ p_1$.

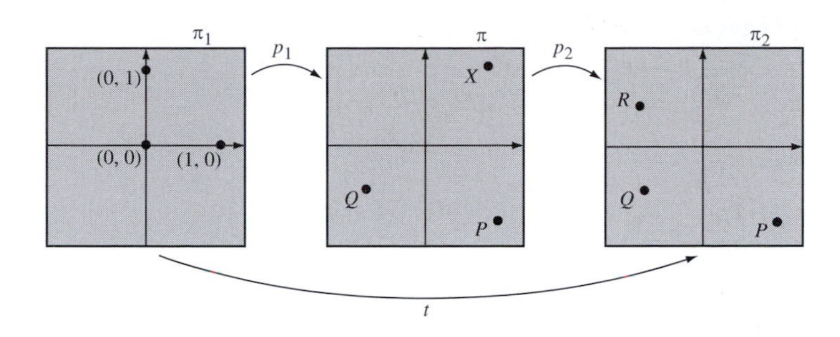

To construct p_1 we embed its domain plane π_1, and its codomain plane π, into $\mathbb{R}^3$ so that the point (0, 0) in π_1 coincides with the point P in π, as shown below. It does not matter how this is done, provided that (1, 0) does not lie in π. We then define p_1 by the family of rays that are parallel to the ray through the point (1, 0) in π_1 and the point Q in π. When defined in this way, p_1 maps (0, 0) to P, (1, 0) to Q, and (0, 1) to some point X, as required.

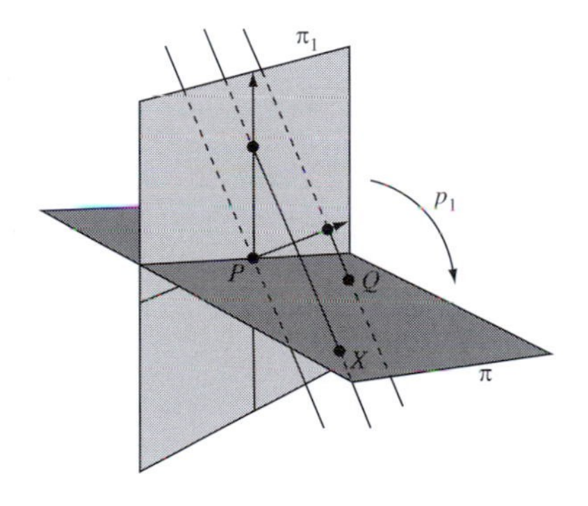

For clarity, we have omitted the axes from the plane π.

To construct p_2 we embed its domain plane π, and its codomain plane π_2, into $\mathbb{R}^3$ so that the points P and Q in π coincide with the points P and Q in π_2, as shown below. Again it does not matter how this is done, provided that X does not lie in π_2. We then define p_2 by the family of rays that are parallel to the ray through the point X in π and the point R in π_2. Then p_2 leaves P and Q fixed and maps X to R.

See the figure below.

Overall, the composite $p_2 \circ p_1$ of the two parallel projections maps (0, 0), (1, 0) and (0, 1) to P, Q and R, respectively. Now p_1 and p_2 are affine transformations, so $p_2 \circ p_1$ is also an affine transformation. Furthermore, $p_2 \circ p_1$ maps

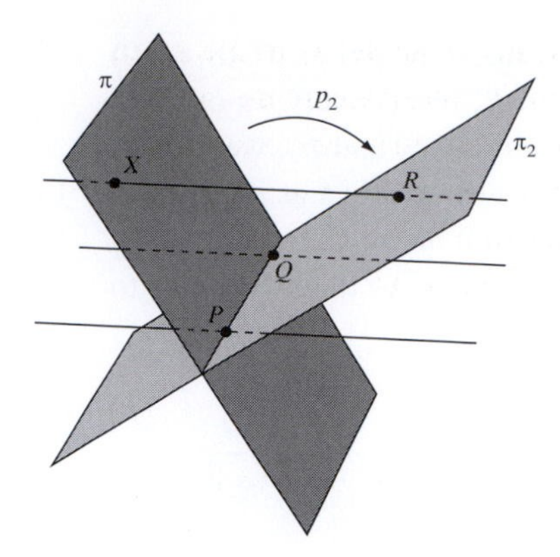

(0, 0), (1, 0) and (0, 1) to the same points as does t. Since such affine transformations are unique, it follows that $t = p_2 \circ p_1$. We have therefore demonstrated the following result.

Theorem 6 An affine transformation can be expressed as the composite of two parallel projections.

An important consequence of this theorem is that all properties of figures that are unchanged by parallel projections must also be unchanged by affine transformations. In particular, the three properties of parallel projections that we met in Subsection 2.2.2 must, in fact, be affine properties.

2.3 Properties of Affine Transformations

In the previous section you saw how parallel projections can be used to explore affine geometry from a visual point of view. In this section we explore some of the same ideas from an algebraic point of view.

2.3.1 Images of Sets Under Affine Transformations

We begin by describing how to find the image of a line under an affine transformation. To do this, recall that an affine transformation is a mapping $t : \mathbb{R}^2 \to \mathbb{R}^2$ given by a formula of the form

$$t(\mathbf{x}) = \mathbf{A}\mathbf{x} + \mathbf{b}, \tag{1}$$

where $\mathbf{A}$ is an invertible 2×2 matrix. The set of such transformations forms a group, in which the transformation inverse to t is given by

$$t^{-1}(\mathbf{x}) = \mathbf{A}^{-1}\mathbf{x} - \mathbf{A}^{-1}\mathbf{b}. \tag{2}$$

Subsection 2.2.1

When equations (1) and (2) are used to find images under t, it is easy to confuse points in the domain plane with points in the codomain plane, as both planes are copies of $\mathbb{R}^2$. To avoid such confusion, we often reserve the symbol $\mathbf{x}$ and the coordinates (x, y) for points in the domain of t, and use the symbol $\mathbf{x}'$ and the coordinates (x', y') to denote the image of $\mathbf{x}$ under t.

With this notation, we may rewrite equations (1) and (2) in the form

$$\mathbf{x}' = \mathbf{A}\mathbf{x} + \mathbf{b}, \tag{3}$$

$$\mathbf{x} = \mathbf{A}^{-1}\mathbf{x}' - \mathbf{A}^{-1}\mathbf{b}. \tag{4}$$

The next example illustrates how these equations can be used to find the image of a line under an affine transformation.

Example 1 Determine the image of the line $y = 2x$ under the affine transformation

$$t(\mathbf{x}) = \begin{pmatrix} 4 & 1 \\ 2 & 1 \end{pmatrix}\mathbf{x} + \begin{pmatrix} 2 \\ -1 \end{pmatrix} \quad \left(\mathbf{x} \in \mathbb{R}^2\right). \tag{5}$$

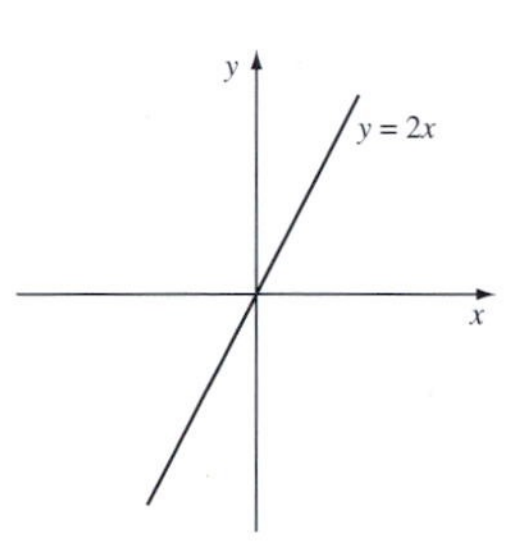

Solution Let (x, y) be an arbitrary point on the line $y = 2x$, and let (x', y') be the image of (x, y) under t. Then

$$\begin{pmatrix} x' \\ y' \end{pmatrix} = \begin{pmatrix} 4 & 1 \\ 2 & 1 \end{pmatrix}\begin{pmatrix} x \\ y \end{pmatrix} + \begin{pmatrix} 2 \\ -1 \end{pmatrix}.$$

Next we use equation (4) to express (x, y) in terms of (x', y'). We have

$$\begin{pmatrix} 4 & 1 \\ 2 & 1 \end{pmatrix}^{-1} = \begin{pmatrix} \frac{1}{2} & -\frac{1}{2} \\ -1 & 2 \end{pmatrix} \quad \text{and} \quad \begin{pmatrix} \frac{1}{2} & -\frac{1}{2} \\ -1 & 2 \end{pmatrix}\begin{pmatrix} 2 \\ -1 \end{pmatrix} = \begin{pmatrix} \frac{3}{2} \\ -4 \end{pmatrix},$$

Recall that the inverse of the invertible matrix $\mathbf{A} = \begin{pmatrix} a & b \\ c & d \end{pmatrix}$ is

$$\mathbf{A}^{-1} = \frac{1}{ad - bc} \times \begin{pmatrix} d & -b \\ -c & a \end{pmatrix}.$$

so

$$\begin{pmatrix} x \\ y \end{pmatrix} = \begin{pmatrix} \frac{1}{2} & -\frac{1}{2} \\ -1 & 2 \end{pmatrix}\begin{pmatrix} x' \\ y' \end{pmatrix} + \begin{pmatrix} -\frac{3}{2} \\ 4 \end{pmatrix}.$$

It follows that under the inverse mapping t^{-1} we have

$$x = \tfrac{1}{2}x' - \tfrac{1}{2}y' - \tfrac{3}{2} \quad \text{and} \quad y = -x' + 2y' + 4.$$

Since x and y are related by the equation $y = 2x$, it follows that x' and y' are related by the equation

$$-x' + 2y' + 4 = 2\left(\tfrac{1}{2}x' - \tfrac{1}{2}y' - \tfrac{3}{2}\right),$$

which simplifies to

$$2x' - 3y' = 7.$$

Dropping the dashes, we see that the image of the line $y = 2x$ under t is the line

$$2x - 3y = 7. \qquad \square$$

Problem 1 Determine the image of the line $3x - y + 1 = 0$ under the affine transformation

$$t(\mathbf{x}) = \begin{pmatrix} \frac{1}{2} & -\frac{1}{2} \\ -1 & 2 \end{pmatrix} \mathbf{x} + \begin{pmatrix} -\frac{3}{2} \\ 4 \end{pmatrix} \quad \left(\mathbf{x} \in \mathbb{R}^2\right).$$

Problem 2 Determine the image of the circle $x^2 + y^2 = 1$ under the affine transformation

$$t(\mathbf{x}) = \begin{pmatrix} \frac{1}{2} & -\frac{1}{2} \\ -1 & 2 \end{pmatrix} \mathbf{x} + \begin{pmatrix} -\frac{3}{2} \\ 4 \end{pmatrix} \quad \left(\mathbf{x} \in \mathbb{R}^2\right).$$

The same technique can be used to find the images of other types of figures, such as other conics. You will meet some examples of this in Section 2.5.

2.3.2 The Fundamental Theorem of Affine Geometry

The algebraic approach can also be used to investigate whether there is an affine transformation which maps one given figure onto another. Recall that if there is such a transformation, then the two figures are said to be **affine-congruent**. This concept of congruence is important because, as we explained in Section 2.1, figures that are affine-congruent to each other share the same affine properties.

In this subsection we prove the remarkable result that *all* triangles are affine-congruent and therefore share the same affine properties. In fact, since a triangle is completely determined by its three vertices, the congruence of triangles follows from the so-called *Fundamental Theorem of Affine Geometry* which states that any three non-collinear points can be mapped to any other three non-collinear points by an affine transformation.

This is very different to Euclidean geometry, where two triangles are congruent only if they have the same shape and size.

First, recall that in Subsection 2.2.3 we described how the points (0, 0), (1, 0) and (0, 1) in $\mathbb{R}^2$ can be mapped to any three non-collinear points P, Q and R by an affine transformation. This transformation is unique in the sense that it is completely determined by the choice of P, Q and R. The following example should remind you of how such transformations are constructed.

There the mapping was constructed in a geometric manner. In this subsection we construct the mapping algebraically.

Example 2 Determine the affine transformation which maps the points (0, 0), (1, 0) and (0, 1) to the points (3, 2), (5, 8) and (7, 3), respectively.

Solution Let t be the affine transformation given by

$$t : \begin{pmatrix} x \\ y \end{pmatrix} \mapsto \begin{pmatrix} a & b \\ c & d \end{pmatrix} \begin{pmatrix} x \\ y \end{pmatrix} + \begin{pmatrix} e \\ f \end{pmatrix}. \tag{6}$$

Since $t(0,0) = (3,2)$, it follows from (6) that $e = 3$ and $f = 2$.

Next, $t(1,0) = (5,8)$, so it follows from (6) that

$$\begin{pmatrix} 5 \\ 8 \end{pmatrix} = \begin{pmatrix} a & b \\ c & d \end{pmatrix} \begin{pmatrix} 1 \\ 0 \end{pmatrix} + \begin{pmatrix} 3 \\ 2 \end{pmatrix} = \begin{pmatrix} a \\ c \end{pmatrix} + \begin{pmatrix} 3 \\ 2 \end{pmatrix}.$$

The first column of the matrix for t is therefore

$$\begin{pmatrix} a \\ c \end{pmatrix} = \begin{pmatrix} 5 \\ 8 \end{pmatrix} - \begin{pmatrix} 3 \\ 2 \end{pmatrix} = \begin{pmatrix} 2 \\ 6 \end{pmatrix}.$$

Finally, $t(0, 1) = (7, 3)$, so that

$$\begin{pmatrix} 7 \\ 3 \end{pmatrix} = \begin{pmatrix} a & b \\ c & d \end{pmatrix}\begin{pmatrix} 0 \\ 1 \end{pmatrix} + \begin{pmatrix} 3 \\ 2 \end{pmatrix} = \begin{pmatrix} b \\ d \end{pmatrix} + \begin{pmatrix} 3 \\ 2 \end{pmatrix}.$$

The second column of the matrix for t is therefore

$$\begin{pmatrix} b \\ d \end{pmatrix} = \begin{pmatrix} 7 \\ 3 \end{pmatrix} - \begin{pmatrix} 3 \\ 2 \end{pmatrix} = \begin{pmatrix} 4 \\ 1 \end{pmatrix}.$$

Hence the desired affine transformation is given by

$$t : \begin{pmatrix} x \\ y \end{pmatrix} \mapsto \begin{pmatrix} 2 & 4 \\ 6 & 1 \end{pmatrix}\begin{pmatrix} x \\ y \end{pmatrix} + \begin{pmatrix} 3 \\ 2 \end{pmatrix}.$$ □

In general, if we want to find an affine transformation t of the form

$$t : \begin{pmatrix} x \\ y \end{pmatrix} \mapsto \begin{pmatrix} a & b \\ c & d \end{pmatrix}\begin{pmatrix} x \\ y \end{pmatrix} + \begin{pmatrix} e \\ f \end{pmatrix} \tag{7}$$

which maps (0, 0) to $\mathbf{p}$, (1, 0) to $\mathbf{q}$ and (0, 1) to $\mathbf{r}$, then we must choose a, b, c, d, e and f so that

$$\begin{aligned} \mathbf{p} &= t(0,0) = (e, f), && \text{so } (e, f) = \mathbf{p}; \\ \mathbf{q} &= t(1,0) = (a,c) + (e, f), && \text{so } (a,c) = \mathbf{q} - \mathbf{p}; \\ \mathbf{r} &= t(0,1) = (b,d) + (e, f), && \text{so } (b,d) = \mathbf{r} - \mathbf{p}. \end{aligned}$$

Notice that any three points $\mathbf{p}$, $\mathbf{q}$ and $\mathbf{r}$ uniquely determine a transformation t of the form (7), but t is *affine* only if the matrix

$$\mathbf{A} = \begin{pmatrix} a & b \\ c & d \end{pmatrix}$$

is invertible. Since the columns of $\mathbf{A}$ correspond to the vectors $\mathbf{q} - \mathbf{p}$ and $\mathbf{r} - \mathbf{p}$, it follows that $\mathbf{A}$ is invertible only if the vectors $\mathbf{q} - \mathbf{p}$ and $\mathbf{r} - \mathbf{p}$ are linearly independent. That is, provided that $\mathbf{p}$, $\mathbf{q}$ and $\mathbf{r}$ are *not* collinear.

So if $\mathbf{p}$, $\mathbf{q}$ and $\mathbf{r}$ are *not* collinear, then we can use the following strategy to find an affine transformation which maps (0, 0) to $\mathbf{p}$, (1, 0) to $\mathbf{q}$ and (0, 1) to $\mathbf{r}$.

Strategy To determine the unique affine transformation $t(\mathbf{x}) = \mathbf{A}\mathbf{x} + \mathbf{b}$ which maps (0, 0), (1, 0) and (0, 1) to the three non-collinear points $\mathbf{p}$, $\mathbf{q}$ and $\mathbf{r}$, respectively:

1. take $\mathbf{b} = \mathbf{p}$;
2. take $\mathbf{A}$ to be the matrix with columns given by $\mathbf{q} - \mathbf{p}$ and $\mathbf{r} - \mathbf{p}$.

Problem 3 Use the strategy to determine the affine transformation which maps the points (0, 0), (1, 0) and (0, 1) to the points (2, 3), (1, 6) and (3, −1), respectively.

We shall use the results of these two problems shortly, in Example 3.

Problem 4 Use the strategy to determine the affine transformation which maps the points (0, 0), (1, 0) and (0, 1) to the points (1, −2), (2, 1) and (−3, 5), respectively.

Notice that the inverse of the transformation in Problem 3 is an affine transformation which maps the points (2, 3), (1, 6) and (3, −1) to the points (0, 0), (1, 0) and (0, 1), respectively. So if, after applying this inverse, we apply the affine transformation in Problem 4, then the overall effect is that of a composite affine transformation which sends the points (2, 3), (1, 6) and (3, −1) to the points (1, −2), (2, 1) and (−3, 5), respectively.

$(2,3) \mapsto (0,0) \mapsto (1,-2)$
$(1,6) \mapsto (1,0) \mapsto (2,1)$
$(3,-1) \mapsto (0,1) \mapsto (-3,5)$

In a similar way, we can find an affine transformation which sends any three non-collinear points to any other three non-collinear points.

Theorem 1 Fundamental Theorem of Affine Geometry
Let **p**, **q**, **r** and **p**′, **q**′, **r**′ be two sets of three non-collinear points in $\mathbb{R}^2$. Then:

(a) there is an affine transformation t which maps **p**, **q** and **r** to **p**′, **q**′ and **r**′, respectively;
(b) the affine transformation t is unique.

Proof

(a) Let t_1 be the affine transformation which maps (0, 0), (1, 0) and (0, 1) to the points **p**, **q** and **r**, respectively, and let t_2 be the affine transformation which maps (0, 0), (1, 0) and (0, 1) to the points **p**′, **q**′ and **r**′, respectively. Then the composite $t = t_2 \circ t_1^{-1}$ is an affine transformation, and it maps **p**, **q** and **r** to **p**′, **q**′ and **r**′, respectively.

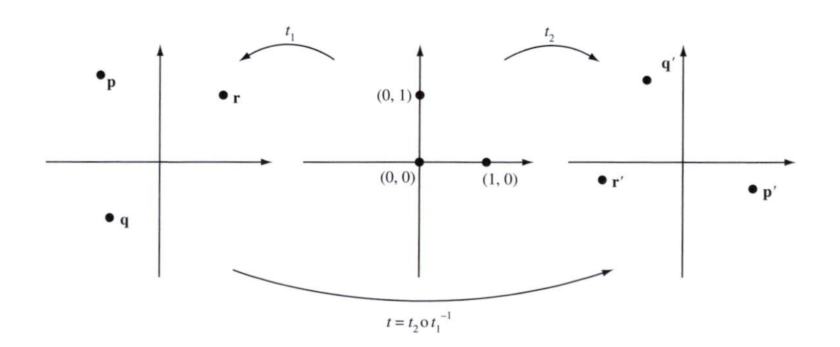

$\mathbf{p} \overset{t_1^{-1}}{\mapsto} (0,0) \overset{t_2}{\mapsto} \mathbf{p}'$
$\mathbf{q} \mapsto (1,0) \mapsto \mathbf{q}'$
$\mathbf{r} \mapsto (0,1) \mapsto \mathbf{r}'$

(b) Suppose that t and s are both affine transformations which map $\mathbf{p}$, $\mathbf{q}$ and $\mathbf{r}$ to $\mathbf{p}'$, $\mathbf{q}'$ and $\mathbf{r}'$, respectively, and let t_1 be the affine transformation defined in part (a). Then the composites $t \circ t_1$ and $s \circ t_1$ are both affine transformations which map (0, 0), (1, 0) and (0, 1) to $\mathbf{p}'$, $\mathbf{q}'$ and $\mathbf{r}'$, respectively. Since an affine transformation is uniquely determined by its effect on the points (0, 0), (1, 0) and (0, 1), it follows that $t \circ t_1 = s \circ t_1$.

If we then compose both $t \circ t_1$ and $s \circ t_1$ on the right with t_1^{-1}, it follows that $t = s$. Thus the mapping t constructed in part (a) is unique. ■

$$(0,0) \overset{t_1}{\mapsto} \mathbf{p} \overset{t \text{ or } s}{\mapsto} \mathbf{p}'$$
$$(1,0) \mapsto \mathbf{q} \mapsto \mathbf{q}'$$
$$(0,1) \mapsto \mathbf{r} \mapsto \mathbf{r}'$$

Now suppose that we are given two arbitrary triangles ΔABC and ΔDEF. By the Fundamental Theorem there is an affine transformation which maps the vertices A, B, C to the vertices D, E, F, respectively. Since this transformation maps straight lines to straight lines, it must map the sides of ΔABC to the sides of ΔDEF, so we have the following important corollary. This will be used extensively in Section 2.4.

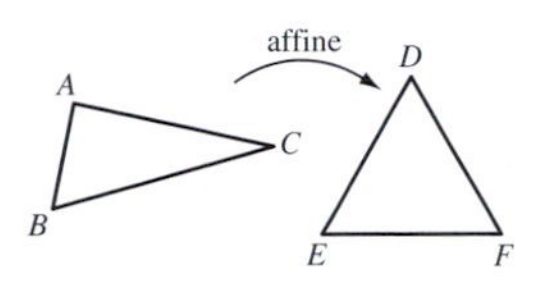

Corollary All triangles are affine-congruent.

In order to find the affine transformation which maps one triangle, vertex to vertex, onto another triangle, we follow the strategy used in part (a) of the proof of the Fundamental Theorem.

Strategy To determine the affine transformation t which maps three non-collinear points $\mathbf{p}$, $\mathbf{q}$ and $\mathbf{r}$ to another three non-collinear points $\mathbf{p}'$, $\mathbf{q}'$ and $\mathbf{r}'$, respectively:

1. determine the affine transformation t_1 which maps (0, 0), (1, 0) and (0, 1) to the points $\mathbf{p}$, $\mathbf{q}$ and $\mathbf{r}$, respectively;
2. determine the affine transformation t_2 which maps (0, 0), (1, 0) and (0, 1) to the points $\mathbf{p}'$, $\mathbf{q}'$ and $\mathbf{r}'$, respectively;
3. calculate the composite $t = t_2 \circ t_1^{-1}$.

Recall that the previous strategy explained how t_1 and t_2 can be determined.

Example 3 Determine the affine transformation which maps the points (2, 3), (1, 6) and (3, −1) to the points (1, −2), (2, 1) and (−3, 5), respectively.

Solution You have already seen in Problem 3 that the affine transformation t_1 which maps the points (0, 0), (1, 0) and (0, 1) to the points (2, 3), (1, 6) and (3, −1), respectively, is given by

$$t_1(\mathbf{x}) = \begin{pmatrix} -1 & 1 \\ 3 & -4 \end{pmatrix} \mathbf{x} + \begin{pmatrix} 2 \\ 3 \end{pmatrix}.$$

Also, in Problem 4 you saw that the affine transformation t_2 which maps the points (0, 0), (1, 0) and (0, 1) to the points (1, −2), (2, 1) and (−3, 5), respectively, is given by

$$t_2(\mathbf{x}) = \begin{pmatrix} 1 & -4 \\ 3 & 7 \end{pmatrix}\mathbf{x} + \begin{pmatrix} 1 \\ -2 \end{pmatrix}.$$

Following the strategy, we need to find the inverse of t_1. We have

$$\begin{pmatrix} -1 & 1 \\ 3 & -4 \end{pmatrix}^{-1} = \begin{pmatrix} -4 & -1 \\ -3 & -1 \end{pmatrix}$$

and

$$\begin{pmatrix} -4 & -1 \\ -3 & -1 \end{pmatrix}\begin{pmatrix} 2 \\ 3 \end{pmatrix} = \begin{pmatrix} -11 \\ -9 \end{pmatrix},$$

so that the inverse of t_1 is given by

$$t_1^{-1}(\mathbf{x}) = \begin{pmatrix} -4 & -1 \\ -3 & -1 \end{pmatrix}\mathbf{x} + \begin{pmatrix} 11 \\ 9 \end{pmatrix}.$$

Thus the affine transformation which maps the points (2, 3), (1, 6) and (3, −1) to the points (1, −2), (2, 1) and (−3, 5), respectively, is given by

$$\begin{aligned} t(\mathbf{x}) &= t_2 \circ t_1^{-1}(\mathbf{x}) \\ &= t_2\left(\begin{pmatrix} -4 & -1 \\ -3 & -1 \end{pmatrix}\mathbf{x} + \begin{pmatrix} 11 \\ 9 \end{pmatrix}\right) \\ &= \begin{pmatrix} 1 & -4 \\ 3 & 7 \end{pmatrix}\left(\begin{pmatrix} -4 & -1 \\ -3 & -1 \end{pmatrix}\mathbf{x} + \begin{pmatrix} 11 \\ 9 \end{pmatrix}\right) + \begin{pmatrix} 1 \\ -2 \end{pmatrix} \\ &= \left(\begin{pmatrix} 8 & 3 \\ -33 & -10 \end{pmatrix}\mathbf{x} + \begin{pmatrix} -25 \\ 96 \end{pmatrix}\right) + \begin{pmatrix} 1 \\ -2 \end{pmatrix} \\ &= \begin{pmatrix} 8 & 3 \\ -33 & -10 \end{pmatrix}\mathbf{x} + \begin{pmatrix} -24 \\ 94 \end{pmatrix}. \end{aligned}$$

□

Problem 5 Determine the affine transformation which maps the points (1, −1), (2, −2) and (3, −4) to the points (8, 13), (3, 4) and (0, −1), respectively.

2.3.3 Proofs of the Basic Properties of Affine Transformations

In Subsection 2.2.2 we used parallel projections to demonstrate that affine transformations have the following basic properties: they map straight lines to straight lines, they map parallel lines to parallel lines, and they preserve ratios of lengths along a given straight line. We now give algebraic proofs of these assertions.

Theorem 2 An affine transformation maps straight lines to straight lines.

Proof

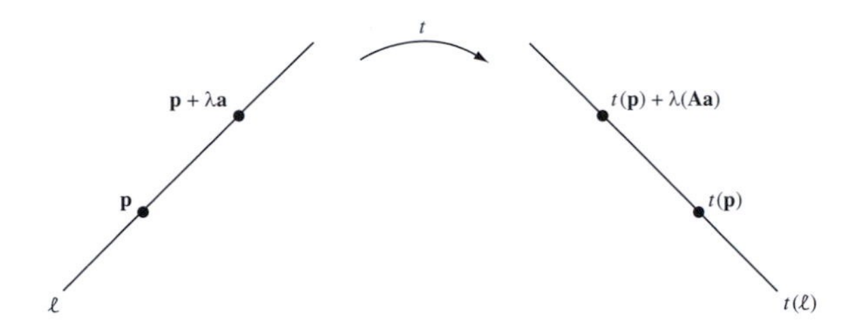

Let ℓ be a line through a point with position vector $\mathbf{p}$, and let the direction of ℓ be that of some vector $\mathbf{a}$. Then

$$\ell = \{\mathbf{p} + \lambda\mathbf{a} : \lambda \in \mathbb{R}\}.$$

Now let $t : \mathbb{R}^2 \to \mathbb{R}^2$ be an affine transformation given by

$$t(\mathbf{x}) = \mathbf{Ax} + \mathbf{b}.$$

We can find the image under t of an arbitrary point $\mathbf{p} + \lambda\mathbf{a}$ on ℓ as follows:

$$\begin{aligned} t(\mathbf{p} + \lambda\mathbf{a}) &= \mathbf{A}(\mathbf{p} + \lambda\mathbf{a}) + \mathbf{b} \\ &= (\mathbf{Ap} + \mathbf{b}) + \lambda\mathbf{Aa} \\ &= t(\mathbf{p}) + \lambda\mathbf{Aa}. \end{aligned}$$

So the image of ℓ is the set

$$t(\ell) = \{t(\mathbf{p}) + \lambda\mathbf{Aa} : \lambda \in \mathbb{R}\},$$

which is a line through $t(\mathbf{p})$ in the direction of the vector $\mathbf{Aa}$. ■

Theorem 3 An affine transformation maps parallel straight lines to parallel straight lines.

Proof

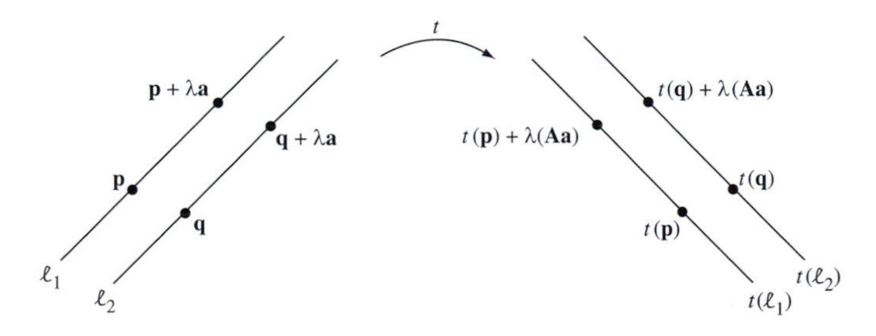

Let ℓ_1 and ℓ_2 be parallel lines through the points with position vectors $\mathbf{p}$ and $\mathbf{q}$, respectively, and let the direction of the lines be that of the vector $\mathbf{a}$. Then

$$\ell_1 = \{\mathbf{p} + \lambda\mathbf{a} : \lambda \in \mathbb{R}\} \quad \text{and} \quad \ell_2 = \{\mathbf{q} + \lambda\mathbf{a} : \lambda \in \mathbb{R}\}.$$

As in the proof of Theorem 2, the images of ℓ_1 and ℓ_2 under the affine transformation $t(\mathbf{x}) = \mathbf{Ax} + \mathbf{b}$ are the sets

$$t(\ell_1) = \{t(\mathbf{p}) + \lambda\mathbf{Aa} : \lambda \in \mathbb{R}\} \quad \text{and} \quad t(\ell_2) = \{t(\mathbf{q}) + \lambda\mathbf{Aa} : \lambda \in \mathbb{R}\}.$$

These sets are straight lines which pass through the image points $t(\mathbf{p})$ and $t(\mathbf{q})$, both in the same direction as that of the vector $\mathbf{Aa}$. Hence the two image lines under t are parallel, as claimed. ∎

Rather than prove that affine transformations preserve ratios of lengths along a given straight line, as in Property 3 of Subsection 2.2.2, we prove the following more general result illustrated in the margin. The original result follows because any line is parallel to itself.

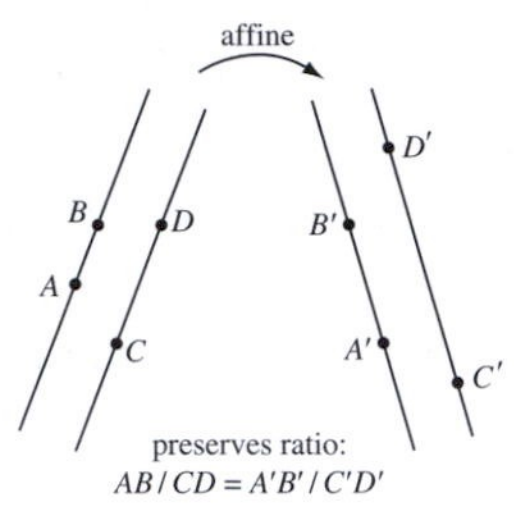

Theorem 4 An affine transformation preserves ratios of lengths along parallel straight lines.

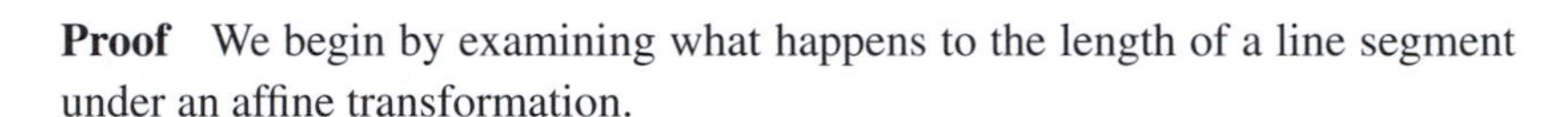

Proof We begin by examining what happens to the length of a line segment under an affine transformation.

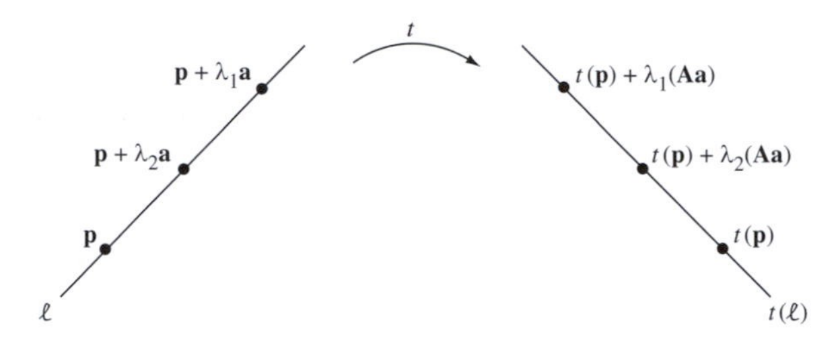

Let ℓ be a line through a point with position vector $\mathbf{p}$, and let the direction of ℓ be that of some *unit* vector $\mathbf{a}$. Then

$$\ell = \{\mathbf{p} + \lambda\mathbf{a} : \lambda \in \mathbb{R}\}.$$

As in the proof of Theorem 2, the image of ℓ under the affine transformation $t(\mathbf{x}) = \mathbf{Ax} + \mathbf{b}$ is the line

$$t(\ell) = \{t(\mathbf{p}) + \lambda\mathbf{Aa} : \lambda \in \mathbb{R}\}.$$

Now consider a segment of ℓ with endpoints $\mathbf{p} + \lambda_1\mathbf{a}$ and $\mathbf{p} + \lambda_2\mathbf{a}$. Since $\mathbf{a}$ is a unit vector, the length of the segment is

$$\|(\mathbf{p} + \lambda_2\mathbf{a}) - (\mathbf{p} + \lambda_1\mathbf{a})\| = |\lambda_2 - \lambda_1| \cdot \|\mathbf{a}\| = |\lambda_2 - \lambda_1|.$$

Recall that $\|\mathbf{a}\|$ means the length of $\mathbf{a}$.

The image of the segment has endpoints $t(\mathbf{p}) + \lambda_1\mathbf{Aa}$ and $t(\mathbf{p}) + \lambda_2\mathbf{Aa}$, so the image of the segment has length

$$\|(t(\mathbf{p}) + \lambda_2\mathbf{Aa}) - (t(\mathbf{p}) + \lambda_1\mathbf{Aa})\| = |\lambda_2 - \lambda_1| \cdot \|\mathbf{Aa}\|.$$

So, in the process of mapping segments along ℓ to segments along $t(\ell)$, lengths are stretched by the factor $\|\mathbf{Aa}\|$. Since this factor is the same for all segments which lie along lines parallel to $\mathbf{a}$, it follows that the *ratios* of lengths along parallel lines are unchanged by t. ∎

2.4 Using the Fundamental Theorem of Affine Geometry

In this section we explain how the Fundamental Theorem of Affine Geometry can be used to deduce the fact that the medians of any triangle are concurrent from the special case that the medians of an equilateral triangle are concurrent. We then use similar methods to prove the classical theorems of Ceva and Menelaus.

These results are named after Giovanni Ceva (Italian mathematician, 1647/48–1734) and Menelaus of Alexandria (Greek geometer, 1st Century AD).

2.4.1 The Median Theorem

Let ΔABC be an arbitrary triangle in the plane. If you join the midpoint of each side of the triangle to the opposite vertex (these lines are called the *medians* of the triangle), these three lines appear to pass through a single point. In fact, no matter what triangle you choose, you find that its medians meet in a single point.

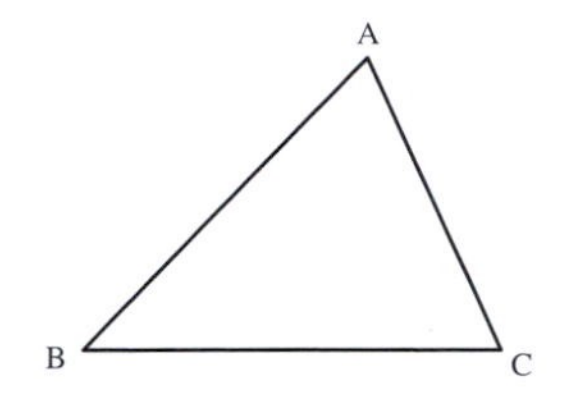

Theorem 1 Median Theorem
The medians of any triangle are concurrent.

We can get some evidence that this theorem holds in general by looking first at a special case where a proof of the theorem is straight-forward – namely, when the triangle is an equilateral triangle.

This technique of looking first to see whether a result holds in a special case is often useful.

To do this, consider an equilateral triangle ΔABC, with medians AP, BQ and CR. Since ΔABC has sides of equal length, it must be symmetric about the line AP. Thus the point at which BQ meets CR must be symmetrically placed with respect to this line – that is, it must actually lie on the line AP. In other words, the lines AP, BQ and CR are concurrent if the triangle is equilateral.

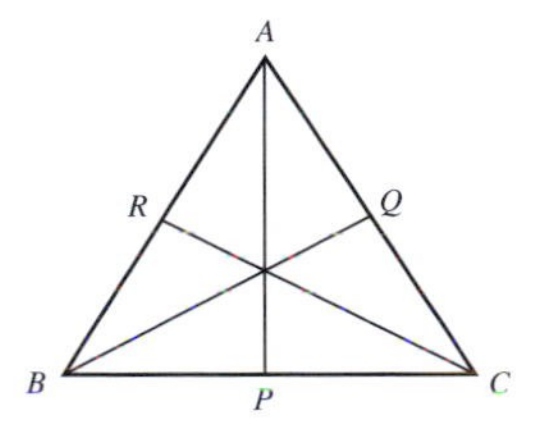

In order to show that the medians of an arbitrary triangle meet at a point, consider an arbitrary triangle ΔABC, and let P, Q and R be the midpoints of the sides BC, CA and AB, respectively. Next, choose a particular *equilateral* triangle $\Delta A'B'C'$, and let P', Q' and R' be the midpoints of the sides $B'C'$, $C'A'$ and $A'B'$, respectively.

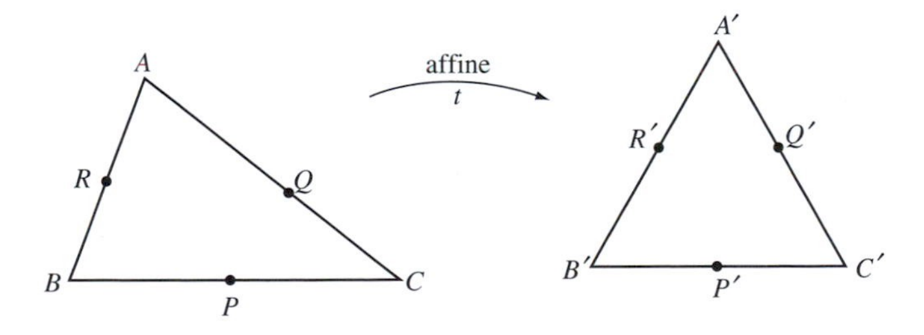

According to the Fundamental Theorem of Affine Geometry there is an affine transformation t which maps ΔABC onto $\Delta A'B'C'$. Moreover, since affine transformations preserve ratios of lengths along lines it follows that t maps the mid-points P, Q and R to the mid-points P', Q' and R', respectively.

From the above discussion we know that the medians of any *equilateral* triangle meet at a point, so in particular we know that $A'P'$, $B'Q'$ and $C'R'$ meet at some point X', say, as shown on the right below.

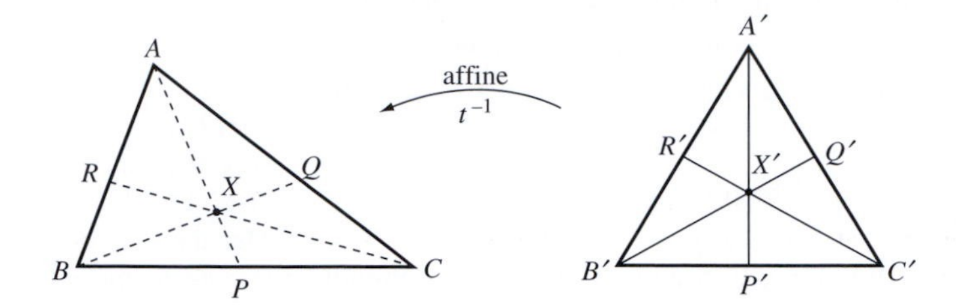

The trick now is to observe that t has an inverse t^{-1} which is also an affine transformation. This inverse maps the medians $A'P'$, $B'Q'$ and $C'R'$ back to the medians AP, BQ and CR of the original triangle ΔABC. Since X' lies on all three of the lines $A'P'$, $B'Q'$ and $C'R'$ it follows that t^{-1} maps X' to some point X which lies on all three of the lines AP, BQ and CR. In other words, the medians of ΔABC are concurrent.

Since ΔABC is an arbitrary triangle we have proved the Median Theorem.

The essence of the above proof is the fact that all triangles are affine-congruent. That powerful result enables us to prove theorems concerning the affine properties of triangles (such as collinearity, lines being parallel, and ratios of lengths along a given line) following a standard pattern. First, we choose a particular type of triangle for which it is easy to prove the result. Then, by asserting the existence of an affine transformation from that triangle to an arbitrary triangle, we deduce that the result holds for all triangles.

The basic affine properties were listed in Subsection 2.2.1.

This is the approach we shall use to prove the theorems of Ceva and Menelaus later in the section.

2.4.2 Ceva's Theorem

We now prove the following theorem due to Ceva.

Theorem 2 Ceva's Theorem
Let ΔABC be a triangle, and let X be a point which does not lie on any of its (extended) sides. If AX meets BC at P, BX meets CA at Q and CX meets BA at R, then

$$\frac{AR}{RB} \cdot \frac{BP}{PC} \cdot \frac{CQ}{QA} = 1.$$

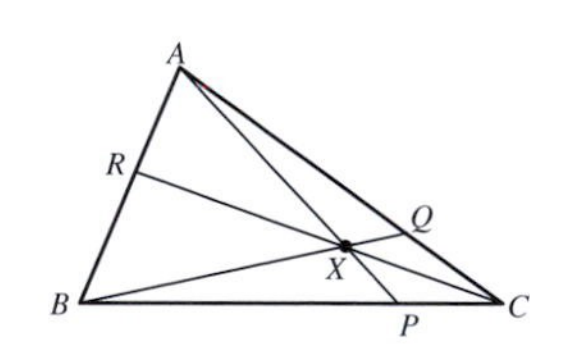

Proof According to the Fundamental Theorem of Affine Geometry there is an affine transformation t which maps the points A, B, C to the points $A' = (0, 1)$, $B' = (0, 0)$, $C' = (1, 0)$, respectively. This transformation maps the triangle ΔABC onto the right-angled triangle $\Delta A'B'C'$, and it maps the point X to some point $X' = (u, v)$.

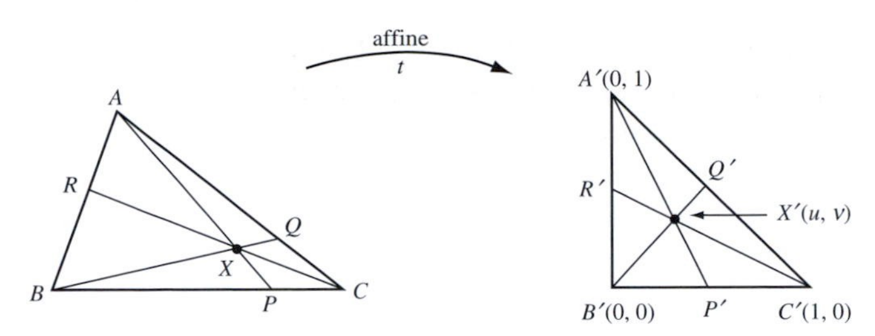

Using coordinate geometry we can calculate the equations of the lines $A'X'$, $B'X'$, $C'X'$ and hence find the coordinates of the point P' where $A'X'$ meets $B'C'$, of the point Q' where $B'X'$ meets $A'C'$, and of the point R' where $C'X'$ meets $A'B'$.

Starting with the point P', we note that the line $B'C'$ has equation $y = 0$.

Also, the line $A'X'$ has slope $\frac{1-v}{0-u}$, so its equation is $y - 1 = \frac{1-v}{0-u}(x - 0)$. Hence, at the point P' where the two lines meet, we must have $y = 0$ and $y - 1 = \frac{1-v}{0-u}(x - 0)$, so

$$P' = \left(\frac{u}{1-v}, 0\right).$$

Similarly, at the point R' we have $x = 0$, and $y - 0 = \frac{0-v}{1-u}(x - 1)$, so

$$R' = \left(0, \frac{v}{1-u}\right).$$

Finally, at Q' we have $x + y = 1$ and $y = \frac{v}{u}x$, so $x = \frac{u}{u+v}$ and $y = \frac{v}{u+v}$. Hence

$$Q' = \left(\frac{u}{u+v}, \frac{v}{u+v}\right).$$

Thus, using the coordinate formulas for calculating ratios we obtain

These formulas are given at the beginning of Appendix 2 just above the Section Formula.

$$\frac{A'R'}{R'B'} = \frac{y_{R'} - y_{A'}}{y_{B'} - y_{R'}} = \frac{\frac{v}{1-u} - 1}{0 - \frac{v}{1-u}} = \frac{u + v - 1}{-v},$$

$$\frac{B'P'}{P'C'} = \frac{x_{P'} - x_{B'}}{x_{C'} - x_{P'}} = \frac{\frac{u}{1-v} - 0}{1 - \frac{u}{1-v}} = \frac{u}{1 - u - v},$$

and

$$\frac{C'Q'}{Q'A'} = \frac{y_{Q'} - y_{C'}}{y_{A'} - y_{Q'}} = \frac{\frac{v}{u+v} - 0}{1 - \frac{v}{u+v}} = \frac{v}{u}.$$

Hence

$$\frac{A'R'}{R'B'} \cdot \frac{B'P'}{P'C'} \cdot \frac{C'Q'}{Q'A'} = 1.$$

Since t^{-1} is an affine transformation, it preserves ratios along a line. It must therefore map P', Q', R' back to the points P, Q, R in such a way that

$$\frac{AR}{RB} \cdot \frac{BP}{PC} \cdot \frac{CQ}{QA} = 1,$$

as required. ■

The next example illustrates how we can use Ceva's Theorem to calculate certain unknown distances along the sides of a triangle. For the method to work correctly, it is important to remember that all the ratios in Ceva's Theorem are *signed* ratios. Thus, if X lies inside the triangle, as in part (a) of the example, then all the ratios are positive. But if X lies outside the triangle, as in part (b), then two of the ratios will be negative.

Example 1

(a) In the figure on the left below, $AR = 1$, $RB = 2$, $BP = 3$, $CQ = 2$ and $QA = 2$. Calculate the distance PC.

(b) For the figure on the right, $AR = 1$, $AB = 3$, $PC = 1$, $CQ = 2$ and $QA = 2$. Calculate the distance BC.

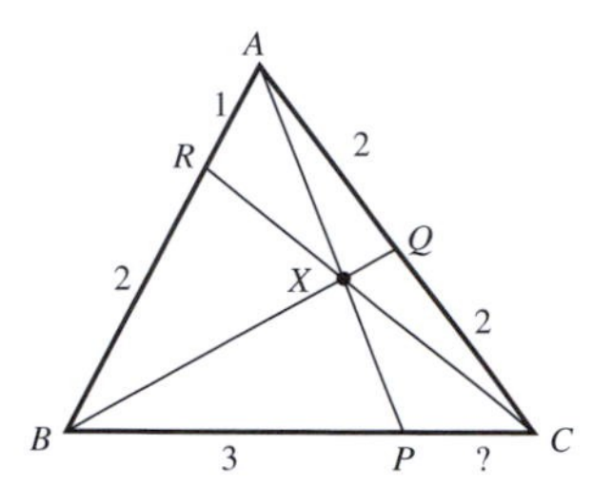

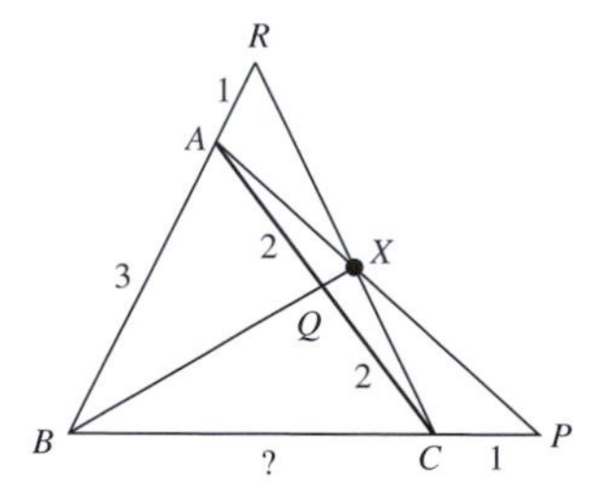

Solution

(a) By Ceva's Theorem, we have

$$\frac{AR}{RB} \cdot \frac{BP}{PC} \cdot \frac{CQ}{QA} = 1;$$

so,

$$\frac{1}{2} \cdot \frac{3}{PC} \cdot \frac{2}{2} = 1.$$

It follows that $PC = \frac{3}{2}$.

(b) By Ceva's Theorem, we have

$$\frac{AR}{RB} \cdot \frac{BP}{PC} \cdot \frac{CQ}{QA} = 1;$$

so,

$$-\frac{1}{4} \cdot \left(-\frac{BC+1}{1}\right) \cdot \frac{2}{2} = 1.$$

It follows that $BC = 3$. □

Problem 1

(a) Determine the ratio $\frac{BP}{PC}$ in the left diagram below, given that

$$\frac{AR}{RB} = \frac{AQ}{QC} = \frac{3}{2}.$$

(b) Determine the ratio $\frac{CQ}{QA}$ in the middle diagram below, given that

$$\frac{AR}{RB} = \frac{1}{2} \quad \text{and} \quad \frac{BP}{PC} = -\frac{2}{7}.$$

(c) Determine the ratio $\frac{AR}{RB}$ in the right diagram below, given that

$$\frac{BP}{PC} = \frac{5}{7} \quad \text{and} \quad \frac{CQ}{QA} = -7.$$

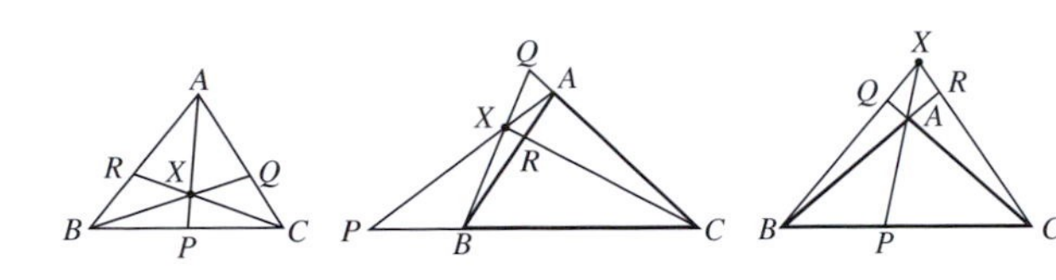

Ceva's Theorem has the following converse, which can be regarded as a generalization of the Median Theorem to configurations where P, Q, R are not all midpoints of sides.

Theorem 3 Converse to Ceva's Theorem
Let P, Q and R be points, other than vertices, on the (possibly extended) sides BC, CA and AB of a triangle ΔABC, such that

$$\frac{AR}{RB} \cdot \frac{BP}{PC} \cdot \frac{CQ}{QA} = 1. \tag{1}$$

Then the lines AP, BQ and CR are concurrent.

In the Median Theorem,

$$\frac{AR}{RB} = 1, \quad \frac{BP}{PC} = 1,$$
$$\frac{CQ}{QA} = 1,$$

so Theorem 3 generalizes the Median Theorem.

Proof Let the lines BQ and CR intersect at a point X, and let the line AX meet BC at some point P'. It is sufficient to prove that $P = P'$.

It follows from Ceva's Theorem that

$$\frac{AR}{RB} \cdot \frac{BP'}{P'C} \cdot \frac{CQ}{QA} = 1. \tag{2}$$

Hence, from equations (1) and (2), we have

$$\frac{BP}{PC} = \frac{BP'}{P'C},$$

so that P and P' must indeed be the same point. ∎

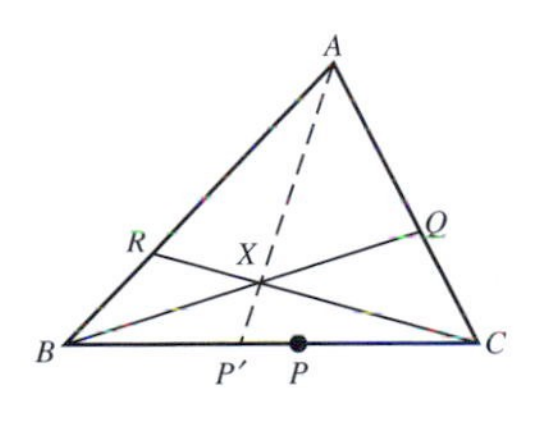

Example 2 The triangle ΔABC has vertices $A(1,3)$, $B(-1,0)$ and $C(4,0)$, and the points $P(0,0)$, $Q\left(\frac{8}{3}, \frac{4}{3}\right)$ and $R\left(-\frac{2}{3}, \frac{1}{2}\right)$ lie on BC, CA and AB, respectively.

(a) Determine the ratios in which P, Q and R divide the sides of the triangle.
(b) Determine whether the lines AP, BQ and CR are concurrent.

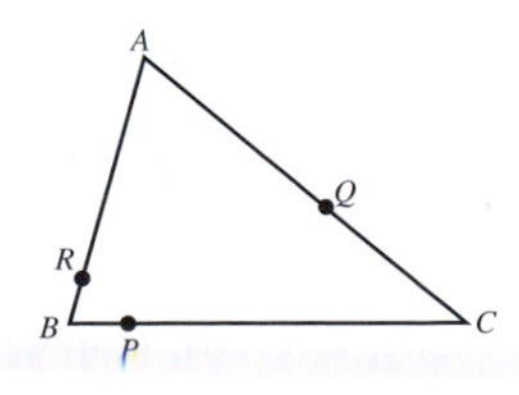

Solution

(a) Using the coordinate formulas for calculating ratios, we obtain

$$\frac{AR}{RB} = \frac{x_R - x_A}{x_B - x_R} = \frac{-\frac{2}{3} - 1}{-1 + \frac{2}{3}} = 5, \quad \frac{BP}{PC} = \frac{x_P - x_B}{x_C - x_P} = \frac{0+1}{4-0} = \frac{1}{4},$$

$$\frac{CQ}{QA} = \frac{x_Q - x_C}{x_A - x_Q} = \frac{\frac{8}{3} - 4}{1 - \frac{8}{3}} = \frac{4}{5}, \tag{3}$$

so that P divides BC in the ratio $1:4$, Q divides CA in the ratio $4:5$ and R divides AB in the ratio $5:1$.

(b) It follows from (3) that the product

$$\frac{AR}{RB} \cdot \frac{BP}{PC} \cdot \frac{CQ}{QA} = 5 \cdot \frac{1}{4} \cdot \frac{4}{5} = 1;$$

so by the converse to Ceva's Theorem the lines AP, BQ and CR must be concurrent. □

Problem 2 The triangle ΔABC has vertices $A(-1,1)$, $B(2,-1)$ and $C(3,2)$, and the points $P\left(\frac{8}{3},1\right)$, $Q\left(2,\frac{7}{4}\right)$ and $R\left(\frac{4}{5},-\frac{1}{5}\right)$ lie on BC, CA and AB, respectively.

(a) Determine the ratios in which P, Q and R divide the sides of the triangle.
(b) Determine whether the lines AP, BQ and CR are concurrent.

2.4.3 Menelaus' Theorem

Ceva's theorem is concerned with lines through the vertices of a triangle that meet at a point. We now use the Fundamental Theorem of Affine Geometry to prove an analogous theorem due to Menelaus which is concerned with points on the sides of a triangle that are collinear.

Theorem 4 Menelaus' Theorem
Let ΔABC be a triangle, and let ℓ be a line that crosses the sides BC, CA, AB at three distinct points P, Q, R, respectively. Then

$$\frac{AR}{RB} \cdot \frac{BP}{PC} \cdot \frac{CQ}{QA} = -1.$$

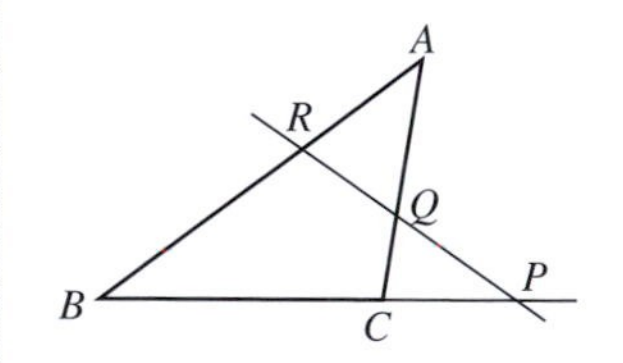

Proof According to the Fundamental Theorem of Affine Geometry there is an affine transformation t which maps the points A, B, C to the points $A'(0,1)$, $B'(0,0)$, $C'(1,0)$, respectively. This transformation maps the triangle ΔABC

onto the right-angled triangle $\Delta A'B'C'$, and it maps the line ℓ to some line ℓ'. Let the equation of ℓ' be $y = mx + c$.

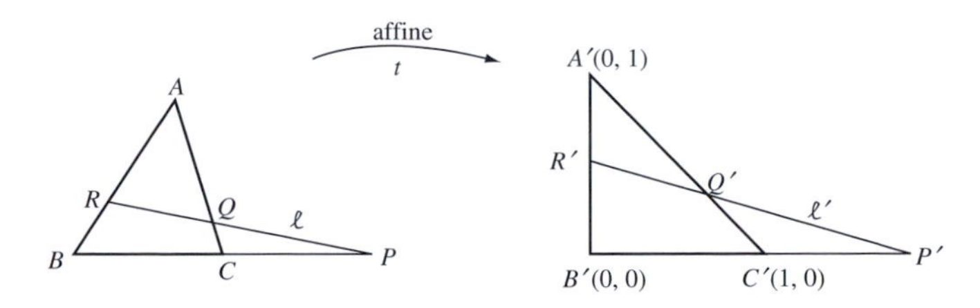

We now calculate the coordinates of the points P', Q' and R' where ℓ' meets the sides $B'C'$, $C'A'$ and $A'B'$, respectively.

At P' we have $y = 0$ and $y = mx + c$. This implies that $x = -\frac{c}{m}$, and hence

$$P' = \left(-\frac{c}{m}, 0\right).$$

At R' we have $x = 0$ and $y = mx + c$. This implies that $y = c$, and hence

$$R' = (0, c).$$

At Q' we have $x + y = 1$ and $y = mx + c$. This implies that $1 - x = mx + c$ so that $x = \frac{1-c}{m+1}$; also $y = m(1 - y) + c$, so that $y = \frac{m+c}{m+1}$; and hence

$$Q' = \left(\frac{1-c}{m+1}, \frac{m+c}{m+1}\right).$$

Using the coordinate formulas for calculating ratios we obtain

$$\frac{A'R'}{R'B'} = \frac{y_{R'} - y_{A'}}{y_{B'} - y_{R'}} = \frac{c-1}{0-c} = \frac{c-1}{-c},$$

$$\frac{B'P'}{P'C'} = \frac{x_{P'} - x_{B'}}{x_{C'} - x_{P'}} = \frac{-\frac{c}{m} - 0}{1 + \frac{c}{m}} = \frac{-c}{m+c},$$

and

$$\frac{C'Q'}{Q'A'} = \frac{x_{Q'} - x_{C'}}{x_{A'} - x_{Q'}} = \frac{\frac{1-c}{m+1} - 1}{0 - \frac{1-c}{m+1}} = \frac{-(m+c)}{c-1}.$$

Hence,

$$\frac{A'R'}{R'B'} \cdot \frac{B'P'}{P'C'} \cdot \frac{C'Q'}{Q'A'} = -1.$$

Since t^{-1} is an affine transformation, it preserves ratios along a line. It must therefore map P', Q', R' back to the points P, Q, R in such a way that

$$\frac{AR}{RB} \cdot \frac{BP}{PC} \cdot \frac{CQ}{QA} = -1,$$

as required. ■

Remark

As for Ceva's Theorem, it is important to remember that all the ratios in Menelaus' Theorem are *signed* ratios. In fact if ℓ passes through the interior of the triangle, then precisely one of the ratios is negative; otherwise all the ratios are negative.

Example 3

(a) In the figure on the left below: $AR = 1$, $RB = 2$, $BC = 2$, $CQ = 1$ and $QA = 1$. Calculate the distance PC.

(b) In the figure on the right below: $AR = 2$, $AB = 1$, $BC = 2$, $CA = 2$ and $BP = 2$. Calculate the distance QA.

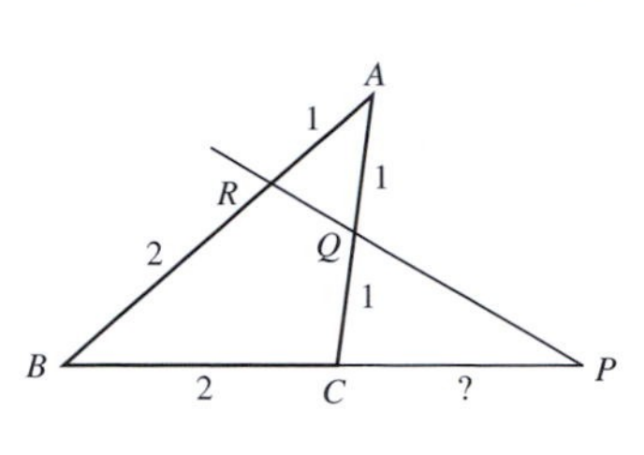

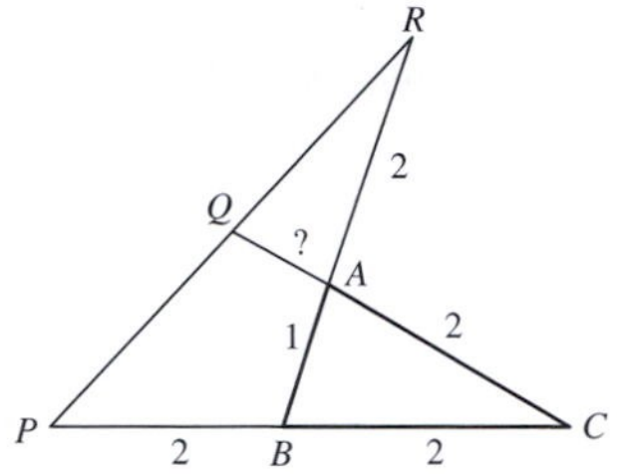

Solution

(a) By Menelaus' Theorem, we have

$$\frac{AR}{RB} \cdot \frac{BP}{PC} \cdot \frac{CQ}{QA} = -1.$$

So

$$\frac{1}{2} \cdot \left(-\frac{2 + PC}{PC}\right) \cdot \frac{1}{1} = -1.$$

It follows that $2 + PC = 2PC$, and hence $PC = 2$.

(b) By Menelaus' Theorem, we have

$$\frac{AR}{RB} \cdot \frac{BP}{PC} \cdot \frac{CQ}{QA} = -1.$$

So

$$\left(-\frac{2}{3}\right) \cdot \left(-\frac{2}{4}\right) \cdot \left(-\frac{2 + QA}{QA}\right) = -1.$$

It follows that $2 + QA = 3QA$, and hence $QA = 1$. □

Problem 3

(a) Determine the ratio $\frac{CQ}{QA}$ in the left diagram below, given that

$$\frac{AR}{RB} = 2 \quad \text{and} \quad \frac{BP}{PC} = -2.$$

(b) Determine the ratio $\frac{CQ}{QA}$ in the right diagram below, given that

$$\frac{AR}{RB} = -\frac{1}{4} \quad \text{and} \quad \frac{BP}{PC} = -2.$$

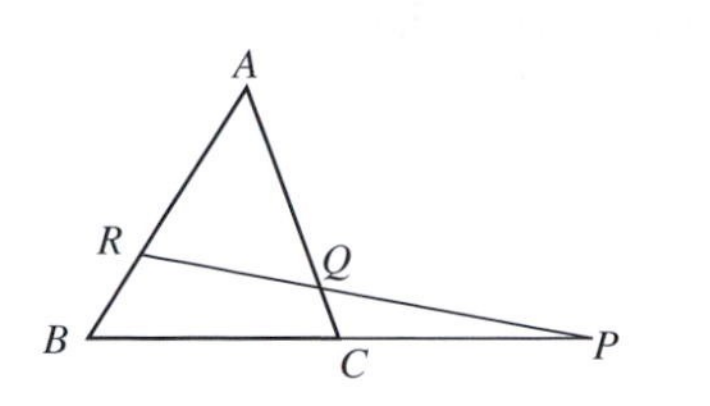

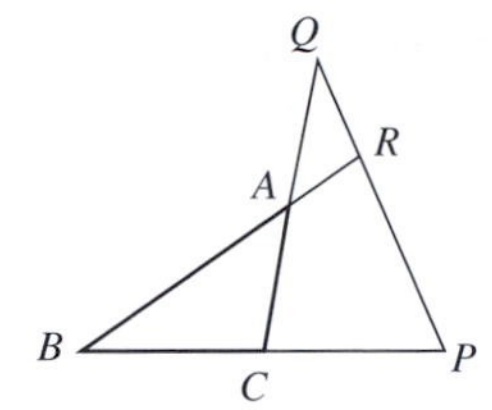

Menelaus' Theorem has a converse that enables us to check whether points on the three sides of a triangle are collinear.

Theorem 5 Converse to Menelaus' Theorem
Let P, Q and R be points other than vertices on the (possibly extended) sides BC, CA and AB of a triangle ΔABC, such that

$$\frac{AR}{RB} \cdot \frac{BP}{PC} \cdot \frac{CQ}{QA} = -1. \tag{3}$$

Then the points P, Q and R are collinear.

Proof Let the line ℓ that passes through Q and R meet BC at some point P'. It is sufficient to prove that $P = P'$.

The strategy of the proof is the same as that of Theorem 3.

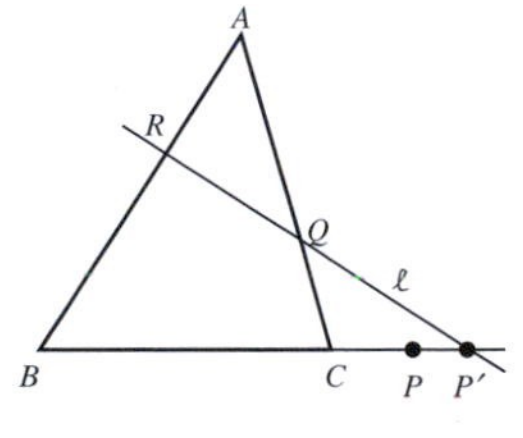

It follows from Menelaus' Theorem that

$$\frac{AR}{RB} \cdot \frac{BP'}{P'C} \cdot \frac{CQ}{QA} = -1. \tag{4}$$

Hence, from equations (3) and (4) we deduce that

$$\frac{BP}{PC} = \frac{BP'}{P'C}.$$

It follows that P and P' must indeed be the same point. ■

Problem 4 The triangle ΔABC has vertices $A(2,4)$, $B(-2,0)$ and $C(1,0)$, and the points $P\left(\frac{5}{2},0\right)$, $Q\left(\frac{3}{2},2\right)$ and $R(1,3)$ lie on BC, CA and AB, respectively.

(a) Determine the ratios in which P, Q and R divide the sides of the triangle.
(b) Hence determine whether the points P, Q and R are collinear.

We end this subsection with two revision problems.

Problem 5 Let ΔABC be a triangle, and let X be a point which does not lie on any of its (extended) sides. Also, let AX meet BC at P, BX meet CA at Q and CX meet BA at R; and let QR and BC meet at T.

Given that $\frac{BP}{PC} = k$, determine $\frac{BT}{TC}$ in terms of k.

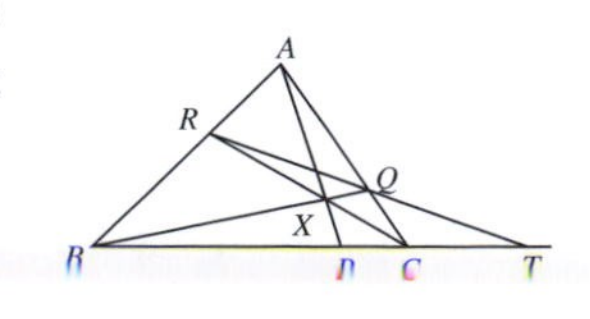

Problem 6 Suppose that P and Q are the midpoints of the sides AB and BC of a parallelogram $ABCD$, and that the lines DP and AQ meet at R.

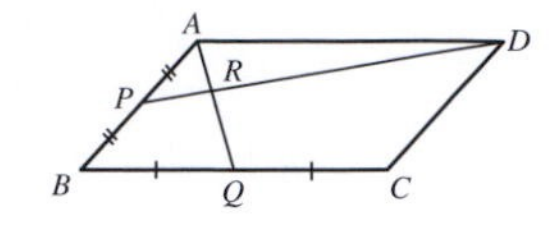

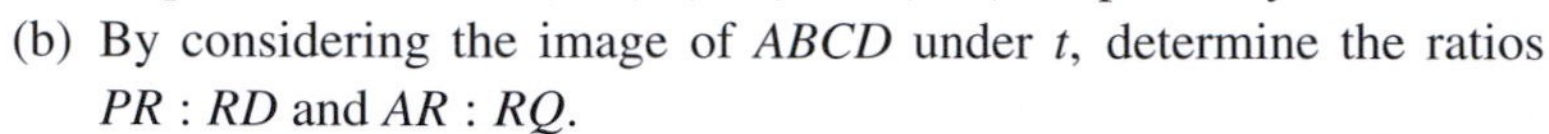

(a) Determine the image of B under the affine transformation t which maps A, D and C to (0, 1), (0, 0) and (1, 0), respectively.
(b) By considering the image of $ABCD$ under t, determine the ratios $PR : RD$ and $AR : RQ$.

2.4.4 Barycentric Coordinates

In this subsection we introduce a new coordinate system, of *barycentric coordinates* with respect to a *triangle of reference*, for points in the plane, which can simplify some calculations. Rather than use two perpendicular axes to determine the coordinates of an arbitrary point in the plane, we use a weighted sum of the coordinates of three non-collinear points – the vertices of the triangle of reference.

Barycentric coordinates were introduced by Möbius in 1827.

Definitions Let $A = (a_1, a_2)$, $B = (b_1, b_2)$ and $C = (c_1, c_2)$ be three non-collinear points in the plane $\mathbb{R}^2$; we will call ΔABC the **triangle of reference**. Then a point (x, y) in the plane has **barycentric coordinates** (ξ, η, ζ) with respect to ΔABC if

$$\left.\begin{aligned} x &= \xi a_1 + \eta b_1 + \zeta c_1, \\ y &= \xi a_2 + \eta b_2 + \zeta c_2, \text{ and} \\ 1 &= \xi \quad + \eta \quad + \zeta. \end{aligned}\right\} \qquad (5)$$

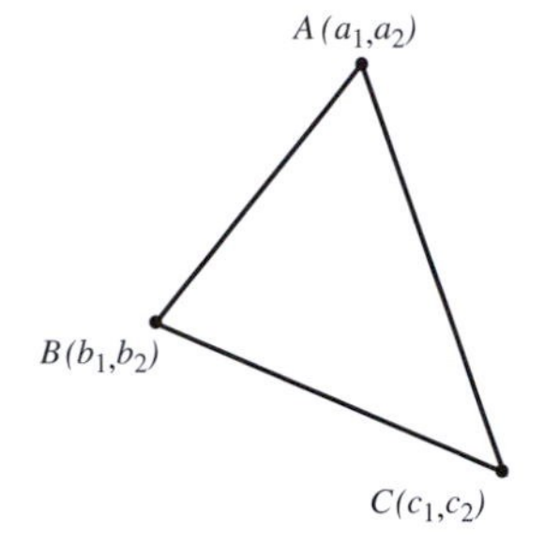

Remark

In particular, notice that the barycentric coordinates of the vertices A, B and C of the triangle of reference ΔABC are (1,0,0), (0,1,0) and (0,0,1), respectively.

Thus, for example, if the triangle of reference ΔABC has vertices $A = (1, 2)$, $B = (-3, 0)$ and $C = (-2, 4)$, then the point with barycentric coordinates $\left(\frac{1}{2}, \frac{3}{4}, -\frac{1}{4}\right)$ has cartesian coordinates

Cartesian coordinates are the standard Euclidean coordinates in the plane.

$$\begin{aligned} &\left(\tfrac{1}{2} \cdot (1) + \tfrac{3}{4} \cdot (-3) - \tfrac{1}{4} \cdot (-2), \tfrac{1}{2} \cdot (2) + \tfrac{3}{4} \cdot (0) - \tfrac{1}{4} \cdot 4\right) \\ &= \left(\frac{2 - 9 + 2}{4}, \frac{4 + 0 - 4}{4}\right) \\ &= \left(-\tfrac{5}{4}, 0\right). \end{aligned}$$

We may express the formula (5) for barycentric coordinates in terms of matrices in the form

$$\begin{pmatrix} x \\ y \\ 1 \end{pmatrix} = \mathbf{M} \begin{pmatrix} \xi \\ \eta \\ \zeta \end{pmatrix}, \quad \text{where } \mathbf{M} = \begin{pmatrix} a_1 & b_1 & c_1 \\ a_2 & b_2 & c_2 \\ 1 & 1 & 1 \end{pmatrix}. \tag{6}$$

Since $A = (a_1, a_2)$, $B = (b_1, b_2)$ and $C = (c_1, c_2)$ are non-collinear, the determinant $\det \mathbf{M}$ is non-zero. Then, since $\mathbf{M}$ is non-singular, we can reformulate the representation (6) as

We do not prove that $\det \mathbf{M} \neq 0$ here.

$$\begin{pmatrix} \xi \\ \eta \\ \zeta \end{pmatrix} = \mathbf{M}^{-1} \begin{pmatrix} x \\ y \\ 1 \end{pmatrix}. \tag{7}$$

Example 4 Determine barycentric coordinates for the point (2,1) with respect to the triangle of reference ΔABC where $A = (1,0)$, $B = (1,-1)$ and $C = (-1,1)$.

Solution The matrix $\mathbf{M}$ for the triangle of reference ΔABC is

$$\mathbf{M} = \begin{pmatrix} 1 & 1 & -1 \\ 0 & -1 & 1 \\ 1 & 1 & 1 \end{pmatrix},$$

whose inverse is

We omit the details of the calculation of this inverse.

$$\begin{pmatrix} 1 & 1 & 0 \\ -\frac{1}{2} & -1 & \frac{1}{2} \\ -\frac{1}{2} & 0 & \frac{1}{2} \end{pmatrix}.$$

It follows from the representation (7) that the point (2,1) has barycentric coordinates with respect to the triangle of reference ΔABC given by

$$\begin{pmatrix} 1 & 1 & 0 \\ -\frac{1}{2} & -1 & \frac{1}{2} \\ -\frac{1}{2} & 0 & \frac{1}{2} \end{pmatrix} \begin{pmatrix} 2 \\ 1 \\ 1 \end{pmatrix} = \begin{pmatrix} 3 \\ -\frac{3}{2} \\ -\frac{1}{2} \end{pmatrix};$$

namely, barycentric coordinates $\left(3, -\frac{3}{2}, -\frac{1}{2}\right)$. □

Problem 7 Determine barycentric coordinates for the point $(-1, 1)$ with respect to the triangle of reference ΔABC where $A = (1, 1)$, $B = (2, 2)$ and $C = (1, 2)$.

Next, we give barycentric versions of the condition for collinearity of three points in the plane and of the equation of a line in the plane.

Theorem 6 The points P, Q and R with barycentric coordinates (ξ_1, η_1, ζ_1), (ξ_2, η_2, ζ_2) and (ξ_3, η_3, ζ_3) are collinear if and only if

$$\begin{vmatrix} \xi_1 & \xi_2 & \xi_3 \\ \eta_1 & \eta_2 & \eta_2 \\ \zeta_1 & \zeta_2 & \zeta_3 \end{vmatrix} = 0.$$

Proof Let the points P, Q and R have cartesian coordinates (x_1, y_1), (x_2, y_2) and (x_3, y_3), respectively. It follows that, if the triangle of reference ΔABC has vertices $A = (a_1, a_2)$, $B = (b_1, b_2)$ and $C = (c_1, c_2)$, then we may apply the formula (5) to each of P, Q and R in turn to obtain

$$x_1 = a_1\xi_1 + b_1\eta_1 + c_1\zeta_1 \text{ and } y_1 = a_2\xi_1 + b_2\eta_1 + c_2\zeta_1,$$
$$x_2 = a_1\xi_2 + b_1\eta_2 + c_1\zeta_2 \text{ and } y_2 = a_2\xi_2 + b_2\eta_2 + c_2\zeta_2,$$
$$x_3 = a_1\xi_3 + b_1\eta_3 + c_1\zeta_3 \text{ and } y_3 = a_2\xi_3 + b_2\eta_3 + c_2\zeta_3.$$

We may write these simultaneous equations in matrix form as

$$\begin{pmatrix} x_1 & x_2 & x_3 \\ y_1 & y_2 & y_3 \\ 1 & 1 & 1 \end{pmatrix} = \begin{pmatrix} a_1 & b_1 & c_1 \\ a_2 & b_2 & c_2 \\ 1 & 1 & 1 \end{pmatrix} \begin{pmatrix} \xi_1 & \xi_2 & \xi_3 \\ \eta_1 & \eta_2 & \eta_3 \\ \zeta_1 & \zeta_2 & \zeta_3 \end{pmatrix}$$
$$= \mathbf{M} \begin{pmatrix} \xi_1 & \xi_2 & \xi_3 \\ \eta_1 & \eta_2 & \eta_3 \\ \zeta_1 & \zeta_2 & \zeta_3 \end{pmatrix}, \qquad (8)$$

The results for P, Q and R are set out as columns 1, 2 and 3, respectively, of the left-hand matrix.

where $\mathbf{M} = \begin{pmatrix} a_1 & b_1 & c_1 \\ a_2 & b_2 & c_2 \\ 1 & 1 & 1 \end{pmatrix}$.

Now, $\mathbf{M}$ is non-singular since the points A, B and C are non-collinear. Then it follows from equation (8) that

$$\begin{vmatrix} \xi_1 & \xi_2 & \xi_3 \\ \eta_1 & \eta_2 & \eta_3 \\ \zeta_1 & \zeta_2 & \zeta_3 \end{vmatrix} = 0 \text{ if and only if } \begin{vmatrix} x_1 & x_2 & x_3 \\ y_1 & y_2 & y_3 \\ 1 & 1 & 1 \end{vmatrix} = 0.$$

But the second determinant equation here is the condition for P, Q and R to be collinear. The desired result then follows immediately. ■

Corollary The line ℓ through the points with barycentric coordinates (ξ_1, η_1, ζ_1) and (ξ_2, η_2, ζ_2) has equation

$$\begin{vmatrix} \xi_1 & \xi_2 & \xi \\ \eta_1 & \eta_2 & \eta \\ \zeta_1 & \zeta_2 & \zeta \end{vmatrix} = 0.$$

That is, an equation in terms of barycentric coordinates.

For example, if the triangle of reference is ΔABC, the equation of the line AB in terms of barycentric coordinates is

$$\begin{vmatrix} 1 & 0 & \xi \\ 0 & 1 & \eta \\ 0 & 0 & \zeta \end{vmatrix} = 0;$$

which simplifies to the equation $\zeta = 0$.

Similarly, BC and CA have equations $\xi = 0$ and $\eta = 0$.

Problem 8 Determine which of the following sets of points, described by their barycentric coordinates with respect to an unspecified triangle of reference, are collinear; for those that are collinear, determine the equation of the line on which they lie.

(a) $(1, 1, -1)$, $(4, -2, -1)$, $\left(\frac{1}{2}, 2, -\frac{3}{2}\right)$
(b) $(1, 1, -1)$, $(2, -2, 1)$, $(-1, 7, -5)$

Next, we meet a version of the Section Formula in terms of barycentric coordinates.

The Section Formula is given in Appendix 2.

Theorem 7 Section Formula
The point R that divides the line ℓ joining the points P and Q with barycentric coordinates (ξ_1, η_1, ζ_1) and (ξ_2, η_2, ζ_2) in the ratio $(1 - \lambda) : \lambda$ has barycentric coordinates

$$(\xi, \eta, \zeta) = \lambda(\xi_1, \eta_1, \zeta_1) + (1 - \lambda)(\xi_2, \eta_2, \zeta_2).$$

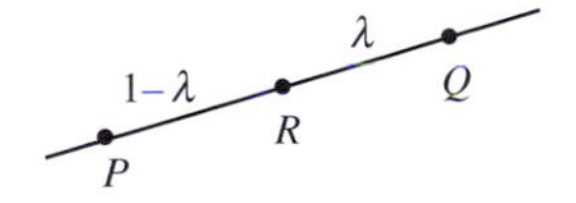

Proof Let P, Q and R have cartesian coordinates (x_1, y_1), (x_2, y_2) and (x, y), respectively. It follows from the cartesian form of the Section Formula that

$$(x, y) = \lambda(x_1, y_1) + (1 - \lambda)(x_2, y_2), \quad \text{for some real number } \lambda;$$

we can rewrite this equation in matrix form as

$$\begin{pmatrix} x \\ y \\ 1 \end{pmatrix} = \lambda \begin{pmatrix} x_1 \\ y_1 \\ 1 \end{pmatrix} + (1 - \lambda) \begin{pmatrix} x_2 \\ y_2 \\ 1 \end{pmatrix}.$$

Multiplying both sides of this equation on the left by the matrix $\mathbf{M}^{-1}$ and using the formula (8), it follows that

$$\begin{pmatrix} \xi \\ \eta \\ \zeta \end{pmatrix} = \lambda \begin{pmatrix} \xi_1 \\ \eta_1 \\ \zeta_1 \end{pmatrix} + (1 - \lambda) \begin{pmatrix} \xi_2 \\ \eta_2 \\ \zeta_2 \end{pmatrix}.$$

This is the desired result. ■

We can now use barycentric coordinate methods to give further proofs of Menelaus's Theorem and Ceva's Theorem.

Theorem 8 Menelaus' Theorem
Let ΔABC be a triangle, and let ℓ be a line that crosses the sides BC, CA and AB at three distinct points P, Q, R, respectively. Then

$$\frac{AR}{RB} \cdot \frac{BP}{PC} \cdot \frac{CQ}{QA} = -1.$$

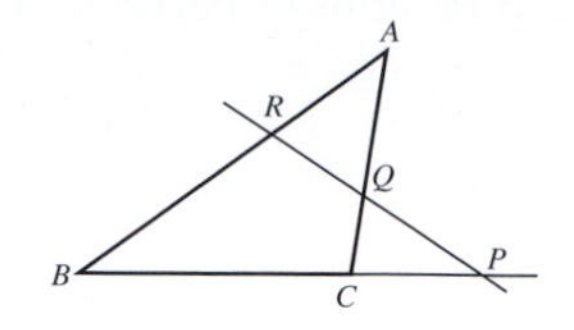

Proof Define λ, μ and ν as follows:

$$\frac{BP}{PC} = \frac{1-\lambda}{\lambda}, \quad \frac{CQ}{QA} = \frac{1-\mu}{\mu}, \quad \frac{AR}{RB} = \frac{1-\nu}{\nu}.$$

Hence, by the Section Formula, P has barycentric coordinates

$$\begin{aligned} P &= \lambda(0,1,0) + (1-\lambda)(0,0,1) \\ &= (0, \lambda, 1-\lambda). \end{aligned}$$

For, P divides BC in the ratio $(1-\lambda) : \lambda$.

Similarly, Q and R have barycentric coordinates $(1-\mu, 0, \mu)$ and $(\nu, 1-\nu, 0)$, respectively.

We omit the details of these calculations.

Then, by Theorem 6 the points P, Q and R are collinear if and only if

$$\begin{vmatrix} 0 & 1-\mu & \nu \\ \lambda & 0 & 1-\nu \\ 1-\lambda & \mu & 0 \end{vmatrix} = 0.$$

Expanding this determinant, we have that P, Q and R are collinear if and only if

$$-(1-\mu)\begin{vmatrix} \lambda & 1-\nu \\ 1-\lambda & 0 \end{vmatrix} + \nu \begin{vmatrix} \lambda & 0 \\ 1-\lambda & \mu \end{vmatrix} = 0;$$

that is, if and only if

$$(1-\lambda)(1-\mu)(1-\nu) + \lambda\mu\nu = 0,$$

or

$$\frac{1-\lambda}{\lambda} \cdot \frac{1-\mu}{\mu} \cdot \frac{1-\nu}{\nu} = -1.$$

From the original definition of λ, μ and ν, it follows that P, Q and R are collinear if and only if

$$\frac{AR}{RB} \cdot \frac{BP}{PC} \cdot \frac{CQ}{QA} = -1.$$

Notice that we have also obtained a proof of the converse of Menelaus' Theorem, since this equality is an 'if and only if' result!

From our assumption that P, Q and R are collinear, the desired result follows. ■

Theorem 9 Ceva's Theorem
Let ΔABC be a triangle, and let X be a point which does not lie on any of its (extended) sides. If AX meets BC at P, BX meets CA at Q and CX meets BA at R, then

$$\frac{AR}{RB} \cdot \frac{BP}{PC} \cdot \frac{CQ}{QA} = 1.$$

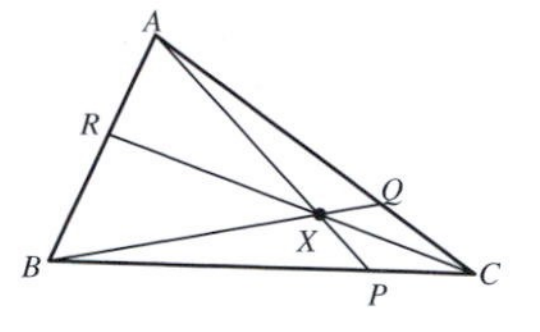

Proof We use the same notation λ, μ and ν as in the previous proof, so that again P, Q and R have homogeneous coordinates $(0, \lambda, 1-\lambda)$, $(1-\mu, 0, \mu)$ and $(\nu, 1-\nu, 0)$, respectively.

Then, using the Corollary to Theorem 6 the equation of AP is

$$\begin{vmatrix} 1 & 0 & \xi \\ 0 & \lambda & \eta \\ 0 & 1-\lambda & \zeta \end{vmatrix} = 0,$$

which simplifies to

$$\begin{vmatrix} \lambda & \eta \\ 1-\lambda & \zeta \end{vmatrix} = 0.$$

This gives that the equation of AP is $\lambda\zeta - (1-\lambda)\eta = 0$, or

$$\zeta = \frac{1-\lambda}{\lambda}\eta.$$

Similarly, we find that the equation of BQ is $\xi = \frac{1-\mu}{\mu}\zeta$.

We omit the details of this calculation.

Since AP and BQ meet at X, it follows that the barycentric coordinates (ξ, η, ζ) of X must satisfy the equations $\zeta = \frac{1-\lambda}{\lambda}\eta$ and $\xi = \frac{1-\mu}{\mu}\zeta$; hence its barycentric coordinates are

$$\left(\frac{1-\lambda}{\lambda} \cdot \frac{1-\mu}{\mu}\eta, \eta, \frac{1-\lambda}{\lambda}\eta\right), \quad \text{for some } \eta \neq 0.$$

Then C, X and R are collinear if and only if

$$\begin{vmatrix} 0 & \frac{1-\lambda}{\lambda} \cdot \frac{1-\mu}{\mu}\eta & \nu \\ 0 & \eta & 1-\nu \\ 1 & \frac{1-\lambda}{\lambda}\eta & 0 \end{vmatrix} = 0;$$

this simplifies to

$$\frac{1-\lambda}{\lambda} \cdot \frac{1-\mu}{\mu} \cdot \frac{1-\nu}{\nu} = 1.$$

It follows that C, X and R are collinear – that is, that AP, BQ and CR are concurrent – if and only if

$$\frac{AR}{RB} \cdot \frac{BP}{PC} \cdot \frac{CQ}{QA} = 1.$$

From our assumption that AP, BQ and CR are concurrent, the desired result follows. ■

Notice that we have also obtained a proof of the converse of Ceva's Theorem, since this equality is an 'if and only if' result!

2.5 Affine Transformations and Conics

2.5.1 Classifying Non-Degenerate Conics in Affine Geometry

In Section 2.2 you saw that under an affine transformation a straight line maps to a straight line. Indeed, it follows from the Fundamental Theorem of Affine Geometry that any straight line can be mapped to any other straight line by some affine transformation. We now explore the corresponding situation for conics.

We discussed conics in Chapter 1.

Recall that a conic is a set in $\mathbb{R}^2$ given by an equation of the form

$$Ax^2 + Bxy + Cy^2 + Fx + Gy + H = 0, \qquad (1)$$

where A, B, C, F, G and H are real numbers, and A, B and C are not all zero. The three types of non-degenerate conic are ellipses, parabolas and hyperbolas. A non-degenerate conic is an ellipse if $B^2 - 4AC < 0$, a parabola if $B^2 - 4AC = 0$, and a hyperbola if $B^2 - 4AC > 0$.

You met the $B^2 - 4AC$ test for conics in Theorem 3 of Section 1.3.

First, consider the case where equation (1) represents an ellipse, as illustrated on the left of the figure below. We can apply a translation to move the centre of the ellipse to the origin, and then a rotation to align its major and minor axes with the directions of the x-axis and y-axis, respectively. After we have applied these two Euclidean transformations, the equation of the ellipse becomes

$$\frac{x^2}{a^2} + \frac{y^2}{b^2} = 1, \quad a \geq b > 0. \qquad (2)$$

If we now apply the affine transformation $t_1 : (x, y) \mapsto (x', y')$, where

$$\begin{pmatrix} x' \\ y' \end{pmatrix} = \begin{pmatrix} 1/a & 0 \\ 0 & 1/b \end{pmatrix} \begin{pmatrix} x \\ y \end{pmatrix},$$

then $x' = x/a$ and $y' = y/b$, so equation (2) becomes

$$(x')^2 + (y')^2 = 1.$$

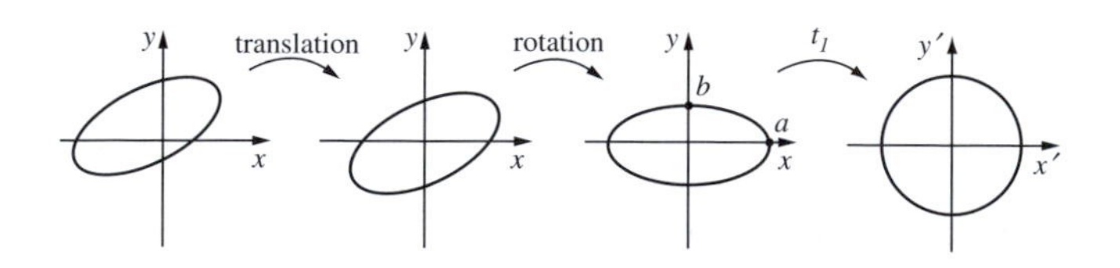

Since the translation, the rotation and the transformation t_1 are all affine, their composite must also be affine. Overall, this shows that each ellipse can be mapped onto the unit circle by an affine transformation. We therefore have the following theorem.

Theorem 1 Every ellipse is affine-congruent to the unit circle with equation $x^2 + y^2 = 1$.

Secondly, consider the case where equation (1) represents a hyperbola, as illustrated on the left of the figure below. Again, we can apply a translation to move the centre of the hyperbola to the origin, and then a rotation to align its major and minor axes with the directions of the x-axis and y-axis, respectively. After we have applied these two transformations, the equation of the hyperbola becomes

$$\frac{x^2}{a^2} - \frac{y^2}{b^2} = 1. \tag{3}$$

Under the affine transformation t_1 defined above, equation (3) becomes

$$(x')^2 - (y')^2 = 1,$$

that is,

$$(x' - y')(x' + y') = 1. \tag{4}$$

Finally, if we apply the affine transformation $t_2 : (x', y') \mapsto (x'', y'')$, where

$$\begin{pmatrix} x'' \\ y'' \end{pmatrix} = \begin{pmatrix} 1 & -1 \\ 1 & 1 \end{pmatrix} \begin{pmatrix} x' \\ y' \end{pmatrix},$$

then equation (4) becomes

$$x''y'' = 1.$$

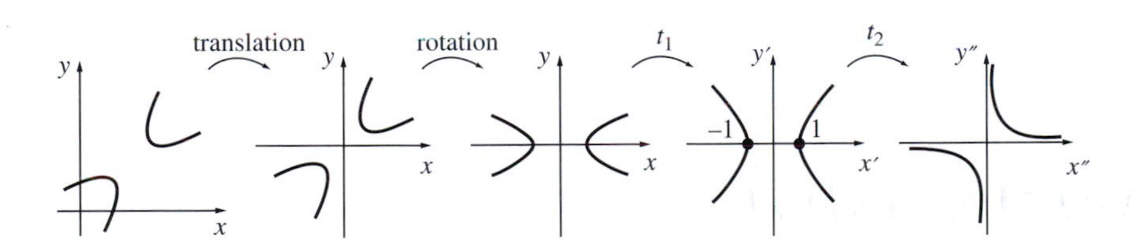

Dropping the dashes from the equation $x''y'' = 1$, we obtain the following theorem.

Theorem 2 Every hyperbola is affine-congruent to the rectangular hyperbola with equation $xy = 1$.

Recall that 'rectangular' means that the asymptotes of the hyperbola are at right angles to each other.

Finally, consider the case where equation (1) represents a parabola, as illustrated on the left of the figure below. We can apply a translation to move the vertex of the parabola to the origin, and then a rotation to align its axis with the (positive) x-axis. After we have applied these two Euclidean transformations, the equation of the parabola becomes

$$y^2 = ax, \tag{5}$$

where a is some positive number which depends on the coefficients in equation (1).

Here we omit the details of the particular transformations involved, and concentrate instead on the principles underlying the successive mappings which are used.

Next, if we apply the affine transformation $t_3 : (x, y) \mapsto (x', y')$, where

$$\begin{pmatrix} x' \\ y' \end{pmatrix} = \begin{pmatrix} 1/a & 0 \\ 0 & 1/a \end{pmatrix} \begin{pmatrix} x \\ y \end{pmatrix},$$

then $x' = x/a$ and $y' = y/a$, so equation (5) becomes $(y'a)^2 = a(x'a)$, or

$$(y')^2 = x'.$$

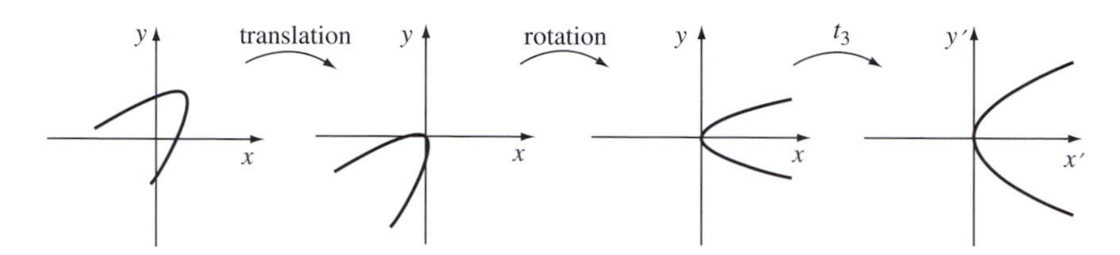

Dropping the dashes, we obtain the following theorem.

Theorem 3 Every parabola is affine-congruent to the parabola with equation $y^2 = x$.

Since all parabolas are affine-congruent to $y^2 = x$, they must be affine-congruent to each other. Similarly, by Theorem 1, all ellipses must be affine-congruent to each other; and, by Theorem 2, all hyperbolas must be affine-congruent to each other.

This raises the question as to whether it is possible for one type of conic (such as an ellipse) to be affine-congruent to another type of conic (such as a hyperbola). The next theorem shows that this cannot happen. In fact, since an affine transformation can be expressed as the composite of two parallel projections, this should not surprise you. After all, no parallel projection can change a bounded curve (such as an ellipse) into an unbounded one (such as a parabola or a hyperbola); nor can it change a curve with two branches (a hyperbola) into a curve with just one branch (an ellipse or a parabola).

Theorem 4 Affine transformations map ellipses to ellipses, parabolas to parabolas, and hyperbolas to hyperbolas.

Remember that a circle is a special type of ellipse.

Proof Consider the non-degenerate conic with equation

$$Ax^2 + Bxy + Cy^2 + Fx + Gy + H = 0, \tag{6}$$

and its image under an affine transformation $t : \mathbf{x} \mapsto \mathbf{x}'$ given by

$$\mathbf{x}' = \mathbf{A}\mathbf{x} + \mathbf{b},$$

where $\mathbf{A}$ is an invertible 2×2 matrix.

The inverse affine transformation $t^{-1} : \mathbf{x}' \mapsto \mathbf{x}$ is given by

$$\mathbf{x} = \mathbf{A}^{-1}\mathbf{x}' - \mathbf{A}^{-1}\mathbf{b},$$

You met this formula for the inverse in Subsection 2.2.1.

which we may write in the form

$$\begin{pmatrix} x \\ y \end{pmatrix} = \begin{pmatrix} p & q \\ r & s \end{pmatrix} \begin{pmatrix} x' \\ y' \end{pmatrix} + \begin{pmatrix} u \\ v \end{pmatrix},$$

for some real numbers p, q, r, s, u and v. It follows that

$$x = px' + qy' + u \quad \text{and} \quad y = rx' + sy' + v. \tag{7}$$

If we now substitute these expressions for x and y into equation (6), then the resulting equation is a second-degree equation in x' and y', so the image of the conic under the affine transformation t must be another conic.

We omit the details of these calculations, as they are complicated and uninformative.

Next we show that this image conic cannot be degenerate. A degenerate image would consist of a pair of lines, a single line, a point, or the empty set. Since the affine transformation t^{-1} maps lines to lines, it would map the degenerate image to another degenerate conic. But this cannot happen since t^{-1} maps the image back to the original non-degenerate conic (6). It follows that the image of (6) cannot be degenerate.

Theorem 2 of Section 2.3

Finally, if we substitute for x and y from equations (7) into equation (6), and keep careful track of the algebra involved, it turns out that the discriminant of the image conic is just

$$(ps - rq)^2(B^2 - 4AC).$$

Recall that the sign of the discriminant of a non-degenerate conic determines the type of the conic.

Here $B^2 - 4AC$ is the discriminant of the original conic. Since $(ps - rq)^2 > 0$, the sign of the discriminant is not changed by an affine transformation of a conic. Hence the type of the conic is also unchanged. ■

Here $ps - rq \neq 0$ since $\mathbf{A}$ is invertible.

We can combine the results of Theorems 1–4 to obtain the following corollary.

Corollary In affine geometry:

(a) all ellipses are congruent to each other;
(b) all hyperbolas are congruent to each other;
(c) all parabolas are congruent to each other.

Non-degenerate conics are congruent only to non-degenerate conics of the same type.

The corollary shows that affine-congruence partitions the set of non-degenerate conics into three disjoint equivalence classes. One class consists of all the ellipses, another class consists of all the hyperbolas, and the third consists of all the parabolas. Each class contains one of the so-called *standard* conics $x^2 + y^2 = 1$, $xy = 1$ and $y^2 = x$.

Just as the Fundamental Theorem of Affine Geometry enables us to deduce a given result about an arbitrary triangle by showing that the result holds for an equilateral triangle, so the corollary enables us to deduce a given result about an arbitrary ellipse, hyperbola or parabola by showing that the result holds for the corresponding standard conic. Of course, this works only if the result is concerned with the affine properties of the conic, so we need to be able to recognize such properties.

The following theorem shows that one such property is the property of being the centre of an ellipse or hyperbola.

Theorem 5 Let t be an affine transformation, and let C be an ellipse or hyperbola with centre R. Then $t(C)$ has centre $t(R)$.

Recall that a parabola does not have a centre.

Proof Let C' and R' be the images of C and R under t. If P' is any point on C', then it must be the image of some point P on C. Since R is the centre of C, we can rotate P about R through an angle π to a point Q which must also lie on C. Hence $Q' = t(Q)$ is a point on C'.

This uses the definition of centre given in Chapter 1.

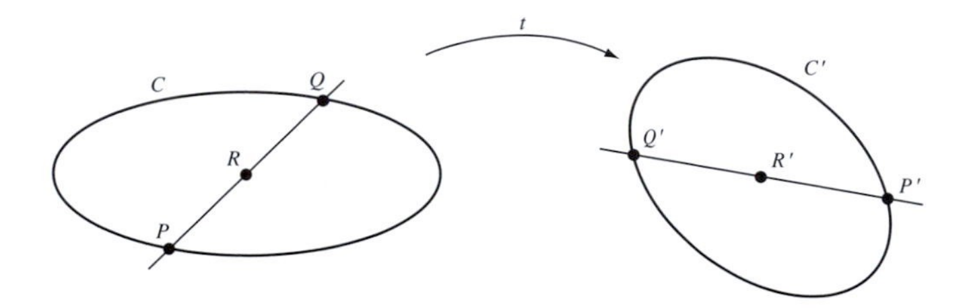

This figure illustrates the proof for an ellipse C, but the proof works equally well for a hyperbola.

Now t preserves ratios of lengths along lines, so the line segment PRQ maps onto the line segment $P'R'Q'$ with $P'R' = R'Q'$. Thus if we rotate P' about R' through an angle π, it must go to Q' on C'. Now, as our choice for P' as a point on C' varies, so do $P = t^{-1}(P')$ and Q, but the point R is always the same point. It follows that the midpoint of $P'Q'$ is always the same point $R' = t(R)$. Hence $R' = t(R)$ is the centre of C', as required. ■

For since $PR/RQ = 1$ it follows that $P'R'/R'Q' = 1$.

Another affine property is the property of being an asymptote of a hyperbola.

Theorem 6 Let t be an affine transformation, and let H be a hyperbola with asymptotes ℓ_1 and ℓ_2. Then $t(H)$ has asymptotes $t(\ell_1)$ and $t(\ell_2)$.

The figure below illustrates that this theorem is plausible for parallel projections.

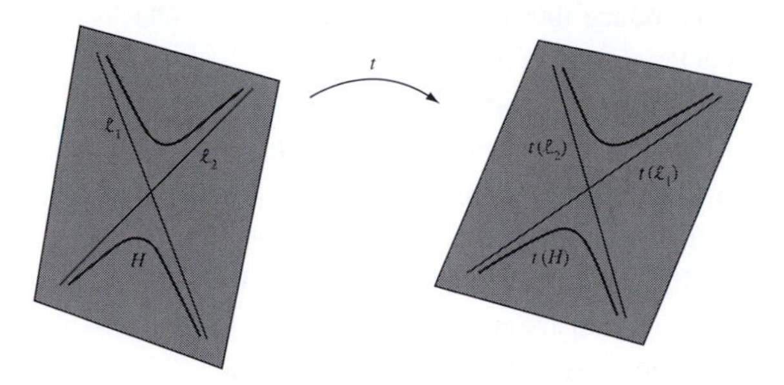

Proof The hyperbola H possesses exactly two (distinct) families of parallel lines each of which fills the plane, with each member of each family meeting H exactly once – apart from one line in each family that is an asymptote of H, and so does not meet H.

The image of H under the affine transformation t is also a hyperbola, $t(H)$. The images under t of the two families of parallel lines are also (distinct) families of parallel lines; within each family, a line that meets H once is mapped onto a line that meets $t(H)$ once, and the single line that does not meet H maps onto a line that does not meet $t(H)$. So the two exceptional lines in the image families must be the asymptotes of the hyperbola $t(H)$.

It follows that the asymptotes of H are mapped by t to the asymptotes of $t(H)$, as required. ■

Many of the problems concerning conics which are particularly amenable to solution using the methods of affine geometry involve tangents.

This is due to the following theorem, which asserts that tangency is an affine property.

Theorem 7 Let t be an affine transformation, and let ℓ be a tangent to a conic C. Then $t(\ell)$ is a tangent to the conic $t(C)$.

The figure below illustrates the theorem for parallel projections.

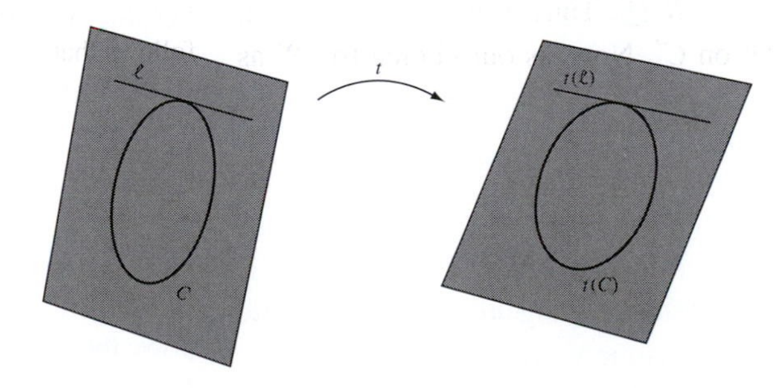

Solution We shall use the fact that a tangent to a conic (whether it is an ellipse, a hyperbola or a parabola) intersects the conic at exactly one point.

However we have to be a little careful. For example, any line parallel to its axis meets a parabola in exactly one point.

First, the image of an ellipse E under an affine transformation t is an ellipse. A tangent to E is a line that intersects E in exactly one point. These properties remain unchanged under an affine projection; hence the image of a tangent to E under an affine transformation t must be a tangent to $t(E)$.

This characterizes tangents to ellipses.

Next, the image of a hyperbola H under an affine transformation t is a hyperbola. A tangent to H is a member of a family of parallel lines that fill the plane such that there are lines in the family that meet H twice, once and not at all; there are exactly two lines in the family that meet H exactly once, and these are tangents to H. The image of the family of lines under t is again a family of parallel lines that fill the plane; it contains lines that meet the parabola $t(H)$ twice and not at all, and exactly two lines that meet H exactly once. These lines are the images of the original tangents to H, and must themselves be tangents to $t(H)$. Hence, the image of a tangent to H under an affine transformation t must be a tangent to $t(H)$.

This characterizes tangents to hyperbolas.

Finally, the image of a parabola P under an affine transformation t is a parabola. A tangent to P is a member of a family of parallel lines that fill the plane such that there are lines in the family that meet P twice, once and not at all; the tangent is the unique member of the family that meets P exactly once. The image of the family of lines under t is again a family of parallel lines that fill the plane; it contains lines that meet the parabola $t(P)$ twice and not at all, and a single line that meets P exactly once. This line is the image of the original tangent to P, and must itself be a tangent to $t(P)$. Hence, the image of a tangent to P under an affine transformation t must be a tangent to $t(P)$.

This characterizes tangents to parabolas.

This completes the proof. ■

In applications we often use the following facts that you met earlier.

Theorem 2, Subsection 1.2.1.

Tangents to Conics in Standard Form The equation of the tangent to a standard conic at the point (x_1, y_1) is as follows.

Conic	*Tangent*
Unit circle $x^2 + y^2 = 1$	$xx_1 + yy_1 = 1$
Rectangular hyperbola $xy = 1$	$xy_1 + yx_1 = 2$
Parabola $y^2 = x$	$2yy_1 = x + x_1$

2.5.2 Applying Affine Geometry to Conics

We are now in a position to apply the methods of affine geometry to the solution of problems involving conics. Of course, affine geometry can be helpful in this task only if the property being investigated is one which is preserved under affine transformations. The underlying idea is that we use an affine transformation to map the original conic onto one of our standard conics, tackle the problem in hand there, and then map back to the original conic.

We used these techniques in Subsection 2.2.3 to prove the Conjugate Diameters Theorem for the ellipse.

Example 1 AB is a diameter of an ellipse. Prove that the tangents to the ellipse at A and B are parallel to the diameter conjugate to AB.

Recall that the diameter conjugate to AB is the set of midpoints of all the chords parallel to AB (see Subsection 2.2.3).

Solution First, map the ellipse onto the unit circle, by an affine transformation t. Since the centre O of the ellipse maps to the centre O' of the circle, the image of the diameter AB is a diameter $A'B'$ of the unit circle.

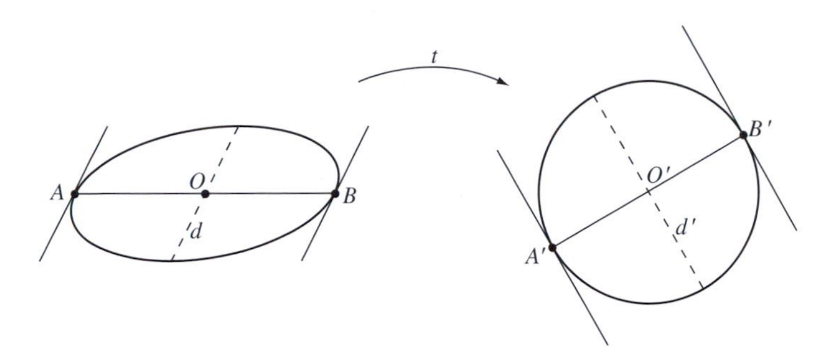

All chords of the circle that are parallel to the tangents at A' and B' are bisected by $A'B'$, and so the diameter through O' is the diameter conjugate to $A'B'$. Since parallel lines map to parallel lines and ratios along parallel lines are preserved under the inverse affine transformation t^{-1}, it follows that all chords of the ellipse that are parallel to the tangents at A and B are bisected by AB, and so the diameter through O that is parallel to the tangents at A and B is the diameter conjugate to AB. □

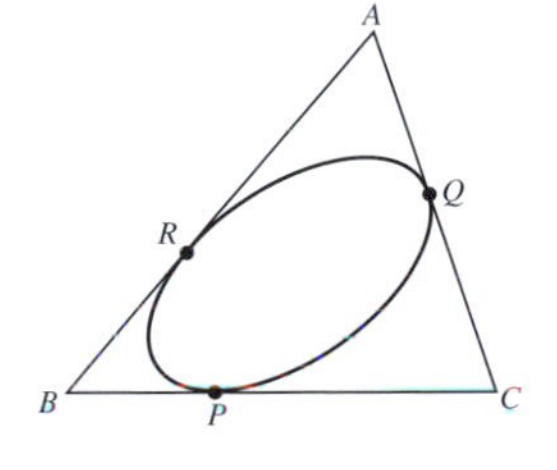

Problem 1 An ellipse touches the sides BC, CA and AB of ΔABC at the points P, Q and R, respectively. Prove that

$$\frac{AR}{RB} \cdot \frac{BP}{PC} \cdot \frac{CQ}{QA} = 1,$$

and deduce that the lines AP, BQ and CR are concurrent.

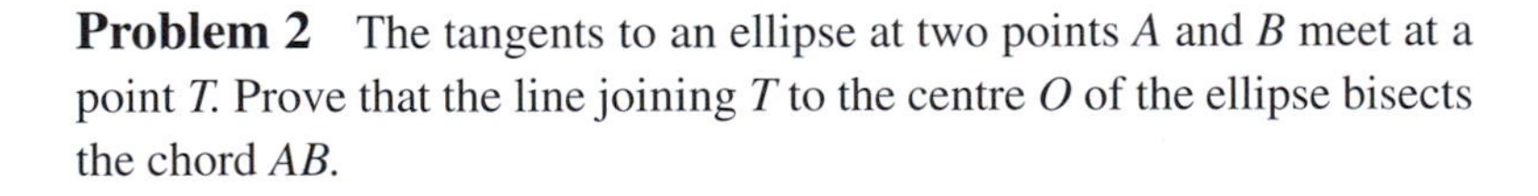

Problem 2 The tangents to an ellipse at two points A and B meet at a point T. Prove that the line joining T to the centre O of the ellipse bisects the chord AB.

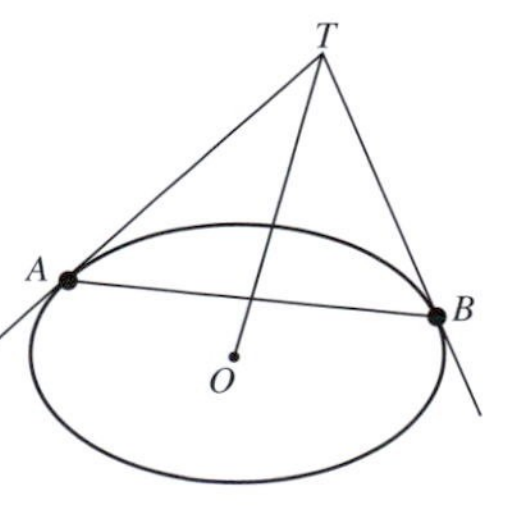

The rectangular hyperbola $H = \{(x, y) : xy = 1\}$ does not possess as much symmetry as does the unit circle; so the fact that every hyperbola is affine-congruent to H may not be sufficient to simplify a given problem. Fortunately, however, we can also arrange for any given point on the original hyperbola to map to the point $(1, 1)$ on H.

To see this, note that for any non-zero number a, the affine transformation

$$t_a : \begin{pmatrix} x \\ y \end{pmatrix} \mapsto \begin{pmatrix} a & 0 \\ 0 & 1/a \end{pmatrix} \begin{pmatrix} x \\ y \end{pmatrix}$$

maps H to itself. For, an arbitrary point on H has coordinates of the form $(x, 1/x)$, $x \neq 0$, and under t_a this is mapped to the point $(ax, 1/ax)$, which also lies on H. As x varies through $\mathbb{R} - \{0\}$, its image $(ax, 1/ax)$ varies over the whole of H, so the image of H under t_a is the whole of H.

So if we start with a given hyperbola and a point P on it, we can map the hyperbola to H by some affine transformation s. The point $s(P)$ will then have coordinates $(b, 1/b)$ for some number $b \in \mathbb{R} - \{0\}$; so if we choose $a = 1/b$, then the affine transformation t_a will map $s(P)$ to (1,1). Overall, the composite $t = t_a \circ s$ is an affine transformation which maps the given hyperbola to H, and maps P to (1, 1). We now state this as a corollary to Theorem 2.

Corollary Given any hyperbola and a point P on it, there is an affine transformation which maps the hyperbola onto the rectangular hyperbola $xy = 1$, and the point P to (1, 1).

Example 2 The tangent at the point P on a hyperbola meets the asymptotes at the points A and B. Prove that $PA = PB$.

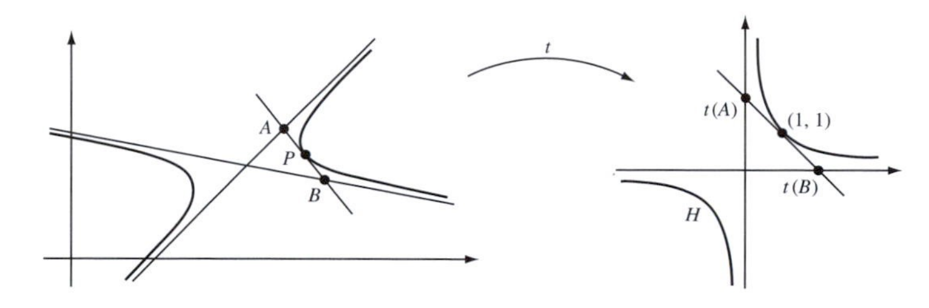

Solution Let t be an affine transformation which maps the hyperbola onto the rectangular hyperbola $H = \{(x, y) : xy = 1\}$ in such a way that $t(P) = (1, 1)$. Then, by Theorem 6 of Subsection 2.5.1, the asymptotes of the hyperbola map to the asymptotes of H; and, by Theorem 7 of Subsection 2.5.1, the tangent at P maps to the tangent at (1,1).

By symmetry, (1, 1) is the midpoint of the line segment from $t(A)$ to $t(B)$. Since midpoints are preserved under the affine transformation t^{-1}, it follows that P is the midpoint of AB. □

Problem 3 P is a point on a hyperbola H with centre O. Prove that there exists a line ℓ through O such that all chords of the hyperbola which are parallel to ℓ are bisected by OP.

This result is an analogue for the hyperbola of the Conjugate Diameters Theorem for the ellipse (Theorem 3 of Subsection 2.2.3).

2.6 Exercises

Section 2.1

1. Let ΔABC be a triangle in which $AB = AC$. Prove that

$$\angle ABC = \angle ACB.$$

Hint: Consider a reflection in the bisector of $\angle BAC$.

2. Determine which of the following transformations $t : \mathbb{R}^2 \to \mathbb{R}^2$ are Euclidean transformations.

(a) $t(\mathbf{x}) = \begin{pmatrix} -\frac{1}{2} & -\frac{\sqrt{3}}{2} \\ -\frac{\sqrt{3}}{2} & \frac{1}{2} \end{pmatrix}\mathbf{x} + \begin{pmatrix} -3 \\ 1 \end{pmatrix}$

(b) $t(\mathbf{x}) = \begin{pmatrix} -\frac{2}{3} & -\frac{1}{3} \\ -\frac{1}{3} & \frac{2}{3} \end{pmatrix}\mathbf{x} + \begin{pmatrix} 3 \\ 2 \end{pmatrix}$

(c) $t(\mathbf{x}) = \begin{pmatrix} -\frac{1}{\sqrt{5}} & \frac{2}{\sqrt{5}} \\ -\frac{2}{\sqrt{5}} & -\frac{1}{\sqrt{5}} \end{pmatrix}\mathbf{x} + \begin{pmatrix} 2 \\ -3 \end{pmatrix}$

3. The Euclidean transformations t_1 and t_2 are given by

$$t_1(\mathbf{x}) = \begin{pmatrix} \frac{1}{\sqrt{5}} & \frac{2}{\sqrt{5}} \\ \frac{2}{\sqrt{5}} & -\frac{1}{\sqrt{5}} \end{pmatrix}\mathbf{x} + \begin{pmatrix} -1 \\ 1 \end{pmatrix}$$

and

$$t_2(\mathbf{x}) = \begin{pmatrix} \frac{1}{\sqrt{5}} & \frac{2}{\sqrt{5}} \\ -\frac{2}{\sqrt{5}} & \frac{1}{\sqrt{5}} \end{pmatrix}\mathbf{x} + \begin{pmatrix} 2 \\ -1 \end{pmatrix}.$$

Determine the composites $t_1 \circ t_2$ and $t_2 \circ t_1$.

4. Determine the inverse of each of the following Euclidean transformations.

(a) $t(\mathbf{x}) = \begin{pmatrix} \frac{5}{13} & -\frac{12}{13} \\ \frac{12}{13} & \frac{5}{13} \end{pmatrix}\mathbf{x} + \begin{pmatrix} -4 \\ 5 \end{pmatrix}$

(b) $t(\mathbf{x}) = \begin{pmatrix} -\frac{12}{13} & -\frac{5}{13} \\ -\frac{5}{13} & \frac{12}{13} \end{pmatrix}\mathbf{x} + \begin{pmatrix} 1 \\ -1 \end{pmatrix}$

5. The Euclidean transformations t_1 and t_2 are given by

$$t_1(\mathbf{x}) = \begin{pmatrix} \frac{1}{\sqrt{2}} & \frac{1}{\sqrt{2}} \\ \frac{1}{\sqrt{2}} & -\frac{1}{\sqrt{2}} \end{pmatrix}\mathbf{x} + \begin{pmatrix} 1 \\ -1 \end{pmatrix}$$

and

$$t_2(\mathbf{x}) = \begin{pmatrix} -\frac{1}{\sqrt{2}} & \frac{1}{\sqrt{2}} \\ -\frac{1}{\sqrt{2}} & -\frac{1}{\sqrt{2}} \end{pmatrix}\mathbf{x} + \begin{pmatrix} 1 \\ 1 \end{pmatrix}.$$

Determine the composite $t_2^{-1} \circ t_1$.

Section 2.2

1. Determine whether or not each of the following transformations $t : \mathbb{R}^2 \to \mathbb{R}^2$ is an affine transformation.

   (a) $t(\mathbf{x}) = \begin{pmatrix} 2 & -2 \\ -3 & 3 \end{pmatrix} \mathbf{x} + \begin{pmatrix} 2 \\ -1 \end{pmatrix}$

   (b) $t(\mathbf{x}) = \begin{pmatrix} 5 & -2 \\ -2 & 5 \end{pmatrix} \mathbf{x} + \begin{pmatrix} -3 \\ -1 \end{pmatrix}$

   (c) $t(\mathbf{x}) = \begin{pmatrix} -1 & 1 \\ -1 & -2 \end{pmatrix} \mathbf{x}$

2. Write down an example (if one exists) of each type of transformation $t : \mathbb{R}^2 \to \mathbb{R}^2$ described below. In each case, justify your answer.

   (a) An affine transformation t which is not a Euclidean transformation
   (b) A Euclidean transformation t which is not an affine transformation
   (c) A transformation t which is both Euclidean and affine
   (d) A transformation t which is one–one, but is neither Euclidean nor affine

3. The affine transformations t_1 and t_2 are given by

$$t_1(\mathbf{x}) = \begin{pmatrix} 2 & -3 \\ 1 & -1 \end{pmatrix} \mathbf{x} + \begin{pmatrix} 1 \\ -1 \end{pmatrix}$$

   and

$$t_2(\mathbf{x}) = \begin{pmatrix} -1 & 2 \\ -1 & 1 \end{pmatrix} \mathbf{x} + \begin{pmatrix} -1 \\ 1 \end{pmatrix}.$$

   Determine the following composites.

   (a) $t_1 \circ t_2$ (b) $t_2 \circ t_1$ (c) $t_1 \circ t_1$

4. Determine the inverse of each of the following affine transformations.

   (a) $t(\mathbf{x}) = \begin{pmatrix} 2 & -3 \\ 3 & -5 \end{pmatrix} \mathbf{x} + \begin{pmatrix} 2 \\ 4 \end{pmatrix}$ (b) $t(\mathbf{x}) = \begin{pmatrix} 3 & 2 \\ 4 & 2 \end{pmatrix} \mathbf{x} + \begin{pmatrix} 1 \\ -2 \end{pmatrix}$

5. Prove that the transformation

$$t(\mathbf{x}) = 3\mathbf{x} \ (\mathbf{x} \in \mathbb{R}^2)$$

   is an affine transformation, but not a parallel projection.

6. Which of the following are affine properties?

   (a) distance (b) collinearity (c) circularity
   (d) magnitude of angle (e) midpoint of line segment

Section 2.3

1. The affine transformation $t : \mathbb{R}^2 \to \mathbb{R}^2$ is given by

$$t(\mathbf{x}) = \begin{pmatrix} 1 & -1 \\ 2 & -3 \end{pmatrix} \mathbf{x} + \begin{pmatrix} 2 \\ -4 \end{pmatrix}.$$

   Determine the image under t of each of the following lines.

   (a) $y = -2x$ (b) $2y = 3x - 1$

2. The affine transformation $t : \mathbb{R}^2 \to \mathbb{R}^2$ is given by

$$t(\mathbf{x}) = \begin{pmatrix} 4 & 5 \\ 1 & 1 \end{pmatrix} \mathbf{x} + \begin{pmatrix} 1 \\ -1 \end{pmatrix}.$$

Determine the image under t of each of the following lines.

(a) $2x - 5y + 3 = 0$ (b) $3x + y - 4 = 0$

3. Determine the affine transformation which maps the points $(0, 0)$, $(1, 0)$ and $(0, 1)$ to the points:
   (a) $(0, -1)$, $(1, 1)$ and $(-1, 1)$, respectively;
   (b) $(-4, -5)$, $(1, 7)$ and $(2, -9)$, respectively.
4. Determine the affine transformation which maps the points $(1, 1)$, $(3, 2)$ and $(4, 1)$ to the points $(0, 1)$, $(1, 2)$ and $(3, 7)$, respectively.
5. Determine the affine transformation which maps the points $(1, -1)$, $(5, -4)$ and $(-2, 1)$ to the points $(1, 1)$, $(4, 0)$ and $(0, 2)$, respectively.
6. Prove that the affine transformation t for which

$$t(\mathbf{x}) = \begin{pmatrix} -1 & 2 \\ 3 & -2 \end{pmatrix} \mathbf{x}$$

maps each point of the line $y = x$ in $\mathbb{R}^2$ onto itself.

7. Determine the matrices $\mathbf{A}$ and $\mathbf{b}$ for the affine transformation

$$t(\mathbf{x}) = \mathbf{A}\mathbf{x} + \mathbf{b},$$

where $\mathbf{A}$ and $\mathbf{b}$ are 2×2 and 2×1 matrices, respectively, given that t maps each point of the line $y = 0$ onto itself and (0,1) onto (2,3). Prove also that t is a parallel projection of $\mathbb{R}^2$ onto itself.

Section 2.4

1. The points P, Q, R and S lie on a line, in that order; the distances between them are 4 units, 2 units and 3 units, respectively. Determine the ratios $PR : RS$ and $PS : SQ$.
2. A point X lies inside a triangle ΔABC, and the lines AX, BX and CX meet the opposite sides of the triangle at P, Q and R, respectively. The ratios $AR : AB$ and $BP : BC$ are $1 : 5$ and $3 : 7$, respectively. Determine the ratio $AC : AQ$.
3. Let ℓ be a line that crosses the sides BC, CA and AB of a triangle ΔABC at three distinct points P, Q and R, respectively. The ratios $BC : CP$ and $CQ : QA$ are $3 : 2$ and $1 : 3$, respectively. Determine the ratio $AR : RB$.
4. $ABCD$ is a parallelogram, and the point P divides AB in the ratio $2 : 1$; the lines AC and DP meet at Q, and the lines BQ and AD meet at R.
   (a) Determine the images of P, Q and R under the affine transformation t which maps A, D and C to $(0, 1)$, $(0, 0)$ and $(1, 0)$, respectively.
   (b) By considering the image of $ABCD$ under t, determine the ratios $BQ : QR$ and $AR : RD$.
5. The triangle ΔABC has vertices $A(-1, 2)$, $B(-3, -1)$ and $C(3, 1)$, and the points $P\left(1, \frac{1}{3}\right)$, $Q\left(1, \frac{3}{2}\right)$ and $R\left(-\frac{5}{3}, 1\right)$ lie on BC, CA and AB, respectively.
   (a) Determine the ratios in which P, Q and R divide the sides of the triangle.
   (b) Determine whether or not the lines AP, BQ and CR are concurrent.

6. The triangle ΔABC has vertices $A(2,0)$, $B(-3,0)$ and $C(3,-3)$, and the points $P(-1,-1)$, $Q(1,3)$ and $R\left(-\frac{1}{4},0\right)$ lie on BC, CA and AB, respectively.
   (a) Determine the ratios in which P, Q and R divide the sides of the triangle.
   (b) Determine whether or not the points P, Q and R are collinear.
7. ΔABC is a triangle, and X a point which does not lie on any of its (extended) sides. Also, AX meets BC at P, BX meets CA at Q and CX meets BA at R. Prove that
$$\frac{AX}{XP} = \frac{AR}{RB} + \frac{AQ}{QC}.$$
(This result is often known as van Aubel's Theorem.)

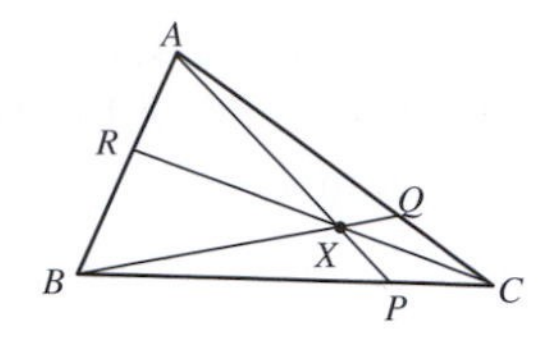

8. ΔABC is a triangle, and X a point which does not lie on any of its (extended) sides. Next, AX meets BC at P, BX meets CA at Q and CX meets BA at R. Also, RQ meets BC at L, PR meets CA at M and PQ meets BA at N. Prove that L, M and N are collinear.
   Hint: Apply the result of Problem 5 in Subsection 2.4.3 to ΔABC and points L, M and N in turn. Then evaluate the product $\frac{BL}{LC} \cdot \frac{CM}{MA} \cdot \frac{AN}{NB}$.

Direct calculation using coordinates is also possible, but very tedious!

9. Three disjoint circles of unequal radii lie in the plane, their centres being non-collinear. Pairs of tangents are drawn to each pair of circles such that the point of intersection of the two tangents to each pair of circles lies beyond the two circles. Prove that the three intersection points are collinear.

Section 2.5

1. An ellipse touches the sides AB, BC, CD, DA of a parallelogram $ABCD$ at the points P, Q, R, S, respectively. Prove that the lengths CQ, QB, BP and CR satisfy the equation
$$\frac{CQ}{QB} = \frac{CR}{BP}.$$

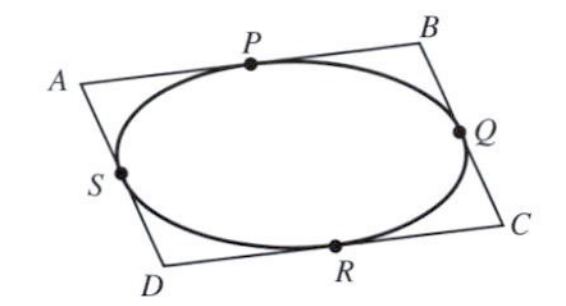

2. Determine the equation of the image of the parabola P with equation $y = x^2$ under the affine transformation $t : \mathbb{R}^2 \mapsto \mathbb{R}^2$ given by
$$t(\mathbf{x}) = \begin{pmatrix} 1 & 0 \\ -2 & 1 \end{pmatrix} \mathbf{x}.$$
Show that the image of the vertex of P is not the vertex of $t(P)$.

This proves that the property of 'being a vertex of a parabola' is not an affine property.

3. Prove that for any triangle ΔABC there exists an ellipse that touches the sides AB, BC and CA at their midpoints.
4. Let $P(a\cos\theta, b\sin\theta)$, where θ is not a multiple of $\pi/2$, be a point on the ellipse $C : \frac{x^2}{a^2} + \frac{y^2}{b^2} = 1$, where $a \geq b > 0$; and $P'(a\cos\theta, a\sin\theta)$ the corresponding point on the 'auxiliary circle' C': $x^2 + y^2 = a^2$. Prove that the tangents at P to C and at P' to C' meet on the x-axis.
   Hint: Write down an affine transformation that maps C to C' and P to P', and that maps each point of the x-axis to itself.
5. Given any two points P and P' on ellipses E and E', respectively, show that there exists an affine transformation that maps E to E' and P to P'.

This result is analogous to that for hyperbolas in the Corollary in Subsection 2.5.2.

6. Find the endpoints of the chord AB of the hyperbola H with equation $xy = 1$ that is bisected by the point $P(2,1)$.

7. E is the ellipse with equation $\frac{x^2}{9} + \frac{y^2}{4} = 1$, and $P\left(\frac{3}{\sqrt{5}}, \frac{2}{\sqrt{5}}\right)$ is a point inside E. AB is a chord of E through P, and O is the centre of E. Find the maximum value of $\frac{AP}{PB}$ as A varies on E.

Summary of Chapter 2

Section 2.1: Geometry and transformations

1. An **isometry** of $\mathbb{R}^2$ is a function which maps $\mathbb{R}^2$ onto $\mathbb{R}^2$ and preserves distances.
   Every isometry of $\mathbb{R}^2$ has one of the following forms:
   - a translation along a line in $\mathbb{R}^2$;
   - a reflection in a line in $\mathbb{R}^2$;
   - a rotation about a point in $\mathbb{R}^2$;
   - a composite of translations, reflections and rotations in $\mathbb{R}^2$.

   The set of all isometries of $\mathbb{R}^2$ forms a group under composition of functions; in particular, the composite of two isometries is an isometry.
2. **Euclidean geometry** is the study of those properties of figures that are unchanged by the group of isometries.
   These properties are called **Euclidean properties**, and include distance, angle, collinearity of points and concurrence of lines.
3. The **Kleinian view of geometry** is the idea that geometry can be thought of in terms of a group of transformations acting on a space.
4. The transformations of $\mathbb{R}^2$ given by

$$\begin{pmatrix} x \\ y \end{pmatrix} \mapsto \begin{pmatrix} \cos\theta & -\sin\theta \\ \sin\theta & \cos\theta \end{pmatrix} \begin{pmatrix} x \\ y \end{pmatrix} + \begin{pmatrix} e \\ f \end{pmatrix}, \text{ and}$$

$$\begin{pmatrix} x \\ y \end{pmatrix} \mapsto \begin{pmatrix} \cos\theta & \sin\theta \\ \sin\theta & -\cos\theta \end{pmatrix} \begin{pmatrix} x \\ y \end{pmatrix} + \begin{pmatrix} e \\ f \end{pmatrix}$$

   are isometries; they represent, respectively, anticlockwise rotation about the origin through an angle θ followed by a translation by (e, f), and reflection in a line through the origin that makes an angle $\theta/2$ with the x-axis followed by a translation by (e, f).
   Every isometry of $\mathbb{R}^2$ is of one or other of these two forms
5. A **Euclidean transformation** of $\mathbb{R}^2$ is a function $t : \mathbb{R}^2 \to \mathbb{R}^2$ of the form $t(\mathbf{x}) = \mathbf{Ux} + \mathbf{a}$, where $\mathbf{U}$ is an orthogonal 2×2 matrix and $\mathbf{a} \in \mathbb{R}^2$. The set of all Euclidean transformations of $\mathbb{R}^2$ is denoted by $E(2)$.
6. Every isometry of $\mathbb{R}^2$ is a Euclidean transformation of $\mathbb{R}^2$, and vice versa.
7. The set of Euclidean transformations of $\mathbb{R}^2$ forms a group under the operation of composition of functions.
8. The inverse of the Euclidean transformation $t(\mathbf{x}) = \mathbf{Ux} + \mathbf{a}$ is given by $t^{-1}(\mathbf{x}) = \mathbf{U}^{-1}\mathbf{x} - \mathbf{U}^{-1}\mathbf{a}$.

9. **Euclidean geometry** is the study of those properties of figures that are preserved by Euclidean transformations of $\mathbb{R}^2$.
10 A figure F_1 is **Euclidean-congruent** to a figure F_2 if there is a Euclidean transformation which maps F_1 onto F_2. Loosely speaking, two figures are congruent if they have the same size and shape.

   Euclidean congruence is an equivalence relation.

   A figure F_1 is ***G*-congruent** to a figure F_2 in some geometry defined by a group G of transformations acting on the space of the geometry if there is a transformation in G which maps F_1 onto F_2. G-congruence is an equivalence relation.

Section 2.2: Affine Transformations and Parallel Projections

1. An **affine transformation** of $\mathbb{R}^2$ is a function $t : \mathbb{R}^2 \to \mathbb{R}^2$ of the form $t(\mathbf{x}) = \mathbf{Ax} + \mathbf{b}$, where $\mathbf{A}$ is an invertible 2×2 matrix and $\mathbf{b} \in \mathbb{R}^2$. The set of all affine transformations of $\mathbb{R}^2$ is denoted by $A(2)$.

   Every Euclidean transformation of $\mathbb{R}^2$ is an affine transformation.
2. The set of affine transformations $A(2)$ forms a group under the operation of composition of functions.

   The inverse of the affine transformation $t(\mathbf{x}) = \mathbf{Ax} + \mathbf{b}$ is $t^{-1}(\mathbf{x}) = \mathbf{A}^{-1}\mathbf{x} - \mathbf{A}^{-1}\mathbf{b}$.
3. **Affine geometry** is the study of those properties (called **affine properties**) of figures in the plane $\mathbb{R}^2$ that are preserved by affine transformations.

   Basic properties of affine transformations

   Affine transformations:
   1. map straight lines to straight lines;
   2. map parallel straight lines to parallel straight lines;
   3. preserve ratios of lengths along a given straight line.
4. A **parallel projection** is a one-one mapping of $\mathbb{R}^2$ to itself defined in the following way. Let π_1 and π_2 be planes in $\mathbb{R}^3$, with parallel rays of light shining through them; then the function p which maps each point P in π_1 to the corresponding point P' in π_2 is a parallel projection from π_1 onto π_2.

   If π_1 and π_2 are parallel, then the parallel projection from π_1 onto π_2 is an isometry.
5. **Basic properties of parallel projections**

   Parallel projections:
   1. map straight lines to straight lines;
   2. map parallel straight lines to parallel straight lines;
   3. preserve ratios of lengths along a given straight line.
6. A **diameter** of an ellipse is a chord of the ellipse that passes through its centre.

   Midpoint Theorem Let ℓ be a chord of an ellipse. Then the midpoints of the chords parallel to ℓ lie on a diameter of the ellipse.
7. **Conjugate Diameters Theorem** Let ℓ be a diameter of an ellipse. Then there is another diameter m of the ellipse such that

(a) the midpoints of all chords parallel to ℓ lie on m;
(b) the midpoints of all chords parallel to m lie on ℓ.

The directions of these two diameters are called **conjugate directions**, and the diameters are called **conjugate diameters**.

8. Given any ellipse, there is a parallel projection which maps the ellipse onto a circle.
9. Under parallel projection certain properties of figures, such as length and angle, are not necessarily preserved. This is one difference between Euclidean geometry and affine geometry.
10. Each parallel projection is an affine transformation.

    However, an affine transformation is not necessarily a parallel projection. For example, the **doubling map** of $\mathbb{R}^2$ to itself given by $t(\mathbf{v}) = 2\mathbf{v}$ is an affine transformation but cannot be modelled by a parallel projection.
11. An affine transformation t is **completely determined** by its effect on the three non-collinear points (0,0), (1,0) and (0,1).

    Hence if we know the points onto which (0,0), (1,0) and (0,1) are mapped by t, we can determine $\mathbf{A}$ and $\mathbf{b}$ in the formula $t(\mathbf{x}) = \mathbf{A}\mathbf{x} + \mathbf{b}$, $\mathbf{x} \in \mathbb{R}^2$.
12. An affine transformation can be expressed as the composite of two parallel projections.

Section 2.3: Properties of Affine Transformations

1. **Strategy** To determine the image of a line or conic in $\mathbb{R}^2$ under an affine transformation $t : \mathbb{R}^2 \to \mathbb{R}^2$ given by $t(\mathbf{x}) = \mathbf{A}\mathbf{x} + \mathbf{b}$, let $\mathbf{x}$ and coordinates (x, y) denote points in the domain copy of $\mathbb{R}^2$, and $\mathbf{x}'$ and coordinates (x', y') denote points in the codomain copy. Then:
   1. express the relationship between $\mathbf{x}$ and $\mathbf{x}'$ in the form $\mathbf{x} = \mathbf{A}^{-1}\mathbf{x}' - \mathbf{A}^{-1}\mathbf{b}$;
   2. determine formulas for x and y in terms of x' and y';
   3. substitute for x and y in the equation of the line or conic;
   4. drop the dashes from x' and y'.

   The resulting equation describes the image under t.
2. **Strategy** To determine the unique affine transformation $t(\mathbf{x}) = \mathbf{A}\mathbf{x} + \mathbf{b}$ which maps (0,0), (1,0) and (0,1) to the three non-collinear points $\mathbf{p}$, $\mathbf{q}$ and $\mathbf{r}$, respectively:
   1. take $\mathbf{b} = \mathbf{p}$;
   2. take $\mathbf{A}$ to be the matrix with columns given by $\mathbf{q} - \mathbf{p}$ and $\mathbf{r} - \mathbf{p}$.

   Warning This Strategy requires that $\mathbf{p}$, $\mathbf{q}$ and $\mathbf{r}$ are non-collinear. If they are collinear, the matrix $\mathbf{A}$ described in the Strategy is non-invertible, and hence the transformation given by the procedure in the Strategy is not an affine transformation.
3. **Fundamental Theorem of Affine Geometry** Let $\mathbf{p}$, $\mathbf{q}$, $\mathbf{r}$ and $\mathbf{p}'$, $\mathbf{q}'$, $\mathbf{r}'$ be two sets of three non-collinear points in $\mathbb{R}^2$. Then:
   (a) there is an affine transformation t which maps $\mathbf{p}$, $\mathbf{q}$ and $\mathbf{r}$ to $\mathbf{p}'$, $\mathbf{q}'$ and $\mathbf{r}'$, respectively;
   (b) the affine transformation t is unique.

Strategy To determine the affine transformation t which maps three non-collinear points $\mathbf{p}$, $\mathbf{q}$ and $\mathbf{r}$ to another three non-collinear points $\mathbf{p}'$, $\mathbf{q}'$ and $\mathbf{r}'$, respectively:

1. determine the affine transformation t_1 which maps (0,0), (1,0) and (0,1) to the points $\mathbf{p}$, $\mathbf{q}$ and $\mathbf{r}$, respectively;
2. determine the affine transformation t_2 which maps (0,0), (1,0) and (0,1) to the points $\mathbf{p}'$, $\mathbf{q}'$ and $\mathbf{r}'$, respectively;
3. calculate the composite $t = t_2 \circ t_1^{-1}$.

4. Two figures are **affine-congruent** if there is an affine transformation which maps one onto the other.

   All triangles are affine-congruent.
5. An affine transformation preserves ratios of lengths along parallel straight lines.

Section 2.4: Using the Fundamental Theorem of Affine Geometry

1. **Median Theorem** The medians of any triangle are concurrent.
2. If a point R divides a line segment PQ in the **ratio** $(1-\lambda):\lambda$, then $\frac{PR}{RQ} = \frac{1-\lambda}{\lambda}$. The **magnitude** of the ratio equals the length of PR divided by the length of RQ, and the ratio is **positive** if $0 < \lambda < 1$ (when $\overrightarrow{PR}$ and $\overrightarrow{RQ}$ lie in the same direction) and **negative** if $\lambda < 0$ or $\lambda > 1$ (when $\overrightarrow{PR}$ and $\overrightarrow{RQ}$ lie in opposite directions).

   If P, Q and R have coordinates (x_P, y_P), (x_Q, y_Q) and (x_R, y_R), respectively, then $\frac{PR}{RQ} = \frac{x_R - x_P}{x_Q - x_R}$ and $\frac{PR}{RQ} = \frac{y_R - y_P}{y_Q - y_R}$. (If the denominator of one of these fractions vanishes, then use the other to determine the ratio.)
3. **Ceva's Theorem** Let ΔABC be a triangle, and let X be a point which does not lie on any of its (extended) sides. If AX meets BC at P, BX meets CA at Q and CX meets BA at R, then
$$\frac{AR}{RB} \cdot \frac{BP}{PC} \cdot \frac{CQ}{QA} = 1.$$
4. **Converse to Ceva's Theorem** Let P, Q and R be points other than vertices on the (possibly extended) sides BC, CA and AB of a triangle ΔABC, such that $\frac{AR}{RB} \cdot \frac{BP}{PC} \cdot \frac{CQ}{QA} = 1$. Then the lines AP, BQ and CR are concurrent.
5. **Menelaus's Theorem** Let ΔABC be a triangle, and let ℓ be a line that crosses the sides BC, CA and AB at three distinct points P, Q and R, respectively. Then $\frac{AR}{RB} \cdot \frac{BP}{PC} \cdot \frac{CQ}{QA} = -1$.
6. **Converse to Menelaus' Theorem** Let P, Q and R be points other than vertices on the (possibly extended) sides BC, CA and AB of a triangle ΔABC, such that $\frac{AR}{RB} \cdot \frac{BP}{PC} \cdot \frac{CQ}{QA} = -1$. Then the points P, Q and R are collinear.

7. Many results about properties of triangles (such as collinearity, lines being parallel, and ratios of lengths along a given line) are preserved under affine transformations, are proved following a standard pattern.

   First, we choose a particular type of triangle for which it is easy to prove the result. Then, by asserting the existence of an affine transformation from that triangle to an arbitrary triangle, we may deduce that the result holds for all triangles.

8. Let $A = (a_1, a_2)$, $B = (b_1, b_2)$ and $C = (c_1, c_2)$ be three non-collinear points in the plane $\mathbb{R}^2$; we call ΔABC the **triangle of reference**. Then a point (x, y) in the plane has **barycentric coordinates** (ξ, η, ζ) with respect to ΔABC if

$$x = \xi a_1 + \eta b_1 + \zeta c_1,$$
$$y = \xi a_2 + \eta b_2 + \zeta c_2, \text{ and}$$
$$1 = \xi + \eta + \zeta.$$

   If we set $\mathbf{M} = \begin{pmatrix} a_1 & b_1 & c_1 \\ a_2 & b_2 & c_2 \\ 1 & 1 & 1 \end{pmatrix}$, then $\mathbf{M}$ is invertible. Also $\begin{pmatrix} x \\ y \\ 1 \end{pmatrix} = \mathbf{M}\begin{pmatrix} \xi \\ \eta \\ \zeta \end{pmatrix}$ and $\begin{pmatrix} \xi \\ \eta \\ \zeta \end{pmatrix} = \mathbf{M}^{-1}\begin{pmatrix} x \\ y \\ 1 \end{pmatrix}$.

9. The points P, Q and R with barycentric coodinates (ξ_1, η_1, ζ_1), (ξ_2, η_2, ζ_2) and (ξ_3, η_3, ζ_3) are collinear if and only if $\begin{vmatrix} \xi_1 & \xi_2 & \xi_3 \\ \eta_1 & \eta_2 & \eta_3 \\ \zeta_1 & \zeta_2 & \zeta_3 \end{vmatrix} = 0.$

   The line ℓ through the points with barycentric coordinates (ξ_1, η_1, ζ_1) and (ξ_2, η_2, ζ_2) has equation $\begin{vmatrix} \xi_1 & \xi_2 & \xi \\ \eta_1 & \eta_2 & \eta \\ \zeta_1 & \zeta_2 & \zeta \end{vmatrix} = 0.$

10. **Section Formula** The point R that divides the line ℓ joining the points P and Q with barycentric coordinates (ξ_1, η_1, ζ_1) and (ξ_2, η_2, ζ_2) in the ratio $(1 - \lambda) : \lambda$ has barycentric coordinates

$$(\xi, \eta, \zeta) = \lambda(\xi_1, \eta_1, \zeta_1) + (1 - \lambda)(\xi_2, \eta_2, \zeta_2).$$

Section 2.5: Affine Transformations and Conics

1. Every ellipse is affine-congruent to the unit circle with equation $x^2 + y^2 = 1$.

   Every hyperbola is affine-congruent to the rectangular hyperbola with equation $xy = 1$.

   Every parabola is affine-congruent to the parabola with equation $y^2 = x$.

2. Affine transformations map ellipses to ellipses, parabolas to parabolas, and hyperbolas to hyperbolas.

   In Affine Geometry:

   (a) all ellipses are congruent to each other;

(b) all hyperbolas are congruent to each other;

(c) all parabolas are congruent to each other.

Non-degenerate conics are congruent only to non-degenerate conics of the same type.

3. Let t be an affine transformation, and let C be an ellipse or hyperbola with centre R. Then $t(C)$ has centre $t(R)$.
4. Let t be an affine transformation, and let H be a hyperbola with asymptotes ℓ_1 and ℓ_2. Then $t(H)$ has asymptotes $t(\ell_1)$ and $t(\ell_2)$.
5. Let t be an affine transformation, and let ℓ be a tangent to a conic C. Then $t(\ell)$ is a tangent to the conic $t(C)$.
6. Affine geometry can be used to tackle problems involving conics when the property being investigated is an affine property. We first use an affine transformation to map the original conic onto one of our standard conics, tackle the problem in hand there, and then map back to the original conic.
7. Given any hyperbola and a point P on it, there is an affine transformation which maps the hyperbola onto the rectangular hyperbola $xy = 1$, and the point P to (1,1).

3 Projective Geometry: Lines

Geometry is one branch of mathematics that has an obvious relevance to the 'real world'. Earlier, we studied some results in Euclidean geometry and we described the group of Euclidean transformations, the isometries. We saw that the Euclidean transformations preserve distances and angles, and have a definite physical significance.

Chapters 1 and 2.

In this chapter we study *projective geometry,* a very different type of geometry, that has important but less obvious applications. It was discovered through artists' attempts over many centuries to paint realistic-looking pictures of scenes composed of objects situated at differing distances from the eye. How can three-dimensional scenes be represented on a two-dimensional canvas? Projective geometry explains how an eye perceives 'the real world', and so explains how artists can achieve realism in their work.

For example, in Computer Graphics and in Art.

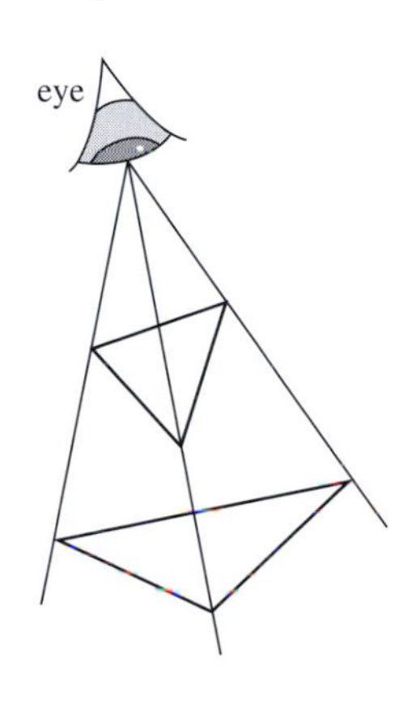

In Section 3.1, we look at the development of perspective in Art and explain the concept of a *perspectivity.* We describe Desargues' Theorem, which concerns a curious property of two triangles whose vertices are in perspective from a single point, and so explain that perspective can play a key role in the statement and the proof of theorems in mathematics.

In Section 3.2, we define the term *projective point* (or *Point*) and call the set of all such Points the *projective plane,* which we denote by $\mathbb{RP}^2$. We also define a *projective line* (or *Line*). To enable us to tackle problems in projective geometry algebraically, we introduce *homogeneous coordinates* to specify the Points in $\mathbb{RP}^2$.

In Section 3.3, we define the projective transformations of $\mathbb{RP}^2$ and use them to define projective geometry. We also prove the *Fundamental Theorem of Projective Geometry,* which states that given two sets of four Points there is a unique projective transformation of $\mathbb{RP}^2$ that maps the Points in one set to the corresponding Points in the other set. This crucial result enables us to apply a preliminary transformation to many geometric problems, thereby simplifying their solution by reducing the arithmetic involved. It turns out that there is a close connection between the idea of perspective in $\mathbb{R}^3$ and projective transformations.

We also require that no three Points in either set lie on a Line.

In Section 3.4, we use the Fundamental Theorem of Projective Geometry to prove several results, including Desargues' Theorem. We also introduce the concept of *duality,* which involves a remarkable relationship between Points and Lines.

Finally, in Section 3.5, we note that the ideas of distance and ratio of distances along a line have no immediate analogues in $\mathbb{RP}^2$; nevertheless, we are able to define a related quantity called the *cross-ratio* of four collinear Points in $\mathbb{RP}^2$. This quantity is very useful in proving various mathematical results, and it has 'real life' applications – such as in aerial photography.

3.1 Perspective

3.1.1 Perspective in Art

The first 'pictures' were probably Cave Art wall paintings: for example, depictions of animals and hunters. Up to the Middle Ages, most pictures were drawn on walls, floors or ceilings of buildings and were intended to convey messages rather than to be accurate illustrations of what an eye might see. For example, Christian religious art portrayed Christ and the Saints, the Bayeux tapestry outlined events such as the Norman Conquest and the Battle of Hastings, and so on.

(Clockwise from left) Hunters below antelopes. Bambata cave, Zimbabwe © M. Jelliffe; Tomb of Rekhmare, Thebes. 1500 BC © Ronald Sheridan; Bayeux Tapestry: The death of Harold. These prints are reproduced by kind permission of A.A. & A. Ancient Art and Architecture Collection.

To the modern eye, the people and animals in these pictures appear to be rather stylized, and the whole scene seems very two-dimensional. The events illustrated do not appear to be properly integrated into the background, even if this is included.

Towards the end of the 13th century, early Renaissance artists began to attempt to portray 'real' situations in a realistic way. For example, people at the back of a group would be drawn higher up than those at the front – a technique known as *terraced perspective.*

Simone Martini 'Maestà' Palazzo Pubblico, Sala del Mappamodo, Siena (su concessione del commune di Siena). Foto LENSINI Siena.

As artists struggled to find better techniques to improve the realism of their work, the idea of *vertical perspective* was developed by the Italian school of artists (including Duccio (1255–1318) and Giotto (1266–1337)). To create an impression of depth in a scene, the artist would represent pairs of parallel lines that are symmetrically placed either side of the scene by lines that meet on the centre line of the picture. The method is not totally realistic, since objects do not appear to recede into the distance in the way that might be expected. The problem of depicting 'distant objects looking smaller', with a properly integrated foreground and background, was tackled by many artists, including notably Ambrogio Lorenzetti (c. 1290–1348).

Giotto is sometimes called the 'Father of Modern Painting'.

'Last Supper' painted by Duccio; Opera del Duomo, Siena. Foto LENSINI Siena.

The modern system of *focused perspective* was discovered around 1425 by the sculptor and architect Brunelleschi (1377–1446), developed by the painter and architect Leone Battista Alberti (1404–1472), and finally perfected by Leonardo da Vinci (1452–1519).

Alberti wrote that the first necessity for a painter is 'to know geometry'.

These artists realized that what the eye actually 'sees' of a scene are the various rays of light travelling from each point in the scene to the eye. An effective way of deciding how to depict a three-dimensional scene on a two-dimensional canvas so as to create a realistic impression is therefore as follows. Imagine a glass screen placed between the eye and the three-dimensional scene. Each line joining the eye to a point of the scene pierces the glass screen at some point. The set of all such points forms an image on the screen known as a *cross-section.* Since the eye cannot distinguish between light rays coming from the points of the actual scene and light rays coming from the corresponding points of the cross-section (since these are in exactly the same direction), the cross-section produces the same impression as the original scene. In other words, the cross-section gives a realistic two-dimensional representation of the three-dimensional scene.

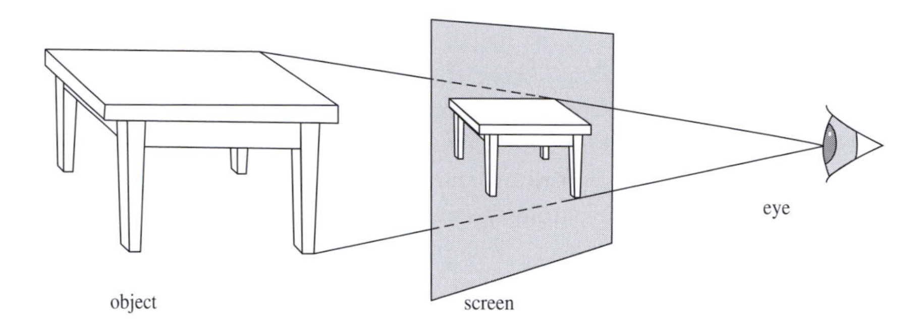

The German artist Albrecht Dürer (1471–1528) introduced the term *perspective* (from the Latin verb meaning 'to see through') to describe this technique, and illustrated it by a series of well-known woodcuts in his book *Underweysung der Messung mit dem Zyrkel und Rychtsscheyed* (1525). The Dürer woodcut below shows an artist peering through a grid on a glass screen to study perspective and the effect of foreshortening.

In English: *Instruction on measuring with compass and straight edge.*
We discuss foreshortening in Subsection 3.1.2.

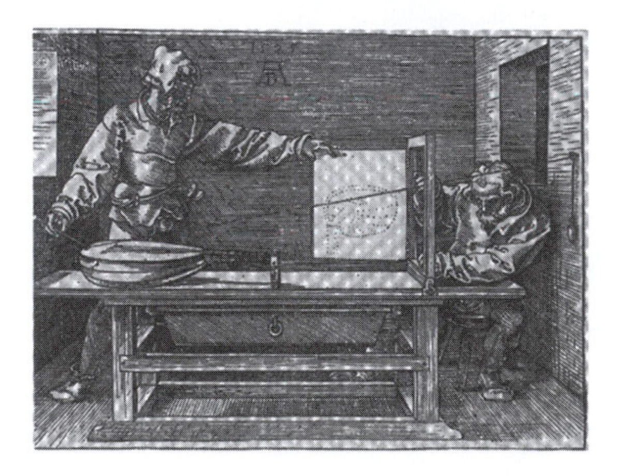

Of course, the picture displayed on the screen is just one representation of the scene. If the screen is placed closer to, or further away from, the eye, the size of the cross-section changes. Also, the screen may be placed at a different angle for a given position of the eye, or the eye itself may be moved to a different position. In each case, a different cross-section is obtained, though they are all related to each other.

3.1.2 Mathematical Perspective

To help us understand the relationship between different representations of a scene, we now look at perspective from a mathematical point of view. In place of an eye and light rays travelling to it, we use the family of all lines in $\mathbb{R}^3$ through a given point. For convenience, this point will often be the origin O. The glass screen is replaced by a plane in $\mathbb{R}^3$ that does not pass through the origin.

In order to compare the cross-sections that appear on different screens, we consider two planes π and π' that do not pass through O. A point P in π and a point Q in π' are said to be *in perspective from* O if there is a straight line through O, P and Q. A *perspectivity from* π *to* π' centred at O is a function that maps a point P of π to a point Q of π' whenever P and Q are in perspective from O. Notice that the planes π and π' may lie on the same side of O as shown on the left below, or they may lie on opposite sides of O as shown on the right.

In terms of O representing an eye, the figure on the right corresponds to the observer having the ability to look simultaneously both forwards and backwards!

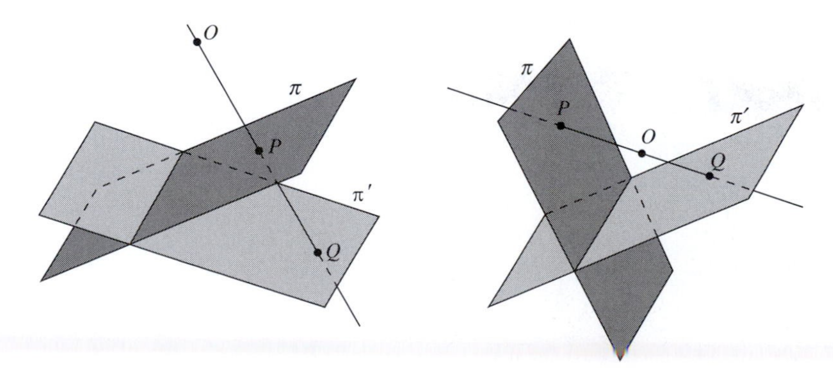

One complication with the above definition of a perspectivity is that the domain of the perspectivity is not necessarily the whole of π. Indeed, if P is any point of π such that OP is parallel to π', as shown in the margin, then P cannot have an image in π', and cannot therefore belong to the domain of the perspectivity. From a mathematical point of view, this need to exclude such exceptional points from the domain of a perspectivity turns out to be rather a nuisance. In Subsection 3.2.3 we shall therefore reformulate the definition of a perspectivity in such a way that these exceptional points can be included in the domain.

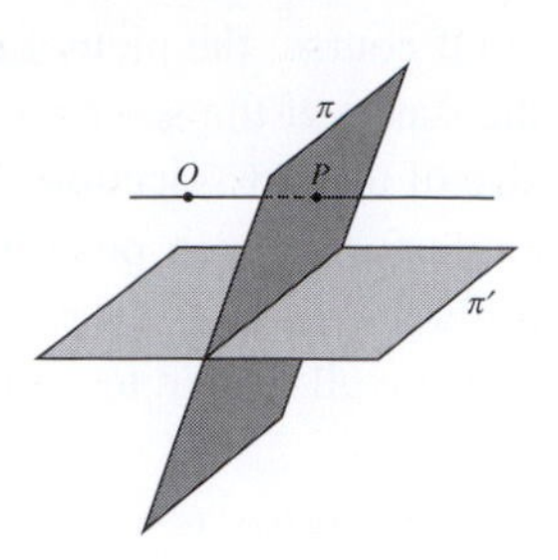

Even with only the preliminary definition of perspectivity given above, it is clear that some features of figures are preserved under a perspectivity, while others are not. For example, the figure on the left below illustrates a particular perspectivity in which a line segment in one plane maps onto a line segment in another plane. This suggests that collinearity is preserved by a perspectivity. On the other hand, the figure on the right illustrates a perspectivity in which a circle in one plane appears to map to a parabolic shape in another plane, which suggests that 'circularity' is not preserved.

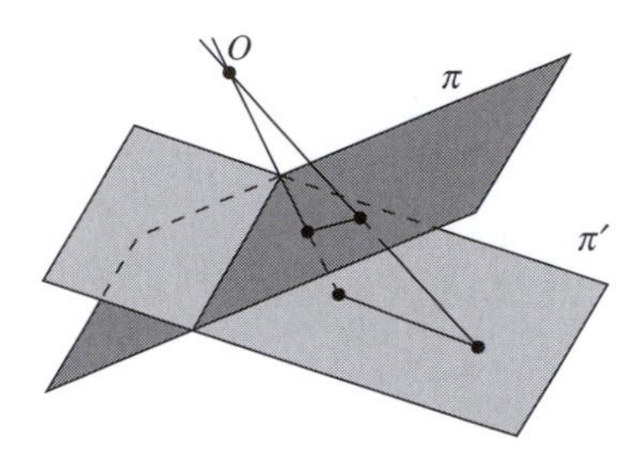

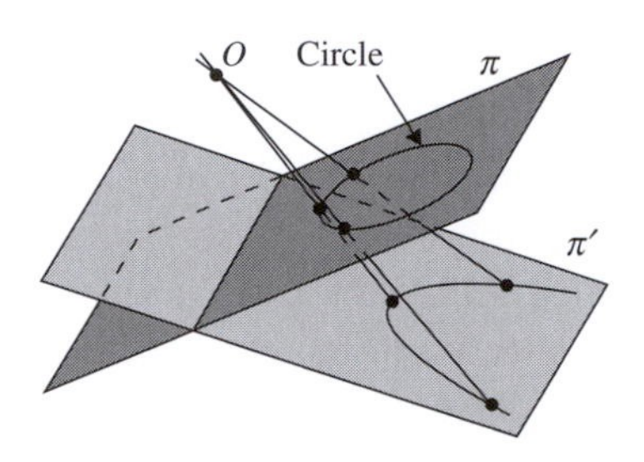

One of our main tasks is to study the images of standard configurations such as lines and conics under perspectivities. This chapter deals with lines; the next chapter deals with conics.

Consider a perspectivity with centre O that maps points in a plane π to points in a plane π'. A convenient way to visualize the image of a line ℓ under the perspectivity is to consider an arbitrary point P on ℓ. As P moves along ℓ, the line OP sweeps out a plane. The line ℓ' where this plane intersects π' is the image of ℓ.

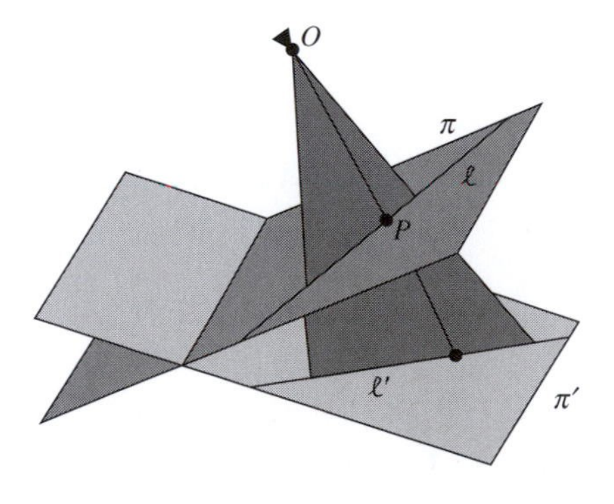

To be specific, consider the perspectivity p with centre O that maps points in a horizontal plane π to points in a vertical plane π', and let L be the line where π and π' intersect. Under p, every line ℓ in π that is parallel to L maps to a horizontal line ℓ' in π'. In particular, L maps to itself. The only exception is the line h that passes through the foot of the perpendicular from O to π. This line does not have an image in π' since the lines joining points of h to O are parallel to π'.

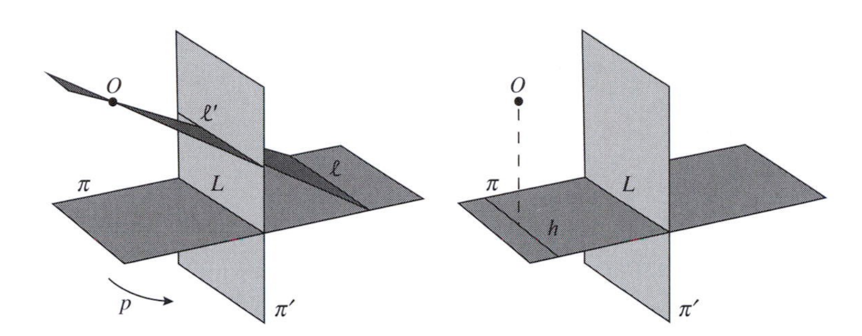

Next, consider the image under the same perspectivity p of a line ℓ in π that is perpendicular to L. To do this, let P denote the foot of the perpendicular from O to the plane π'. Although P is not the image of any point of π, the plane through O and ℓ meets π' in some line ℓ' that passes through P. It follows that the image of ℓ under p is some line ℓ' through P, with the point P itself omitted.

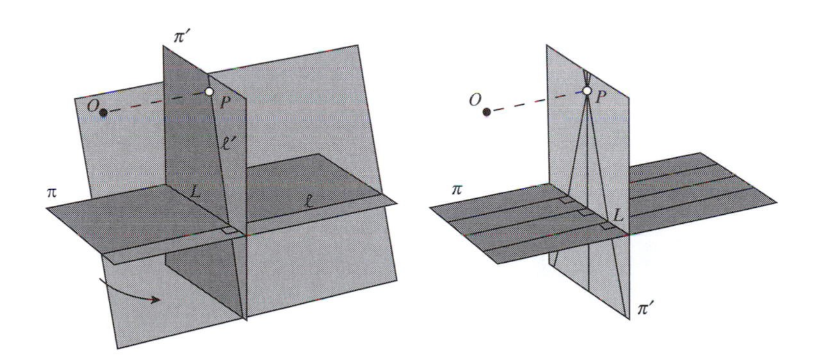

The above argument works for *any* line in π that is perpendicular to L. All such lines are mapped by the perspectivity p to lines in π' that pass through P, and that omit the point P itself.

We may combine our observations concerning lines in π that are parallel to L or perpendicular to L in the following way. Let $ABCD$ be a rectangle in π on the opposite side of L from O, with sides AB and CD that lie on lines ℓ_1 and ℓ_2, perpendicular to L. Then AD and BC both map onto horizontal lines in π' between L and P. As the side BC recedes from L, its image $B'C'$ under the perspectivity p moves further up π' towards P, becoming shorter as it moves.

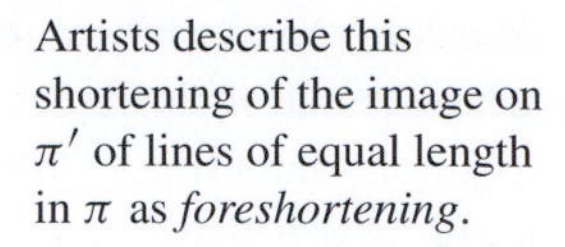
Artists describe this shortening of the image on π' of lines of equal length in π as *foreshortening*.

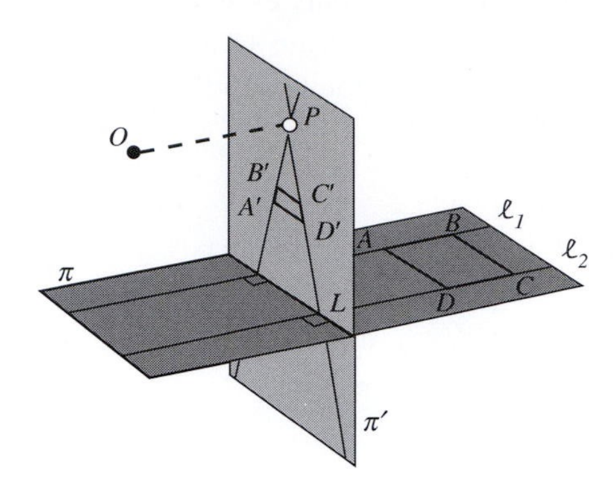

To an observer whose eye is located at O, the lines ℓ_1 and ℓ_2 *appear* to meet 'at infinity', and this corresponds to their images under p *appearing* to meet at P. The point P is called the *principal vanishing point* of the perspectivity p because the images in π' of all lines in π perpendicular to L appear to vanish there.

You can think of ℓ_1 and ℓ_2 as a pair of railroad lines disappearing into the distance.

In fact, a perspectivity has many vanishing points. For instance, let ℓ be any line in π that intersects L at an angle of $\pi/4$. Now let h' be the horizontal line in π' through P, and let D be the point on h' such that OD is parallel to ℓ. Then the plane through O and ℓ meets π' in some line ℓ' that passes through D. It follows that the image of ℓ under p is a line through D, with the point D itself omitted.

Here the symbol π is being used in two different ways: as a label for the embedding plane, and as an angle.

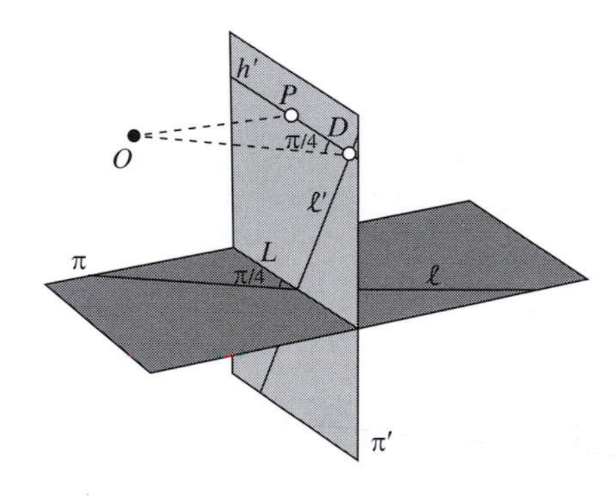

The point D is called a *diagonal vanishing point* of the perspectivity. All lines in the plane π that are parallel to the given line ℓ have images in π' that are lines through D, with the point D itself omitted.

That is, the images appear to *vanish* at D.

In the same way, each point of the horizontal line h' in π' through P is a *vanishing point* for the images of all lines in π in some direction; hence the line h'

is called the *vanishing line*. It corresponds to the 'horizon' in the plane – in other words, to the points 'at infinity' towards which an observer's eye is pointing when looking in a horizontal direction.

3.1.3 Desargues' Theorem

The idea that information in three dimensions can be related to information in two dimensions, and vice versa, plays an important role in mathematics just as it does in Art. For example, consider the following three-dimensional figure that consists of two triangles ΔABC and $\Delta A'B'C'$ which are in perspective from a point U. For the moment we shall assume that no pair of corresponding sides BC and $B'C'$, CA and $C'A'$, and AB and $A'B'$, are parallel.

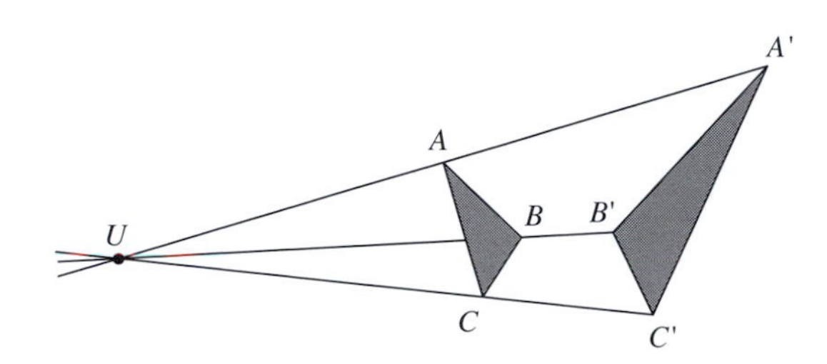

We shall show that this three-dimensional figure has the property that BC and $B'C'$, CA and $C'A'$, AB and $A'B'$ meet at P, Q, R, respectively, where P, Q and R are collinear. This will enable us to formulate an equivalent two-dimensional result, known as *Desargues' Theorem.*

Girard Desargues (1593–1662) was a French engineer and architect.

To prove the three-dimensional result, observe that both BC and $B'C'$ lie in the plane that passes through the points U, B and C. Since BC and $B'C'$ are coplanar but not parallel, they must meet at some point P.

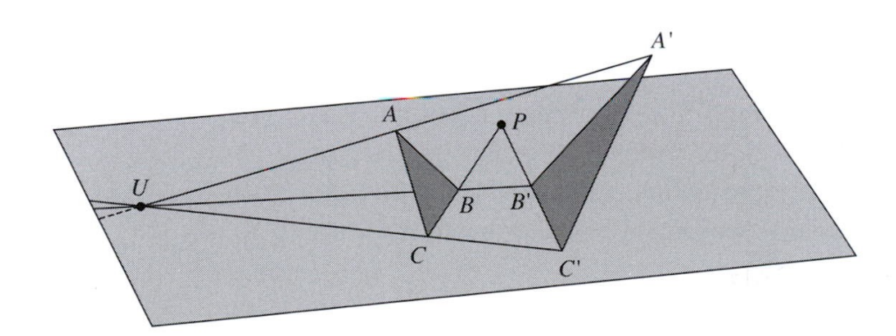

Similarly, the sides CA and $C'A'$ meet at some point Q, and the sides AB and $A'B'$ meet at some point R.

Since the points P, Q and R lie both on the plane which contains the triangle ΔABC and on the plane which contains the triangle $\Delta A'B'C'$, they must lie on the line ℓ where the two planes meet. It follows that P, Q and R are collinear.

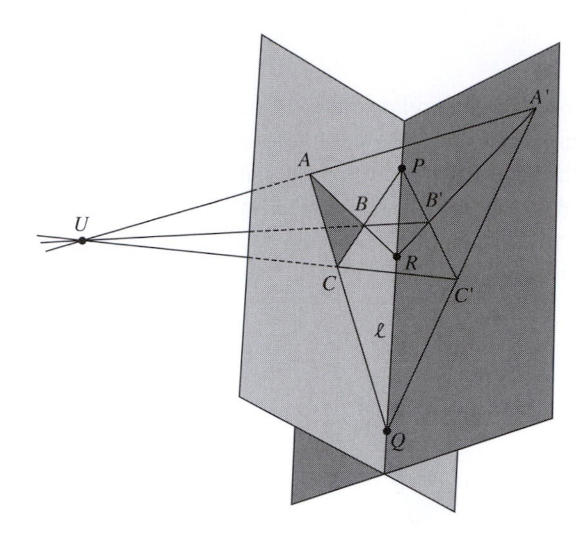

To obtain the equivalent two-dimensional result, imagine that you are viewing the three-dimensional configuration through a transparent screen. Since this viewing process will not alter the collinearity of points or the coincidence of lines, we may reinterpret the three-dimensional result in terms of the image on the screen to obtain the following theorem.

We give a rigorous proof of Desargues' Theorem in Theorem 1, Subsection 3.4.1.

Theorem 1 Desargues' Theorem
Let ΔABC and $\Delta A'B'C'$ be triangles in $\mathbb{R}^2$ such that the lines AA', BB' and CC' meet at a point U. Let BC and $B'C'$ meet at P, CA and $C'A'$ meet at Q, and AB and $A'B'$ meet at R. Then P, Q and R are collinear.

Strictly speaking, we have not proved this theorem since it is not immediately obvious that ΔABC and $\Delta A'B'C'$ can be obtained as images of triangles in $\mathbb{R}^3$ which have corresponding sides that are not parallel. Nevertheless, the above argument does provide reasonably convincing evidence that the theorem is true.

One remarkable feature of the above argument is the way in which the geometry of the figure on the transparent screen is characterized by the rays of light that enter an eye. Thus a point on the screen corresponds to a single ray of light that enters the eye, a line on the screen corresponds to a plane of rays of light that enter the eye, and so on. The geometry of the figure can be investigated entirely in terms of these rays of light. The screen is needed only to interpret the result in terms of a two-dimensional figure.

In the rest of Chapters 3 and 4, we introduce a geometry known as *projective geometry* that enables us to work with figures on a plane (a screen) as if they correspond to rays of light that enter an eye in the way described above.

3.2 The Projective Plane $\mathbb{RP}^2$

You have already met the Kleinian view that a geometry consists of a group of transformations acting on a space of points. In this section we begin our discussion of projective geometry by investigating its space of points. The group of transformations is discussed in Section 3.3.

Introductory remarks to Chapter 2

3.2.1 Projective Points

Imagine an eye situated at the origin of $\mathbb{R}^3$ looking at a fixed screen. As we mentioned in Subsection 3.1.1, each point of the screen corresponds to the ray of light that enters the eye from the point. This correspondence between points of the screen and rays of light through the origin is the clue that we need to define a space of points for our new geometry.

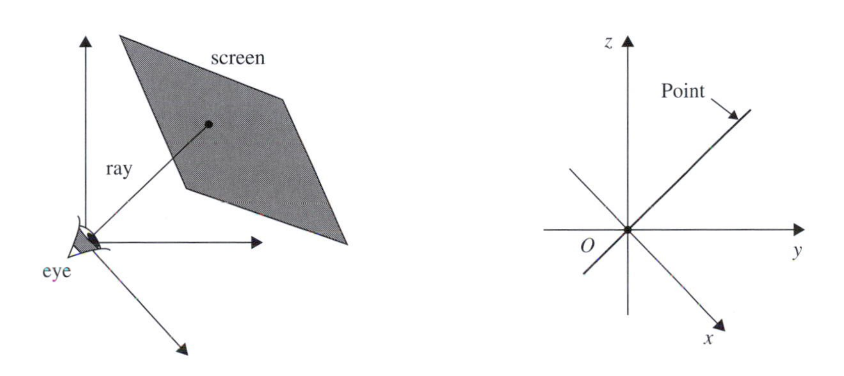

Rather than use the points of the screen directly, we use the rays of light that enable an eye to 'see' the points from the origin. We can express this idea mathematically by defining a *projective point* to be a Euclidean line in $\mathbb{R}^3$ that passes through the origin. In order to avoid confusion with Euclidean points of $\mathbb{R}^3$, we write Point with a capital P whenever we mean a projective point.

It is important that you use the capital letter P in 'Point'.

Definitions **A Point** (or **projective point**) is a line in $\mathbb{R}^3$ that passes through the origin of $\mathbb{R}^3$. The **real projective plane** $\mathbb{RP}^2$ is the set of all such Points.

In order to prove results in projective geometry algebraically, we need to have an algebraic notation that can be used to specify the Points of $\mathbb{RP}^2$. To do this, we use the fact that a line ℓ through the origin O in $\mathbb{R}^3$ is uniquely determined once we have specified a Euclidean point (other than O) that lies on ℓ. For example, there is a unique line ℓ in $\mathbb{R}^3$ through O and the point with Euclidean coordinates (4, 2, 6), so we can use these coordinates to specify a projective point. When doing this we write the coordinates in the form [4, 2, 6], with square brackets to indicate that the coordinates refer to a projective point.

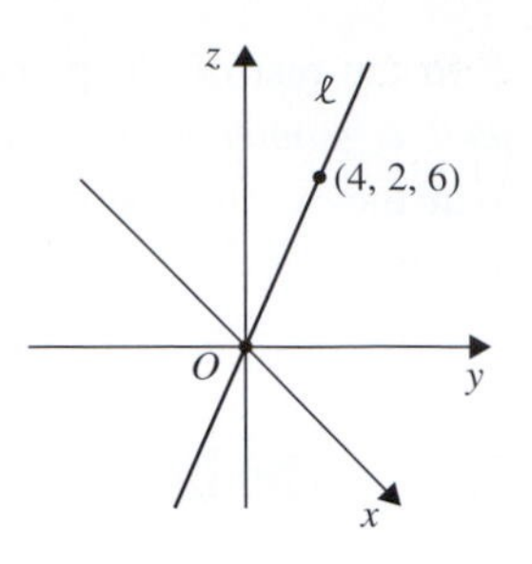

Definition The expression $[a, b, c]$, in which the numbers a, b, c are not all zero, represents the Point P in $\mathbb{RP}^2$ which consists of the unique line in $\mathbb{R}^3$ that passes through (0, 0, 0) and (a, b, c). We refer to $[a, b, c]$ as **homogeneous coordinates** of P. If (a, b, c) has position vector $\mathbf{v}$, then we often denote P by $[\mathbf{v}]$ and we say that P can be **represented** by $\mathbf{v}$.

Note that [0, 0, 0] is *not* defined.

Remark

Often we abuse our notation slightly, by talking about 'the Point $[a, b, c]$' when strictly speaking we should say 'the Point with homogeneous coordinates $[a, b, c]$'.

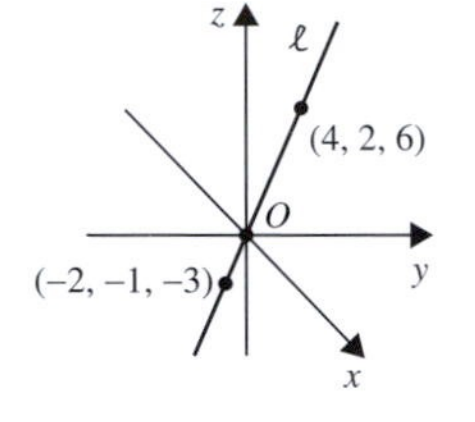

Notice that the homogeneous coordinates of a Point are not unique. For example, the Point with homogeneous coordinates [4, 2, 6] consists of a line that passes through (0, 0, 0) and (4, 2, 6). But this line also passes through $(-2, -1, -3)$, so [4, 2, 6] and $[-2, -1, -3]$ both represent the same Point.

In general, if (a, b, c) is any point on a line through the origin, and λ is any real number, then $(\lambda a, \lambda b, \lambda c)$ also lies on the line. Moreover, if (a, b, c) is not at the origin and $\lambda \neq 0$, then $(\lambda a, \lambda b, \lambda c)$ is not at the origin either. It follows that $[a, b, c]$ and $[\lambda a, \lambda b, \lambda c]$ both represent the same Point, for any $\lambda \neq 0$. We express this by writing

$$[a, b, c] = [\lambda a, \lambda b, \lambda c], \qquad \text{for any } \lambda \neq 0. \tag{1}$$

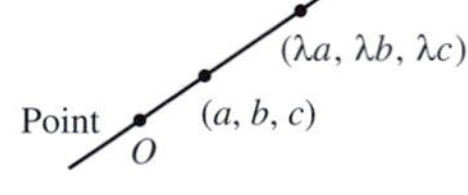

Conversely, if there is no non-zero real number λ such that

$$(a', b', c') = (\lambda a, \lambda b, \lambda c),$$

then (a, b, c) and (a', b', c') cannot lie on the same line through the origin, and so the homogeneous coordinates $[a, b, c]$ and $[a', b', c']$ must represent different Points in $\mathbb{RP}^2$.

Example 1 Which of the following homogeneous coordinates represent the same Point in $\mathbb{RP}^2$ as [6, 3, 2]?

(a) [18, 9, 6] (b) $[12, -6, 4]$ (c) $\left[1, \frac{1}{2}, \frac{1}{3}\right]$ (d) [1, 2, 3]

Solution

Throughout the solution we use equation (1):

$$[a, b, c] = [\lambda a, \lambda b, \lambda c],$$

for any $\lambda \neq 0$.

(a) This represents the same Point as [6, 3, 2], for if $\lambda = 3$, then

$$[18, 9, 6] = [6\lambda, 3\lambda, 2\lambda] = [6, 3, 2].$$

(b) This represents a Point different from [6, 3, 2], for there is no λ that satisfies the simultaneous equations

$$12 = 6\lambda, -6 = 3\lambda, 4 = 2\lambda.$$

(c) This represents the same Point as [6, 3, 2], for if $\lambda = \frac{1}{6}$, then

$$\left[1, \tfrac{1}{2}, \tfrac{1}{3}\right] = [6\lambda, 3\lambda, 2\lambda] = [6, 3, 2].$$

(d) This represents a Point different from [6, 3, 2], for there is no λ that satisfies the simultaneous equations

$$1 = 6\lambda, \quad 2 = 3\lambda, \quad 3 = 2\lambda.$$

□

Problem 1 Which of the following homogeneous coordinates represent the same Point in $\mathbb{RP}^2$ as [1, 2, 3]?

(a) [2, 4, 6] (b) $[1, 2, -3]$ (c) $[-1, -2, -3]$ (d) [11, 12, 13]

At first sight it may seem rather unsatisfactory that the coordinates of a Point are not unique. However, this ambiguity can often be turned to our advantage. For example, if a calculation yields a Point of $\mathbb{RP}^2$ with fractional homogeneous coordinates such as $\left[1, \frac{1}{2}, \frac{1}{3}\right]$, then the rest of the calculation may be simpler if we 'clear' the fractions and represent the Point by the integer homogeneous coordinates [6, 3, 2] instead.

Problem 2 For each of the following homogeneous coordinates, find integer homogeneous coordinates which represent the same Point.

(a) $\left[\frac{3}{4}, \frac{1}{2}, -\frac{1}{8}\right]$ (b) $\left[0, 4, \frac{2}{3}\right]$ (c) $\left[\frac{1}{6}, -\frac{1}{3}, -\frac{1}{2}\right]$

Given a collection of homogeneous coordinates, it is not always easy to spot those that represent the same Point. In such cases it is sometimes possible to rewrite the coordinates in a form that makes the comparison easier.

Example 2 Determine homogeneous coordinates of the form $[a, b, 1]$ for the Points

$$[2, -1, 4], \quad [4, 2, 8], \quad [2\pi, -\pi, 4\pi],$$

$$[200, 100, 400], \quad \left[-\frac{1}{2}, -\frac{1}{4}, -1\right], \quad [6, -9, -12].$$

Hence decide which homogeneous coordinates represent the same Points.

Solution According to equation (1), a Point of $\mathbb{RP}^2$ is unchanged if its homogeneous coordinates are multiplied (or divided) by any non-zero real number. Since the third coordinate of each Point is non-zero, we may divide by this third coordinate to obtain homogeneous coordinates of the form $[a, b, 1]$ as follows:

For, dividing by a non-zero number λ is equivalent to multiplying by the non-zero number $1/\lambda$.

$$[2,-1,4]=\left[\tfrac{1}{2},-\tfrac{1}{4},1\right]; \qquad [4,2,8]=\left[\tfrac{1}{2},\tfrac{1}{4},1\right];$$

$$[2\pi,-\pi,4\pi]=\left[\tfrac{1}{2},-\tfrac{1}{4},1\right]; \qquad [200,100,400]=\left[\tfrac{1}{2},\tfrac{1}{4},1\right];$$

$$\left[-\tfrac{1}{2},-\tfrac{1}{4},-1\right]=\left[\tfrac{1}{2},\tfrac{1}{4},1\right]; \qquad [6,-9,-12]=\left[-\tfrac{1}{2},\tfrac{3}{4},1\right].$$

Since $[a,b,1]=[a',b',1]$ if and only if $a=a'$ and $b=b'$, it follows that:

$[2,-1,4]$ and $[2\pi,-\pi,4\pi]$ represent the same Point;

$[4,2,8]$, $[200,100,400]$ and $\left[-\frac{1}{2},-\frac{1}{4},-1\right]$ represent the same Point;

$[6,-9,-12]$ represents none of the other Points. □

Notice that the method used in Example 2 works only if the third coordinates of all the Points are non-zero. If this is not the case, then you may still be able to apply the technique using the first or second coordinates.

Problem 3 Determine homogeneous coordinates of the form $[1, b, c]$ for the Points

$$[2,3,-5], \qquad [-8,-12,20], \qquad \left[\sqrt{2},\sqrt{3},-\sqrt{5}\right],$$

$$[4,-6,10], \qquad [-20,-30,50], \qquad [74,148,0].$$

Hence decide which homogeneous coordinates represent the same Points.

Having defined projective points, we are now in a position to define a *projective figure.* Just as a figure in Euclidean geometry is defined to be a subset of $\mathbb{R}^2$, so figures in projective geometry are defined to be subsets of $\mathbb{RP}^2$.

Definition **A projective figure** is a subset of $\mathbb{RP}^2$.

Projective figures are just sets of lines in $\mathbb{R}^3$ that pass through the origin. Thus a double cone with a vertex at O, and a double square pyramid with a vertex at O, are both examples of projective figures, for they can both be formed from sets of lines that pass through the origin of $\mathbb{R}^3$.

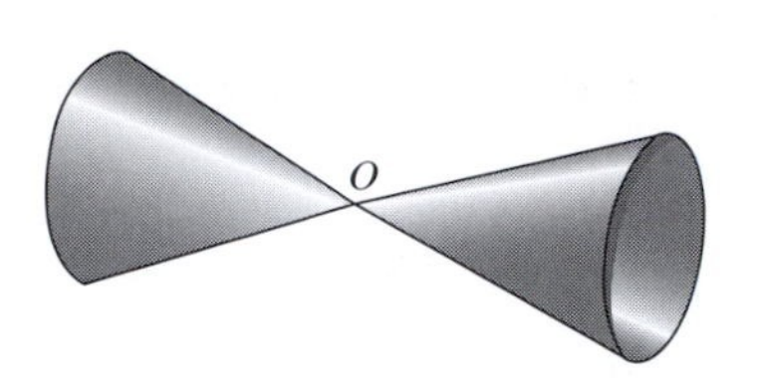

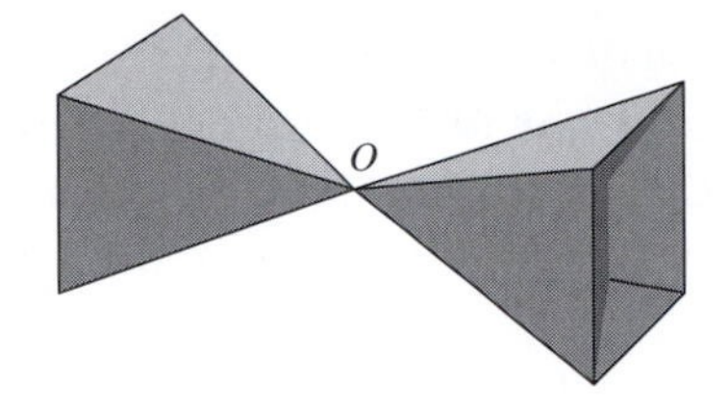

3.2.2 Projective Lines

A particularly simple type of projective figure is a plane through the origin. Such a plane is a projective figure because it can be formed from the set of all Points (lines through the origin of $\mathbb{R}^3$) that lie on the plane. Since all but one of these Points can be thought of as rays of light that come from a line on a screen, it seems reasonable to define any plane through the origin to be a *projective line*.

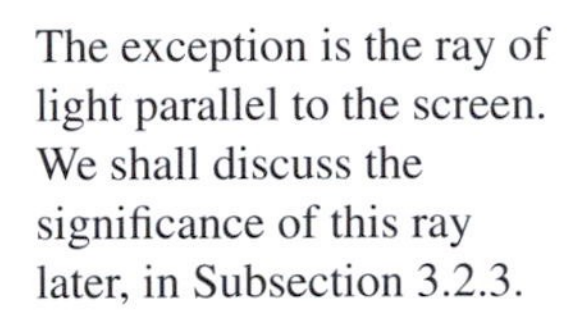
The exception is the ray of light parallel to the screen. We shall discuss the significance of this ray later, in Subsection 3.2.3.

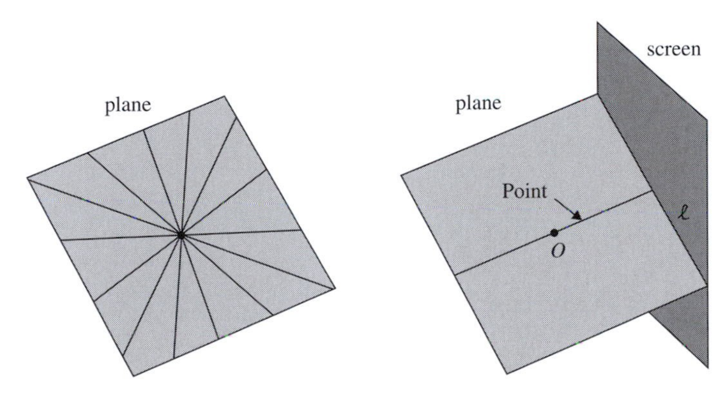

Just as we use 'Point' to refer to a 'projective point', so we use 'Line' to refer to a 'projective line'. The use of a capital L avoids any confusion with lines in $\mathbb{R}^3$.

Definitions **A Line** (or **projective line**) in $\mathbb{RP}^2$ is a plane in $\mathbb{R}^3$ that passes through the origin. Points in $\mathbb{RP}^2$ are **collinear** if they lie on a Line.

Since a Line in $\mathbb{RP}^2$ is simply a plane in $\mathbb{R}^3$ that passes through the origin, it must consist of the set of Euclidean points (x, y, z) that satisfy an equation of the form

$$ax + by + cz = 0,$$

where a, b and c are real and not all zero. We can interpret this fact in terms of $\mathbb{RP}^2$ as follows.

Theorem 1 The general equation of a Line in $\mathbb{RP}^2$ is

$$ax + by + cz = 0, \tag{2}$$

where a, b, c are real and not all zero.

Remark

1. The equation of a Line is not unique, for, if $\lambda \neq 0$, then $\lambda ax+\lambda by+\lambda cz=0$ is also an equation for the Line. We can use this fact to 'clear fractions' from the coefficients just as we did for the homogeneous coordinates of a Point.
2. From the figure in the margin it is clear that a Point lies on a Line, or a Line passes through a Point, if and only if the Point has homogeneous coordinates $[x, y, z]$ which satisfy the equation of the Line. For example, $[1, -1, 1]$ lies on the Line $3x + y - 2z = 0$, but $[0, 1, 3]$ does not.

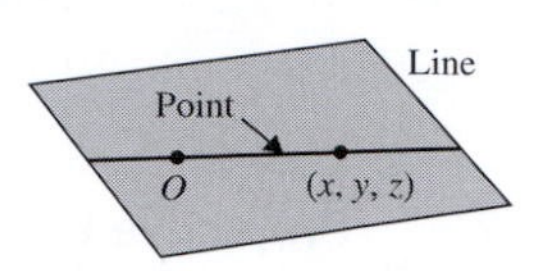

In Euclidean geometry there is a unique line that passes through any two distinct points, as illustrated on the left of the figure below. Similarly, in projective geometry two distinct Points (lines through the origin) lie on a unique Line (plane through the origin).

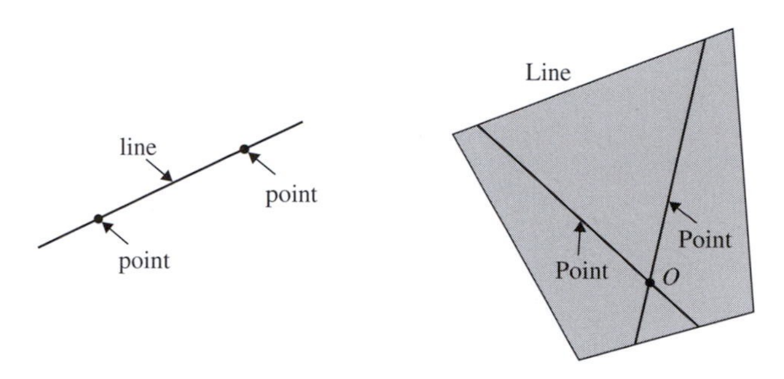

We express this observation in the form of a theorem, as follows.

Theorem 2 Collinearity Property of $\mathbb{RP}^2$
Any two distinct Points of $\mathbb{RP}^2$ lie on a unique Line.

It is sometimes possible to find an equation for the Line that passes through two distinct Points of $\mathbb{RP}^2$ simply by spotting an equation of the form (2) that is satisfied by the homogeneous coordinates of both Points.

Example 3 For each of the following pairs of Points, write down an equation for the Line that passes through them.

(a) $[3, 2, 0]$ and $[3, 4, 0]$ (b) $[1, 2, 1]$ and $[3, 0, 3]$
(c) $[1, 0, 0]$ and $[0, 0, 1]$

Solution

(a) Both the Points have a z-coordinate equal to 0, so the homogeneous coordinates must satisfy the equation $z = 0$. This equation is of the form (2) with $a = 0, b = 0$ and $c = 1$, so it must be the required equation for the Line.

The equation $x = 3$ is not of the form (2), and so is not the equation of a Line.

(b) The homogeneous coordinates of both Points satisfy $x = z$. This equation is of the form (2) with $a = 1$, $b = 0$ and $c = -1$. It must therefore be the required equation for the Line.

(c) The homogeneous coordinates of both Points satisfy $y = 0$. This equation is of the form (2) with $a = 0$, $b = 1$ and $c = 0$, so it must be the required equation for the Line. □

Problem 4 For each of the following pairs of Points, write down an equation for the Line that passes through them.

(a) [0, 1, 0] and [0, 0, 1] (b) [2, 2, 3] and [3, 3, 7]

But how do we find an equation for a Line through two given Points in cases where it cannot be found by inspection? As an example, consider the Points $[2, -1, 4]$ and $[1, -1, 1]$. We could certainly substitute the values $x = 2$, $y = -1$, $z = 4$ and $x = 1$, $y = -1$, $z = 1$ into equation (2), to obtain the pair of simultaneous equations

$$2a - b + 4c = 0,$$
$$a - b + c = 0.$$

Then subtracting twice the second equation from the first, we obtain $b = -2c$. So from the second equation it follows that $a = -3c$. If we set $c = -1$, say, then $a = 3$ and $b = 2$, so an equation for the Line is

$$3x + 2y - z = 0.$$

Of course, we could set c to have any non-zero value, but $c = -1$ keeps the calculation simple.

In this case the calculations are fairly straightforward, but there is an alternative method that is often simpler. Notice that the Line in $\mathbb{RP}^2$ through the Points $[2, -1, 4]$ and $[1, -1, 1]$ is the Euclidean plane in $\mathbb{R}^3$ that contains the position vectors of the points $(2, -1, 4)$ and $(1, -1, 1)$ in $\mathbb{R}^3$. A point (x, y, z) lies in this plane if and only if the vector (x, y, z) is a linear combination of the vectors $(2, -1, 4)$ and $(1, -1, 1)$; in other words, if and only if the vectors (x, y, z), $(2, -1, 4)$ and $(1, -1, 1)$ are linearly dependent.

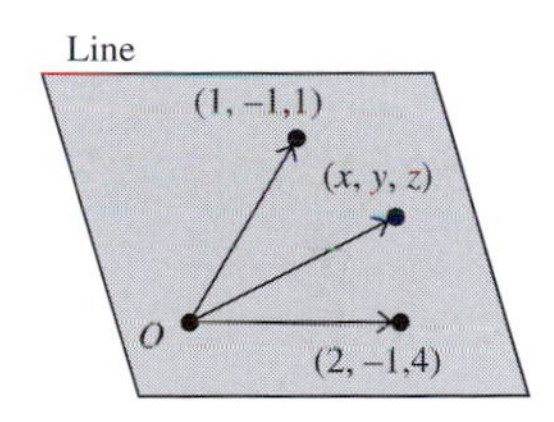

But three vectors in $\mathbb{R}^3$ are linearly dependent if and only if the 3×3 determinant that has these vectors as its rows is zero. It follows that (x, y, z) lies in the plane containing the position vectors $(2, -1, 4)$ and $(1, -1, 1)$ if and only if

$$\begin{vmatrix} x & y & z \\ 2 & -1 & 4 \\ 1 & -1 & 1 \end{vmatrix} = 0.$$

Translating this statement back into a statement concerning $\mathbb{RP}^2$, we deduce that the Point $[x, y, z]$ lies on the Line through the Points $[2, -1, 4]$ and $[1, -1, 1]$ if and only if

$$\begin{vmatrix} x & y & z \\ 2 & -1 & 4 \\ 1 & -1 & 1 \end{vmatrix} = 0.$$

Expanding this determinant in terms of the entries in its first row, we obtain

$$\begin{vmatrix} x & y & z \\ 2 & -1 & 4 \\ 1 & -1 & 1 \end{vmatrix} = x\begin{vmatrix} -1 & 4 \\ -1 & 1 \end{vmatrix} - y\begin{vmatrix} 2 & 4 \\ 1 & 1 \end{vmatrix} + z\begin{vmatrix} 2 & -1 \\ 1 & -1 \end{vmatrix}$$

$$= 3x + 2y - z.$$

Hence an equation for the required Line in $\mathbb{RP}^2$ is

$$3x + 2y - z = 0. \qquad (3)$$

Remark

It is always sensible to check your arithmetic by checking that the two given Points actually lie on the Line that you have found. For instance, the answer above is correct, since equation (3) is a homogeneous linear equation in x, y and z, and the equation is satisfied by $x = 2$, $y = -1$, $z = 4$ and by $x = 1$, $y = -1$, $z = 1$.

We may summarize the above method in the form of a strategy, as follows.

Strategy To determine an equation for the Line in $\mathbb{RP}^2$ through the Points $[d, e, f]$ and $[g, h, k]$:

1. write down the equation

$$\begin{vmatrix} x & y & z \\ d & e & f \\ g & h & k \end{vmatrix} = 0;$$

2. expand the determinant in terms of the entries in its first row to obtain the required equation in the form $ax + by + cz = 0$.

Example 4 Find an equation for the Line that passes through the Points $[1, 2, 3]$ and $[2, -1, 4]$.

Solution An equation for the Line is

$$\begin{vmatrix} x & y & z \\ 1 & 2 & 3 \\ 2 & -1 & 4 \end{vmatrix} = 0.$$

Now

$$\begin{vmatrix} x & y & z \\ 1 & 2 & 3 \\ 2 & -1 & 4 \end{vmatrix} = x\begin{vmatrix} 2 & 3 \\ -1 & 4 \end{vmatrix} - y\begin{vmatrix} 1 & 3 \\ 2 & 4 \end{vmatrix} + z\begin{vmatrix} 1 & 2 \\ 2 & -1 \end{vmatrix}$$

$$= 11x + 2y - 5z.$$

An equation for the Line is therefore

$$11x + 2y - 5z = 0. \qquad \square$$

You can easily check that the Points [1, 2, 3] and [2, −1, 4] lie on this Line.

Problem 5 Determine an equation for each of the following Lines in $\mathbb{RP}^2$:

(a) the Line through the Points [2, 5, 4] and [3, 1, 7];
(b) the Line through the Points [−2, −4, 5] and [3, −2, −4].

A similar technique can be used to check whether three given Points are collinear. Indeed, three Points $[a, b, c]$, $[d, e, f]$, $[g, h, k]$ are collinear if and only if the position vectors of the points (a, b, c), (d, e, f), (g, h, k) are linearly dependent; that is, if and only if

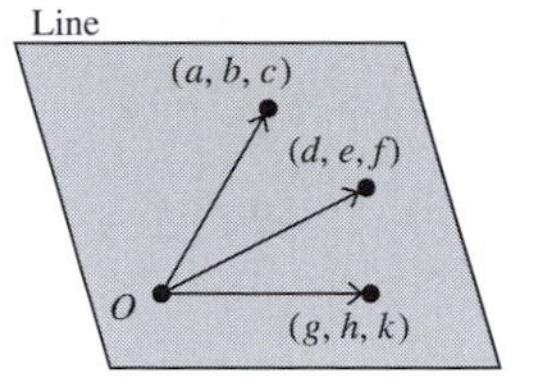

$$\begin{vmatrix} a & b & c \\ d & e & f \\ g & h & k \end{vmatrix} = 0.$$

Example 5 Determine whether the Points [2, 1, 3], [1, 2, 1] and [−1, 4, −3] are collinear.

Solution We have

$$\begin{aligned}\begin{vmatrix} 2 & 1 & 3 \\ 1 & 2 & 1 \\ -1 & 4 & -3 \end{vmatrix} &= 2\begin{vmatrix} 2 & 1 \\ 4 & -3 \end{vmatrix} - 1\begin{vmatrix} 1 & 1 \\ -1 & -3 \end{vmatrix} + 3\begin{vmatrix} 1 & 2 \\ -1 & 4 \end{vmatrix} \\ &= 2(-6-4) - (-3+1) + 3(4+2) \\ &= -20 + 2 + 18 \\ &= 0.\end{aligned}$$

Since this is zero it follows that [2, 1, 3], [1, 2, 1] and [−1, 4, −3] are collinear. □

We summarize the method of Example 5 in the following strategy.

Strategy To determine whether three Points $[a, b, c]$, $[d, e, f]$, $[g, h, k]$ are collinear:

1. evaluate the determinant $\begin{vmatrix} a & b & c \\ d & e & f \\ g & h & k \end{vmatrix}$;
2. the Points $[a, b, c]$, $[d, e, f]$, $[g, h, k]$ are collinear if and only if this determinant is zero.

Problem 6 Determine whether the following sets of Points are collinear.

(a) $[1, 2, 3], [1, 1, -2], [2, 1, -9]$ (b) $[1, 2, -1], [2, 1, 0], [0, -1, 3]$

Before rushing to solve a problem using determinants, you should always stop to see if you can solve the problem more easily by inspection. For example, suppose that you are asked to check whether the Points [1, 0, 0], [0, 1, 0], [1, 1, 1] are collinear. Clearly, [1, 0, 0] and [0, 1, 0] lie on the Line $z = 0$, whereas [1, 1, 1] does not, so the Points are not collinear.

Problem 7 Verify that no three of the Points [1, 0, 0], [0, 1, 0], [0, 0, 1] and [1, 1, 1] are collinear.

The Points that you considered in Problem 7 play an important part in our development of the theory of projective geometry, so we give them special names.

Definitions The Points [1, 0, 0], [0, 1, 0], [0, 0, 1] are known as the **triangle of reference.** The Point [1, 1, 1] is called the **unit Point.**

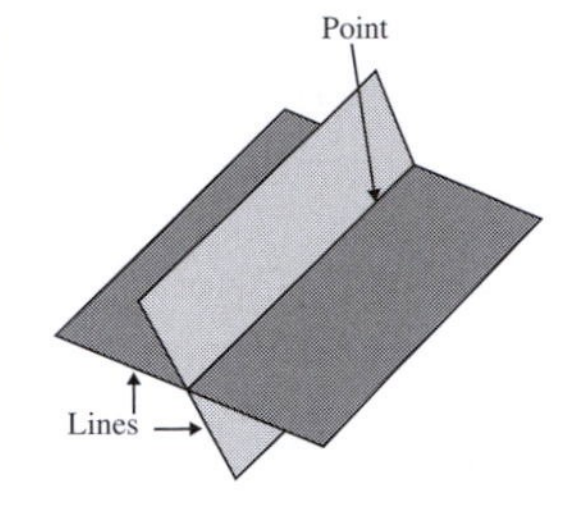

Next, observe that any two distinct Lines necessarily meet at a unique Point. Indeed, a Line in $\mathbb{RP}^2$ is simply a plane in $\mathbb{R}^3$ that passes through the origin, and two distinct planes through the origin of $\mathbb{R}^3$ must intersect in a unique Euclidean line through the origin; that is, in a Point. This is very different to the situation in Euclidean geometry where parallel lines do not meet.

Theorem 3 Incidence Property of $\mathbb{RP}^2$
Any two distinct Lines in $\mathbb{RP}^2$ intersect in a unique Point of $\mathbb{RP}^2$.

This result neatly complements Theorem 2, the *Collinearity Property* of $\mathbb{RP}^2$.

We can determine the Point of intersection of two Lines simply by solving the equations of the two Lines as a pair of simultaneous equations.

Example 6 Determine the Point of intersection of the Lines in $\mathbb{RP}^2$ with equations $x + 6y - 5z = 0$ and $x - 2y + z = 0$.

We know that there is a unique Point of intersection, by Theorem 3.

Solution At the Point of intersection $[x, y, z]$ of the two Lines, we have

$$x + 6y - 5z = 0,$$
$$x - 2y + z = 0.$$

Subtracting the second equation from the first, we obtain

$$8y - 6z = 0,$$

so that $y = \frac{3}{4} z$. Substituting this into the second equation, we obtain $x = \frac{1}{2} z$.

It follows that the Point of intersection has homogeneous coordinates $\left[\frac{1}{2}z, \frac{3}{4}z, z\right]$ which we can rewrite in the form $\left[\frac{1}{2}, \frac{3}{4}, 1\right]$ or [2, 3, 4]. □

Note that $z \neq 0$, since [0, 0, 0] are not allowed as homogeneous coordinates.

Problem 8 Determine the Point of intersection of each of the following pairs of Lines in $\mathbb{RP}^2$:

(a) the Lines with equations $x - y - z = 0$ and $x + 5y + 2z = 0$;
(b) the Lines with equations $x + 2y - z = 0$ and $2x + y - 4z = 0$.

Problem 9 Determine the Point of $\mathbb{RP}^2$ at which the Line through the Points $[1, 2, -3]$ and $[2, -1, 0]$ meets the Line through the Points $[1, 0, -1]$ and [1, 1, 1].

In some cases we can write down the Point at which two Lines intersect without having to solve any equations at all. For example, the Lines with equations $x = 0$ and $y = 0$ clearly meet at the Point [0, 0, 1].

Problem 10 Determine the Point of $\mathbb{RP}^2$ at which the Line through the Points [1, 0, 0] and [0, 1, 0] meets the Line through the Points [0, 0, 1] and [1, 1, 1].

3.2.3 Embedding Planes

So far we have used three-dimensional space to develop the theory of projective geometry. In practice, however, we want to use projective geometry to study two-dimensional figures in a plane. In order to do this, we now investigate a way of associating figures in a plane with figures in $\mathbb{RP}^2$, and vice versa.

Suppose that a plane π contains a figure F. We can place π into $\mathbb{R}^3$, making sure that it does not pass through the origin, and then construct a corresponding projective figure by drawing in all the Points of $\mathbb{RP}^2$ that pass through the points of F. For example, if F is the triangle shown on the left below, then the corresponding projective figure is a double triangular pyramid. Note that if we change the position of π in $\mathbb{R}^3$, we obtain a different projective figure corresponding to F.

It is a *double* pyramid because the Points which make up the pyramid are lines that emerge from the origin in *both* directions.

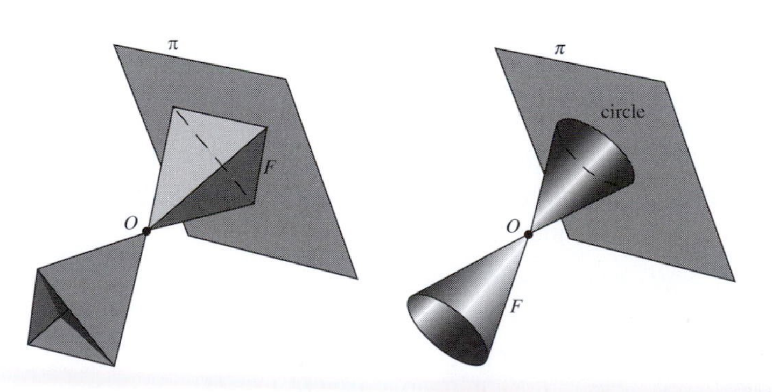

Conversely, suppose that we start with a projective figure F. The corresponding Euclidean figure in π consists of the Euclidean points where the Points of F pierce π. For example, if F is a double cone whose axis is at right angles to the embedding plane, as shown on the right above, then the corresponding Euclidean figure is a circle. Note that if we change the position of π in $\mathbb{R}^3$, we obtain a different plane figure corresponding to F.

This correspondence between projective figures and Euclidean figures works well provided that each Point of the projective figure pierces the plane π, as shown in the margin. Unfortunately, any Point of $\mathbb{RP}^2$ that consists of a line through the origin *parallel* to π does not pierce π, and so cannot be associated with a point of π. Such a Point is called an **ideal Point** for π.

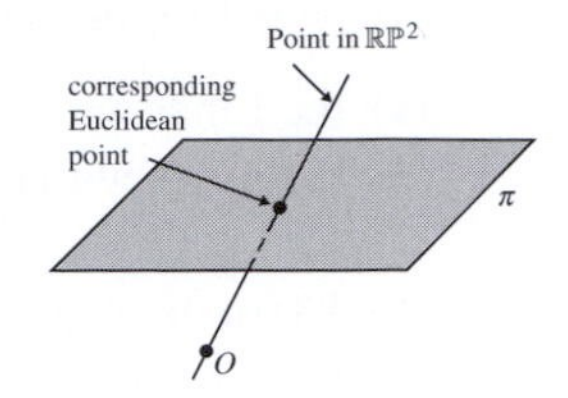

All the ideal Points for π lie on a plane through O parallel to π. This plane is a projective line known as the **ideal Line** for π.

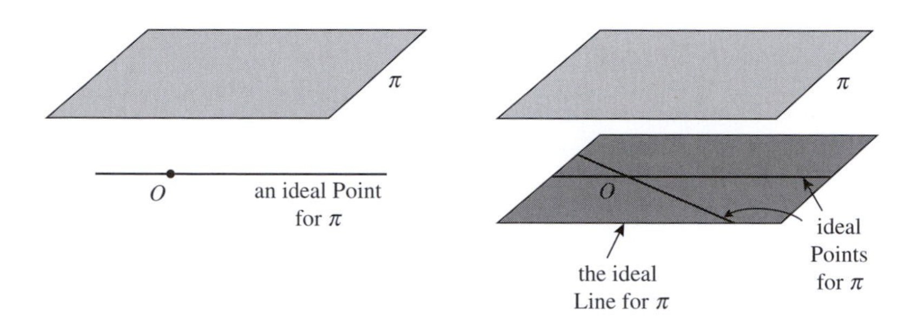

How can we represent a projective figure on π if the figure includes some of the ideal Points for π? As a simple example, consider the Line illustrated in the margin. This is a projective figure which intersects π in a line ℓ. Every Point of the Line pierces the embedding plane at a point of ℓ except for the ideal Point P which cannot be represented on π. In order to represent the Line completely, we need not only the line ℓ but also the ideal Point P. In other words, the Line is represented by $\ell \cup \{P\}$.

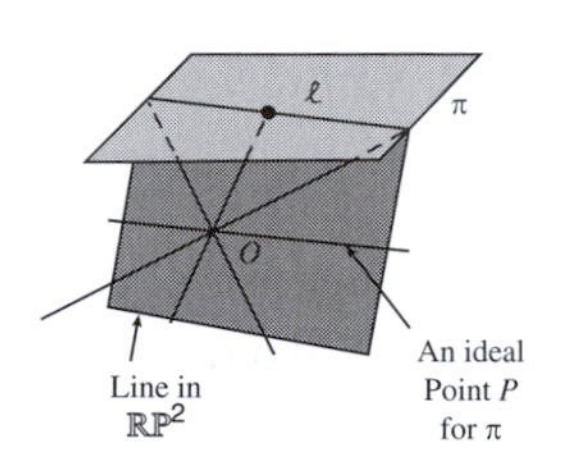

In general, a projective figure can be represented by a figure in π provided that we are prepared to include a subset of Points taken from the ideal Line for π. In order to allow for these additional ideal Points, we introduce the concept of an *embedding plane*.

Definitions An **embedding plane** is a plane, π, which does not pass through the origin, together with the set of all ideal Points for π. The plane in $\mathbb{R}^3$ with equation $z = 1$ is called the **standard embedding plane**. The mapping of $\mathbb{RP}^2$ into the standard embedding plane is called the **standard embedding** of $\mathbb{RP}^2$.

We frequently use π to denote both a plane and an embedding plane, but no confusion should arise.

We may summarize the above discussion by saying that for a given embedding plane, every projective figure in $\mathbb{RP}^2$ corresponds to a figure in the embedding plane, and vice versa. The figure in the embedding plane may include some ideal Points but is otherwise a Euclidean figure.

If two embedding planes are parallel to each other, the same Points of $\mathbb{RP}^2$ correspond to ideal Points of the embeddings; whereas, if the embedding planes are not parallel, different Points of $\mathbb{RP}^2$ correspond to ideal Points of the two embedding planes.

Once we have represented a projective figure in an embedding plane, we can investigate the relationship between its Points and Lines without having to refer to three-dimensional space at all. For example, consider the representation of the triangle of reference and unit Point on the embedding plane $x + y + z = 1$, shown on the left below. If we extract the embedding plane from $\mathbb{R}^3$, as shown on the right, we can use the algebraic theory developed earlier to write down an equation for the Line through any two given Points, without reference to $\mathbb{R}^3$.

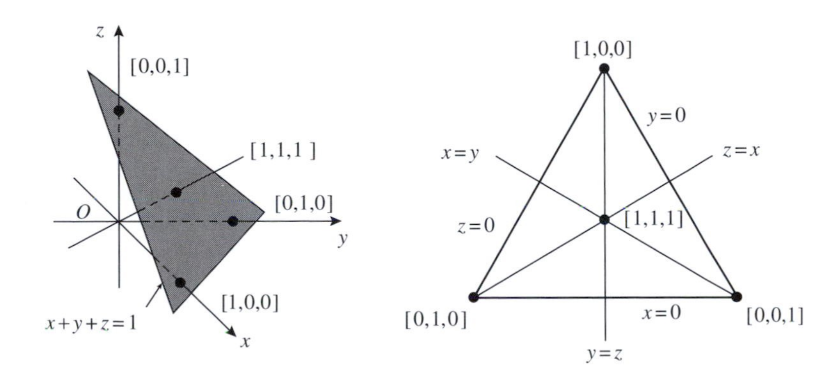

Similarly, we can use the algebraic techniques to calculate the homogeneous coordinates of the Point of intersection of any two given Lines.

Problem 11 On the right-hand diagram above, insert the homogeneous coordinates of the Points where the Lines through [1, 1, 1] meet the sides of the triangle of reference.

Any plane may be used as an embedding plane provided that it does not pass through the origin. For example, if we take π to be the plane $z = -1$, then the ideal Line for π has equation $z = 0$, and the ideal Points are Points of the form $[a, b, 0]$, where a and b are not both zero. Any other Point $[a, b, c]$ has $c \neq 0$ and can therefore be represented in π by the Euclidean point$(-a/c, -b/c, -1)$.

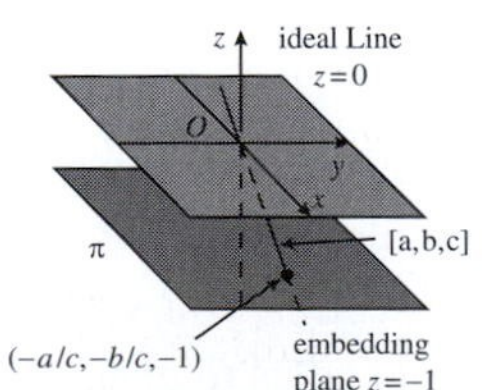

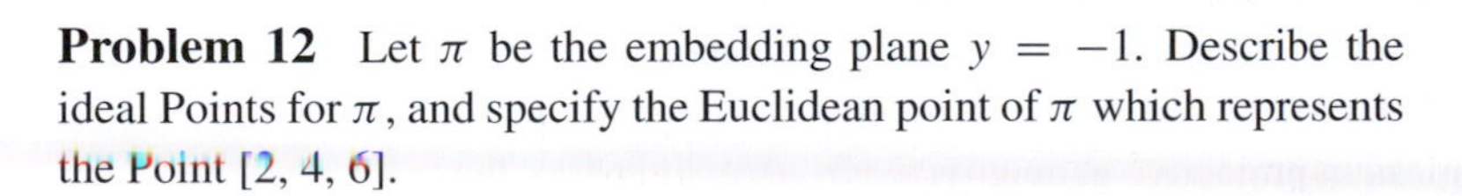
Problem 12 Let π be the embedding plane $y = -1$. Describe the ideal Points for π, and specify the Euclidean point of π which represents the Point [2, 4, 6].

Although we can choose any embedding plane to represent figures of $\mathbb{RP}^2$, the representation does depend on the choice. For example, suppose that π_1 is the embedding plane $y = -1$, and that π_2 is the embedding plane $z = -1$. Now consider the projective figure which consists of two Lines ℓ_1 and ℓ_2 with equations $x = -z$ and $x = z$, respectively. These Lines intersect at the Point $[0, 1, 0]$.

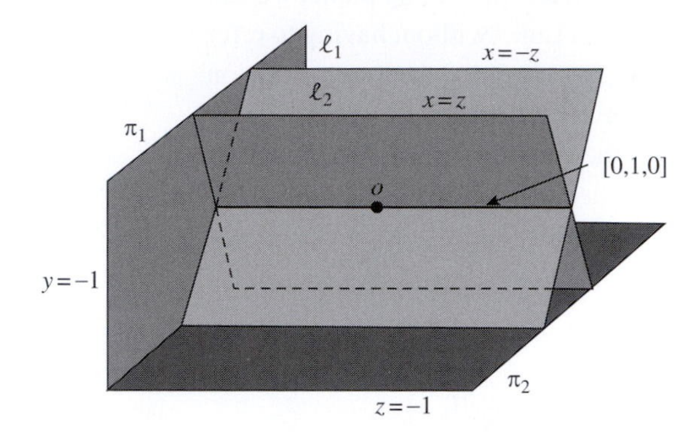

On the embedding plane π_1 the Lines ℓ_1 and ℓ_2 are represented by two lines that can be seen to meet at the point corresponding to $[0, 1, 0]$. However, on the embedding plane π_2 the Point of intersection $[0, 1, 0]$ is an ideal Point and so the Lines ℓ_1 and ℓ_2 are represented by parallel lines that do not appear to meet. The contrast between the two representations of ℓ_1 and ℓ_2 is particularly striking if we extract the two embedding planes from $\mathbb{R}^3$ and lay them side by side, as follows.

This is the mathematical fact which explains why artists sometimes draw parallel lines (such as railroad lines) as intersecting lines.

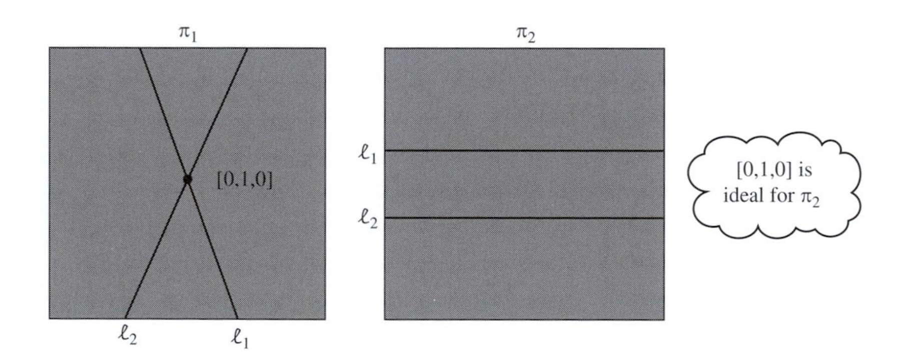

This example illustrates that Lines which appear to be parallel in one embedding plane may not appear to be parallel in another embedding plane. In the next section you will see that the transformations of projective geometry are chosen so as to ensure that the projective properties of a figure are unaffected by the choice of embedding plane. Since parallelism does depend on the choice of embedding plane, it cannot be a projective property, so the concept of parallel Lines is meaningless in projective geometry.

3.2.4 An equivalent definition of Projective Geometry

In our work on projective geometry, we have used Euclidean points in a plane in $\mathbb{R}^3$ to construct the projective points (Points) of the geometry $\mathbb{RP}^2$, homogeneous coordinates for those Points, and projective lines (Lines).

Equivalently, we could have defined $\mathbb{RP}^2$ as the set of ordered triples $[a, b, c]$, where a, b, c are real and not all zero, with the convention that we regard $[\lambda a, \lambda b, \lambda c]$ and $[a, b, c]$ (where $\lambda \neq 0$) as the same Point in the geometry. We would then have defined projective lines (Lines) as the set of points $[x, y, z]$ in $\mathbb{RP}^2$ that satisfy an equation of the form $ax + by + cz = 0$, where a, b, c are real and not all zero, Then we would continue to develop the theory of projective geometry in the same way as we have done here.

However, we chose to start our work by looking at a model of $\mathbb{RP}^2$ obtained by using an embedding plane π in $\mathbb{R}^3$ that does not pass through the origin. We modeled the projective points $[a, b, c]$ by the Euclidean lines through the origin and the corresponding Euclidean points (a, b, c), plus 'points at infinity' (the ideal Points); and we modeled the projective lines by Euclidean planes through the origin, For convenience, we chose often to use Euclidean points (a, b, c) on a given embedding plane to describe the Euclidean model.

Any plane that does not pass through the origin in $\mathbb{R}^3$ will serve as an embedding plane.

The formal method of defining projective geometry, though, is less intuitive than the description motivated by the $\mathbb{R}^3$ model!

3.3 Projective Transformations

3.3.1 The Group of Projective Transformations

By now you should be familiar with the idea that a geometry consists of a space of points together with a group of transformations which act on that space.

Having introduced the space of projective points $\mathbb{RP}^2$ in Section 3.2, we are now in a position to describe the transformations of $\mathbb{RP}^2$. First we shall define the transformations algebraically, then we give a geometrical interpretation of the transformations using the ideas of perspectivity introduced in Section 3.1, and finally meet the Fundamental Theorem of Projective Geometry.

Subsection 3.2.1

Recall that a point of $\mathbb{R}^3$ (other than the origin) on an embedding plane π (that does not pass through the origin) has coordinates $\mathbf{x} = (x, y, z)$ with respect to the standard basis of $\mathbb{R}^3$, and homogeneous coordinates of the corresponding Point $[\mathbf{x}]$ in $\mathbb{RP}^2$ (which represents the points $\{\lambda \mathbf{x} : \lambda \in \mathbb{R}\}$) are $[\lambda x, \lambda y, \lambda z]$ for some real $\lambda \neq 0$. Since the Points of $\mathbb{RP}^2$ are just lines through the origin of $\mathbb{R}^3$, we need a group of transformations that map the lines through the origin of $\mathbb{R}^3$ onto the lines through the origin of $\mathbb{R}^3$. Suitable transformations of $\mathbb{R}^3$ that do this are the invertible linear transformations.

If $\mathbf{A}$ is the matrix of an invertible linear transformation of $\mathbb{R}^3$ to itself, the transformation maps points $\mathbf{x} = (x, y, z)$ of $\mathbb{R}^3$ to points $\mathbf{Ax}$ of $\mathbb{R}^3$; then the *projective transformation* with matrix $\mathbf{A}$ maps Points $[\mathbf{x}]$ of $\mathbb{RP}^2$ to Points $[\mathbf{Ax}]$ of $\mathbb{RP}^2$. This suggests that we define the transformations of projective geometry as follows.

In fact, any 'continuous' transformation of $\mathbb{RP}^2$ to itself that maps Lines to Lines and that preserves incidences of Lines corresponds to an invertible linear transformation of $\mathbb{R}^3$. We omit a proof of this fact.

Definitions A **projective transformation** of $\mathbb{RP}^2$ is a function $t : \mathbb{RP}^2 \to \mathbb{RP}^2$ of the form

$$t : [\mathbf{x}] \mapsto [\mathbf{Ax}],$$

where $\mathbf{A}$ is an invertible 3×3 matrix. We say that $\mathbf{A}$ is a matrix **associated** with t. The set of all projective transformations of $\mathbb{RP}^2$ is denoted by $P(2)$.

Example 1 Show that the function $t : \mathbb{RP}^2 \to \mathbb{RP}^2$ defined by

$$t : [x, y, z] \mapsto [2x + z, -x + 2y - 3z, x - y + 5z]$$

is a projective transformation, and find the image of $[1, 2, 3]$ under t.

Solution The transformation t has the form $t : [\mathbf{x}] \mapsto [\mathbf{Ax}]$, where $\mathbf{x} = (x, y, z)$ and

$$\mathbf{A} = \begin{pmatrix} 2 & 0 & 1 \\ -1 & 2 & -3 \\ 1 & -1 & 5 \end{pmatrix}.$$

Now

$$\begin{aligned} \det \mathbf{A} &= \begin{vmatrix} 2 & 0 & 1 \\ -1 & 2 & -3 \\ 1 & -1 & 5 \end{vmatrix} \\ &= 2(10 - 3) - 0 + (1 - 2) \\ &= 13 \neq 0. \end{aligned}$$

So $\mathbf{A}$ is invertible. It follows that t is a projective transformation.

We have

$$t([1, 2, 3]) = [2 + 3, -1 + 4 - 9, 1 - 2 + 15] = [5, -6, 14]. \quad \square$$

Problem 1 Decide which of the following functions t from $\mathbb{RP}^2$ to itself are projective transformations. For those that are projective transformations, write down a matrix associated with t.

(a) $t : [x, y, z] \mapsto [-2y + 3z, -x + 5y - z, -3x]$
(b) $t : [x, y, z] \mapsto [x - 7y + 4z, -x + 5y - z, x - 9y + 7z]$
(c) $t : [x, y, z] \mapsto [x - 1 + z, 2y - 4z + 5, 2x]$

Problem 2 Let t be the projective transformation associated with the matrix.

$$\mathbf{A} = \begin{pmatrix} 1 & 1 & -1 \\ -1 & -2 & 1 \\ 4 & -3 & 4 \end{pmatrix}.$$

Determine the image under t of each of the following Points.

(a) $[1, 2, -1]$ (b) $[1, 0, 0]$ (c) $[0, 1, 0]$
(d) $[0, 0, 1]$ (e) $[1, 1, 1]$

Since we can multiply the homogeneous coordinates of Points in $\mathbb{RP}^2$ by any non-zero real number λ without altering the Point itself, it follows that if $\mathbf{A}$ is a matrix associated with a particular projective transformation then so is the matrix $\lambda\mathbf{A}$, provided that $\lambda \neq 0$. For example, another matrix associated with the transformation in Example 1 is

$$\mathbf{B} = \begin{pmatrix} -4 & 0 & -2 \\ 2 & -4 & 6 \\ -2 & 2 & -10 \end{pmatrix},$$

for we have $\mathbf{B} = -2\mathbf{A}$.

Problem 3 Write down a matrix with top left-hand entry $\frac{1}{2}$ which is associated with the transformation in Example 1.

Before we can use the projective transformations to define projective geometry, we must first check that they form a group.

Theorem 1 The set of projective transformations $P(2)$ forms a group under the operation of composition of functions.

Recall that a similar result holds for affine transformations.

Proof We check that the four group axioms hold.

G1 CLOSURE Let t_1 and t_2 be projective transformations defined by

$$t_1 : [\mathbf{x}] \mapsto [\mathbf{A}_1\mathbf{x}] \quad \text{and} \quad t_2 : [\mathbf{x}] \mapsto [\mathbf{A}_2\mathbf{x}],$$

where $\mathbf{A}_1$ and $\mathbf{A}_2$ are invertible 3×3 matrices. Then

$$\begin{aligned} t_1 \circ t_2([\mathbf{x}]) &= t_1(t_2([\mathbf{x}])) \\ &= t_1([\mathbf{A}_2\mathbf{x}]) \\ &= [(\mathbf{A}_1\mathbf{A}_2)\mathbf{x}]. \end{aligned}$$

Since $\mathbf{A}_1$ and $\mathbf{A}_2$ are invertible, it follows that $\mathbf{A}_1\mathbf{A}_2$ is invertible. So by definition $t_1 \circ t_2$ is a projective transformation.

G2 IDENTITY Let $i : \mathbb{RP}^2 \to \mathbb{RP}^2$ be the transformation defined by

$$i : [\mathbf{x}] \mapsto [\mathbf{I}\mathbf{x}],$$

where $\mathbf{I}$ is the 3×3 identity matrix; this is a projective transformation, since $\mathbf{I}$ is invertible.

Let t: $\mathbb{RP}^2 \to \mathbb{RP}^2$ be an arbitrary projective transformation, defined by t :$[\mathbf{x}] \mapsto [\mathbf{Ax}]$, for some invertible 3 × 3 matrix **A**. Then for any $[\mathbf{x}] \in \mathbb{RP}^2$,

$$t \circ i([\mathbf{x}]) = [\mathbf{A}(\mathbf{Ix})] = [\mathbf{Ax}]$$

and

$$i \circ t([\mathbf{x}]) = [\mathbf{I}(\mathbf{Ax})] = [\mathbf{Ax}].$$

Thus $t \circ i = i \circ t = t$. Hence i is the identity transformation.

G3 INVERSES Let $t : \mathbb{RP}^2 \to \mathbb{RP}^2$ be an arbitrary projective transformation defined by

$$t\colon [\mathbf{x}] \mapsto [\mathbf{Ax}],$$

for some invertible 3 × 3 matrix **A**. Then we can define another projective transformation $t' : \mathbb{RP}^2 \to \mathbb{RP}^2$ by

$$t'\colon [\mathbf{x}] \mapsto [\mathbf{A}^{-1}\mathbf{x}].$$

Now, for each $[\mathbf{x}] \in \mathbb{RP}^2$, we have

$$t \circ t'([\mathbf{x}]) = \mathbf{t}([\mathbf{A}^{-1}\mathbf{x}]) = [\mathbf{A}(\mathbf{A}^{-1}\mathbf{x})] = [\mathbf{x}]$$

and

$$t' \circ t([\mathbf{x}]) = t'([\mathbf{Ax}]) = [\mathbf{A}^{-1}(\mathbf{Ax})] = [\mathbf{x}].$$

Thus t' is an inverse for t.

G4 ASSOCIATIVITY Composition of functions is always associative.

It follows that the set of projective transformations $P(2)$ forms a group. ■

The above proof shows that if t_1 and t_2 are projective transformations with associated matrices $\mathbf{A}_1$ and $\mathbf{A}_2$, respectively, then $t_1 \circ t_2$ is a projective transformation with an associated matrix $\mathbf{A}_1\mathbf{A}_2$. We therefore have the following strategy for composing projective transformations.

Strategy To compose two projective transformations t_1 and t_2:

1. write down matrices $\mathbf{A}_1$ and $\mathbf{A}_2$ associated with t_1 and t_2;
2. calculate $\mathbf{A}_1\mathbf{A}_2$;
3. write down the composite $t_1 \circ t_2$ with which $\mathbf{A}_1\mathbf{A}_2$ is associated.

The proof also shows that if t is a projective transformation with an associated matrix **A**, then t^{-1} is a projective transformation with associated matrix $\mathbf{A}^{-1}$. We therefore have the following strategy for calculating the inverse of a projective transformation.

See Appendix 2 for one method to calculate $\mathbf{A}^{-1}$.

Strategy To find the inverse of a projective transformation t:

1. write down a matrix $\mathbf{A}$ associated with t;
2. calculate $\mathbf{A}^{-1}$;
3. write down the inverse t^{-1} with which $\mathbf{A}^{-1}$ is associated.

Example 2 Let t_1 and t_2 be projective transformations defined by

$$t_1 : [x, y, z] \mapsto [x + z, x + y + 3z, -2x + z],$$

$$t_2 : [x, y, z] \mapsto [2x, x + y + z, 4x + 2y].$$

Determine the projective transformations $t_2 \circ t_1$ and t_1^{-1}.

Solution The transformations t_1 and t_2 have associated matrices

$$\mathbf{A}_1 = \begin{pmatrix} 1 & 0 & 1 \\ 1 & 1 & 3 \\ -2 & 0 & 1 \end{pmatrix} \quad \text{and} \quad \mathbf{A}_2 = \begin{pmatrix} 2 & 0 & 0 \\ 1 & 1 & 1 \\ 4 & 2 & 0 \end{pmatrix},$$

respectively. It follows that $t_2 \circ t_1$ has an associated matrix

$$\mathbf{A}_2\mathbf{A}_1 = \begin{pmatrix} 2 & 0 & 0 \\ 1 & 1 & 1 \\ 4 & 2 & 0 \end{pmatrix} \begin{pmatrix} 1 & 0 & 1 \\ 1 & 1 & 3 \\ -2 & 0 & 1 \end{pmatrix} = \begin{pmatrix} 2 & 0 & 2 \\ 0 & 1 & 5 \\ 6 & 2 & 10 \end{pmatrix},$$

so

$$t_2 \circ t_1 : [x, y, z] \mapsto [2x + 2z, y + 5z, 6x + 2y + 10z].$$

Next, t_1^{-1} has an associated matrix $\mathbf{A}_1^{-1}$ given by

$$\mathbf{A}_1^{-1} = \begin{pmatrix} \frac{1}{3} & 0 & -\frac{1}{3} \\ -\frac{7}{3} & 1 & -\frac{2}{3} \\ \frac{2}{3} & 0 & \frac{1}{3} \end{pmatrix};$$

a simpler matrix associated with t_1^{-1} is then

$$\begin{pmatrix} 1 & 0 & -1 \\ -7 & 3 & -2 \\ 2 & 0 & 1 \end{pmatrix},$$

so

$$t_1^{-1} : [x, y, z] \mapsto [x - z, -7x + 3y - 2z, 2x + z]. \qquad \square$$

Problem 4 Let t_1 and t_2 be projective transformations defined by

$$t_1 : [x, y, z] \mapsto [2x + y, -x + z, y + z],$$

$$t_2 : [x, y, z] \mapsto [5x + 8y, 3x + 5y, 2z].$$

Determine the projective transformations $t_1 \circ t_2$ and t_1^{-1}.

Having shown that the set of projective transformations forms a group under composition of functions, we can now define **projective geometry** to be the study of those properties of figures in $\mathbb{RP}^2$ that are preserved by projective transformations. Those properties that are preserved by projective transformations are known as **projective properties.**

3.3.2 Some Properties of Projective Transformations

We now check two important properties of projective transformations, namely, that they preserve collinearity and incidence.

A Line in $\mathbb{RP}^2$ is a plane in $\mathbb{R}^3$ that passes through the origin. It therefore consists of the set of points (x, y, z) of $\mathbb{R}^3$ that satisfy an equation of the form

$$ax + by + cz = 0,$$

where a, b and c are not all zero. We can write this condition equivalently in the matrix form $\mathbf{Lx} = 0$, where $\mathbf{L}$ is the non-zero row matrix $(a\ b\ c)$ and $\mathbf{x} = (x\ y\ z)^T$.

Now let t be a projective transformation defined by $t : [\mathbf{x}] \mapsto [\mathbf{Ax}]$, where $\mathbf{A}$ is an invertible 3×3 matrix, and let $[\mathbf{x}]$ be an arbitrary Point on the Line $\mathbf{Lx} = 0$. Then the image of $[\mathbf{x}]$ under t is a Point $[\mathbf{x}']$ where $\mathbf{x}' = \mathbf{Ax}$. Since $\mathbf{x}$ satisfies the equation $\mathbf{Lx} = 0$, it follows that $\mathbf{x}'$ satisfies $\mathbf{L}(\mathbf{A}^{-1}\mathbf{x}') = 0$, or $(\mathbf{LA}^{-1})\,\mathbf{x}' = 0$. Dropping the dash, we conclude that the image of the Line $\mathbf{Lx} = 0$ under t is the Line with equation

$$\left(\mathbf{LA}^{-1}\right)\mathbf{x} = 0.$$

Here, $\mathbf{LA}^{-1}$ is non-zero, for if

$$\mathbf{LA}^{-1} = \mathbf{0},$$

then

$$\mathbf{0} = \left(\mathbf{LA}^{-1}\right)\mathbf{A} = \mathbf{L}\left(\mathbf{A}^{-1}\mathbf{A}\right) = \mathbf{L},$$

which is not the case.

Since the image of a Line in $\mathbb{RP}^2$ is a Line, it follows that collinearity is preserved under a projective transformation.

Notice that if $\mathbf{B}$ is any matrix associated with t^{-1}, then $\mathbf{B} = \lambda\mathbf{A}^{-1}$ for some non-zero real number λ, and so $\left(\mathbf{LA}^{-1}\right)\mathbf{x} = 0$ if and only if $(\mathbf{LB})\mathbf{x} = 0$. It follows that the image of the Line can equally well be written as $(\mathbf{LB})\mathbf{x} = 0$. (For instance, since $\mathbf{A}^{-1} = \operatorname{adj}(\mathbf{A})/\det(\mathbf{A})$ so that t^{-1} also has $\operatorname{adj}(\mathbf{A})$ as an associated matrix, we can express the image of the Line as $(\mathbf{L}\operatorname{adj}(\mathbf{A}))\mathbf{x} = 0$.)

We therefore summarize the above discussion in the form of a strategy, as follows.

Strategy To find the image of a Line

$$ax + by + cz = 0$$

under a projective transformation $t : [\mathbf{x}] \mapsto [\mathbf{AX}]$:

1. write the equation of the Line in the form $\mathbf{Lx} = 0$, where $\mathbf{L}$ is the matrix $(a\,b\,c)$;
2. find a matrix $\mathbf{B}$ associated with t^{-1};
3. write down the equation of the image as $(\mathbf{LB})\mathbf{x} = 0$.

Example 3 Find the image of the Line $2x + y - 3z = 0$ under the projective transformation t_1 defined by

$$t_1 : [x, y, z] \mapsto [x + z, x + y + 3z, -2x + z].$$

Solution The equation of the Line can be written in the form $\mathbf{Lx} = 0$, where

$$\mathbf{L} = (2 \quad 1 \quad -3).$$

In Example 2 we showed that ${t_1}^{-1}$ has an associated matrix

$$\mathbf{B} = \begin{pmatrix} 1 & 0 & -1 \\ -7 & 3 & -2 \\ 2 & 0 & 1 \end{pmatrix}.$$

So

$$\mathbf{LB} = (2 \quad 1 - 3) \begin{pmatrix} 1 & 0 & -1 \\ -7 & 3 & -2 \\ 2 & 0 & 1 \end{pmatrix} = (-11 \quad 3 - 7).$$

It follows that the required image has equation

$$-11x + 3y - 7z = 0.$$ □

Problem 5 Find the image of the Line $x + 2y - z = 0$ under the projective transformation t_1 defined by

$$t_1 : [x, y, z] \mapsto [2x + y, -x + z, y + z].$$

Next, we consider the incidence property. If two Lines intersect at the Point P, then P lies on both Lines. So if t is a projective transformation, then $t(P)$ lies on the images of both Lines. It follows that the image under t of the Point of intersection of the two Lines is the Point of intersection of the images of the two Lines. In other words, incidence is also preserved under a projective transformation.

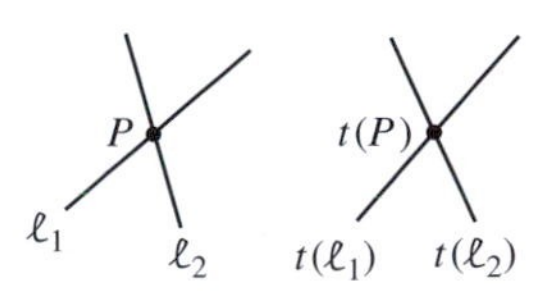

Theorem 2 Collinearity and incidence are both projective properties.

3.3.3 Fundamental Theorem of Projective Geometry

In Chapter 2 we discussed the Fundamental Theorem of Affine Geometry which states that given any two sets of three non-collinear points of $\mathbb{R}^2$ there is a unique affine transformation which maps the points in one set to the corresponding points in the other set. So an affine transformation is uniquely determined by its effect on any given triangle.

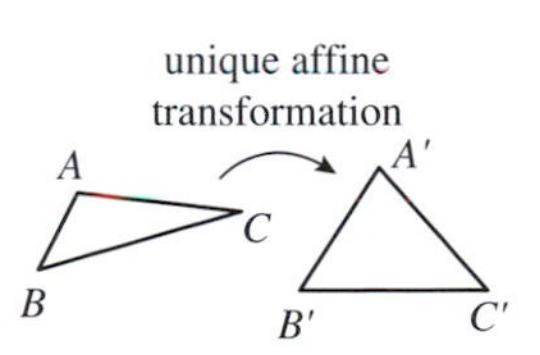

In this subsection we explore an analogous result for projective geometry known as the *Fundamental Theorem of Projective Geometry.* We begin by asking you to tackle the following problem.

Problem 6 Let t_1 and t_2 be the projective transformations with associated matrices

$$\mathbf{A}_1 = \begin{pmatrix} -4 & -1 & 1 \\ -3 & -2 & 1 \\ 4 & 2 & -1 \end{pmatrix} \text{ and } \mathbf{A}_2 = \begin{pmatrix} -8 & -6 & -2 \\ -3 & 4 & 7 \\ 6 & 0 & -4 \end{pmatrix},$$

respectively. Find the images of the Points $[1, -1, 1]$, $[1, -2, 2]$ and $[-1, 2, -1]$ under t_1 and t_2.

You should have found that both of the projective transformations t_1 and t_2 map the Points $[1, -1, 1]$, $[1, -2, 2]$ and $[-1, 2, -1]$ to the Points $[-2, 0, 1]$, $[0, 3, -2]$ and $[1, -2, 1]$, respectively. Notice, however, that t_1 and t_2 are not the same projective transformation, since their matrices are not multiples of each other. It follows that, unlike affine transformations, projective transformations are not uniquely determined by their effect on three (non-collinear) Points.

This raises the question as to whether it is possible to specify how many Points *are* required to determine a projective transformation. According to the Fundamental Theorem of Projective Geometry, the answer is four. In fact the theorem states that given any two sets of four Points, no three of which are collinear, there is a *unique* projective transformation that maps the Points in one set to the corresponding Points in the second set. Thus, in projective geometry a transformation is uniquely determined by its effect on a quadrilateral.

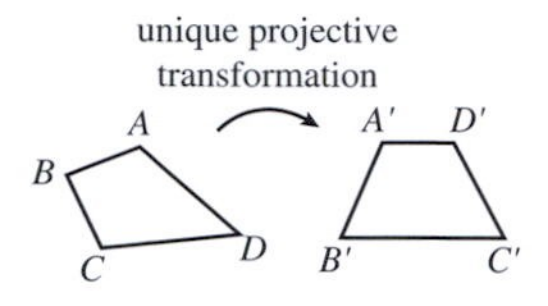

To understand why a triangle is insufficient to determine a projective transformation uniquely, consider what happens when we look for a projective transformation that maps the triangle of reference to three given non-collinear Points.

Recall that the Points [1, 0, 0], [0, 1, 0] and [0, 0, 1] are known as the *triangle of reference.*

Example 4 Find a projective transformation t that maps the Points $[1, 0, 0]$, $[0, 1, 0]$ and $[0, 0, 1]$ to the non-collinear Points $[1, -1, 1], [1, -2, 2]$ and $[-1, 2, -1]$, respectively.

Solution Let $\mathbf{A}$ be a matrix associated with t, and let the first column of $\mathbf{A}$ be $\begin{pmatrix} a \\ b \\ c \end{pmatrix}$. Then since

$$\left[\begin{pmatrix} a & * & * \\ b & * & * \\ c & * & * \end{pmatrix}\begin{pmatrix} 1 \\ 0 \\ 0 \end{pmatrix}\right] = \left[\begin{pmatrix} a \\ b \\ c \end{pmatrix}\right] = \left[\begin{pmatrix} 1 \\ -1 \\ 1 \end{pmatrix}\right],$$

Here the asterisks * denote unspecified numbers.

it follows that we may take $\begin{pmatrix} 1 \\ -1 \\ 1 \end{pmatrix}$ as the first column of $\mathbf{A}$.

Similarly, since

$$\left[\begin{pmatrix} * & d & * \\ * & e & * \\ * & f & * \end{pmatrix}\begin{pmatrix} 0 \\ 1 \\ 0 \end{pmatrix}\right] = \left[\begin{pmatrix} d \\ e \\ f \end{pmatrix}\right] = \left[\begin{pmatrix} 1 \\ -2 \\ 2 \end{pmatrix}\right]$$

and

$$\left[\begin{pmatrix} * & * & g \\ * & * & h \\ * & * & k \end{pmatrix}\begin{pmatrix} 0 \\ 0 \\ 1 \end{pmatrix}\right] = \left[\begin{pmatrix} g \\ h \\ k \end{pmatrix}\right] = \left[\begin{pmatrix} -1 \\ 2 \\ -1 \end{pmatrix}\right],$$

it follows that a suitable transformation is given by $t : [\mathbf{x}] \mapsto [\mathbf{A}x]$ where

$$\mathbf{A} = \begin{pmatrix} 1 & 1 & -1 \\ -1 & -2 & 2 \\ 1 & 2 & -1 \end{pmatrix}.$$

□

Notice that because the Points $[1, -1, 1]$, $[1, -2, 2]$ and $[-1, 2, -1]$ are not collinear it follows that the columns of $\mathbf{A}$ are linearly independent, so that $\mathbf{A}$ is invertible.

This example illustrates the fact that we can always find a projective transformation $t : [\mathbf{x}] \mapsto [\mathbf{Ax}]$ which maps the triangle of reference to three non-collinear Points simply by writing the homogeneous coordinates of the Points as the columns of $\mathbf{A}$. Notice, however, that the transformation we obtain is not unique. Indeed, if the Points $[1, -1, 1]$, $[1, -2, 2]$ and $[-1, 2, -1]$ in Example 4 are rewritten in the form $[u, -u, u]$, $[v, -2v, 2v]$ and $[-w, 2w, -w]$, for some non-zero real numbers u, v, w, then the matrix becomes

$$\mathbf{A} = \begin{pmatrix} u & v & -w \\ -u & -2v & 2w \\ u & 2v & -w \end{pmatrix}.$$

The corresponding transformation $t : [\mathbf{x}] \mapsto [\mathbf{Ax}]$ still maps the triangle of reference to the Points $[1, -1, 1]$, $[1, -2, 2]$ and $[-1, 2, -1]$, as required, but the effect that t has on the other Points of $\mathbb{RP}^2$ depends on the numbers u, v and w.

So if we wish to specify t uniquely we need to assign particular values to u, v and w. We can do this by specifying the effect that t has on a fourth Point $[1, 1, 1]$.

Recall that the Point $[1, 1, 1]$ is known as the *unit Point*.

Example 5 Find the projective transformation t which maps the Points $[1, 0, 0]$, $[0, 1, 0]$, $[0, 0, 1]$ and $[1, 1, 1]$ to the Points $[1, -1, 1]$, $[1, -2, 2]$, $[-1, 2, -1]$ and $[0, 1, 2]$, respectively.

Solution If $\mathbf{A}$ is the matrix associated with t, then its columns must be multiples of the homogeneous coordinates $[1, -1, 1]$, $[1, -2, 2]$, $[-1, 2, -1]$; that is,

$$\mathbf{A} = \begin{pmatrix} u & v & -w \\ -u & -2v & 2w \\ u & 2v & -w \end{pmatrix}.$$

Also, to ensure that t maps [1, 1, 1] to [0, 1, 2] we must choose u, v and w so that

$$\left[\begin{pmatrix} u & v & -w \\ -u & -2v & 2w \\ u & 2v & -w \end{pmatrix}\begin{pmatrix} 1 \\ 1 \\ 1 \end{pmatrix}\right] = \left[\begin{pmatrix} 0 \\ 1 \\ 2 \end{pmatrix}\right].$$

We can do this by solving the equations

$$u + v - w = 0,$$
$$-u - 2v + 2w = 1,$$
$$u + 2v - w = 2.$$

Adding the second and third equations we obtain $w = 3$. If we then subtract the first equation from the third we obtain $v = 2$. Finally, if we substitute v and w into the first equation we obtain $u = 1$. The required projective transformation is therefore given by $t : [\mathbf{x}] \mapsto [\mathbf{Ax}]$, where

$$\mathbf{A} = \begin{pmatrix} 1 & 2 & -3 \\ -1 & -4 & 6 \\ 1 & 4 & -3 \end{pmatrix}.$$

□

The columns of **A** are still linearly independent because they are non-zero multiples of the linearly independent vectors $(1, -1, 1)$, $(1, -2, 2)$ and $(-1, 2, -1)$.

It is natural to ask whether the method used in this example can be adapted to find a projective transformation which maps the triangle of reference and unit Point to *any* four given Points. The answer is usually yes, but since collinearity is a projective property, and since no three of the Points [1, 0, 0], [0, 1, 0], [0, 0, 1], [1, 1, 1] are collinear, the method must fail if three of the four given Points lie on a Line. Provided we exclude this possibility, the answer is yes!

We explain why the method works in the Remark that follows the strategy.

Strategy To find the projective transformation which maps

$[1, 0, 0]$ to $[a_1, a_2, a_3]$,

$[0, 1, 0]$ to $[b_1, b_2, b_3]$,

$[0, 0, 1]$ to $[c_1, c_2, c_3]$,

$[1, 1, 1]$ to $[d_1, d_2, d_3]$,

where no three of $[a_1, a_2, a_3], [b_1, b_2, b_3], [c_1, c_2, c_3], [d_1, d_2, d_3]$ are collinear:

1. find u, v, w such that

$$\begin{pmatrix} a_1u & b_1v & c_1w \\ a_2u & b_2v & c_2w \\ a_3u & b_3v & c_3w \end{pmatrix}\begin{pmatrix} 1 \\ 1 \\ 1 \end{pmatrix} = \begin{pmatrix} d_1 \\ d_2 \\ d_3 \end{pmatrix};$$

2. write down the required projective transformation in the form $t : [\mathbf{x}] \mapsto [\mathbf{Ax}]$, where $\mathbf{A}$ is any non-zero real multiple of the matrix

$$\begin{pmatrix} a_1u & b_1v & c_1w \\ a_2u & b_2v & c_2w \\ a_3u & b_3v & c_3w \end{pmatrix}.$$

The non-zero multiple can be used to clear fractions from the entries of the matrix $\mathbf{A}$.

Remark

To see why this strategy always works, notice that we can rewrite the equation from Step 1 in the form

$$u\begin{pmatrix} a_1 \\ a_2 \\ a_3 \end{pmatrix} + v\begin{pmatrix} b_1 \\ b_2 \\ b_3 \end{pmatrix} + w\begin{pmatrix} c_1 \\ c_2 \\ c_3 \end{pmatrix} = \begin{pmatrix} d_1 \\ d_2 \\ d_3 \end{pmatrix}.$$

From this we can make the following observations.

(a) The equation in Step 1 must have a unique solution for u, v, w because the required values of u, v and w are simply the coordinates of (d_1, d_2, d_3) with respect to the basis of $\mathbb{R}^3$ formed from the three linearly independent vectors $(a_1, a_2, a_3), (b_1, b_2, b_3), (c_1, c_2, c_3)$.

(b) The values *of* u, v and w must all be non-zero, because otherwise three of the vectors $(a_1, a_2, a_3), (b_1, b_2, b_3), (c_1, c_2, c_3), (d_1, d_2, d_3)$ would be linearly dependent.

(c) Since the columns of $\mathbf{A}$ are non-zero, multiples of the linearly independent vectors $(a_1, a_2, a_3), (b_1, b_2, b_3), (c_1, c_2, c_3)$ it follows that $\mathbf{A}$ is invertible, and hence that t is a projective transformation.

Since no three of the Points $[a_1, a_2, a_3]$, $[b_1, b_2, b_3]$, $[c_1, c_2, c_3]$, $[d_1, d_2, d_3]$ are collinear, it follows that *any three* of the vectors (a_1, a_2, a_3), (b_1, b_2, b_3), (c_1, c_2, c_3), (d_1, d_2, d_3) must be linearly independent.

There is no need to check whether any three of the four given Points are collinear because any failure of this condition will emerge in the process of applying the strategy. Indeed, if the equation in Step 1 fails to yield unique non-zero values for u, v and w, then it must be because three of the Points $(a_1, a_2, a_3), (b_1, b_2, b_3), (c_1, c_2, c_3), (d_1, d_2, d_3)$ lie on a Line.

Problem 7 Use the above strategy to find the projective transformation which maps the Points [1, 0, 0], [0, 1, 0], [0, 0, 1] and [1, 1, 1] to the Points:

(a) $[-1, 0, 0], [-3, 2, 0], [2, 0, 4]$ and $[1, 2, -5]$, respectively;
(b) [1, 0, 0], [0, 0, 1], [0, 1, 0] and [3, 4, 5], respectively;
(c) [2, 1, 0], $[1, 0, -1], [0, 3, -1]$ and $[3, -1, 2]$, respectively.

Now consider the transformation t_1 in Problem 7(a). The inverse of this, t_1^{-1}, is a projective transformation which maps the Points $[-1, 0, 0], [-3, 2, 0]$, [2, 0, 4] and $[1, 2, -5]$ back to the triangle of reference and unit Point. So if, after applying this inverse, we apply the projective transformation t_2 in Problem 7(c), then the overall effect of the composite $t_2 \circ t_1^{-1}$ is that of a projective transformation which sends the Points $[-1, 0, 0], [-3, 2, 0]$, [2, 0, 4] and $[1, 2, -5]$ directly to the Points [2, 1, 0], $[1, 0, -1], [0, 3, -1]$ and $[3, -1, 2]$, respectively.

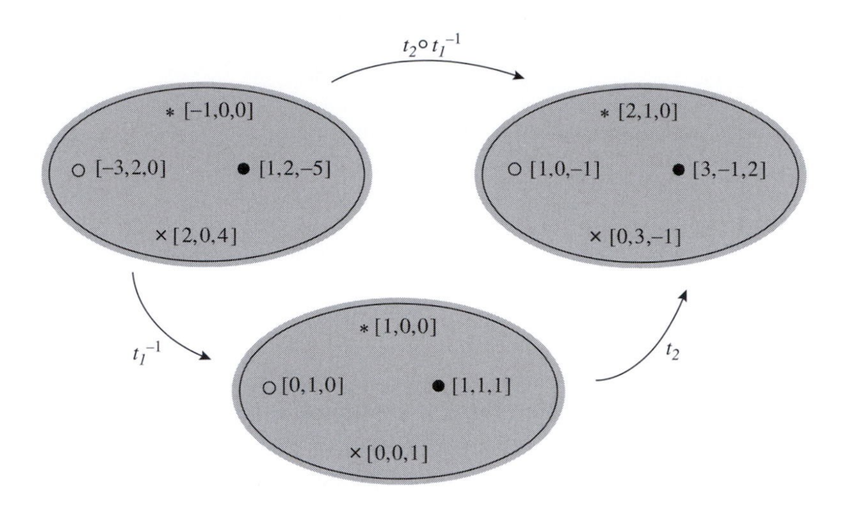

In a similar way we can find a projective transformation which maps any set of four Points to any other set of four Points. The only constraint is that no three of the Points in either set can be collinear. In the following statement of the Fundamental Theorem we express this constraint by requiring that each of the four sets of Points lie at the vertices of some quadrilateral, where a *quadrilateral* is defined as follows. A *quadrilateral* is a set of four Points A, B, C and D (no three of which are collinear), together with the Lines AB, BC, CD and DA.

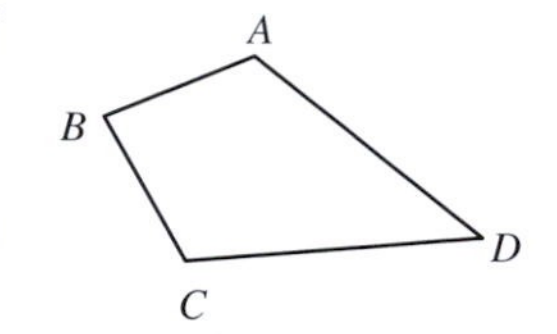

Theorem 3 The Fundamental Theorem of Projective Geometry
Let $ABCD$ and $A'B'C'D'$ be two quadrilaterals in $\mathbb{RP}^2$. Then:

(a) there is a projective transformation t which maps

$$A \text{ to } A', B \text{ to } B', C \text{ to } C', D \text{ to } D';$$

(b) the projective transformation t is unique.

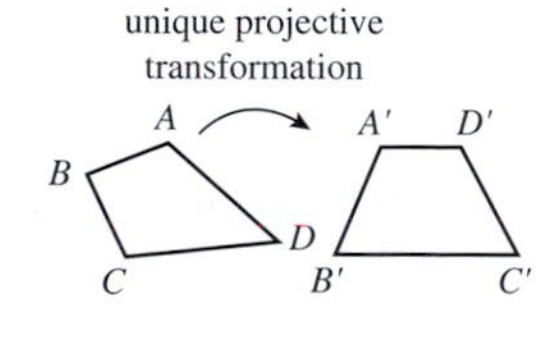

Proof According to the strategy above, there is a projective transformation t_1 which maps the Points [1, 0, 0], [0, 1, 0], [0, 0, 1], [1, 1, 1] to the Points A, B, C, D, respectively. Similarly, there is a projective transformation t_2 which maps the Points [1, 0, 0], [0, 1, 0], [0, 0, 1], [1, 1, 1] to the Points A', B', C', D', respectively.

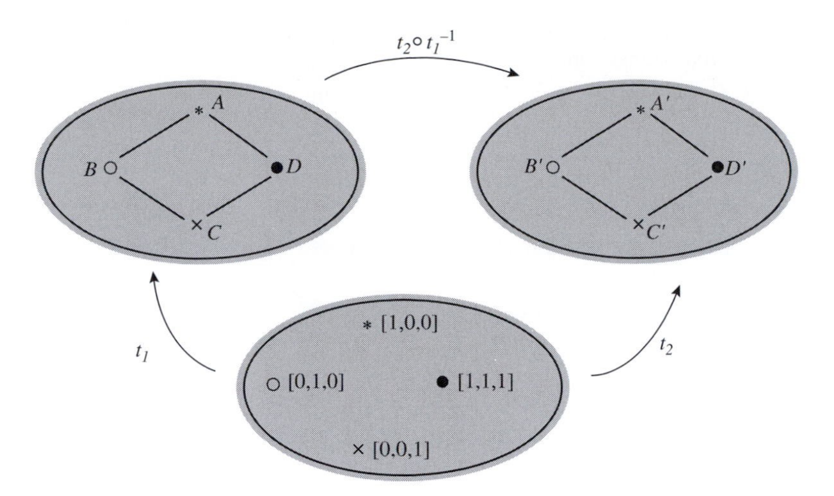

(a) The composite $t = t_2 \circ t_1^{-1}$ is then a projective transformation which maps A to A', B to B', C to C', D to D'.

(b) To check uniqueness of t, we first check that the identity transformation is the only projective transformation which maps each of the Points [1, 0, 0], [0, 1, 0], [0, 0, 1], [1, 1, 1] to themselves. In fact any projective transformation with this property must have an associated matrix which is some non-zero multiple of the matrix

$$\begin{pmatrix} u & 0 & 0 \\ 0 & v & 0 \\ 0 & 0 & w \end{pmatrix}, \quad \text{where} \quad \begin{pmatrix} u & 0 & 0 \\ 0 & v & 0 \\ 0 & 0 & w \end{pmatrix} \begin{pmatrix} 1 \\ 1 \\ 1 \end{pmatrix} = \begin{pmatrix} 1 \\ 1 \\ 1 \end{pmatrix}.$$

Such a matrix must be (a non-zero multiple of) the identity matrix, and so the transformation must indeed be the identity.

This follows from the discussion leading to the strategy above.

Next suppose that t and t' are two projective transformations which satisfy the conditions of the theorem. Then the composites $t_2^{-1} \circ t \circ t_1$ and $t_2^{-1} \circ t' \circ t_1$ must both be projective transformations which map each of the Points [1, 0, 0], [0, 1, 0], [0, 0, 1], [1, 1, 1] to themselves. Since this implies that both composites are equal to the identity, we deduce that

$$t_2^{-1} \circ t \circ t_1 = t_2^{-1} \circ t' \circ t_1.$$

If we now compose both sides of this equation with t_2 on the left and with t_1^{-1} on the right, then we obtain $t = t'$, as required. ■

The Fundamental Theorem tells us that there is a projective transformation which maps any given quadrilateral onto any other given quadrilateral. So we have the following corollary.

Corollary All quadrilaterals are projective-congruent.

By *projective-congruent* we mean that there is a projective transformation that maps any quadrilateral onto any other quadrilateral.

If we actually need to find the projective transformation which maps one given quadrilateral onto another given quadrilateral, we simply follow the strategy used to prove part (a) of the Fundamental Theorem.

Strategy To determine the projective transformation t which maps the vertices of the quadrilateral $ABCD$ to the corresponding vertices of the quadrilateral $A'B'C'D'$:

1. find the projective transformation t_1 which maps the triangle of reference and unit Point to the Points A, B, C, D, respectively;
2. find the projective transformation t_2 which maps the triangle of reference and unit Point to the Points A', B', C', D', respectively;
3. calculate $t = t_2 \circ t_1^{-1}$.

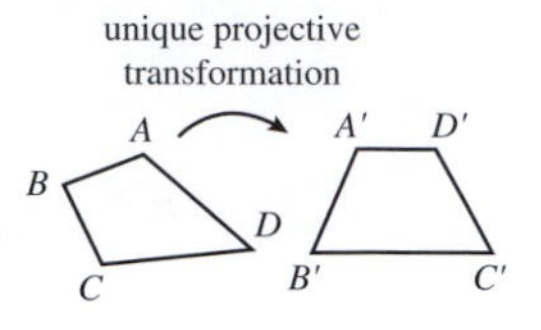

Example 6 Find the projective transformation t which maps the Points $[1, -1, 2]$, $[1, -2, 1]$, $[5, -1, 2]$, $[1, 0, 1]$ to the Points $[-1, 3, -2]$, $[-3, 7, -5]$, $[2, -5, 4]$, $[-3, 8, -5]$, respectively.

Solution We follow the steps in the above strategy.

(a) Any matrix associated with the projective transformation t_1 which maps the Points $[1, 0, 0]$, $[0, 1, 0]$, $[0, 0, 1]$, $[1, 1, 1]$ to the Points $[1, -1, 2]$, $[1, -2, 1]$, $[5, -1, 2]$, $[1, 0, 1]$, respectively, must be a multiple of the matrix

$$\begin{pmatrix} u & v & 5w \\ -u & -2v & -w \\ 2u & v & 2w \end{pmatrix}, \quad \text{where} \begin{pmatrix} u & v & 5w \\ -u & -2v & -w \\ 2u & v & 2w \end{pmatrix}\begin{pmatrix} 1 \\ 1 \\ 1 \end{pmatrix} = \begin{pmatrix} 1 \\ 0 \\ 1 \end{pmatrix}.$$

Solving the equations

$$u + v + 5w = 1,$$
$$-u - 2v - w = 0,$$
$$2u + v + 2w = 1,$$

we obtain $u = \frac{1}{2}, v = -\frac{1}{3}, w = \frac{1}{6}$. So a suitable choice of matrix for t_1 is

$$\begin{pmatrix} \frac{1}{2} & -\frac{1}{3} & \frac{5}{6} \\ -\frac{1}{2} & \frac{2}{3} & -\frac{1}{6} \\ 1 & -\frac{1}{3} & \frac{1}{3} \end{pmatrix}, \quad \text{or more simply } \mathbf{A}_1 = \begin{pmatrix} 3 & -2 & 5 \\ -3 & 4 & -1 \\ 6 & -2 & 2 \end{pmatrix}.$$

It is simpler to multiply the first matrix by 6 to obtain integer entries. This does not alter the projective transformation t_1 with which the matrix is associated.

(b) Any matrix associated with the projective transformation t_2 which maps the Points $[1, 0, 0]$, $[0, 1, 0]$, $[0, 0, 1]$, $[1, 1, 1]$ to the Points $[-1, 3, -2]$, $[-3, 7, -5]$, $[2, -5, 4]$, $[-3, 8, -5]$, respectively, must be a multiple of the matrix

$$\begin{pmatrix} -u & -3v & 2w \\ 3u & 7v & -5w \\ -2u & -5v & 4w \end{pmatrix}, \text{ where} \begin{pmatrix} -u & -3v & 2w \\ 3u & 7v & -5w \\ -2u & -5v & 4w \end{pmatrix}\begin{pmatrix} 1 \\ 1 \\ 1 \end{pmatrix} = \begin{pmatrix} -3 \\ 8 \\ -5 \end{pmatrix}.$$

Solving the equations

$$-u - 3v + 2w = -3,$$
$$3u + 7v - 5w = 8,$$
$$-2u - 5v + 4w = -5,$$

we obtain $u = 2, v = 1, w = 1$. So a suitable choice of matrix for t_2 is

$$\mathbf{A}_2 = \begin{pmatrix} -2 & -3 & 2 \\ 6 & 7 & -5 \\ -4 & -5 & 4 \end{pmatrix}.$$

(c) A matrix associated with the inverse, t_1^{-1}, of t_1 is $\mathbf{A}_1^{-1}$, which we can calculate to be

$$\mathbf{A}_1{}^{-1} = \begin{pmatrix} -\frac{1}{12} & \frac{1}{12} & \frac{1}{4} \\ 0 & \frac{1}{3} & \frac{1}{6} \\ \frac{1}{4} & \frac{1}{12} & -\frac{1}{12} \end{pmatrix};$$

then a simpler matrix associated with t_1^{-1} is

$$\mathbf{B} = \begin{pmatrix} 1 & -1 & -3 \\ 0 & -4 & -2 \\ -3 & -1 & 1 \end{pmatrix}.$$

The required projective transformation is therefore $t : [\mathbf{x}] \mapsto [\mathbf{A}x]$, where

$$\mathbf{A} = \mathbf{A}_2\mathbf{B} = \begin{pmatrix} -2 & -3 & 2 \\ 6 & 7 & -5 \\ -4 & -5 & 4 \end{pmatrix} \begin{pmatrix} 1 & -1 & -3 \\ 0 & -4 & -2 \\ -3 & -1 & 1 \end{pmatrix}$$
$$= \begin{pmatrix} -8 & 12 & 14 \\ 21 & -29 & -37 \\ -16 & 20 & 26 \end{pmatrix}.$$

□

Problem 8 Find the projective transformation t that maps the Points $[-1, 0, 0]$, $[-3, 2, 0]$, $[2, 0, 4]$, $[1, 2, -5]$ to the Points $[2, 1, 0]$, $[1, 0, -1]$, $[0, 3, -1]$, $[3, -1, 2]$, respectively.

Problem 9 Find the projective transformation t that maps the Points $[1, 0, -3]$, $[1, 1, -2]$, $[3, 3, -5]$, $[6, 4, -13]$ to the Points $[3, -5, 3]$, $[\frac{1}{2}, -1, 0]$, $[3, -5, 6]$, $[8, -13, 12]$, respectively.

3.3.4 Geometrical Interpretation of Projective Transformations

You may omit this subsection at a first reading, as it is quite hard going.

In this subsection we discuss the relationship between projective transformations and the perspectivities introduced in Section 3.1.

Starting from the geometric definition of perspective in Subsection 3.1.1, we will define the term *perspective transformation*, show how a perspective transformation may be interpreted as a projective transformation, and finally prove that any projective transformation can be expressed as the composite of at most three perspective transformations.

Recall that a *projective transformation* of $\mathbb{RP}^2$ is a map $[\mathbf{x}] \to [\mathbf{Ax}]$, where $\mathbf{A}$ is an invertible 3×3 matrix.

So, let π and π' be two embedding planes in $\mathbb{R}^3$ that do not pass through the origin O in $\mathbb{R}^3$, and let C ($\neq O$) be another point in $\mathbb{R}^3$ such that OC is not parallel to either π or π'. Let C have position vector $\mathbf{c}$ (based at O). Also, let σ denote an arbitrary perspectivity from the point C that maps the plane π to the plane π'.

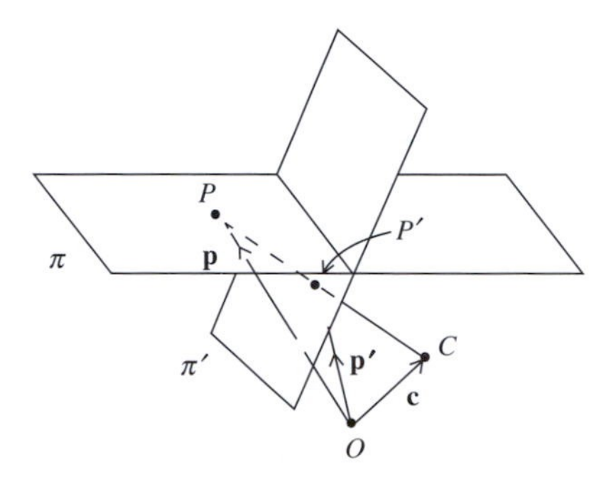

Note that O, C, P and P' are coplanar.

Now, the perspectivity σ will map any point P (with position vector $\mathbf{p}$) in π onto some point P' (with position vector $\mathbf{p}'$) in π', so long as the vector $\mathbf{p} - \mathbf{c}$ is not parallel to the plane π'. We then define the perspective transformation associated with σ to be the mapping of $\mathbb{R}^3$ to itself that maps the line $[\mathbf{p} - \mathbf{c}]$ onto the line $[\mathbf{p}' - \mathbf{c}]$. But since C, P and P' are collinear, it follows that the vectors $\mathbf{p} - \mathbf{c}$ and $\mathbf{p}' - \mathbf{c}$ must be parallel (equivalently, that $[\mathbf{p} - \mathbf{c}] = [\mathbf{p}' - \mathbf{c}]$); hence there is some real number t such that

Recall that a line through the origin in $\mathbb{R}^2$ is a 1-dimensional vector space.

The value of t will depend on the particular point P under discussion.

$$\mathbf{p}' - \mathbf{c} = t(\mathbf{p} - \mathbf{c}).$$

We can then rewrite this formula in the form

$$\mathbf{p}' = t(\mathbf{p} - \mathbf{c}) + \mathbf{c},$$

or

$$\mathbf{p}' = t\mathbf{p} + (1 - t)\mathbf{c},$$

so that

$$[\mathbf{p}] \mapsto [t\mathbf{p} + (1 - t)\mathbf{c}].$$

In this way, the perspectivity σ gives a one-one mapping from π onto π', except that it is not defined on the line ℓ in π where π cuts the plane through C parallel to π'; also, there are no points of π that map onto points of π' on the line ℓ' where π' cuts the plane through C parallel to π. We use the fact that the ideal Points for a plane correspond to the directions of lines in the plane, rather than actual points in the plane, to extend our definition of the map.

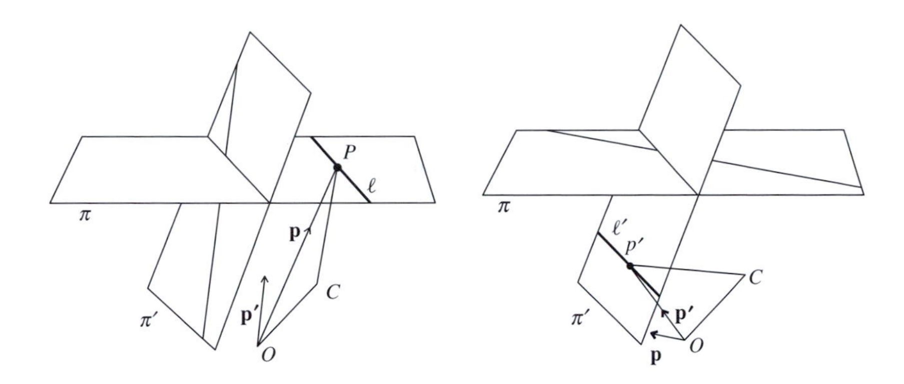

So, first, let P be a point of π that lies on the line ℓ; then the points O, C, P are not collinear, since OC is not parallel to the plane π'. Denote the position vector $\overrightarrow{OP}$ by $\mathbf{p}$. Let the plane through O, C and P meet the plane π' in a line, and let $\mathbf{p}'$ be a position vector based at O that is parallel to this line. Then we specify that our (extended) map σ maps the line $[\mathbf{p}]$ through O onto the line $[\mathbf{p}']$ through O.

Similarly, let P' be a point of π' that lies on the line ℓ'; then the points O, C, P' are not collinear, since OC is not parallel to the plane π. Denote the position vector $\overrightarrow{OP'}$ by $\mathbf{p}'$. Let the plane through O, C and P' meet the plane π in a line, and let $\mathbf{p}$ be a position vector based at O that is parallel to this line. Then we specify that our (further extended) map σ maps the line $[\mathbf{p}]$ through O onto the line $[\mathbf{p}']$ through O.

Finally, we specify that the extended map σ maps the line through O that is parallel to the line of intersection of π and π' onto itself.

In this way, we have constructed a transformation that is a one-one mapping of $\pi \cup \{\text{the ideal Points for } \pi\}$ onto $\pi' \cup \{\text{the ideal Points for } \pi'\}$ associated in a natural way with the given perspectivity σ, and we call it the associated **perspective transformation**. This maps the family of Euclidean lines through O onto itself, in other words $\mathbb{RP}^2$ onto itself.

We now explain why we can think of this perspective transformation as a projective transformation.

First, we consider a perspectivity in $\mathbb{R}^2$ as this will prove useful later in our discussion. So, let ℓ and ℓ' be two lines in $\mathbb{R}^2$ that do not pass through the origin O in $\mathbb{R}^2$, let L be the common point of the two lines, and let $C(\neq O)$ be another point in $\mathbb{R}^2$ such that OC is not parallel to either ℓ or ℓ'. Consider any perspectivity σ, with centre some point C, say, that maps ℓ 'onto' ℓ'.

In our discussion we shall omit discussion of 'the exceptional points', for simplicity.

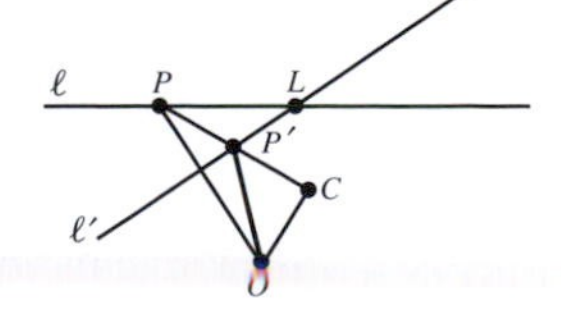

We are interested in the map τ that sends lines through O to lines through O that is obtained from σ as follows. Let σ map the point P on ℓ onto the point P' on ℓ'. Then we define $\tau(OP) = OP'$.

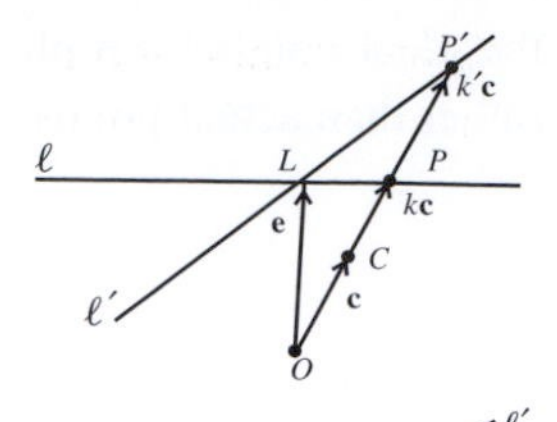

It is clear that $\tau(OL) = OL$, because L is fixed by σ.

It is also clear that $\tau(OC) = OC$, because if the line OC meets ℓ at P and the line ℓ' at P' then O, C, P and P' are collinear and so $\tau(OC) = OC$.

We therefore choose to take as basis vectors in $\mathbb{R}^2$ the vector $\mathbf{e} = \overrightarrow{OL}$ and the vector $\mathbf{c} = \overrightarrow{OC}$.

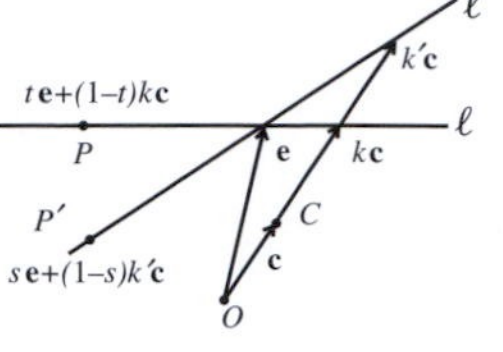

We now find the effect of the map τ on a line through the origin O. We shall suppose that the line OC meets the line ℓ at the point with position vector $k\mathbf{c}$ and the line ℓ' at the point with position vector $k'\mathbf{c}$. Any point P on the line ℓ then has position vector $t\mathbf{e} + (1-t)k\mathbf{c}$, for some real number t, and any point P' on the line ℓ' has position vector $s\mathbf{e} + (1-s)k'\mathbf{c}$, for some real number s.

Next, the line OP consists of the points with coordinates $u(\mathbf{e} + m\mathbf{c}) = u\mathbf{e} + mu\mathbf{c}$ for a fixed value of m and varying values of u. This gives two (equivalent) expressions for the position of the point P, namely $t\mathbf{e}+(1-t)k\mathbf{c}$ and $u\mathbf{e}+mu\mathbf{c}$. It follows that we must have

$$u = t \quad \text{and} \quad mu = (1-t)k.$$

Dividing the second equation by the first, we get

$$m = \frac{1-t}{t}k$$

so that

$$tm = k - kt,$$

which yields the formula

$$t = \frac{k}{m+k}.$$

Similarly, the line OP' consisting of points $u'(\mathbf{e} + m'\mathbf{c}) = u'\mathbf{e} + m'u'\mathbf{c}$, for a fixed value of m' and varying values of u', meets the line ℓ' at the point P' where

$$s = \frac{k'}{m'+k'}.$$

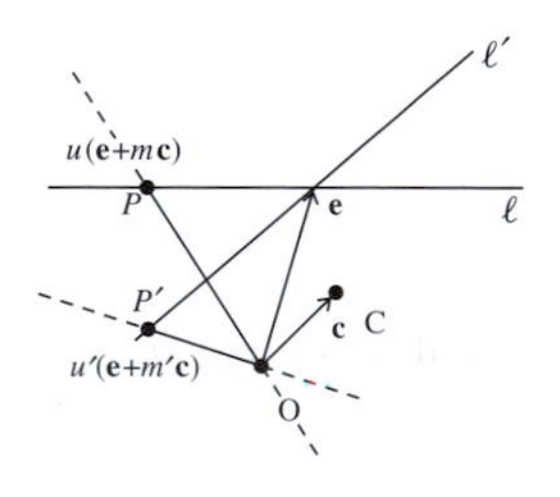

Now, we have $\tau(OP) = OP'$ if and only if the points C, P and P' are collinear; that is, if and only if there is a real number r such that

$$r(t\mathbf{e} + (1-t)k\mathbf{c} - \mathbf{c}) = s\mathbf{e} + (1-s)k'\mathbf{c} - \mathbf{c}.$$

This is the case if and only if

$$rt = s \quad \text{and} \quad r((1-t)k - 1) = (1-s)k' - 1.$$

We can eliminate r by dividing the second equation by the first equation, so obtaining

$$\frac{(1-t)k-1}{t} = \frac{(1-s)k'-1}{s}.$$

Then, if we substitute $t = \frac{k}{m+k}$ and $s = \frac{k'}{m'+k'}$ into this equation, after some manipulation we obtain the remarkable result that

$$m' = m\frac{(k-1)k'}{(k'-1)k}.$$

It follows that, in terms of the basis elements **e** and **c**, the map τ of the family of lines through O to itself given by

$$\left[\begin{pmatrix}1\\m\end{pmatrix}\right] \mapsto \left[\begin{pmatrix}1\\m'\end{pmatrix}\right]$$

can be represented by the matrix

$$\begin{pmatrix}1 & 0\\ 0 & \frac{(k-1)k'}{(k'-1)k}\end{pmatrix}.$$

For convenience, we write this matrix in the form

$$\begin{pmatrix}1 & 0\\ 0 & r\end{pmatrix},$$

where r is a fixed number that depends only on the geometry of the two lines ℓ and ℓ' and on our choice of the point C. Furthermore, $r \neq 0$ since $k \neq 1$ (for $k = 1$ implies that C lies on π) and $k' \neq 0$ (for $k' = 0$ implies that O lies on π').

Now we consider the situation in $\mathbb{R}^3$, in relation to $\mathbb{RP}^2$. Consider an arbitrary perspectivity σ, with centre some point C, say, that maps a plane π 'onto' a plane π', with neither O nor C lying on π or π'. As before, we shall suppose that the line OC meets the plane π at a point with position vector $k\mathbf{c}$ and the plane π' at a point with position vector $k'\mathbf{c}$.

Once again, we are interested in the map τ that sends lines (in $\mathbb{R}^3$) through O to lines through O that is obtained from σ as follows. Let σ map a point P in π to a point P' in π'. Then we define $\tau(OP) = OP'$.

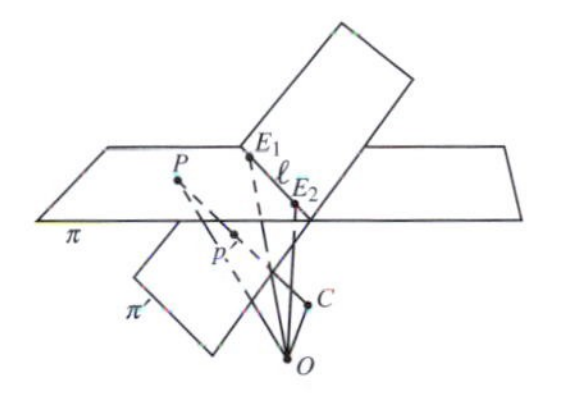

Let E_1 and E_2 be any two points on the common line ℓ of the two planes. Then, clearly, $\tau(OE_1) = OE_1$ and $\tau(OE_2) = OE_2$ because every point of ℓ is fixed by σ.

It is also clear that $\tau(OC) = OC$, because if the line OC meets the plane π at Q and the plane π' at Q', then O, C, Q and Q' are collinear, and so $\tau(OC) = OC$.

We therefore now take as basis vectors in $\mathbb{R}^3$ the vectors $\mathbf{e}_1 = \overrightarrow{OE_1}$, $\mathbf{e}_2 = \overrightarrow{OE_2}$, and $\mathbf{c} = \overrightarrow{OC}$.

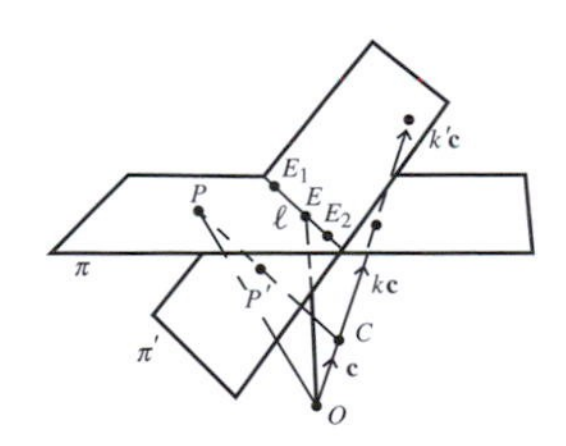

We now find the effect of τ on a line in $\mathbb{R}^3$ through the origin O. We can simplify our task by observing that the lines OC, OP and OP' all lie in a plane, π'' say; let this plane meet the line common to the given planes π and π' at a point E. We then define $\mathbf{e} = \overrightarrow{OE}$.

We can now apply our earlier discussion of the planar case to the restriction of the mapping τ to the plane π''. If we denote the lines OP and OP' by the parametrizations

$$OP = u(\mathbf{e} + m\mathbf{c}) = u\mathbf{e} + um\mathbf{c}, \quad \text{where } m \text{ is fixed and } u \text{ varies,}$$

$$OP' = u'(\mathbf{e} + m'\mathbf{c}) = u'\mathbf{e} + u'm'\mathbf{c}, \quad \text{where } m' \text{ is fixed and } u' \text{ varies,}$$

it follows from our earlier discussion that m' can be expressed in the form rm, where r is a fixed number that depends only on the geometry of the two planes π and π' and on our choice of the point C. Hence we can rewrite the parametrization of OP' as

$$OP' = u'(\mathbf{e} + rm\mathbf{c}) = u'\mathbf{e} + u'rm\mathbf{c}, \quad \text{where } r,\ m \text{ are fixed and } u' \text{ varies.}$$

Next, we can express the position vector $\mathbf{e}$ of the point E in the form

$$\mathbf{e} = t\mathbf{e}_1 + (1-t)\mathbf{e}_2, \quad \text{for some number } t.$$

Then we have that the mapping τ maps the line

$$OP = u(t\mathbf{e}_1 + (1-t)\mathbf{e}_2 + m\mathbf{c}) = ut\mathbf{e}_1 + u(1-t)\mathbf{e}_2 + um\mathbf{c}$$

onto the line

$$OP' = u'(t\mathbf{e}_1 + (1-t)\mathbf{e}_2 + rm\mathbf{c}) = u't\mathbf{e}_1 + u'(1-t)\mathbf{e}_2 + u'rm\mathbf{c}.$$

It follows analogously to the two-dimensional situation that, in terms of the basis elements $\mathbf{e}_1$, $\mathbf{e}_2$, $\mathbf{c}$, the mapping τ of the family of lines in $\mathbb{R}^3$ to itself can be represented by the matrix

$$\begin{pmatrix} 1 & 0 & 0 \\ 0 & 1 & 0 \\ 0 & 0 & r \end{pmatrix},$$

Notice that this maps the line represented by the points $[t, 1-t, 0]$ to itself pointwise, as it should.

where the (non-zero) constant r depends only on the position of the planes π and π'. Since this is an invertible 3×3 matrix, this is our required description of the perspective transformation as a projective transformation.

We now go the other way round, and obtain a projective transformation as a sequence of three perspective transformations. We have a lot of freedom, because we can choose the centres of perspectivities and the planes.

Suppose we are given a projective transformation τ and two embedding planes π and π'. Let $[\mathbf{a}]$, $[\mathbf{b}]$, $[\mathbf{c}]$, $[\mathbf{d}]$ be any four non-collinear Points, and $[\mathbf{a}']$, $[\mathbf{b}']$, $[\mathbf{c}']$, $[\mathbf{d}']$ be any other four non-collinear Points; we can represent these Points by (Euclidean) points A, B, C, D in π and A', B', C', D' in π', respectively.

We will use in our discussion the existence and uniqueness parts of the Fundamental Theorem of Projective Geometry: namely, that there is one and only one projective transformation mapping four non-collinear Points to any other four non-collinear Points. This means that if we can find a composite of three perspective transformations that maps $[\mathbf{a}]$, $[\mathbf{b}]$, $[\mathbf{c}]$, $[\mathbf{d}]$ to $[\mathbf{a}']$, $[\mathbf{b}']$, $[\mathbf{c}']$, $[\mathbf{d}']$, respectively, then this composite must be a projective transformation (by the existence part) and it must equal the given projective transformation τ (by the uniqueness part).

Subsection 3.3.3, Theorem 3

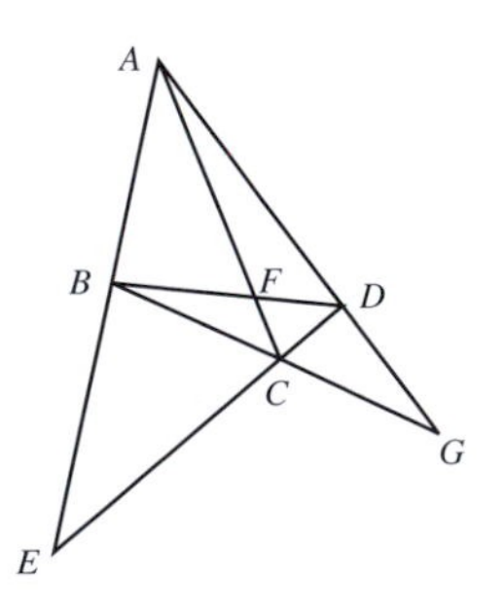

We now exhibit a sequence of three perspectivities the composite of which maps A, B, C, D in π to A', B', C', D' in π'. In our discussion it is convenient to let AB and CD meet at E, AC and BD meet at F, and AD and BC meet at G, with analogous definitions of E', F' and G'. (You may find the figures below helpful to follow through the argument; though, for simplicity, we have omitted the initial plane π and the points F', F'' and F'''.)

1. The first perspectivity is from the plane π to a plane π'' that passes through A'. The centre of this perspectivity is an arbitrary point P_1 on the line AA' (if $A = A'$, then P_1 can be chosen to lie anywhere not on π or π'). This perspectivity maps A to A' and B, C, D, E, F, G to B'', C'', D'', E'', F'', G'', say, respectively. We can assume, by suitably varying π'', that $B'B''$ and $E'E''$ are not parallel.

$A \mapsto A'$

$B \mapsto B''$, $C \mapsto C''$,

$D \mapsto D''$, $E \mapsto E''$,

$F \mapsto F''$, $G \mapsto G''$

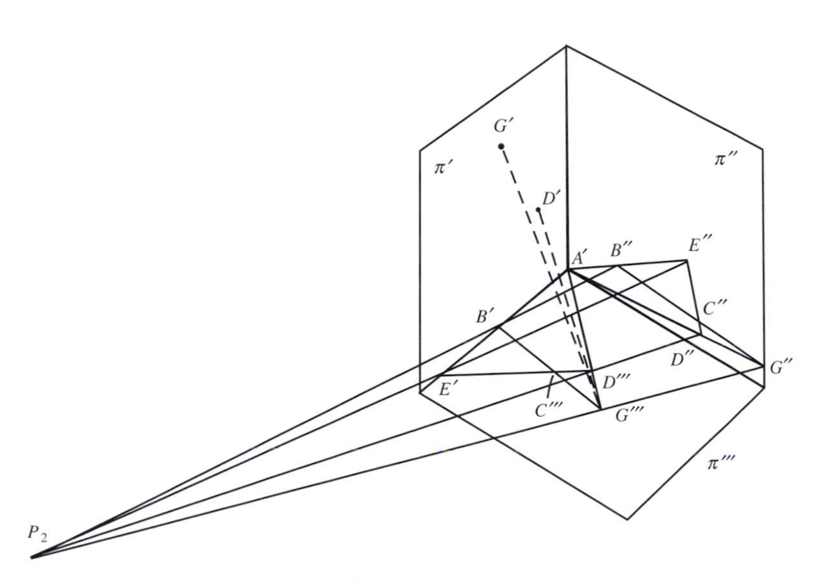

2. Now, by the definition of E' the points A', B', E' lie on a line through A', and since the points A, B, E are collinear the points A', B'', E'' lie on a line through A'; so these five points lie in a plane, and because the lines $B'B''$ and $E'E''$ are not parallel they meet in a point, P_2 say. We then pick a plane π''' through the line $A'B'E'$ and map the plane π'' onto π''' by the perspectivity with centre P_2. This sends A', B'', E'' to A', B', E', respectively, and the points C'', D'', F'', G'' to, say, C''', D''', F''', G''', respectively. As before, we can assume that $D'D'''$ and $G'G'''$ are not parallel.

A' fixed;
$B'' \mapsto B'$, $E'' \mapsto E'$

$C'' \mapsto C'''$, $D'' \mapsto D'''$,

$F'' \mapsto F'''$, $G'' \mapsto G'''$

3. Now, since A, D, G are collinear, the points A', D', G' lie on a line through A' and the points A', D''', G''' lie on a line through A', so these five points lie in a plane; and because the lines $D'D'''$ and $G'G'''$ are not parallel they meet in a point, P_3 say. We then map the plane π''' onto the plane π' by the perspectivity with centre P_3. This map sends A', B', D''', E''', G''' to A', B', D', E', G', respectively.

   This third perspectivity sends the line $B'G'''$ to the line $B'G'$ and the line $D'''E'$ to the line $D'E'$, so it maps the point C''' to the point C'.

Refer here to the figure below.

A', B', E' fixed

$D''' \mapsto D'$, $G''' \mapsto G'$

Thus the composite of these three perspectivities maps the points A, B, C, D in π to A', B', C', D' in π', as required.

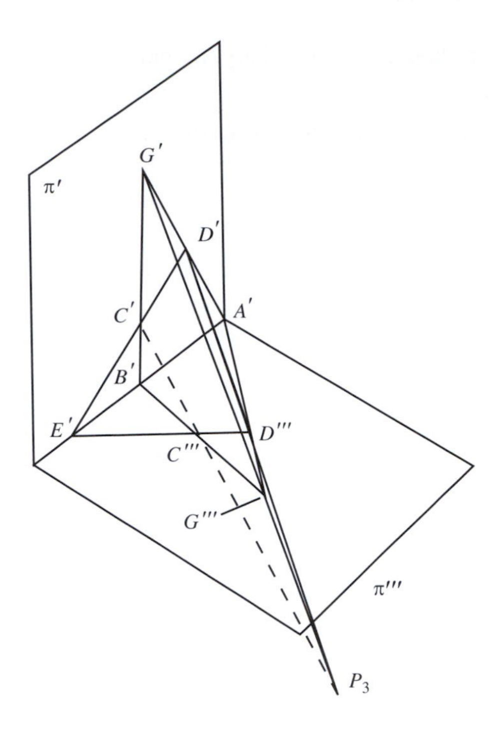

Theorem 4 Perspectivity Theorem
Every projective transformation can be expressed as the composite of three perspective transformations.

3.4 Using the Fundamental Theorem of Projective Geometry

In Section 3.2 we described how an embedding plane π can be used to represent projective space $\mathbb{RP}^2$. The Points of $\mathbb{RP}^2$ are represented by Euclidean points in π and the Lines of $\mathbb{RP}^2$ are represented by Euclidean lines in π.

In general, any Euclidean figure in an embedding plane corresponds to a projective figure in $\mathbb{RP}^2$, and visa versa. This correspondence enables us to compare Euclidean theorems about a figure in an embedding plane with projective theorems about the corresponding projective figure. Provided that the theorems are concerned exclusively with projective properties, such as collinearity and incidence, then a Euclidean theorem will hold if and only if the corresponding projective theorem holds.

The Euclidean figure may have some ideal Points attached to it.

3.4.1 Desargues' Theorem and Pappus' Theorem

The advantage of interpreting a Euclidean theorem as a projective theorem in this way is that we can often obtain a much simpler proof of the theorem than would be possible using Euclidean geometry directly. We illustrate this by using projective geometry to prove the theorem of Desargues.

We introduced Desargues' Theorem in Subsection 3.1.3.

Theorem 1 Desargues' Theorem
Let ΔABC and $\Delta A'B'C'$ be triangles in $\mathbb{R}^2$ such that the lines AA', BB' and CC' meet at a point U. Let BC and $B'C'$ meet at P, CA and $C'A'$ meet at Q, and AB and $A'B'$ meet at R. Then P, Q and R are collinear.

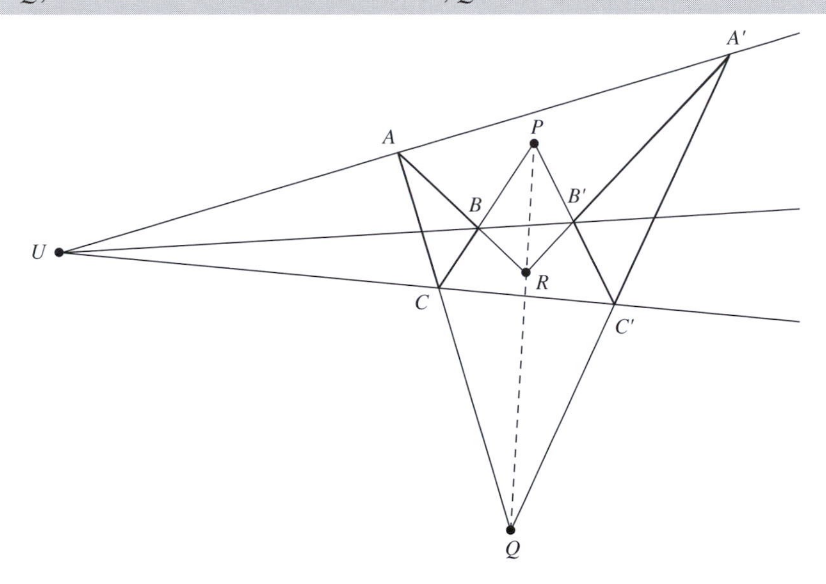

Proof Because this theorem is concerned exclusively with the projective properties of collinearity and incidence we can interpret it as a projective theorem in $\mathbb{RP}^2$. Moreover, by the Fundamental Theorem of Projective Geometry we know that any configuration of the theorem is projective-congruent to a configuration of the theorem in which $A = [1, 0, 0]$, $B = [0, 1, 0]$, $C = [0, 0, 1]$ and $U = [1, 1, 1]$. If we can prove the theorem in this special case then we can use the fact that projective-congruence preserves projective properties to deduce that the theorem holds in general.

To prove the special case we use the algebraic techniques described in Section 3.2. First observe that the Line AU passes through the Points $[1, 0, 0]$ and $[1, 1, 1]$, and therefore has equation $y = z$. Since A' is a Point on AU, it must have homogeneous coordinates of the form $[a, b, b]$, for some real numbers a and b. Now, $b \neq 0$, since $A \neq A'$; so we may write the homogeneous coordinates of A' in the form $[p, 1, 1]$ (where $p = a/b$).

For $[a, 0, 0] = [1, 0, 0]$.

Similarly, the homogeneous coordinates of the Points B' and C' may be written in the form $[1, q, 1]$ and $[1, 1, r]$, respectively, for some real numbers q and r.

We omit the details of the calculations.

We now find the Point P where BC and $B'C'$ intersect. The Line BC has equation $x = 0$. Since the Line $B'C'$ passes through the Points $B' = [1, q, 1]$ and $C' = [1, 1, r]$, it must have equation

$$\begin{vmatrix} x & y & z \\ 1 & q & 1 \\ 1 & 1 & r \end{vmatrix} = 0,$$

which we may rewrite in the form

$$(qr-1)x-(r-1)y+(1-q)z=0.$$

It follows that at the Point P of intersection of the Lines BC and $B'C'$ we must have $x=0$ and $(r-1)y=(1-q)z$, so that P has homogeneous coordinates $[0,1-q,r-1]$.

Similarly, the Points Q and R have homogeneous coordinates $[1-p, 0, r-1]$ and $[1-p, q-1, 0]$, respectively.

We omit the details of the calculations.

Now, the Points P, Q and R are collinear if

$$\begin{vmatrix} 0 & 1-q & r-1 \\ 1-p & 0 & r-1 \\ 1-p & q-1 & 0 \end{vmatrix} = 0.$$

But

$$\begin{aligned}
&\begin{vmatrix} 0 & 1-q & r-1 \\ 1-p & 0 & r-1 \\ 1-p & q-1 & 0 \end{vmatrix} \\
&= -(1-q)\begin{vmatrix} 1-p & r-1 \\ 1-p & 0 \end{vmatrix} + (r-1)\begin{vmatrix} 1-p & 0 \\ 1-p & q-1 \end{vmatrix} \\
&= -(1-q)(1-p)(1-r)+(r-1)(1-p)(q-1) \\
&= 0.
\end{aligned}$$

It follows that P, Q and R are collinear, as asserted. The general result now holds, by projective-congruence. ■

When using the Fundamental Theorem to simplify proofs of results in projective geometry, we do not usually refer to projective-congruence. Instead, *so long as the properties involved are projective properties*, we content ourselves with an initial remark of the type: 'By the Fundamental Theorem of Projective Geometry, we may choose the four Points..., no three of which are collinear, to be the triangle of reference and the unit Point; that is, to have homogeneous coordinates [1, 0, 0], [0, 1, 0], [0, 0, 1] and [1, 1, 1], respectively'.

Problem 1 Let ΔABC be a triangle in $\mathbb{R}^2$, and let U be any point of $\mathbb{R}^2$ that is not collinear with any two of the points A, B and C. Let the lines AU, BU and CU meet the lines BC, CA and AB at the points A', B' and C', respectively. Next, let the lines BC and $B'C'$ meet at P, AC and $A'C'$ meet at Q, and AB and $A'B'$ meet at R. Prove that P, Q and R are collinear.

Hint: Let A, B, C be the vertices of the triangle of reference, and let U be the unit Point. Then determine the homogeneous coordinates of the Points A', B' and C'.

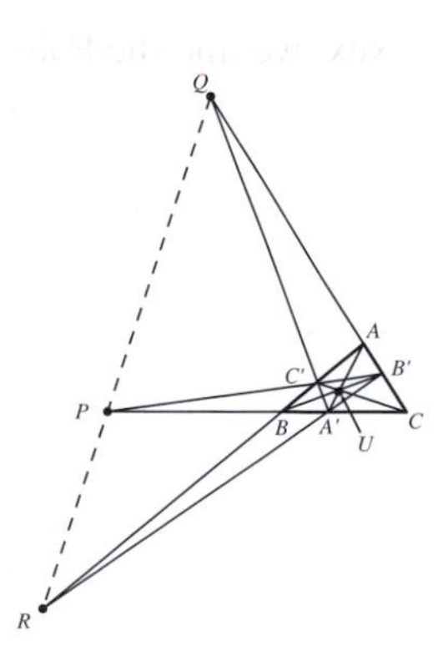

Next we use the Fundamental Theorem of Projective Geometry to prove Pappus' Theorem.

Theorem 2 Pappus' Theorem
Let A, B and C be three points on a line in $\mathbb{R}^2$, and let A', B' and C' be three points on another line. Let BC' and $B'C$ meet at P, CA' and $C'A$ meet at Q, and AB' and $A'B$ meet at R. Then P, Q, R are collinear.

This theorem is named after Pappus, a Greek mathematician who discovered it in the 3rd century AD.

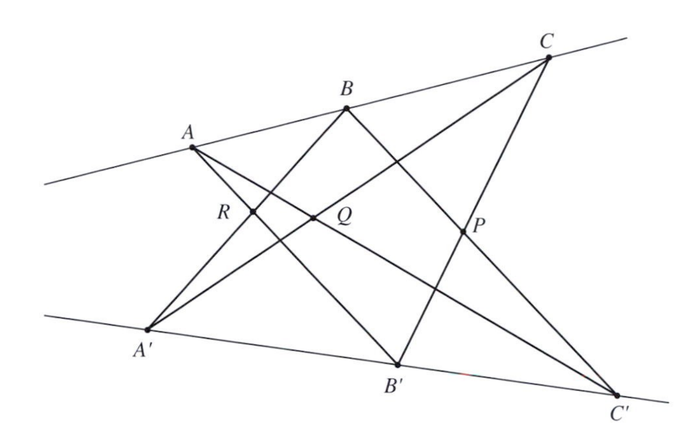

Proof We interpret the theorem as a projective theorem, so: by the Fundamental Theorem of Projective Geometry we may choose the four Points A, A', P, R, no three of which are collinear, to be the triangle of reference and the unit Point; that is, to have homogeneous coordinates [1, 0, 0], [0, 1, 0], [0, 0, 1] and [1, 1, 1], respectively.

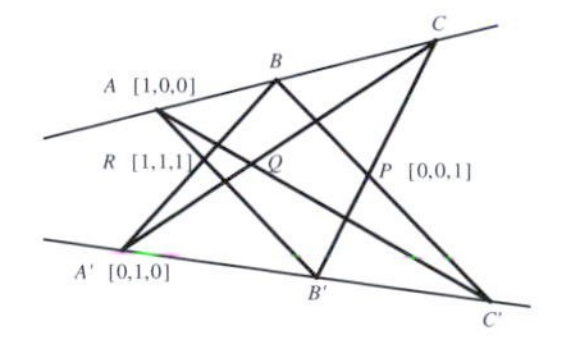

First observe that the Line AR passes through the Points [1, 0, 0] and [1, 1, 1], and must therefore have equation $y = z$. Since B' is a Point on AR, it must have homogeneous coordinates of the form $[a, b, b]$ for some real numbers a and b. Now, $b \neq 0$ since $A \neq B'$, so we may write the homogeneous coordinates of B' in the form $[r, 1, 1]$ (where $r = a/b$).

Similarly, the Point B lies on the Line $x = z$ through the Points $A' = [0, 1, 0]$ and $R = [1, 1, 1]$, so it must have homogeneous coordinates of the form $[1, s, 1]$.

Next we find the Point C where the Line AB intersects the Line $B'P$. Since the Line AB passes through the Points $A = [1, 0, 0]$ and $B = [1, s, 1]$, it must have equation $y = sz$. Also since the Line $B'P$ passes through the Points $B' = [r, 1, 1]$ and $P = [0, 0, 1]$ it must have equation $x = ry$. At the Point C where AB meets $B'P$ we have $y = sz$ and $x = ry$, so $C = [rs, s, 1]$.

Similarly, C' is the point where the Line BP intersects the Line $A'B'$. Since $B = [1, s, 1]$ and $P = [0, 0, 1]$, BP has equation $y = sx$; and, since $A' = [0, 1, 0]$ and $B' = [r, 1, 1]$, $A'B'$ has equation $x = rz$. It follows that $C' = [r, rs, 1]$.

Finally we find the point Q where AC' intersects $A'C$. Since the Line AC' passes through the Points $A = [1, 0, 0]$ and $C' = [r, rs, 1]$ it must have equation $y = rsz$. Also the Line $A'C$ passes through the Points $A' = [0, 1, 0]$ and $C = [rs, s, 1]$ so it must have equation $x = rsz$. At the Point Q where AC' intersects A'C we have $y = rsz$ and $x = rsz$, so $Q = [rs, rs, 1]$.

To complete the proof we simply observe that the Points $R = [1, 1, 1]$, $Q = [rs, rs, 1]$ and $P = [0, 0, 1]$ all lie on the Line $x = y$. It follows that P, Q and R are collinear. ■

A fortunate choice of Points for the triangle of reference and unit Point meant that we did not have to use the determinant criterion for collinearity at the final stage of the argument.

Although we can sometimes simplify the proof of a Euclidean theorem by using projective geometry, there is another more subtle reason for interpreting a Euclidean theorem as a projective theorem. By doing so we can often avoid having to make special provision for exceptional cases, such as when two lines are parallel. In projective geometry, Lines which correspond to a pair of parallel lines in an embedding plane actually meet and are therefore no different to any other Lines.

As an example, consider the diagram in the margin. This illustrates the situation that occurs in Pappus' Theorem when the Point of intersection R of $A'B$ and AB' is an ideal Point for the embedding plane. The above proof of Pappus' Theorem is able to cope with this situation because it uses arguments from $\mathbb{RP}^2$! Our interpretation of the theorem on an embedding plane in this situation is that the Points P and Q must be collinear with the ideal Point R at which $A'B$ and AB' meet. That is, PQ must be parallel to both $A'B$ and AB'.

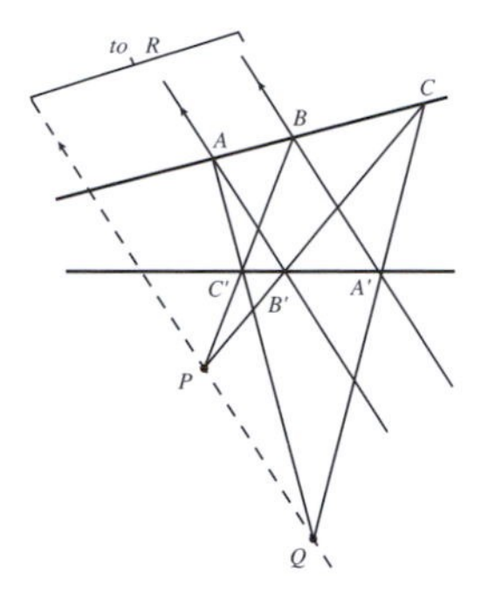

Problem 2 Give a Euclidean interpretation of Desargues' Theorem on an embedding plane π in the case where Q is an ideal Point for π.

3.4.2 Duality

Recall that two key projective properties that we have met so far have a certain symmetry between them.

Collinearity Property	Incidence Property
Any two distinct *Points* lie on a unique *Line.*	Any two distinct *Lines* meet in a unique *Point.*

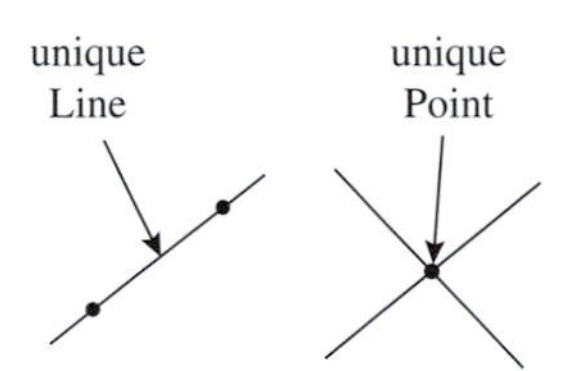

We can obtain one property from the other simply by interchanging the words 'Point' and 'Line', and making whatever other changes are needed to ensure that the sentence makes sense. We say that this interchanging process *dualizes* one statement into the other, and that each statement is the *dual* of the other.

For example, 'a family of Points on a Line' becomes 'a family of Lines through a Point' under dualization. Similarly, 'a triangle' or 'a family of three non-collinear Points and the three Lines joining them' dualizes to 'a family of three non-concurrent Lines and the three Points where they meet', which is again a triangle. Since a triangle is thus dual to a triangle, we say that triangles are *self-dual* figures.

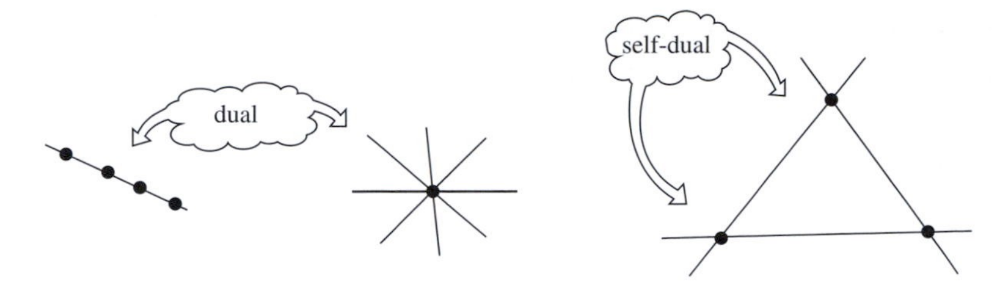

The dualization process is particularly interesting when applied to theorems. We shall illustrate this in the context of Pappus' Theorem. In order to do this, it is helpful to rephrase Pappus' Theorem using the term *hexagon.* As you would expect, a **hexagon** in $\mathbb{RP}^2$ consists of six Points joined by six Lines. The figure below illustrates (Euclidean) hexagons in an embedding plane; the corresponding hexagons in $\mathbb{RP}^2$ are the corresponding six Points and six Lines — that we model as six lines and six planes in $\mathbb{R}^3$.

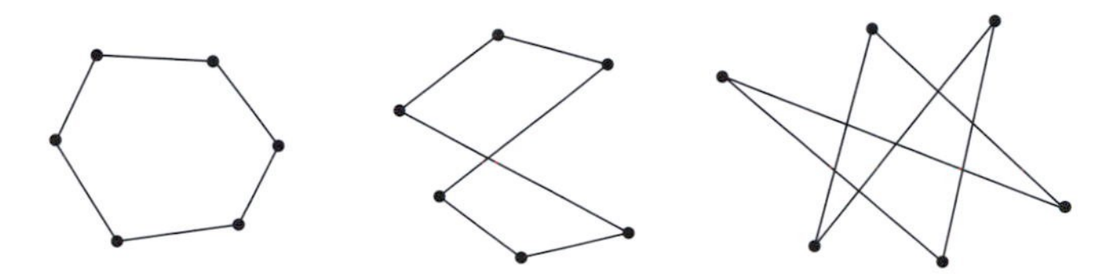

We can now rephrase Pappus' Theorem in the following form.

Theorem 3 Pappus' Theorem (rephrased)
Let the vertices A, B', C, A', B and C' of a hexagon lie alternately on two different Lines. Then the Points of intersection of opposite sides $B'C$ and BC', CA' and $C'A$, AB' and $A'B$, are collinear.

You should compare this formulation with that in Subsection 3.4.1.

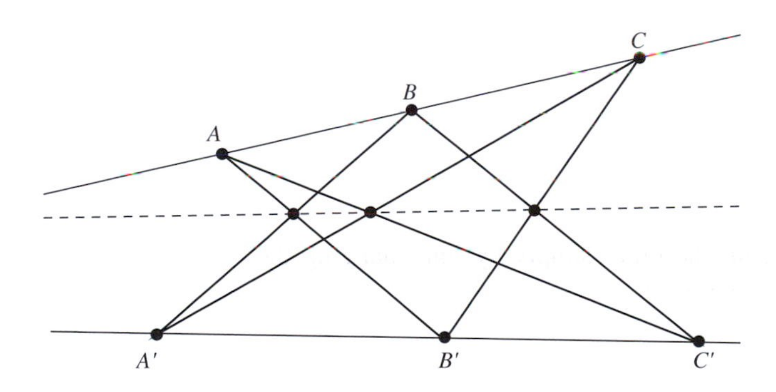

If we dualize this theorem, we obtain the following theorem.

Theorem 4 Brianchon's Theorem
Let the sides $AB', B'C, CA', A'B, BC', C'A$ of a hexagon pass alternately through two (different) Points P and Q in $\mathbb{RP}^2$. Then the Lines joining opposite vertices A and A′, B and B', C and C', are concurrent.

Charles J. Brianchon (1785–1864) was one of many distinguished French geometers who studied under Gaspard Monge (1746–1818).

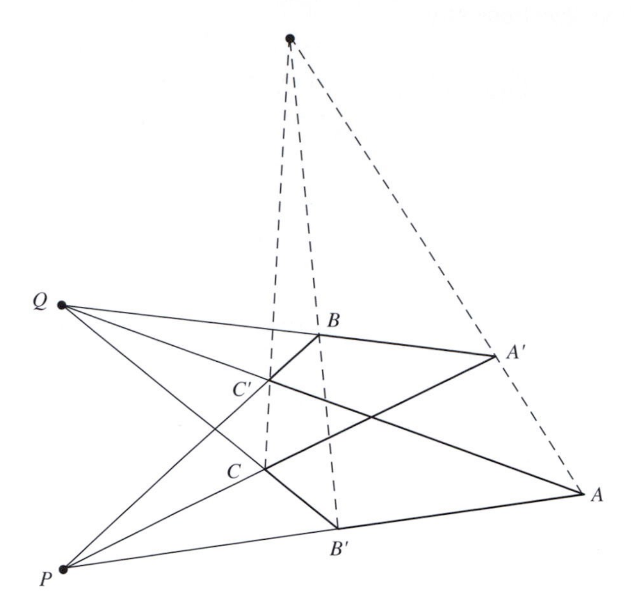

Problem 3 Prove Brianchon's Theorem.
Hint: Let P, C, Q, C' be the Points [1, 0, 0], [0, 1, 0], [0, 0, 1], [1, 1, 1], respectively.

It turns out that the dual of any true statement concerning Points, Lines and their projective properties remains true after dualization; that is, if we dualize any theorem in projective geometry, then the statement that we obtain is itself a theorem.

We do not prove this assertion, as it would take us beyond the scope of this book.

Problem 4 Earlier you saw that 'three Points $[a, b, c]$, $[d, e, f]$, $[g, h, k]$ are collinear if and only if $\begin{vmatrix} a & b & c \\ d & e & f \\ g & h & k \end{vmatrix} = 0$'. Write down the dual result of this statement.

Subsection 3.2.2, Strategy

We end this subsection by forming the dual of Desargues' Theorem, as follows.

Desargues' Theorem	Dual Theorem
Let two *triangles* be such that the *Lines* joining corresponding *vertices* meet at a *Point*. Then the *Points* of intersection of the corresponding *sides* of the two *triangles* are *collinear*.	Let two *triangles* be such that the *Points* through which corresponding *sides* pass are *collinear*. Then the *Lines* through the corresponding *vertices* of the two *triangles* are *concurrent*.

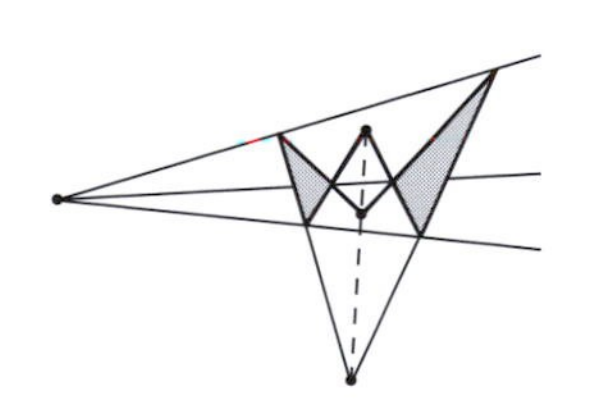

Note that the dual theorem is simply the converse result for Desargues' Theorem! Thus the Principle of Duality enables us to deduce that the converse of Desargues' Theorem holds.

3.5 Cross-Ratio

3.5.1 Another Projective Property

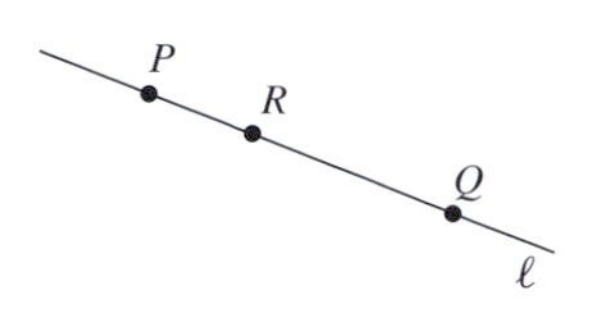

Earlier, in Subsection 2.2.1, we noted that *ratio of lengths along a line* is an affine property. Thus, in affine geometry, if we are given two points P and Q on a line ℓ, then we can locate the position of a third point R along ℓ by specifying the ratio $PR : RQ$. In particular, it is possible to talk about the point midway between P and Q.

In projective geometry it is meaningless to talk about the Point midway between two other Points. In one embedding plane π a Point R may appear to be midway between the Points P and Q, whereas in another embedding plane π' the ratio $PR : RQ$ may be very different.

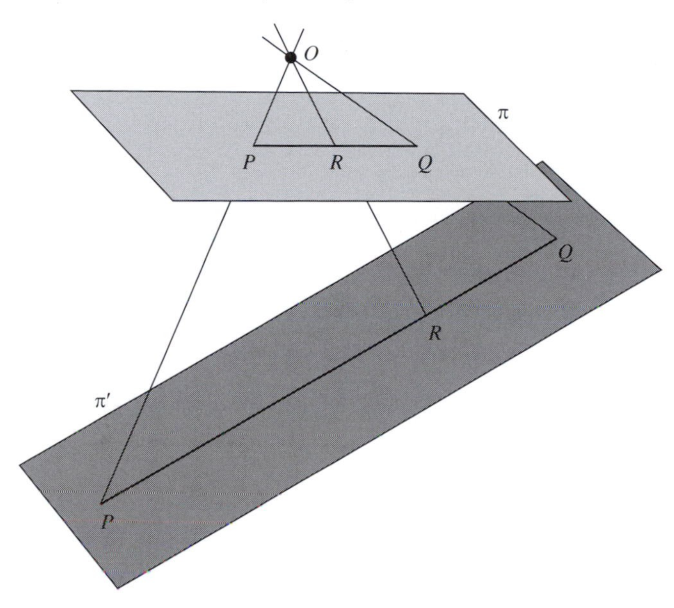

This ambiguity arises from the fact that perspectivities do not preserve the ratio of lengths along a line, so: *ratio of lengths along a line* is not a projective property.

In some embedding planes, such as the plane π' illustrated in the margin, the Point R does not even appear to lie between P and Q, so *betweenness* is not a projective property either!

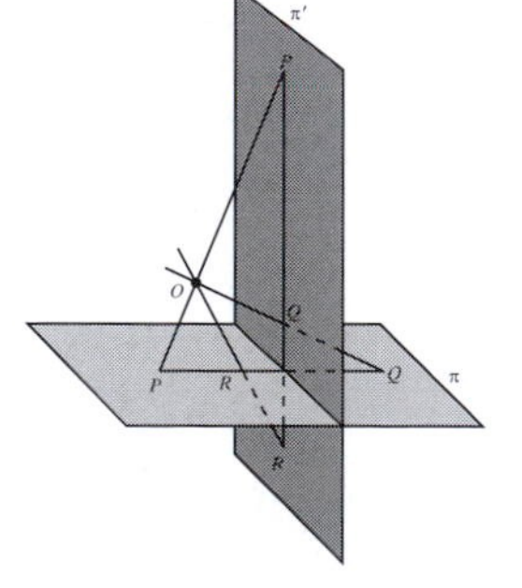

Fortunately, there is a quantity, known as *cross-ratio,* that is preserved under all projective transformations. To see how this is defined, consider four collinear Points $A = [\mathbf{a}]$, $B = [\mathbf{b}]$, $C = [\mathbf{c}]$, $D = [\mathbf{d}]$ in $\mathbb{RP}^2$. We can express the fact that A, B, C, D are collinear by writing $\mathbf{c}$ and $\mathbf{d}$ as linear combinations of $\mathbf{a}$ and $\mathbf{b}$. Thus we can write

$$\mathbf{c} = \alpha\mathbf{a} + \beta\mathbf{b} \quad \text{and} \quad \mathbf{d} = \gamma\mathbf{a} + \delta\mathbf{b},$$

for suitable real numbers $\alpha, \beta, \gamma, \delta$.

The cross-ratio is then defined to be the ratio of the ratios $\frac{\beta}{\alpha}$ and $\frac{\delta}{\gamma}$.

Definition Let A, B, C, D be four collinear Points in $\mathbb{RP}^2$ represented by position vectors **a**, **b**, **c**, **d**, and let

$$\mathbf{c} = \alpha\mathbf{a} + \beta\mathbf{b} \quad \text{and} \quad \mathbf{d} = \gamma\mathbf{a} + \delta\mathbf{b}.$$

Then the **cross-ratio** of A, B, C, D is

$$(ABCD) = \frac{\beta}{\alpha} \Big/ \frac{\delta}{\gamma}.$$

Equivalently, we can write $(ABCD) = \frac{\beta\gamma}{\alpha\delta}$.

Of course, before we can be sure that this definition makes sense, we must ensure that it does not depend on the particular choice of position vectors **a**, **b**, **c**, **d** that are used to represent the Points A, B, C, D. We shall check this shortly, but first we illustrate how cross-ratios are calculated.

Example 1 Let $A = [1, 2, 3]$, $B = [1, 1, 2]$, $C = [3, 5, 8]$, $D = [1, -1, 0]$ be Points of $\mathbb{RP}^2$. Calculate the cross-ratio $(ABCD)$.

Solution First, we have to find real numbers α and β such that the following vector equation holds:

$$(3, 5, 8) = \alpha(1, 2, 3) + \beta(1, 1, 2).$$

Note that we have not verified that A, B, C, D are collinear; but if they were not, the equations for $\alpha, \beta, \gamma, \delta$ could not be solved.

Comparing corresponding coordinates on both sides of this vector equation, we deduce that

$$3 = \alpha + \beta, \quad 5 = 2\alpha + \beta \quad \text{and} \quad 8 = 3\alpha + 2\beta.$$

Solving these equations gives $\alpha = 2$, $\beta = 1$.

Next, we find real numbers γ and δ such that the vector equation

$$(1, -1, 0) = \gamma(1, 2, 3) + \delta(1, 1, 2)$$

holds. Comparing corresponding coordinates on both sides of this vector equation, we deduce that

$$1 = \gamma + \delta, \quad -1 = 2\gamma + \delta \quad \text{and} \quad 0 = 3\gamma + 2\delta.$$

Solving these equations gives $\gamma = -2$, $\delta = 3$.

It follows from the definition of cross-ratio that

$$(ABCD) = \frac{\beta}{\alpha} \Big/ \frac{\delta}{\gamma} = \frac{1}{2} \Big/ \frac{3}{-2} = -\frac{1}{3}.$$ □

Problem 1 Calculate the cross-ratio $(ABCD)$ for each of the following sets of collinear Points in $\mathbb{RP}^2$.

(a) $A = [1, -1, -1]$, $B = [1, 3, -2]$, $C = [3, 5, -5]$, $D = [1, -5, 0]$
(b) $A = [1, 2, 3]$, $B = [2, 2, 4]$, $C = [-3, -5, -8]$, $D = [3, -3, 0]$

You may have noticed that the Points A, B, C, D in Problem 1 (b) are the same as those which appear in Example 1. The only difference is that different homogeneous coordinates are used to represent the Points in each case. As we mentioned after the definition of cross-ratio, the value of the cross-ratio *(ABCD)* does not depend on the homogeneous coordinates that are used to represent A, B, C, D, so it is not surprising that the cross-ratio turned out to have the value $-\frac{1}{3}$ in both cases.

Theorem 1 The cross-ratio *(ABCD)* is independent of the homogeneous coordinates that are used to represent the collinear Points A, B, C, D.

Proof Suppose that $A = [\mathbf{a}]$, $B = [\mathbf{b}]$, $C = [\mathbf{c}]$, $D = [\mathbf{d}]$, and let

$$\mathbf{c} = \alpha\mathbf{a} + \beta\mathbf{b} \quad \text{and} \quad \mathbf{d} = \gamma\mathbf{a} + \delta\mathbf{b}. \tag{1}$$

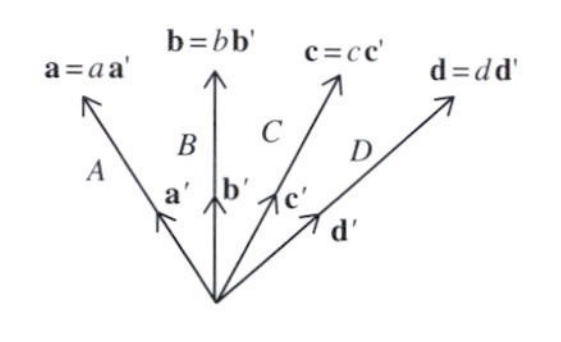

Now suppose that $A = [\mathbf{a}']$, $B = [\mathbf{b}']$, $C = [\mathbf{c}']$, $D = [\mathbf{d}']$. Then

$$\mathbf{a} = a\mathbf{a}', \quad \mathbf{b} = b\mathbf{b}', \quad \mathbf{c} = c\mathbf{c}', \quad \mathbf{d} = d\mathbf{d}',$$

where a, b, c, d are some non-zero real numbers.

By substituting these expressions into the equations (1), we obtain

$$c\mathbf{c}' = \alpha a\mathbf{a}' + \beta b\mathbf{b}' \quad \text{and} \quad d\mathbf{d}' = \gamma a\mathbf{a}' + \delta b\mathbf{b}',$$

which we can rewrite in the form

$$\mathbf{c}' = \alpha'\mathbf{a}' + \beta'\mathbf{b}' \quad \text{and} \quad \mathbf{d}' = \gamma'\mathbf{a}' + \delta'\mathbf{b}', \tag{2}$$

where $\alpha' = \alpha a/c, \beta' = \beta b/c, \gamma' = \gamma a/d, \delta' = \delta b/d$.

We can now check that equations (1) and (2) yield the same value for the cross-ratio:

$$\begin{aligned}\frac{\beta'}{\alpha'}\Big/\frac{\delta'}{\gamma'} &= \frac{\beta b/c}{\alpha a/c}\Big/\frac{\delta b/d}{\gamma a/d} \\ &= \frac{\beta b}{\alpha a}\Big/\frac{\delta b}{\gamma a} \\ &= \frac{\beta}{\alpha}\Big/\frac{\delta}{\gamma}.\end{aligned}$$

So, as expected, the cross-ratio is independent of the choice of homogeneous coordinates. ■

The next problem illustrates that although the value of the cross-ratio *(ABCD)* is independent of the choice of homogeneous coordinates that are used to represent A, B, C, D, the value of the cross-ratio does depend on the order in which the Points A, B, C, D appear.

Problem 2 Calculate the cross-ratios *(BACD)* and *(ACBD)* for the four Points used in Problem 1(a).

When answering Problem 2 you may have noticed that *(BACD)* is the reciprocal of the value which we obtained for *(ABCD)* in Problem 1(a). Also, *(ACBD)* is equal to 1 – *(ABCD)*. The next result shows that this is not simply chance!

Note the way in which the Points are changed from their original ordering in the various cross-ratios. We take the reciprocal when swapping the first or last pair of Points, and we subtract from 1 when swapping the inner or outer pair of Points.

Theorem 2 Let A, B, C, D be four distinct collinear Points in $\mathbb{RP}^2$, and let $(ABCD) = k$. Then

$$(BACD) = (ABDC) = 1/k,$$

$$(ACBD) = (DBCA) = 1 - k.$$

Proof Let **a**, **b**, **c**, **d** be any position vectors in $\mathbb{R}^3$ in the directions of the Points A, B, C, D, respectively, of $\mathbb{RP}^2$, and let $\alpha, \beta, \gamma, \delta$ be real numbers such that

$$\mathbf{c} = \alpha\mathbf{a} + \beta\mathbf{b} \quad \text{and} \quad \mathbf{d} = \gamma\mathbf{a} + \delta\mathbf{b}.$$

Then, by definition of cross-ratio, the cross-ratio $(ABCD)$ of the four Points A, B, C, D is the quantity

$$(ABCD) = \frac{\beta}{\alpha} \bigg/ \frac{\delta}{\gamma} = \frac{\beta\gamma}{\alpha\delta} = k, \text{ say.}$$

To determine (*BACD*), we interchange the roles of *A* and *B* in the evaluation of *ABCD* above; it follows that, since

$$\mathbf{c} = \beta\mathbf{b} + \alpha\mathbf{a} \quad \text{and} \quad \mathbf{d} = \delta\mathbf{d} + \gamma\mathbf{a},$$

the cross-ratio (*BACD*) is the quantity

$$(BACD) = \frac{\alpha}{\beta} \bigg/ \frac{\gamma}{\delta} = \frac{\alpha\delta}{\beta\gamma} = \frac{1}{k}.$$

To determine (*ABDC*), we interchange the roles of *C* and *D* in the evaluation of (*ABCD*) above; it follows that, since

$$\mathbf{d} = \gamma\mathbf{a} + \delta\,\mathbf{b} \quad \text{and} \quad \mathbf{c} = \alpha\mathbf{a} + \beta\mathbf{b},$$

the cross-ratio (*ABDC*) is the quantity

$$(ABDC) = \frac{\delta}{\gamma} \bigg/ \frac{\beta}{\alpha} = \frac{\alpha\delta}{\beta\gamma} = \frac{1}{k}.$$

This completes the first part of the proof.

To evaluate (*ACBD*), we use the equations

$$\mathbf{c} = \alpha\mathbf{a} + \beta\mathbf{b} \quad \text{and} \quad \mathbf{d} = \gamma\mathbf{a} + \delta\mathbf{b} \tag{3}$$

to express **b** and **d** in terms of **a** and **c**, as follows.

From the first equation in (3) we have

$$\begin{aligned}\mathbf{b} &= (\mathbf{c} - \alpha\mathbf{a})/\beta \\ &= (-\alpha/\beta)\mathbf{a} + (1/\beta)\mathbf{c}. \qquad (4)\end{aligned}$$

β cannot be zero, for if it were then we would have $\mathbf{c} = \alpha\mathbf{a}$; this cannot happen since A and C are distinct Points.

If we then substitute this expression for $\mathbf{b}$ into the second equation in (3), we obtain

$$\begin{aligned}\mathbf{d} &= \gamma\mathbf{a} + \delta((-\alpha/\beta)\mathbf{a} + (1/\beta)\mathbf{c}) \\ &= ((\beta\gamma - \alpha\delta)/\beta)\mathbf{a} + (\delta/\beta)\mathbf{c}. \qquad (5)\end{aligned}$$

It follows from the coefficients of $\mathbf{a}$ and $\mathbf{c}$ in equations (4) and (5) that

$$\begin{aligned}(ACBD) &= \frac{1/\beta}{-\alpha/\beta}\bigg/\frac{\delta/\beta}{(\beta\gamma - \alpha\delta)/\beta} \\ &= -\left(\frac{\beta\gamma - \alpha\delta}{\alpha\delta}\right) \\ &= 1 - \frac{\beta\gamma}{\alpha\delta} \\ &= 1 - k.\end{aligned}$$

Finally, we can use the previous parts of the proof to evaluate $(DBCA)$, as follows:

This avoids the algebra involved in expressing $\mathbf{c}$ and $\mathbf{a}$ in terms of $\mathbf{d}$ and $\mathbf{b}$.

$$\begin{aligned}(DBCA) &= 1/(BDCA) && \text{(swap first two Points)} \\ &= (BDAC) && \text{(swap last two Points)} \\ &= 1 - (BADC) && \text{(swap middle two Points)} \\ &= 1 - 1/(ABDC) && \text{(swap first two Points)} \\ &= 1 - (ABCD) && \text{(swap last two Points)} \\ &= 1 - k.\end{aligned}$$

■

Example 1

Earlier, we showed that the cross-ratio $(ABCD)$ of the four collinear Points $A = [1, 2, 3]$, $B = [1, 1, 2]$, $C = [3, 5, 8]$, $D = [1, -1, 0]$ in $\mathbb{RP}^2$ is $-\frac{1}{3}$. Theorem 2 enables us to deduce that

$$\begin{aligned}(BACD) &= -3, \qquad & (ABDC) &= -3, \\ (ACBD) &= \tfrac{4}{3}, \qquad & (DBCA) &= \tfrac{4}{3}.\end{aligned}$$

Problem 3 Let the Points $A = [1, -1, -1]$, $B = [1, 3, -2]$, $C = [3, 5, -5]$, $D = [1, -5, 0]$ be collinear Points of $\mathbb{RP}^2$. By applying Theorem 2 to the solution of Problem 1(a), determine the values of the cross-ratios $(ABDC)$, $(DBCA)$ and $(ACBD)$.

The next theorem confirms that cross-ratio is preserved by projective transformations.

Theorem 3 Let t be a projective transformation, and let A, B, C, D be any four collinear Points in $\mathbb{RP}^2$. If $A' = t(A), B' = t(B), C' = t(C), D' = t(D)$, then

$$(ABCD) = (A'B'C'D').$$

Proof Let t be the projective transformation $t : [\mathbf{x}] \mapsto [\mathbf{Ax}]$, where $\mathbf{A}$ is an invertible 3×3 matrix. If $A = [\mathbf{a}], B = [\mathbf{b}], C = [\mathbf{c}], D = [\mathbf{d}]$, and

$$\mathbf{a}' = \mathbf{Aa},\ \mathbf{b}' = \mathbf{Ab},\ \mathbf{c}' = \mathbf{Ac},\ \mathbf{d}' = \mathbf{Ad},$$

then $A' = [\mathbf{a}'],\ B' = [\mathbf{b}'],\ C' = [\mathbf{c}'],\ D' = [\mathbf{d}']$.

Since A, B, C, D are collinear, we can write

$$\mathbf{c} = \alpha\mathbf{a} + \beta\mathbf{b} \quad \text{and} \quad \mathbf{d} = \gamma\mathbf{a} + \delta\mathbf{b}, \tag{6}$$

so

$$(ABCD) = \frac{\beta}{\alpha}\Big/\frac{\delta}{\gamma}.$$

Multiplying each equation in (6) through by $\mathbf{A}$, we obtain

$$\mathbf{c}' = \alpha\mathbf{a}' + \beta\mathbf{b}' \quad \text{and} \quad \mathbf{d}' = \gamma\mathbf{a}' + \delta\mathbf{b}',$$

so that

$$(A'B'C'D') = \frac{\beta}{\alpha}\Big/\frac{\delta}{\gamma}.$$

It follows that

$$(A'B'C'D') = (ABCD). \qquad \blacksquare$$

We now use Theorem 3 to prove that if four distinct Points on a Line are in perspective with four distinct Points on another Line, then the cross-ratios of the four Points on each Line are equal.

Theorem 4 Let A, B, C, D be four distinct Points on a Line, and let A', B', C', D' be four distinct Points on another Line such that AA', BB', CC', DD' all meet at a Point U. Then

$$(ABCD) = (A'B'C'D').$$

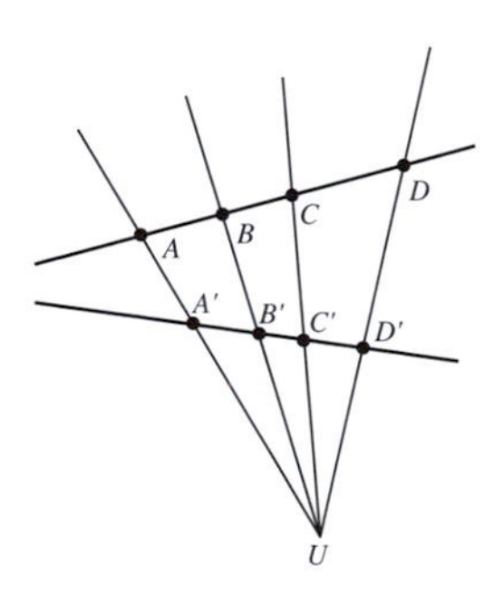

Proof By the Fundamental Theorem of Projective Geometry, there is a unique projective transformation t which maps B to B', C to C', B' to B, and C' to C. We shall show that $t(A) = A'$ and $t(D) = D'$, and hence by Theorem 3 it follows that $(ABCD) = (A'B'C'D')$.

First observe that the composite $t \circ t$ fixes the Points B, C, B' and C'. By the Fundamental Theorem of Projective Geometry, the only projective transformation which does this is the identity transformation, so $t \circ t = i$ and $t = t^{-1}$.

Next observe that t maps the Line BC onto the Line $B'C'$, and vice versa; so the Point T at which BC and $B'C'$ intersect must be fixed by t. Also, t maps the Lines BB' and CC' onto themselves, so their Point of intersection U must be fixed by t.

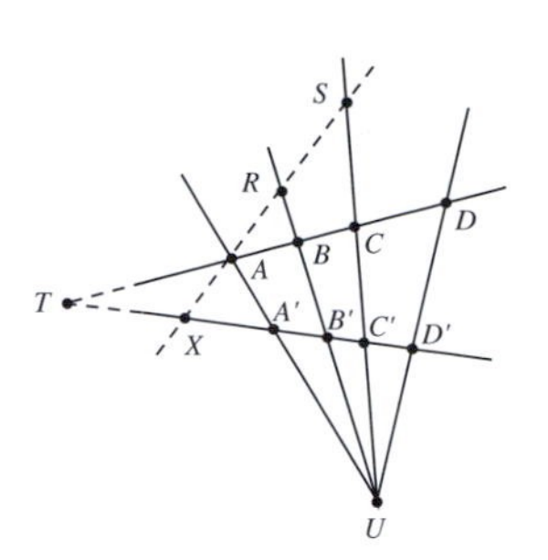

Now let X be the image of A under t. Then X lies on $B'C'$. We want to show that $X = A'$.

Suppose that $X \neq A'$; then AX cannot pass through U so it must intersect BB' at R and CC' at S, where R, S and U are distinct Points.

Since t is self-inverse, it maps X back to A and therefore maps AX onto itself. But this implies that t fixes the four Points R, S, T, U; so by the Fundamental Theorem of Projective Geometry t must be the identity transformation. This is a contradiction with the hypothesis that the Lines $ABCD$ and $A'B'C'D'$ are different. It follows that we must conclude that $X = A'$, that is, $t(A) = A'$. A similar argument shows that $t(D) = D'$.

Finally, it follows by Theorem 3 that $(ABCD) = (A'B'C'D')$, as required. ■

In affine geometry, if we are given two points A and B, then the ratio AC/CB uniquely determines a third point C on the line AB. We now explore the analogous result for projective geometry, namely that if we are given any three collinear Points A, B, C in $\mathbb{RP}^2$, then the value of the cross-ratio $(ABCD)$ uniquely determines a fourth Point D.

Theorem 5 Unique Fourth Point Theorem
Let A, B, C, X, Y be collinear Points in $\mathbb{RP}^2$ such that

$$(ABCX) = (ABCY).$$

Then $X = Y$.

Proof Let $A = [\mathbf{a}]$, $B = [\mathbf{b}]$, $C = [\mathbf{c}]$, $X = [\mathbf{x}]$, $Y = [\mathbf{y}]$. Since A, B, C, X, Y are collinear, it follows that there are real numbers α, β, γ, δ, λ, μ such that

$$\mathbf{c} = \alpha\mathbf{a} + \beta\mathbf{b}, \quad \mathbf{x} = \gamma\mathbf{a} + \delta\mathbf{b} \quad \text{and} \quad \mathbf{y} = \lambda\mathbf{a} + \mu\mathbf{b}. \tag{7}$$

Then

$$(ABCX) = \frac{\beta\gamma}{\alpha\delta} \quad \text{and} \quad (ABCY) = \frac{\beta\lambda}{\alpha\mu}.$$

Since $(ABCX) = (ABCY)$, it follows that

This is the hypothesis of the theorem.

$$\frac{\gamma}{\delta} = \frac{\lambda}{\mu},$$

so $\lambda = \gamma\mu/\delta$. If we substitute this value of λ into the expression for $\mathbf{y}$ in equation (7), we obtain

$$\mathbf{y} = (\gamma\mu/\delta)\mathbf{a} + \mu\mathbf{b} = (\mu/\delta)(\gamma\mathbf{a} + \delta\mathbf{b}) = (\mu/\delta)\mathbf{x}.$$

Since $\mathbf{y}$ is a scalar multiple of $\mathbf{x}$, it follows that $X = Y$, as required. ■

In Theorem 4 we showed that the cross-ratios $(ABCD)$ and $(A'B'C'D')$ are equal if the Points A', B', C', D' are in perspective with the Points A, B, C, D. Our next result is a partial converse of this result.

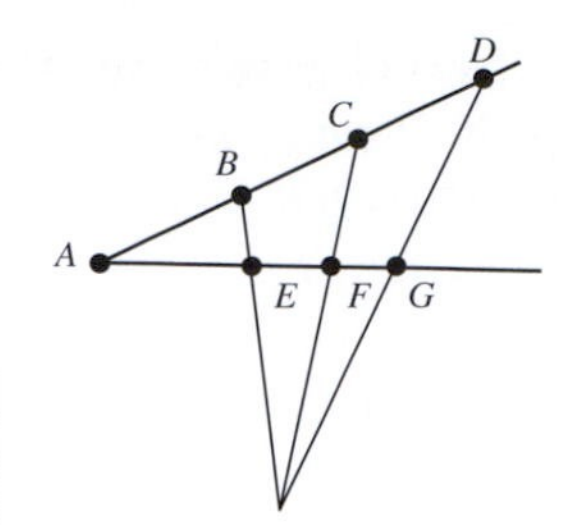

Theorem 6 Let A, B, C, D and A, E, F, G be two sets of collinear Points (on different Lines in $\mathbb{RP}^2$) such that the cross-ratios $(ABCD)$ and $(AEFG)$ are equal. Then the Lines BE, CF and DG are concurrent.

Proof Let P be the Point at which the Lines BE and CF meet, and let X be the Point at which the Line PG meets the Line $ABCD$. Then the Points A, B, C and X are in perspective from P with the Points A, E, F and G, so that

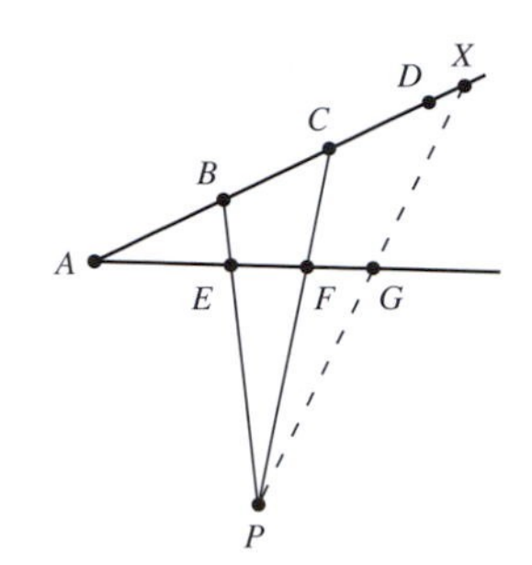

$$(ABCX) = (AEFG).$$

Since we know that $(AEFG) = (ABCD)$, it follows that

$$(ABCX) = (ABCD).$$

By Theorem 5, we must therefore have $X = D$. Hence the Points A, B, C, D and the Points A, E, F, G are in perspective from P. ■

We can now use Theorem 6 together with the other properties of cross-ratio to give a second proof of Pappus' Theorem.

Theorem 7 Pappus' Theorem
Let A, B and C be three Points on a Line in $\mathbb{RP}^2$, and let A', B' and C' be three Points on another Line. Let BC' and $B'C$ meet at P, CA' and $C'A$ meet at Q, and AB' and $A'B$ meet at R. Then P, Q and R are collinear.

You met this Theorem earlier, in Subsection 3.4.1.

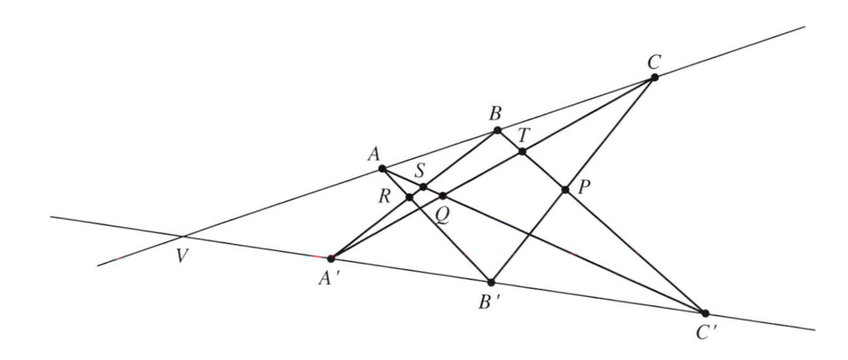

Proof Let V be the Point of intersection of the two given Lines. Also let the Lines BA' and AC' meet at the Point S, and the Lines BC' and CA' meet at the Point T.

Now, the Points V, A', B', C' are in perspective from A with the Points B, A', R, S, so that

$$(VA'B'C') = (BA'RS). \quad (8)$$

By Theorem 4

Similarly, the Points V, A', B', C' are in perspective from C with the Points B, T, P, C', so that

$$(VA'B'C') = (BTPC'). \quad (9)$$

It follows from equations (8) and (9) that

$$(BA'RS) = (BTPC'),$$

so that by Theorem 6 the Lines $A'T, RP, SC'$ are concurrent.

We may rephrase this statement as follows: the Line RP passes through the Point where $A'T$ meets SC'; that is, the Line RP passes through Q. In other words, P, Q and R are collinear. ■

3.5.2 Cross-Ratio on Embedding Planes

So far, we have calculated a given cross-ratio $(ABCD)$ by applying the definition of cross-ratio directly to the Points A, B, C, D. However, it is sometimes convenient to evaluate the cross-ratio by examining the representation of the Points on some embedding plane.

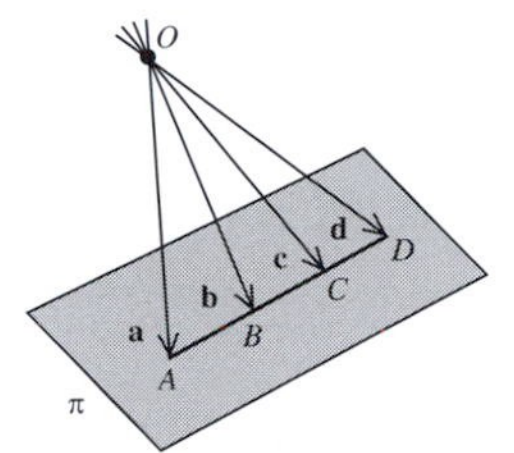

Suppose that four collinear Points of $\mathbb{RP}^2$ pierce an embedding plane π at the points A, B, C, D with position vectors $\mathbf{a}$, $\mathbf{b}$, $\mathbf{c}$, $\mathbf{d}$, respectively.

According to the Section Formula, we can write $\mathbf{c}$ and $\mathbf{d}$ in the form

See Appendix 2.

$$\mathbf{c} = \lambda\mathbf{a} + (1-\lambda)\mathbf{b} \quad \text{and} \quad \mathbf{d} = \mu\mathbf{a} + (1-\mu)\mathbf{b},$$

where $(1-\lambda) : \lambda$ is the ratio $AC : CB$, and $(1-\mu) : \mu$ is the ratio $AD : DB$. Then from the definition of cross-ratio

$$(ABCD) = \frac{1-\lambda}{\lambda} \bigg/ \frac{1-\mu}{\mu},$$

so

$$(ABCD) = \frac{AC}{CB} \bigg/ \frac{AD}{DB}. \quad (10)$$

Example 2 In an embedding plane, the points A, B, C, D lie in order along a line with the distances AB, BC, CD being 1 unit, 3 units and 2 units, respectively. Determine the cross-ratios $(ABCD)$, $(BACD)$ and $(ACBD)$.

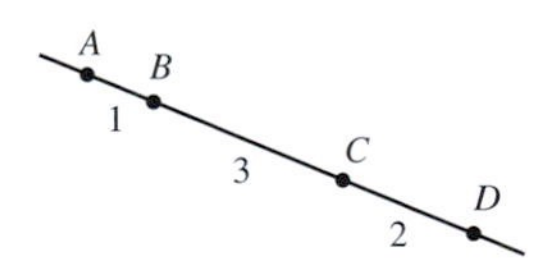

Solution Using equation (10) and the sign convention for ratios, we have

$$(ABCD) = \frac{AC}{CB} \bigg/ \frac{AD}{DB} = \left(-\frac{4}{3}\right) \bigg/ \left(-\frac{6}{5}\right) = \frac{10}{9},$$

$$(BACD) = \frac{BC}{CA} \bigg/ \frac{BD}{DA} = \left(-\frac{3}{4}\right) \bigg/ \left(-\frac{5}{6}\right) = \frac{9}{10}$$

and

$$(ACBD) = \frac{AB}{BC} \Big/ \frac{AD}{DC} = \left(\frac{1}{3}\right) \Big/ \left(-\frac{6}{2}\right) = -\frac{1}{9}. \qquad \square$$

Problem 4 The points A, B, C, D lie in order along a line with the distances AB, BC, CD being 2 units, 1 unit and 3 units, respectively. Determine the cross-ratios $(ABCD)$ and $(DBCA)$.

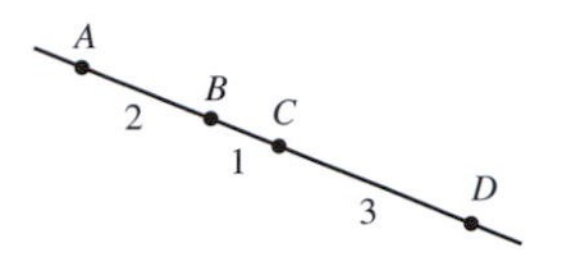

Sometimes one of the Points whose cross-ratio we are trying to find turns out to be an ideal Point for the embedding plane. In such cases, formula (10) cannot be used since some of the distances in the formula will not be defined.

To be specific, suppose that the Points A, B, C, D are collinear, but that A is an ideal Point for the embedding plane π, as shown in the margin. As before, we can let **b**, **c**, **d** be the position vectors of the points B, C, D on π, but we take **a** to be a unit vector along A. Then

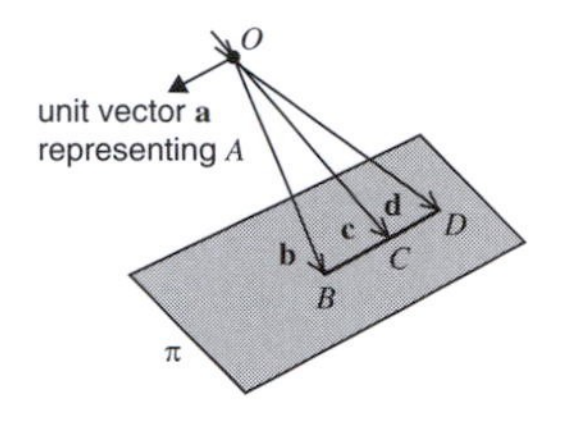

$$\mathbf{c} = -(CB)\mathbf{a} + \mathbf{b} \quad \text{and} \quad \mathbf{d} = -(DB)\mathbf{a} + \mathbf{b}.$$

From the definition of cross-ratio, it follows that

$$(ABCD) = \frac{1}{-CB} \Big/ \frac{1}{-DB} = \frac{DB}{CB}. \tag{11}$$

We can now obtain the corresponding formulas for the cases where B, C or D is an ideal Point, by applying Theorem 2. For example, if B is an ideal Point, then

$$\begin{aligned} (ABCD) &= \frac{1}{(BACD)} && \text{(swap first two terms)} \\ &= (BADC) && \text{(swap last two terms)} \\ &= \frac{CA}{DA} && \text{by equation (11).} \end{aligned}$$

Problem 5 Use Theorem 2 To prove that:

(a) $(ABCD) = \frac{AC}{BC}$ if D is an ideal Point;

(b) $(ABCD) = \frac{BD}{AD}$ if C is an ideal Point.

We now summarize the various formulas for cross-ratio in the form of a strategy, as follows.

Strategy To use an embedding plane to calculate the cross-ratio of four collinear Points:

1. if the four Points pierce the embedding plane at A, B, C, D, then

$$(ABCD) = \frac{AC}{CB} \Big/ \frac{AD}{DB};$$

2. if one of the Points is an ideal Point for the embedding plane, then

$$(ABCD) = \frac{DB}{CB} \text{ if } A \text{ is ideal,}$$

$$(ABCD) = \frac{CA}{DA} \text{ if } B \text{ is ideal,}$$

$$(ABCD) = \frac{BD}{AD} \text{ if } C \text{ is ideal,}$$

$$(ABCD) = \frac{AC}{BC} \text{ if } D \text{ is ideal.}$$

Example 3 Determine $(ABCD)$ for the collinear points A, B, C, D illustrated in the margin, where C is an ideal Point.

Solution Since C is an ideal Point, we have

$$(ABCD) = \frac{BD}{AD} = \frac{4}{1} = 4.$$

□

Problem 6 Determine $(ABCD)$ for the collinear points A, B, C, D illustrated in the margin, where B is an ideal Point.

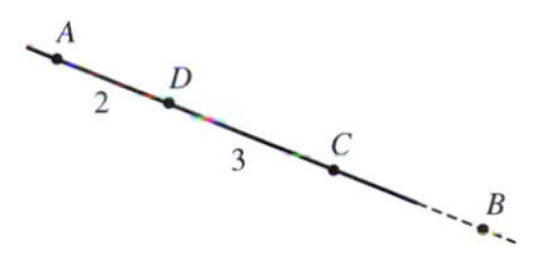

3.5.3 An Application of Cross-Ratio

Earlier, we described how projective geometry can be used to obtain two-dimensional representations of three-dimensional scenes. We now describe how cross-ratios can be used to obtain information about a three-dimensional scene from a two-dimensional representation of the scene. We do this in the context of aerial photography.

Subsection 3.1.1

For simplicity, consider an aerial camera that takes pictures on a flat film behind its lens, L, of features on a flat piece of land in front of L. Since a point on the ground lies on the same line through L as its image on the film, we can regard the process of taking a photograph as a perspectivity centred at L.

Since collinearity is invariant under a perspectivity, the image of any line ℓ on the ground is a line on the film. Moreover, the cross-ratio of any four collinear points is invariant under a perspectivity, so the cross-ratio of any four points on ℓ must be equal to the cross-ratio of their images on the film.

Section 3.3, Theorem 2

Theorem 3

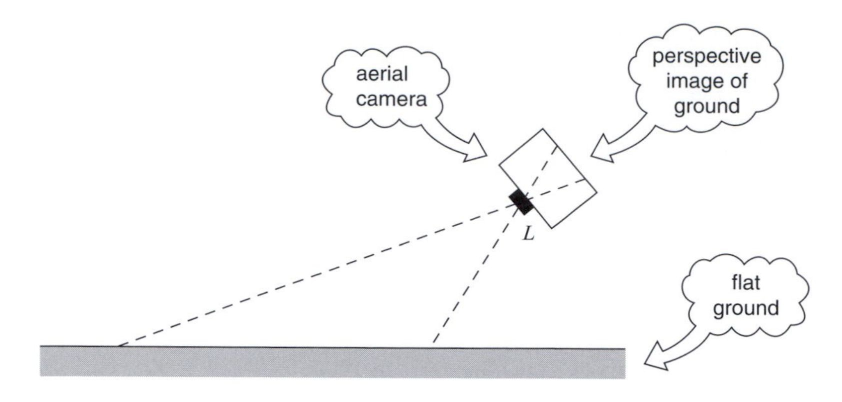

Example 4 An aerial camera photographs a car travelling along a straight road on flat ground towards a junction. Before the junction there are two warning signs at distances of 4 km and 2 km from the junction. On the film the signs are 1 cm and 3 cm from the junction, and the car is $\frac{3}{7}$ cm from the junction. How far is the car from the junction on the ground?

Strictly speaking, the car is not in line with the two signposts. Consequently, the distances marked on the photograph are approximations measured along the line of the left-hand kerb of the road.

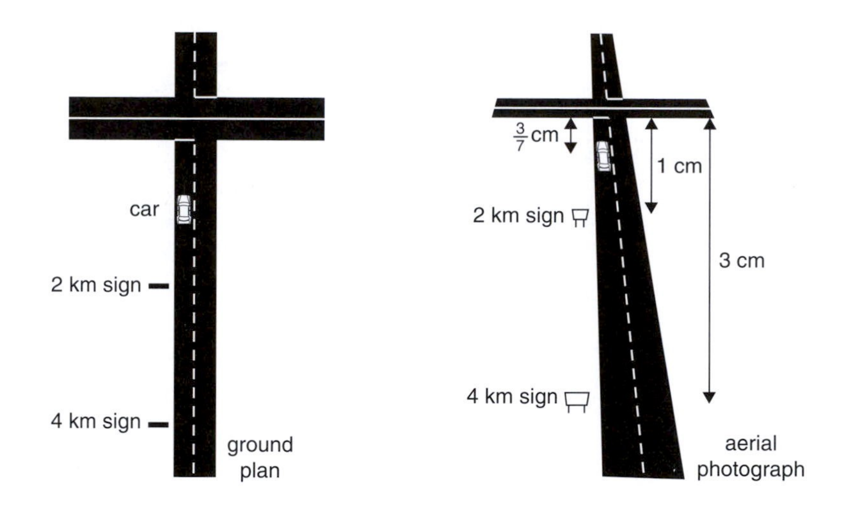

Solution Let A and B denote the signs, C denote the car, and D denote the junction, and let A', B', C', D' be their images on the film. Then

$$\begin{aligned}(A'B'C'D') &= \frac{A'C'}{C'B'} \Big/ \frac{A'D'}{D'B'} \\ &= \left(-\frac{18/7}{4/7}\right) \Big/ \left(-\frac{3}{1}\right) \\ &= \frac{3}{2}.\end{aligned}$$

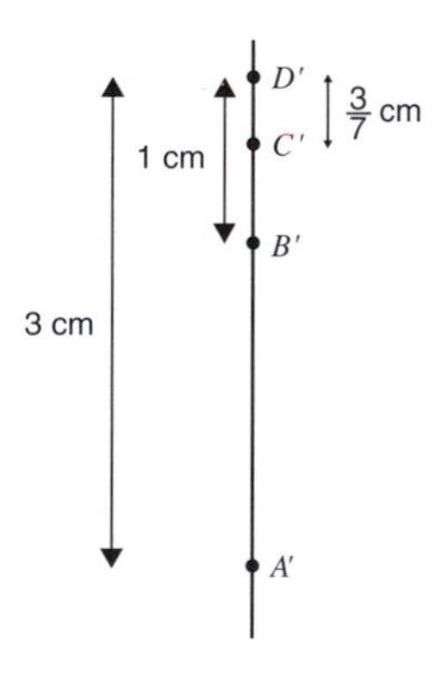

Now let the car be n km from the junction. Then

$$
\begin{aligned}
(ABCD) &= \frac{AC}{CB} \bigg/ \frac{AD}{DB} \\
&= \left(-\frac{4-n}{2-n}\right) \bigg/ \left(-\frac{4}{2}\right) \\
&= \frac{4-n}{2(2-n)}.
\end{aligned}
$$

Since $(ABCD)$ and $(A'B'C'D')$ must be equal, it follows that

$$\frac{4-n}{2(2-n)} = \frac{3}{2}.$$

Hence

$$4 - n = 3(2 - n).$$

and so $n = 1$. That is, the car is 1 km from the junction. □

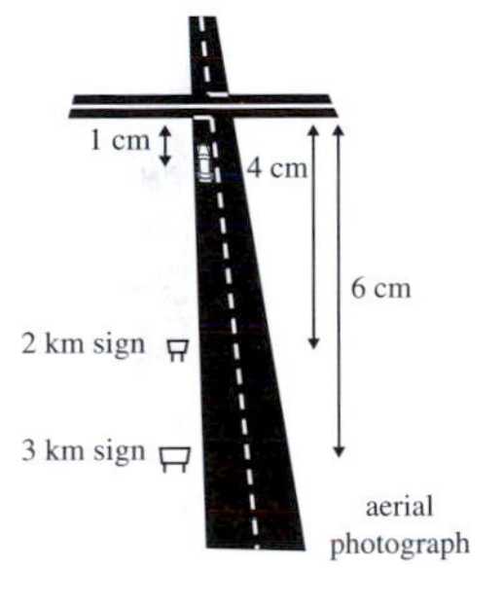

Problem 7 An aerial camera photographs a car travelling along a straight road on flat ground towards a junction. Before the junction there are two warning signs, at distances of 2 km and 3 km from the junction. On the film the signs are 4 cm and 6 cm from the junction, and the car is 1 cm from the junction. How far is the car from the junction on the ground?

If two lines that are known to be parallel on the ground appear to meet on the film, then the point of intersection on the film corresponds to the ideal Point where the 'parallel lines meet'. We can therefore use the above technique even when one of the Points is ideal, for we can use the second part of the strategy in Subsection 3.5.2 to calculate the cross-ratio whenever one of the Points is ideal.

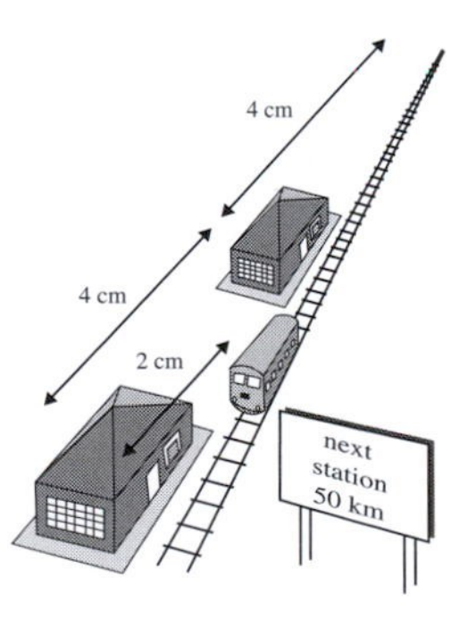

Problem 8 An aerial camera photographs a train travelling between two stations along a straight track on flat ground. The stations are 50 km apart. When the film is inspected, the stations are 4 cm apart, the train is midway between the stations, and the rails appear to meet (or vanish) 4 cm beyond the station towards which the train is travelling. How far has the train to travel to the next station?

3.6 Exercises

Section 3.2

1. (a) Write down numbers a, b, c and d such that

$$[1, a, b] = \left[-\tfrac{1}{2}, 3, 4\right] \quad \text{and} \quad [c, d, 2] = [3, 0, 1].$$

   (b) Which of the following homogeneous coordinates represent the same Point of $\mathbb{RP}^2$ as $[4, -8, 2]$?

   (i) $[1, 4, -2]$ (ii) $\left[\frac{1}{4}, -\frac{1}{2}, \frac{1}{8}\right]$ (iii) $\left[-\frac{1}{2}, -2, 1\right]$

   (iv) $[-2, 4, -1]$ (v) $\left[-\frac{1}{8}, -\frac{1}{2}, \frac{1}{4}\right]$

2. Determine an equation for each of the following Lines in $\mathbb{RP}^2$:
   (a) the Line through the Points $[1, 2, 3]$ and $[3, 0, -2]$;
   (b) the Line through the Points $[1, -1, -1]$ and $[2, 1, -3]$.
3. Determine whether each of the following sets of Points are collinear:
   (a) $[1, -1, 0]$, $[1, 0, -1]$ and $[2, -1, -1]$;
   (b) $[1, 0, 1]$, $[0, 1, 2]$ and $[1, 2, 3]$.
4. Determine the Point of intersection of each of the following pairs of Lines in $\mathbb{RP}^2$:
   (a) the Lines with equations $x - 2y + z = 0$ and $x - y - z = 0$;
   (b) the Lines with equations $x + 2y + 5z = 0$ and $3x - y + z = 0$.
5. Determine the Point of $\mathbb{RP}^2$ at which the Line through the Points $[8, -1, 2]$ and $[1, -2, -1]$ meets the Line through the Points $[0, 1, -1]$ and $[2, 3, 1]$.
6. Determine the Point of $\mathbb{RP}^2$ at which the Line through the Points $[1, 2, 2]$ and $[2, 3, 3]$ meets the Line through the Points $[0, 1, 2]$ and $[0, 1, 3]$.

Section 3.3

In these exercises, you may find the following list of matrices and their inverses useful.

$$\mathbf{A}: \begin{pmatrix} 2 & 1 & 0 \\ -1 & 0 & 1 \\ 0 & 1 & 1 \end{pmatrix} \begin{pmatrix} -2 & 0 & 1 \\ 0 & 3 & -2 \\ 1 & -3 & 1 \end{pmatrix} \begin{pmatrix} 0 & 3 & -1 \\ 2 & 0 & -1 \\ 0 & 0 & 1 \end{pmatrix} \begin{pmatrix} 0 & 3 & 4 \\ -1 & 3 & 2 \\ 3 & -3 & 3 \end{pmatrix}$$

$$\mathbf{A}^{-1}: \begin{pmatrix} 1 & 1 & -1 \\ -1 & -2 & 2 \\ 1 & 2 & -1 \end{pmatrix} \begin{pmatrix} -1 & -1 & -1 \\ -\frac{2}{3} & -1 & -\frac{4}{3} \\ -1 & -2 & -2 \end{pmatrix} \begin{pmatrix} 0 & \frac{1}{2} & \frac{1}{2} \\ \frac{1}{3} & 0 & \frac{1}{3} \\ 0 & 0 & 1 \end{pmatrix} \begin{pmatrix} 5 & -7 & -2 \\ 3 & -4 & -\frac{4}{3} \\ -2 & 3 & 1 \end{pmatrix}$$

1. Determine which of the following transformations t of $\mathbb{RP}^2$ are projective transformations. For those that are projective transformations, write down a matrix associated with t.

(a) $t : [x, y, z] \mapsto [2x, y + 3z, 1]$
(b) $t : [x, y, z] \mapsto [x, x - y + 3z, x + y]$
(c) $t : [x, y, z] \mapsto [2y, y - 4z, x]$
(d) $t : [x, y, z] \mapsto [x + y - z, y + 3z, x + 2y + 2z]$

2. Determine the images of the Points [1, 2, 3], [0, 1, 0] and [1, −1, 1] under the projective transformation t associated with the matrix

$$\mathbf{A} = \begin{pmatrix} 2 & 0 & 1 \\ -1 & 1 & 0 \\ 0 & 1 & 1 \end{pmatrix}.$$

3. Let

$$t_1 : [x, y, z] \mapsto [2x + y, -x + z, y + z],$$

$$t_2 : [x, y, z] \mapsto [x + y, 3x - z, 4y - 2z]$$

be projective transformations from $\mathbb{RP}^2$ to $\mathbb{RP}^2$.
(a) Write down matrices associated with each of t_1 and t_2.
(b) Determine formulas for $t_2 \circ t_1$ and $t_2 \circ t_1^{-1}$.

4. Find the image of the Line $x + 2y + 3z = 0$ under the projective transformation t_1 defined in Exercise 3.

5. Determine matrices for the projective transformations which map the Points [1, 0, 0], [0, 1, 0], [0, 0, 1] and [1, 1, 1] onto the following Points:
(a) $[-2, 0, 1]$, $[0, 1, -1]$, $[-1, 2, -1]$ and $[-1, 1, -1]$;
(b) [0, 1, 0], [1, 0, 0], $[-1, -1, 1]$ and [2, 1, 1];
(c) $[0, 1, -3]$, $[1, 1, -1]$, [4, 2, 3] and [7, 4, 3].

6. Use the results of Exercise 5 to determine the projective transformations that map:
(a) the Points

$$[-2, 0, 1],\ [0, 1, -1],\ [-1, 2, -1],\ [-1, 1, -1]$$

to the Points

$$[0, 1, 0],\ [1, 0, 0],\ [-1, -1, 1],\ [2, 1, 1],$$

respectively;
(b) the Points

$$[0, 1, 0],\ [1, 0, 0],\ [-1, -1, 1],\ [2, 1, 1]$$

to the Points

$$[0, 1, -3],\ [1, 1, -1],\ [4, 2, 3],\ [7, 4, 3],$$

respectively;
(c) the Points

$$[0, 1, -3],\ [1, 1, -1],\ [4, 2, 3],\ [7, 4, 3]$$

to the Points

$$[-2, 0, 1],\ [0, 1, -1],\ [-1, 2, -1],\ [-1, 1, -1],$$

respectively.

Section 3.4

1. For which of the following configurations of Points A, B, C and D in $\mathbb{RP}^2$ is there a projective transformation sending A, B, C to the triangle of reference and D to the unit Point?

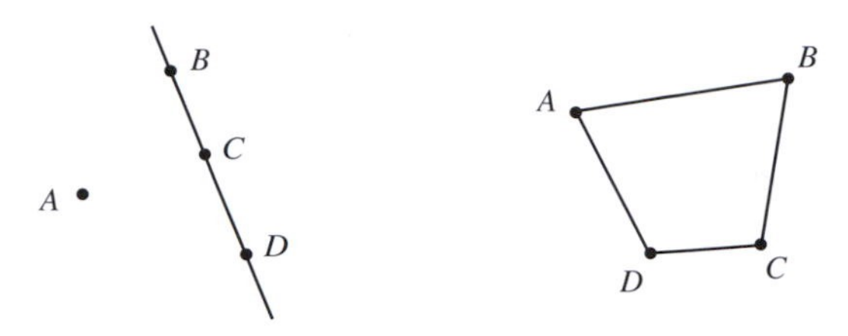

2. Let ΔABC be a triangle in $\mathbb{R}^2$, and let U be any point of $\mathbb{R}^2$ that is not collinear with any two of the points A, B, C. Let the Lines BC and AU meet at P, CA and BU meet at Q, and AB and CU meet at R. Prove that P, Q, R cannot be collinear.

Section 3.5

1. For each of the following sets of Points A, B, C, D, calculate the cross-ratio $(ABCD)$.
   (a) $A = [2, 1, 3],\ B = [1, 2, 3],\ C = [8, 1, 9],\ D = [4, -1, 3]$
   (b) $A = [2, 1, 1],\ B = [-1, 1, -1],\ C = [1, 2, 0],\ D = [-1, 4, -2]$
   (c) $A = [-1, 1, 1],\ B = [0, 0, 2],\ C = [5, -5, 3],\ D = [-3, 3, 7]$
2. For the Points A, B, C, D in Exercise 1(a), determine the cross-ratios $(BACD)$, $(BDCA)$ and $(ADBC)$.
3. For each set of collinear points A, B, C, D illustrated below, calculate the cross-ratio $(ABCD)$.

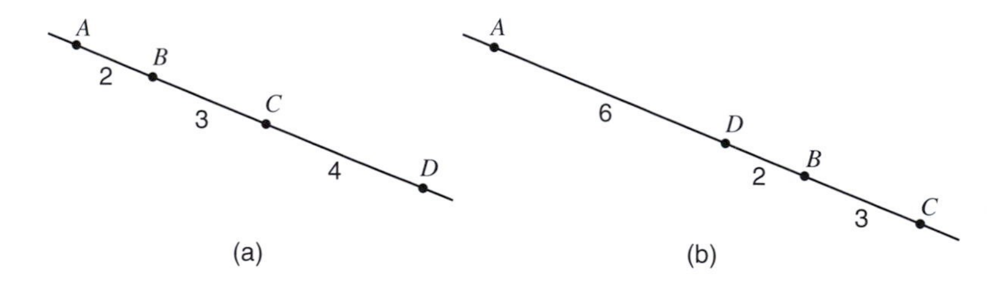

4. Calculate the cross-ratio $(ABCD)$ for the collinear points A, B, C, D illustrated below, where D is an ideal Point.

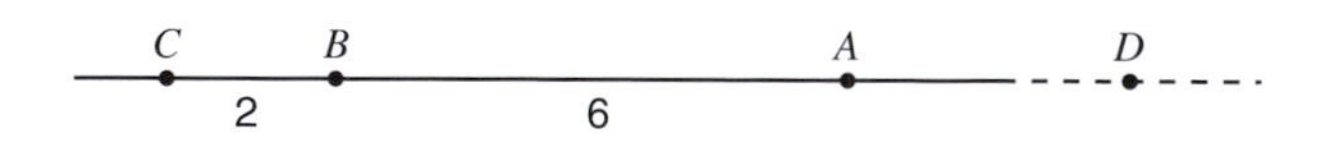

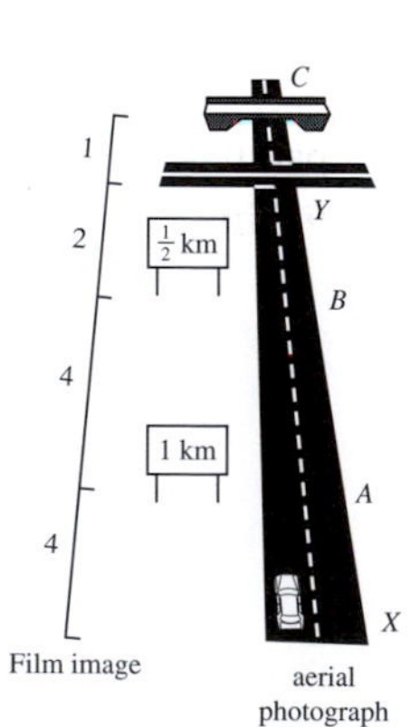

5. The diagram in the margin represents an aerial photograph of a straight road on flat ground. At A there is a sign 'Junction 1 km', at B a sign 'Junction

$\frac{1}{2}$ km', and C is the road junction. Also, a police patrol car is at X, and a bridge is at Y. The distances marked on the left of the diagram are measured in cm from the photograph.

Calculate the actual distances (in km) of the patrol car and the bridge from the junction.

Summary of Chapter 3

Section 3.1: Perspective

1. Renaissance artists used **terraced perspective** and later **vertical perspective** in an attempt to portray 'real' scenes in a realistic way. The modern system of **focused perspective** was discovered by Brunelleschi and finally perfected by Leonardo da Vinci; it is well illustrated by the woodcuts of Albrecht Dürer.

   The family of lines joining an eye to each point of a scene meets a screen in front of the eye, and the image on the screen is called a **cross-section** (or **section**). The cross-section gives a realistic two-dimensional representation of the three-dimensional scene.
2. For two planes π and π' that do not pass through the origin O in $\mathbb{R}^3$, points P in π and Q in π' are **in perspective from O** if there is a straight line through O, P and Q.

   A **perspectivity from π to π'** centred at O is a function that maps a point P of π to a point Q of π' whenever P and Q are in perspective from O. (The planes π and π' may lie on the same or on opposite sides of O.)
3. The domain of a perspectivity may not be the whole of π; for, if P is any point of π such that OP is parallel to to π', then P cannot have an image in π'.

   The image of a line ℓ under a perspectivity is another line, possibly minus one point.

   Foreshortening is the effect under a perspectivity of lines of equal lengths at different distances from a screen corresponding to different lengths on the screen.
4. Two parallel lines in a horizontal plane π appear to an observer to meet at a **vanishing point** on a vertical screen π'; this is the **principal vanishing point** if the lines are perpendicular to the line of intersection of π and π', and a **diagonal vanishing point** otherwise.

   The family of vanishing points is a line, the **vanishing line**, that corresponds to the horizon line in a picture.
5. **Desargues' Theorem** Let ΔABC and $\Delta A'B'C'$ be triangles in $\mathbb{R}^2$ such that the lines AA', BB' and CC' meet at a point U. Let BC and $B'C'$meet at P, CA and $C'A'$meet at Q, and AB and $A'B'$meet at R. Then P, Q and R are collinear.

Section 3.2: The Projective Plane $\mathbb{RP}^2$

1. A **Point** (or **projective point**) is a line in $\mathbb{R}^3$ that passes through the origin of $\mathbb{R}^3$.

   The **real projective plane** $\mathbb{RP}^2$ is the set of all such Points.
2. The expression $[a, b, c]$, in which the numbers a, b and c are not all zero, represents the Point P of $\mathbb{RP}^2$ which consists of the unique line in $\mathbb{R}^3$ that passes through (0,0,0) and (a, b, c). We refer to $[a, b, c]$ as **homogeneous coordinates** of P.

   If (a, b, c) has position vector $\mathbf{v}$, then we often denote P by $[\mathbf{v}]$ and we say that P can be **represented** by $\mathbf{v}$.

   It makes NO sense to write the expression [0,0,0], since not all of a, b and c can be zero.
3. The homogeneous coordinates $[a, b, c]$ and $[\lambda a, \lambda b, \lambda c]$ (where $\lambda \neq 0$) represent the same Point of $\mathbb{RP}^2$; that is, $[a, b, c] = [\lambda a, \lambda b, \lambda c]$, for any $\lambda \neq 0$.

   If there is no non-zero real number λ such that $[a, b, c] = \left[\lambda a', \lambda b', \lambda c'\right]$, then the homogeneous coordinates $[a, b, c]$ and $\left[a', b', c'\right]$ represent different Points of $\mathbb{RP}^2$.

   Further, $\left[a', b', 1\right] = \left[a'', b'', 1\right]$ if and only if $a' = a''$ and $b' = b''$.
4. A **projective figure** is a subset of $\mathbb{RP}^2$.
5. A **Line** (or **projective line**) in $\mathbb{RP}^2$ is a plane in $\mathbb{R}^3$ that passes through the origin. Points of $\mathbb{RP}^2$ are **collinear** if they lie on a Line.
6. The general equation of a Line in $\mathbb{RP}^2$ is $ax + by + cz = 0$, where a, b, c are real and not all zero.
7. **Collinearity Property of** $\mathbb{RP}^2$ Any two distinct Points of $\mathbb{RP}^2$ lie on a unique Line.

   Strategy To determine an equation for the Line in $\mathbb{RP}^2$ through the Points $\left[d, e, f\right]$ and $\left[g, h, k\right]$:
   1. write down the equation $\begin{vmatrix} x & y & z \\ d & e & f \\ g & h & k \end{vmatrix} = 0$;
   2. expand the determinant in terms of the entries in its first row to obtain the required equation in the form $ax + by + cz = 0$.

   Sometimes it is possible to 'spot' the equation of a Line through two Points without using the determinant.
8. **Strategy** To determine whether three Points $[a, b, c]$, $[d, e, f]$ and $[g, h, k]$ are collinear:
   1. evaluate the determinant $\begin{vmatrix} a & b & c \\ d & e & f \\ g & h & k \end{vmatrix}$;
   2. the Points $[a, b, c]$, $[d, e, f]$ and $[g, h, k]$ are collinear if and only if this determinant is zero.
9. The Points [1,0,0], [0,1,0], [0,0,1] are known as the **triangle of reference**. The Point [1,1,1] is called the **unit Point**.

10. **Incidence Property of $\mathbb{RP}^2$** Any two distinct Lines in $\mathbb{RP}^2$ intersect in a unique Point of $\mathbb{RP}^2$.
11. Let π be any plane in $\mathbb{R}^3$ that does not pass through the origin O. Then there is a one-one correspondence between the points of π and those Points of $\mathbb{RP}^2$ which pierce π. Those Points of $\mathbb{RP}^2$ which do not pierce π are called **ideal Points** for π.

    The set of ideal Points for π is a plane through O parallel to π, called the **ideal Line** for π.
12. An **embedding plane** is a plane, π, which does not pass through the origin, together with the set of all ideal Points for π.

    The plane in $\mathbb{R}^3$ with equation $z = 1$ is called the **standard embedding plane**. The mapping of $\mathbb{RP}^2$ into the standard embedding plane is called the **standard embedding** of $\mathbb{RP}^2$.
13. Parallelism is not a projective property.

Section 3.3: Projective Transformations

1. A **projective transformation** of $\mathbb{RP}^2$ is a function t: $\mathbb{RP}^2 \to \mathbb{RP}^2$ of the form $t : [\mathbf{x}] \mapsto [\mathbf{Ax}]$, where $\mathbf{A}$ is an invertible 3×3 matrix. We say that $\mathbf{A}$ is a matrix **associated** with t. The set of all projective transformations is denoted by $P(2)$.

    If $\mathbf{A}$ is a matrix **associated** with t, then so is $\lambda\mathbf{A}$ for any non-zero number λ.
2. The set of projective transformations $P(2)$ forms a group under the operation of composition of functions. In particular, if t_1 and t_2 are projective transformations with associated matrices $\mathbf{A}_1$ and $\mathbf{A}_2$, respectively, then $t_1 \circ t_2$ and t_1^{-1} are projective transformations with associated matrices $\mathbf{A}_1\mathbf{A}_2$ and $\mathbf{A}_1^{-1}$.

    Strategy To compose two projective transformations t_1 and t_2:
    1. write down matrices $\mathbf{A}_1$ and $\mathbf{A}_2$ associated with t_1 and t_2;
    2. calculate $\mathbf{A}_1\mathbf{A}_2$;
    3. write down the composite $t_1 \circ t_2$ with which $\mathbf{A}_1\mathbf{A}_2$ is associated.

    Strategy To find the inverse of a projective transformation t:
    1. write down a matrix $\mathbf{A}$ associated with t;
    2. calculate $\mathbf{A}^{-1}$;
    3. write down the inverse t^{-1} with which $\mathbf{A}^{-1}$ is associated.
3. **Strategy** To find the image of a Line $ax + by + cz = 0$ under a projective transformation $t : [\mathbf{x}] \mapsto [\mathbf{Ax}]$:
    1. write the equation of the Line in the form $\mathbf{Lx} = 0$, where $\mathbf{L}$ is the matrix $(a\,b\,c)$;
    2. find a matrix $\mathbf{B}$ associated with t^{-1};
    3. write down the equation of the image as $(\mathbf{LB})\,\mathbf{x} = 0$.
4. **Projective geometry** is the study of those properties of figures in $\mathbb{RP}^2$ that are preserved by projective transformations. Collinearity and incidence are both projective properties.

5. A **quadrilateral** is a set of four Points A, B, C and D (no three of which are collinear), together with the Lines AB, BC, CD and DA.
   All quadrilaterals are projective-congruent.
6. **Strategy** To find the projective transformation which maps [1,0,0] to $[a_1, a_2, a_3]$, [0,1,0] to $[b_1, b_2, b_3]$, [0,0,1] to $[c_1, c_2, c_3]$, [1,1,1] to $[d_1, d_2, d_3]$, where no three of $[a_1, a_2, a_3]$, $[b_1, b_2, b_3]$, $[c_1, c_2, c_3]$ and $[d_1, d_2, d_3]$ are collinear:
   1. find u, v, w for which $\begin{pmatrix} a_1u & b_1v & c_1w \\ a_2u & b_2v & c_2w \\ a_3u & b_3v & c_3w \end{pmatrix} \begin{pmatrix} 1 \\ 1 \\ 1 \end{pmatrix} = \begin{pmatrix} d_1 \\ d_2 \\ d_3 \end{pmatrix}$;
   2. write down the required projective transformation in the form t: $[\mathbf{x}] \mapsto [\mathbf{Ax}]$, where $\mathbf{A}$ is any non-zero real multiple of the matrix $\begin{pmatrix} a_1u & b_1v & c_1w \\ a_2u & b_2v & c_2w \\ a_3u & b_3v & c_3w \end{pmatrix}$.
7. **Fundamental Theorem of Projective Geometry** Let $ABCD$ and $A'B'C'D'$ be two quadrilaterals in $\mathbb{RP}^2$. Then:
   (a) there is a projective transformation t which maps A to A', B to B', C to C', D to D';
   (a) the projective transformation t is unique.
8. **Strategy** To determine the projective transformation t which maps the vertices of the quadrilateral $ABCD$ to the corresponding vertices of the quadrilateral $A'B'C'D'$:
   1. find the projective transformation t_1 which maps the triangle of reference and unit Point to the Points A, B, C, D, respectively;
   2. find the projective transformation t_2 which maps the triangle of reference and unit Point to the Points A', B', C', D', respectively;
   3. calculate $t = t_2 \circ t_1^{-1}$.
9. With any given perspectivity σ we can construct an associated **perspective transformation** that is a one-one mapping of $\pi \cup \{\text{the ideal Points for } \pi\}$ onto $\pi' \cup \{\text{the ideal Points for } \pi'\}$. This maps $\mathbb{RP}^2$ onto itself.
   Every projective transformation can be expressed as the composite of three perspective transformations.

Section 3.4: Using the Fundamental Theorem of Projective Geometry

1. **Desargues' Theorem** Let ΔABC and $\Delta A'B'C'$ be triangles in $\mathbb{R}^2$ such that the lines AA', BB' and CC' meet at a point U. Let BC and $B'C'$ meet at P, CA and $C'A'$ meet at Q, and AB and $A'B'$ meet at R. Then P, Q and R are collinear.
2. The Fundamental Theorem is often used to simplify proofs of results in projective geometry, where the properties involved are projective properties. Generally, we do not explicitly refer to the corresponding auxiliary projective transformation t concerned, but simply comment that "By the

Fundamental Theorem of Projective Geometry, we may choose the four Points ... (no three of which are collinear) to be the triangle of reference and the unit Point; that is, to have homogeneous coordinates [1,0,0], [0,1,0], [0,0,1] and [1,1,1], respectively."

3. **Pappus' Theorem** Let A, B and C be three points on a line in $\mathbb{R}^2$, and let A', B' and C' be three points on another line. Let BC' and $B'C$ meet at P, CA' and $C'A$ meet at Q, and AB' and $A'B$ meet at R. Then P, Q and R are collinear.
4. Any Euclidean figure in an embedding plane corresponds to a projective figure in $\mathbb{RP}^2$. It follows that a Euclidean theorem concerned with projective properties (such as collinearity and coincidence) holds if and only if the corresponding projective theorem holds.
5. The **dual** of a statement about the projective properties of some figure in $\mathbb{RP}^2$ is the corresponding statement about $\mathbb{RP}^2$ in which the terms 'Point' and 'Line' are interchanged, and such other changes are made that ensure that the sentence makes sense.

   A triangle (three non-collinear Points and the Lines joining them) is self-dual.
6. A **hexagon** in $\mathbb{RP}^2$ consists of six Points joined by the six Lines joining them in turn.

   Pappus' Theorem (rephrased) Let the vertices A, B', C, A', B and C' of a hexagon lie alternately on two different Lines. Then the Points of intersection of opposite sides $B'C$ and BC', CA' and $C'A$, and AB' and $A'B$, are collinear.
7. **Brianchon's Theorem** (the dual of Pappus' Theorem) Let the sides $AB', B'C, CA', A'B, BC', C'A$ of a hexagon pass alternately through two (different) Points P and Q in $\mathbb{RP}^2$. Then the Lines joining opposite vertices A and A', B and B', C and C', are concurrent.
8. **Converse of Desargues' Theorem** (also its dual) Let two triangles be such that the Points through which corresponding sides pass are collinear. Then the Lines through the corresponding vertices of the two triangles are concurrent.

Section 3.5: Cross-Ratio

1. Let A, B, C, D be four collinear Points in $\mathbb{RP}^2$ represented by the position vectors **a**, **b**, **c**, **d**, and let $\mathbf{c} = \alpha\mathbf{a} + \beta\mathbf{b}$ and $\mathbf{d} = \gamma\mathbf{a} + \delta\mathbf{b}$. Then the **cross-ratio** of A, B, C, D is $(ABCD) = \frac{\beta}{\alpha}\Big/\frac{\delta}{\gamma}$.

   The cross-ratio $(ABCD)$ is independent of the homogeneous coordinates that are used to represent the collinear Points A, B, C, D.
2. Let A, B, C, D be four distinct collinear Points in $\mathbb{RP}^2$, and let $(ABCD) = k$. Then $(BACD) = (ABDC) = 1/k$ and $(ACBD) = (DBCA) = 1 - k$.
3. Let t be a projective transformation, and let A, B, C, D be any four collinear Points in $\mathbb{RP}^2$. If $A' = t(A)$, $B' = t(B)$, $C' = t(C)$, $D' = t(D)$, then $(ABCD) = (A'B'C'D')$.

4. Let A, B, C, D be four distinct Points on a Line, and let A', B', C', D' be four distinct Points on another Line such that AA', BB', CC', DD' all meet at a Point U. Then $(ABCD) = \left(A'B'C'D'\right)$.
5. **Unique Fourth Point Theorem** Let A, B, C, X, Y be collinear Points in $\mathbb{RP}^2$ such that $(ABCX) = (ABCY)$. Then $X = Y$.
6. Let A, B, C, D and A, E, F, G be two sets of collinear Points (on different Lines in $\mathbb{RP}^2$) such that the cross-ratios $(ABCD)$ and $(AEFG)$ are equal. Then the Lines BE, CF and DG are concurrent.
7. Let four collinear Points of $\mathbb{RP}^2$ pierce an embedding plane at the points A, B, C, D with position vectors $\mathbf{a}$, $\mathbf{b}$, $\mathbf{c}$, $\mathbf{d}$, respectively. Then, if we can write $\mathbf{c}$ and $\mathbf{d}$ in the form $\mathbf{c} = \lambda\mathbf{a} + (1-\lambda)\,\mathbf{b}$ and $\mathbf{d} = \mu\mathbf{a} + (1-\mu)\,\mathbf{b}$, we have
$$(ABCD) = \frac{1-\lambda}{\lambda} \Big/ \frac{1-\mu}{\mu} = \frac{AC}{CB} \Big/ \frac{AD}{DB}.$$
8. **Strategy** To use an embedding plane to calculate the cross-ratio of four collinear Points:
   1. if the four Points pierce the embedding plane at A, B, C, D, then $(ABCD) = \frac{AC}{CB} \Big/ \frac{AD}{DB}$;
   2. if one of the Points is an ideal Point for the embedding plane, then
$$(ABCD) = \frac{DB}{CB} \quad \text{if } A \text{ is ideal,}$$
$$(ABCD) = \frac{CA}{DA} \quad \text{if } B \text{ is ideal,}$$
$$(ABCD) = \frac{BD}{AD} \quad \text{if } C \text{ is ideal,}$$
$$(ABCD) = \frac{AC}{BC} \quad \text{if } D \text{ is ideal.}$$
9. Cross-ratios can be used to measure distances on the ground from aerial photographs, since the cross-ratio of any four points on a line on the ground equals the cross-ratio of their images on the film.

4 Projective Geometry: Conics

In the previous chapter, we used projective geometry to prove Pappus' Theorem.

Subsection 3.4.1, Theorem 2

Pappus' Theorem Let A, B and C be three points on a line in $\mathbb{R}^2$, and let A', B' and C' be three points on another line. Let BC' and $B'C$ meet at P, CA' and $C'A$ meet at Q, and AB' and $A'B$ meet at R. Then P, Q and R are collinear.

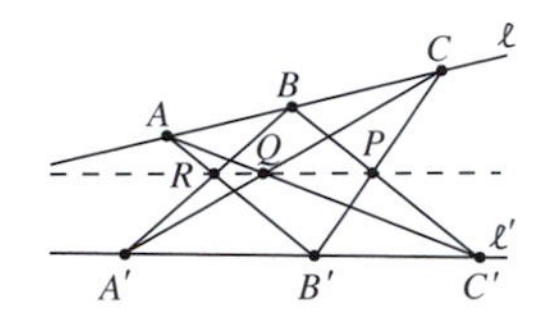

In fact, we could also show that if the six points A, B, C, A', B' and C' lie not on two lines but on an ellipse, a parabola or a hyperbola, then the three points of intersection P, Q and R are still collinear.

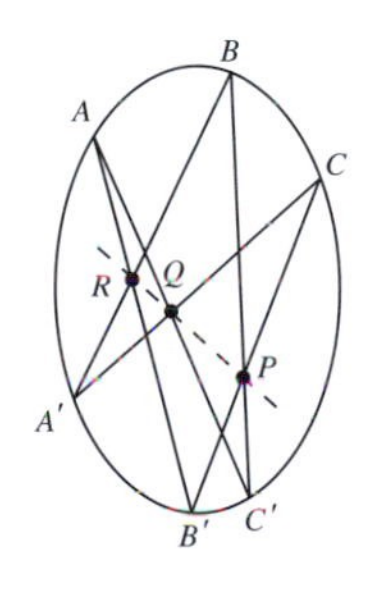

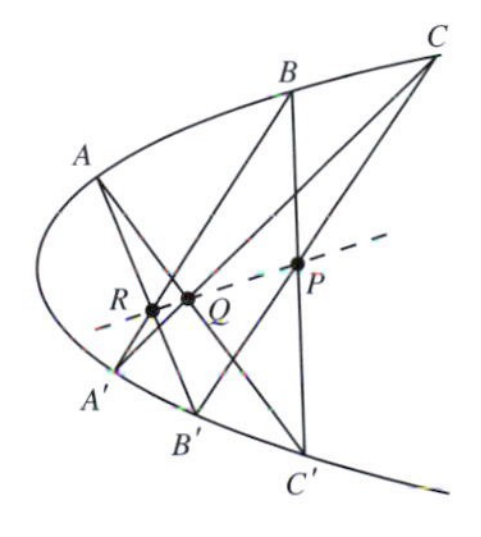

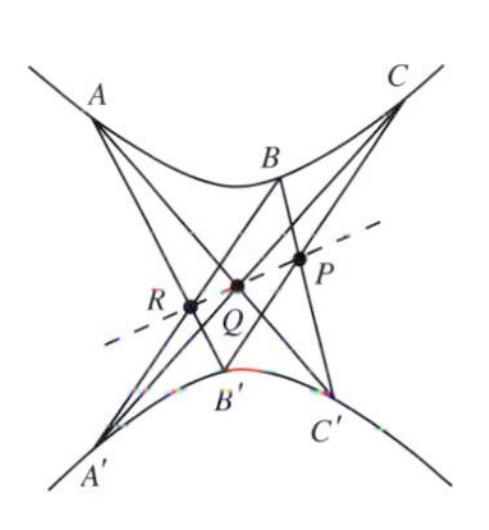

The similarity of these three results is remarkable! It suggests that, instead of being three new theorems, perhaps these results are particular instances of a general theorem about conics. We shall show that this is indeed the case. The general theorem, known as *Pascal's Theorem*, is proved in this chapter along with many other such results about conics.

But why should a result like Pascal's Theorem apply to different types of conic? A clue is provided by our discussion of conics in Chapter 1. There we explained that conics are so called because they are the shapes that we obtain when we take plane sections through a double cone. In particular, the *non-degenerate conics* are the *ellipses, parabolas* and *hyperbolas*, and they arise when we slice through a double cone with a plane that does not pass through the cone's vertex v (see the figure below).

Subsections 1.1.1 and 1.1.5

This has an exciting implication for projective geometry. Given any two non-degenerate conics, such as the circle and the parabola shown in the margin below, there is a perspectivity centred at v that maps one onto the other. It follows that any two non-degenerate conics are projective-congruent. Consequently, any result involving the projective properties of collinearity and concurrence that holds for ellipses, for example, necessarily holds also for parabolas and hyperbolas.

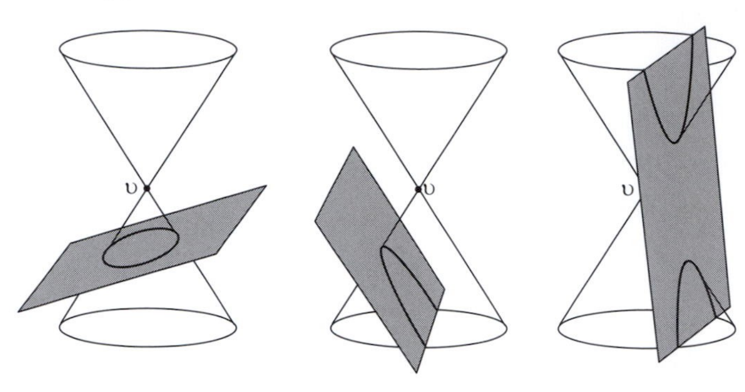

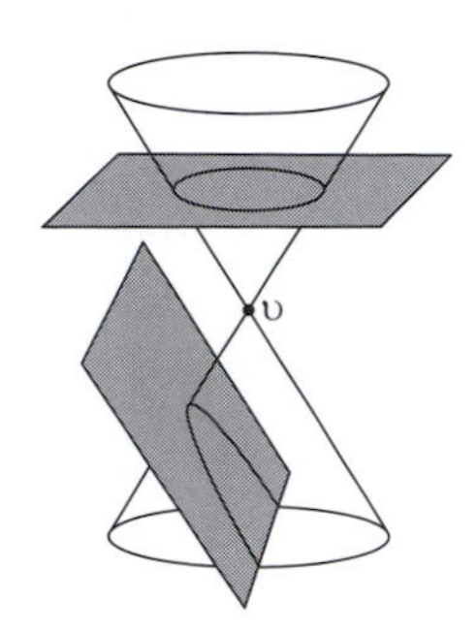

We saw earlier that in affine geometry all ellipses are affine-congruent, all parabolas are affine-congruent, and all hyperbolas are affine-congruent. One of the exciting features of projective geometry is that all non-degenerate conics are projective-congruent. Thus there is no distinction between ellipses, parabolas and hyperbolas in this geometry, so we simply call them *projective conics*.

In Section 4.1 we define the surfaces in $\mathbb{R}^3$ that are called projective conics, we see that between any two non-degenerate plane conics there is a perspectivity, and we note that (analogously) all projective conics are projective-congruent.

In Section 4.2 we observe that tangency is a projective property. We introduce a compact notation due to Joachimsthal for projective conics, and use it to find formulas for tangents, tangent pairs and polars for projective conics.

In Section 4.3 we introduce two standard forms for the equation of a projective conic, and use these to prove theorems such as Pascal's Theorem.

In Section 4.4 we prove that all non-degenerate projective conics are projective-congruent, using Linear Algebra methods.

Finally, Section 4.5 discusses *duality* in the context of projective conics.

4.1 Projective Conics

4.1.1 What is a Projective Conic?

In the previous chapter we described how, given any (Euclidean) figure F in an embedding plane π, we can obtain the corresponding projective figure by drawing in all the Points of $\mathbb{RP}^2$ that pass through points of F.

Subsection 3.2.1

For example, if F is the circle $\{(x, y, z) : x^2 + y^2 = 1, z = 1\}$ which lies in the embedding plane π with equation $z = 1$, then the corresponding projective figure is a right circular cone. If a Point $[x', y', z']$ of $\mathbb{RP}^2$ lies on this cone, then it pierces the embedding plane at the point $(x'/z', y'/z', 1)$. Since this point lies on F, it follows that

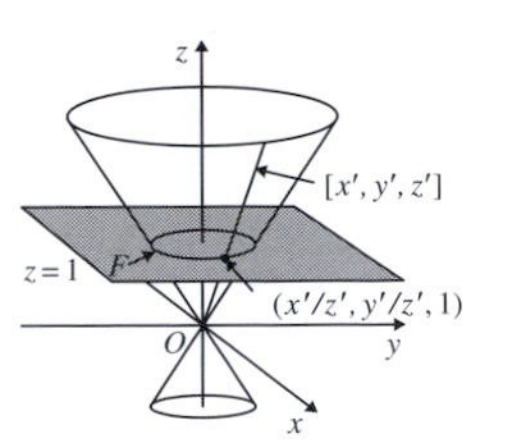

$$\left(\frac{x'}{z'}\right)^2 + \left(\frac{y'}{z'}\right)^2 = 1.$$

Multiplying by $(z')^2$, and dropping the dashes, we obtain the following equation for the cone:

$$x^2 + y^2 = z^2. \tag{1}$$

Conversely, suppose that we start with a projective figure. Then we can represent the projective figure in an embedding plane π. Of course, the representation that we obtain depends on the embedding plane that we use. For example, if the projective figure is the hollow cone $x^2 + y^2 = z^2$, then the representation can be bounded, unbounded, or even in two 'bits' depending on the position of π.

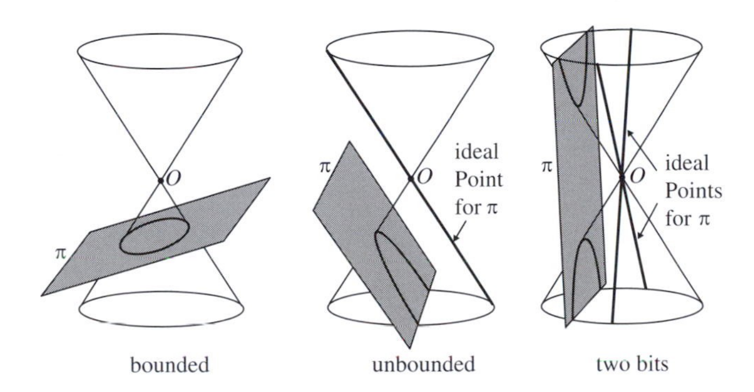

In Subsection 4.1.4 we show that these representations of the cone are, in fact, ellipses, parabolas and hyperbolas, respectively. Notice, however, that in the case of the parabola and the hyperbola, the representation is incomplete unless we include additional ideal Points for the embedding plane concerned. The parabola requires one ideal Point, and the hyperbola requires two. These are indicated by the thick lines in the diagram above.

Let us now concentrate on just one embedding plane π, and consider the kinds of projective figures that correspond to conics in π. To keep the algebra simple, we shall choose π to be the so-called **standard embedding plane** $z = 1$. We have already explained how to find the equation of the projective figure which corresponds to the circle $\{(x, y, z) : x^2 + y^2 = 1, z = 1\}$ in π, so let us now consider what happens when we find the equation of the projective figure which corresponds to an unbounded conic, such as the hyperbola

$$\{(x, y, z) : y^2 - 4x^2 = 1, z = 1\}.$$

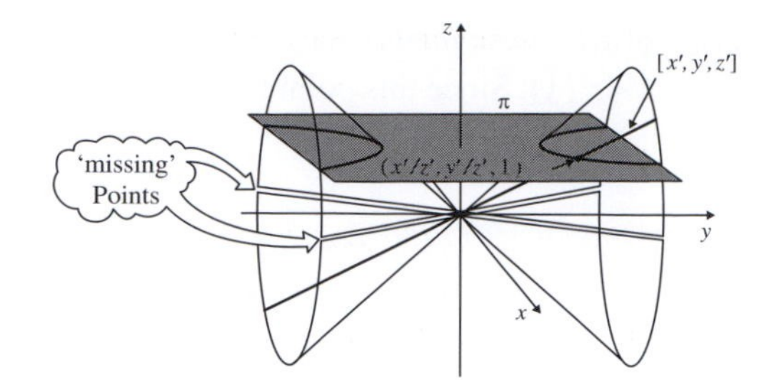

Any Point $[x', y', z']$ on the corresponding projective figure must pierce π at a point $(x'/z', y'/z', 1)$ on the hyperbola, and so

$$\left(\frac{y'}{z'}\right)^2 - 4\left(\frac{x'}{z'}\right)^2 = 1.$$

Since z' is non-zero, we can multiply by $(z')^2$ and drop the dashes, to obtain the equivalent equation

If z' were zero, then $[x', y', z']$ would not pierce π.

$$y^2 - 4x^2 = z^2, \quad z \neq 0.$$

As in the case of the circle, this is the equation of a cone-like family of Points in $\mathbb{RP}^2$ with vertex at the origin, as shown above. Notice, however, that because $z \neq 0$, two Points are missing from the family. These must be Points of the form $[x, y, 0]$ which satisfy the equation $y^2 - 4x^2 = 0$. Since this equation implies that $y = \pm 2x$, it follows that the missing Points are $[1, 2, 0]$ and $[1, -2, 0]$.

But should these Points really be omitted from the projective figure? After all, both Points are ideal Points for the standard embedding plane, and figures in an embedding plane often have ideal Points associated with them. In projective geometry we take the view that the ideal Points $[1, 2, 0]$ and $[1, -2, 0]$ *are* associated with the hyperbola $y^2 - 4x^2 = 1$, and hence that the corresponding projective figure consists of the entire cone-like family of Points $[x, y, z]$ that satisfy the equation $y^2 - 4x^2 = z^2$.

Problem 1 Find an equation for the projective figure in $\mathbb{RP}^2$ which corresponds to the parabola $\{(x, y, z) : y = x^2, z = 1\}$ in the standard embedding plane. Which ideal Points should be associated with the parabola?

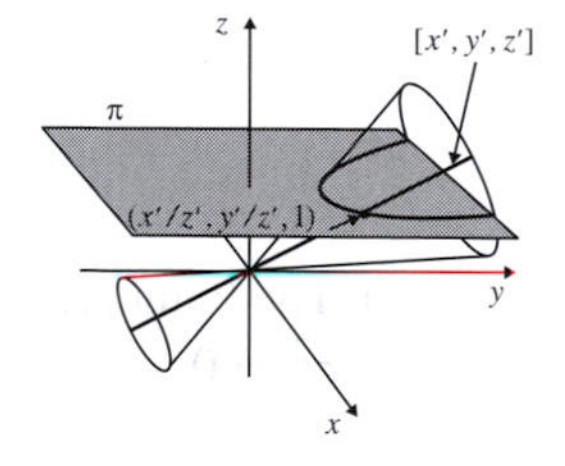

In general, any conic in the standard embedding plane can be expressed in the form

$$\{(x, y, z) : Ax^2 + Bxy + Cy^2 + Fx + Gy + H = 0, z = 1\}.$$

Since any Point $[x', y', z']$ on the corresponding projective figure must pierce the standard embedding plane at a point $(x'/z', y'/z', 1)$ on the conic, it follows that

$$A\left(\frac{x'}{z'}\right)^2 + B\left(\frac{x'}{z'}\right)\left(\frac{y'}{z'}\right) + C\left(\frac{y'}{z'}\right)^2 + F\left(\frac{x'}{z'}\right) + G\left(\frac{y'}{z'}\right) + H = 0.$$

Multiplying by $(z')^2$, and dropping the dashes, we obtain the equivalent equation

$$Ax^2 + Bxy + Cy^2 + Fxz + Gyz + Hz^2 = 0, \quad z \neq 0.$$

If we drop the constraint that $z \neq 0$, then we can include those ideal Points for the standard embedding plane that should be associated with the plane conic. The corresponding projective figure (including the additional ideal Points) is known as a *projective conic*.

The additional Points have homogeneous coordinates of the form $[x, y, 0]$ where

$$Ax^2 + Bxy + Cy^2 = 0.$$

Definition A **projective conic** in $\mathbb{RP}^2$ is a set of Points whose homogeneous coordinates satisfy a second-degree equation of the form

$$Ax^2 + Bxy + Cy^2 + Fxz + Gyz + Hz^2 = 0. \qquad (2)$$

For example, $xy + xz + yz = 0$ defines a projective conic, because it has the form of equation (2) with $A = C = H = 0$ and $B = F = G = 1$. However, $x^2 + y^2 - 3y + z^2 = 0$ does not define a projective conic, because it includes a linear term in y.

Problem 2 Which of the following equations define projective conics?

(a) $x^2 + xy - 3y^2 + 4x - 3y + z^2 = 0$
(b) $x^2 + xy + y^2 + yz = 0$
(c) $y^2 = xz$
(d) $x^2 + y^2 + z^2 = 2$

We say that a Point P **lies on** a projective conic, or a projective conic **passes through** a Point P, if the homogeneous coordinates of P satisfy the equation of the projective conic. For example, the Point $[3, 4, 5]$ lies on the projective conic $x^2 + y^2 = z^2$, since $3^2 + 4^2 = 5^2$; however, $x^2 + y^2 = z^2$ does not pass through the Point $[1, 1, 1]$, since $1^2 + 1^2 \neq 1^2$.

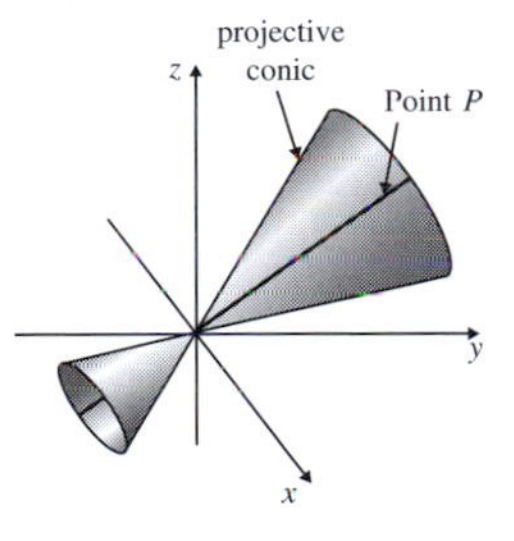

Problem 3 Which of the following statements are true?

(a) The projective conic $xy + xz + yz = 0$ passes through the Point $[1, 0, 0]$.
(b) The Point $[1, 2, 0]$ lies on the projective conic $2x^2 - y^2 + xy + xz + z^2 = 0$.
(c) The projective conic $3x^2 + 2y^2 - z^2 = 0$ passes through the Point $[1, 2, 3]$.

In this chapter we concentrate on those projective conics that can be represented by non-degenerate conics in the standard embedding plane.

It follows that a degenerate projective conic consists of a pair of Lines, a single Line, a Point, or the empty set (that is, 'no Points').

Definition A projective conic is **non-degenerate** if it can be represented by a non-degenerate conic in the standard embedding plane.

For example, the projective conic with equation

$$x^2 + 2y^2 - z^2 = 0$$

is non-degenerate because it is represented in the standard embedding plane by the ellipse

$$\{(x, y, z) : x^2 + 2y^2 = 1, z = 1\}.$$

On the other hand, the projective conic with equation

$$27x^2 + 30xy - 8y^2 + 14yz - 3z^2 = 0$$

is degenerate because

$$27x^2 + 30xy - 8y^2 + 14yz - 3z^2 = (9x - 2y + 3z)(3x + 4y - z),$$

so the projective conic intersects the standard embedding plane in the degenerate conic which consists of the following pair of lines:

$$\{(x, y, z) : 9x - 2y + 3 = 0, z = 1\}$$

and

$$\{(x, y, z) : 3x + 4y - 1 = 0, z = 1\}.$$

The following theorem shows that non-degenerate projective conics are preserved by projective transformations.

Theorem 1 Let t be a projective transformation, and let E be a non-degenerate projective conic. Then $t(E)$ is a non-degenerate projective conic.

Proof Let E have equation

$$Ax^2 + Bxy + Cy^2 + Fxz + Gyz + Hz^2 = 0. \tag{2}$$

Then any Point that lies on E has homogeneous coordinates $[x, y, z]$ which satisfy equation (2). If $[x', y', z']$ is the image of $[x, y, z]$ under t, then under the inverse transformation we have $[x, y, z] = t^{-1}([x', y', z'])$. It follows that if

$$\begin{pmatrix} a & b & c \\ d & e & f \\ g & h & k \end{pmatrix}$$

is a matrix associated with t^{-1}, then

$$x = ax' + by' + cz', \quad y = dx' + ey' + fz', \quad z = gx' + hy' + kz'.$$

Substituting these expressions for x, y and z into equation (2), we obtain a second-degree equation in x', y' and z'. It follows that the image of E under t is a projective conic.

Next we show that this image cannot be degenerate. A degenerate image would consist of a pair of Lines, a single Line, a Point, or the empty set (that is 'no Points'). Since the projective transformation t^{-1} maps Lines to Lines and Points to Points, it would map the degenerate image to another degenerate projective conic. But this cannot happen since t^{-1} maps the image back to the original non-degenerate projective conic E. It follows that $t(E)$ cannot be degenerate. ■

4.1.2 Tangents to Projective Conics

Let E be any non-degenerate projective conic, and let ℓ be a Line in $\mathbb{RP}^2$. Then ℓ is a plane in $\mathbb{R}^3$ which passes through O, and E is a surface in $\mathbb{R}^3$ which consists of a 'cone-like' family of lines through the origin O. It follows that there are three possibilities:

ℓ can meet E at a pair of Points;
ℓ can meet E at a single Point;
ℓ can meet E at no Points.

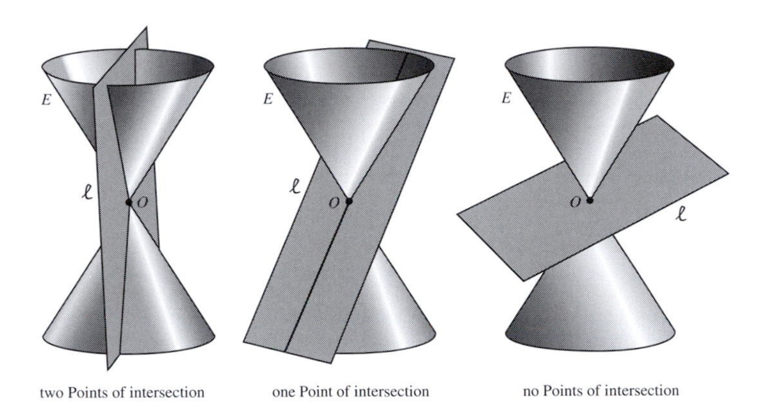

In the second case, ℓ just 'touches' E along a 'Point of contact' P. This suggests that in such a case we define ℓ to be the tangent to E at P.

Definition Let E be a non-degenerate projective conic. Then a Line ℓ is a **tangent to** E **at** P if ℓ meets E at a Point P, and at no other Point.

We can also define whether a Point lies inside or outside a projective conic.

Definitions Let E be a non-degenerate projective conic. A Point Q lies **inside** E if every Line through Q meets E at two distinct Points. A Point R lies **outside** E if there is a Line through R that meets E at no Points.

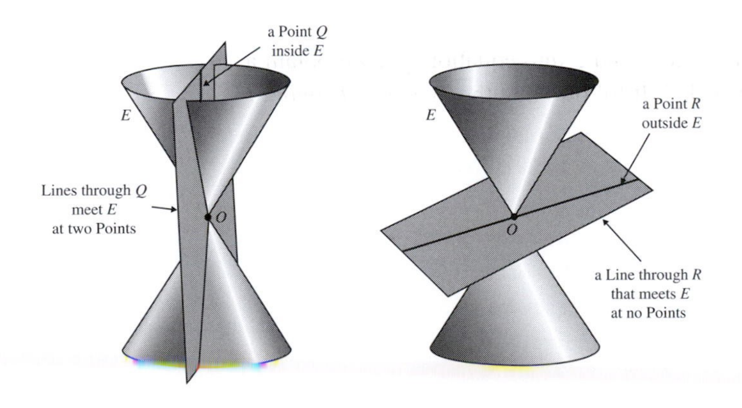

The following theorem shows that 'tangency' and 'lying inside or outside a projective conic' are projective properties.

Theorem 2 Let t be a projective transformation, and let the Line ℓ be a tangent to a non-degenerate projective conic E at a Point P. Then $t(\ell)$ is a tangent to $t(E)$ at $t(P)$. Also, if Q is a Point inside E, then $t(Q)$ lies inside $t(E)$; and if R is a Point outside E, then $t(R)$ lies outside $t(E)$.

Proof By the definition of tangent, P is the only Point that ℓ and E have in common. Since t is a one–one map of $\mathbb{RP}^2$ onto itself, it follows that $t(P)$ is the only Point that $t(\ell)$ and $t(E)$ have in common. In other words, $t(\ell)$ is a tangent to $t(E)$ at $t(P)$.

Also, if Q lies inside E, then any Line ℓ' through $t(Q)$ must meet $t(E)$ at two distinct Points, for otherwise the Line $t^{-1}(\ell')$ through Q would not meet E in two distinct Points. It follows that $t(Q)$ lies inside $t(E)$.

Again, if R lies outside E, then there is a Line ℓ through R that does not meet E. It follows that $t(\ell)$ is a Line through $t(R)$ that does not meet $t(E)$, and so $t(R)$ lies outside $t(E)$. ■

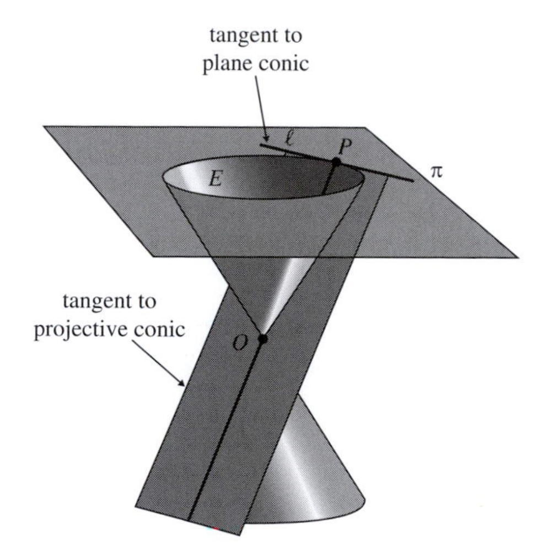

The figure above illustrates that, in an embedding plane, tangents to projective conics correspond to tangents to plane conics, and vice versa, and so the two notions of tangency are consistent.

Moreover, since a projective transformation preserves the property of being a tangent, it follows that we can use projective geometry to tackle problems involving tangents to plane conics. You will meet several examples of this later, but first we must show that all non-degenerate projective conics are projective-congruent.

4.1.3 Some Preliminaries

In our proof that all non-degenerate projective conics are projective-congruent we need a number of facts concerning plane conics. In order to concentrate on the key ideas in the proof then, we deal with these facts now.

Theorem 5, Subsection 4.1.4

First, in our proof we shall need to calculate the eccentricity of conics which are symmetrical about the v-axis in the (u, v)-plane, and which have a focus that lies on the v-axis. We do this by using the following result.

Theorem 3 Eccentricity Formula
Let E be a non-degenerate plane conic with equation

$$u^2 + Cv^2 + Gv + H = 0.$$

If E has a focus on the v-axis, then the eccentricity e of E is given by the formula

$$e^2 = 1 - C.$$

We use this result in Subsection 4.1.4.

We use (u, v) as our coordinate system here rather than (x, y) to match our notation in later discussion.

For example, the eccentricity e of the ellipse with equation

$$u^2 + \tfrac{1}{2}v^2 - 7v + 4 = 0$$

is given by $e^2 = 1 - \frac{1}{2} = \frac{1}{2}$, so that $e = 1/\sqrt{2}$.

Proof First suppose that E is a circle. Then $C = 1$, and by convention $e = 0$, so $e^2 = 1 - C$, as required.

Next suppose that E is not a circle. Since the equation of E has no terms that involve u or uv, it follows that E is symmetrical about the v-axis. Also, since E has a focus F on the v-axis, it follows, by symmetry, that the directrix which corresponds to F is perpendicular to the v-axis. Hence F has coordinates $(0, r)$ for some real number r, and the directrix has equation $v = s$, for some real number s.

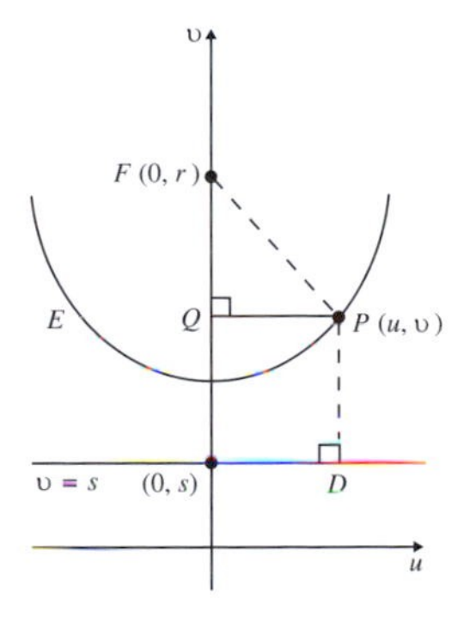

Let $P(u, v)$ be an arbitrary point on the conic, and let PQ be the perpendicular from P to the v-axis. Then, by Pythagoras' Theorem, we have

$$PF^2 = FQ^2 + QP^2 = (v - r)^2 + u^2.$$

Now let PD be the perpendicular from P to the directrix. Then, by the focus-directrix property, we have $PF = e \cdot PD = e \cdot |v - s|$, so

$$e^2(v - s)^2 = PF^2 = (v - r)^2 + u^2.$$

Expanding the brackets and collecting terms, we obtain

$$u^2 + (1 - e^2)v^2 - 2(r - e^2 s)v + (r^2 - e^2 s^2) = 0.$$

Comparing this with the equation of E in the statement of the theorem, we see that $C = (1 - e^2)$, and hence that $e^2 = 1 - C$. ∎

Problem 4 Determine the eccentricity of the hyperbola with equation

$$2u^2 - 6v^2 + 5v - 1 = 0.$$

In our work in the following subsection we shall also need an understanding of the relationship between conics which have the same eccentricity. So, first, observe that all non-degenerate conics with a given distance d between the focus and the directrix, and a given eccentricity e, are Euclidean-congruent to each other. This means that the size and shape of a non-degenerate conic are completely determined by the numbers e and d.

Euclidean-congruence was explained in Subsection 2.1.2.

If we now fix e and vary the size of d, then the shape of the conic remains the same but the size of the conic changes. The following figure shows the effect that an increase in d has on each type of conic.

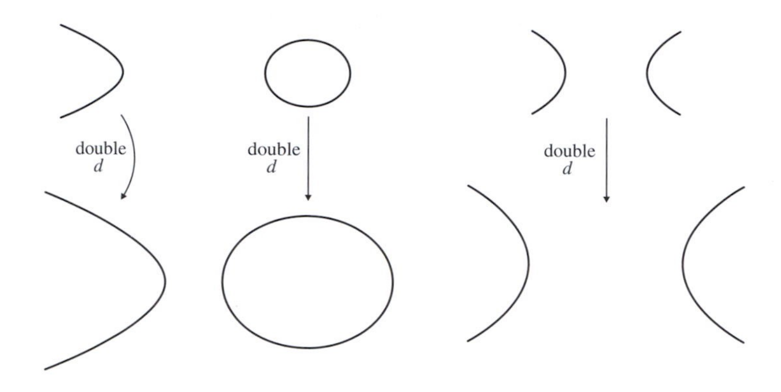

By allowing d to vary throughout the interval $(0, \infty)$, we obtain every size of conic with a given eccentricity e. We can therefore obtain any non-degenerate conic by first choosing a conic with the correct eccentricity and then adjusting its size by varying d.

Recall that $(0, \infty)$ denotes the interval $\{x : x > 0\}$.

4.1.4 Conics in Perspective

We now demonstrate that, for any two given plane conics, there is a perspectivity between them. Indeed, we can draw a right circular cone such that each of the given plane conics arises as the intersection of the cone with a suitable plane. It follows that there is a projective transformation that maps any given projective conic onto any other, so the property of 'being a conic' is a projective property. This enables us to prove quite surprising results about plane conics.

This fact about the intersection was known to the ancient Greek mathematicians around 300 BC.

Plane Sections of a Right Circular Cone

Earlier, we asserted that we can construct all conics by taking different plane sections through double cones. We now justify that claim, and describe its relevance to projective geometry.

Subsection 1.1.1

First, we generate a hollow right circular cone by taking a line through the origin in $\mathbb{R}^3$ at some angle ϕ to the (x, y)-plane, and rotating that line around the z-axis. We call the line a **generator** of the cone. Next, we cut the cone with planes (none of which passes through the origin) at various angles, and make the following observations illustrated below. (a) When the plane is horizontal,

Here $0 < \phi < \pi/2$

the section is a circle. (b) As the plane tilts, its curve of intersection with the cone looks like an ellipse. (c) When the plane becomes parallel to one of the generators of the cone, the curve of intersection looks like a parabola. (d) As the plane tilts further, it meets both portions of the cone, and the curve of intersection looks like a hyperbola. (e) When the plane is vertical we obtain the fattest possible hyperbola that can be obtained from *this* cone. The asymptotes of this hyperbola are parallel to a pair of lines on the cone that pass through the origin.

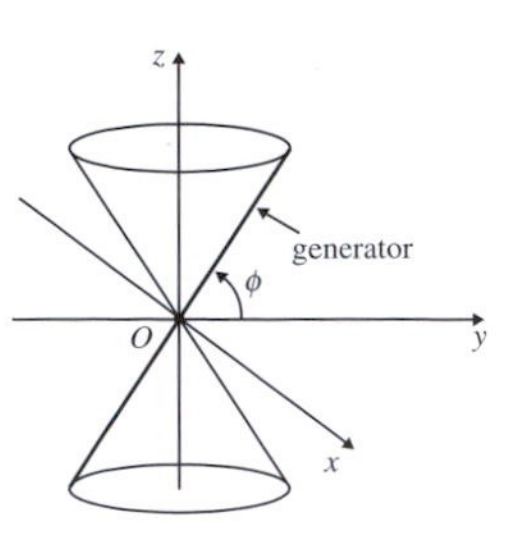

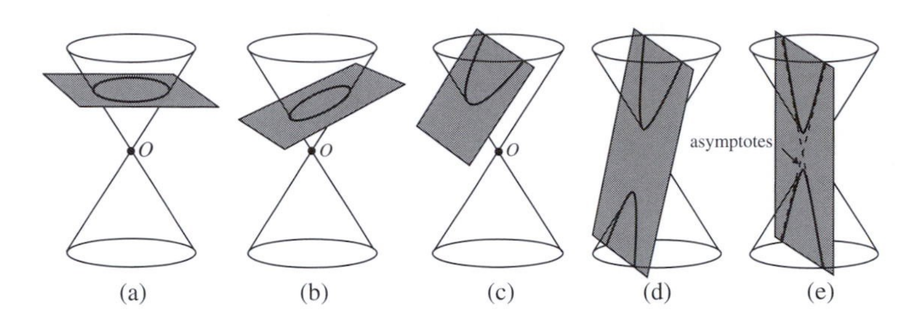

To prove that these observations are correct, we do some algebra.

Since the generator of the cone through the first quadrant of the (y, z) -plane has slope $\tan\phi$, this generator has equation $\{(x, y, y) : z = \tan\phi \cdot y,\ x = 0\}$, or $\left\{(x, y, z) : y = z\big/\tan\phi,\ x = 0\right\}$. Rotating this generator around the z-axis gives the whole of the cone, which must therefore have equation

$$x^2 + y^2 = \frac{z^2}{\tan^2\phi}. \tag{3}$$

This equation describes *both* the upper part *and* the lower part of the cone.

Next, let π be the plane in which we are interested that cuts the cone, and let θ be the angle between the (x, y)-plane and π. We shall assume for convenience that the x- and y-axes have been chosen so that π intersects the (x, y)-plane in a line parallel to the x-axis. Then if we rotate the (x, y)-plane about the x-axis through an angle θ, it becomes parallel to π.

Here $0 \leq \theta < \pi/2$.

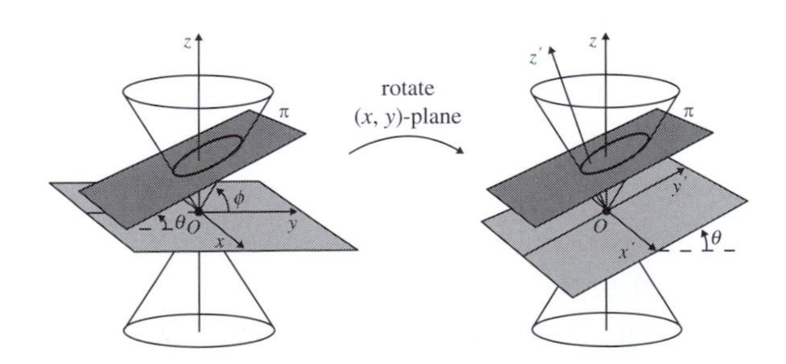

This rotation moves the x-, y- and z-axes into new positions, which we call the x'-, y'- and z'-axes. If a given point in $\mathbb{R}^3$ has coordinates (x, y, z) and

(x', y', z') with respect to these two sets of axes, then the connection between these coordinates is given by the matrix equation

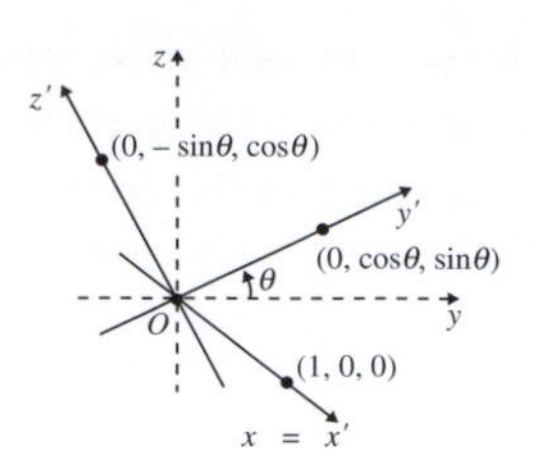

$$\begin{pmatrix} x \\ y \\ z \end{pmatrix} = \begin{pmatrix} 1 & 0 & 0 \\ 0 & \cos\theta & -\sin\theta \\ 0 & \sin\theta & \cos\theta \end{pmatrix} \begin{pmatrix} x' \\ y' \\ z' \end{pmatrix}. \qquad (4)$$

Notice that here the columns of the 3×3 matrix of the rotation comprise the coordinates (with respect to the initial x-, y-, z-axes) of the points 1 unit along each of the final x'-, y'-, z'-axes (as shown in the margin).

Next, we translate the (x', y')-plane through some distance d parallel to the z'-axis until it coincides with the plane π. This translation sends the x'-, y'- and z'-axes into new positions, which we call the u-, v- and w-axes.

Here d does not denote a focus-directrix distance, but an arbitrary positive number.

If a given point in $\mathbb{R}^3$ has coordinates (x', y', z') and (u, v, w) with respect to these two sets of axes, then the coordinates are related by

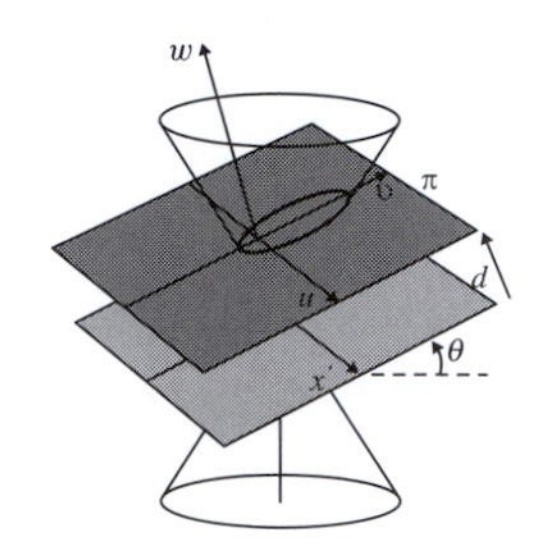

$$\begin{pmatrix} x' \\ y' \\ z' \end{pmatrix} = \begin{pmatrix} u \\ v \\ w \end{pmatrix} + \begin{pmatrix} 0 \\ 0 \\ d \end{pmatrix}. \qquad (5)$$

Overall, it follows from equations (4) and (5) that if a given point has coordinates (x, y, z) with respect to the x-, y- and z-axes, and coordinates (u, v, w) with respect to the u-, v- and w-axes, then

$$\begin{aligned} \begin{pmatrix} x \\ y \\ z \end{pmatrix} &= \begin{pmatrix} 1 & 0 & 0 \\ 0 & \cos\theta & -\sin\theta \\ 0 & \sin\theta & \cos\theta \end{pmatrix} \left(\begin{pmatrix} u \\ v \\ w \end{pmatrix} + \begin{pmatrix} 0 \\ 0 \\ d \end{pmatrix} \right) \\ &= \begin{pmatrix} 1 & 0 & 0 \\ 0 & \cos\theta & -\sin\theta \\ 0 & \sin\theta & \cos\theta \end{pmatrix} \begin{pmatrix} u \\ v \\ w \end{pmatrix} + \begin{pmatrix} 0 \\ -d\sin\theta \\ d\cos\theta \end{pmatrix}. \end{aligned}$$

Now, for points on the curve where π intersects the cone, we have $w = 0$. So at these points,

$$\begin{pmatrix} x \\ y \\ z \end{pmatrix} = \begin{pmatrix} 1 & 0 & 0 \\ 0 & \cos\theta & -\sin\theta \\ 0 & \sin\theta & \cos\theta \end{pmatrix} \begin{pmatrix} u \\ v \\ 0 \end{pmatrix} + \begin{pmatrix} 0 \\ -d\sin\theta \\ d\cos\theta \end{pmatrix},$$

that is,

$$\left.\begin{aligned} x &= u, \\ y &= v\cos\theta - d\sin\theta, \\ z &= v\sin\theta + d\cos\theta. \end{aligned}\right\} \qquad (6)$$

Since the points of intersection lie on the cone, their coordinates (x, y, z) must satisfy the equation of the cone given in equation (3). Hence, if we substitute for x, y and z from equations (6) into equation (3), we obtain

$$u^2 + (v\cos\theta - d\sin\theta)^2 = \frac{(v\sin\theta + d\cos\theta)^2}{\tan^2\phi}.$$

After rearranging the terms in this equation, we obtain an equation of the form

$$u^2 + Cv^2 + Gv + H = 0, \tag{7}$$

where C, G and H are expressions involving θ and ϕ. In particular,

$$C = \cos^2\theta - \frac{\sin^2\theta}{\tan^2\phi} = \cos^2\theta\left(1 - \frac{\tan^2\theta}{\tan^2\phi}\right). \tag{8}$$

Since equation (7) is a second-degree equation in u and v, the curve of intersection is certainly a plane conic. Moreover, the conic is clearly non-degenerate with a focus on the v-axis. But which type of conic is it?

First, suppose that $\theta < \phi$, so that π is less steep than the generators of the cone. Then, it follows from equation (8) that $C > 0$. Hence, if we apply the '$B^2 - 4AC$ test' to equation (7), we find that $B^2 - 4AC = -4 \cdot 1 \cdot C < 0$, and so the curve of intersection is an ellipse.

This test was described in Section 1.3.

It follows from equations (7) and (8) and the Eccentricity Formula that the eccentricity e of the ellipse is given by

Theorem 3, Subsection 4.1.3

$$\begin{aligned} e^2 = 1 - C &= 1 - \cos^2\theta\left(1 - \frac{\tan^2\theta}{\tan^2\phi}\right) \\ &= 1 - \cos^2\theta + \frac{\sin^2\theta}{\tan^2\phi} \\ &= \sin^2\theta + \frac{\sin^2\theta}{\tan^2\phi} \\ &= \sin^2\theta(1 + \cot^2\phi) \\ &= \frac{\sin^2\theta}{\sin^2\phi}, \end{aligned}$$

so that

$$e = \frac{\sin\theta}{\sin\phi}.$$

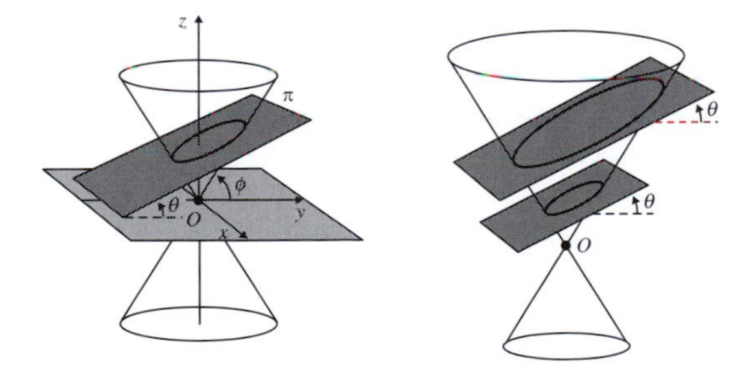

As θ increases from 0 to ϕ, e increases from 0 to 1. Thus, by tilting π through a suitable angle θ in the interval $[0, \phi)$, we can obtain an ellipse with any desired eccentricity between 0 and 1. If we then move π parallel to itself, the angle θ remains the same, and so the eccentricity of the ellipse remains the same, but the size of the ellipse can be adjusted by any desired dilatation factor.

As θ increases, the ellipse becomes longer and thinner.

It follows that we can obtain ellipses of all possible eccentricities and sizes by choosing the intersecting plane π to be at the appropriate angle and at the appropriate distance from the origin.

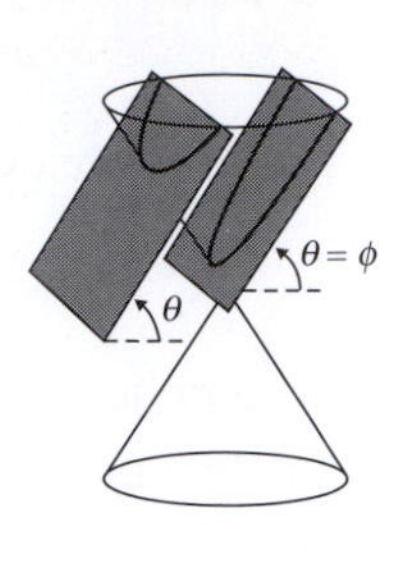

Next, suppose that $\theta = \phi$, so that π is parallel to a generator of the cone. Then, it follows from equation (8) that $C = 0$. Hence, if we apply the '$B^2 - 4AC$ test' to equation (7), we find that $B^2 - 4AC = -4 \cdot 1 \cdot C = 0$, and so the curve of intersection is a parabola.

As the plane moves further from the origin, the size of the parabola increases. It follows that we can obtain parabolas of all possible sizes by choosing the intersecting plane to be at the appropriate distance from the origin.

Finally, suppose that $\theta > \phi$, so that π is steeper than the generators of the cone. Then, it follows from equation (8) that $C < 0$. Hence, if we apply the '$B^2 - 4AC$ test' to equation (7), we find that $B^2 - 4AC = -4 \cdot 1 \cdot C > 0$, and so the curve of intersection is a hyperbola.

But, as we saw earlier, the eccentricity e of the hyperbola is given by

$$e = \frac{\sin\theta}{\sin\phi}.$$

As θ increases from ϕ to $\pi/2$, e increases from 1 to cosec ϕ.

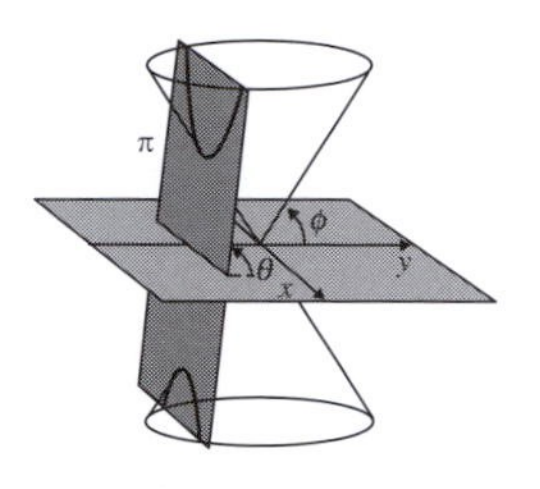

In particular, for each given value of θ, the eccentricities of the hyperbolas obtained from all planes of that slope are equal. And as the plane moves further from the origin, the size of the hyperbolas increases.

Thus, by tilting the plane of intersection through a suitable angle θ in the interval $[\phi, \pi/2]$, we can obtain a hyperbola with any eccentricity in the interval $[1, \text{cosec}\,\phi]$, and by moving the plane parallel to itself, we can adjust the size of the hyperbola by any desired factor. But how can we obtain a hyperbola with eccentricity greater than cosec ϕ?

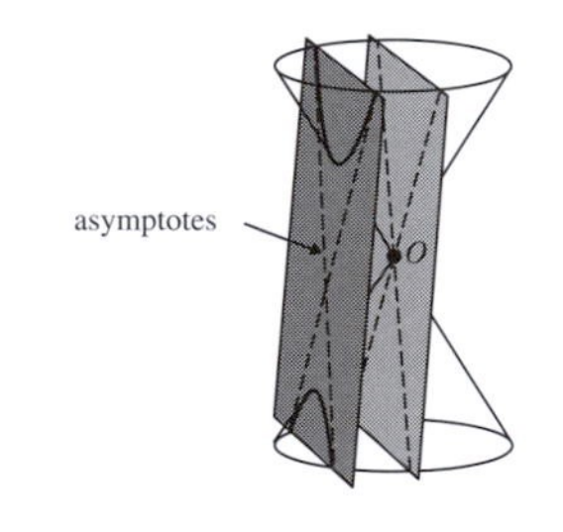

Well, notice that the asymptotes of each hyperbola are parallel to the two lines of intersection of the cone with a plane through the origin that is parallel to the intersecting plane. It follows that the angle between the asymptotes of the hyperbola can be any angle from 0 to the angle between the lines of intersection of the cone with a vertical plane through the origin. In other words, they can be no further apart than two opposite generators of the cone. So not every hyperbola can be found in every cone! Hence, in order to obtain any given hyperbola as a curve of intersection of a plane with a cone, we need to choose a 'fat enough' cone (that is, a cone with a sufficiently small angle ϕ).

This completes the proof of the following fact.

Theorem 4 Every non-degenerate plane conic can be found as the curve of intersection of a suitable right circular cone with a suitable plane.

We can now use this theorem to illustrate that there is a perspectivity between any two non-degenerate plane conics.

First, we illustrate why there is a perspectivity between any two ellipses E_1 and E_2. Choose any right circular cone with vertex at the origin O. Then it follows from the details in the proof of Theorem 4 that there are two planes π_1 and π_2 whose curves of intersection with the cone are E_1 and E_2. The required perspectivity is simply the point-to-point mapping of E_1 onto E_2 along the generators of the cone.

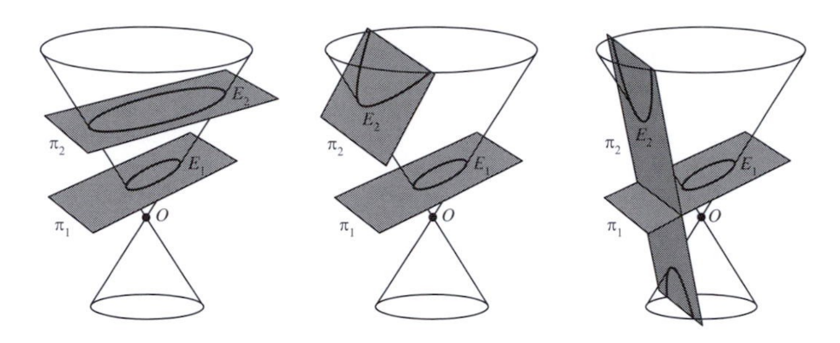

Similarly, there is a perspectivity between any ellipse E_1 and any parabola E_2. For, given any right circular cone with vertex at the origin O, it follows from the earlier explanations that there are two planes π_1 and π_2 whose curves of intersection with the cone are E_1 and E_2.

Here we are assuming that π_2 is an embedding plane, and hence that its intersection E_2 with the cone includes the ideal Point associated with the parabola.

In the same way, there is a perspectivity between any ellipse E_1 and any hyperbola E_2. Again, the perspectivity maps each pair of curves point-to-point along generators of the cone—but this time we need to choose a sufficiently fat cone in the first place, so that *some* plane intersects the cone in the hyperbola E_2.

Here E_2 includes two ideal Points.

In general, there is a perspectivity which maps any given non-degenerate plane conic onto any other given non-degenerate plane conic. It can be realized as a point-to-point map along the generators of a cone that is fat enough to yield both conics as sections through the cone.

This map is one–one and onto provided that the conics include their associated ideal Points.

Next, just as there is a perspectivity between any two non-degenerate plane conics there is a projective transformation that maps any non-degenerate projective conic onto any other non-degenerate projective conic.

Theorem 5 All non-degenerate projective conics are projective-congruent.

We postpone the proof of Theorem 5 to Section 4.4, to concentrate on our main story-line.

Three Tangents Theorem

The correspondence between plane conics in an embedding plane and projective conics in $\mathbb{RP}^2$ enables us to match Euclidean theorems about plane conics with projective theorems about projective conics. Provided that the theorems are concerned exclusively with projective properties, then a Euclidean theorem will hold if and only if the corresponding projective theorem holds.

Later we shall meet a whole range of applications of the projective-congruence of all non-degenerate projective conics. However we end this

Section 4.3

section with the following striking result, to give you a 'taster' of things to come!

Theorem 6 Three Tangents Theorem
Let a non-degenerate plane conic touch the sides BC, CA and AB of a triangle $\triangle ABC$ in $\mathbb{R}^2$ at the points P, Q and R, respectively. Then AP, BQ and CR are concurrent.

The following figures illustrate the Three Tangents Theorem for an ellipse and a parabola.

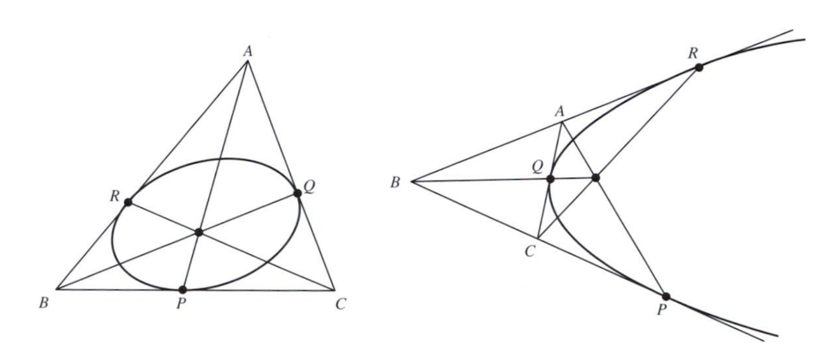

Proof The theorem concerns a *non-degenerate conic*, its *tangents*, and *concurrency of lines*. Since all of these properties are projective properties, it is sufficient to prove the result for *any* non-degenerate plane conic. For simplicity, therefore, we take the plane conic to be a circle.

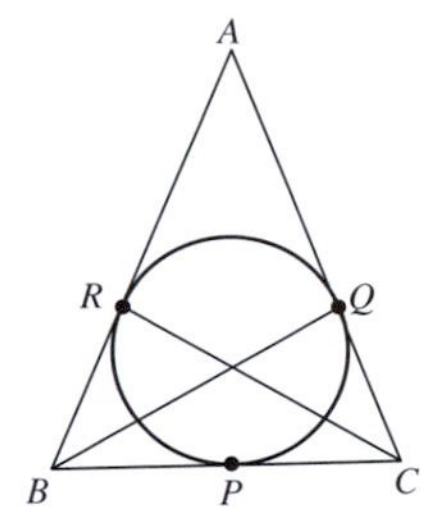

Subsection 2.4.2

Since, by symmetry, the two tangents from a point to a circle are of equal length, it follows that $AQ = AR, BP = BR$ and $CP = CQ$, so that, in particular,

$$\frac{AR}{RB} \cdot \frac{BP}{PC} \cdot \frac{CQ}{QA} = 1.$$

By applying the converse to Ceva's Theorem to this equation, it follows that the lines AP, BQ and CR are concurrent. ■

4.2 Tangents

Many results about plane conics and projective conics involve properties of their tangents and polars. In this section we introduce a notation due to Joachimsthal that can be used to write down the equations of such tangents and polars.

Such results appear in Section 4.3.

Ferdinand Joachimsthal (1818–1861) was a distinguished German geometer, noted for his mature, polished exposition.

4.2.1 Tangents to Plane Conics

Let P be a point on a non-degenerate plane conic E, and let ℓ be the tangent at P to E. Next, let P' be a point close to P on E, and let ℓ' be the line through

P and P'. If we let P' approach P along the curve E, the direction of the line ℓ' approaches the direction of the tangent ℓ. We may phrase this rather loosely as follows: 'the direction of the tangent is the limiting direction of the chords'.

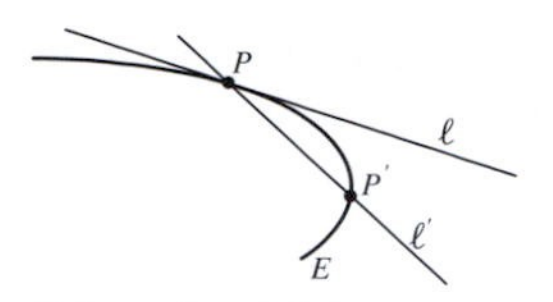

Subsection 1.2.1

We begin by recalling that a non-degenerate plane conic has an equation of the form

$$Ax^2 + Bxy + Cy^2 + Fx + Gy + H = 0.$$

If we denote the expression on the left-hand side of this equation by the symbol s, then the equation of the conic can be written very simply as

$$s = 0.$$

Joachimsthal's approach is to investigate the equations of tangents and polars to the conic by systematically attaching subscripts to the symbol s. For example, to check whether a point $P_1 = (x_1, y_1)$ lies on the conic it is necessary to replace the variables x and y in s by x_1 and y_1, respectively. This yields a number which we denote by the symbol s_{11}; in other words,

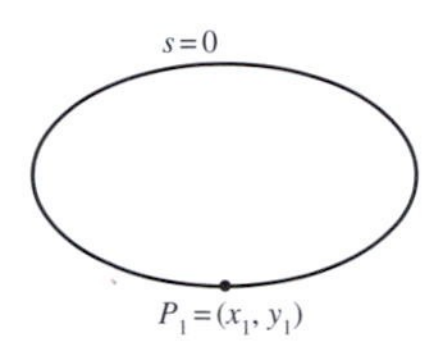

$$s_{11} = Ax_1^2 + Bx_1y_1 + Cy_1^2 + Fx_1 + Gy_1 + H.$$

If $s_{11} = 0$, then we can conclude that P_1 lies on the conic; and if $s_{11} \neq 0$, that P_1 does not lie on the conic.

Similarly, the point $P_2 = (x_2, y_2)$ lies on the conic if and only if the number s_{22} defined by

$$s_{22} = Ax_2^2 + Bx_2y_2 + Cy_2^2 + Fx_2 + Gy_2 + H$$

is equal to zero.

We may define a number s_{33} associated with a point $P_3 = (x_3, y_3)$ in a similar way.

Joachimsthal's notation can also be used to obtain a kind of 'average' number when we 'mix' the subscripts 1 and 2 to define a number s_{12} associated with two given points $P_1 = (x_1, y_1)$ and $P_2 = (x_2, y_2)$ to be

$$s_{12} = Ax_1x_2 + B\left(\frac{x_1y_2 + x_2y_1}{2}\right) + Cy_1y_2 + F\left(\frac{x_2 + x_1}{2}\right) + G\left(\frac{y_1 + y_2}{2}\right) + H.$$

We could define s_{13} and s_{23} in a similar way.

The reason for defining s_{12} in this way will become apparent shortly. Notice, for the moment, that the definition of s_{12} is symmetrical in the sense that the value of s_{12} is unaltered if we interchange the subscripts 1 and 2. In other words $s_{12} = s_{21}$, a fact which we will use later.

So far we have attached double subscripts to the symbol s, and this has always produced a *number*. Notice, however, that if we drop the second subscript in the definition of s_{12}, then the resulting expression is a linear expression in x and y, defined by

$$s_1 = Ax_1x + B\left(\frac{x_1y + xy_1}{2}\right) + Cy_1y + F\left(\frac{x + x_1}{2}\right) + G\left(\frac{y_1 + y}{2}\right) + H.$$

Here we temporarily use the term *linear expression in x and y* rather loosely, to mean an expression of the form $ax + by + c$, for some real numbers a, b and c.

Once again the reason for defining s_1 in this way will become apparent shortly. A clue is provided by the fact that in plane geometry a line can be defined by setting a linear expression in x and y equal to zero. It will turn out that the line defined by $s_1 = 0$ plays a particularly important part in our discussion of tangents.

We may give a general summary of Joachimsthal's notation in which we use the symbols i and j to stand for *arbitrary* subscripts, each of which can take the values 1, 2 or 3, as follows.

Joachimsthal's Notation for Plane Conics

Let a plane conic have equation $s = 0$, where

$$s = Ax^2 + Bxy + Cy^2 + Fx + Gy + H,$$

and let $P_1 = (x_1, y_1)$, $P_2 = (x_2, y_2)$ and $P_3 = (x_3, y_3)$ be points of $\mathbb{R}^2$. Then we define

$$s_i = Ax_ix + B\frac{x_iy + xy_i}{2} + Cy_iy + F\frac{x_i + x}{2} + G\frac{y_i + y}{2} + H,$$

$$s_{ii} = Ax_i^2 + Bx_iy_i + Cy_i^2 + Fx_i + Gy_i + H,$$

and

$$s_{ij} = Ax_ix_j + B\frac{x_iy_j + x_jy_i}{2} + Cy_iy_j + F\frac{x_i + x_j}{2} + G\frac{y_i + y_j}{2} + H,$$

where i and j can each take the values 1, 2, or 3.

In general it is simpler NOT to try to remember these formulas, but to remember the pattern for obtaining the expressions s_i, s_{ii} and s_{ij} from s.

The following example illustrates how to work out these expressions in Joachimsthal's notation for conics.

Example 1 Determine s_{11}, s_{22}, s_{12} and s_1 for the hyperbola with equation

$$3x^2 - 2xy - y^2 + 5x - y - 4 = 0$$

at the points $P_1 = (3, 2)$ and $P_2 = (-5, -2)$. Hence determine whether either P_1 or P_2 lies on the hyperbola.

Solution The equation of the conic may be written in Joachimsthal's notation as $s = 0$, where

$$s = 3x^2 - 2xy - y^2 + 5x - y - 4.$$

Since here we have $x_1 = 3, y_1 = 2, x_2 = -5$ and $y_2 = -2$, we deduce that

$$s_{11} = 3 \cdot 3^2 - 2 \cdot 3 \cdot 2 - 2^2 + 5 \cdot 3 - 2 - 4 = 20,$$

$$s_{22} = 3 \cdot (-5)^2 - 2 \cdot (-5) \cdot (-2) - (-2)^2 + 5 \cdot (-5) - (-2) - 4 = 24,$$

$$s_{12} = 3 \cdot 3 \cdot (-5) - 2\frac{3 \cdot (-2) + (-5) \cdot 2}{2} - 2 \cdot (-2) + 5\frac{3 - 5}{2}$$
$$- \frac{2 - 2}{2} - 4 = -34,$$

and

$$s_1 = 3 \cdot 3 \cdot x - 2\frac{3y + x \cdot 2}{2} - 2 \cdot y + 5\frac{3 + x}{2} - \frac{2 + y}{2} - 4$$
$$= \frac{19}{2}x - \frac{11}{2}y + \frac{5}{2}.$$

Since s_{11} and s_{22} are both non-zero, it follows that neither of the points P_1 or P_2 lies on the hyperbola. □

Problem 1 Determine s_{11}, s_{22}, s_{12} and s_1 for the plane conic with equation

$$2x^2 + 3xy - y^2 + x + 2y + 1 = 0$$

at the points $P_1 = (1, 0)$ and $P_2 = (2, 1)$.

Having introduced Joachimsthal's notation we now turn our attention to finding the equations of tangents to a conic $s = 0$. Since a tangent is a line which intersects the conic at two coincident points, we first describe how to determine the points where a given line ℓ meets $s = 0$.

Recall that every point P on the line ℓ through two given points $P_1 = (x_1, y_1)$ and $P_2 = (x_2, y_2)$ divides the segment P_1P_2 in the ratio $k : 1$, for some real number k, and so has coordinates that may be written in the form

Section Formula, Appendix 2

$$\left(\frac{kx_2 + x_1}{k + 1}, \frac{ky_2 + y_1}{k + 1}\right).$$

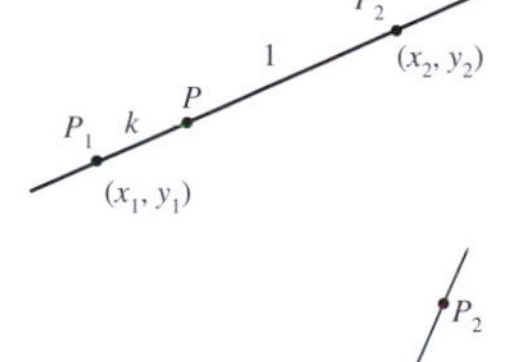

It follows that the line through the points P_1 and P_2 meets the conic with equation

$$(s =)\ Ax^2 + Bxy + Cy^2 + Fx + Gy + H = 0$$

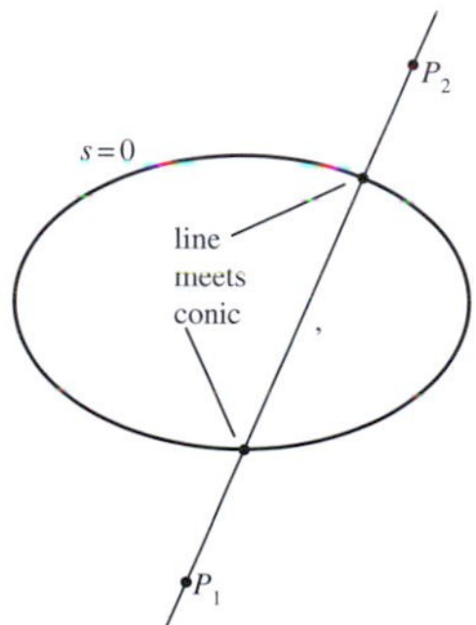

at points which divide the segment P_1P_2 in the ratio $k : 1$, where

$$A\left(\frac{kx_2 + x_1}{k + 1}\right)^2 + B\left(\frac{kx_2 + x_1}{k + 1}\right)\left(\frac{ky_2 + y_1}{k + 1}\right) + C\left(\frac{ky_2 + y_1}{k + 1}\right)^2$$
$$+ F\frac{kx_2 + x_1}{k + 1} + G\frac{ky_2 + y_1}{k + 1} + H = 0.$$

If we multiply both sides of this equation by $(k + 1)^2$ and collect the coefficients of the terms involving k^2, k and the terms independent of k, it turns out that we can rewrite this equation in terms of Joachimsthal's notation in the marvelously simple form

We omit the unedifying details.

$$s_{22}k^2 + 2s_{12}k + s_{11} = 0.$$

This equation occurs so frequently in our work that we give it a special name, **Joachimsthal's Section Equation.**

Since Joachimsthal's Section Equation is a quadratic equation in k, the line through P_1 and P_2 meet the conic at two distinct points, at one repeated point, or not at all depending on whether the quadratic equation has two distinct real roots, one repeated real root or no real roots, respectively.

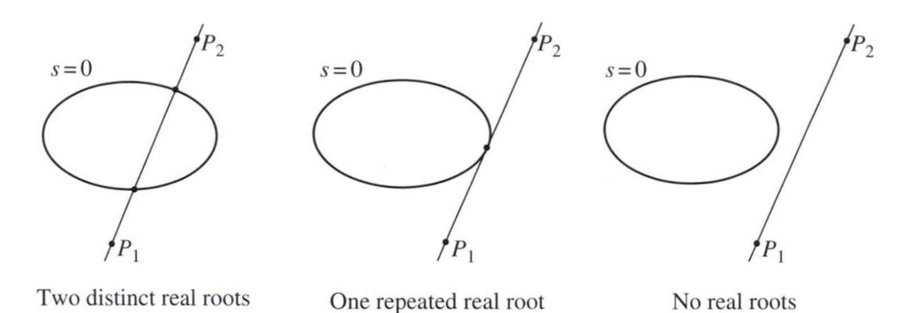

Example 2 Determine the ratios in which the hyperbola with equation

$$3x^2 - 2xy - y^2 + 5x - y - 4 = 0$$

divides the line segment from $P_1 = (3, 2)$ to $P_2 = (-5, -2)$.

Solution First observe that the hyperbola and the points P_1 and P_2 are the same as those used in Example 1, so we can use the values $s_{11} = 20, s_{22} = 24$ and $s_{12} = -34$ calculated there. It follows that we can rewrite Joachimsthal's Section Equation in this case as

$$24k^2 - 68k + 20 = 0,$$

or

$$6k^2 - 17k + 5 = 0,$$

so that

$$(3k - 1)(2k - 5) = 0.$$

Thus $k = \frac{1}{3}$ or $k = \frac{5}{2}$. Thus, the hyperbola divides the line segment P_1P_2 at two distinct points, in the ratios $\frac{1}{3} : 1$ and $\frac{5}{2} : 1$; that is, in the ratios 1:3 and 5:2, respectively. □

Problem 2 Determine the ratios in which the hyperbola with equation

$$2x^2 + 3xy - y^2 + x + 2y + 1 = 0$$

divides the line segment from $P_1 = (1, 0)$ to $P_2 = (2, 1)$.
Hint: Use your results from Problem 1.

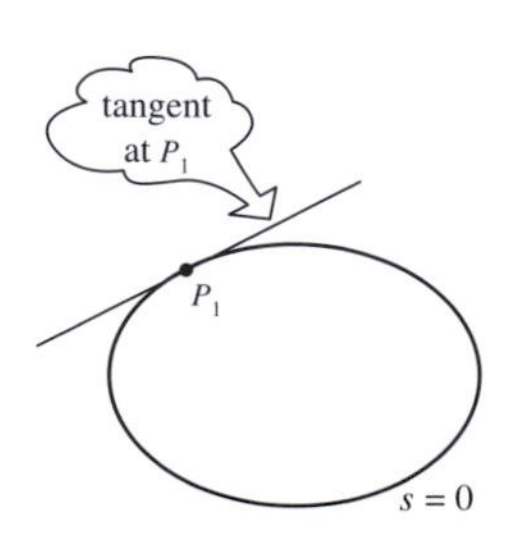

We are now in a position to find the equation of the tangent to the plane conic with equation $s = 0$ at a point $P_1 = (x_1, y_1)$ on the conic. To do this, let ℓ be a chord of the conic which passes through P_1, and let $P_2 = (x_2, y_2)$ be a point on ℓ which lies outside the conic. Since the chord through P_1 and P_2 meets the conic at two distinct points, it follows that Joachimsthal's Section Equation has two distinct (real) roots k.

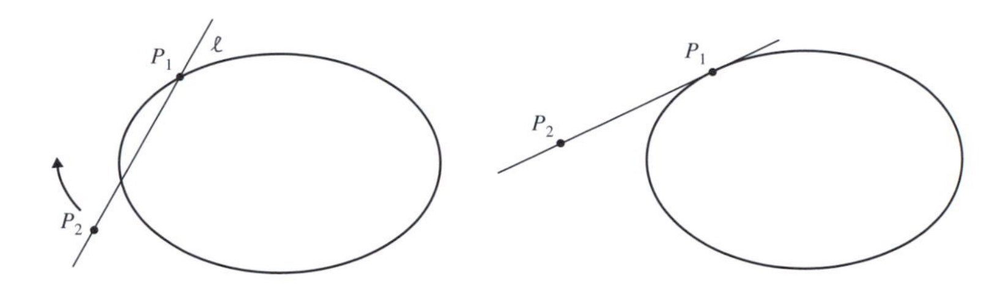

Now let P_2 move so that the chord through P_1 and P_2 becomes a tangent to the conic at P_1. This tangent is a line that meets the conic at the point P_1 alone, and so the corresponding Joachimsthal's Section Equation must have a repeated real root. Hence it follows that we must have

$$(2s_{12})^2 = 4s_{11}s_{22},$$

Recall that the condition for a quadratic equation

$$ax^2 + bx + c = 0$$

to have a repeated root is just $b^2 = 4ac$.

or, equivalently,

$$(s_{12})^2 = s_{11}s_{22}.$$

Since P_1 lies on the conic, we know that $s_{11} = 0$; so we must have

$$s_{12} = 0.$$

It follows from this equation that the point $P_2 = (x_2, y_2)$ must satisfy the equation $s_1 = 0$. But P_2 is an arbitrary point on the tangent to the conic at P_1, and so the tangent at P_1 must have the equation $s_1 = 0$.

Remember that s_1 is obtained from s_{12} by dropping the subscript 2.

Theorem 1 Let $P_1 = (x_1, y_1)$ be a point on a non-degenerate plane conic E with equation $s = 0$. Then the equation of the tangent to E at P_1 is $s_1 = 0$.

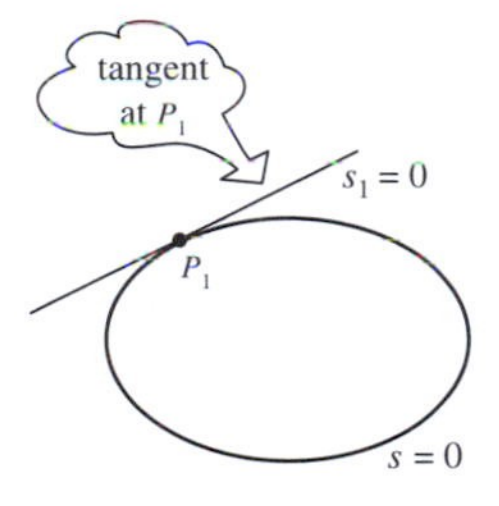

Example 3 Determine the equation of the tangent at $P_1 = (1, 1)$ to the hyperbola with equation

$$3x^2 - 2xy - y^2 + 5x - y - 4 = 0.$$

Solution The equation of the conic may be written in Joachimsthal's notation as $s = 0$, where

$$s = 3x^2 - 2xy - y^2 + 5x - y - 4.$$

Since here we have $x_1 = 1$ and $y_1 = 1$, we deduce that

$$\begin{aligned} s_1 &= 3 \cdot 1 \cdot x - 2\frac{1 \cdot y + x \cdot 1}{2} - 1 \cdot y + 5\frac{1 + x}{2} - \frac{1 + y}{2} - 4 \\ &= \tfrac{9}{2}x - \tfrac{5}{2}y - 2. \end{aligned}$$

The equation of the tangent at (1, 1) to the hyperbola is therefore

$$9x - 5y - 4 = 0. \qquad \square$$

Problem 3 Determine the equation of the tangent to each of the following non-degenerate plane conics at the given point:

(a) $x^2 - xy + 2y - 7 = 0$ at the point $(-1, 2)$;
(b) $3x^2 + 2xy - y^2 + x - 2y - 3 = 0$ at the point (1, 1).

Tangent Pair from a Point to a Conic

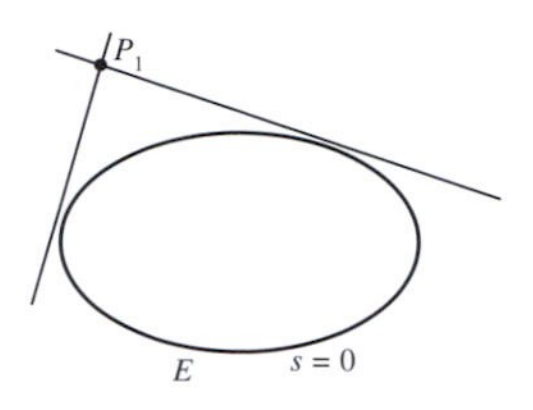

Now suppose that we wish to find the equation of a tangent to a plane conic E from a point P_1 outside the conic. Intuitively we would expect there to be *two* such tangents, as shown in the diagram in the margin. We now check this algebraically. The technique is similar to the case where P_1 lies on the conic, as we discussed above.

First we consider a chord of E that passes through the given point P_1 and another point P_2 that is also outside the conic; then we move P_2 so that the line becomes a tangent. As before, this occurs when Joachimsthal's Section Equation has a repeated real root, that is, when

$$(S_{12})^2 = S_{22} \cdot S_{11}.$$

Here $s_{11} \neq 0$, since P_1 does not lie on E.

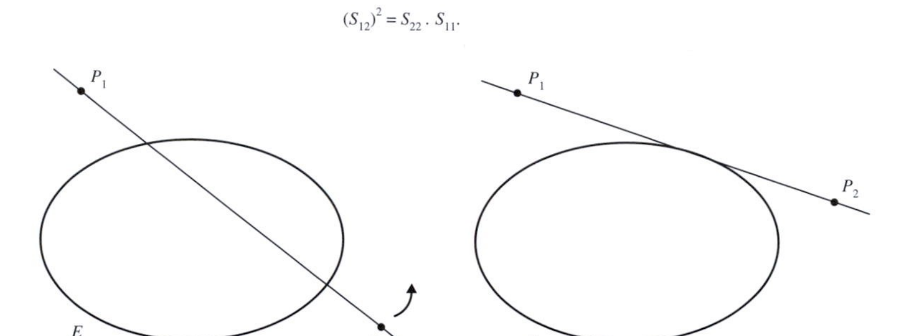

Then, since the point P_2 is an arbitrary point on the tangent, we can drop the subscript 2 to obtain the equation

$$(s_1)^2 = s \cdot s_{11}.$$

This is a second-degree equation in x and y; it factorizes into two linear equations which represent the *pair* of tangents from P_1 to E.

Theorem 2 Let P_1 be a point outside a non-degenerate plane conic E with equation $s = 0$. Then the equation of the tangent pair from P_1 to E is

$$(s_1)^2 = s \cdot s_{11}.$$

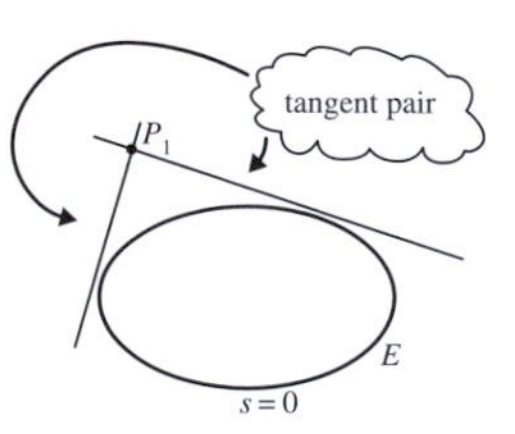

Example 4 Find the equations of the tangents from the point (1, 1) to the ellipse with equation $x^2 + 2y^2 = 1$.

Solution The equation of the ellipse may be written in Joachimsthal's notation as $s = 0$, where

$$s = x^2 + 2y^2 - 1.$$

Since here we have $x_1 = 1$ and $y_1 = 1$, we deduce that

$$s_{11} = 1^2 + 2 \cdot 1^2 - 1 = 2$$

and

$$s_1 = 1 \cdot x + 2 \cdot 1 \cdot y - 1 = x + 2y - 1.$$

The equation of the tangent pair is therefore

That is, $(s_1)^2 = s \cdot s_{11}$.

$$(x + 2y - 1)^2 = (x^2 + 2y^2 - 1) \cdot 2.$$

Multiplying out both sides and rearranging terms, we can rewrite this equation in the form

$$x^2 - 4xy + 2x + 4y - 3 = 0,$$

which we can then factorize as

$$(x - 1)(x - 4y + 3) = 0.$$

Hence the equations of the two tangents from the point (1, 1) to the ellipse are $x - 1 = 0$ and $x - 4y + 3 = 0$. □

Problem 4 One of the two tangents from the point (2, 1) to the hyperbola $4xy + 1 = 0$ has the equation $y = x - 1$. Find the equation of the other.

Poles and Polars

In Subsection 1.2.1 we defined the *polar* of a point with respect to the unit circle. We now extend this definition to a general non-degenerate plane conic.

Let $P_1 = (x_1, y_1)$ be a point outside a non-degenerate plane conic E with equation $s = 0$ (in Joachimsthal's notation), and suppose that $P_2 = (x_2, y_2)$ and $P_3 = (x_3, y_3)$ are the points where the tangents through P_1 meet E. These tangents have equations $s_2 = 0$ and $s_3 = 0$, respectively. Since P_1 lies on both tangents, it follows that

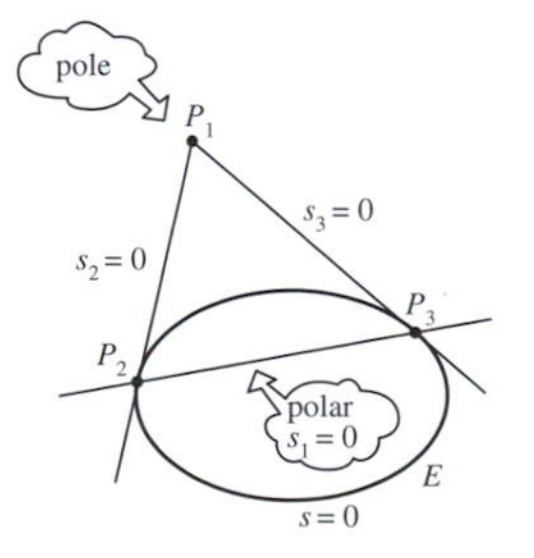

$$s_{12} = 0 \quad \text{and} \quad s_{13} = 0.$$

It follows from these equations that the points P_2 and P_3 both lie on the line with equation $s_1 = 0$, the so-called *chord of contact* or *polar* of P_1 with respect to the conic E.

Definitions Let E be a non-degenerate plane conic with equation $s = 0$, and let $P_1 = (x_1, y_1)$ be an arbitrary point of $\mathbb{R}^2$. Then the **polar of P_1 with respect to E** is the line with equation $s_1 = 0$. The point P_1 is called the **pole of the line $s_1 = 0$ with respect to E.**

This definition is analogous to the definition of polar with respect to a circle in Subsection 1.2.1.

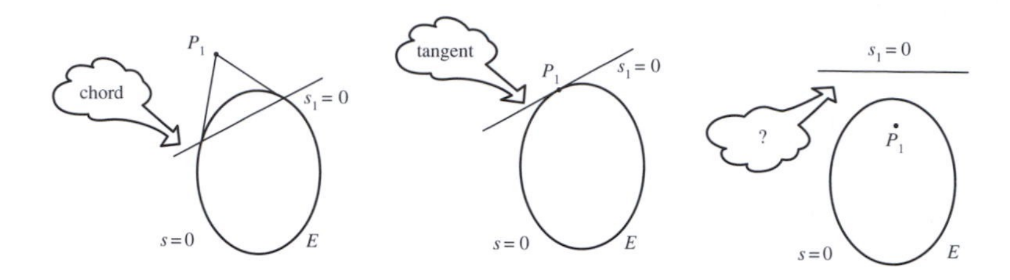

We have seen that if P_1 lies *outside* the conic E, then the polar $s_1 = 0$ is the chord which passes through the points where the tangent pair from P_1 touch E. Also, if P_1 is a point *on* the conic $s = 0$, then it follows from Theorem 1 that the polar $s_1 = 0$ is simply the tangent to E at P_1. If P_1 lies *inside* the conic E, then the polar is simply a particular line outside E that corresponds to P_1.

We will see in Theorem 3 below that this definition for the case that P_1 lies *inside* E results in theorems about poles and polars where we do not need to worry whether a point in the plane lies outside, on or inside a given conic.

Example 5 Determine the polar of $P_1 = (2, 2)$ with respect to the hyperbola E with equation

$$3x^2 - 2xy - y^2 + 5x - y - 4 = 0.$$

Solution The equation of the hyperbola E may be written in Joachimsthal's notation as $s = 0$, where

You met this hyperbola E previously, in Example 3.

$$s = 3x^2 - 2xy - y^2 + 5x - y - 4.$$

Since here we have $x_1 = 2$ and $y_1 = 2$, we deduce that

$$\begin{aligned} s_1 &= 3 \cdot 2 \cdot x - 2\frac{2 \cdot y + x \cdot 2}{2} - 2 \cdot y + 5\frac{2+x}{2} - \frac{2+y}{2} - 4 \\ &= \tfrac{13}{2}x - \tfrac{9}{2}y. \end{aligned}$$

It follows that the equation of the polar of P_1 with respect to E has equation $\frac{13}{2}x - \frac{9}{2}y = 0$, or $y = \frac{13}{9}x$. □

Problem 5 Determine the polar of $(1, -1)$ with respect to the hyperbola E with equation $2x^2 + xy - 3y^2 + x - 6 = 0$.

We are now able to state and prove a stunningly beautiful result concerning poles and polars.

Theorem 3 La Hire's Theorem
Let E be a non-degenerate plane conic, and let p_1 be the polar of a point P_1 in $\mathbb{R}^2$. Then each point of p_1 has a polar which passes through P_1.

Proof Let $P_1 = (x_1, y_1)$ be a point in $\mathbb{R}^2$. Then, by definition, the polar p_1 of P_1 with respect to E has the equation $s_1 = 0$.

Let $P_2 = (x_2, y_2)$ be any point on p_1, so that in particular we have

$$s_{12} = 0.$$

But from the definition of polar, we know that the polar of P_2 with respect to E must have equation $s_2 = 0$. It then follows from the equation $s_{12} = 0$ that the point P_1 must lie on the polar of P_2 with respect to E. This completes the proof. ■

4.2.2 Tangents to Projective Conics

Earlier, we defined a Line ℓ to be a tangent to a projective conic E if ℓ meets E at precisely one Point P. We then explained the connection between tangents to projective conics and tangents to plane conics. If ℓ is a tangent to a projective conic E, then the line ℓ' which represents ℓ in an embedding plane π is a tangent to the plane conic E' which represents E in π.

Subsection 4.1.2

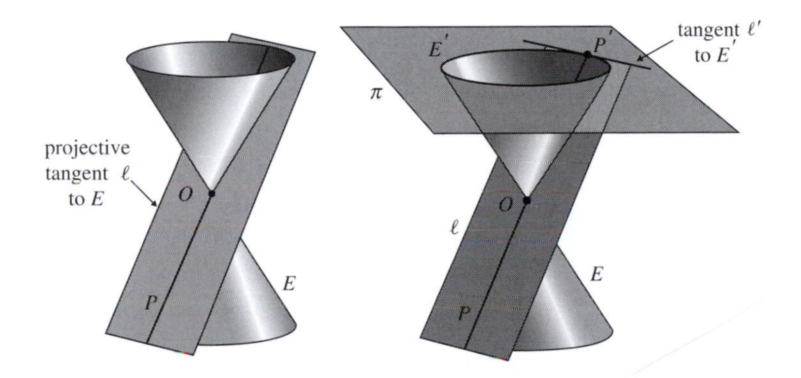

Because of this correspondence, we are able to use results about figures in $\mathbb{RP}^2$ to deduce results about figures in an embedding plane, and vice versa. In this subsection we use this correspondence to deduce Joachimsthal's formulas for projective conics from the corresponding formulas for plane conics. In preparation for this, we first extend the concepts of a *tangent pair* and a *polar* to a projective conic in $\mathbb{RP}^2$.

Let E be a non-degenerate projective conic, and let P be a Point in $\mathbb{RP}^2$ which lies outside E. If π is an embedding plane, then E is represented in π by a non-degenerate plane conic E', and P is represented by a Euclidean point P' in π. Now, in the embedding plane π we can draw a pair of tangents ℓ_1 and ℓ_2 from P' to E'. Back in $\mathbb{RP}^2$, the planes which pass through the origin

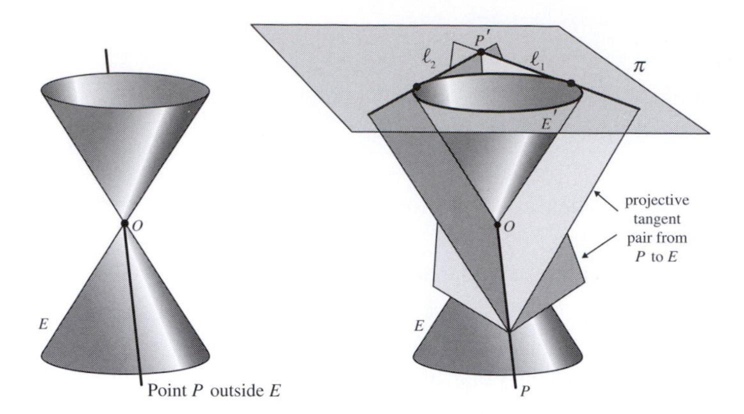

and the lines ℓ_1 and ℓ_2 are projective tangents to E. These tangents meet at the Point P, so they are called the **projective tangent pair** (or, simply, the **tangent pair**) from the Point P to the projective conic E.

Now let Q' and R' be the points in the embedding plane π at which ℓ_1 and ℓ_2 touch E'. Then the polar of P' with respect to E' is the line ℓ through Q' and R'. Back in $\mathbb{RP}^2$, the plane which passes through the origin and ℓ is a Line which we call the **projective polar** (or, simply, **polar**) of the Point P with respect to the projective conic E.

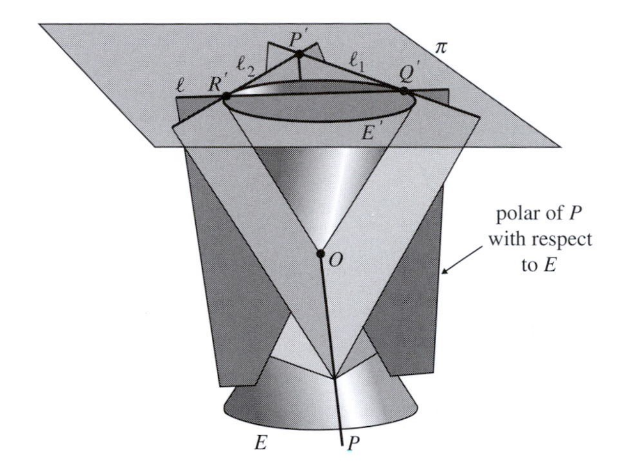

In order to see how Joachimsthal's notation can be extended from $\mathbb{R}^2$ to $\mathbb{RP}^2$, suppose that the work in Subsection 4.2.1 had all been carried out in the embedding plane $z = 1$. To illustrate the ideas involved, recall that Joachimsthal's equation for a plane conic is $s = 0$, where

$$s = Ax^2 + Bxy + Cy^2 + Fx + Gy + H.$$

Earlier, in Subsection 4.1.1, we described how the equation for a conic in $\mathbb{RP}^2$ can be obtained from the equation of the corresponding plane conic by replacing x by x/z and y by y/z, and then multiplying by z^2 to clear the fractions. The equation is still $s = 0$, but now

$$s = Ax^2 + Bxy + Cy^2 + Fxz + Gyz + Hz^2.$$

In a similar way, Joachimsthal's expressions, such as $s_1 = 0$ for a polar, or $s_1^2 = s \cdot s_{11}$ for a tangent pair, still hold in $\mathbb{RP}^2$ provided that we amend the expressions for s_1 and s_{11} by replacing x, y, x_1, y_1 by $x/z, y/z, x_1/z_1, y_1/z_1$, respectively, and then multiplying by powers of z and z_1 to clear fractions. The resulting amendments to s_1, s_{11}, etc., are as specified in the following notation.

Joachimsthal's Notation for Projective Conics
Let a projective conic have equation $s = 0$, where

$$s = Ax^2 + Bxy + Cy^2 + Fxz + Gyz + Hz^2,$$

and let $[x_1, y_1, z_1]$ and $[x_2, y_2, z_2]$ be Points of $\mathbb{RP}^2$. Then we define

$$\begin{aligned} s_1 &= Ax_1x + \tfrac{1}{2}B(x_1y + xy_1) + Cy_1y \\ &\quad + \tfrac{1}{2}F(x_1z + xz_1) + \tfrac{1}{2}G(y_1z + yz_1) + Hz_1z, \\ s_{11} &= Ax_1^2 + Bx_1y_1 + Cy_1^2 + Fx_1z_1 + Gy_1z_1 + Hz_1^2, \\ s_{12} &= Ax_1x_2 + \tfrac{1}{2}B(x_1y_2 + x_2y_1) + Cy_1y_2 + \tfrac{1}{2}F(x_1z_2 + x_2z_1) \\ &\quad + \tfrac{1}{2}G(y_1z_2 + y_2z_1) + Hz_1z_2. \end{aligned}$$

Example 6 Determine s_1, s_{11} and s_{12} for the projective conic

$$4x^2 + xy - 2y^2 - 8xz - 2yz + 4z^2 = 0$$

at the Points $[x_1, y_1, z_1] = [1, 0, 2]$ and $[x_2, y_2, z_2] = [1, 2, -1]$.

Solution Using Joachimsthal's notation with $A = 4, B = 1, C = -2$, $F = -8, G = -2$ and $H = 4$, we deduce that

$$\begin{aligned} s_1 &= 4x + \tfrac{1}{2}(y + 0) - 2 \cdot 0 - 4(z + 2x) - (0 + 2y) + 4 \cdot 2z \\ &= -4x - \tfrac{3}{2}y + 4z; \\ s_{11} &= 4 \cdot 1 + 0 - 0 - 8 \cdot 2 - 0 + 4 \cdot 4 = 4; \\ s_{12} &= 4 \cdot 1 + \tfrac{1}{2}(2 + 0) - 0 - 4(-1 + 2) - (0 + 4) + 4 \cdot 2 \cdot (-1) \\ &= -11. \end{aligned}$$

□

With the changes to Joachimsthal's notation described above, all of the formulas for polars, tangents and tangent pairs carry over from $\mathbb{R}^2$ to $\mathbb{RP}^2$. We therefore have the following theorem.

We omit the details.

Theorem 4 Let a projective conic E in $\mathbb{RP}^2$ have equation $s = 0$.

(a) If $P = [x_1, y_1, z_1]$ lies on E, then the tangent to E at P has equation $s_1 = 0$.
(b) If $P = [x_1, y_1, z_1]$ lies outside E, then the pair of tangents to E from P are given by the equation $s_1^2 = s \cdot s_{11}$.
(c) If $P = [x_1, y_1, z_1]$ is any Point in $\mathbb{RP}^2$, then the polar of P with respect to E is the Line with equation $s_1 = 0$.

In fact, if P lies inside or on E, then we *define* the **polar** of P to be the Line with equation $s_1 = 0$.

Example 7 The projective conic E has equation

$$4x^2 + xy - 2y^2 - 8xz - 2yz + 4z^2 = 0.$$

(a) Determine the equation of the tangent to E at the Point $[0, 1, 1]$.
(b) Determine the equations of the two tangents to E that pass through the Point $[1, 0, 2]$.
(c) Determine the polar of the Point $[1, 0, 2]$ with respect to E.

Solution

(a) Let $s = 4x^2 + xy - 2y^2 - 8xz - 2yz + 4z^2$ and $[x_1, y_1, z_1] = [0, 1, 1]$. Then

$$\begin{aligned} s_1 &= 0 + \tfrac{1}{2}(0 + x) - 2y - 4(0 + x) - (z + y) + 4z \\ &= -\tfrac{7}{2}x - 3y + 3z. \end{aligned}$$

Hence the equation of the tangent to E at $[0, 1, 1]$ is

$$-\tfrac{7}{2}x - 3y + 3z = 0,$$

or

$$7x + 6y - 6z = 0.$$

(b) The pair of tangents from the Point $[1, 0, 2]$ to E are given by the equation $s_1^2 = s \cdot s_{11}$, where, by Example 6, $s_1 = -4x - \tfrac{3}{2}y + 4z$ and $s_{11} = 4$. Thus the equation of the tangent pair is

$$\left(-4x - \tfrac{3}{2}y + 4z\right)^2 = (4x^2 + xy - 2y^2 - 8xz - 2yz + 4z^2) \cdot 4,$$

or

$$\begin{aligned} &16x^2 + \tfrac{9}{4}y^2 + 16z^2 + 12xy - 32xz - 12yz \\ &\quad = 16x^2 + 4xy - 8y^2 - 32xz - 8yz + 16z^2. \end{aligned}$$

After some rearrangement, this becomes

$$\tfrac{41}{4}y^2 + 8xy - 4yz = 0,$$

or

$$y\left(\tfrac{41}{4}y + 8x - 4z\right) = 0.$$

Thus the equations of the two tangents to the projective conic are

$$y = 0 \quad \text{and} \quad \tfrac{41}{4}y + 8x - 4z = 0.$$

(c) The polar of the Point $[1, 0, 2]$ is given by the equation $s_1 = 0$, where, by Example 6,

$$s_1 = -4x - \tfrac{3}{2}y + 4z.$$

Thus the equation of the polar of $[1, 0, 2]$ with respect to E is

$$-4x - \tfrac{3}{2}y + 4z = 0,$$

or

$$8x + 3y - 8z = 0. \qquad \square$$

Problem 6 The projective conic E has equation

$$y^2 + z^2 + 2xy - 4yz + zx = 0.$$

(a) Determine the equation of the tangent to E at the Point $[1, 0, 0]$. Verify that $[0, 1, -2]$ lies on this tangent.
(b) Determine the equations of the two tangents to E that pass through the Point $[0, 1, -2]$.
(c) Determine the polar of the Point $[0, 1, -2]$ with respect to E.

4.3 Theorems

In this section we use the fact that all non-degenerate (projective) conics are projective-congruent, together with the Fundamental Theorem of Projective Geometry, to prove many interesting results about projective conics. It then follows that the corresponding results hold for plane conics too.

Subsection 3.3.4, Theorem 4

4.3.1 Points on Projective Conics

Recall that the general form of the equation of a projective conic in $\mathbb{RP}^2$ is

Subsection 4.1.1

$$Ax^2 + Bxy + Cy^2 + Fxz + Gyz + Hz^2 = 0.$$

Although this equation involves six arbitrary constants A, B, C, F, G and H, it is only their five ratios that matter. In fact, it takes exactly five Points to determine a projective conic.

Example 1 Determine the equation of the projective conic which passes through the Points [1, 0, 0], [0, 1, 0], [0, 0, 1], [1, 1, 1] and [1, 2, 3].

Solution Let the projective conic have equation

$$Ax^2 + Bxy + Cy^2 + Fxz + Gyz + Hz^2 = 0.$$

Since [1, 0, 0] lies on the projective conic, we must have $A = 0$. Similarly, since [0, 1, 0] and [0, 0, 1] lie on the projective conic, we must also have $C = 0$ and $H = 0$. Thus the equation of the projective conic reduces to the form

$$Bxy + Fxz + Gyz = 0.$$

Since [1, 1, 1] and [1, 2, 3] both satisfy this equation, we deduce that

$$B + F + G = 0 \quad (1)$$

and

$$2B + 3F + 6G = 0. \quad (2)$$

Subtracting equation (2) from twice equation (1), we deduce that $-F - 4G = 0$ so that $F = -4G$; and subtracting equation (2) from three times equation (1), we deduce that $B - 3G = 0$ so that $B = 3G$. It follows that the equation of the projective conic must be of the form

$$3Gxy - 4Gxz + Gyz = 0,$$

The equation of a projective conic is of second degree, and so $G \neq 0$.

or

$$3xy - 4xz + yz = 0. \qquad \square$$

Problem 1 Determine the equation of the projective conic which passes through the Points [1, 0, 0], [0, 1, 0], [0, 0, 1], [1,1,1] and $[-2, 3, 1]$.

We now use the approach in Example 1 to prove the following result.

Theorem 1 Five Points Theorem
There is a unique non-degenerate projective conic through any given set of five Points, no three of which are collinear. In particular, if the five Points are $[1, 0, 0]$, $[0, 1, 0]$, $[0, 0, 1]$, $[1, 1, 1]$ and $[a, b, c]$, then the equation of the conic is

$$c(a - b)xy + b(c - a)xz + a(b - c)yz = 0.$$

Proof By the Fundamental Theorem of Projective Geometry, there is a projective transformation t which maps four of the Points to [1, 0, 0], [0, 1, 0], [0, 0, 1] and [1, 1, 1]. Let $[a, b, c]$ be the image of the fifth Point under t. Since t^{-1} preserves collinearity, it follows that no three of the Points [1, 0, 0],

[0, 1, 0], [0, 0, 1], [1, 1, 1] and $[a, b, c]$ are collinear. This observation enables us to deduce that the numbers a, b and c are all different and non-zero.

For example, $b \neq c$, for otherwise $[a, b, c]$, [1, 0, 0] and [1, 1, 1] would all lie on the Line $y = z$. Similarly, $a \neq b$ and $c \neq a$.

Also, $c \neq 0$, for otherwise $[a, b, c]$, [1, 0, 0] and [0, 1, 0] would all lie on the Line $z = 0$. Similarly, $a \neq 0$ and $b \neq 0$.

Since t is a one-one transformation which preserves non-degenerate projective conics, the theorem holds if and only if there is a unique non-degenerate projective conic through the Points [1, 0, 0], [0, 1, 0], [0, 0, 1], [1, 1, 1] and $[a, b, c]$. In fact, since no *degenerate* projective conic can pass through [1, 0, 0], [0, 1, 0], [0, 0, 1], [1, 1, 1] and $[a, b, c]$, it is sufficient to show that there is a unique projective conic (with the desired equation) through these Points.

Subsection 4.1.1, Theorem 1

A degenerate projective conic consists of a pair of Lines, a single Line, a Point, or 'no Points'. Such a projective conic cannot pass through five Points without three of the Points being collinear.

Now any projective conic has equation

$$Ax^2 + Bxy + Cy^2 + Fxz + Gyz + Hz^2 = 0,$$

and if it passes through the Point [1, 0, 0], then $A = 0$. Similarly, if it passes through the Points [0, 1, 0] and [0, 0, 1], then $C = 0$ and $H = 0$. It follows that any projective conic which passes through [1, 0, 0], [0, 1, 0] and [0, 0, 1] must have an equation of the form

$$Bxy + Fxz + Gyz = 0, \qquad (3)$$

for some real numbers B, F and G.

If the projective conic also passes through the Points [1, 1, 1] and $[a, b, c]$, then

$$B + F + G = 0 \qquad (4)$$

and

$$Bab + Fac + Gbc = 0. \qquad (5)$$

We may regard equations (4) and (5) as simultaneous equations in B and F. If we subtract equation (5) from ab times equation (4), we obtain

$$F(ab - ac) + G(ab - bc) = 0,$$

so

$$F = -G\frac{ab - bc}{ab - ac}; \qquad (6)$$

Note that $ab - ac \neq 0$ since $a \neq 0$ and $b \neq c$.

and if we subtract equation (5) from ac times equation (4), we obtain

$$B(ac - ab) + G(ac - bc) = 0,$$

so

$$B = -G\frac{ac - bc}{ac - ab}. \qquad (7)$$

It follows from equations (3), (6) and (7) that any projective conic through the Points [1, 0, 0], [0, 1, 0], [0, 0, 1], [1, 1, 1] and $[a, b, c]$ must have an equation of the form

$$-G\frac{ac - bc}{ac - ab}xy - G\frac{ab - bc}{ab - ac}xz + Gyz = 0,$$

The equation of a projective conic is of second degree, so $G \neq 0$.

or

$$c(a-b)xy + b(c-a)xz + a(b-c)yz = 0.$$

Since a, b and c are uniquely determined (up to a multiple) by the fifth Point, it follows that the projective conic is unique. ■

If we multiply a, b and c by a constant λ, then the equation of the projective conic is multiplied by λ^2; however, this does not change the projective conic.

Now any theorem which is concerned exclusively with the projective properties of Points, Lines and projective conics can be interpreted as a theorem about the corresponding points, lines and plane conics in an embedding plane. For example, if we are given any set of five points in an embedding plane, no three of which are collinear, then Theorem 1 tells us that there is a unique plane conic which passes through the points. In particular, this result remains true if we require that none of the five points is an ideal Point for the embedding plane, so we have the following result about plane conics.

Some of the points may be ideal Points for the embedding plane.

Corollary 1 There is a unique plane conic through any given set of five points, no three of which are collinear.

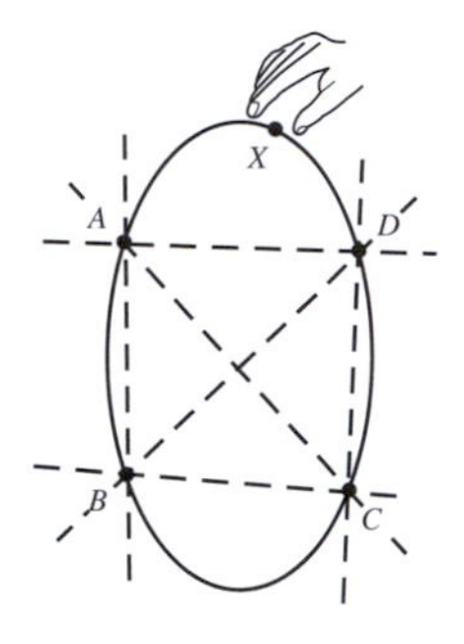

Now consider any four distinct Points A, B, C, D, no three of which are collinear. If X is a Point in $\mathbb{RP}^2$ that does not lie on any of the various Lines through A, B, C, D, then Theorem 1 tells us that there is a unique non-degenerate projective conic through A, B, C, D and X. If we now move X around $\mathbb{RP}^2$ (avoiding the various Lines through A, B, C, D) we obtain an infinite family of non-degenerate projective conics through A, B, C, D. We therefore have the following corollary of Theorem 1.

Corollary 2 There are infinitely many non-degenerate projective conics through any given set of four Points, no three of which are collinear.

Problem 2 Find the equations of two different (non-degenerate) projective conics through the Points [1, 0, 0], [0, 1, 0], [0, 0, 1] and [1, 2, 3].

Warning
We now use this corollary to warn you about a mistake that is frequently made in projective geometry.

Warning

The mistake is to assume that there exists a projective transformation t which maps one projective conic E_1 onto another projective conic E_2 in such a way that four given Points on E_1 are mapped to four given Points on E_2.

Of course, there is certainly a projective transformation t_1 which maps E_1 onto E_2, and by the Fundamental Theorem of Projective Geometry there is certainly a projective transformation t_2 which maps the four Points on E_1 to the four Points on E_2. The trouble is that t_1 may not be the same transformation as t_2.

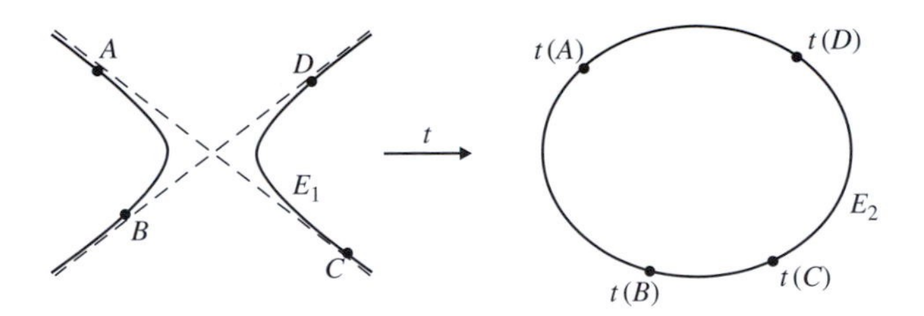

For example, consider two different projective conics E_1 and E_2 through the Points $A = [1, 0, 0], B = [0, 1, 0], C = [0, 0, 1], D = [1, 1, 1]$. (This is possible by Corollary 2.) By the Fundamental Theorem of Projective Geometry the only projective transformation which maps each of the Points A, B, C, D to itself is the identity transformation, and this certainly cannot map E_1 onto E_2.

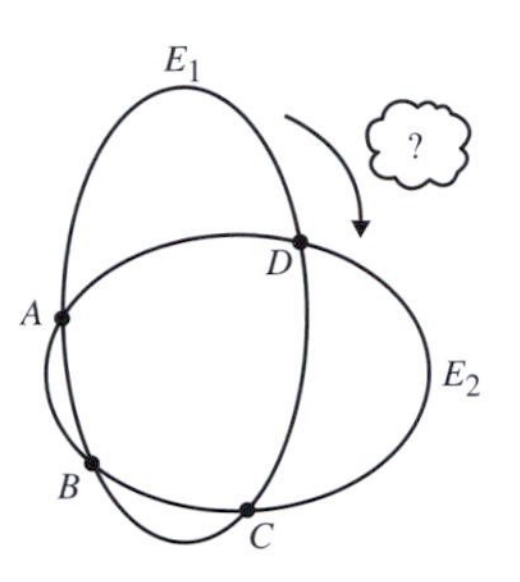

The following theorem shows that the situation is very different if, instead of having to map four given Points on one projective conic to four given Points on another, we have to map just three.

Theorem 2 Three Points Theorem
Let E_1 and E_2 be non-degenerate projective conics which pass through the Points P_1, Q_1, R_1 and P_2, Q_2, R_2, respectively. Then there is a projective transformation t which maps E_1 onto E_2 in such a way that

$$t(P_1) = P_2, \quad t(Q_1) = Q_2, \quad t(R_1) = R_2.$$

Proof First, let t' be any projective transformation that maps P_1, Q_1, R_1 to [1, 0, 0], [0, 1, 0], [0, 0, 1], respectively. Then t' maps E_1 to a non-degenerate projective conic E' which passes through the triangle of reference. If E' has equation

$$Ax^2 + Bxy + Cy^2 + Fxz + Gyz + Hz^2 = 0,$$

then the fact that the Point [1, 0, 0] lies on E' forces A to be zero, and the fact that the Points [0, 1, 0] and [0, 0, 1] lie on E' forces C and H to be zero. It follows that the equation of E' can be written in the form

$$Bxy + Fxz + Gyz = 0,$$

for some non-zero real numbers B, F and G. By dividing by BFG we can rewrite the equation of E' in the form

If B were zero, then the equation of the projective conic could be written as

$$(Fx + Gy)z = 0,$$

which is the equation of a degenerate projective conic consisting of two Lines. It follows that $B \neq 0$. Similarly, $F \neq 0$ and $G \neq 0$.

$$\frac{x}{G} \cdot \frac{y}{F} + \frac{x}{G} \cdot \frac{z}{B} + \frac{y}{F} \cdot \frac{z}{B} = 0.$$

Now define the projective transformation t'' by $t''([x, y, z]) = [x', y', z']$, where

$$\begin{pmatrix} x' \\ y' \\ z' \end{pmatrix} = \begin{pmatrix} 1/G & 0 & 0 \\ 0 & 1/F & 0 \\ 0 & 0 & 1/B \end{pmatrix} \begin{pmatrix} x \\ y \\ z \end{pmatrix}.$$

Then t'' maps E' to the conic with equation $x'y' + x'z' + y'z' = 0$, or, after dropping the dashes, to the conic with equation $xy + xz + yz = 0$.

Since t'' leaves the triangle of reference unchanged, it follows that the composite projective transformation $t_1 = t'' \circ t'$ maps E_1 to the projective conic with equation

For example,
$t''([1,0,0]) = [1/G,0,0]$
$= [1,0,0]$.

$$xy + xz + yz = 0$$

in such a way that $t_1(P_1) = [1,0,0], t_1(Q_1) = [0,1,0]$ and $t_1(R_1) = [0,0,1]$.

Similarly, there is a projective transformation t_2 which maps E_2 onto the projective conic with equation

$$xy + xz + yz = 0$$

in such a way that $t_2(P_2) = [1,0,0], t_2(Q_2) = [0,1,0]$ and $t_2(R_2) = [0,0,1]$.

It follows that the composite projective transformation $t = t_2^{-1} \circ t_1$ maps E_1 onto E_2 in such a way that $t(P_1) = P_2, t(Q_1) = Q_2, t(R_1) = R_2$, as required. ■

Example 2 The Points [1, 1, 1], [1, 2, 2], [1, 2, 1] lie on the projective conic E with equation $2x^2 + 2xy - y^2 + yz - 5xz + z^2 = 0$.

(a) Verify that the projective transformation $t_1 : [\mathbf{x}] \mapsto [\mathbf{x}']$ with associated matrix $\mathbf{A} = \begin{pmatrix} 2 & -1 & 0 \\ -1 & 0 & 1 \\ 0 & 1 & -1 \end{pmatrix}$ maps the Points [1, 1, 1], [1, 2, 2], [1, 2, 1] to the Points [1, 0, 0], [0, 1, 0], [0, 0,1], respectively, and maps E onto the projective conic E' with equation $x'y' - x'z' + y'z' = 0$.

Hint: The inverse of $\mathbf{A}$ is $\mathbf{A}^{-1} = \begin{pmatrix} 1 & 1 & 1 \\ 1 & 2 & 2 \\ 1 & 2 & 1 \end{pmatrix}$.

(b) Determine the equation of the projective conic E'' that is the image of E' under the projective transformation $t_2 : [\mathbf{x}'] \mapsto [\mathbf{x}'']$ with associated matrix $\mathbf{B} = \begin{pmatrix} 1 & 0 & 0 \\ 0 & -1 & 0 \\ 0 & 0 & 1 \end{pmatrix}$.

(c) Hence determine a matrix associated with a projective transformation that maps E onto the projective conic with equation $xy + yz + zx = 0$.

Solution

(a) Let $\mathbf{x}' = \mathbf{A}\mathbf{x}$, so that

$$\begin{pmatrix} x' \\ y' \\ z' \end{pmatrix} = \begin{pmatrix} 2 & -1 & 0 \\ -1 & 0 & 1 \\ 0 & 1 & -1 \end{pmatrix} \begin{pmatrix} x \\ y \\ z \end{pmatrix}.$$

Since

$$\begin{pmatrix} 2 & -1 & 0 \\ -1 & 0 & 1 \\ 0 & 1 & -1 \end{pmatrix} \begin{pmatrix} 1 \\ 1 \\ 1 \end{pmatrix} = \begin{pmatrix} 1 \\ 0 \\ 0 \end{pmatrix},$$

$$\begin{pmatrix} 2 & -1 & 0 \\ -1 & 0 & 1 \\ 0 & 1 & -1 \end{pmatrix} \begin{pmatrix} 1 \\ 2 \\ 2 \end{pmatrix} = \begin{pmatrix} 0 \\ 1 \\ 0 \end{pmatrix}.$$

and

$$\begin{pmatrix} 2 & -1 & 0 \\ -1 & 0 & 1 \\ 0 & 1 & -1 \end{pmatrix} \begin{pmatrix} 1 \\ 2 \\ 1 \end{pmatrix} = \begin{pmatrix} 0 \\ 0 \\ 1 \end{pmatrix},$$

it follows that the images under t_1 of [1, 1, 1], [1, 2, 2], [1, 2, 1] are [1, 0, 0], [0, 1, 0], [0, 0, 1], respectively.
Next, $\mathbf{x} = \mathbf{A}^{-1}\mathbf{x}'$ so that

$$\begin{pmatrix} x \\ y \\ z \end{pmatrix} = \begin{pmatrix} 1 & 1 & 1 \\ 1 & 2 & 2 \\ 1 & 2 & 1 \end{pmatrix} \begin{pmatrix} x' \\ y' \\ z' \end{pmatrix};$$

thus

$$x = x' + y' + z',$$
$$y = x' + 2y' + 2z',$$

and

$$z = x' + 2y' + z'.$$

It follows that t_1 maps the given projective conic onto the projective conic with equation

$$\begin{aligned} &2(x' + y' + z')^2 + 2(x' + y' + z')(x' + 2y' + 2z') \\ &\quad - (x' + 2y' + 2z')^2 + (x' + 2y' + 2z')(x' + 2y' + z') \\ &\quad - 5(x' + y' + z')(x' + 2y' + z') + (x' + 2y' + z')^2 = 0. \end{aligned}$$

After some simplification, this becomes $x'y' - x'z' + y'z' = 0$, as required. We omit the details.

(b) Under the projective transformation $t_2 : \mathbf{x}' \mapsto \begin{pmatrix} 1 & 0 & 0 \\ 0 & -1 & 0 \\ 0 & 0 & 11 \end{pmatrix} \mathbf{x}' = \mathbf{x}''$,

we have

$$x'' = x', y'' = -y' \quad \text{and} \quad z'' = z'.$$

so that

$$x' = x'', y' = -y'' \quad \text{and} \quad z' = z''.$$

Hence the image under t_2 of the projective conic $x'y' - x'z' + y'z' = 0$ is the projective conic with equation

$$-x''y'' - x''z'' - y''z'' = 0,$$

or

$$x''y'' + x''z'' + y''z'' = 0.$$

(c) It follows from parts (a) and (b) that the projective transformation $t_2 \circ t_1$, with matrix

$$\mathbf{BA} = \begin{pmatrix} 1 & 0 & 0 \\ 0 & -1 & 0 \\ 0 & 0 & 1 \end{pmatrix} \begin{pmatrix} 2 & -1 & 0 \\ -1 & 0 & 1 \\ 0 & 1 & -1 \end{pmatrix}$$

$$= \begin{pmatrix} 2 & -1 & 0 \\ -1 & 0 & -1 \\ 0 & 1 & -1 \end{pmatrix}$$

maps E onto the projective conic with equation $xy + yz + zx = 0$, as required. □

We can use a similar approach to find a projective transformation that maps any given projective conic onto the standard projective conic $xy+yz+zx = 0$.

We explain the designation 'standard projective conic' in the next subsection.

Strategy To determine a projective transformation t that maps a given projective conic E onto the standard projective conic $xy + yz + zx = 0$:

1. choose three Points P, Q, R on E;
2. determine a matrix $\mathbf{A}$ associated with a projective transformation that maps P, Q, R onto [1, 0, 0], [0, 1, 0], [0, 0, 1], respectively;
3. determine the equation $Bx'y' + Fx'z' + Gy'z' = 0$ of $t(E)$, for some real numbers B, F and G;
4. then a matrix associated with t is $\mathbf{BA}$, where

$$\mathbf{B} = \begin{pmatrix} 1/G & 0 & 0 \\ 0 & 1/F & 0 \\ 0 & 0 & 1/B \end{pmatrix}.$$

Problem 3 The Points $[-2, 0, 1], [0, -3, 2], [1, -2, 1]$ lie on the projective conic E with equation $17x^2 + 47xy + 32y^2 + 67xz + 92yz + 66z^2 = 0$.

(a) Verify that the projective transformation t with an associated matrix

$$\mathbf{A} = \begin{pmatrix} 1 & 2 & 3 \\ 2 & 3 & 4 \\ 3 & 4 & 6 \end{pmatrix},$$

maps $[-2, 0, 1]$, $[0, -3, 2]$, $[1, -2, 1]$ to [1, 0, 0], [0, 1, 0], [0, 0,1], respectively.

(b) Verify that the inverse of $\mathbf{A}$ is $\mathbf{A}^{-1} = \begin{pmatrix} -2 & 0 & 1 \\ 0 & 3 & -2 \\ 1 & -2 & 1 \end{pmatrix}$.

(c) Determine the equation of the projective conic $t(E)$.

(d) Hence determine a matrix associated with a projective transformation that maps E onto the projective conic with equation $xy + yz + zx = 0$.

4.3.2 The Standard Form $xy + yz + zx = 0$

In the previous chapter you saw that we can often simplify problems about Points and Lines by mapping certain Points to the triangle of reference and the unit Point. In a similar way, we can often simplify problems about a projective conic by mapping it onto another projective conic that has a simpler equation.

Subsection 3.4.1

Since all non-degenerate projective conics are projective-congruent, it follows that there is a projective transformation which maps any given non-degenerate projective conic in $\mathbb{RP}^2$ onto the projective conic with equation $xy + yz + zx = 0$. This equation turns out to be particularly useful for tackling a large number of problems about projective conics, so we give it a special name.

Subsection 4.1.4, Theorem 5

The existence of such a projective transformation was also proved in Theorem 2 above, and the subsequent Strategy.

Definitions The equation

$$xy + yz + zx = 0$$

is called a **standard form** for the equation of a projective conic.

The conic defined by this standard form is called a **standard projective conic.**

The following diagram illustrates this standard projective conic together with its representation in the standard embedding plane $z = 1$. The representation has equation $xy + y + x = 0$, or $(x + 1)(y + 1) = 1$, and is therefore a rectangular hyperbola with asymptotes $x = -1$ and $y = -1$.

Recall that a *rectangular hyperbola* is a hyperbola whose asymptotes meet at right angles.

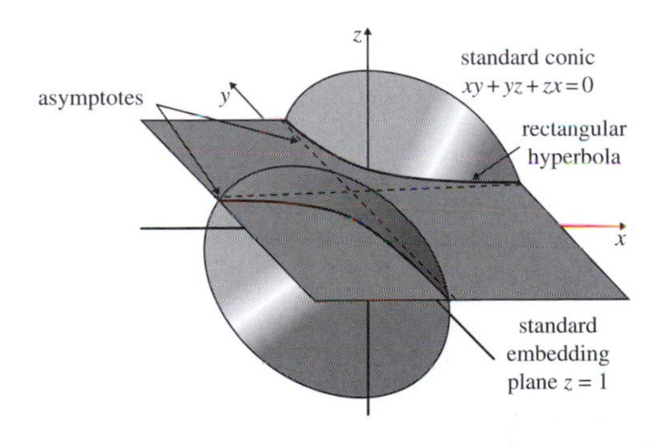

Notice the unconventional orientation of the axes in this figure. This is done to give a better view of the cone and its intersection with the embedding plane.

Since this standard projective conic is defined by the equation $xy + yz + zx = 0$, it must pass through the triangle of reference $[1, 0, 0]$, $[0, 1, 0]$, $[0, 0, 1]$. This fact can often be used to simplify calculations involving projective conics. Other Points on the projective conic that appear in such calculations may then be expressed in terms of a single real parameter.

Recall that the sides of the triangle of reference in $\mathbb{RP}^2$ are the x-, y- and z-axes in $\mathbb{R}^3$.

Theorem 3 Parametrization Theorem
Let E be a projective conic with equation in the standard form

$$xy + yz + zx = 0.$$

Then each Point on E, other than [1, 0, 0], has homogeneous coordinates of the form $[t^2+t, t+1, -t]$, where $t \in \mathbb{R}$. Moreover, each such Point lies on E.

Proof Let $[x, y, z]$ be any Point on E. If $x = 0$, then we must have $yz = 0$, so either $y = 0$ (in which case the Point has homogeneous coordinates $[0, 0, z] = [0, 0, 1]$) or $z = 0$ (in which case the Point has homogeneous coordinates $[0, y, 0] = [0, 1, 0]$). A similar discussion of the possibilities when y or z is zero shows that the only Points on E for which one of the homogeneous coordinates vanishes are the three Points [1, 0, 0], [0, 1, 0] and [0, 0,1].

We cannot have x, y and z all zero for Points $[x, y, z]$ in $\mathbb{RP}^2$.

Suppose next that $[x, y, z]$ is a Point on E for which none of the coordinates vanishes, and let $t = x/y$. Then $x = ty$ and so

Notice that $t \neq 0$ since $x \neq 0$.

$$(ty)y + yz + z(ty) = 0,$$

that is, $ty^2 + (t+1)yz = 0$. Since $y \neq 0$, it follows that

$$ty + (t+1)z = 0.$$

Thus $y = -\left(\frac{t+1}{t}\right)z$, and so $x = -(t+1)z$. It follows that the Point $[x, y, z]$ has homogeneous coordinates $\left[-(t+1)z, -\left(\frac{t+1}{t}\right)z, z\right]$; and, since $z \neq 0$ and $t \neq 0$, we may rewrite these coordinates in the form $[t(t+1), t+1, -t]$.

Also, notice that we can obtain the Point [0, 1, 0] by choosing $t = 0$, and the Point [0, 0, 1] by choosing $t = -1$; so every Point on E, other than [1, 0, 0], can be written in the form $[t^2+t, t+1, -t]$ for some $t \in \mathbb{R}$.

Conversely, every Point of the form $[t^2+t, t+1, -t]$, where $t \in \mathbb{R}$, lies on E because

$$(t^2+t)(t+1) + (t+1)(-t) + (-t)(t^2+t) = 0.$$

This completes the proof. ■

Using a preliminary projective transformation, we can transform any problem involving Points on a projective conic to a problem involving Points on the standard projective conic with equation $xy + yz + zx = 0$. By the Three Points Theorem, we can assume that three of the Points are [1, 0, 0], [0, 1, 0] and [0, 0,1], and by the Parametrization Theorem we may express any remaining Points in the form $[t^2+t, t+1, -t]$ for some real number t.

We illustrate this technique by proving Pascal's Theorem.

Theorem 4 Pascal's Theorem
Let A, B, C, A', B' and C' be six distinct Points on a non-degenerate projective conic. Let BC and $B'C$ intersect at P, CA' and $C'A$ intersect at Q, and AB' and $A'B$ intersect at R. Then P, Q and R are collinear.

Blaise Pascal (1623–1662) proved this theorem while still a schoolboy aged 16.

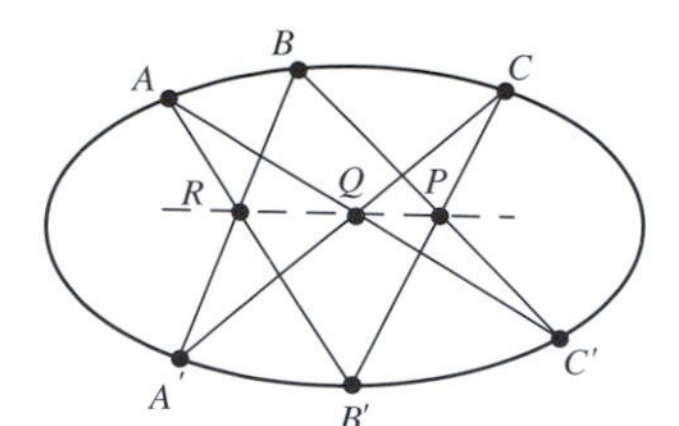

Notice that in geometric illustrations of results in projective geometry, we usually draw the projective conic as a plane ellipse in $\mathbb{R}^2$. We draw it in $\mathbb{R}^2$ simply for convenience, since the page is part of $\mathbb{R}^2$! We draw it as an ellipse to avoid having to cope with ideal Points, etc.

Proof By the Three Points Theorem we can let the equation of the projective conic be in the standard form $xy + yz + zx = 0$, with $A = [1,0,0], B = [0,1,0], C = [0,0,1]$. Also, by the Parametrization Theorem, we can let $A' = [a^2 + a, a+1, -a], B' = [b^2 + b, b+1, -b], C' = [c^2 + c, c+1, -c]$, for some real numbers a, b, c.

First we find the Point P. The Line BC' joins the Points $[0,1,0]$ and $\left[c^2 + c, c+1, -c\right]$, and so clearly has equation $x = -(c+1)z$ since both Points lie on this Line. The Line $B'C$ joins the Points $\left[b^2 + b, b+1, -b\right]$ and $[0,0,1]$, and so clearly has equation $x = by$ since both Points lie on this Line.

The point P lies on both BC' and $B'C$, so that its homogeneous coordinates $\left[x, y, z\right]$ must satisfy the two equations $x = -(c+1)z$ and $x = by$. It follows that P has homogeneous coordinates $[b(c+1), c+1, -b]$.

Similar arguments show that the Lines CA' and $C'A$ have equations $x = ay$ and $cy = -(c+1)z$, so that their Point of intersection Q has homogeneous coordinates $[a(c+1), c+1, -c]$. Also, the Lines AB' and $A'B$ have equations $by = -(b+1)z$ and $x = -(a+1)z$, so that their Point of intersection R has homogeneous coordinates $[b(a+1), b+1, -b]$.

Finally, P, Q and R are collinear since

Here we use the determinant criterion for collinearity given in Subsection 3.2.2.

$$\begin{vmatrix} b(c+1) & c+1 & -b \\ a(c+1) & c+1 & -c \\ b(a+1) & b+1 & -b \end{vmatrix}$$

$$= \begin{vmatrix} bc+b & c+1 & -b \\ ac+a & c+1 & -c \\ ab+b & b+1 & -b \end{vmatrix}$$

$$= \begin{vmatrix} bc-ab & c-b & 0 \\ ac+a & c+1 & -c \\ ab+b & b+1 & -b \end{vmatrix} \quad (row\ 1 - row\ 3)$$

Elementary row operations do not affect the linear independence of the rows.

$$= b(c-a)\begin{vmatrix} c+1 & -c \\ b+1 & -b \end{vmatrix} - (c-b)\begin{vmatrix} ac+a & -c \\ ab+b & -b \end{vmatrix}$$

$$= b(c-a)(-b+c) - (c-b)(-ab+bc)$$

$$= 0.$$

This completes the proof of Pascal's Theorem. ■

By representing the configuration of Pascal's Theorem in an embedding plane, we can obtain a version of the theorem that holds in $\mathbb{R}^2$. We state this in the form of a corollary, as follows.

Corollary 3 Let A, B, C, A', B' and C' be six distinct points on a non-degenerate plane conic, with BC' and $B'C$ intersecting at P, CA' and $C'A$ intersecting at Q, and AB' and $A'B$ intersecting at R. Then P, Q and R are collinear.

Notice that because this corollary is stated as a result in $\mathbb{R}^2$, certain configurations are excluded. For example, the lines BC' and $B'C$ cannot be parallel since P is assumed to lie in $\mathbb{R}^2$. Such cases have to be treated separately.

Problem 4 Give an interpretation in $\mathbb{R}^2$ of Pascal's Theorem for which the lines AB' and $A'B$ are parallel (but AC' meets $A'C$ and BC' meets $B'C$).

4.3.3 Converse of Pascal's Theorem

Pascal's Theorem states that if six distinct Points A, B, C, A', B' and C' lie on a non-degenerate projective conic with BC' and $B'C$ intersecting at P, CA' and $C'A$ intersecting at Q, and AB' and $A'B$ intersecting at R, then P, Q and R are collinear. The converse states that if the intersection Points P, Q and R are collinear, then the six Points A, B, C, A', B' and C' lie on a non-degenerate projective conic. Again, our proof is algebraic, and depends on a suitable choice of coordinates.

Subsection 4.3.2, Theorem 4

Theorem 5 Converse of Pascal's Theorem
Let A, B, C, A', B' and C' be six Points, no three of which are collinear, with BC' and $B'C$ intersecting at P, CA' and $C'A$ intersecting at Q, and AB' and $A'B$ intersecting at R. If P, Q and R are collinear, then the Points A, B, C, A', B' and C' lie on a non-degenerate projective conic.

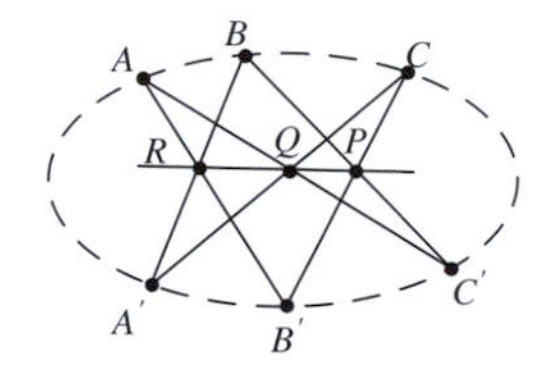

Proof Since no three of the Points A, B, C, A', B' and C' are collinear, we can (by a preliminary projective transformation, if necessary) assume that the points A, B, C and A' have homogeneous coordinates [1, 0, 0], [0, 1, 0], [0, 0, 1] and [1,1,1], respectively. Suppose that B' and C' have homogeneous coordinates $[a, b, c]$ and $[r, s, t]$, respectively.

Here we are using the Fundamental Theorem of Projective Geometry.

By Theorem 1 we know that there is a unique non-degenerate projective conic through the five Points A, B, C, A' and B', and its equation is

Subsection 4.3.1

$$c(a-b)xy + b(c-a)xz + a(b-c)yz = 0. \quad (8)$$

We must verify that the Point C' also lies on this projective conic. We do this by calculating the homogeneous coordinates of the Points P, Q and R, and using the determinant condition that these three Points are collinear.

Subsection 3.2.2

First, the Line BC' passes through the Points [0, 1, 0] and $[r, s, t]$, and therefore has equation $tx = rz$. Similarly, the Line $B'C$ passes through the Points $[a, b, c]$ and [0, 0, 1], and therefore has equation $bx = ay$. It follows that the Point P has homogeneous coordinates $[ar, br, at]$.

Next, the line CA' passes through the Points [0, 0, 1] and [1, 1, 1], and therefore has equation $x = y$. Similarly, the line $C'A$ passes through the points $[r, s, t]$ and $[1, 0, 0]$, and therefore has equation $ty = sz$. It follows that the Point Q has homogeneous coordinates $[s, s, t]$.

Finally, the Line AB' passes through the Points [1, 0, 0] and $[a, b, c]$, and therefore has equation $cy = bz$. Similarly, the Line $A'B$ passes through the Points [1, 1, 1] and [0, 1, 0], and therefore has equation $x = z$. It follows that the Point R has homogeneous coordinates $[c, b, c]$.

Then it follows from the fact that P, Q and R are collinear that

Recall that the collinearity of P, Q and R is a hypothesis of the theorem.

$$\begin{aligned}0 &= \begin{vmatrix} ar & br & at \\ s & s & t \\ c & b & c \end{vmatrix} \\ &= ar\begin{vmatrix} s & t \\ b & c \end{vmatrix} - br\begin{vmatrix} s & t \\ c & c \end{vmatrix} + at\begin{vmatrix} s & s \\ c & b \end{vmatrix} \\ &= ar(sc - bt) - br(sc - ct) + at(sb - cs).\end{aligned}$$

By rearranging the terms in this equation we get

$$c(a - b)rs + b(c - a)rt + a(b - c)st = 0. \tag{9}$$

By comparing equations (8) and (9), we observe that equation (8) holds for the Point $C' = [r, s, t]$. In other words, the Point C' lies on the protective conic through the Points A, B, C, A' and B', which has equation (8). This shows that A, B, C, A', B' and C' lie on the same (non-degenerate) projective conic, as required. ■

The following corollary gives a version of the converse of Pascal's Theorem that holds in $\mathbb{R}^2$.

Corollary 4 Let A, B, C, A', B' and C' be six points in $\mathbb{R}^2$, no three of which are collinear, with BC' and $B'C$ intersecting at P, CA' and $C'A$ intersecting at Q, and AB' and $A'B$ intersecting at R. If P, Q and R are collinear, then the points A, B, C, A', B' and C' lie on a non-degenerate plane conic.

We conclude this subsection by showing how this corollary can be used to construct further points on the unique plane conic which passes through five given points A, B, C, A' and B', no three of which are collinear.

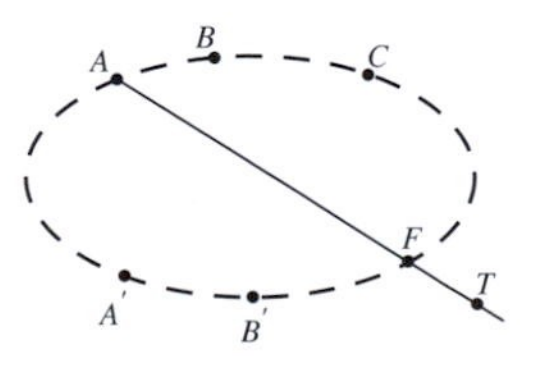

Let T be any point in $\mathbb{R}^2$ such that AT does not pass through B, C, A' or B'. Then the following construction, using only a straight edge and pencil, determines a sixth point, F say, where the conic meets AT.

1. Draw the lines AB' and $A'B$, and so determine the point of intersection R of these lines.

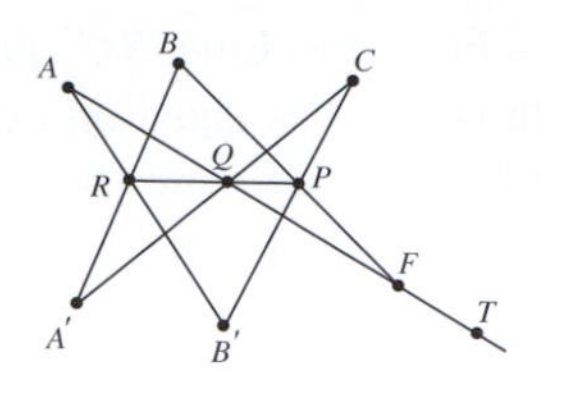

2. Draw the lines CA' and TA, and so determine the point of intersection Q of these lines.
3. Draw the lines QR and $B'C$, and so determine the point of intersection P of these lines.
4. Then the point of intersection F of the lines BP and AT lies on the conic.

We can see easily why this construction works. The points A, B, C, A', B' and F are such that the point R where AB' meets $A'B$, the point Q where CA' meets FA, and the point P where BF meets $B'C$ are all collinear. It follows from the converse of Pascal's Theorem that the points A, B, C, A', B' and F all lie on the same plane conic.

Thus F is the point where AT meets the conic which passes through the five points A, B, C, A' and B'.

4.3.4 The Standard Form $x^2 + y^2 = z^2$

Since all non-degenerate projective conics are projective-congruent, there is a projective transformation that maps any non-degenerate projective conic in $\mathbb{RP}^2$ onto the projective conic with equation $x^2 + y^2 = z^2$. This is usually the simplest equation to use for problems that involve tangents and polars of non-degenerate projective conics.

Subsection 4.1.4, Theorem 5

Definitions The equation

$$x^2 + y^2 = z^2$$

is called **a standard form** for the equation of a projective conic.

The conic defined by this standard form is called a **standard projective conic**.

This is the second standard form that we have discussed (see Subsection 4.3.2 for the other).

This standard projective conic is a right circular cone in $\mathbb{R}^3$, as shown in the margin. It meets the embedding plane $z = 1$ in a circle of unit radius.

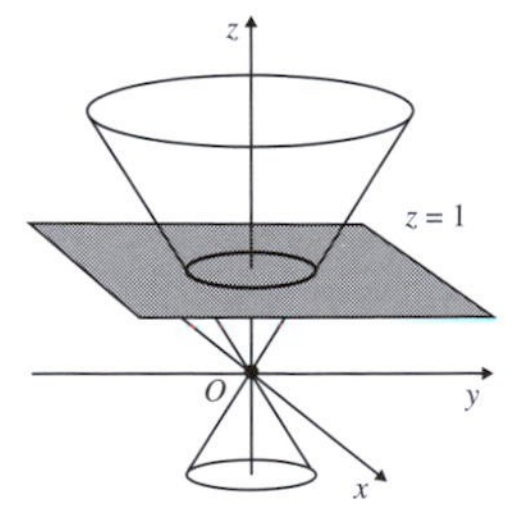

The reason why this standard projective conic is so useful for studying polars and tangents is that the equations of its polars and tangents have particularly simple forms.

Theorem 6 Let E be the projective conic with equation

$$x^2 + y^2 = z^2,$$

and let $P = [a, b, c]$ be any Point in $\mathbb{RP}^2$. Then:

(a) if $P \in E$, then $ax + by - cz = 0$ is the equation of the tangent to E at P;
(b) if $P \notin E$, then $ax + by - cz = 0$ is the equation of the polar of P with respect to E.

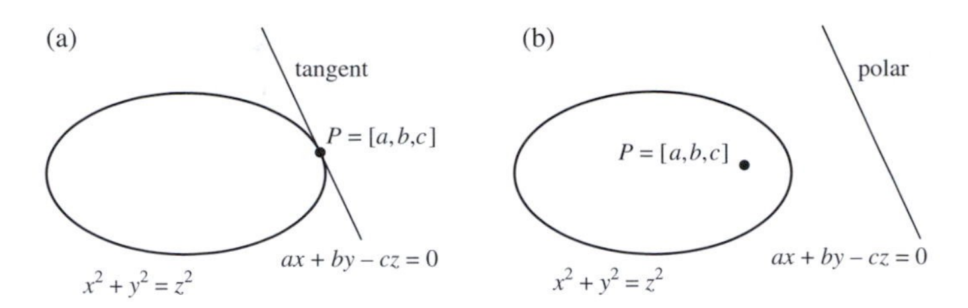

Proof In Joachimsthal's notation, the standard form becomes $s = 0$, where $s = x^2 + y^2 - z^2$. So at any Point $P = [x_1, y_1, z_1]$ of $\mathbb{RP}^2$, we have $s_1 = xx_1 + yy_1 - zz_1$.

Subsection 4.2.2

Now recall that when P lies on the projective conic, the equation $s_1 = 0$ gives the equation of the tangent to the projective conic at P, and when P does not lie on the projective conic, it gives the equation of the polar of P. The result follows by noting that at the Point $P = [a, b, c]$ the equation $s_1 = 0$ becomes $ax + by - cz = 0$. ■

Example 3 Determine whether each of the following Lines touches the projective conic with equation $x^2 + y^2 = z^2$. For each Line that does, state the Point of tangency.

(a) $3x - 5y + 4z = 0$ (b) $3x - 4y + 5z = 0$

Solution If a Line is the tangent to the projective conic $x^2 + y^2 - z^2 = 0$ at some Point $P = [a, b, c]$, say, then its equation must be $ax + by - cz = 0$ (or some multiple of this).

(a) Comparing the equations $3x - 5y + 4z = 0$ and $ax + by - cz = 0$, we see that P must have homogeneous coordinates $[3, -5, -4]$. However, since

$$(3)^2 + (-5)^2 - (-4)^2 = 9 + 25 - 16 = 18 \neq 0,$$

the Point $[3, -5, -4]$ cannot lie on the projective conic; hence the Line cannot be a tangent to the projective conic.

(b) Comparing the equations $3x - 4y + 5z = 0$ and $ax + by - cz = 0$, we see that P must have homogeneous coordinates $[3, -4, -5]$. Since

$$(3)^2 + (-4)^2 - (-5)^2 = 9 + 16 - 25 = 0,$$

the Line is a tangent to the projective conic at the Point $[3, -4, -5]$. □

Problem 5 Determine whether each of the following Lines touches the projective conic $x^2 + y^2 = z^2$. For each Line that does, state the Point of tangency.

(a) $91x - 60y - 109z = 0$ (b) $4x + 5y + 3z = 0$

We can use these ideas to provide an alternative proof of La Hire's Theorem concerning polars.

We first proved a version of La Hire's Theorem in $\mathbb{R}^2$, in Subsection 4.2.1.

Theorem 7 La Hire's Theorem
Let E be a non-degenerate projective conic, and let P be any Point of $\mathbb{RP}^2$, with polar p with respect to E. Then the polar of any Point Q on p passes through P.

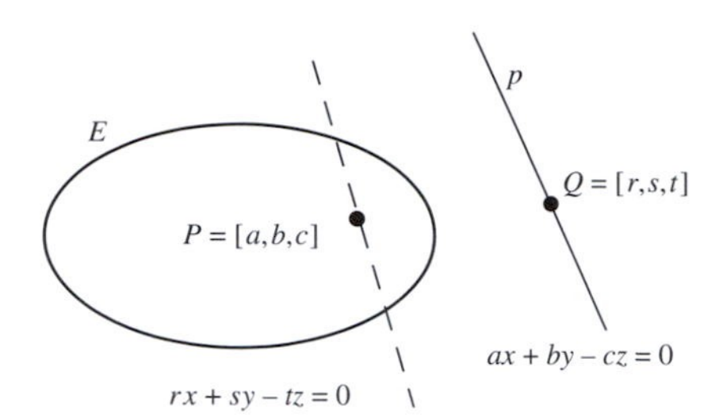

Proof We may assume that the equation of the projective conic is in the standard form $x^2 + y^2 - z^2 = 0$, and that P has homogeneous coordinates $[a, b, c]$.

Then, by Theorem 6, the equation of p is

$$ax + by - cz = 0.$$

So if $Q = [r, s, t]$ is any Point on p, then

$$ar + bs - ct = 0. \tag{10}$$

Now, by applying Theorem 6 again, we know that the polar of $Q = [r, s, t]$ with respect to the projective conic $x^2 + y^2 - z^2 = 0$ has equation

$$rx + sy - tz = 0,$$

so, by equation (10), it passes through P. This completes the proof. ■

We now prove the following interesting result as another application of Theorem 6.

Theorem 8 Three Tangents and Three Chords Theorem
Let a non-degenerate plane conic touch the sides of a triangle at the points P, Q and R, respectively, and let the tangents at P, Q and R meet the extended chords QR, RP and PQ at the points A, B and C, respectively. Then A, B and C are collinear.

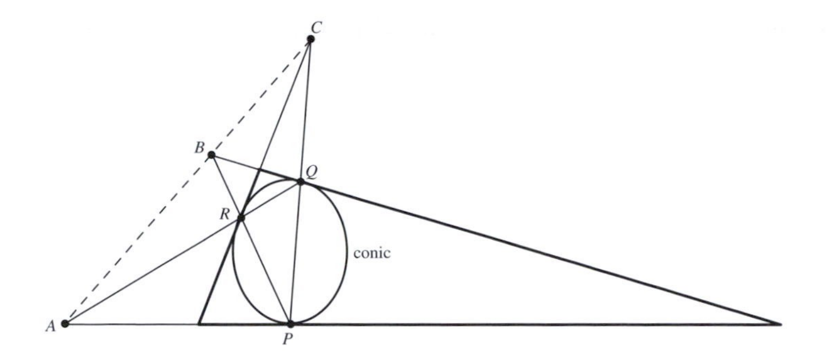

The initial triangle is drawn in the diagram with bold lines.

Proof Since the theorem is concerned exclusively with projective properties, we shall prove it as a projective theorem about a projective conic. Since this projective conic is non-degenerate, we may assume that its equation is in the standard form $x^2 + y^2 - z^2 = 0$, and that the Points P, Q and R have homogeneous coordinates $[1, 0, 1]$, $[1, 0, -1]$ and $[0, 1, 1]$, respectively.

The result then follows as an interpretation in an embedding plane.

Here we are using the Three Points Theorem (Subsection 4.3.1, Theorem 2).

By Theorem 6, part (a), the equation of the tangent to the projective conic at P is

$$x - z = 0. \tag{11}$$

Also, the equation of the chord QR is

$$\begin{vmatrix} x & y & z \\ 1 & 0 & -1 \\ 0 & 1 & 1 \end{vmatrix} = 0,$$

which we can rewrite in the form

$$x - y + z = 0. \tag{12}$$

It follows that at the Point A, both equations (11) and (12) must hold, so that $x = z$ and $y = x + z = 2z$. Hence A must have homogeneous coordinates $[z, 2z, z]$ or, equivalently, $[1, 2, 1]$.

Similar arguments show that the tangent to the projective conic at Q has equation $x + z = 0$, and the chord PR has equation $x + y - z = 0$, so their Point of intersection B is $[-1, 2, 1]$, Similarly, the tangent to the projective conic at R has equation $y - z = 0$, and the chord PQ has equation $y = 0$, so their Point of intersection C is $[1, 0, 0]$.

We omit the details.

Finally, A, B and C are collinear since

$$\begin{vmatrix} 1 & 2 & 1 \\ -1 & 2 & 1 \\ 1 & 0 & 0 \end{vmatrix} = 1 \cdot \begin{vmatrix} 2 & 1 \\ 0 & 0 \end{vmatrix} - 2 \cdot \begin{vmatrix} -1 & 1 \\ 1 & 0 \end{vmatrix} + 1 \cdot \begin{vmatrix} -1 & 2 \\ 1 & 0 \end{vmatrix}$$

$$= 1 \cdot (0) - 2 \cdot (-1) + 1 \cdot (-2) = 0. \quad \blacksquare$$

Problem 6 Let E_1 be the projective conic with equation $xy + yz + zx = 0$, and E_2 the projective conic with equation $x^2 + y^2 - z^2 = 0$.

(a) Verify that the Points $[1, 0, 0]$, $[0, 1, 0]$, $[0, 0, 1]$, $[2, 2, -1]$ and $[2, -1, 2]$ lie on E_1.

(b) Determine the images of the Points in part (a) under the projective transformation t_1 with associated matrix $\mathbf{A} = \begin{pmatrix} 1 & -1 & 0 \\ 0 & 0 & 2 \\ 1 & 1 & 2 \end{pmatrix}$.

(c) Use the results of parts (a) and (b) to write down a matrix associated with the projective transformation that maps E_1 onto E_2.

(d) Hence verify that $\mathbf{B} = \begin{pmatrix} -1 & 1 & -1 \\ 1 & 1 & -1 \\ 0 & -1 & 0 \end{pmatrix}$ is a matrix associated with the projective transformation t_2 that maps E_2 onto E_1.

Problem 7 Using the results of Problem 3, part (b), and Problem 6, part (c), determine a matrix associated with a projective transformation that maps the projective conic with equation

$$17x^2 + 47xy + 32y^2 + 67xz + 92yz + 66z^2 = 0$$

onto the projective conic with equation $x^2 + y^2 - z^2 = 0$.

Parametrization of the Projective Conic $x^2 + y^2 = z^2$

Sometimes it is useful to have a convenient parametrization of Points on the projective conic E with equation $x^2 + y^2 = z^2$.

Earlier we described a parametrization of Points on the projective conic with equation $xy + yz + zx = 0$.

First, notice that we cannot have $z = 0$ for Points $[x, y, z]$ on E; for then we would also have $x = y = 0$, which is impossible. So, since the coordinates $[x, y, z]$ are homogeneous coordinates, we may assume temporarily that $z = 1$; in other words, we consider the intersection of E with the embedding plane $z = 1$.

We may parametrize points of the unit circle $x^2 + y^2 = 1$ in the embedding plane as $\{(\cos\theta, \sin\theta, 1) : \theta \in (-\pi, \pi]\}$. Putting $t = \tan\frac{1}{2}\theta$, for $\theta \in (-\pi, \pi)$ we obtain the parametrization

$$\left[\frac{1-t^2}{1+t^2}, \frac{2t}{1+t^2}, 1\right] = \left[1-t^2, 2t, 1+t^2\right], \quad t \in \mathbb{R}.$$

We may multiply each coordinate by $(1 + t^2)$ since they are homogeneous coordinates.

As θ varies over $(-\pi, \pi)$, $\frac{1}{2}\theta$ varies over $(-\frac{1}{2}\pi, \frac{1}{2}\pi)$. In particular, as $\frac{1}{2}\theta$ varies over $(-\frac{1}{2}\pi, \frac{1}{2}\pi)$, $t = \tan\frac{1}{2}\theta$ takes all values in $\mathbb{R}$ exactly once. Finally, when $\theta = \pi$, the parametrization $(\cos\theta, \sin\theta, 1)$ gives the Point $[-1, 0, 1]$.

We can summarize the above discussion as follows:

Theorem 9 Parametrization Theorem
Each Point of the projective conic with equation $x^2 + y^2 = z^2$, other than the Point $[-1, 0, 1]$, has homogeneous coordinates of the form $\left[1-t^2, 2t, 1+t^2\right]$, where $t \in \mathbb{R}$.

Thus, for example, the Point P with homogeneous coordinates $\left[1, -\sqrt{3}, 2\right]$ on the projective conic E with equation $x^2 + y^2 = z^2$ also has homogeneous coordinates $\left[\frac{1}{2}, -\frac{\sqrt{3}}{2}, 1\right]$. It follows that at P we may take

$$\frac{1-t^2}{1+t^2} = \frac{1}{2} \quad \text{and} \quad \frac{2t}{1+t^2} = -\frac{\sqrt{3}}{2}.$$

From the first equation we have $2(1-t^2) = 1 + t^2$, so that $1 = 3t^2$ or $t = \pm 1/\sqrt{3}$. Since (from the second equation displayed above) $2t/(1+t^2)$ is negative, it follows that we must have t negative. Hence, at the Point P, $t = -1/\sqrt{3}$.

Problem 8 Determine the value of the parameter t at the Point $[1, 2\sqrt{2}, 3]$ in the parametrization $[1-t^2, 2t, 1+t^2]$ of the projective conic with equation $x^2 + y^2 = z^2$.

This parametrization is often useful, as the following problem illustrates.

Problem 9 Let E be the projective conic with equation $x^2 + y^2 = z^2$, and let $A = [0,1,1], B = [1,0,-1], C = [0,1,-1], D = [1,0,1]$ and $P = [1-t^2, 2t, 1+t^2]$, where $t \in \mathbb{R}$, be Points on E. Let the tangents to E at A, B, C and D meet the tangent to E at P at the Points A', B', C' and D', respectively.

(a) Determine the equations of the tangents to E at A, B, C, D and P.
(b) Determine homogeneous coordinates for the Points A', B', C' and D'.
(c) Determine the value of the cross-ratio $(A'B'C'D')$. (Cross-ratio was defined in Subsection 3.5.1)

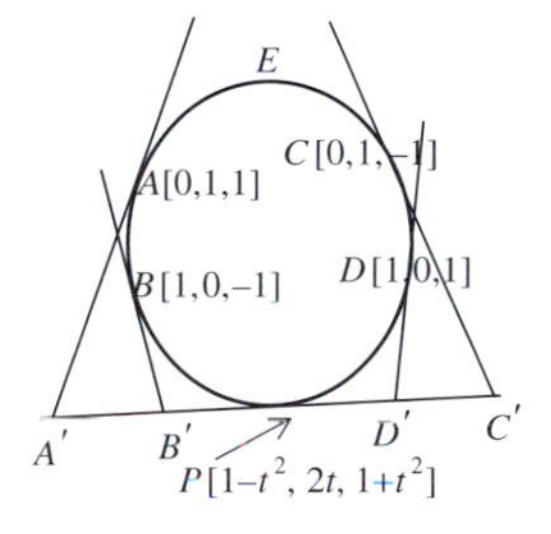

4.3.5 Some General Remarks

One of the principal features of our approach to projective geometry is the use of *algebraic methods* (via homogeneous coordinates) to prove *geometric results.* This is analogous to the use of Cartesian coordinates to prove geometric results in Euclidean geometry.

In projective geometry we can often use the Fundamental Theorem of Projective Geometry, or the Three Points Theorem for projective conics, to assign homogeneous coordinates to particular Points in some convenient way that makes the algebra as simple as possible.

Once we have made the choice of coordinates, the proofs of many geometric theorems are then simply a question of ploughing carefully through algebraic calculations. In a way this is a great advantage, since it is often much simpler to do routine algebra to obtain a proof than to sit and wait for geometrical inspiration! Unfortunately, the routine nature of the algebraic work hides the geometric ideas behind the results themselves. For example, when we discussed Desargues' Theorem in the previous chapter, you probably gained a deeper understanding of the theorem from the geometric discussion in Subsection 3.1.3 than from the algebraic proof in Subsection 3.4.1.

4.4 Applying Linear Algebra to Projective Conics

We now prove that all non-degenerate projective conics are projective-congruent. In order to do this, we first describe how to express the general equation of a projective conic in matrix form.

Let E be a projective conic with equation

$$Ax^2 + Bxy + Cy^2 + Fxz + Gyz + Hz^2 = 0,$$

and let

$$\mathbf{A} = \begin{pmatrix} A & \frac{1}{2}B & \frac{1}{2}F \\ \frac{1}{2}B & C & \frac{1}{2}G \\ \frac{1}{2}F & \frac{1}{2}G & H \end{pmatrix} \quad \text{and} \quad \mathbf{x} = \begin{pmatrix} x \\ y \\ z \end{pmatrix},$$

so that $\mathbf{x}^T = (x\ y\ z)$. It follows that

$$\begin{aligned}\mathbf{x}^T\mathbf{A}\mathbf{x} &= (x \quad y \quad z)\begin{pmatrix} A & \frac{1}{2}B & \frac{1}{2}F \\ \frac{1}{2}B & C & \frac{1}{2}G \\ \frac{1}{2}F & \frac{1}{2}G & H \end{pmatrix}\begin{pmatrix} x \\ y \\ z \end{pmatrix} \\ &= \Big(Ax + \tfrac{1}{2}By + \tfrac{1}{2}Fz \quad \tfrac{1}{2}Bx + Cy + \tfrac{1}{2}Gz \\ &\qquad \tfrac{1}{2}Fx + \tfrac{1}{2}Gy + Hz\Big)\begin{pmatrix} x \\ y \\ z \end{pmatrix} \\ &= Ax^2 + Bxy + Cy^2 + Fxz + Gyz + Hz^2.\end{aligned}$$

We may therefore write the equation of the projective conic in the form

$$\mathbf{x}^T\mathbf{A}\mathbf{x} = 0.$$

This suggests that we make the following definition.

Definition Let E be a projective conic with equation

$$Ax^2 + Bxy + Cy^2 + Fxz + Gyz + Hz^2 = 0.$$

Then

$$\mathbf{A} = \begin{pmatrix} A & \frac{1}{2}B & \frac{1}{2}F \\ \frac{1}{2}B & C & \frac{1}{2}G \\ \frac{1}{2}F & \frac{1}{2}G & H \end{pmatrix}$$

is a **matrix associated** with E.

Note that if $\mathbf{A}$ is a matrix associated with a projective conic E, so also is $\lambda\mathbf{A}$ for any non-zero real number λ.

For example, a matrix associated with the projective conic with equation

$$17x^2 + 47xy + 32y^2 + 67xz + 92yz + 66z^2 = 0$$

is

$$\mathbf{A} = \begin{pmatrix} 17 & \frac{47}{2} & \frac{67}{2} \\ \frac{47}{2} & 32 & 46 \\ \frac{67}{2} & 46 & 66 \end{pmatrix}.$$

Problem 1 Write down a matrix associated with the projective conic given by each of the following equations.

(a) $x^2 - xy + 3y^2 - 2xz + 3yz - \frac{1}{2}z^2 = 0$
(b) $2x^2 - y^2 + 4z^2 - xy + yz - 3zx = 0$

We may summarize the above discussion in the form of a theorem, as follows.

Theorem 1 Let E be a projective conic with an associated matrix $\mathbf{A}$. Then E has an equation of the form $\mathbf{x}^T\mathbf{A}\mathbf{x} = 0$.

Having expressed the general equation of a projective conic in matrix form, we can prove the result we have been seeking.

Theorem 2 All non-degenerate projective conics are projective-congruent.

Proof It is sufficient to show that all non-degenerate projective conics are projective-congruent to the standard projective conic with equation $x^2 + y^2 = z^2$.

Let E be any non-degenerate projective conic with equation $\mathbf{x}^T\mathbf{A}\mathbf{x} = 0$, where $\mathbf{A}$ is a matrix associated with E. Then by definition $\mathbf{A}$ is a symmetric matrix.

Recall that a matrix $\mathbf{A}$ is *symmetric* if $\mathbf{A}^T = \mathbf{A}$.

It follows that $\mathbf{A}$ has three orthonormal eigenvectors $\mathbf{v}_1, \mathbf{v}_2$ and $\mathbf{v}_3$, with eigenvalues λ_1, λ_2 and λ_3, respectively. If $\mathbf{P}$ is the matrix whose columns are the coordinates of $\mathbf{v}_1, \mathbf{v}_2$ and $\mathbf{v}_3$, and in this order, then

$$\mathbf{D} = \begin{pmatrix} \lambda_1 & 0 & 0 \\ 0 & \lambda_2 & 0 \\ 0 & 0 & \lambda_3 \end{pmatrix},$$

then $\mathbf{P}$ is an orthogonal matrix and $\mathbf{P}^T\mathbf{A}\mathbf{P} = \mathbf{D}$.

A matrix $\mathbf{P}$ is *orthogonal* if $\mathbf{P}^T\mathbf{P} = \mathbf{I}$; or, equivalently $\mathbf{P}^{-1} = \mathbf{P}^T$.

Now, the transformation t_1 of coordinates given by $\mathbf{x} = \mathbf{P}\mathbf{x}'$ or $\mathbf{x}' = \mathbf{P}^T\mathbf{x}$ transforms the projective conic with equation $\mathbf{x}^T\mathbf{A}\mathbf{x} = 0$ into a projective conic with equation $(\mathbf{P}\mathbf{x}')^T\mathbf{A}(\mathbf{P}\mathbf{x}') = 0$. We can rewrite this equation in the form $(\mathbf{x}')^T(\mathbf{P}^T\mathbf{A}\mathbf{P})\mathbf{x}' = 0$ or $(\mathbf{x}')^T\mathbf{D}\mathbf{x}' = 0$; in other words, as

$$\lambda_1(x')^2 + \lambda_2(y')^2 + \lambda_3(z')^2 = 0. \quad (1)$$

Geometrically, this transformation corresponds to a rotation of the coordinate axes around the origin, keeping the origin fixed and the axes at right angles to each other, possibly composed with a reflection in one of the coordinate planes.

Since the projective conic is non-degenerate, we cannot have all the λ's positive or all the λ's negative, since then equation (1) describes only the origin in $\mathbb{R}^3$.

Also, none of the λ's in equation (1) can be zero. For example, if $\lambda_3 = 0$, then λ_1 and λ_2 must be of opposite sign and equation (1) can be written as the equation of a pair of Lines in $\mathbb{RP}^2$:

$$\sqrt{|\lambda_1|}x' = \pm\sqrt{|\lambda_2|}y'.$$

The two Lines are coincident if $\lambda_1 = 0$ or $\lambda_2 = 0$.

The only other possibility is that two of the λ's, say λ_1 and λ_2, are of the same sign, positive say, and the sign of the third, λ_3, is negative. Then equation (1) can be rewritten in the form

If necessary, we can re-order the eigenvalues and eigenvectors to ensure that this is the case.

$$|\lambda_1|(x')^2 + |\lambda_2|(y')^2 = |\lambda_3|(z')^2.$$

This is the equation of a cone in $\mathbb{R}^3$ whose axis is along the z'-axis and whose horizontal cross-sections are ellipses. This 'elliptical' cone can be mapped onto the 'circular' cone with equation $(x'')^2 + (y'')^2 = (z'')^2$ by means of a transformation of coordinates $\mathbf{x}' \mapsto \mathbf{x}''$ given by $\mathbf{x}'' = \mathbf{B}\mathbf{x}'$, where

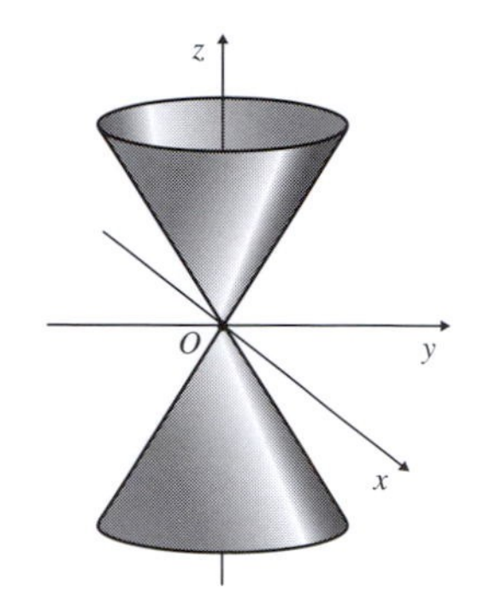

$$\mathbf{B} = \begin{pmatrix} \sqrt{|\lambda_1|} & 0 & 0 \\ 0 & \sqrt{|\lambda_2|} & 0 \\ 0 & 0 & \sqrt{|\lambda_3|} \end{pmatrix}.$$

After dropping the dashes, it follows that we can map E onto the projective conic with equation $x^2 + y^2 = z^2$ by using the projective transformation $t : [\mathbf{x}] \mapsto [\mathbf{B}\mathbf{P}^T\mathbf{x}]$. From the remark at the start of the proof, we conclude that all projective conics are projective-congruent. ■

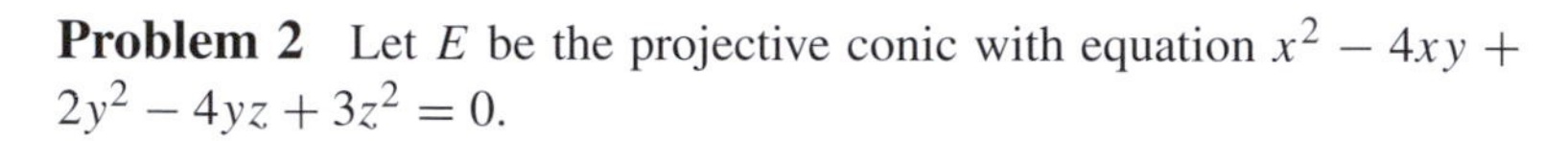

Problem 2 Let E be the projective conic with equation $x^2 - 4xy + 2y^2 - 4yz + 3z^2 = 0$.

(a) Write down a matrix $\mathbf{A}$ associated with E.
(b) Find an orthogonal matrix $\mathbf{P}$ such $\mathbf{P}^T\mathbf{A}\mathbf{P}$ is a diagonal matrix.
(c) Find a matrix associated with a projective transformation that maps E onto the standard projective conic $x^2 + y^2 = z^2$.

4.5 Duality and Projective Conics

You saw earlier that in projective geometry there is a *duality* between Points and Lines. For example, the Collinearity Property states that any two Points lie on a unique Line, and the Incidence Property states that any two Lines meet in a unique Point. In general, the *Principle of Duality* states that any true statement about Points, Lines and their projective properties remains true after dualization.

Subsection 3.4.2

But why should the Principle of Duality hold? A clue is provided by La Hire's Theorem. Suppose that E is any projective conic. Then with respect to this projective conic, every Point P in $\mathbb{RP}^2$ can be associated with a Line p, namely the polar of P with respect to E. By La Hire's Theorem, any Points Q and R on p have polars q and r which pass through the Point P. It follows that the association of Points with their polars changes collinearity into concurrence, and vice versa. This is precisely what is required for the Duality Principle.

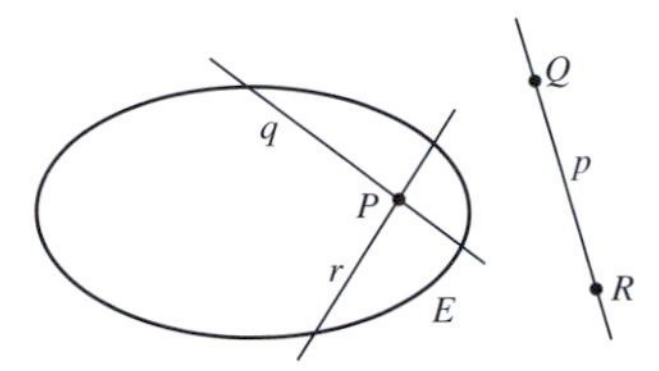

We now explore the dual of a projective conic E. If P is any point on E, then its polar with respect to E is the tangent to E at P. The discussion above suggests that we should be able to dualize the projective conic E by replacing each Point on E by the tangent to E at that Point. If we do that, then we obtain a collection of Lines which forms an envelope around E, as shown in the following diagram.

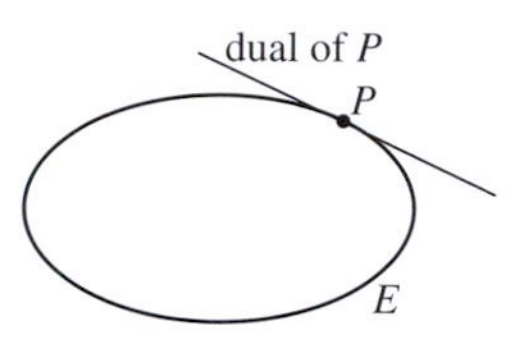

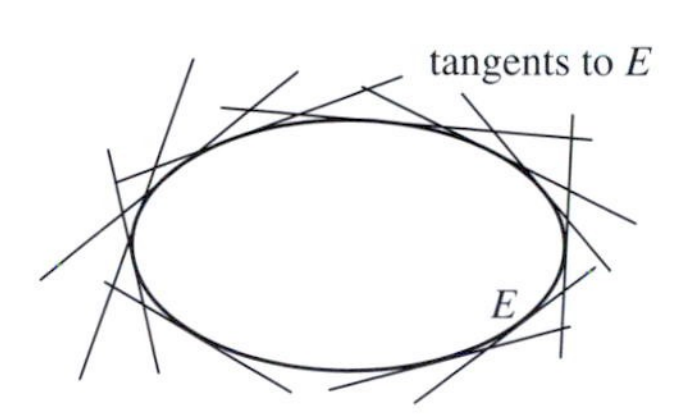

At first sight this collection of tangents appears to be rather an unwieldy object. Notice, however, that although we usually define a projective conic by specifying its family of Points, we could equally well define the projective conic by specifying its family of tangents. Both definitions uniquely determine the projective conic. When a projective conic is defined by its family of Points, we refer to the projective conic as a **Point conic**. A projective conic which is defined by its envelope of tangents is called a **Line conic**. Of course, E is the same projective conic however it is defined, so we say that E is **self-dual**.

We have already seen that the Principle of Duality is a powerful tool for discovering 'new theorems from old' when the theorems involve Lines. We now investigate an example of its use with projective conics, by dualizing Pascal's Theorem.

Subsection 3.4.2

Subsection 4.3.2, Theorem 4

Pascal's Theorem Let A, B, C, A', B' and C' be six distinct Points on a non-degenerate projective conic. Let BC' and $B'C$ intersect at P, CA' and $C'A$ intersect at Q, and AB' and $A'B$ intersect at R. Then P, Q and R are collinear.

We proceed by making the appropriate modifications to the statement of Pascal's Theorem. For clarity and convenience, we reword Pascal's Theorem slightly and set out the theorem and its dual side by side.

Pascal's Theorem	*Its dual*
Let A, B, C, A', B' and C' be six distinct Points on a non-degenerate projective conic.	Let a, b, c, a', b' and c' be six distinct tangents to a non-degenerate projective conic.
Let the Lines through B and C' and B' and C meet at a Point P, the Lines through C and A' and C' and A meet at a Point Q, and the Lines through A and B' and A' and B meet at a Point R.	Let the Points of intersection of b and c' and b' and c lie on a Line p, the Points of intersection of c and a' and c' and a lie on a Line q, and the Points of intersection of a and b' and a' and b lie on a Line r.
Then P, Q and R are collinear.	Then p, q and r are concurrent.

The dual result is known as Brianchon's Theorem, after its discoverer, and can be reworded rather more memorably, as follows.

Brianchon's Theorem The diagonals joining opposite vertices of a (projective) hexagon circumscribed around a non-degenerate projective conic are concurrent.

Pascal's Theorem can be worded similarly as follows: the opposite sides of a (projective) hexagon inscribed in a non-degenerate projective conic meet in three collinear Points.

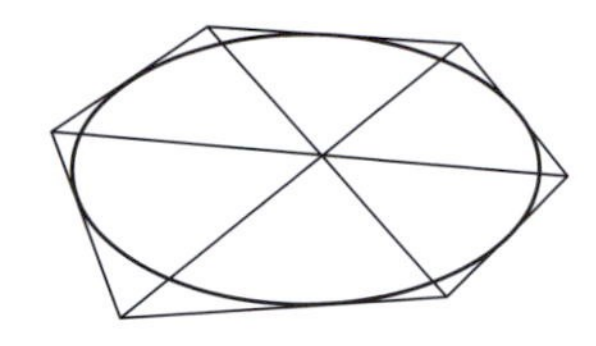

We may interpret this result in the plane $\mathbb{R}^2$ as follows: if we circumscribe a hexagon around a (non-degenerate) plane conic, then the lines joining the opposite vertices of the hexagon meet in a single point.

Here we are describing the intersection of the projective conic in Brianchon's Theorem with an embedding plane.

Problem 1 Write down the result dual to the Five Points Theorem; namely, that '*There is a unique non-degenerate projective conic through any given set of five Points, no three of which are collinear.*'

Subsection 4.3.1, Theorem 1

Many other beautiful results concerning projective and plane conics can also be discovered using the Principle of Duality.

4.6 Exercises

Section 4.1

1. Find the equation for the projective conic in $\mathbb{RP}^2$ which corresponds to each of the following plane conics in the standard embedding plane $z = 1$.

$$\left\{(x, y, z) : 9x^2 + 6xy + y^2 + x - 3y - 4 = 0,\ z = 1\right\}$$

$$\left\{(x, y, z) : 3x^2 - 9xy - 12y^2 - 30x - 64y + 1 = 0,\ z = 1\right\}$$

In each case, which ideal Points should be associated with the plane conic?

2. Which of the following equations define projective conics?
   (a) $2x^2 + 3z^2 - y^2 + xy + 4xz - 3yz = 0$
   (b) $x^2 = 4yz$
   (c) $x^2 + x - y + y^2 + z = 0$
   (d) $x^2 + z^2 + y^2 - 2 = 0$
3. Which of the following statements are true?
   (a) The projective conic $xy + xz + yz = 0$ passes through the Point $[1, 1, 1]$.
   (b) The Point $[1, -2, 1]$ lies on the projective conic $y^2 = 4xz$.
4. Determine the eccentricity of the ellipse E in $\mathbb{R}^2$ with equation

$$8u^2 + 5v^2 + 2v - 6 = 0.$$

Section 4.2

Exercises 1–4 concern the parabola E in $\mathbb{R}^2$ with equation

$$x^2 + 2xy + y^2 + 2x - y - 3 = 0.$$

1. Determine the ratio in which the parabola E divides the line segment from $(0, 0)$ to $(1, 2)$.
2. Determine the equations of the tangents to E at the points $(1, -1)$ and $(-3, 0)$.
3. One of the two tangents from $(2, 1)$ to E has equation $y = 2x - 3$. Determine the equation of the other.
4. (a) Determine the equation of the polar of $(0, \frac{3}{5})$ with respect to E. Verify that $(2, 1)$ lies on this polar.
   (b) Determine the equation of the polar of $(2, 1)$ with respect to E. Verify that $(0, \frac{3}{5})$ lies on this polar.
5. This question concerns the projective conic E with equation

$$x^2 + y^2 - 2z^2 + 2xy - yz + 4zx = 0.$$

   (a) Determine the equation of the tangent to E at the Point $[0, 1, -1]$. Verify that $[3, 0, 2]$ lies on this tangent.
   (b) Determine the equation of the other tangent to E that passes through the Point $[3, 0, 2]$.
   (c) Determine the polar of the Point $[3, 0, 2]$ with respect to E.

Section 4.3

Hint for Exercises 2, 3 *and* 7 Take the equation of the projective conic to be $xy + yz + zx = 0$, and (by the Three Points Theorem) the Points A, B and C to have homogeneous coordinates $[1, 0, 0]$, $[0, 1, 0]$ and $[0, 0, 1]$, respectively.

1. Determine the equation of the projective conic through the Points $[1, 0, 0]$, $[0, 1, 0]$, $[0, 0, 1]$, $[1, -1, 1]$ and $[4, -1, -3]$.

2. Let E be a projective conic through the vertices of a quadrilateral $ABCD$. Let AB meet CD at P, AC meet BD at Q, and AD meet BC at R. The triangle $\triangle PQR$ is called the *diagonal triangle* of $ABCD$.
   (a) Prove that the tangents to E at A and B intersect at a Point on QR.
   (b) Prove that the Line PQ is the polar of R.

In fact, each pair of tangents to E at vertices of $ABCD$ intersect on its diagonal triangle, and each vertex of the diagonal triangle has its opposite side as polar.

3. Let E be a projective conic through the vertices of a quadrilateral $ABCD$, and let the tangents to E at A and C meet at P on the Line BD. Show that the tangents to E at B and D meet at a Point Q on AC.
4. Let A, B, C, D be the Points $[1, 0, 0]$, $[0, 1, 0]$, $[2, 2, -1]$, $[0, 0, 1]$, respectively, on the projective conic E with equation $xy + yz + zx = 0$. Let T be any other Point on E, and let TB and TC meet AD at B' and C', respectively. Determine the value of the cross-ratio $(AB'C'D)$.

Cross-ratio was defined in Subsection 3.5.1.

5. Let A, B, C, D and P be the Points $[0, 1, 1]$, $[0, 1, -1]$, $[3, 4, 5]$, $[5, 12, 13]$ and $[1, 0, 1]$, respectively, on the projective conic E with equation $x^2 + y^2 = z^2$, and let the tangents to E at A, B, C and D meet the tangent to E at P at the Points A', B', C' and D', respectively.
   Determine the value of the cross-ratio $\left(A'B'C'D'\right)$.
6. The Points $P = [1, -1, 1]$, $Q = [1, -2, 2]$ and $R = [1, -2, 1]$ lie on the projective conic E with equation

$$-2x^2 + 3xy + 3y^2 + 6xz + 6yz + 2z^2 = 0.$$

   (a) Verify that the projective transformation $t : [\mathbf{x}] \mapsto [\mathbf{x}']$, where $\mathbf{x}' = \mathbf{A}\mathbf{x}$ and $\mathbf{A} = \begin{pmatrix} 2 & 1 & 0 \\ -1 & 0 & 1 \\ 0 & 1 & 1 \end{pmatrix}$ maps P, Q and R to $[1, 0, 0]$, $[0, 1, 0]$ and $[0, 0, 1]$, respectively.
   (b) Verify that the inverse of $\mathbf{A}$ is $\mathbf{A}^{-1} = \begin{pmatrix} 1 & 1 & -1 \\ -1 & -2 & 2 \\ 1 & 2 & -1 \end{pmatrix}$.
   (c) Determine the equation of the projective conic $t(E)$.
   (d) Determine a matrix associated with a projective transformation that maps E onto the projective conic with equation $xy + yz + zx = 0$.
   (e) Hence determine a matrix associated with a projective transformation that maps E onto the projective conic with equation $x^2 + y^2 = z^2$.

In tackling part (e), you should use your result from Problem 6, part (c), in Subsection 4.3.4.

7. Let A and B be Points on a projective conic E, and let P be the Point in $\mathbb{RP}^2$ with polar AB. The Line ℓ through P meets AB at Q, and E at C and D. Prove that $(PQCD) = -1$.

We shall use this result in Section 8.3.

Section 4.4

1. Let E be the projective conic with equation

$$x^2 + 2xy + 3y^2 + 6xz + 2yz + z^2 = 0.$$

   (a) Write down a matrix $\mathbf{A}$ associated with E.

(b) Find an orthogonal matrix $\mathbf{P}$ such $\mathbf{P}^T\mathbf{AP}$ is a diagonal matrix.
(c) Find a matrix associated with a projective transformation that maps E onto the standard projective conic $x^2 + y^2 = z^2$.

Summary of Chapter 4

Section 4.1: Projective conics

1. A **projective conic** in $\mathbb{RP}^2$ is a set of Points whose homogeneous coordinates satisfy a second degree equation of the form $Ax^2 + Bxy + Cy^2 + Fxz + Gyz + Hz^2 = 0$.

   A projective conic is **non-degenerate** if it can be represented by a non-degenerate conic in the standard embedding plane (which has equation $z = 1$).

   A degenerate projective conic consists of a pair of Lines, a single Line, a Point, or 'no Points'.
2. Let E be a projective conic, and let π be an embedding plane for $\mathbb{RP}^2$. Then E is represented in π by a plane conic E', where each Point of E either pierces π at a Point of E' or is an ideal Point for π associated with E'. The number of associated ideal Points depends on the type of the conic E': there are two such Points if E' is a hyperbola, one if $E'$is a parabola, and none if E' is an ellipse.
3. Let t be a projective transformation, and let E be a non-degenerate projective conic. Then $t(E)$ is a non-degenerate projective conic.
4. Let E be a non-degenerate projective conic. Then a Line ℓ is a **tangent to *E* at *P*** if ℓ meets E at a Point P, and at no other Point.
5. Let E be a non-degenerate projective conic. A Point Q lies **inside** E if every Line through Q meets E at two distinct Points. A Point R lies **outside** E if there is a Line through R that meets E at no Points.
6. Let t be a projective transformation, and let the Line ℓ be a tangent to a non-degenerate projective conic E at a Point P. Then $t(\ell)$ is a tangent to $t(E)$ at $t(P)$. Also, if Q is a Point inside E, then $t(Q)$ lies inside $t(E)$; and if R is a Point outside E, then $t(R)$ lies outside E.

   Tangency and 'lying inside or outside a projective conic' are projective properties. Hence we can use projective transformations to tackle problems involving tangents to plane conics.

   In an embedding plane tangents to projective conics correspond to tangents to plane conics, and vice versa.
7. **Eccentricity Formula** Let E be a non-degenerate plane conic with equation $u^2 + Cv^2 + Gv + H = 0$. If E has a focus on the v-axis, then the eccentricity e of E is given by the formula $e^2 = 1 - C$.
8. Any line on the surface of a right circular cone passes through the vertex of the cone, and is a **generator** of the cone; that is, the cone can be obtained

by rotating that line about a fixed line (the axis of the cone) through the vertex.

9. Every non-degenerate plane conic can be found as the curve of intersection of a suitable right circular cone with a suitable plane.
10. Every ellipse and every parabola occurs as the intersection of any right circular cone and a suitable intersecting plane.

    The intersection of a right circular cone and a suitable intersecting plane is a hyperbola, with the property that the angle between its asymptotes is less than the angle between two opposite generators of the cone. Every hyperbola occurs as the intersection of a sufficiently 'fat' right circular cone and a suitable intersecting plane
11. All non-degenerate projective conics are projective-congruent.
12. **Three Tangents Theorem** Let a non-degenerate plane conic touch the sides *BC*, *CA* and *AB* of a triangle ΔABC in $\mathbb{R}^2$ at the points P, Q and R, respectively. Then *AP*, *BQ* and *CR* are concurrent.

Section 4.2: Tangents

1. **Joachimsthal's notation for plane conics** Let a plane conic have equation $s = 0$, where $s = Ax^2 + Bxy + Cy^2 + Fx + Gy + H = 0$, and let $P_1 = (x_1, y_1)$, $P_2 = (x_2, y_2)$ and $P_3 = (x_3, y_3)$ be points of $\mathbb{R}^2$. Then we define
$$s_i = Ax_ix + B\frac{x_iy + xy_i}{2} + Cy_iy + F\frac{x_i + x}{2} + G\frac{y_i + y}{2} + H,$$
$$s_{ii} = Ax_i^2 + Bx_iy_i + Cy_i^2 + Fx_i + Gy_i + H,$$
$$s_{ij} = Ax_ix_j + B\frac{x_iy_j + x_jy_i}{2} + Cy_iy_j + F\frac{x_i + x_j}{2} + G\frac{y_i + y_j}{2} + H,$$
where i and j can each take the values 1, 2 or 3.
2. **Joachimsthal's Section Equation** The point $\left(\frac{kx_2+x_1}{k+1}, \frac{ky_2+y_1}{k+1}\right)$ divides the line segment from the point (x_1, y_1) to the point (x_2, y_2) in the ratio k:1; and lies on the plane conic with equation $s = 0$ if $s_{22}k^2 + 2s_{12}k + s_{11} = 0$.
3. The **equation of the tangent** at the point (x_1, y_1) to the non-degenerate plane conic with equation $s = 0$ is $s_1 = 0$.
4. The equation of the **tangent pair** from the point (x_1, y_1) to the non-degenerate plane conic with equation $s = 0$ is $(s_1)^2 = s \cdot s_{11}$.
5. The **polar** (or **polar line**) of a point $P\,(x_1, y_1)$ with respect to a non-degenerate plane conic E with equation $s = 0$ is the line with equation $s_1 = 0$. P is the pole of this line with respect to E.

    If P lies outside E, then the polar is the line through the two points at which the tangents from P meet E. If P lies on E, then the polar of P is the tangent to E at P.
6. **La Hire's Theorem** Let E be a non-degenerate plane conic, and let p_1 be the polar of a point P_1 in $\mathbb{R}^2$. Then each point of p_1 has a polar which passes through P_1.

7. **Joachimsthal's notation for projective conics** Let a projective conic have equation $s = 0$, where $s = Ax^2 + Bxy + Cy^2 + Fxz + Gyz + Hz^2$, and let $[x_1, y_1, z_1]$ and $[x_2, y_2, z_2]$ be Points of $\mathbb{RP}^2$. Then we define

$$\begin{aligned} s_1 &= Ax_1x + \tfrac{1}{2}B(x_1y + xy_1) + Cy_1y \\ &\quad + \tfrac{1}{2}F(x_1z + xz_1) + \tfrac{1}{2}G(y_1z + yz_1) + Hz_1z, \\ s_{11} &= Ax_1^2 + Bx_1y_1 + Cy_1^2 + Fx_1z_1 + Gy_1z_1 + Hz_1^2, \\ s_{12} &= Ax_1x_2 + \tfrac{1}{2}B(x_1y_2 + x_2y_1) + Cy_1y_2 \\ &\quad + \tfrac{1}{2}F(x_1z_2 + x_2z_1) + \tfrac{1}{2}G(y_1z_2 + y_2z_1) + Hz_1z_2. \end{aligned}$$

8. The two tangents to a projective conic E that pass through a Point P outside E are called the **projective tangent pair** (or **tangent pair**) from P to E.
9. The **projective polar** (or **polar**) of a Point P outside a projective conic E with respect to E is the Line through the two Points at which the tangents from P to E meet E.
10. Let a projective conic E in $\mathbb{RP}^2$ have equation $s = 0$.
    (a) If $P = [x_1, y_1, z_1]$ lies on E, then the tangent to E at P has equation $s_1 = 0$.
    (b) If $P = [x_1, y_1, z_1]$ lies outside E, then the pair of tangents to E from P are given by the equation $s_1^2 = s \cdot s_{11}$.
    (c) If $P = [x_1, y_1, z_1]$ is any Point in $\mathbb{RP}^2$, then the polar of P with respect to E is the Line with equation $s_1 = 0$.

Section 4.3: Theorems

1. **Five Points Theorem** There is a unique non-degenerate projective conic through any given set of five Points, no three of which are collinear. In particular, if the five Points are $[1, 0, 0]$, $[0, 1, 0]$, $[0, 0, 1]$, $[1, 1, 1]$ and $[a, b, c]$, then the equation of the conic is $c(a - b)xy + b(c - a)xz + a(b - c)yz = 0$.
2. There is a unique plane conic through any given set of five points, no three of which are collinear.
3. There are infinitely many non-degenerate projective conics through any given set of four Points, no three of which are collinear.
4. **Three Points Theorem** Let E_1 and E_2 be non-degenerate projective conics which pass through the Points P_1, Q_1, R_1 and P_2, Q_2, R_2, respectively. Then there is a projective transformation t which maps E_1 onto E_2 in such a way that $t(P_1) = P_2, t(Q_1) = Q_2, t(R_1) = R_2$.

   It is not always possible to map one projective conic E_1 onto another projective conic E_2 in such a way that four given Points of E_1 are mapped to four given Points of E_2.

5. **Strategy** To determine a projective transformation t that maps a given projective conic onto the projective conic $xy + yz + zx = 0$:
   1. choose three Points P, Q, R on E;
   2. determine a matrix $\mathbf{A}$ associated with a projective transformation that maps P, Q, R onto $[1, 0, 0]$, $[0, 1, 0]$, $[0, 0, 1]$, respectively;
   3. determine the equation $Bx'y' + Fx'z' + Gy'z' = 0$ of $t(E)$, for some real numbers B, F and G;
   4. then a matrix associated with t is $\mathbf{BA}$, where

$$\mathbf{B} = \begin{pmatrix} 1/G & 0 & 0 \\ 0 & 1/F & 0 \\ 0 & 0 & 1/B \end{pmatrix}.$$

6. The equation $xy+yz+zx = 0$ is called a **standard form** for the equation of a projective conic. The conic defined by this standard form is called a **standard projective conic**.
7. **Parametrization Theorem** Let E be a projective conic with equation in the standard form $xy + yz + zx = 0$. Then each Point of E, other than $[1, 0, 0]$, has homogeneous coordinates of the form $\left[t^2 + t, t + 1, -t\right]$, where $t \in \mathbb{R}$. Moreover, each such Point lies on E.
8. **Pascal's Theorem** Let A, B, C, A', B' and C' be six distinct Points on a non-degenerate projective conic. Let BC' and $B'C$ intersect at P, CA' and $C'A$ intersect at Q, and AB' and $A'B$ intersect at R. Then P, Q and R are collinear.

   Corollary Let A, B, C, A', B' and C' be six distinct points on a non-degenerate plane conic, with BC' and $B'C$ intersecting at P, CA' and $C'A$ intersecting at Q, and AB' and $A'B$ intersecting at R. Then P, Q and R are collinear.
9. **Converse of Pascal's Theorem** Let A, B, C, A', B' and C' be six Points, no three of which are collinear, with BC' and $B'C$ intersecting at P, CA' and $C'A$ intersecting at Q, and AB' and $A'B$ intersecting at R. If P, Q and R are collinear, then the Points A, B, C, A', B' and C' lie on a non-degenerate projective conic.

   Corollary Let A, B, C, A', B' and C' be six points in $\mathbb{R}^2$, no three of which are collinear, with BC' and $B'C$ intersecting at P, CA' and $C'A$ intersecting at Q, and AB' and$A'B$ intersecting at R. If P, Q and R are collinear, then the points A, B, C, A', B' and C' lie on a non-degenerate plane conic.
10. To locate, using only a straight edge and a pencil, further points on the unique plane conic through five given points A, B, C, A' andB', no three of which are collinear:
    1. draw the lines AB' and $A'B$, and so determine the point of intersection R of these lines;
    2. for any point T in $\mathbb{R}^2$ such that AT does not pass through B, C, A' orB', draw the lines CA' and TA, and so determine the point of intersection Q of these lines;

3. draw the lines QR and $B'C$, and so determine the point of intersection P of these lines;
4. then the point of intersection F of the lines BP and AT lies on the conic.

11. The equation $x^2 + y^2 = z^2$ is called a **standard form** for the equation of a projective conic. The conic defined by this standard form is called a **standard projective conic**. It is a right circular cone in $\mathbb{R}^3$.
12. Let E be the projective conic with equation $x^2 + y^2 = z^2$, and let $P = [a, b, c]$ be any Point in $\mathbb{RP}^2$. Then:
    (a) if $P \in E$, then $ax + by - cz = 0$ is the equation of the tangent to E at P;
    (b) if $P \notin E$, then $ax + by - cz = 0$ is the equation of the polar of P with respect to E.
13. **La Hire's Theorem** Let E be a non-degenerate projective conic, and let P be any Point of $\mathbb{RP}^2$, with polar p with respect to E. Then the polar of any Point Q on p passes through P.
14. **Three Tangents and Three Chords Theorem** Let a non-degenerate plane conic touch the sides of a triangle at the points P, Q and R, respectively, and let the tangents at P, Q and R meet the extended chords QR, RP and PQ at the points A, B and C, respectively. Then A, B and C are collinear.
15. If E_1 and E_2 are the projective conics with equations $xy + yz + zx = 0$ and $x^2 + y^2 = z^2$, respectively, then matrices associated with projective transformations that map E_1 onto E_2 and E_2 onto E_1 are $\begin{pmatrix} 1 & -1 & 0 \\ 0 & 0 & 2 \\ 1 & 1 & 2 \end{pmatrix}$ and $\begin{pmatrix} -1 & 1 & -1 \\ 1 & 1 & -1 \\ 0 & -1 & 0 \end{pmatrix}$, respectively.
16. **Parametrization Theorem** Each Point of the projective conic with equation $x^2 + y^2 = z^2$, other than the Point $[-1, 0, 1]$ has homogeneous coordinates of the form $\left[1 - t^2, 2t, 1 + t^2\right]$, where $t \in \mathbb{R}$.

Section 4.4: Applying Linear Algebra to Projective Conics

1. Let E be a projective conic with equation $Ax^2 + Bxy + Cy^2 + Fxz + Gyz + Hz^2 = 0$. Then $\mathbf{A} = \begin{pmatrix} A & \frac{1}{2}B & \frac{1}{2}F \\ \frac{1}{2}B & C & \frac{1}{2}G \\ \frac{1}{2}F & \frac{1}{2}G & H \end{pmatrix}$ is a matrix **associated** with E.
   The equation of E can be written in the form $\mathbf{x}^T \mathbf{A} \mathbf{x} = 0$.
2. All non-degenerate projective conics are projective-congruent.

Section 4.5: Duality and projective conics

1. **Principle of Duality** Any true statement about Points, Lines and their projective properties remains true after dualization.

2. When a projective conic is defined by its family of Points, we refer to it as a **Point conic**; when we define a projective conic by its envelope of tangents, we refer to it as a **Line conic**.
3. Every projective conic is **self-dual** in the sense that the dual of a Point conic is the corresponding Line conic, and vice versa.
4. **Brianchon's Theorem** The diagonals joining opposite vertices of a projective hexagon circumscribed around a non-degenerate projective conic are concurrent.

   This is the dual of Pascal's Theorem for projective conics.

5 Inversive Geometry

In this chapter we introduce a geometry known as *inversive geometry* and use it to investigate circles and lines.

We also prove the *Apollonian Circles Theorem*: namely, that if A and B are two given points in the plane, then all the points P in the plane such that $PA : PB$ is a fixed ratio $k : 1$, for some positive real number $k \neq 1$, lie on a circle. Notice that if $k = 1$, the locus of points P such that $PA : PB = 1 : 1$ is simply the set of points equidistant from A and B, namely the perpendicular bisector of the segment AB.

This surprising fact was proved by the Greek geometer Apollonius of Perga (2nd century BC).

So for each pair of points A, B there is a family of circles known as the *circles of Apollonius*. Small values of k yield circles close to A, whereas large values yield circles close to B. It is sometimes helpful to regard the points A and B as point circles, corresponding to k being zero and 'infinity', respectively.

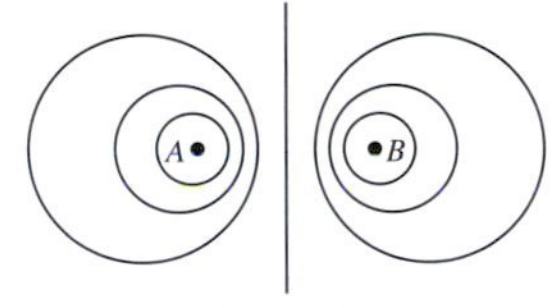

Apollonian circles

The key tool in our work is *inversion*, which is a generalization of the notion of reflection of points in a line. Just as reflection in a line maps points on one side of the line to points on the other, so inversion in a circle maps points inside the circle to points outside, and vice versa.

One difficulty that arises when we tackle problems like the Apollonian Circles Theorem is the presence of a line amongst what would otherwise be a family of circles. Clearly, it would be convenient if we could tackle such problems without having to treat the line as a special case. An ingenious way to do this is to think of the line as a circle of infinite radius in which the 'ends' of the line have been 'joined' by an additional 'point at infinity'. This interpretation enables us to introduce the term *generalized circle* to mean either a line or a circle.

To make the idea of a point at infinity precise, we introduce a space known as the *extended plane*. This consists of the ordinary plane $\mathbb{R}^2$ together with an additional point that we define to be the *point at infinity*. This extended plane is the space in which we study inversive geometry. The transformations of this geometry, known as *inversive transformations*, are defined to be the composites of inversions.

Inversive transformations have the remarkable properties that they preserve the magnitude of angles between intersecting curves, and they map generalized circles onto generalized circles. These properties enable us to use inversive geometry to prove results like the Apollonian Circles Theorem.

In Section 5.1, we define inversion and study its basic properties.

In Section 5.2, we introduce the *extended plane*, and show how certain transformations of this plane can be represented using complex numbers. We then introduce the idea of a *generalized circle*, and show that every inversion preserves the magnitude of angles, and maps generalized circles onto generalized circles. We end the section by giving an interpretation of the extended plane in terms of the so-called *Riemann sphere*. On this sphere all generalized circles actually look like ordinary circles.

In Section 5.3, we formally define what is meant by an *inversive transformation* and *inversive geometry*. The group of all inversive transformations has a subgroup which consists of the so-called *Möbius transformations*. This subgroup is analogous to the subgroup of direct isometries in Euclidean geometry.

In Section 5.4, we prove the *Fundamental Theorem of Inversive Geometry:* namely, that any three points in the extended plane can be mapped onto any other three points by an inversive transformation. We use this result to show that, in inversive geometry, all generalized circles are congruent.

Finally, in Section 5.5 we use inversive geometry to prove the Apollonian Circles Theorem and to study various families of circles.

5.1 Inversion

5.1.1 Reflection and Inversion

We begin our exploration of inversive geometry by introducing a type of transformation known as *inversion*. These transformations will be used in Section 5.3 to define inversive geometry.

Roughly speaking, an inversion is a transformation of the plane that generalizes reflection in a line. Instead of mapping points from one side of a line to the other, an inversion maps points inside a circle to points outside the circle, and vice versa.

Recall that under reflection in a line ℓ, a point A is mapped to an image point A' that lies an equal distance from ℓ, but on the opposite side of ℓ. In order to generalize this notion of a reflection, we shall reformulate it in a way that provides us with a sensible analogue when the line ℓ is replaced by a circle C.

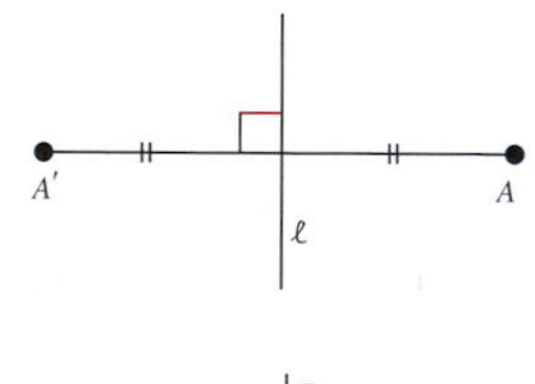

To do this, let m be a line parallel to AA' that crosses ℓ at some point P. Under reflection in ℓ, $\angle PAA'$ maps to $\angle PA'A$, so these two angles must be equal. But since the lines m and AA' are parallel, the angle $\angle PA'A$ is equal to the angle between PA' and m, so $\angle PAA'$ is equal to the angle between PA' and m. This is the clue that we need to generalize reflection.

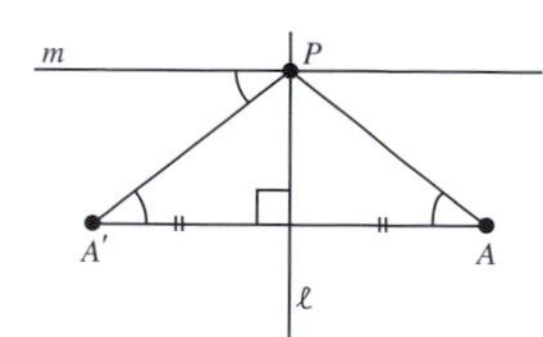

By a stretch of the imagination, we can think of ℓ as an infinitely large circle with m lying along the radius through P. If we replace ℓ by a circle C of finite radius, and we replace m by a radial line that meets C at some point P, then, by analogy with reflection, we can define the image of A to be the point A' on the line segment OA for which $\angle OPA'$ is equal to $\angle PAO$. We say that A' is the point *inverse to A with respect to the circle C*.

Of course, for this definition to work, we must check that for a given point A, the position of the point A' is independent of P. To do this, observe that the triangles $\Delta POA'$ and ΔAOP are similar, for they have a common angle at O and $\angle OPA' = \angle OAP$. It follows that

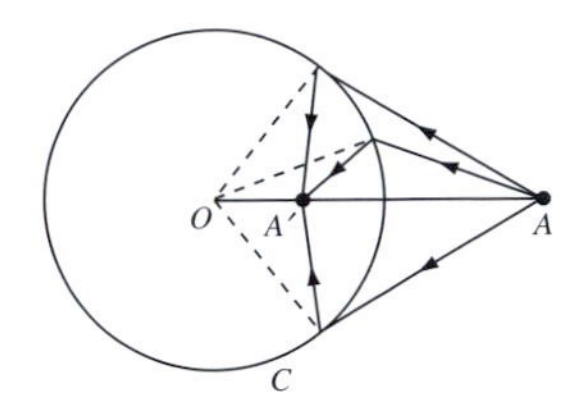

$$\frac{OA'}{OP} = \frac{OP}{OA},$$

so

$$OA \cdot OA' = OP^2.$$

But OP is equal to the radius r of C, so

$$OA \cdot OA' = r^2. \qquad (1)$$

Since there is only one point A' on the line segment OA that satisfies equation (1), and since the equation does not depend on P, it follows that the position of A' does not depend on the choice of P.

Although the above construction illustrates how inversion can be defined as a generalization of reflection, it is worth noting that equation (1) is all that we need to determine the point A' that corresponds to a given point A. For simplicity, we shall therefore use equation (1) as our formal definition of inversion.

Definitions Let C be a circle with centre O and radius r, and let A be any point other than O. If A' is the point on the line OA that lies on the same side of O as A and satisfies the equation

$$OA \cdot OA' = r^2, \qquad (2)$$

then we call A' the **inverse** of A with respect to the circle C. The point O is called the **centre of inversion**, and C is called the **circle of inversion**. The transformation t defined by

$$t(A) = A' \quad \left(A \in \mathbb{R}^2 - \{O\}\right)$$

is known as **inversion in** C.

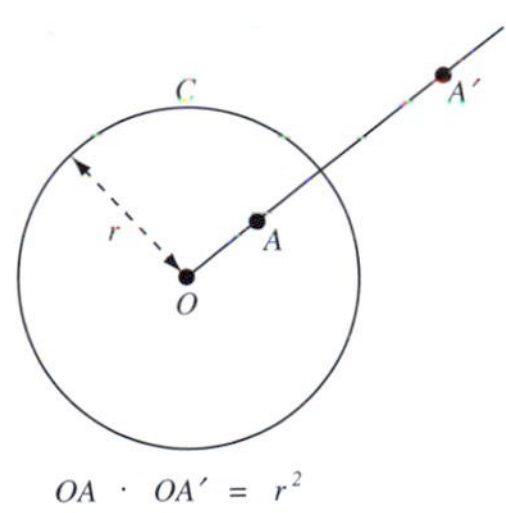

$OA \cdot OA' = r^2$

Remark

Since $OA \cdot OA' = r^2$ is non-zero, neither OA nor OA' can be zero, and so neither A nor A' can coincide with O. It is for this reason that O is excluded from the domain of the transformation t since there is no point to which O can be mapped. Likewise, there is no point that maps to O.

We can sometimes write down the inverse of a point directly from the above definition. For example, if C is the unit circle $\{(x, y) : x^2 + y^2 = 1\}$, then the inverse of (0, 2) with respect to C is the point $\left(0, \frac{1}{2}\right)$, and the inverse of $\left(-\frac{1}{3}, 0\right)$ is the point $(-3, 0)$.

Problem 1 Write down the inverse of each of the following points with respect to the circle of unit radius, centred at the origin.

(a) $(4, 0)$ (b) $(0, 1)$ (c) $\left(0, -\frac{1}{3}\right)$ (d) $\left(\frac{1}{4}, 0\right)$

Inversion distorts the plane considerably, for it maps points inside a circle C to points outside C, and vice versa. Indeed, if $OA < r$, then $OA' = r^2/OA > r$, whereas, if $OA > r$, then $OA' = r^2/OA < r$. Any point that lies on C maps to itself.

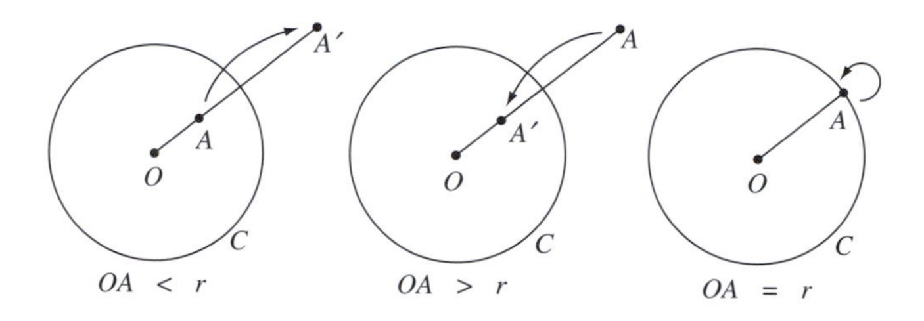

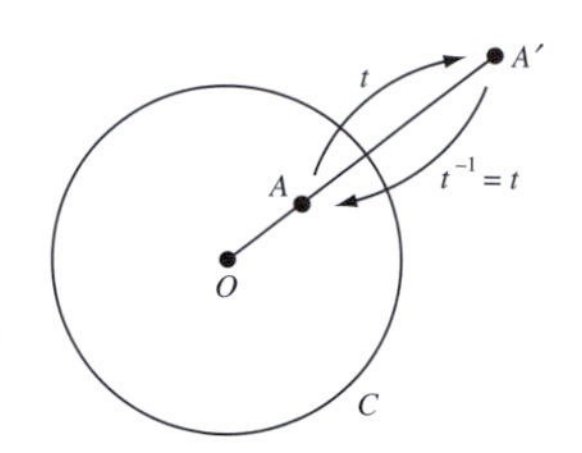

Note that if A' is the inverse of A, then A must be the inverse of A', for if $OA \cdot OA' = r^2$ then clearly $OA' \cdot OA = r^2$; we say that A and A' are **inverse points** with respect to C. In this sense, inversion is like reflection; if we reflect a point in a line and then reflect the reflection, we obtain the original point back again. Any transformation t that has this property is said to be *self-inverse* because it shows that t^{-1} exists and is equal to t.

Theorem 1 Inversion in a circle is a self-inverse transformation.

Since any transformation that has an inverse is one–one, it follows that every inversion is a one–one transformation. Remember, however, that since O is excluded from its domain, an inversion is a one–one transformation of $\mathbb{R}^2 - \{O\}$ onto itself.

Earlier, we mentioned that inversion can be regarded as a generalization of reflection. To see how this follows from the definition of inversion, we examine what happens to the inverse of a point A as we increase the radius of the circle.

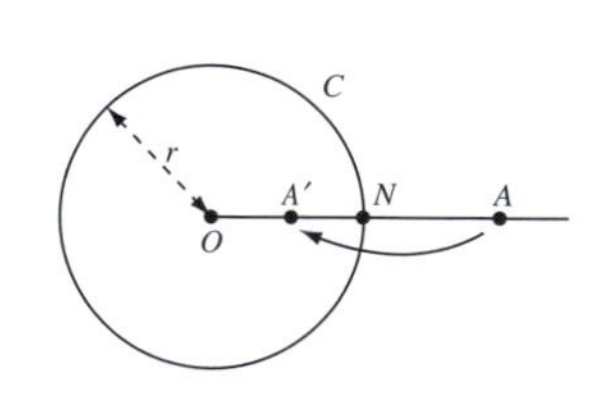

Let A be a point outside a circle C with centre O and radius r, let A' be the inverse of A with respect to C, and let the line segment AA' meet the circle at N. Then $OA = r + AN$ and $OA' = r - A'N$, so the equation $OA \cdot OA' = r^2$ becomes

$$(r + AN)(r - A'N) = r^2.$$

After expanding the brackets, cancelling the r^2 terms, and solving for $A'N$, we obtain

$$A'N = \frac{AN \cdot r}{r + AN} = \frac{AN}{1 + AN/r}.$$

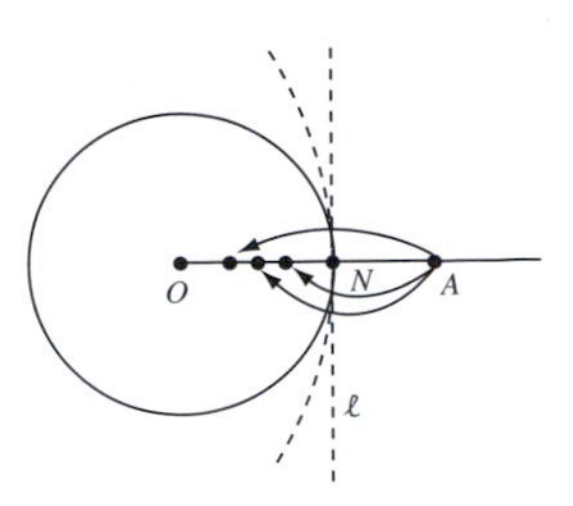

Now fix A and N, and let the radius r of the circle tend to infinity. As it does so, it follows from the above equation that the length of $A'N$ tends to the length of AN. In other words, reflection in a line can be regarded as the limiting case of inversion in circles of increasing radii. For this reason, we adopt the following useful convention.

Convention We use the term **inversion** to mean either reflection in a line or inversion in a circle.

The following gives a geometric method for constructing inverse points with respect to a given circle.

Let A be a point outside a circle C with centre O and radius r, let AP and AQ be the two tangents from A to C, and let A' be the point of intersection of OA and PQ. Then A and A' are inverse with respect to C.

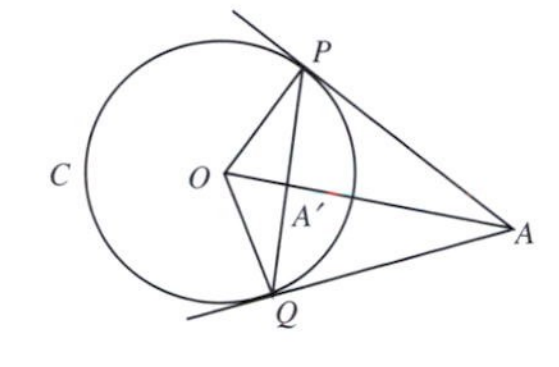

For, in the triangles $\Delta OPA'$ and ΔOAP, the angles at O are equal and the two angles $\angle OA'P$ and $\angle OPA$ are equal, since they are both right angles; hence all the angles in the two triangles are equal. Thus the triangles are similar; hence, in particular, we must have that

$$\frac{OA'}{OP} = \frac{OP}{OA},$$

so that

$$OA \cdot OA' = OP^2 = r^2.$$

Thus A and A' are inverse with respect to C, as claimed.

We end this subsection by showing that an inversion in a circle is a different type of transformation from those that arise in Euclidean and affine geometry.

First, an inversion in a circle is not an affine transformation since it does not map lines to lines. For example, let C be the unit circle and ℓ be the line with equation $x = 2$. All the points on ℓ lie outside the unit circle, and so inversion in C must map all the points of ℓ to points that lie inside the unit circle. In particular, the image of ℓ cannot be a line. Thus, since the image of a line under any affine transformation is itself a line, it follows that inversion cannot be an affine transformation.

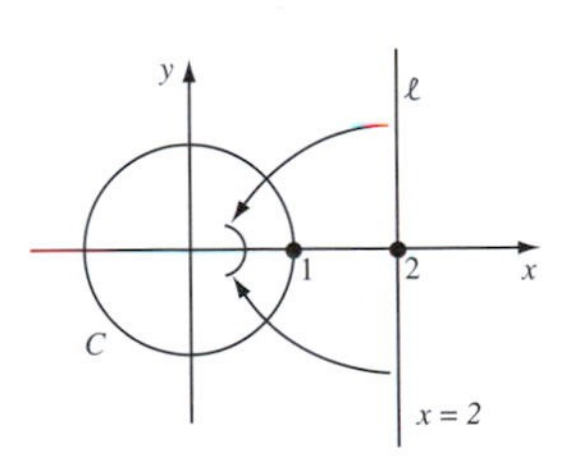

An inversion in a circle is not a Euclidean transformation either, for like affine transformations, Euclidean transformations map lines to lines. Indeed, a Euclidean transformation is just a special type of affine transformation.

5.1.2 The Effect of Inversion on Lines and Circles

In the previous subsection you saw how to construct the point A' that is the inverse of a given point A with respect to a given circle C. However, in order to study many of the properties of inversion, we require an *algebraic* formula that relates the coordinates of A and A'.

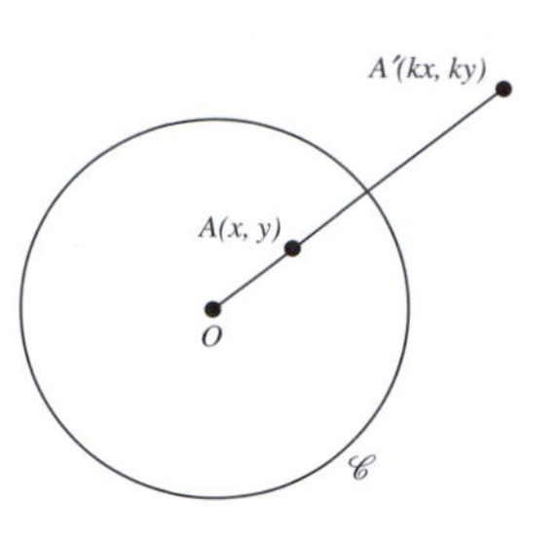

For our present purposes it is sufficient to derive the formula for the case where C is the unit circle $\{(x, y) : x^2 + y^2 = 1\}$. This circle will occur so frequently in our work that we denote it by the special symbol $\mathscr{C}$.

Let A be a point $(x, y) \in \mathbb{R}^2 - \{O\}$, and let A' be its image under inversion in the unit circle $\mathscr{C}$. Since A' lies on the same half-line from the origin as does A, it follows that A' must have coordinates (kx, ky), for some positive number k.

Since the radius of $\mathscr{C}$ is 1, we must have $OA \cdot OA' = 1$. Thus $OA^2 \cdot OA'^2 = 1$, and so

$$\left(x^2 + y^2\right)\left(k^2x^2 + k^2y^2\right) = 1.$$

It follows that

$$k^2 = \frac{1}{\left(x^2 + y^2\right)^2},$$

and hence that

$$k = \frac{1}{x^2 + y^2}.$$

Thus A' is the point $\left(\frac{x}{x^2+y^2}, \frac{y}{x^2+y^2}\right)$. We therefore have the following algebraic description of inversion in the unit circle.

Theorem 2 Inversion in the unit circle $\mathscr{C}$ is the function

$$t : (x, y) \mapsto \left(\frac{x}{x^2 + y^2}, \frac{y}{x^2 + y^2}\right) \qquad \left((x, y) \in \mathbb{R}^2 - \{O\}\right).$$

We may use this theorem to find the image of any non-zero point of $\mathbb{R}^2$ under inversion in the unit circle $\mathscr{C}$. For example, the image of $(3, -2)$ is the point

$$\left(\frac{3}{3^2 + (-2)^2}, \frac{-2}{3^2 + (-2)^2}\right) = \left(\tfrac{3}{13}, -\tfrac{2}{13}\right).$$

Problem 2 Determine the image of each of the following points under inversion in $\mathscr{C}$.

(a) $(4,1)$ (b) $\left(\frac{1}{2}, -\frac{1}{4}\right)$

Now let (x', y') be the image of (x, y) under inversion in $\mathscr{C}$. Since inversion is self-inverse, it follows that

$$(x, y) = \left(\frac{x'}{(x')^2 + (y')^2}, \frac{y'}{(x')^2 + (y')^2}\right).$$

Using this relationship between (x, y) and (x', y'), we can find the image of a curve under inversion in $\mathscr{C}$ in much the same way as we found the image of a curve in affine geometry. The main difference is that we must be careful to remember that inversion is not defined at the origin (the centre of inversion) and that no point is mapped to the origin.

We shall return to the question of what happens at the origin later, in Subsection 5.2.3.

Example 1 Determine the image under inversion in $\mathscr{C}$ of the line $2x+4y=1$.

Solution Let (x, y) be an arbitrary point on the line $2x+4y=1$, and let (x', y') be the image of (x, y) under inversion in $\mathscr{C}$. Then

$$(x, y) = \left(\frac{x'}{(x')^2+(y')^2}, \frac{y'}{(x')^2+(y')^2}\right).$$

Since x and y are related by the equation $2x+4y=1$, it follows that x' and y' are related by the equation

$$\frac{2x'}{(x')^2+(y')^2} + \frac{4y'}{(x')^2+(y')^2} = 1.$$

Multiplying by $(x')^2+(y')^2$, we obtain $2x'+4y' = (x')^2+(y')^2$, and by completing the square we may write this as

$$(x'-1)^2+(y'-2)^2 = 5.$$

Dropping the dashes, we see that points on the image must have (x, y)-coordinates which satisfy the equation

$$(x-1)^2+(y-2)^2 = 5.$$

This is the equation of a circle with centre $(1, 2)$ and radius $\sqrt{5}$. Since this passes through the origin, and since the origin cannot be part of the image under the inversion, the required image must be the circle with the origin removed. □

For every point on $(x'-1)^2+(y'-1)^2 = 5$ other than $(0, 0)$ is the image of some point on $2x+4y=1$.

In this example we used dashes to distinguish the coordinates of a point (x, y) from the coordinates of its image (x', y'). However, once you have understood the method you may prefer to adopt the strategy below. This uses the same method as in the example, but it drops the dashes throughout.

Strategy To determine an equation for the image of a curve under inversion in the unit circle $\mathscr{C}$:

1. write down an equation that relates the x- and y-coordinates of points on the curve;
2. replace x by $\frac{x}{x^2+y^2}$ and y by $\frac{y}{x^2+y^2}$, and simplify the resulting equation.

When using this strategy to find the image of a curve that passes through the origin, we must first remove the origin from the curve.

We may also have to exclude the origin from the image.

Example 2 Determine the image under inversion in $\mathscr{C}$ of the line $y = x$ (with the origin removed).

Solution Replacing x by $\frac{x}{x^2+y^2}$ and y by $\frac{y}{x^2+y^2}$, we obtain

$$\frac{y}{x^2+y^2} = \frac{x}{x^2+y^2}.$$

Hence

$$y = x.$$

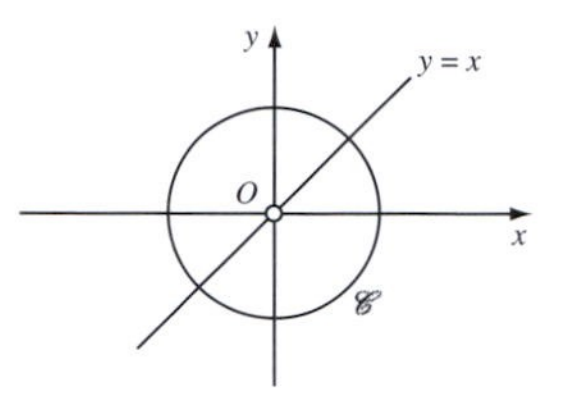

This is the line we started with. Just as the origin had to be excluded from that line before we could find its image, so the origin has to be excluded from the image. Hence the line $y = x$ (with the origin removed) maps to itself under inversion in $\mathscr{C}$. □

We shall use the term **punctured** to describe a line or circle that has one of its points removed. More specifically, if A is the point removed from a line or circle, then the line or circle is said to be **punctured at** A.

Problem 3 Determine the image under inversion in $\mathscr{C}$ of the line $y = 2x$ punctured at the origin.

Problem 4 Let ℓ be the line with equation $x + y = 1$.

(a) Determine the image of ℓ under inversion in $\mathscr{C}$.
(b) Explain why the points $(1, 0)$, $(0, 1)$ lie both on ℓ and on its image.
(c) Sketch ℓ and its image on a single diagram.

The solutions to these problems and the preceding examples illustrate the following general result.

Theorem 3 Images of Lines under Inversion
Under inversion in a circle with centre O:

(a) a line that does not pass through O maps onto a circle punctured at O;
(b) a line punctured at O maps onto itself.

Proof First choose a pair of coordinate axes with origin at O, and choose a unit of length equal to the radius of the circle in which we are inverting. Then the circle in which we are inverting becomes the unit circle $\mathscr{C}$.

(a) If ℓ is a line that does not pass through the origin, then it has an equation of the form

$$ax + by + c = 0,$$

where c is non-zero. Using the above strategy, we know that the image of ℓ under inversion in $\mathscr{C}$ has equation

$$\frac{ax}{x^2+y^2}+\frac{by}{x^2+y^2}+c=0.$$

Since c is non-zero, we may rewrite this in the form

$$x^2+y^2+(a/c)x+(b/c)y=0.$$

This is the equation of a circle C through the origin. If the origin is removed from this circle, then each remaining point A' is the image of the point A at which OA' intersects ℓ. It follows that the image of ℓ is the whole of the punctured circle $C-\{O\}$.

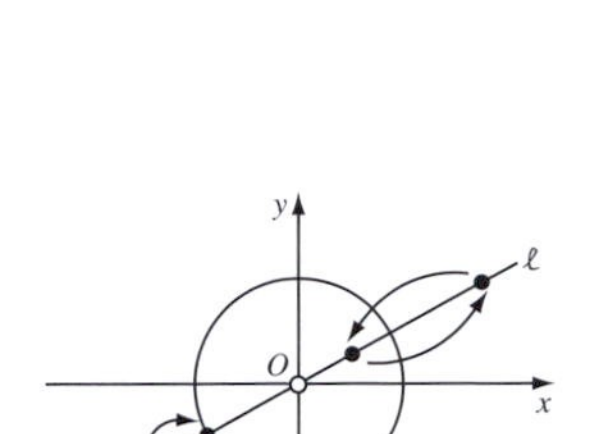

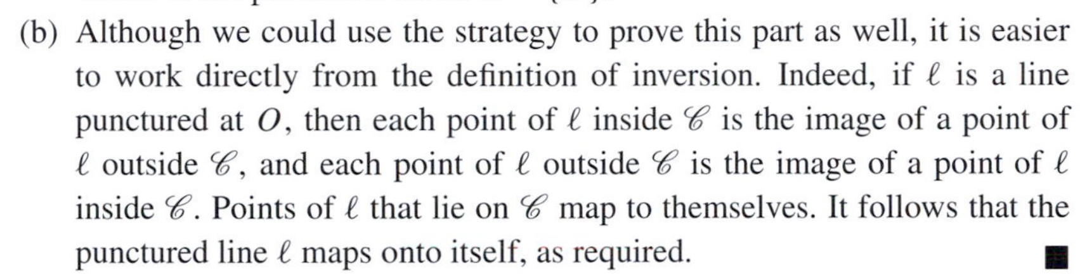

(b) Although we could use the strategy to prove this part as well, it is easier to work directly from the definition of inversion. Indeed, if ℓ is a line punctured at O, then each point of ℓ inside $\mathscr{C}$ is the image of a point of ℓ outside $\mathscr{C}$, and each point of ℓ outside $\mathscr{C}$ is the image of a point of ℓ inside $\mathscr{C}$. Points of ℓ that lie on $\mathscr{C}$ map to themselves. It follows that the punctured line ℓ maps onto itself, as required. ∎

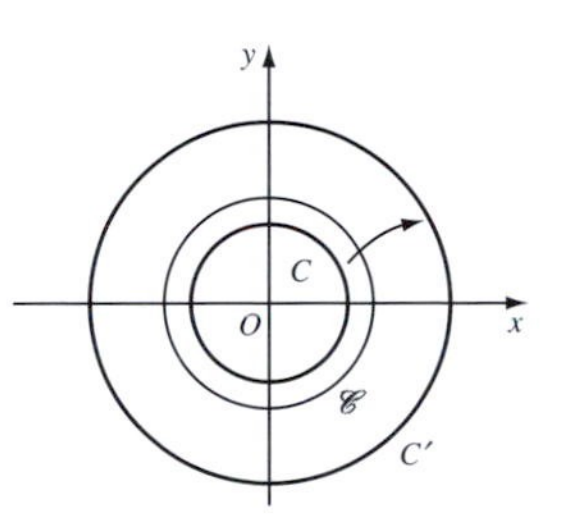

Next, we consider the images of circles under inversion. Since points on the circle of inversion map to themselves, the image of that circle is also a circle. Also, any circle C with centre the centre of inversion O must map onto another circle C' with centre O, for, by symmetry, every point of C is mapped an equal distance along a radial line.

This raises the question as to whether an inversion always maps circles to circles.

Example 3 Use the above strategy to determine the image under inversion in $\mathscr{C}$ of the circle C with centre (2, 0) and radius 1.

Solution The circle C has equation $(x-2)^2+y^2=1$, which we may rewrite in the form

$$x^2+y^2-4x+3=0.$$

Using the above strategy, we deduce that the image of C under inversion in $\mathscr{C}$ has equation

Note that the origin does not lie on C.

$$\left(\frac{x}{x^2+y^2}\right)^2+\left(\frac{y}{x^2+y^2}\right)^2-\frac{4x}{x^2+y^2}+3=0.$$

We may add together the first two terms of this equation to obtain

$$\frac{1}{x^2+y^2}-\frac{4x}{x^2+y^2}+3=0,$$

which we may rearrange in the form

$$x^2 + y^2 - \tfrac{4}{3}x + \tfrac{1}{3} = 0.$$

By completing the square we obtain

$$\left(x - \tfrac{2}{3}\right)^2 + y^2 = \left(\tfrac{1}{3}\right)^2.$$

This is the equation of a circle with centre $\left(\frac{2}{3}, 0\right)$ and radius $\frac{1}{3}$. □

So in this example the circle C does indeed map to another circle. Notice, however, that the centre $(2, 0)$ of C maps to $(\frac{1}{2}, 0)$ which is *not* the centre of the image of C. It follows that even if an inversion maps one circle onto another, it may not map the centres to each other.

Problem 5 Determine the image under inversion in $\mathscr{C}$ of the circle with centre $(2, 2)$ and radius 1.

The next example illustrates what happens when we use the strategy to find the image of a circle that passes through the origin.

Example 4 Let C be the circle with centre $(-2, 0)$ and radius 2, punctured at the origin. Determine the image of C under inversion in $\mathscr{C}$.

Solution The circle C has equation $(x + 2)^2 + y^2 = 2^2$, which we may rewrite in the form

$$x^2 + y^2 + 4x = 0.$$

Using the above strategy, we deduce that the image of C under inversion in $\mathscr{C}$ has equation

$$\left(\frac{x}{x^2 + y^2}\right)^2 + \left(\frac{y}{x^2 + y^2}\right)^2 + \frac{4x}{x^2 + y^2} = 0.$$

Adding together the first two terms of this equation, we obtain

$$\frac{1}{x^2 + y^2} + \frac{4x}{x^2 + y^2} = 0,$$

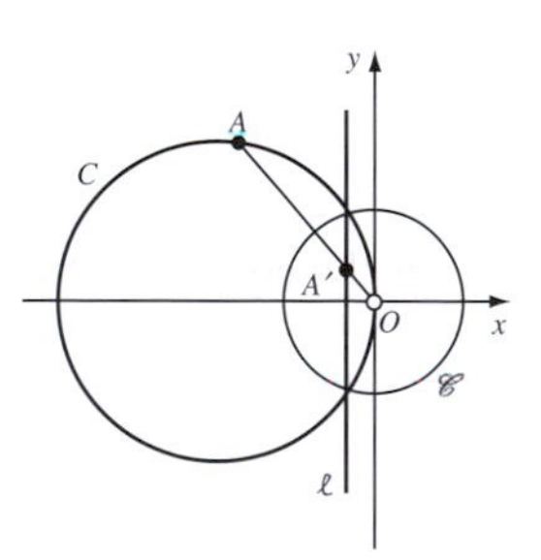

which we may rearrange in the form $1 + 4x = 0$.

It follows that the image of the punctured circle C is the line ℓ with equation $x = -\frac{1}{4}$. From the figure in the margin it is clear that every point of ℓ is the image of some point on C, so there is no need to puncture ℓ. □

Problem 6 Let C be the circle with centre $\left(0, -\frac{1}{4}\right)$ and radius $\frac{1}{4}$, punctured at the origin. Determine the image of C under inversion in $\mathscr{C}$.

The conclusions of Examples 3 and 4, and of Problems 5 and 6, suggest the following result.

Theorem 4 Images of Circles under Inversion
Under inversion in a circle with centre O:

(a) a circle that does not pass through O maps onto a circle;
(b) a circle punctured at O maps onto a line that does not pass through O.

Proof As for Theorem 3, we choose a pair of coordinate axes that makes the circle in which we are inverting the unit circle $\mathscr{C}$.

Now let C be an arbitrary circle with centre (a,b) and radius r. This has equation

$$(x-a)^2+(y-b)^2=r^2,$$

which we may rewrite in the form

$$x^2+y^2-2ax-2by+c=0,$$

Note that this passes through O if and only if $c=0$.

where $c=a^2+b^2-r^2$. Using the strategy for determining the images of curves under inversion, we deduce that the image of C under inversion has equation

$$\left(\frac{x}{x^2+y^2}\right)^2+\left(\frac{y}{x^2+y^2}\right)^2-\frac{2ax}{x^2+y^2}-\frac{2by}{x^2+y^2}+c=0.$$

Adding together the first two terms of this equation, we obtain

$$\frac{1}{x^2+y^2}-\frac{2ax}{x^2+y^2}-\frac{2by}{x^2+y^2}+c=0.$$

By multiplying this equation by $\left(x^2+y^2\right)$, we may rearrange it in the form

$$1-2ax-2by+c\left(x^2+y^2\right)=0. \qquad (3)$$

This is the equation of either a line or a circle, depending on whether or not C passes through O.

(a) If C does not pass through O, then c is non-zero. We can therefore divide equation (3) by c to obtain

$$x^2+y^2-2\tfrac{a}{c}x-2\tfrac{b}{c}y+\tfrac{1}{c}=0.$$

This is the equation of a circle on which the image of C must lie.

(b) If C does pass through O, then $c = 0$, so equation (3) becomes

$$1 - 2ax - 2by = 0.$$

This is the equation of a line ℓ that does not pass through O. ■

The following box summaries the results of Theorem 3 and 4.

Under inversion with respect to a circle with centre O:

a line punctured at O	*maps onto*	the same line punctured at O;
a line not through O	*maps onto*	a circle punctured at O;
a circle punctured at O	*maps onto*	a line not through O;
a circle not through O	*maps onto*	a circle not through O.

If the circle of inversion is the unit circle $\mathscr{C}$, then O is the origin.

There is no need to remember the details of this summary since its predictions can be recalled intuitively as follows. First, points on a line or circle through the origin can be chosen arbitrarily close to the origin. The images of such points can therefore be chosen arbitrarily far from the origin, and must therefore lie on a line. Secondly, points on a line can be chosen arbitrarily far from the origin. The images of these points can be chosen arbitrarily close to the origin and must therefore lie on a circle or line punctured at the origin.

With a little practice it is easy to use the summary to simplify the work needed to determine the image of a circle or line under an inversion.

Example 5 Determine the image of each of the following under inversion in the unit circle $\mathscr{C}$:

(a) the line ℓ with equation $x = 2$;
(b) the circle C with centre (0, 2) and radius 1.

Solution

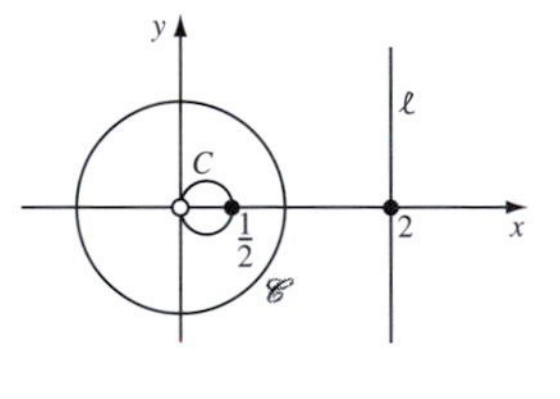

(a) From the summary, we know that ℓ maps to a circle C punctured at the origin. This circle passes through the point $\left(\frac{1}{2}, 0\right)$, since $\left(\frac{1}{2}, 0\right)$ is the image of the point (2, 0) on the line. Since ℓ is symmetrical about the x-axis, it follows that the image circle must also be symmetrical about the x-axis. The only circle C that fulfils all these criteria is the circle with radius $\frac{1}{4}$ and centre $\left(\frac{1}{4}, 0\right)$.

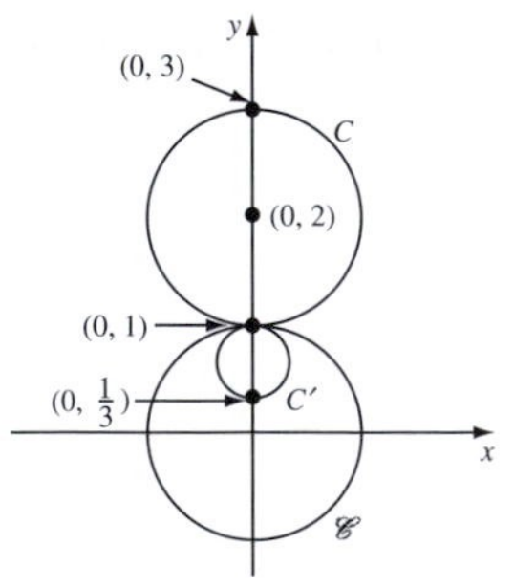

(b) From the summary, the image C' of C is a circle that does not pass through the origin. It must be symmetrical about the y-axis (because C is), and it must pass through the points $\left(0, \frac{1}{3}\right)$ and (0, 1) (the images of (0, 3) and (0, 1), respectively). The only circle C' that fulfils all these criteria is the circle with radius $\frac{1}{3}$ and centre $\left(0, \frac{2}{3}\right)$. □

Problem 7 Determine the image of each of the following under inversion in the unit circle $\mathscr{C}$:

(a) the line ℓ with equation $y = 1$;
(b) the circle C with centre (0, 1) and radius 1 (punctured at the origin).

5.1.3 The Effect of Inversion on Angles

In this subsection we shall show that inversion preserves the magnitude of the angle at which two curves meet. First, however, we must clarify what it means to measure the angle between two intersecting curves.

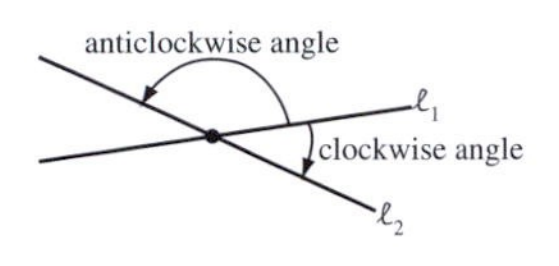

In the case of two intersecting lines ℓ_1 and ℓ_2 there are two ways in which we can measure the angle from ℓ_1 to ℓ_2. We can measure it either in a clockwise direction, or in an anticlockwise direction, as shown in the margin. Clearly, the magnitude of the angle depends on the direction we choose, so when specifying an angle we must give both its magnitude and direction.

In the case of two intersecting *curves*, we define the angle between the curves by using tangents, as follows.

Definitions Let c_1 and c_2 be two curves that intersect at the point A, and let the tangents to the curves at A be ℓ_1 and ℓ_2, respectively. Then the **anticlockwise angle** from c_1 to c_2 at A is the anticlockwise angle from ℓ_1 to ℓ_2, and the **clockwise angle** from c_1 to c_2 at A is the clockwise angle from ℓ_1 to ℓ_2.

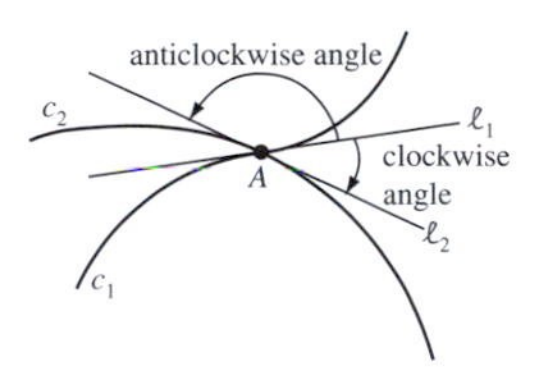

To examine what happens to the angles between two curves under an inversion, it is sufficient to examine what happens to the angles between the corresponding tangents.

For the moment let us concentrate on what happens to a single line ℓ under inversion in a circle centred at O. We know that ℓ maps onto a circle C punctured at O. But we can say more, for if m is the line through O that is perpendicular to ℓ, then ℓ is symmetrical about m. It follows that the circle C is symmetrical about m, and so ℓ is parallel to the tangent to C at O. We state this result as the Symmetry Lemma.

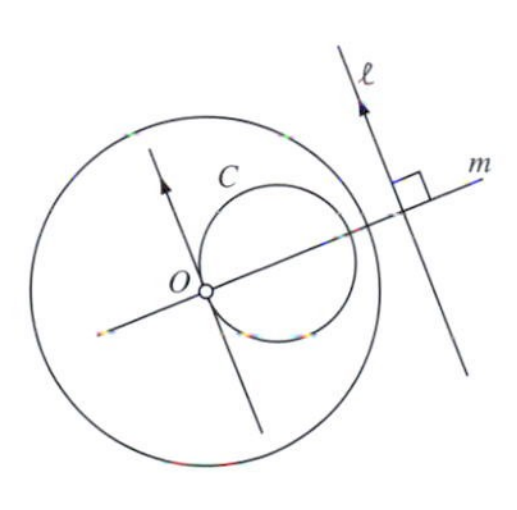

Lemma 1 Symmetry Lemma
Let ℓ be a line that does not pass through the point O. Then under inversion in a circle with centre O, ℓ maps to a circle C (punctured at O), and the tangent to C at O is parallel to ℓ.

Now consider what happens to the angle between two lines ℓ_1 and ℓ_2 which intersect at some point A other than O, as shown below. For the moment we shall assume that neither line passes through O, so that under the inversion the lines map to punctured circles C_1 and C_2, respectively. These punctured circles meet at the point A', where A' is the image of A. We want to compare

We ask you to investigate what happens when one of the lines passes through O in Problem 8.

the angle from ℓ_1 to ℓ_2 at A with the angle from C_1 to C_2 at A'. The trick we use is to compare both angles with the angle from C_1 to C_2 at O.

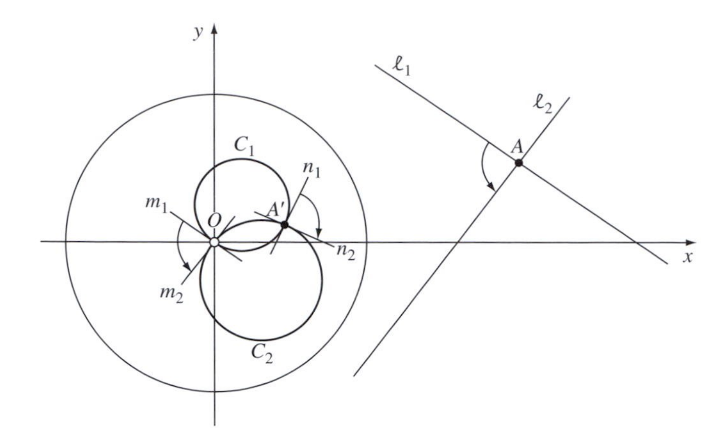

By the Symmetry Lemma, ℓ_1 is parallel to the tangent m_1 to C_1 at O, and ℓ_2 is parallel to the tangent m_2 to C_2 at O. It follows that the angle from ℓ_1 to ℓ_2 must be equal in magnitude and direction to the angle we have shown from m_1 to m_2.

Next observe that the reflection in the line through the centres of C_1 and C_2 sends the tangents m_1, m_2 at O to the tangents n_1, n_2 at A'. Since the reflection preserves the magnitude of an angle but changes its orientation, we conclude that the angle from n_1 to n_2 at A' must be equal in magnitude but *opposite* in orientation to the angle from m_1 to m_2 at O.

Overall, we have shown that the angle from C_1 to C_2 at A' must be equal in magnitude but opposite in orientation to the angle from ℓ_1 to ℓ_2 at A.

We sometimes abbreviate this by saying that the angle at A' is *equal but opposite* to the angle at A.

Theorem 5 Angle Theorem
An inversion in any circle preserves the magnitude of angles between curves but reverses their orientation.

The next problem asks you to complete our proof of this theorem.

Problem 8 Prove the Angle Theorem in the case where one of the lines passes through the centre of inversion.

The Angle Theorem provides us with a very powerful tool for locating the images of two or more circles or lines under inversion.

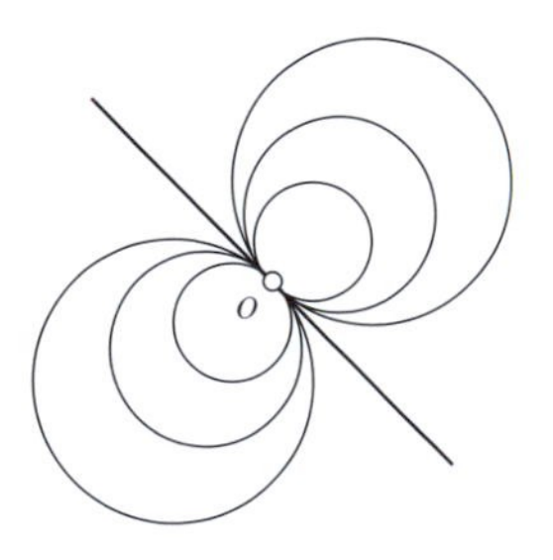

Example 6 A family of circles shares a common tangent at the origin O, as shown in the margin. Describe the effect of inverting the family of circles (all punctured at O) in the unit circle $\mathscr{C}$. Illustrate your answer with a sketch.

Solution Let ℓ be the common tangent to the circles, as shown on the left below. If d is the punctured line through O perpendicular to ℓ, then each of the circles crosses d at right angles. Since all the circles are punctured at O, their images under the inversion must be straight lines, as shown on the right. By the Angle Theorem, these straight lines must cross the image of d at right angles. But d maps onto itself under the inversion, so the punctured circles map to a family of parallel lines perpendicular to d.

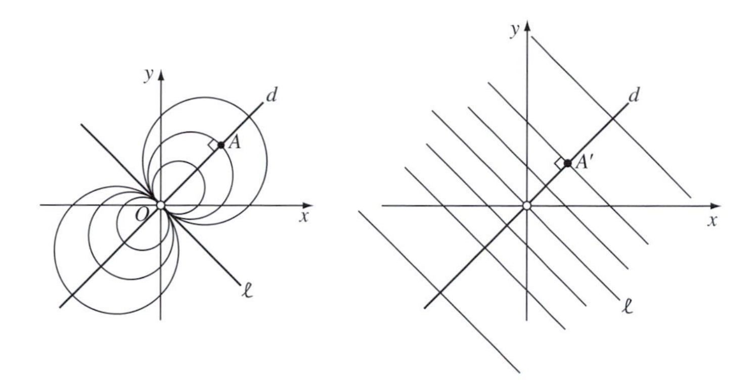

In fact, apart from ℓ, every line perpendicular to d must be the image of a punctured circle in the family. For if a perpendicular line meets d at some point A', then it must be the image of the punctured circle that passes through the point A which is mapped to A' under the inversion. □

Problem 9 One family of circles touches the x-axis at the origin O, and another family of circles touches the y-axis at O. Describe the effect of inverting the two families of circles (all punctured at O) in the unit circle $\mathscr{C}$. Illustrate your answer with a sketch.

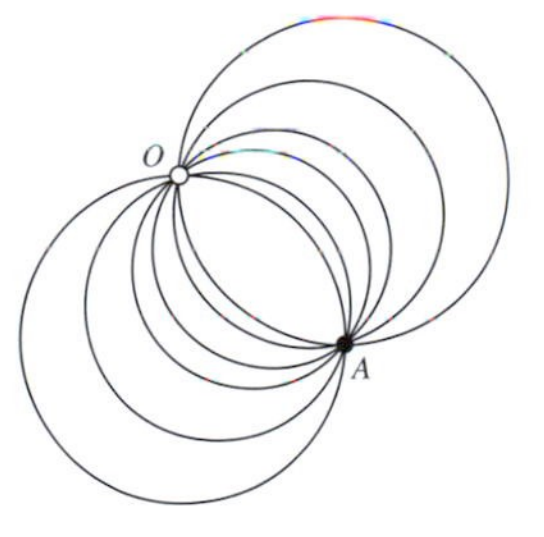

Problem 10 A family of circles intersects at the origin O and at another point A, as shown in the margin. Describe the effect of inverting the family of circles (all punctured at O) in the unit circle $\mathscr{C}$. Illustrate your answer with a sketch.

Problem 11 Let C_1, C_2 and C_3 be circles in the plane such that C_1 and C_2 touch at the origin O, C_3 and C_1 touch at another point A, and C_2 and C_3 touch at the further point B. Describe the effect of inverting the circles C_1, C_2 (both punctured at O) and C_3 in the unit circle $\mathscr{C}$. Illustrate your answer with a sketch.

These problems illustrate that, by carefully choosing centres of inversion, we can transform some of the circles in a figure to straight lines. Since straight lines are often easier to deal with than are circles, we can use such transformed figures to investigate those properties of the original figures that are preserved by inversions. In essence, this is the idea that underlies inversive geometry, and you will see several example of its use in Section 5.5.

5.2 Extending the Plane

In order to define a geometry in which we can study the properties of circles, lines and angles, we need a group of transformations that preserve these properties. Among the transformations that do this are the Euclidean transformations, and the inversions introduced in the previous section. In this section we describe how such transformations can be represented in terms of complex numbers. This will enable us to manipulate the transformations by using the algebra of complex numbers.

5.2.1 Transformations of the Complex Plane

We begin by reminding you of some facts concerning complex numbers.

First, there is a one–one correspondence between points (x, y) in the plane $\mathbb{R}^2$ and complex numbers $z = x+iy$ in the complex plane $\mathbb{C}$; we call x and y the **real part** and the **imaginary part** of the complex number z, and denote them by the symbols 'Re z' and 'Im z', respectively. All the arithmetic operations may be carried out in $\mathbb{C}$ as for real numbers, except that we replace i^2 by -1 wherever i^2 occurs.

Recall that $x + iy$ is the *Cartesian* form of the complex number.

If z is the complex number $x + iy$, then its **conjugate** $\bar{z}$ is defined by

$$\bar{z} = x - iy,$$

and its **modulus** $|z|$ is defined by

$$|z| = \sqrt{x^2 + y^2}.$$

Recall that $|z|^2 = z\bar{z}$.

Problem 1 Let $z_1 = 2 - 3i$ and $z_2 = -3 + 4i$.

(a) Determine each of the following complex numbers in Cartesian form.

(i) $z_1 + z_2$ (ii) $z_1 - z_2$ (iii) $z_1 z_2$
(iv) z_1/z_2 (v) $\overline{z_1}$ (vi) $\overline{z_2}$

(b) Determine $|z_1|$ and $|z_2|$.

If a non-zero complex number $z = x + iy$ has modulus r, and if the position vector of the point (x, y) lies at an angle θ to the positive x-axis, then we can express z in the form

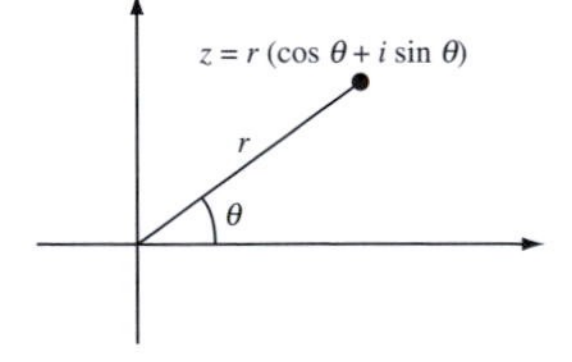

$$r(\cos\theta + i\sin\theta).$$

Such an expression is known as a **polar form** of z, and the angle θ is known as an **argument** of z, written $\arg z$. Polar forms and arguments are not unique, since the same z could equally well be expressed in the form

$$r(\cos(\theta + 2\pi n) + i \sin(\theta + 2\pi n))$$

for any integer n. The **principal argument** of z, written $\operatorname{Arg} z$, is the unique value of $\arg z$ that lies in the interval $(-\pi, \pi]$.

We can obtain the Cartesian form of a complex number from any of its polar forms by using the equations

$$x = r\cos\theta \quad \text{and} \quad y = r\sin\theta,$$

and we can obtain a polar form of a complex number from its Cartesian form by using the equations

$$r = \sqrt{x^2 + y^2}, \quad \cos\theta = \frac{x}{r} \quad \text{and} \quad \sin\theta = \frac{y}{r}.$$

Problem 2 Determine the polar forms of the complex numbers $z_1 = 1 - i$ and $z_2 = -\sqrt{3} + i$ in terms of their principal arguments.

The following strategy can be used to multiply and divide complex numbers given in polar form.

Strategy To multiply two complex numbers given in polar form, multiply their moduli and add their arguments.

To divide two complex numbers given in polar form, divide their moduli and subtract their arguments.

Although this strategy gives an argument for a product or quotient, you may require the principal argument, in which case you will need to adjust the argument by adding an appropriate multiple of 2π.

Problem 3 Let $z_1 = 1 - i$ and $z_2 = -\sqrt{3} + i$. Determine the polar forms of the following complex numbers in terms of their principal arguments.

(a) $z_1 z_2$ (b) z_1/z_2

We can now use the algebra of complex numbers described above to represent some of the basic Euclidean transformations of the plane.

Translations

First, consider the transformation

$$t(z) = z + c \quad (z \in \mathbb{C}), \tag{1}$$

where $c = a + ib$. This maps an arbitrary point $x + iy \in \mathbb{C}$ to the point

$$(x + a) + i(y + b),$$

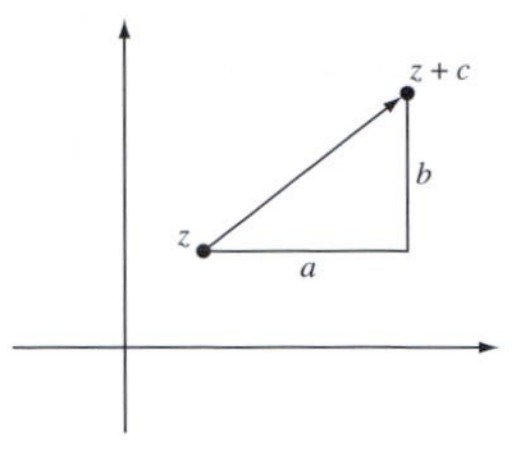

and therefore corresponds to a translation through the vector (a, b). Clearly, such transformations preserve angles, and map circles and lines to circles and lines.

Reflection in the x-Axis

Next, consider the transformation

$$t(z) = \bar{z} \quad (z \in \mathbb{C}). \tag{2}$$

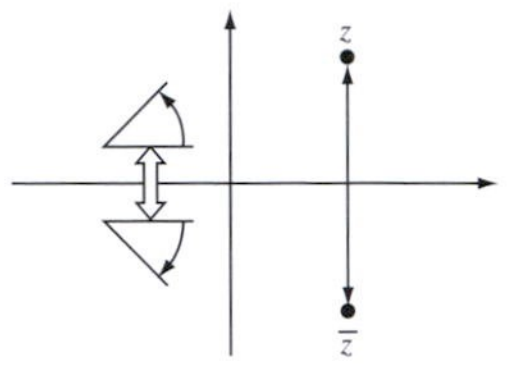

This maps an arbitrary point $x + iy \in C$ to the point $x - iy$, and therefore corresponds to a *reflection* in the x-axis. It maps circles and lines to circles and lines, and it preserves the magnitude of angles; however, the orientation of angles is reversed.

Rotation about the Origin

Now, consider the transformation

$$t(z) = az \quad (z \in \mathbb{C}), \tag{3}$$

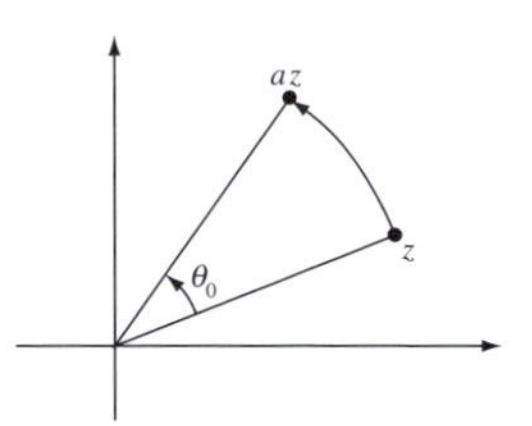

where $|a| = 1$. Since $|a| = 1$ we can write $a = \cos\theta_0 + i\sin\theta_0$, where $\theta_0 = \operatorname{Arg} a$. From the above strategy, t maps an arbitrary point $r(\cos\theta + i\sin\theta)$ in $\mathbb{C}$ to the point

$$r(\cos(\theta + \theta_0) + i\sin(\theta + \theta_0)),$$

and therefore corresponds to a rotation through the angle $\theta_0 = \operatorname{Arg} a$ about the origin. The rotation is clockwise if $\operatorname{Arg} a < 0$ and anticlockwise if $\operatorname{Arg} a > 0$.

Arbitrary Isometries

All the other isometries can be represented in the complex plane as composites of the basic transformations described above. In fact, we have the following result.

Theorem 1 Each isometry t of the plane can be represented in the complex plane by one of the functions

$$t(z) = az + b \quad \text{or} \quad t(z) = a\bar{z} + b,$$

where $a, b \in \mathbb{C}$, $|a| = 1$. Conversely, all such functions represent isometries.

Proof The converse is easy to prove, because every function of the type described is a composite of the basic isometries described above.

So let t be an isometry of the complex plane, and let $t(0) = b$, $t(1) = c$. If we denote $c - b$ by a, then $|a|$ is the distance between $t(0)$ and $t(1)$, and since t is an isometry it follows that $|a| = 1$. Now let s be the isometry defined by $s(z) = az + b$; then

$$s(0) = b = t(0) \quad \text{and} \quad s(1) = a + b = c = t(1).$$

Thus $s^{-1} \circ t$ is an isometry that fixes 0 and 1, and so, since it is an isometry, it must fix each point of the x-axis.

Then using the fact that $s^{-1} \circ t$ is an isometry (that is, it does not alter distances) that leaves each point of the x-axis unaltered, we may deduce that $s^{-1} \circ t$ is EITHER the identity transformation OR a reflection in the x-axis. If it is the identity, then $s^{-1} \circ t(z) = z$, in which case $t(z) = s(z) = az + b$. If it is a reflection, then $s^{-1} \circ t(z) = \bar{z}$, in which case $t(z) = s(\bar{z}) = a\bar{z} + b$. ■

Theorem 1 may be interpreted as saying that every isometry of the complex plane can be obtained as a rotation (through an angle Arg a) followed by a translation (through the vector (Re b, Im b)), possibly all preceded by a reflection in the real axis.

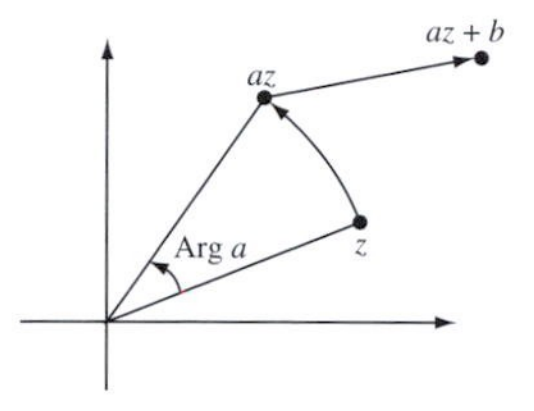

Slightly more surprising is the fact that every isometry of the plane can be expressed as a composite of reflections alone. This is because each rotation and each translation can be expressed as a composite of two reflections.

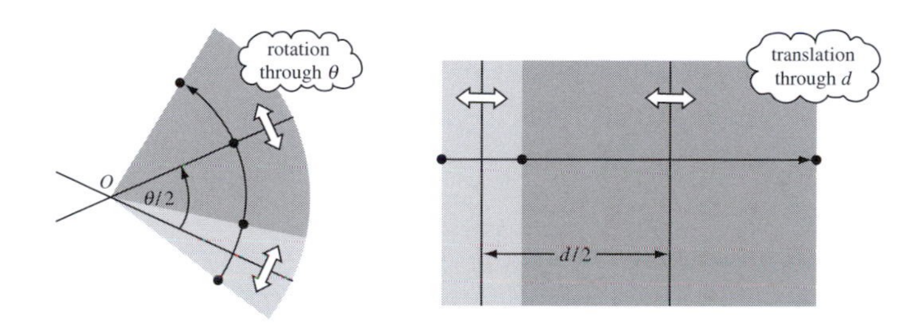

Thus a rotation about a point O can be expressed as a reflection in any line through O followed by a reflection in a second line through O, where the angle from the first line to the second line is half the desired angle rotation.

Similarly, a translation can be expressed as a reflection in any line perpendicular to the direction of the translation followed by a reflection in a second line that is parallel to the first and in the direction of the translation relative to the first line. The distance between the lines is half the distance of the translation.

So we have the following result.

Theorem 2 Every isometry can be expressed as a composite of reflections.

Example 1 Let t be the isometry defined by

$$t(z) = i\bar{z} + 4 + 2i \quad (z \in \mathbb{C}).$$

(a) Show that t represents an isometry.
(b) Interpret t as the composite of a reflection, a rotation and a translation.
(c) Interpret t as a composite of reflections.

Solution

(a) The coefficient of $\bar{z}$ is i, which has modulus $|i| = 1$. By Theorem 1, it follows that t is an isometry.

(b) In the formula that defines t, the conjugation corresponds to a reflection in the x-axis, the multiplication by i corresponds to an anticlockwise rotation through $\pi/2$, and the addition of $4+2i$ corresponds to a translation through the vector (4, 2).

(c) First, let r be the reflection in the x-axis that corresponds to the conjugation.
Next, observe that the anticlockwise rotation through $\pi/2$ can be interpreted as the composite $r_1 \circ r$, where r is the reflection in the x-axis again, and r_1 is the reflection in the line $y = x$ through the origin that makes an angle $\pi/4$ with the x-axis.
Finally, observe that the translation through the vector (4, 2) can be interpreted as the composite $r_3 \circ r_2$, where r_2 is the reflection in the line $4x + 2y = 0$ through the origin that is perpendicular to the vector (4, 2), and r_3 is the reflection in the parallel line $4x + 2y = 10$ that passes through $\frac{1}{2}(4, 2) = (2, 1)$.
Overall, we have $t = r_3 \circ r_2 \circ r_1 \circ r \circ r$ or, since r is its own inverse, $t = r_3 \circ r_2 \circ r_1$. □

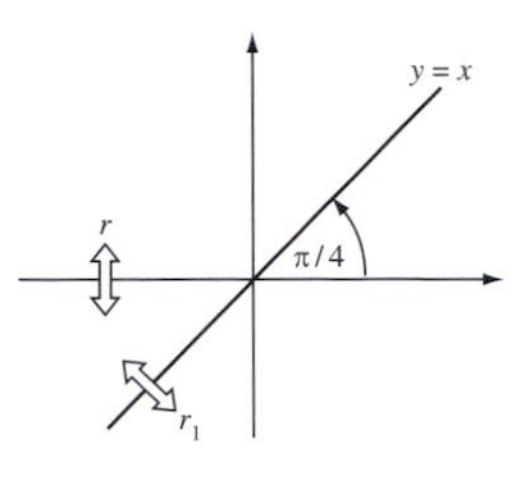

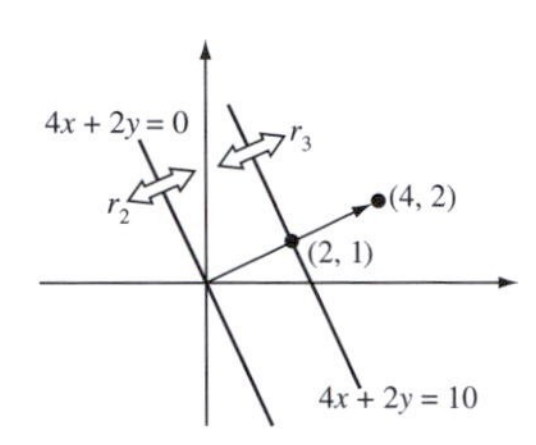

The decompositions illustrated above generalize as follows.

Problem 4 Let t be the transformation defined by

$$t(z) = -iz + 6 - 4i \quad (z \in \mathbb{C}).$$

(a) Show that t represents an isometry.
(b) Interpret t as the composite of a rotation and a translation.
(c) Interpret t as a composite of reflections.
Hint: Use the decompositions described in the margin.

To *rotate* the plane $\mathbb{R}^2$ through an angle θ: first reflect in the line $\{z : \text{Arg } z = 0\}$, then reflect in the line $\{z : \text{Arg } z = \frac{1}{2}\theta\}$.

To *translate* the plane $\mathbb{R}^2$ through a vector (a, b): first reflect in the line $\{(x, y) : ax + by = 0\}$, then reflect in the line $\{(x, y) : ax + by = \frac{1}{2}(a^2 + b^2)\}$.

Having discussed isometries, we now turn our attention to two other transformations of the complex plane that preserve the magnitude of angles and map circles and lines to circles and lines.

Scalings

The transformation defined by

$$t(z) = kz \quad (z \in \mathbb{C}), \tag{4}$$

where k is real and positive, multiplies the modulus of each complex number by a factor k but leaves its argument unchanged. It is therefore a *scaling* by the factor k. Clearly, scalings preserve angles, and map circles and lines to circles and lines.

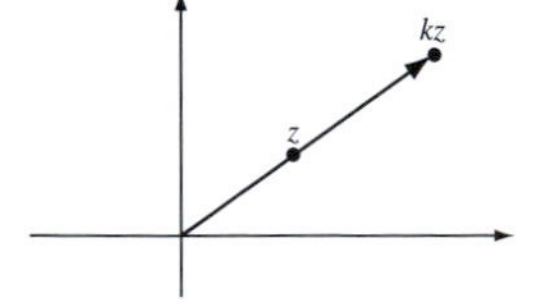

Inversions

The primary reason for introducing complex numbers into our discussion is that they provide a particularly convenient way in which to represent the effect of an inversion on points in the plane.

Theorem 3 An inversion in a circle C of radius r with centre (a, b) may be represented in the complex plane by the transformation

$$t(z) = \frac{r^2}{\overline{z - c}} + c \quad (z \in \mathbb{C} - \{c\}),$$

where $c = a + ib$.

Proof We first consider the case where C is the unit circle $\mathscr{C}$. The image under inversion in $\mathscr{C}$ of the point (x, y) is the point $\left(\frac{x}{x^2+y^2}, \frac{y}{x^2+y^2}\right)$. We may reformulate this expression in terms of complex numbers, by using the fact that the modulus $|z|$ of a complex number $z = x + iy$ satisfies the identity

By Theorem 2, Subsection 5.1.2

$$x^2 + y^2 = |z|^2 = z\bar{z}.$$

Thus the image under inversion of the point $z = x + iy$ is the point

$$\frac{x}{x^2 + y^2} + i\frac{y}{x^2 + y^2} = \frac{x + iy}{x^2 + y^2} = \frac{z}{z\bar{z}} = \frac{1}{\bar{z}}.$$

Next we consider the general case where C is a circle of radius r with centre (a, b). In this case the inversion in C can be expressed as the composite $t = t_3 \circ t_2 \circ t_1$, where

$t_1(z) = \frac{z-c}{r}$ is a translation and a scaling that sends C to the unit circle;

$t_2(z) = 1/\bar{z}$ is the inversion in the unit circle;

$t_3(z) = rz + c$ is the inverse of t_1 and sends the unit circle back to C.

Then

$$t(z) = t_3 \circ t_2 \circ t_1(z) = \frac{r^2}{\overline{z - c}} + c,$$

as required. ■

The representation of inversion provided by Theorem 3 has a particularly simple form in the case where C is the unit circle $\mathscr{C}$. In that case, $r = 1$ and $c = 0$ so the inversion is represented by the transformation

$$t(z) = \frac{1}{\bar{z}} \quad (z \in \mathbb{C} - \{O\}).$$

You met this formula in the proof of Theorem 3 above.

For example, the image of the point $1 - i$ under inversion in the unit circle $\mathscr{C}$ is the point

$$\frac{1}{\overline{1-i}} = \frac{1}{1+i} = \frac{1-i}{(1+i)(1-i)} = \tfrac{1}{2}(1-i).$$

Note that the technique of multiplying both numerator and denominator by the conjugate of the denominator is often useful.

Problem 5 Determine, in Cartesian form, the image under inversion in the unit circle $\mathscr{C}$ of each of the following points.

(a) $-\sqrt{3}+i$ (b) $-3-4i$

Problem 6 Let C be the circle of radius 2 with centre at the origin. Write down the inversion in C as a transformation of $\mathbb{C} - \{O\}$.

We know that inversions preserve the magnitude of angles. But what about their effect on circles and lines? Unfortunately, it is not strictly correct to say that an inversion maps circles and lines to circles and lines since some of the circles and lines may have to be punctured before the map can be carried out. We describe how to overcome this complication later, in Subsection 5.2.3.

By Theorem, Subsection 5.1.3

5.2.2 Linear and Reciprocal Functions

We can use the 'basic' complex functions described in the previous subsection to give a geometric interpretation of many other complex functions. We illustrate this for the so-called *linear* and *reciprocal functions*.

Definition A **linear function** is a function of the form

$$t(z) = az + b \quad (z \in \mathbb{C}),$$

where $a, b \in \mathbb{C}$ and $a \neq 0$.

Every linear function $t(z) = az + b$ can be decomposed into a composite $t_2 \circ t_1$ where

t_1 is the scaling $t_1(z) = |a|z$,
t_2 is the isometry $t_2(z) = (a/|a|)z + b$.

The geometrical interpretation of the linear function t depends on how we choose to interpret the isometry t_2. We can say that the linear function is a scaling by the factor $|a|$, followed by a rotation through the angle Arg $(a/|a|)$, followed by a translation through the vector $(\operatorname{Re} b, \operatorname{Im} b)$. Alternatively, we can use Theorem 2 and say that the linear function is a scaling composed with a number of reflections.

Either way, since both the scaling and the isometry preserve angles and map circles and lines to circles and lines, it follows that the same must be true of all linear functions.

Next, we consider the *reciprocal function*.

Definition The **reciprocal function** is defined by

$$t(z) = \frac{1}{z} \quad (z \in \mathbb{C} - \{O\}).$$

This function can be decomposed into the composite $t_2 \circ t_1$ where

t_1 is the inversion $t_1(z) = \frac{1}{\bar{z}}$,
t_2 is the conjugation $t_2(z) = \bar{z}$.

It follows that, geometrically, the reciprocal function can be interpreted as an inversion in the unit circle $\mathscr{C}$ followed by a reflection in the real axis, as shown below.

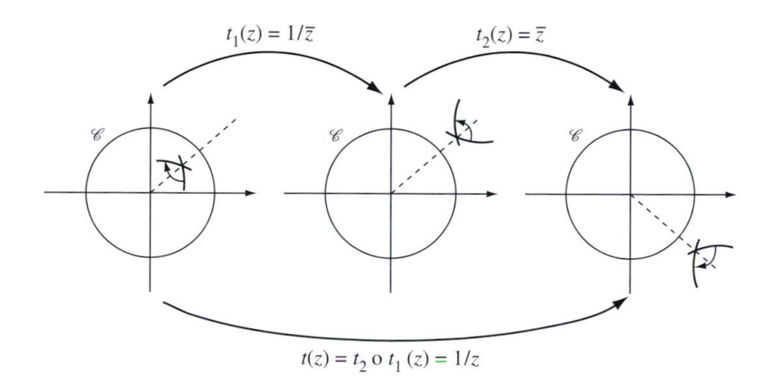

Since the inversion and the conjugation are both one–one functions that preserve the magnitude of angles and reverse their orientation, it follows that the reciprocal function must be a one–one function that preserves both the magnitude and the orientation of angles.

As in the case of inversions, it is not strictly correct to say that the reciprocal function maps circles and lines to circles and lines, because some of the circles and lines may have to be punctured before the reciprocal function can be applied. We next show how we can overcome this complication by extending the complex plane.

5.2.3 The Extended Plane

So far, our discussion of the effect of inversion on lines and circles has been complicated by the need to puncture those lines and circles that pass through the centre of inversion. Also, the need to distinguish between those images that are circles and those that are lines makes the description of the inversion process somewhat cumbersome. We can deal with both these difficulties in a very elegant way by adding an additional point to the plane to obtain the so-called *extended complex plane*.

To illustrate the ideas involved, consider the line ℓ with equation $x = 2$. Recall that inversion in the unit circle $\mathscr{C}$ maps ℓ to a circle C with radius $\frac{1}{4}$ and centre $(\frac{1}{4}, 0)$, punctured at the origin.

Example 5, part (a), Subsection 5.1.2

The point (2, 0) is mapped to $(\frac{1}{2}, 0)$, (2, 1) is mapped to $(\frac{2}{5}, \frac{1}{5})$, (2, 2) is mapped to $(\frac{1}{4}, \frac{1}{4})$, (2, 3) is mapped to $(\frac{2}{13}, \frac{3}{13})$ and, in general, the point $(2, y)$ is mapped to the point $(\frac{2}{4+y^2}, \frac{y}{4+y^2})$ on C. As the point $(2, y)$ moves from (2, 0) *up* the line ℓ, its image under inversion moves from $(\frac{1}{2}, 0)$ around the circle C towards the origin in an *anticlockwise* direction. Similarly, as the point $(2, y)$ moves from (2, 0) *down* the line ℓ, its image under inversion moves from $(\frac{1}{2}, 0)$ around the circle C towards the origin in a *clockwise* direction. The 'gap' in the circle C at the origin arises because there is no point on ℓ, or indeed any point in $\mathbb{R}^2$, that is inverted to the origin. To 'fill the gap' we attach an additional *point at infinity* to the plane.

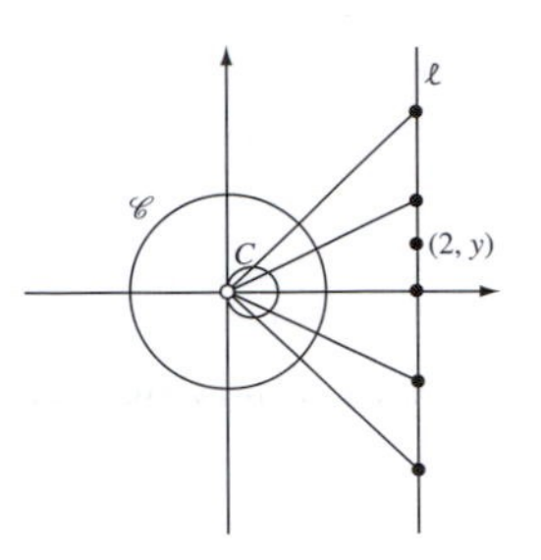

Definitions The **extended plane** is the union of the Euclidean plane $\mathbb{R}^2$ and one extra point, the **point at infinity**, denoted by the symbol ∞. When we wish to consider the plane as the complex plane $\mathbb{C}$, then we call the extended plane the **extended complex plane** and denote it by the symbol $\hat{\mathbb{C}}$; thus $\hat{\mathbb{C}} = \mathbb{C} \cup \{\infty\}$.

$\hat{\mathbb{C}}$ is read as 'C hat'.

Remarks

1. The *extended plane* and the *extended complex plane* both consist of the ordinary plane together with the point ∞. We use both terms interchangeably, depending on whether we wish to think of the 'ordinary' points in the plane as being represented by pairs of real numbers or by complex numbers.
2. The symbol ∞ does *not* represent a complex number and so it should *not* be used in association with arithmetic operations that act on complex numbers. For example, $\infty + 3i$ is a meaningless expression.

When we wish to think of points as complex numbers, we often denote them by lower-case letters such as z or a rather than the upper case letters usually used for points.

Having extended the plane in this way, we now extend the definition of inversion in a circle with centre O. We simply define the image of O to be ∞, and the image of ∞ to be O. Other points are mapped as specified by the definition of inversion given in Subsection 5.1.1.

A more formal definition of inversion in $\hat{\mathbb{C}}$ is given below.

With this extended definition of inversion, the point O corresponds to ∞, and so the circle C in the above discussion now corresponds under inversion in $\mathscr{C}$ to the set $\ell \cup \{\infty\}$. This certainly fills the 'gap' in C, but how can the set $\ell \cup \{\infty\}$ be interpreted?

As a point A on the circle C moves anticlockwise towards O, its image A' under inversion moves up ℓ. When A reaches O, A' reaches ∞. As A continues around C below O, its image returns up ℓ from below. You can think of the point ∞ as 'linking' the two ends of the line ℓ, thereby enabling points to travel 'round and round' the line. With some stretch of your imagination, you

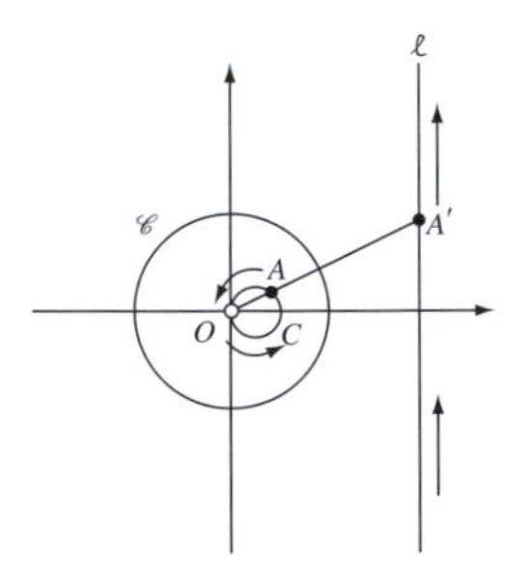

can therefore think of the line ℓ as a circle of infinite radius, where the 'gap at infinity' has been filled by the point ∞.

Any line ℓ in the plane may have its 'gap at infinity' filled by forming the set $\ell \cup \{\infty\}$. Such a set is called an **extended line**. Since an extended line can be thought of as a circle of infinite radius, we make the following definition.

Definition A **generalized circle** in the extended plane is a set that is either a circle or an extended line.

Remark

With this definition, you can think of an extended line as a generalized circle which passes through ∞, and you can think of an ordinary line as a generalized circle which has been punctured at ∞.

Recall that in Subsection 5.1.1 we adopted the convention that an inversion is either a reflection in a line, or an inversion with respect to a circle. We can now regard both of these as inversions in generalized circles.

Definition Let C be a generalized circle in the extended complex plane. Then an **inversion of the extended plane with respect to** C is a function t defined by one of the following rules:

(a) if C is a circle of radius r with centre O, then

$$t(A) = \begin{cases} \text{the inverse of } A \text{ with respect to } C, & \text{if } A \in \mathbb{C} - \{O\}, \\ \infty, & \text{if } A = O, \\ O, & \text{if } A = \infty; \end{cases}$$

(b) if C is an extended line $\ell \cup \{\infty\}$, then

$$t(A) = \begin{cases} \text{the reflection of } A \text{ in } \ell, & \text{if } A \in \mathbb{C}, \\ \infty, & \text{if } A = \infty. \end{cases}$$

Remember that A' is the inverse of A with respect to C if it lies on the same radial half-line from O as A, and $OA \cdot OA' = r^2$.

Remark

Note that any inversion in an extended line fixes the point at infinity. Conversely, every inversion that fixes ∞ must be an inversion in an extended line.

The above definition of inversion ensures that generalized circles map to generalized circles. Indeed, we already know from Subsection 5.1.2 that this is true if we allow the circles to be punctured, but we still need to check that the gap in a punctured circle is mapped to the gap in its image circle.

In the case of inversion with respect to a circle with centre O, we know that a circle or line punctured at O maps to a line, and this is consistent with the fact that O maps to ∞. Also, a line maps to a circle or line that is punctured at O, and this is consistent with the fact that ∞ maps to O.

In the case of inversion with respect to an extended line, we know that lines reflect to lines, which is consistent with the fact that ∞ maps to ∞. Ordinary circles are not a problem since they reflect onto circles.

We therefore have the following important result.

Theorem 4 Inversions of the extended plane map generalized circles onto generalized circles.

We shall sometimes find it convenient to write inversions of the extended plane as inversions of $\hat{\mathbb{C}}$. For example, by Theorem 3, we can write the inversion of the extended plane with respect to the unit circle $\mathscr{C}$ in the form

Subsection 5.2.1

$$t(z) = \begin{cases} \dfrac{1}{\bar{z}}, & \text{if } z \in \mathbb{C} - \{O\}, \\ 0, & \text{if } z = \infty, \\ \infty, & \text{if } z = 0. \end{cases}$$

Problem 7 Write down each of the following inversions of the extended plane as a transformation of $\hat{\mathbb{C}}$:

(a) the inversion with respect to the circle of radius 2 with centre the origin;
(b) the inversion with respect to the extended real axis.

The inversion of the extended plane that we asked you to write down in Problem 7, part (b), is particularly important because it provides us with a natural way of extending the conjugation function from $\mathbb{C}$ to $\hat{\mathbb{C}}$.

Definition The function $t : \hat{\mathbb{C}} \to \hat{\mathbb{C}}$ defined by

$$t(z) = \begin{cases} \bar{z}, & \text{if } z \in \mathbb{C}, \\ \infty, & \text{if } z = \infty, \end{cases}$$

is called the **extended conjugation function**.

This function occurs so frequently that we shall introduce a notation for the images of the points in its domain. Since we already have the notation $\bar{z}$ for the conjugate of a complex number z, we simply adopt the convention that $\overline{\infty} = \infty$.

Inversions are not the only transformations that can be extended to $\hat{\mathbb{C}}$. Indeed, most of the transformations discussed at the beginning of this section can be extended in a natural way to $\hat{\mathbb{C}}$.

Definitions

(a) The function $t : \hat{\mathbb{C}} \to \hat{\mathbb{C}}$ defined by

$$t(z) = \begin{cases} \frac{1}{z}, & \text{if } z \in \mathbb{C} - \{O\}, \\ \infty, & \text{if } z = 0, \\ 0, & \text{if } z = \infty, \end{cases}$$

is called the **extended reciprocal function**.

(b) A function $t : \hat{\mathbb{C}} \to \hat{\mathbb{C}}$ of the form

$$t(z) = \begin{cases} az + b, & \text{if } z \in \mathbb{C}, \\ \infty, & \text{if } z = \infty, \end{cases}$$

where $a, b \in \mathbb{C}$ and $a \neq 0$, is called an **extended linear function**.

Since scalings, rotations and translations are all special types of linear function, they too are extended to $\hat{\mathbb{C}}$ by this definition.

The solution to the next example shows that the extended reciprocal function can be expressed as a composite of two inversions.

Example 2 Find the composite $t = t_2 \circ t_1$ where

t_1 is the inversion in the unit circle $\mathscr{C}$,
t_2 is the extended conjugation function.

Solution Since

$$t_1(z) = \begin{cases} \frac{1}{\bar{z}}, & \text{if } z \in \mathbb{C} - \{O\}, \\ \infty, & \text{if } z = 0, \\ 0, & \text{if } z = \infty, \end{cases} \quad \text{and} \quad t_2(z) = \begin{cases} \bar{z}, & \text{if } z \in \mathbb{C}, \\ \infty, & \text{if } z = \infty, \end{cases}$$

we have

$$t(\infty) = t_2 \circ t_1(\infty) = t_2(0) = 0,$$
$$t(0) = t_2 \circ t_1(0) = t_2(\infty) = \infty.$$

For the remaining values of $z \in \mathbb{C} - \{O\}$ we have

$$t(z) = t_2 \circ t_1(z) = t_2\left(\frac{1}{\bar{z}}\right) = \overline{\left(\frac{1}{\bar{z}}\right)} = \frac{1}{z}.$$

It follows that $t = t_2 \circ t_1$ is the extended reciprocal function. □

Next we describe how an extended linear function can be expressed as a composite of inversions. In preparation for this, we ask you to tackle the following problem.

Problem 8 Find the composite $t = t_2 \circ t_1$ where

t_1 is the inversion with respect to the unit circle $\mathscr{C}$,
t_2 is the inversion with respect to the circle of radius 2 with centre 0.

Give an interpretation of the composite function t. Also describe what would happen if the circle of radius 2 were replaced by a circle of radius $\sqrt{k}$ with centre 0.

Earlier, we noted that every linear function is a composite $t \circ s$ of a scaling s followed by an isometry t. Since every isometry t is a composite of reflections, it follows that every linear function is a composite $r_n \circ \ldots \circ r_2 \circ r_1 \circ s$ of a scaling s followed by a number of reflections $r_1, r_2, \ldots, r_n$.

Now the only difference between a linear function and an *extended* linear function is that the latter contains an additional point at infinity in its domain. Since this additional point maps to itself, it follows that an extended linear function is a composite of a scaling that also fixes ∞, followed by a number of reflections that also fix ∞.

But a reflection that fixes ∞ is just an inversion in an extended line. Also, by the solution to Problem 8, a scaling that also fixes ∞ is a composite of two inversions. It follows that every extended linear function is a composite of inversions.

Combining this with the result of Example 2 above we have the following theorem.

Theorem 5 The extended reciprocal function and the extended linear functions can be decomposed into composites of inversions.

Since inversions of the extended plane map generalized circles onto generalized circles, we have the following corollary of Theorem 5.

Corollary The extended linear functions and the extended reciprocal function map generalized circles onto generalized circles.

Example 3 Let t be the extended linear function defined by

$$t(z) = \begin{cases} 2\left(-1 + \sqrt{3}i\right) z + (4 - 2i), & \text{if } z \in \mathbb{C}, \\ \infty, & \text{if } z = \infty. \end{cases}$$

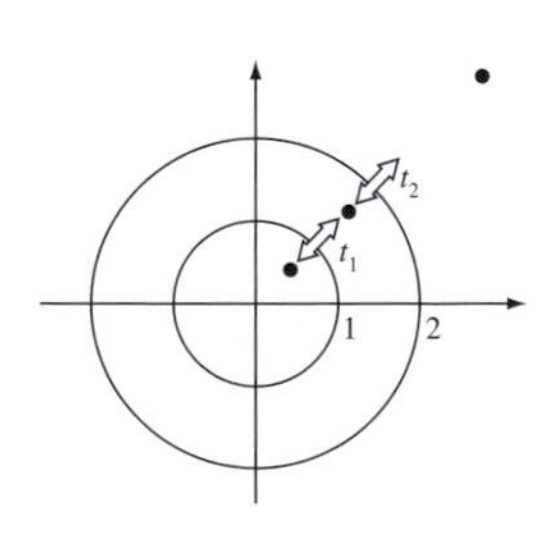

Express t as a composite of inversions of the extended complex plane.

Solution In addition to mapping ∞ to ∞, the transformation t *scales* the complex plane by the factor $|2(-1 + \sqrt{3}i)| = 4$, *rotates* it through the angle $\operatorname{Arg}(2(-1 + \sqrt{3}i)) = \frac{2\pi}{3}$, and then *translates* it through the vector $(4, -2)$.

The *scaling* by the factor 4 can be decomposed into the composite $t_2 \circ t_1$, where

t_1 is the inversion in the unit circle $\mathscr{C}$,
t_2 is the inversion in the circle of radius $\sqrt{4} = 2$ centred at the origin.

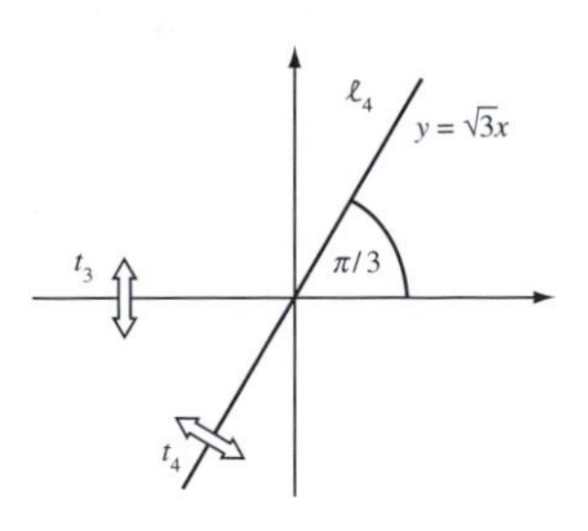

The *rotation* through the angle $2\pi/3$ can be decomposed into the composite $t_4 \circ t_3$, where

t_3 is the inversion in the extended real axis,
t_4 is the inversion in the extended line $\ell_4 \cup \{\infty\}$, where ℓ_4 is the line $y = \sqrt{3}x$.

(Here ℓ_4 is the line through the origin that makes an angle $\pi/3$ with the x-axis.)

The *translation* through the vector $(4, -2)$ can be decomposed into the composite $t_6 \circ t_5$, where

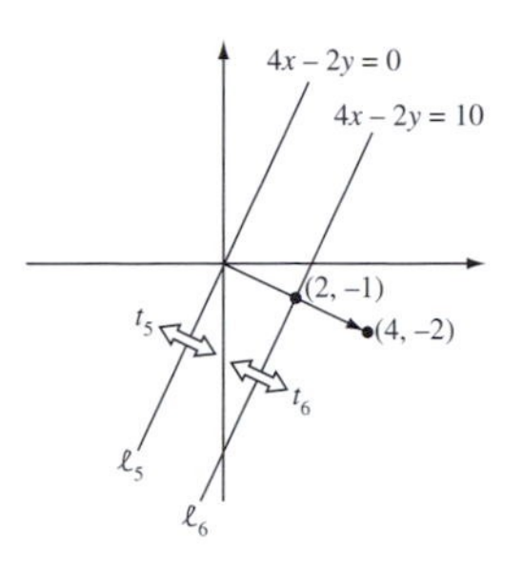

t_5 is the inversion in the extended line $\ell_5 \cup \{\infty\}$, where ℓ_5 is the line $4x - 2y = 0$,
t_6 is the inversion in the extended line $\ell_6 \cup \{\infty\}$, where ℓ_6 is the line $4x - 2y = 10$.

(Note that ℓ_5 is the line through the origin that is perpendicular to the vector $(4, -2)$, and ℓ_6 is the line that passes through $(2, -1)$ parallel to ℓ_5.)

Since t_6, t_5, t_4, t_3 and $t_2 \circ t_1$ all map ∞ to itself, it follows that

$$t = t_6 \circ t_5 \circ t_4 \circ t_3 \circ t_2 \circ t_1. \qquad \square$$

Problem 9 Let t be the extended linear function defined by

$$t(z) = \begin{cases} -9z + (6 - 10i), & \text{if } z \in \mathbb{C}, \\ \infty, & \text{if } z = \infty. \end{cases}$$

Express t as a composite of inversions of the extended complex plane.
Hint: Use the decompositions described in the margin.

The decompositions illustrated above generalize as follows.

To *scale* $\hat{\mathbb{C}}$ by a factor k: first invert in the circle $\{z : |z| = 1\}$, then invert in the circle $\{z : |z| = \sqrt{k}\}$.

To *rotate* $\hat{\mathbb{C}}$ through an angle θ: first invert in the extended line $\{z : \operatorname{Arg} z = 0\} \cup \{\infty\}$, then invert in the extended line $\{z : \operatorname{Arg} z = \frac{1}{2}\theta\} \cup \{\infty\}$.

To *translate* $\hat{\mathbb{C}}$ through a vector (a, b): first invert in the extended line $\{(x, y) : ax + by = 0\} \cup \{\infty\}$, then invert in the extended line $\{(x, y) : ax + by = \frac{1}{2}(a^2 + b^2)\} \cup \{\infty\}$.

5.2.4 The Riemann Sphere

In the previous subsection, we introduced the point at infinity and the extended complex plane in order to provide a simplified explanation of the effect of inversion on lines and circles. However, it may seem unsatisfactory to have to visualize a straight line as a (generalized) circle, since the 'ends' of the line appear to be infinitely far apart. Also, it is difficult to visualize where the point at infinity should be placed relative to $\mathbb{C}$, other than to think of it as smeared in some vague way around the 'outer edge' of $\mathbb{C}$.

Fortunately, there is a model of the extended (complex) plane in which the point at infinity appears as an actual point! Consider the complex plane $\mathbb{C}$ as lying in $\mathbb{R}^3$ with the real and imaginary axes aligned along the x-axis and y-axis, respectively. Then each complex number $x + iy$ may be represented by the point $(x, y, 0)$ in the (x, y)-plane.

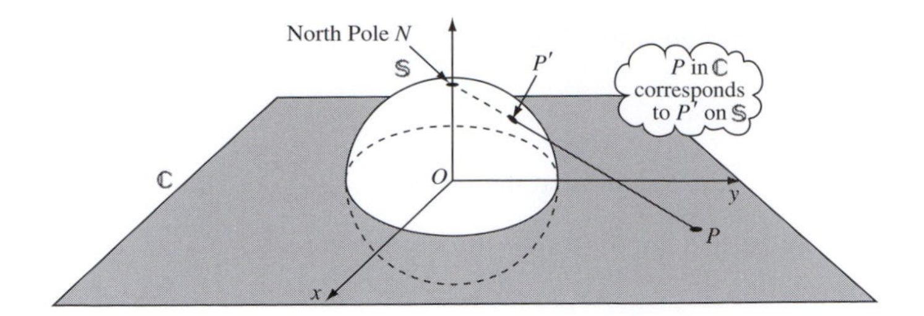

Next, draw the sphere $\mathbb{S}$ with centre at the origin and with radius 1; we call this the **Riemann sphere**. By analogy with the Earth, we refer to the point $N = (0, 0, 1)$ at the top of the sphere as the **North Pole** of $\mathbb{S}$, and the point $S = (0, 0, -1)$ at the bottom of the sphere as the **South Pole** of $\mathbb{S}$.

Each line joining a point P in the complex plane to the North Pole intersects the Riemann sphere at some point P', and vice versa. In this way, we obtain a one–one correspondence between all points P in the complex plane and all but one of the points P' on the sphere. The one point on the sphere that cannot be associated with a point in the complex plane is the North Pole N.

As the point P' on the sphere moves closer to N, the corresponding point P in the plane moves further away from the origin O. This suggests that we associate the North Pole N with the point ∞ in the extended complex plane.

The function $\pi : \mathbb{S} \to \hat{\mathbb{C}}$, which maps points on the Riemann sphere to the associated points in the extended complex plane, is called **stereographic projection**. Since π is one–one and onto, it follows that we can use the Riemann sphere as a convenient visualization of the extended complex plane $\hat{\mathbb{C}}$.

In fact, we can find an explicit formula for stereographic projection of the point (X, Y, Z) onto the point $z = x + iy$, and vice versa.

We use capital letters to denote coordinates of points in $\mathbb{R}^3$, and small letters to denote points or coordinates in the complex plane.

Theorem 6 Let π denote the mapping of the Riemann sphere $\mathbb{S}$ onto $\hat{\mathbb{C}}$ given by stereographic projection. Then the stereographic projection of the point (X, Y, Z) of $\mathbb{S}$ onto the point $z = x + iy$ of $\hat{\mathbb{C}}$ is given by

$$\pi(X, Y, Z) = \frac{X}{1 - Z} + i\frac{Y}{1 - Z}.$$

Also, the inverse mapping is given by

$$\pi^{-1}(x + iy) = \left(\frac{2x}{x^2 + y^2 + 1}, \frac{2y}{x^2 + y^2 + 1}, \frac{x^2 + y^2 - 1}{x^2 + y^2 + 1}\right).$$

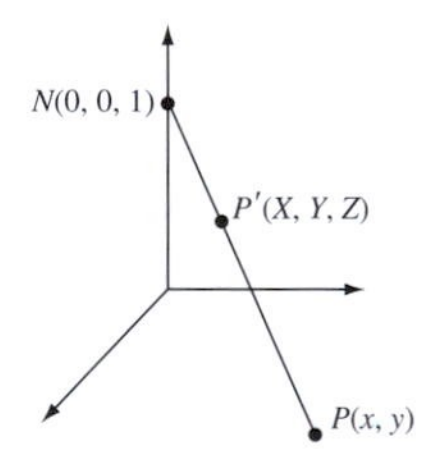

Proof Let the point $P'(X, Y, Z)$ be a point on $\mathbb{S}$ (other than N), and $P(x, y)$ the point in the plane that corresponds to P' under stereographic projection. Project the line $NP'P$ perpendicularly onto the (Y, Z)-plane, with P' and P projecting onto the points Q' and Q; then Q' and Q have coordinates $(0, Y, Z)$ and $(0, y, 0)$ in $\mathbb{R}^3$. Draw the perpendicular $Q'R$ from Q' to the Z-axis.

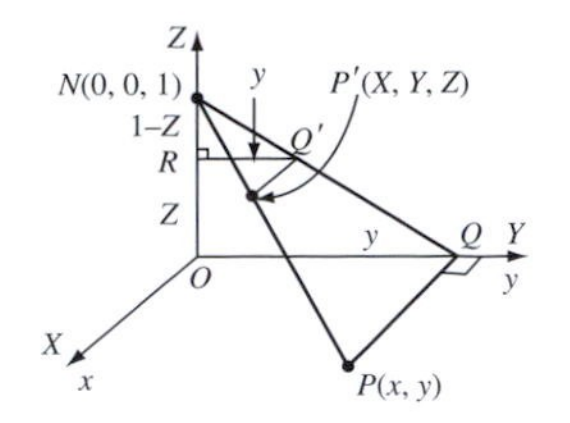

Since RQ' is parallel to OQ, the triangles $\Delta NRQ'$ and ΔNOQ are similar. It follows that $NR : RQ' = NO : OQ$, or

$$\frac{1-Z}{Y} = \frac{1}{y},$$

from which we obtain that

$$y = \frac{Y}{1-Z}.$$

By projecting the line $NP'P$ onto the (X,Z)-plane and using a similar argument, we can show that

We omit the details.

$$x = \frac{X}{1-Z}.$$

It then follows that the mapping π is given by

$$\pi(X,Y,Z) = x + iy = \frac{X}{1-Z} + i\frac{Y}{1-Z}.$$

Clearly this formula also holds when N is mapped to ∞.

For, here $Z = 1$.

To find the formula for π^{-1}, we use the fact that $X^2 + Y^2 + Z^2 = 1$. Substituting the values of X and Y from the above formulas into the equation $X^2 + Y^2 + Z^2 = 1$ and doing some manipulation, we deduce that

We omit the details.

$$x^2 + y^2 + 1 = \frac{2}{1-Z}. \qquad (5)$$

It follows from equation (5) and the earlier equations for X and Y that

$$X = x(1-Z) = \frac{2x}{x^2+y^2+1}$$

and

$$Y = y(1-Z) = \frac{2y}{x^2+y^2+1}.$$

Also, we may rearrange equation (5) in the form

$$Z = \frac{x^2+y^2-1}{x^2+y^2+1}.$$

We can then combine these formulas to give the required formula for π^{-1}:

$$\pi^{-1}(x+iy) = (X,Y,Z) = \left(\frac{2x}{x^2+y^2+1}, \frac{2y}{x^2+y^2+1}, \frac{x^2+y^2-1}{x^2+y^2+1}\right). \qquad \blacksquare$$

Notice that stereographic projection distorts the distances between points. For example, two points that are close together on $\mathbb{S}$ may project onto points that are close together in $\mathbb{C}$, or onto points that are far from each other in $\mathbb{C}$.

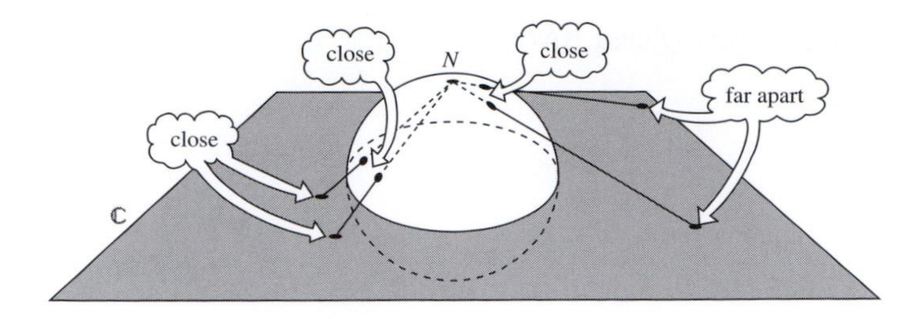

We can use the Riemann sphere to visualize extended lines as generalized circles in a very natural way, as follows. Consider a point P on a line ℓ in $\mathbb{C}$.

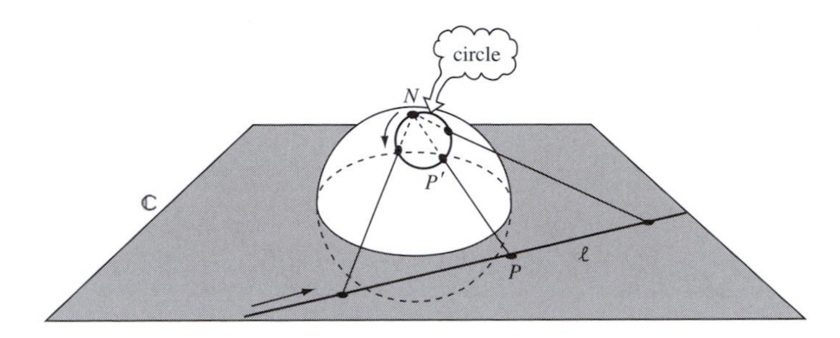

As P moves along ℓ, the line that joins P to the North Pole sweeps out the plane through ℓ and N; this intersects the sphere in a circle. Hence, as P moves along ℓ, the corresponding point P' on the sphere traces out a circle. As P moves further out along the line, the point P' moves towards the North Pole. The point P' never actually reaches the North Pole, since the North Pole corresponds to the 'gap' in the line ℓ that we mentioned earlier.

Subsection 5.2.3

We filled the 'gap' in ℓ by attaching ∞ to ℓ to obtain the extended line $\ell \cup \{\infty\}$ in $\hat{\mathbb{C}}$. On the sphere, the point N fills the corresponding gap in the circle, so that the extended line $\ell \cup \{\infty\}$ corresponds to an *actual* circle on the sphere – in fact, a circle through N.

This fact is a particular instance of the following general result.

Theorem 7 Under stereographic projection, circles on the Riemann sphere map onto generalized circles in $\hat{\mathbb{C}}$.

In particular, circles on the sphere that pass through N map onto extended lines in $\hat{\mathbb{C}}$, and circles on the sphere that do not pass through N map onto ordinary circles in $\hat{\mathbb{C}}$.

Proof A circle on the sphere is the intersection of the sphere $X^2 + Y^2 + Z^2 = 1$ with some plane $aX + bY + cZ + d = 0$, where a, b and c are not all zero. It follows from substituting the expressions

You met these equations in the proof of Theorem 6 above.

$$X = \frac{2x}{x^2 + y^2 + 1}, \quad Y = \frac{2y}{x^2 + y^2 + 1} \quad \text{and} \quad Z = \frac{x^2 + y^2 - 1}{x^2 + y^2 + 1},$$

for X, Y and Z into the equation $X^2 + Y^2 + Z^2 = 1$ that

$$\frac{2ax + 2by + c\left(x^2 + y^2 - 1\right)}{x^2 + y^2 + 1} + d = 0.$$

This equation may be rewritten in the form

$$2ax + 2by + c\left(x^2 + y^2 - 1\right) + d\left(x^2 + y^2 + 1\right) = 0,$$

or

$$(c + d)x^2 + (c + d)y^2 + 2ax + 2by + (d - c) = 0.$$

For its equation is that of a circle or of an (extended) line.

It follows that the image of a circle on the sphere is a generalized circle in the (extended) plane.

The circle on the sphere passes through $N(0, 0, 1)$ if the plane $aX + bY + cZ + d = 0$ passes through N, and so if $c + d = 0$.

It follows that if the circle on the sphere passes through N, its image in the extended plane has an equation of the form

$$2ax + 2by + (d - c) = 0,$$

and is an extended line.

On the other hand, if the circle on the sphere does not pass through N, then $c + d \neq 0$ and its image is an ordinary circle in the plane. ∎

Problem 10 Determine the images under stereographic projection onto $\hat{\mathbb{C}}$ of the following circles on $\mathbb{S}$.

(a) The circle $\{(X, Y, Z) : X^2 + Y^2 + Z^2 = 1, X = \frac{1}{2}\}$
(b) The circle of intersection of $\mathbb{S}$ with the plane $3X + 2Y + Z = 1$

As an illustration of Theorem 7, we describe what happens to the 'lines of latitude and longitude' of the Riemann sphere under stereographic projection.

A 'line of longitude' on the Riemann sphere is a circle on $\mathbb{S}$ that passes through the North Pole and the South Pole of $\mathbb{S}$; this projects onto a line through the origin in the plane. Similarly, a 'line of latitude' on the Riemann sphere is a circle at a constant height above the (x, y)-plane, so, by symmetry, this projects onto a circle centred at the origin in the plane.

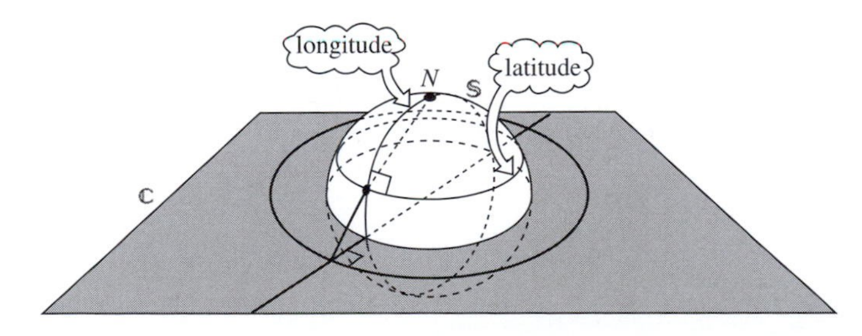

Just as the lines of latitude and longitude meet at right angles on the sphere, so do their projections. This is a consequence of the following remarkable result.

Theorem 8 Stereographic projection preserves the magnitude of angles.

Recall that inversion has a similar property (see Theorem 5, Subsection 5.1.3).

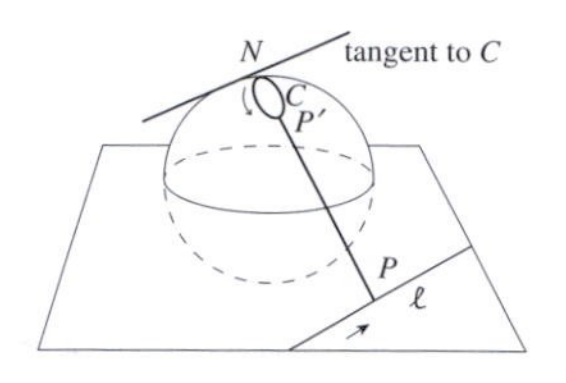

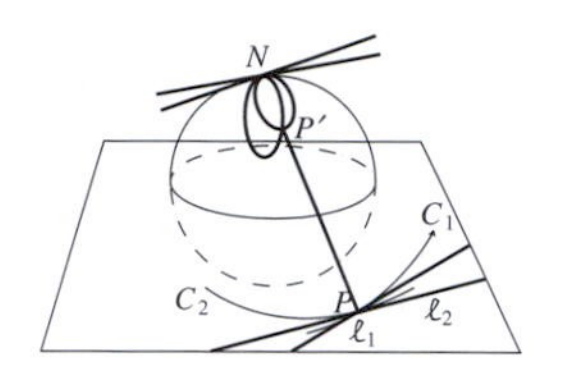

Proof First, we make the following crucial observation. Recall that if a point P moves along a line ℓ in the complex plane, then the corresponding point P' on the Riemann sphere moves round a circle through N. Indeed, as P moves out 'towards ∞', P' approaches N; and the line NP' (in $\mathbb{R}^3$) approaches the tangent at N to the circle on the sphere. It follows that the tangent at N to the circle on the sphere must be parallel to the original line ℓ.

Let C_1 and C_2 be curves in the complex plane that intersect at some point P, and let ℓ_1 and ℓ_2 be the tangents to C_1 and C_2 at P. (Recall that we define the angle between C_1 and C_2 at P to be the angle between ℓ_1 and ℓ_2 at P.)

The curves on the Riemann sphere that correspond to ℓ_1 and ℓ_2 are circles through the North Pole N and the point P' on the sphere that corresponds to the point P in the plane. We have to show that the angle between these circles at P' (that is, the angle between their tangents at P') is equal to the angle between the lines ℓ_1 and ℓ_2 at P.

But, by symmetry, the angle between the circles at P' is equal to the angle between the same circles when they meet again at N. And, as we saw above, the lines ℓ_1 and ℓ_2 through P are parallel to the corresponding tangent lines through N. Thus the angle between the circles at N must be equal to the original angle between ℓ_1 and ℓ_2 at P.

Hence, stereographic projection does preserve the magnitude of angles, as asserted. ■

Earlier, we showed that inversion preserves the magnitude of angles but reverses their orientation. Having introduced the Riemann sphere, we can now show that there is a sense in which this is true even for angles which have a vertex at the centre of inversion, or at ∞.

To see this, let ℓ_1 and ℓ_2 be two lines that intersect at the centre of inversion A. The corresponding extended lines $\ell_1 \cup \{\infty\}$ and $\ell_2 \cup \{\infty\}$ intersect again at the point ∞. On the Riemann sphere these extended lines become circles C_1 and C_2 that intersect at the North Pole and at the point A' corresponding to A.

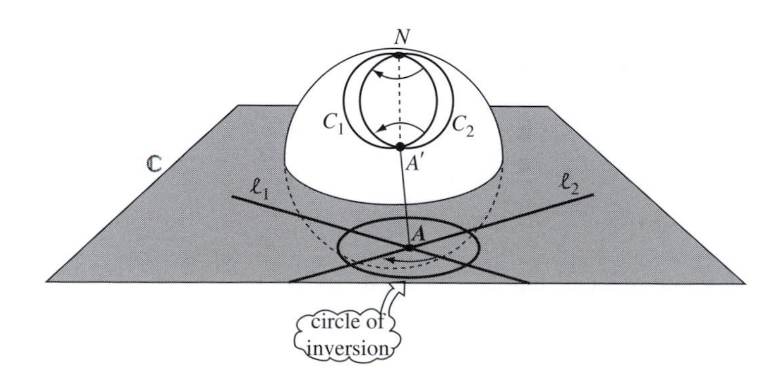

Under the inversion, each extended line maps to itself, so C_1 and C_2 also map to themselves. Notice, however, that the points A' and N swap over, and so the angles at A' and N swap over as well. By symmetry, these angles are equal in magnitude but opposite in direction. So, provided we interpret the angle between curves at ∞ as the corresponding angle on the Riemann sphere, we can conclude that the fact that inversion preserves the magnitude of angles holds throughout the whole of $\hat{\mathbb{C}}$.

5.3 Inversive Geometry

5.3.1 Inversive Geometry

We now come to the main purpose of this chapter, which is to introduce a geometry, known as *inversive geometry*, that we can use to study the properties of circles, lines and angles.

Recall that according to Klein, a geometry has the following ingredients:

Introduction to Chapter 2, before Subsection 2.1.1

a space consisting of a set of points;
a group of transformations that act on the space.

Each geometry is used to study those properties of figures in its space that are preserved by its transformations. For example, Euclidean geometry is used to study those properties of figures in $\mathbb{R}^2$, such as angle and distance, that are preserved by the isometries of $\mathbb{R}^2$.

Since each isometry of $\mathbb{R}^2$ can be decomposed into a composite of reflections, we can think of the group associated with Euclidean geometry as the group of all possible composites of reflections.

This provides the clue that we need to define the transformations of our new geometry. Rather than consider composites of reflections, we consider composites of inversions.

Definition A transformation $t : \hat{\mathbb{C}} \to \hat{\mathbb{C}}$ is an **inversive transformation** if it can be expressed as a composite of inversions.

It is important to distinguish between inversions and inversive transformations.

For example, the extended reciprocal function is an inversive transformation because it can be expressed as a composite $t_2 \circ t_1$ where t_1 is the inversion in the unit circle and t_2 is the inversion in the real axis. Similarly, all extended linear functions can be expressed as composites of inversions, so they too are inversive transformations.

Theorem 1 The extended reciprocal function and the extended linear functions are inversive transformations.

Since every inversion preserves the magnitude of angles and maps generalized circles to generalized circles, the same must be true of all composites of inversions. We therefore have the following result.

> **Theorem 2** Inversive transformations preserve the magnitude of angles, and map generalized circles to generalized circles.

Before we can use the inversive transformations to define a geometry, we must first check that they form a group.

> **Theorem 3** The set of inversive transformations forms a group under the operation of composition of functions.

Proof We check that the four group axioms hold.

G1 CLOSURE Let r and s be inversive transformations. Then we can write

$$r = t_1 \circ t_2 \circ \ldots \circ t_k$$

and

$$s = t_{k+1} \circ t_{k+2} \circ \ldots \circ t_n,$$

where $t_1, t_2, \ldots, t_n$ are inversions. Thus

$$r \circ s = (t_1 \circ t_2 \circ \ldots \circ t_k) \circ (t_{k+1} \circ t_{k+2} \circ \ldots \circ t_n)$$

is a composite of inversions, and is therefore an inversive transformation.

G2 IDENTITY The identity for composition of functions is the identity transformation given by

$$t(z) = z \quad (z \in \hat{\mathbb{C}}).$$

This is an inversive transformation since $t = s \circ s$, where s is the inversion in the unit circle.

Here we could have chosen any other inversion for s since all inversions are self-inverse.

G3 INVERSES If t is an inversive transformation, then we can write

$$t = t_1 \circ t_2 \circ \ldots \circ t_n,$$

where $t_1, t_2, \ldots t_n$ are inversions. It follows that t has inverse

$$t^{-1} = t_n^{-1} \circ t_{n-1}^{-1} \circ \ldots \circ t_1^{-1} = t_n \circ t_{n-1} \circ \ldots \circ t_1,$$

which is an inversive transformation.

G4 ASSOCIATIVITY Composition of functions is always associative.

Since all four group properties hold, it follows that the set of inversive transformations forms a group under composition of functions. ■

Having shown that the inversive transformations form a group, we can use them to define a geometry. But what space should we use for the geometry? Since the inversive transformations act on $\hat{\mathbb{C}}$, the space will have to be $\hat{\mathbb{C}}$. But what do figures look like in this space? Just as figures in $\mathbb{R}^2$ are defined to be subsets of $\mathbb{R}^2$, so **figures in** $\hat{\mathbb{C}}$ are defined to be subsets of $\hat{\mathbb{C}}$. Thus circles and extended lines are both examples of figures in $\hat{\mathbb{C}}$. We can now make the following definition.

Definition **Inversive geometry** is the study of those properties of figures in $\hat{\mathbb{C}}$ that are preserved by inversive transformations.

We have already met the following inversive properties: generalized circles, magnitude of angles, and tangency (since zero angles are preserved).

5.3.2 Relationship with Other Geometries

Although the point at infinity is crucial to the theory of inversive geometry, when we come to interpret results about figures we often choose to confine our attention to points in $\mathbb{C}$. By ignoring the point ∞ in this way, the remaining points in $\hat{\mathbb{C}}$ can be interpreted as points of $\mathbb{R}^2$ and any extended lines in the figure can be interpreted as ordinary lines.

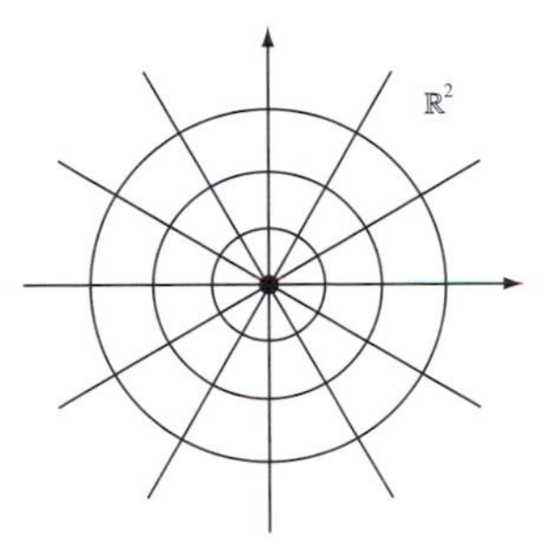

In this sense figures in $\mathbb{R}^2$ can be interpreted as figures in $\hat{\mathbb{C}}$, and vice versa. For example, the figure in the margin is a figure in $\mathbb{R}^2$. It consists of a family of circles centred at the origin, and a family of lines that intersect at the origin. If we add the point ∞ to the figure, then all the lines become generalized circles and the figure becomes a figure in $\hat{\mathbb{C}}$.

Earlier, you saw that every Euclidean transformation can be expressed as a composite of reflections. Since a reflection can be interpreted as an inversion that fixes ∞, it follows that we can interpret every Euclidean transformation as a composite of inversions, and hence as an inversive transformation.

We could make this interpretation precise by adopting the convention that a Euclidean transformation maps ∞ to ∞ when it acts on $\hat{\mathbb{C}}$.

An immediate consequence of the above observations is that when figures in $\hat{\mathbb{C}}$ are interpreted as figures in $\mathbb{R}^2$, all their inversive properties become Euclidean properties. This is because any property that is preserved by all inversive transformations must also be preserved by all Euclidean transformations. For example, the magnitude of angles is both an inversive and a Euclidean property.

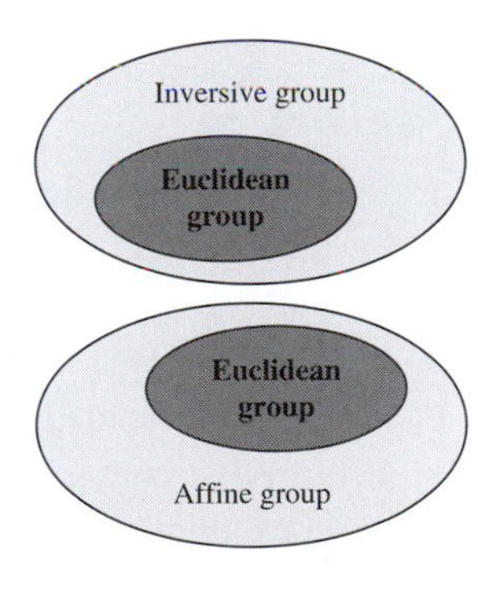

A pictorial representation of the relationship between inversive and Euclidean geometry is given in the margin. Provided that we ignore the point ∞, the group of Euclidean transformations can be regarded as a subgroup of the group of inversive transformations. Because the Euclidean group is smaller than the inversive group, it follows that Euclidean geometry has more properties than does inversive geometry. For example, length is a Euclidean property but it is not an inversive property.

How does affine geometry fit into the scheme? In Subsection 2.2.1 we showed that every Euclidean transformation is an affine transformation. But what is the relationship between affine transformations and inversive transformations?

Certainly, some affine transformations are inversive without being Euclidean. For example, earlier you saw that the 'doubling map',

See Subsection 2.2.3

$$t(z) = 2z \quad (z \in \mathbb{C}),$$

is an affine transformation. This transformation is not a Euclidean transformation, and yet it can be decomposed into the composite $t_2 \circ t_1$, where t_1 is the inversion in the unit circle $\mathscr{C}$, and t_2 is the inversion in the circle $\{|z| = \sqrt{2} : z \in \hat{\mathbb{C}}\}$. So, with the usual proviso about ∞, t is an inversive transformation.

You saw how a scaling can be decomposed into a composite of inversions in Problem 8, Subsection 5.2.3.

However, not all affine transformations are inversive transformations. For example, the transformation

$$t(\mathbf{x}) = \begin{pmatrix} 2 & 0 \\ 0 & 1 \end{pmatrix} \mathbf{x} \quad \left(\mathbf{x} \in \mathbb{R}^2\right)$$

represents a horizontal shear. This cannot be an inversive transformation, since it does not preserve angles.

It follows that the affine group of transformations contains the Euclidean group but overlaps the inversive group, as illustrated in the margin.

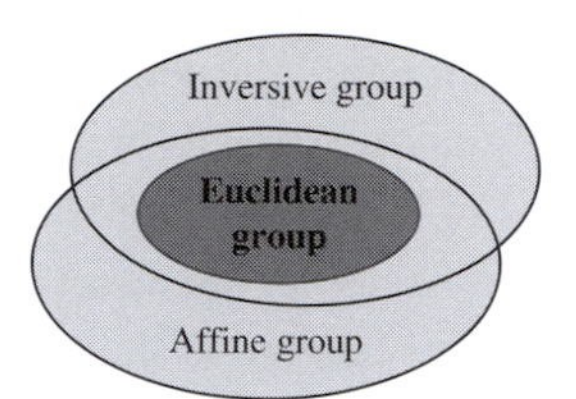

5.3.3 Möbius Transformations

To enable us to tackle problems in inversive geometry algebraically, we need an algebraic representation of the inversive transformations.

We shall show that each inversive transformation has either the form $t(z) = M(z)$, or the form $t(z) = M(\bar{z})$, where M is a so-called *Möbius transformation*.

Definition A **Möbius transformation** is a function $M : \hat{\mathbb{C}} \to \hat{\mathbb{C}}$ of the form

$$M(z) = \frac{az + b}{cz + d},$$

where $a, b, c, d \in \mathbb{C}$ and $ad - bc \neq 0$.

If $c = 0$, then we adopt the convention that $M(\infty) = \infty$; otherwise, we adopt the convention that $M(-d/c) = \infty$ and $M(\infty) = a/c$.

Remarks

1. Every (extended) linear function is a Möbius transformation, as can be seen by setting $c = 0$, $d = 1$. Also, the (extended) reciprocal function is a Möbius transformation with $a = d = 0$ and $b = c = 1$.
2. If $c = 0$, then the formula for M reduces to $M(z) = (a/d)z + (b/d)$. This defines an extended linear function, because the condition $ad - bc \neq 0$ ensures that both a and d are non-zero, and so a/d is also non-zero. Also, the convention that $M(\infty) = \infty$ complies with the definition of an extended linear function.
3. The condition $ad - bc \neq 0$ is equivalent to the statement that the ratios $a : c$ and $b : d$ are unequal. This is necessary to ensure that the numerator is not simply a multiple of the denominator, for if it were, M would be a constant function.

For example,
$$6i \times 2 - 4i \times 3 = 0,$$
so that
$$\frac{6iz + 4i}{3z + 2} = 2i, \text{ a constant.}$$

Problem 1 Which of the following formulas define a Möbius transformation?

(a) $M(z) = \dfrac{5}{z}$ (b) $M(z) = \dfrac{-z + 2i}{3z - 4i}$

(c) $M(z) = -3z + \dfrac{i}{z}$ (d) $M(z) = 1 + \dfrac{5}{z + 2i}$

For those formulas that do define a Möbius transformation, state the image of ∞ under M.

Before we look at the properties of Möbius transformations, we first verify that a Möbius transformation is indeed an inversive transformation.

Theorem 4 Every Möbius transformation is an inversive transformation.

Proof Let M be the Möbius transformation defined by the formula

$$M(z) = \frac{az + b}{cz + d}.$$

If $c = 0$, then M is an extended linear function, and is therefore an inversive transformation.

If $c \neq 0$, then for $z \in \mathbb{C} - \{-d/c\}$ we can write

$$\begin{aligned} M(z) &= \frac{-ad + bc + a(cz + d)}{c(cz + d)} \\ &= -\left(\frac{ad - bc}{c}\right) \cdot \left(\frac{1}{cz + d}\right) + \frac{a}{c}. \end{aligned}$$

It follows that M may be expressed as the composite $t_3 \circ t_2 \circ t_1$, where t_2 is the extended reciprocal function, and t_1 and t_3 are the extended linear functions

$$t_1(z) = \begin{cases} cz + d, & \text{if } z \neq \infty, \\ \infty, & \text{if } z = \infty, \end{cases}$$

and

$$t_3(z) = \begin{cases} -((ad - bc)/c)z + (a/c), & \text{if } z \neq \infty, \\ \infty, & \text{if } z = \infty. \end{cases}$$

Next, we check that the transformations $t_3 \circ t_2 \circ t_1$ and M agree also at the exceptional points ∞ and $-d/c$, as follows:

$$t_3 \circ t_2 \circ t_1(\infty) = t_3 \circ t_2(\infty) = t_3(0) = a/c = M(\infty);$$

$$t_3 \circ t_2 \circ t_1(-d/c) = t_3 \circ t_2(0) = t_3(\infty) = \infty = M(-d/c).$$

Since the extended reciprocal function and the extended linear functions are inversive transformations, it follows that M is an inversive transformation. ∎

Since Möbius transformations are inversive transformations, they must preserve the magnitude of angles and map generalized circles to generalized circles. In fact, we can say slightly more than this, for in the proof of Theorem 4 we showed that every Möbius transformation is either an extended linear function or a composite of two extended linear functions and an extended reciprocal function. Since the extended linear functions and the extended reciprocal function preserve both the magnitude and *orientation* of angles, the same must be true of Möbius transformations.

Theorem 5 Möbius transformations preserve the magnitude and orientation of angles, and map generalized circles onto generalized circles.

5.3.4 Matrix Representation of Möbius Transformations

In order to explore some of the other properties of Möbius transformations, it is helpful to establish a correspondence between Möbius transformations and matrices.

Recall that, if $a, b, c, d, e, f, g, h \in \mathbb{C}$, then 2×2 matrices have the following properties:

$$\begin{pmatrix} a & b \\ c & d \end{pmatrix} \cdot \begin{pmatrix} e & f \\ g & h \end{pmatrix} = \begin{pmatrix} ae + bg & af + bh \\ ce + dg & cf + dh \end{pmatrix},$$

and

$$\begin{pmatrix} a & b \\ c & d \end{pmatrix}^{-1} = \frac{1}{ad - bc} \begin{pmatrix} d & -b \\ -c & a \end{pmatrix}, \quad \text{if } ad - bc \neq 0.$$

Here, the condition $ad - bc \neq 0$ for the matrix to be invertible is reminiscent of the condition $ad - bc \neq 0$ that appears in the definition of a Möbius transformation. This suggests the following fruitful connection between Möbius transformations and 2×2 invertible matrices.

Definition Let M be a Möbius transformation defined by

$$M(z) = \frac{az + b}{cz + d}, \tag{1}$$

where $a, b, c, d \in \mathbb{C}$. Then

$$\mathbf{A} = \begin{pmatrix} a & b \\ c & d \end{pmatrix}$$

is a matrix **associated** with M.

Remark

1. Every matrix $\begin{pmatrix} a & b \\ c & d \end{pmatrix}$ associated with a Möbius transformation $M(z) = \frac{az+b}{cz+d}$ is invertible because $ad - bc \neq 0$.

2. A matrix associated with a Möbius transformation M is not unique, because we can multiply the numerator and denominator of the formula (1) by the same non-zero constant without altering the transformation. For example, both
$$M(z) = \frac{2z + i}{3z + 2i} \quad \text{and} \quad M(z) = \frac{2iz - 1}{3iz - 2}$$
specify the same Möbius transformation M, and so both the matrices
$$\begin{pmatrix} 2 & i \\ 3 & 2i \end{pmatrix} \quad \text{and} \quad \begin{pmatrix} 2i & -1 \\ 3i & -2 \end{pmatrix}$$
are associated with M.

In general, if $\mathbf{A}$ is a matrix associated with a Möbius transformation M, then for any non-zero $c \in \mathbb{C}$, $c\mathbf{A}$ is also a matrix associated with M. In fact, every matrix associated with M has the form $c\mathbf{A}$ for some $c \in \mathbb{C} - \{0\}$.

Example 1 Decide which, if any, of the matrices
$$\mathbf{A}_1 = \begin{pmatrix} 0 & 4 \\ i & 0 \end{pmatrix}, \quad \mathbf{A}_2 = \begin{pmatrix} 8 & 0 \\ -2i & -8 \end{pmatrix}, \quad \mathbf{A}_3 = \begin{pmatrix} -4i & 0 \\ 0 & 1 \end{pmatrix}$$
are associated with each of the following Möbius transformations M.

(a) $M(z) = \dfrac{-3iz + 2}{z - 3i}$ (b) $M(z) = \dfrac{-4i}{z}$ (c) $M(z) = \dfrac{4iz}{z - 4i}$

Solution

(a) Every matrix associated with this M is a non-zero multiple of the matrix
$$\begin{pmatrix} -3i & 2 \\ 1 & -3i \end{pmatrix}.$$
Hence none of the three given matrices is associated with M.

(b) Every matrix associated with this M is a non-zero multiple of the matrix
$$\mathbf{A} = \begin{pmatrix} 0 & -4i \\ 1 & 0 \end{pmatrix}.$$
Since $\mathbf{A}_1 = i\mathbf{A}$, it follows that $\mathbf{A}_1$ is a matrix associated with M.

(c) Every matrix associated with this M is a non-zero multiple of the matrix
$$\mathbf{A} = \begin{pmatrix} 4i & 0 \\ 1 & -4i \end{pmatrix}.$$
Since $\mathbf{A}_2 = -2i\mathbf{A}$, it follows that $\mathbf{A}_2$ is a matrix associated with M. □

Problem 2 Decide which, if any, of the matrices
$$\mathbf{A}_1 = \begin{pmatrix} 0 & 1 \\ -\frac{1}{2}i & 0 \end{pmatrix}, \quad \mathbf{A}_2 = \begin{pmatrix} 0 & 2 \\ 1 & -2i \end{pmatrix}, \quad \mathbf{A}_3 = \begin{pmatrix} 2i & 0 \\ 1 & -i \end{pmatrix}$$

are associated with each of the following Möbius transformations M.

(a) $M_1(z) = \dfrac{2i}{iz+2}$ (b) $M_2(z) = \dfrac{2i}{z}$ (c) $M_3(z) = \dfrac{iz+2}{2z-i}$

A particularly important transformation of $\hat{\mathbb{C}}$ is the *identity* function defined by

$$t(z) = z \quad (z \in \hat{\mathbb{C}}).$$

This is a Möbius transformation because it can be written in the form

$$t(z) = \frac{z+0}{0z+1}.$$

A matrix associated with this Möbius transformation is the identity matrix

$$\mathbf{I} = \begin{pmatrix} 1 & 0 \\ 0 & 1 \end{pmatrix}.$$

5.3.5 Composing Möbius Transformations

The next example illustrates what happens when two Möbius transformations are composed.

Example 2 Determine the composite $M_1 \circ M_2$, where M_1 and M_2 are the Möbius transformations defined by

$$M_1(z) = \frac{iz+1}{2z-2} \quad \text{and} \quad M_2(z) = \frac{z+i}{2z-1}.$$

Solution Since M_1 and M_2 are one–one mappings of $\hat{\mathbb{C}}$ onto $\hat{\mathbb{C}}$, the same must be true of $M_1 \circ M_2$. A formula for $M_1 \circ M_2(z)$ is

$$\begin{aligned} M_1 \circ M_2(z) &= M_1\left(\frac{z+i}{2z-1}\right) \\ &= \frac{i\left(\frac{z+i}{2z-1}\right)+1}{2\left(\frac{z+i}{2z-1}\right)-2} \\ &= \frac{i\,(z+i)+(2z-1)}{2\,(z+i)-2\,(2z-1)} \\ &= \frac{(2+i)\,z-2}{-2z+(2+2i)}. \end{aligned}$$

Strictly speaking, we should check our convention for the exceptional points separately. For example, by convention $M_2(\infty) = \frac{1}{2}$, so that

$$\begin{aligned} M_1 \circ M_2(\infty) &= M_1\left(\tfrac{1}{2}\right) \\ &= \frac{\frac{1}{2}i+1}{1-2} \\ &= \frac{2+i}{-2}. \end{aligned}$$

This agrees with the value obtained when we apply our convention to the formula for $M_1 \circ M_2$.

We can check that this formula defines a Möbius transformation by noting that

$$(2+i)(2+2i) - (-2)(-2) = -2+6i \neq 0.$$

Thus $M_1 \circ M_2$ is a Möbius transformation given by

$$M_1 \circ M_2(z) = \frac{(2+i)z - 2}{-2z + (2+2i)}. \qquad \square$$

Problem 3 Write down matrices $\mathbf{A}_1$ and $\mathbf{A}_2$ associated with the Möbius transformations M_1 and M_2 defined in Example 2. Calculate the product $\mathbf{A}_1\mathbf{A}_2$, and compare it with the Möbius transformation $M_1 \circ M_2$.

In Example 2, the composite of two Möbius transformations M_1 and M_2 turned out to be another Möbius transformation. The solution to Problem 3 demonstrates that the product of matrices associated with M_1 and M_2 is a matrix associated with $M_1 \circ M_2$. The following theorem confirms that this is always the case.

Theorem 6 Composition of Möbius Transformations
Let M_1 and M_2 be Möbius transformations with associated matrices $\mathbf{A}_1$ and $\mathbf{A}_2$, respectively. Then $M_1 \circ M_2$ is a Möbius transformation with an associated matrix $\mathbf{A}_1\mathbf{A}_2$.

Proof Let M_1 and M_2 be defined by

$$M_1(z) = \frac{az+b}{cz+d} \quad \text{and} \quad M_2(z) = \frac{ez+f}{gz+h}.$$

Since M_1 and M_2 are one–one mappings of $\hat{\mathbb{C}}$ onto $\hat{\mathbb{C}}$, the same must be true of $M_1 \circ M_2$. We can obtain a formula for $M_1 \circ M_2(z)$ as follows:

$$\begin{aligned} M_1 \circ M_2(z) &= M_1\left(\frac{ez+f}{gz+h}\right) \\ &= \frac{a\left(\frac{ez+f}{gz+h}\right)+b}{c\left(\frac{ez+f}{gz+h}\right)+d} \\ &= \frac{a(ez+f)+b(gz+h)}{c(ez+f)+d(gz+h)} \\ &= \frac{(ae+bg)z+(af+bh)}{(ce+dg)z+(cf+dh)}. \end{aligned}$$

Strictly speaking, we should check the exceptional cases separately. However, we shall omit this checking from here onwards.

Since $M_1 \circ M_2$ is one–one, this formula cannot remain constant as z varies, so $(ae+bg)(cf+dh) - (af+bh)(ce+dg) \neq 0$. The formula must therefore define a Möbius transformation.

A matrix associated with $M_1 \circ M_2$ is

$$\begin{pmatrix} ae+bg & af+bh \\ ce+dg & cf+dh \end{pmatrix},$$

which is the product of the matrices

$$\begin{pmatrix} a & b \\ c & d \end{pmatrix} \quad \text{and} \quad \begin{pmatrix} e & f \\ g & h \end{pmatrix}$$

associated with M_1 and M_2, respectively. ■

Theorem 6 enables us to calculate the composite of two Möbius transformations by using the following strategy.

Strategy To compose two Möbius transformations M_1 and M_2:

1. write down matrices $\mathbf{A}_1$ and $\mathbf{A}_2$ associated with M_1 and M_2;
2. calculate $\mathbf{A}_1\mathbf{A}_2$;
3. write down the Möbius transformation $M_1 \circ M_2$ with which $\mathbf{A}_1\mathbf{A}_2$ is associated.

Example 3 Use the strategy to determine the composite $M_1 \circ M_2$ of the Möbius transformations

$$M_1(z) = \frac{3z+1}{iz-2} \quad \text{and} \quad M_2(z) = \frac{2iz+3}{z-2}.$$

Solution The Möbius transformations M_1 and M_2 have associated matrices

$$\mathbf{A}_1 = \begin{pmatrix} 3 & 1 \\ i & -2 \end{pmatrix} \quad \text{and} \quad \mathbf{A}_2 = \begin{pmatrix} 2i & 3 \\ 1 & -2 \end{pmatrix},$$

respectively. It follows that a matrix associated with $M_1 \circ M_2$ is

$$\mathbf{A}_1\mathbf{A}_2 = \begin{pmatrix} 3 & 1 \\ i & -2 \end{pmatrix}\begin{pmatrix} 2i & 3 \\ 1 & -2 \end{pmatrix} = \begin{pmatrix} 1+6i & 7 \\ -4 & 4+3i \end{pmatrix}.$$

The composite $M_1 \circ M_2$ is therefore the Möbius transformation defined by

$$M_1 \circ M_2(z) = \frac{(1+6i)z+7}{-4z+(4+3i)}. \quad \square$$

Problem 4 Let M_1 and M_2 be the Möbius transformations defined by

$$M_1(z) = \frac{3z+1}{iz-2} \quad \text{and} \quad M_2(z) = \frac{2iz+3}{z-2}.$$

Use the strategy to determine each of the following composites.

(a) $M_2 \circ M_1$ (b) $M_1 \circ M_1$

Problem 5 Let M_1 and M_2 be the Möbius transformations defined by

$$M_1(z) = \frac{z-i}{iz+2} \quad \text{and} \quad M_2(z) = \frac{2z+i}{-iz+1}.$$

Use the strategy to determine the composite $M_1 \circ M_2$.

5.3.6 Inverting Möbius Transformations

In the solution to Problem 5 you saw that the composite $M_1 \circ M_2$ of the Möbius transformations

$$M_1(z) = \frac{z-i}{iz+2} \quad \text{and} \quad M_2(z) = \frac{2z+i}{-iz+1}$$

is the identity function on $\hat{\mathbb{C}}$. This shows that M_2 is the inverse function of M_1. In terms of matrices, this is equivalent to saying that there are matrices associated with M_1 and M_2 whose product is the identity matrix $\mathbf{I}$. For example,

$$\begin{pmatrix} 1 & -i \\ i & 2 \end{pmatrix} \begin{pmatrix} 2 & i \\ -i & 1 \end{pmatrix} = \begin{pmatrix} 1 & 0 \\ 0 & 1 \end{pmatrix}.$$

We can use this idea to find the inverse of any given Möbius transformation M. The inverse function M^{-1} certainly exists, since M is a one–one transformation from $\hat{\mathbb{C}}$ onto $\hat{\mathbb{C}}$. To find the inverse, let $\mathbf{A}$ be a matrix associated with M. Since $\mathbf{A}$ is invertible,

$$\mathbf{A}\mathbf{A}^{-1} = \mathbf{I} = \mathbf{A}^{-1}\mathbf{A},$$

so the Möbius transformation associated with the matrix $\mathbf{A}^{-1}$ must be the inverse function M^{-1}.

Now, if $M(z) = \frac{az+b}{cz+d}$, we can take $\mathbf{A} = \begin{pmatrix} a & b \\ c & d \end{pmatrix}$, so that

$$\mathbf{A}^{-1} = \frac{1}{ad-bc} \begin{pmatrix} d & -b \\ -c & a \end{pmatrix}.$$

But any non-zero multiple of $\mathbf{A}^{-1}$ is also a matrix associated with M^{-1}, so we shall usually use the matrix $\begin{pmatrix} d & -b \\ -c & a \end{pmatrix}$ as a matrix for M^{-1}. Then

$$M^{-1}(z) = \frac{dz-b}{-cz+a}.$$

We summarize the result of this discussion in the following theorem.

Theorem 7 Inverse of a Möbius Transformation
The inverse of the Möbius transformation

$$M(z) = \frac{az + b}{cz + d}$$

is also a Möbius transformation, and it may be written in the form

$$M^{-1}(z) = \frac{dz - b}{-cz + a}.$$

For example, the inverse of the Möbius transformation

$$M(z) = \frac{iz + 1}{2z - 2}$$

is given by

$$M^{-1}(z) = \frac{-2z - 1}{-2z + i} = \frac{2z + 1}{2z - i}.$$

Problem 6 Determine the inverse of each of the following Möbius transformations.

(a) $M_1(z) = \frac{-3iz+2}{z-3i}$ (b) $M_2(z) = \frac{-4i}{z}$ (c) $M_3(z) = \frac{4iz}{z-4i}$

5.3.7 The Inversive Group

We now have all the information we need to prove the following theorem.

Theorem 8 The set of all Möbius transformations forms a group under composition of functions.

Proof We show that the four group axioms hold.

G1 CLOSURE By Theorem 6, the composite of two Möbius transformations is itself a Möbius transformation.

G2 IDENTITY The identity is the Möbius transformation given by $M(z) = \frac{1z+0}{0z+1}$.

G3 INVERSES By Theorem 7, every Möbius transformation has an inverse.

G4 ASSOCIATIVITY Composition of functions is always associative.

It follows that the set of all Möbius transformations forms a group under composition of functions. ■

Having shown that the set of Möbius transformations forms a group, we now investigate its relationship with the group of all inversive transformations. We know from Theorem 4 that every Möbius transformation is an inversive transformation, but is every inversive transformation a Möbius transformation? Clearly, the answer is no. For example, an inversion cannot be a Möbius transformation since it *reverses* the orientation of angles, whereas all Möbius transformations *preserve* the orientation of angles.

Although inversions are not Möbius transformations, there is a close connection between inversions and Möbius transformations.

Theorem 9 Every inversion t has the form $t(z) = M(\bar{z})$, where M is a Möbius transformation.

Proof If t is an inversion of $\hat{\mathbb{C}}$ in an extended line, then by Theorem 1 of Section 5.2 it must have the form $t(z) = a\bar{z} + b$, with $t(\infty) = \infty$. It follows that $t(z) = M(\bar{z})$, where M is the Möbius transformation

It cannot have the form $t(z) = az + b$, since t reverses the orientation of angles.

$$M(z) = \frac{az + b}{0z + 1}.$$

On the other hand, if t is an inversion of $\hat{\mathbb{C}}$ in a circle of radius r with centre c, then by Theorem 3 of Section 5.2,

$$t(z) = \frac{r^2}{\overline{z - c}} + c = \frac{r^2 + c(\bar{z} - \bar{c})}{\bar{z} - \bar{c}} = \frac{c\bar{z} + (r^2 - c\bar{c})}{\bar{z} - \bar{c}}.$$

So once again t has the form $t(z) = M(\bar{z})$, where M is the Möbius transformation

$$M(z) = \frac{cz + \left(r^2 - c\bar{c}\right)}{z - \bar{c}}. \qquad \blacksquare$$

Notice that

$$c \cdot (-\bar{c}) - (r^2 - c\bar{c}) \cdot 1$$
$$= -r^2 \neq 0.$$

We can now show that every inversive transformation t has the form $t(z) = M(z)$ or $t(z) = M(\bar{z})$, where M is a Möbius transformation.

Theorem 10 Every inversive transformation t can be represented in $\hat{\mathbb{C}}$ by one of the formulas

$$t(z) = \frac{az + b}{cz + d} \quad \text{or} \quad t(z) = \frac{a\bar{z} + b}{c\bar{z} + d},$$

where $a, b, c, d \in \mathbb{C}$ and $ad - bc \neq 0$.

Proof We first show that the composite of two inversions t_1 and t_2 is a Möbius transformation. By Theorem 9 above, we can write $t_1(z) = M_1(\bar{z})$ and $t_2(z) = M_2(\bar{z})$, where M_1 and M_2 are Möbius transformations. Thus

$$t_1 \circ t_2(z) = t_1(M_2(\bar{z})) = M_1(\overline{M_2(\bar{z})}).$$

But if

$$M_2(z) = \frac{az + b}{cz + d},$$

then we can define a Möbius transformation M_3 by

$$M_3(z) = \overline{M_2(\bar{z})} = \overline{\left(\frac{a\bar{z}+b}{c\bar{z}+d}\right)} = \frac{\bar{a}z+\bar{b}}{\bar{c}z+\bar{d}}.$$

Since $t_1 \circ t_2(z) = M_1(\overline{M_2(\bar{z})}) = M_1(M_3(z))$, it follows that $t_1 \circ t_2$ is a composite of two Möbius transformations, and is therefore a Möbius transformation.

Next, let t be an arbitrary inversive transformation, and write

$$t = t_1 \circ t_2 \circ \ldots \circ t_n,$$

where $t_1, t_2, \ldots, t_n$ are inversions. If n is even, then we can rewrite t as a composite of Möbius transformations by pairing together the inversions in the form

$$t = (t_1 \circ t_2) \circ (t_3 \circ t_4) \circ \ldots \circ (t_{n-1} \circ t_n).$$

It follows that t is a Möbius transformation, M say, so we can write $t(z) = M(z)$.

To deal with the case where n is odd, let r be the extended conjugation function. Since r is its own inverse, we can write

$$t = (t_1 \circ t_2 \circ \ldots \circ t_n \circ r) \circ r.$$

Here the composite in the bracket involves an even number of inversions, so it must be a Möbius transformation, M say. Hence $t(z) = M \circ r(z) = M(\bar{z})$.

■

Theorem 10 provides us with an insight into the structure of the group G of all inversive transformations. Those inversive transformations that can be written in the form

$$t(z) = \frac{az+b}{cz+d}$$

are Möbius transformations, and we know from Theorem 8 that they form a subgroup of G. We refer to these transformations as **direct** inversive transformations because they preserve the orientation of angles.

The remaining inversive transformations are of the form

$$t(z) = \frac{a\bar{z}+b}{c\bar{z}+d}.$$

These are Möbius transformations composed with the extended conjugation function, and so they reverse the orientation of angles. For this reason we shall refer to them as **indirect** inversive transformations.

5.3.8 Images of Generalized Circles

We have shown that an inversive transformation maps generalized circles onto generalized circles. However, apart from the special case of inversion in the unit circle, we have not yet given a strategy for finding the image of a generalized circle under an inversive transformation.

One such strategy is based on the fact that every generalized circle in $\mathbb{C}$ is uniquely determined by any three points lying on it. Indeed, if the three points lie in $\mathbb{C}$, and are non-collinear, then they determine a unique circle in $\mathbb{C}$. On the other hand, if the three points are collinear, or if one of the points is ∞, then they determine a unique (extended) line.

Strategy To determine the image of a generalized circle C under an inversive transformation t:

1. write down three points z_1, z_2, z_3 on C;
2. determine the images $t(z_1), t(z_2), t(z_3)$;
3. the image $t(C)$ is the (unique) generalized circle through $t(z_1)$, $t(z_2)$, $t(z_3)$.

Example 4 Use the strategy to find the image of the unit circle $\mathscr{C}$ under the inversive transformation defined by

$$t(z) = \frac{\bar{z} + i}{\bar{z} - 1}.$$

Solution We first pick three distinct points on the unit circle. There is no definite rule about which points should be chosen, so to keep the calculations simple, we pick the points 1, i and -1, as shown below. Now

$$t(1) = \infty,\ t(i) = \frac{\bar{i} + i}{\bar{i} - 1} = 0 \quad \text{and}$$

$$t(-1) = \frac{\overline{-1} + i}{\overline{-1} - 1} = \frac{1}{2}(1 - i).$$

If you can pick a point that maps to ∞ then you know immediately that the image is an extended line.

So the image of $\mathscr{C}$ is the generalized circle through the points ∞, 0 and $\frac{1}{2}(1 - i)$. This is an extended line through the origin with slope -1. □

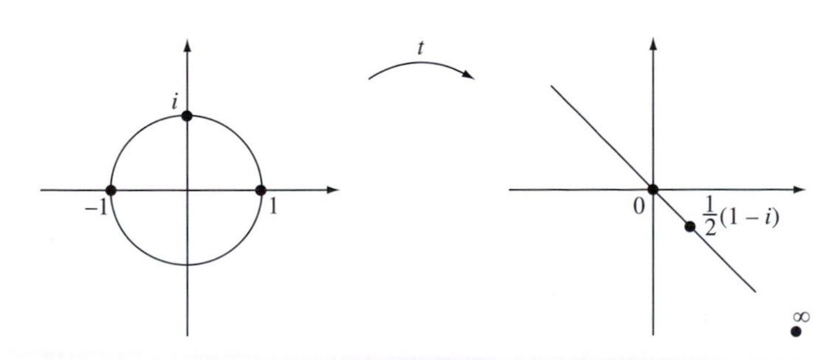

Problem 7 Let t be the inversive transformation defined by

$$t(z) = \frac{z - 2i}{z + 2}.$$

Use the strategy to determine the image of each of the following generalized circles under t:

(a) the extended line $\ell \cup \{\infty\}$, where ℓ is the line with equation $x+y = 2$;

(b) the circle with equation $(x + 1)^2 + y^2 = 1$.

Here $x + iy = z$, as usual.

5.4 Fundamental Theorem of Inversive Geometry

5.4.1 Comparison with Affine Geometry

In Subsection 2.3.2 we introduced the Fundamental Theorem of Affine Geometry, which states that, given any three non-collinear points in the plane, there is always an affine transformation that maps the points to another three given non-collinear points. Since a triangle is uniquely determined by its three (non-collinear) vertices, and since affine transformations map triangles to triangles, we were able to use the Fundamental Theorem of Affine Geometry to show that all triangles are affine-congruent.

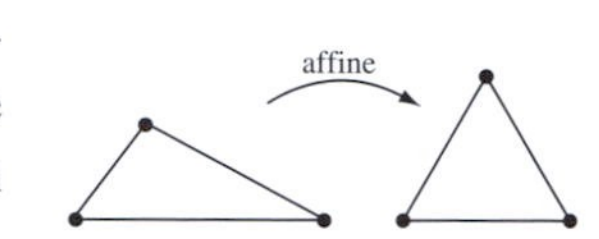

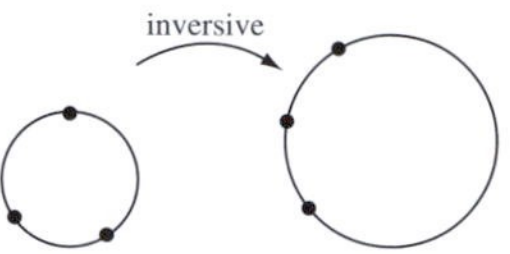

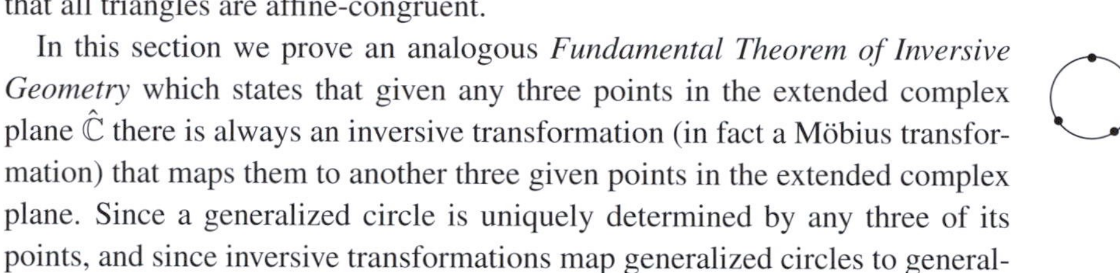

In this section we prove an analogous *Fundamental Theorem of Inversive Geometry* which states that given any three points in the extended complex plane $\hat{\mathbb{C}}$ there is always an inversive transformation (in fact a Möbius transformation) that maps them to another three given points in the extended complex plane. Since a generalized circle is uniquely determined by any three of its points, and since inversive transformations map generalized circles to generalized circles, we can show that all generalized circles are inversive-congruent. That is, given any two generalized circles C_1 and C_2 in the extended plane, it is always possible to find an inversive transformation (in fact a Möbius transformation) that maps C_1 onto C_2.

5.4.2 Mapping Three Points to Three Points

The proof of the Fundamental Theorem of Inverse Geometry is very similar to the proof of the Fundamental Theorem of Affine Geometry which we gave in Subsection 2.3.2. There we described how to construct an affine transformation which maps one given set of three (non-collinear) points in $\mathbb{R}^2$ onto another. We did this by forming a composite of two affine transformations which map the points via the auxiliary points $(0, 0)$, $(1, 0)$ and $(0, 1)$.

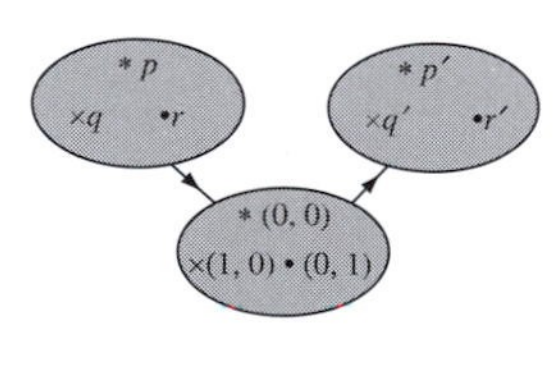

A similar idea works for the Fundamental Theorem of Inversive Geometry. In this geometry the space is $\hat{\mathbb{C}}$ rather than $\mathbb{R}^2$, and we map via the auxiliary points 0, 1 and ∞.

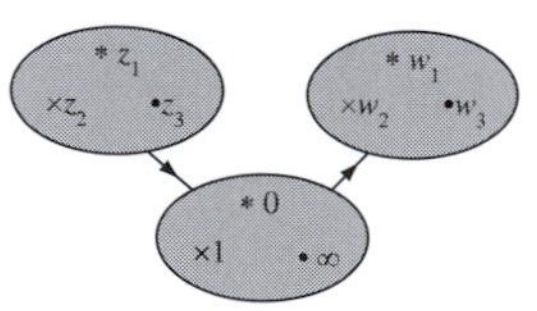

So let us start by considering how to construct a Möbius transformation that maps three given points z_1, z_2, z_3 in $\hat{\mathbb{C}}$ to the points $0, 1, \infty$, respectively.

First observe that if z_1, z_2, z_3 belong to $\mathbb{C}$, then any Möbius transformation of the form

$$M(z) = K\frac{z - z_1}{z - z_3}$$

Here K can be any complex number.

maps z_1 to 0 and z_3 to ∞. We can therefore obtain the required Möbius transformation by choosing K so that $M(z_2) = 1$. The following example illustrates how this is done. It also shows how the method can be modified to deal with cases where one of z_1, z_2, z_3 is ∞.

Example 1 For each set of three points given below, determine a Möbius transformation M that maps the points to $0, 1, \infty$, respectively.

(a) $\frac{1}{2}, -1, 3$ (b) $\infty, i, 2$ (c) $i, \infty, 3$ (d) $5, 2, \infty$

Solution

(a) To ensure that $M\left(\frac{1}{2}\right) = 0$ and $M(3) = \infty$ we let M have the form

$$M(z) = K\frac{z - \frac{1}{2}}{z - 3},$$

This has the form

$$M(z) = \frac{az + b}{cz + d}$$

with $a = K, b = -\frac{1}{2}K$, $c = 1$ and $d = -3$.

for some complex number K. Since $M(-1) = 1$, we must have

$$1 = K\frac{-1 - \frac{1}{2}}{-1 - 3},$$

so that $K = \frac{8}{3}$. The required Möbius transformation is therefore

$$M(z) = \frac{8z - 4}{3z - 9}.$$

(b) To ensure that $M(\infty) = 0$ and $M(2) = \infty$ we let M have the form

$$M(z) = \frac{K}{z - 2}.$$

This has the form

$$M(z) = \frac{az + b}{cz + d}$$

with $a = 0, b = K, c = 1$ and $d = -2$. Since $c \neq 0$, it follows from the definition of a Möbius transformation that $M(\infty) = a/c = 0$, as required.

Since $M(i) = 1$, we must have

$$1 = \frac{K}{i - 2},$$

so that $K = i - 2$. It follows that

$$M(z) = \frac{i - 2}{z - 2}.$$

(c) To ensure that $M(i) = 0$ and $M(3) = \infty$ we let M have the form

$$M(z) = \frac{z - i}{z - 3}.$$

This has the form

$$M(z) = \frac{az+b}{cz+d}$$

with $a = 1, b = -i$, $c = 1$ and $d = -3$. Since $c \neq 0$, we have $M(\infty) = a/c = 1$, as required.

In this case there is no need to include a constant K because $M(\infty)$ is already equal to 1.

(d) Here we require $M(5) = 0$ and $M(\infty) = \infty$, so M is an extended linear function of the form

$$M(z) = K(z - 5).$$

Since $M(2) = 1$, we must have

$$1 = K(2 - 5),$$

so that $K = -\frac{1}{3}$. It follows that

$$M(z) = -\tfrac{1}{3}(z - 5). \qquad \square$$

This has the form $M(z) = \frac{az+b}{cz+d}$ with $a = K$, $b = -5K$, $c = 0$ and $d = 1$. Since $c = 0$, it follows from the definition of a Möbius transformation that $M(\infty) = \infty$.

Guided by the above example we can summarize all the possible cases to obtain the following general strategy.

Strategy To determine the Möbius transformation M which maps three given points z_1, z_2, z_3 onto the points $0, 1, \infty$ respectively:

1. choose the appropriate form of mapping from the following formulas for M:
   - for $z_1, z_2, z_3 \longmapsto 0, 1, \infty$ use $M(z) = K\frac{z-z_1}{z-z_3}$;
   - for $\infty, z_2, z_3 \longmapsto 0, 1, \infty$ use $M(z) = \frac{K}{z-z_3}$;
   - for $z_1, \infty, z_3 \longmapsto 0, 1, \infty$ use $M(z) = \frac{z-z_1}{z-z_3}$;
   - for $z_1, z_2, \infty \longmapsto 0, 1, \infty$ use $M(z) = K(z - z_1)$;
2. find the complex number K for which $M(z_2) = 1$.

Problem 1 For each set of three points given below, determine a Möbius transformation M that maps the points to $0, 1, \infty$, respectively.

(a) $-1, -3, 0$ (b) $\frac{3}{2}, 2, 1$ (c) $\infty, -3, 2$ (d) $\frac{3}{2}, 2, \infty$

We are now in a position to prove the main theorem of this section.

Theorem 1 The Fundamental Theorem of Inversive Geometry
Let z_1, z_2, z_3 and w_1, w_2, w_3 be two sets of three points in the extended complex plane $\hat{\mathbb{C}}$. Then there is a unique Möbius transformation M which maps z_1 to w_1, z_2 to w_2, and z_3 to w_3.

Proof According to the above strategy there is a Möbius transformation M_1 which maps the points z_1, z_2, z_3 to the points $0, 1, \infty$, respectively. Similarly, there is a Möbius transformation M_2 which maps the points w_1, w_2, w_3 to the points $0, 1, \infty$, respectively.

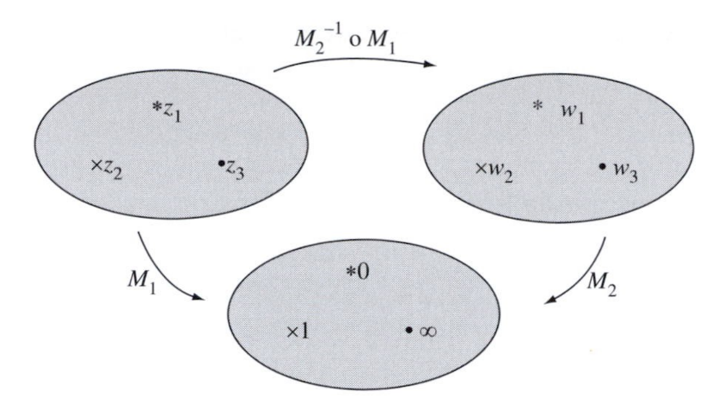

The composite $M = M_2^{-1} \circ M_1$ is therefore the required Möbius transformation which maps z_1, z_2, z_3 to the points w_1, w_2, w_3, respectively.

To check uniqueness we first observe that the identity is the only Möbius transformation which maps each of the points 0, 1, ∞ to themselves. Indeed, if

$$M(z) = \frac{az + b}{cz + d}$$

is a Möbius transformation which maps ∞ to itself, then $c = 0$. Also, if M maps 0 to itself, then $M(0) = b/d = 0$, so $b = 0$. It follows that $M(z) = (a/d)z$. But if M maps 1 to itself, then $M(1) = a/d = 1$, which implies that $a = d$. Putting all this together we conclude that M is the identity $M(z) = z$, as required.

Next suppose that M and M' are two Möbius transformations which satisfy the conditions of the theorem.

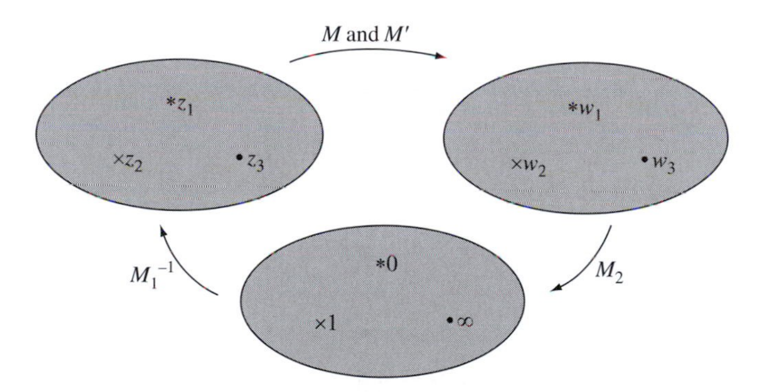

Then the composites $M_2 \circ M \circ M_1^{-1}$ and $M_2 \circ M' \circ M_1^{-1}$ must both be Möbius transformations which map each of the points 0, 1, ∞ to themselves. Since this implies that both composites are equal to the identity, we can write

$$M_2 \circ M \circ M_1^{-1} = M_2 \circ M' \circ M_1^{-1}.$$

If we now compose both sides of this equation with M_2^{-1} on the left and with M_1 on the right, then we obtain $M = M'$, as required. ■

If we actually need to find the Möbius transformation which maps one set of three points onto another set of three points we simply follow the strategy used to prove part (a) of the Fundamental Theorem.

Strategy To determine the Möbius transformation M which maps the points z_1, z_2, z_3 to the points w_1, w_2, w_3, respectively:

1. find the Möbius transformation M_1 which maps the points z_1, z_2, z_3 to the points $0, 1, \infty$, respectively;
2. find the Möbius transformation M_2 which maps the points w_1, w_2, w_3 to the points $0, 1, \infty$, respectively;
3. calculate $M = M_2^{-1} \circ M_1$.

The following example illustrates how to implement this strategy.

Example 2 Find the Möbius transformation M which maps the points $i, \infty, 3$ to the points $\frac{1}{2}, -1, 3$, respectively.

Solution We follow the steps in the above strategy.

1. From part (c) of Example 1, we know that the Möbius transformation M_1 which maps the points $i, \infty, 3$ to the points $0, 1, \infty$, respectively, is given by

$$M_1(z) = \frac{z - i}{z - 3}.$$

2. Also, from part (a) of Example 1, we know that the Möbius transformation M_2 which maps the points $\frac{1}{2}, -1, 3$ to the points $0, 1, \infty$, respectively, is given by

$$M_2(z) = \frac{8z - 4}{3z - 9}.$$

3. Matrices associated with M_1 and M_2 are

$$\begin{pmatrix} 1 & -i \\ 1 & -3 \end{pmatrix} \quad \text{and} \quad \begin{pmatrix} 8 & -4 \\ 3 & -9 \end{pmatrix};$$

also, by Theorem 7 of Subsection 5.3.6, a matrix associated with the inverse of M_2 is

$$\begin{pmatrix} -9 & 4 \\ -3 & 8 \end{pmatrix}.$$

Hence, a matrix associated with $M = M_2^{-l} \circ M_1$ is given by

$$\begin{pmatrix} -9 & 4 \\ -3 & 8 \end{pmatrix} \begin{pmatrix} 1 & -i \\ 1 & -3 \end{pmatrix} = \begin{pmatrix} -5 & -12 + 9i \\ 5 & -24 + 3i \end{pmatrix}.$$

The required Möbius transformation is therefore

$$M(z) = \frac{-5z - 12 + 9i}{5z - 24 + 3i}. \qquad \square$$

As a check, you can verify that $M(i) = \frac{1}{2}$, $M(\infty) = -1$ and $M(3) = 3$.

Problem 2

(a) Find the Möbius transformation which maps the points -1, i, 1 to the points -1, -3, 0, respectively.
(b) Find the Möbius transformation which maps the points 3, ∞, -2 to the points 3, $\frac{7}{3}$, 1, respectively.

5.4.3 Circles through Four Points

We now explore an application of the Fundamental Theorem which enables us to determine whether four given points z_1, z_2, z_3 and z_4 lie on some generalized circle.

If there is a generalized circle that passes through z_1, z_2, z_3 and z_4, then it must be the unique generalized circle C that passes through z_1, z_2 and z_3. We have to decide whether C also passes through z_4. To do this, consider the Möbius transformation M that maps z_1, z_2, z_3 to 0, 1, ∞, respectively. Under M the image of C is the extended real axis, for this is the only generalized circle that passes through 0, 1 and ∞. If z_4 lies on C, then its image under M must lie on the real axis; whereas if z_4 does not lie on C, then its image under M cannot lie on the real axis.

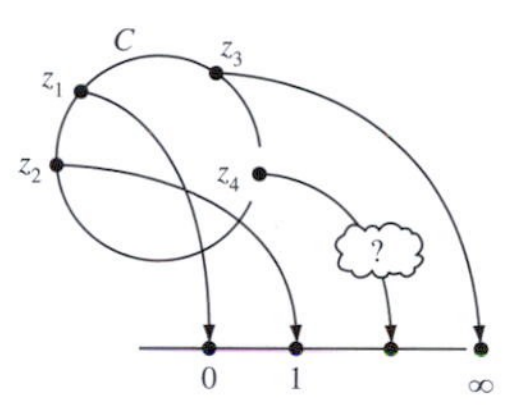

Example 3 Determine whether the four points i, $1 + 4i$, 3, $4 + 3i$ lie on a generalized circle.

Solution First, we determine the Möbius transformation M that maps i, $1 + 4i$, 3 to 0, 1, ∞, respectively. Following the strategy in the previous subsection, we observe that this transformation must be of the form

$$M(z) = K\frac{z - i}{z - 3}$$

for some complex number K. Since $M(1 + 4i) = 1$, we must have

$$1 = K\frac{(1 + 4i) - i}{(1 + 4i) - 3} = K\frac{1 + 3i}{-2 + 4i},$$

so

$$K = \frac{-2 + 4i}{1 + 3i} = \frac{(-2 + 4i)(1 - 3i)}{10} = 1 + i.$$

You met this method of simplifying a quotient by multiplying both numerator and denominator by the conjugate of the denominator earlier, in Subsection 5.2.1.

Thus the transformation M is given by

$$M(z) = \frac{(z - i)(1 + i)}{z - 3}.$$

It follows that

$$M(4+3i) = \frac{(4+2i)(1+i)}{1+3i} = \frac{2+6i}{1+3i} = 2.$$

Since this is real, it follows that i, $1+4i$, 3, $4+3i$ do all lie on a generalized circle. □

In this example we showed that the points i, $1+4i$, 3, $4+3i$ all lie on a generalized circle; but suppose we want to show that the points lie on an *ordinary* circle. To do this we need to check that the generalized circle that passes through the points is not an extended line. If it were an extended line, then it would pass through ∞, and so $M(\infty)$ would be real. Since $M(\infty) = 1+i$ is not real, it follows that the points do indeed lie on an ordinary circle.

$M(\infty)$ cannot be ∞, as M is one–one and $M(3) = \infty$.

Strategy To determine whether z_1, z_2, z_3 and z_4 lie on a circle:

1. find the Möbius transformation M which maps z_1, z_2, z_3 to 0, 1, ∞, respectively;
2. the points z_1, z_2, z_3, z_4, lie on a generalized circle if and only if $M(z_4)$ is real;
3. the generalized circle in Step 2 is a circle provided that $M(\infty)$ is not real.

If $M(z_4)$ and $M(\infty)$ are both real, then z_1, z_2, z_3, z_4 lie on a line.

Problem 3 Determine whether each of the following sets of four points lies on a circle.

(a) $0, -4, -2i, -1-3i$ (b) $-1, -i, i, 2-i$

5.4.4 Congruence of Generalized Circles

We end this section by stating an important consequence of the Fundamental Theorem of Inversive Geometry.

Theorem 2 Let C_1 and C_2 be generalized circles in the extended complex plane. Then there is a Möbius transformation that maps C_1 onto C_2.

Proof Let a, b, c be any three points on C_1 and let d, e, f be any three points on C_2. By the Fundamental Theorem of Inversive Geometry, there is a Möbius transformation M that maps a, b, c to d, e, f, respectively.

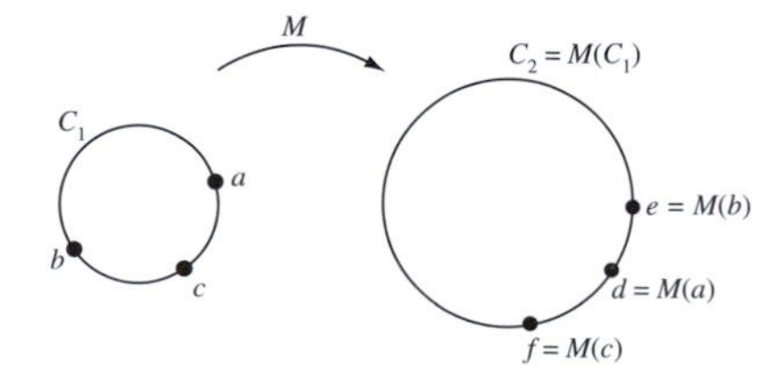

Since M maps generalized circles to generalized circles, it must map C_1 to a generalized circle through d, e and f. But a generalized circle is uniquely determined by any three of its points, so C_2 is the only generalized circle that passes through d, e and f. It follows that M maps C_1 onto C_2. ■

One of the remarkable consequences of Theorem 2 is that all generalized circles are inversive-congruent. In particular, an ordinary circle is inversive-congruent to any given (extended) line. Since lines are often easier to investigate than circles, we can sometimes simplify a problem that involves the inversive properties of a figure by using an inversive transformation to map one or more of the circles in a figure onto (extended) lines. You will see several examples of this in the next section.

5.5 Coaxal Families of Circles

We now prove some lovely theorems about circles, where the beauty of the final results is matched by the beauty of the proofs themselves. Inversion is our key tool.

Sadly this is not always the case in Mathematics!

5.5.1 Apollonian Circles Theorem

We start with the proof of the Apollonian Circles Theorem, which we stated at the beginning of the chapter.

Theorem 1 Apollonian Circles Theorem
Let A and B be two distinct points in the plane, and let k be a positive real number other than 1. Then the locus of points P that satisfy $PA : PB = k : 1$ is a circle whose centre lies on the line through A and B.

Recall that if $k = 1$, then P must lie on the perpendicular bisector of AB.

We give two proofs of this result. The first uses methods from Euclidean geometry, and has the advantage of providing us with equations for the circles in terms of k. The second proof uses the methods of inversive geometry, and has the advantage of providing us with a deeper insight into the geometry of the circles.

The second proof is given later.

First Proof

To keep the algebra simple, we introduce x- and y-axes into the plane such that A and B have coordinates $(-a, 0)$ and $(a, 0)$, respectively, where $a > 0$.

Now fix a value of $k > 0$, $k \neq 1$, and let C be the locus of points P that satisfy $PA : PB = k : 1$. Then a point $P(x, y)$ belongs to C if and only if $PA = k \cdot PB$, and since $k > 0$ this is equivalent to the equation

$$PA^2 = k^2 \cdot PB^2.$$

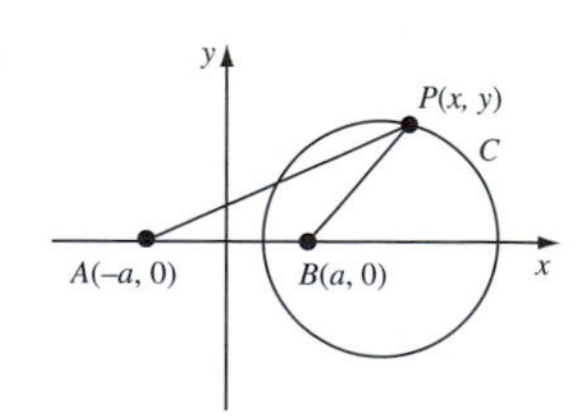

Using the Euclidean formula for distance between points, we see that this last equation holds if and only if

$$(x + a)^2 + y^2 = k^2((x - a)^2 + y^2).$$

Multiplying out the brackets and collecting terms in x^2, y^2 and x, we see that this is equivalent to

$$x^2(1-k^2)+y^2(1-k^2)+2ax\,(1+k^2)+a^2(1-k^2)=0.$$

Since $k \neq 1$, this holds if and only if

$$x^2+y^2+2a\left(\frac{1+k^2}{1-k^2}\right)x+a^2=0.$$

It follows that the locus C is a circle with centre c and radius r where

$$c=\left(-a\left(\frac{1+k^2}{1-k^2}\right),0\right)$$

and

$$r=\sqrt{a^2\left(\frac{1+k^2}{1-k^2}\right)^2-a^2}=\frac{2ak}{|1-k^2|}.$$

Here we are using the formulas for centre and radius given in Theorem 2, Subsection 1.1.2.

In particular, the centre lies on the x-axis, which by our choice of axes is the line through A and B. ■

Notice that the centre of the circle is *not* at either of the points A or B.

If $k=1$, then the locus of points that satisfy $PA = k \cdot PB$ is the line ℓ that bisects AB at right angles. If we adopt the convention that this locus includes the point ∞, then we can think of this locus as the extended line $\ell \cup \{\infty\}$. With this convention, every positive value of k gives rise to a generalized circle known as a **circle of Apollonius**. The family of all such circles is known as the **Apollonian family of circles** defined by the points A and B.

A 'circle of Apollonius' is often referred to as an 'Apollonian circle'.

As k increases through the interval $(0,\infty)$, the corresponding circles range through the Apollonian family.

When $0<k<1$, we have that $PA<PB$, and so P is closer to A than it is to B. The circles that correspond to these values of k therefore lie on the same side of ℓ as does A. For values of k close to 0, $PA = k \cdot PB$ is small, and so the corresponding circles are close to A. As k tends to 1, the circles 'grow' and become ever closer to ℓ.

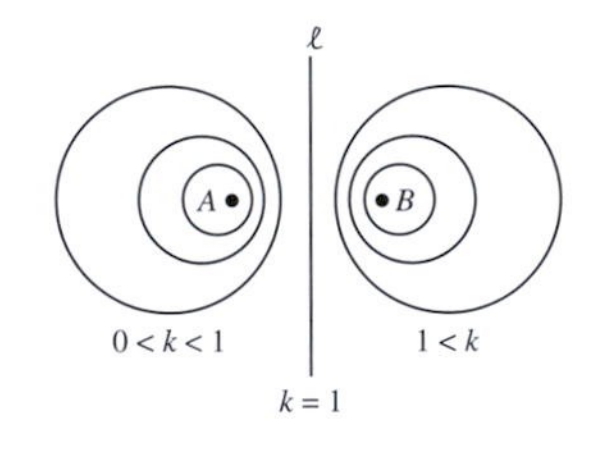

When $k=1$, the corresponding 'circle' is the extended line $\ell \cup \{\infty\}$.

When $k>1$, we have that $PA>PB$, so that P is closer to B than it is to A. The circles that correspond to these values of k therefore lie on the same side of ℓ as does B. As k increases from 1, $PB=(1/k)\cdot PA$ becomes smaller, and the corresponding circles 'shrink' and become ever closer to B.

In light of the above discussion, it is sometimes convenient to refer to the points A and B as **point circles** corresponding to the cases $k=0$ and '$k=\infty$', respectively. With this convention, every point of the plane belongs to a unique 'circle' associated with the Apollonian family.

Example 1 For the Apollonian family defined by the point circles $(-1,0)$ and $(1,0)$, determine the equation of the circle in the family that passes through the point $(2, 1)$.

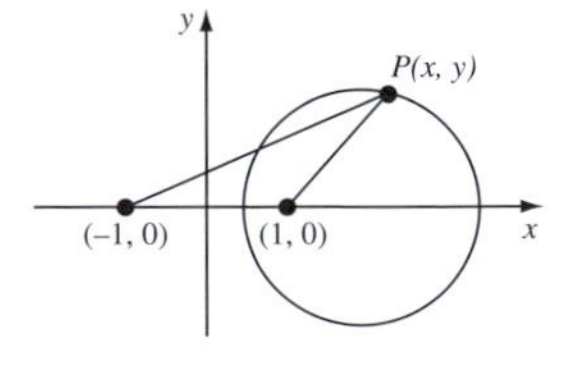

Solution Let P be a point (x, y) in the plane whose distance from the point $(-1, 0)$ is k times its distance from the point $(1, 0)$. Then, if we use the Euclidean formula for distance between points in the plane, we obtain

$$(x+1)^2 + y^2 = k^2((x-1)^2 + y^2).$$

For each value of k this yields an equation for the corresponding Apollonian circle. The point $(2, 1)$ lies on the Apollonian circle for which

$$(2+1)^2 + 1^2 = k^2((2-1)^2 + 1^2),$$

that is, $k = \sqrt{5}$.

The equation of the Apollonian circle through the point $(2, 1)$ is therefore

$$(x+1)^2 + y^2 = 5((x-1)^2 + y^2),$$

which simplifies to

$$x^2 + y^2 - 3x + 1 = 0. \qquad \square$$

Problem 1 For the Apollonian family defined by the point circles $(0, -1)$ and $(0, 2)$, determine the equation of the circle in the family that passes through the point $(1, 1)$.

Problem 2 The circles C_1 and C_2 in an Apollonian family of circles have the segments $[-18, -2]$ and $[3, 12]$, respectively, as diameter. Determine the point circles in the family, and hence the equation of the Apollonian circle in the family that passes through the point $(6, 9)$.

We now give a second proof of the Apollonian Circles Theorem, using inversive geometry.

Theorem 1 Apollonian Circles Theorem
Let A and B be two given points in the plane, and let k be a positive real number other than 1. Then the locus of points P that satisfy $PA : PB = k : 1$ is a circle whose centre lies on the line through A and B.

Second Proof

Let C be the locus of points P that satisfy $PA = k \cdot PB$, and let t be the inversion in the circle with centre A and radius 1. We shall show that $C' = t(C)$ is a circle, and hence that $C = t^{-1}(C')$ is a generalized circle.

We specify radius 1 here simply for definiteness; any positive radius would serve equally well.

Let P be an arbitrary point in the extended plane, and let $t(B) = B'$, $t(P) = P'$. Then by the definition of inversion,

$$AB \cdot AB' = 1 \quad \text{and} \quad AP \cdot AP' = 1, \tag{1}$$

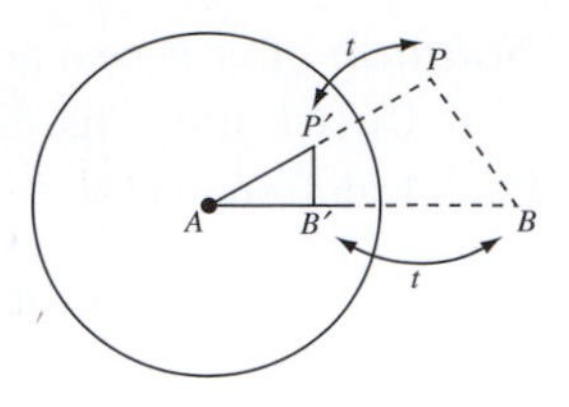

so that

$$\frac{AB}{AP'} = \frac{AP}{AB'}.$$

This shows that the sides AB and AP of $\triangle APB$ are proportional to the sides AP' and AB' of $\triangle AB'P'$. Since $\angle BAP$ is the same as $\angle P'AB'$, it follows that $\triangle APB$ is similar to $\triangle AB'P'$. Hence

$$\frac{B'P'}{PB} = \frac{AP'}{AB}.$$

But, from equation (1), $AP' = 1/AP$, so P is related to P' by the equation

$$B'P' = \frac{PB}{AB \cdot AP}. \qquad (2)$$

Now if P lies on the locus C, then AP $(= PA) = k \cdot PB$, so that

$$B'P' = \frac{1}{k \cdot AB}.$$

Thus P' lies on a circle C' of radius $1/(k \cdot AB)$ and with centre B'.

Conversely, if P' lies on C', then $B'P' = 1/(k \cdot AB)$, so from equation (2),

$$\frac{1}{k \cdot AB} = \frac{PB}{AB \cdot AP}.$$

Thus $AP = k \cdot PB$, and so P lies on the locus C.

It follows that the inversion t maps the locus C onto the circle C'. But $t^{-1} = t$ maps generalized circles to generalized circles, so $C = t^{-1}(C')$ is a generalized circle. ■

This proof sets up a one–one correspondence between the family of Apollonian circles and a family of concentric circles. Each Apollonian circle corresponds to a value of k, and this in turn corresponds to a circle of radius $1/(k \cdot AB)$ with centre $B' = t(B)$.

In particular, the extended line in the Apollonian family corresponds to $k = 1$, which corresponds to the circle of radius $l/AB = AB'$, with centre B'. As you would expect, this circle passes through the centre A of the inversion t.

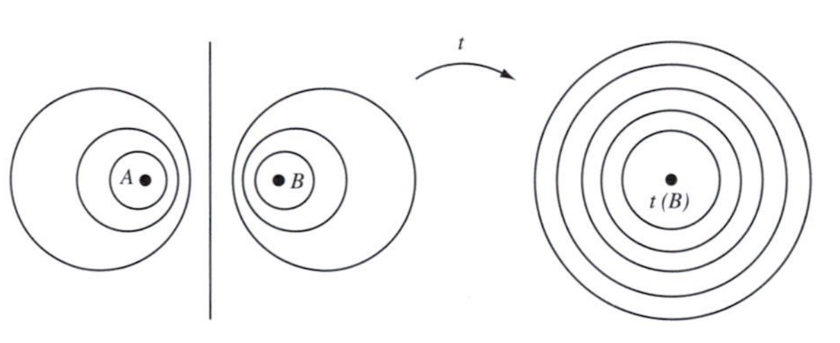

The importance of this correspondence is that it enables us to characterize Apollonian families of circles in terms of inversive transformations. Such a characterization is ideal for tackling problems in inversive geometry so we state it in the form of a theorem.

Theorem 2 Let A and B be distinct points in the plane, and let t be the inversion in the circle with centre A and radius 1. Then the Apollonian family of circles defined by the point circles A and B is mapped by t to the family of all concentric circles with centre $t(B)$; and the family of all concentric circles with centre $t(B)$ is mapped by t to the Apollonian family of circles defined by the point circles A and B.

For t is self-inverse.

A remarkable consequence of this theorem is that in inversive geometry all Apollonian and concentric families of circles are congruent to each other. Indeed the theorem shows that every Apollonian family of circles is congruent to a family of concentric circles, so it is sufficient to show that all families of concentric circles are congruent to each other. This is easily achieved by noting that any family of concentric circles can be mapped onto any other family of concentric circles by a translation that makes their centres coincide.

Recall that a translation is an inversive transformation provided we adopt the convention that ∞ maps to ∞.

5.5.2 Families of Circles

Any two (distinct) generalized circles in the extended plane are related in precisely one of the following ways:

1. they may not intersect;
2. they may intersect in precisely *one* point;
3. they may intersect in *two* distinct points.

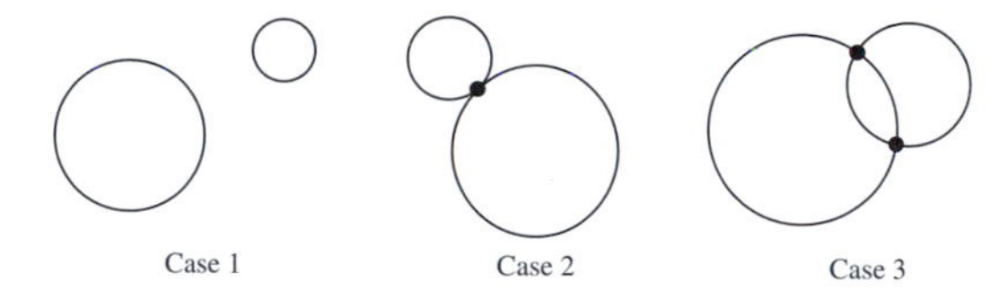

Case 1 Case 2 Case 3

In the previous subsection you met Apollonian families of circles, in which no two circles intersect; in the following subsection we will show that any two non-intersecting circles determine an Apollonian family of circles. We now consider the two other families, namely families of generalized circles that intersect at precisely one point and families of generalized circles that intersect at precisely two points.

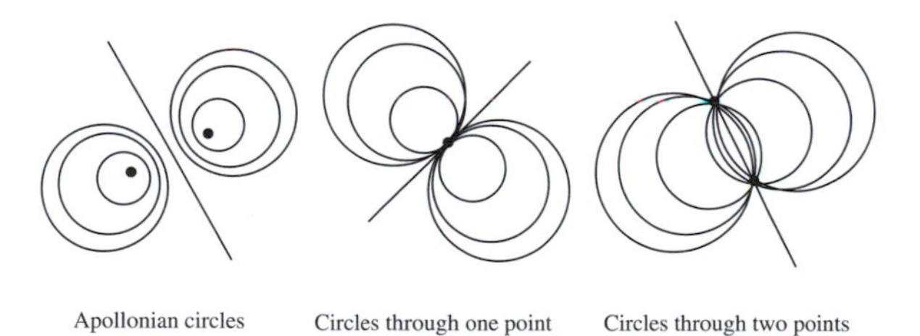

Apollonian circles Circles through one point Circles through two points

Each of these three families contains precisely one extended line known as the *radical axis* of the family. The centres of the circles in each family lie on a line, or *axis*, that is perpendicular to the radical axis. Since all the circles in each family are symmetrical about the axis through their centres, the families are known as *coaxal families*.

Definition A **coaxal family of circles** in the plane is a family of (generalized) circles of one of the following types:

1. an Apollonian family, with particular point circles;
2. a family that intersect at one particular point;
3. a family that intersect at two particular points.

The extended line in each family is called the **radical axis** of the family.

Given the Apollonian family of circles $\mathscr{F}$ defined by point circles A and B, there is a corresponding coaxal family of circles $\mathscr{G}$ that pass through A and B. Conversely, given a coaxal family of circles $\mathscr{G}$ that pass through distinct points A and B, there is a corresponding Apollonian family of circles $\mathscr{F}$ defined by the point circles A and B.

The following theorem describes a remarkable relationship between the two families of circles, $\mathscr{F}$ and $\mathscr{G}$.

Theorem 3 Coaxal Circles Theorem
Let A and B be distinct points in the plane. Let $\mathscr{F}$ be the Apollonian family defined by the point circles A and B, and let $\mathscr{G}$ be the family of all generalized circles through A and B. Then every member of $\mathscr{F}$ is orthogonal to every member of $\mathscr{G}$.

Two circles are *orthogonal* if they meet at right angles.

Proof Let t be the inversion in the unit circle with centre A. By Theorem 2, the Apollonian family $\mathscr{F}$ is mapped by t to the family of concentric circles with centre $t(B)$.

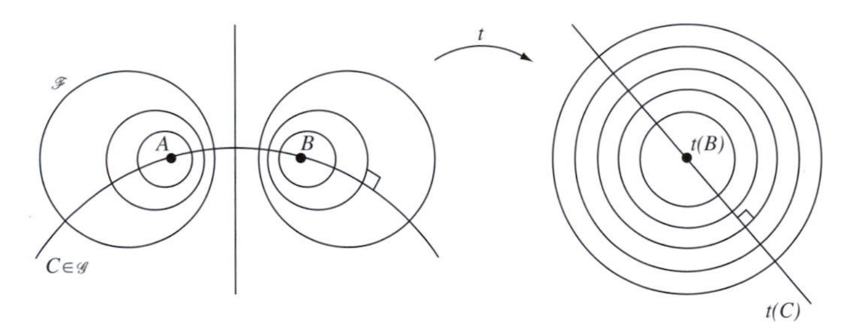

Now let C be an arbitrary member of $\mathscr{G}$. Since C is a generalized circle that passes through A and B, it follows that $t(C)$ is a generalized circle that passes through $t(A) = \infty$ and $t(B)$. In other words, $t(C)$ is an extended line through $t(B)$. Clearly, this line $t(C)$ intersects each of the circles with centre $t(B)$ at right angles. Since inversion preserves the magnitude of angles, it follows that C meets each of the Apollonian circles in $\mathscr{F}$ at right angles. ■

For an Apollonian family of circles defined by point circles A and B, the Coaxal Circles Theorem states that every generalized circle through A and B meets each of the Apollonian circles at right angles. The next theorem enables us to use this fact to deduce that A and B are inverse points with respect to each of the Apollonian circles.

Theorem 4 Two points A and B in the extended complex plane are inverse points with respect to a generalized circle C if and only if every generalized circle through A and B meets C at right angles.

Proof We first show that A and B are inverse points if every generalized circle through A and B meets C at right angles.

Let C_1 be any generalized circle through A and B that meets C at right angles, at points R and S. Now invert the figure C_1 in the generalized circle C. Since R and S remain fixed, C_1 is mapped to a generalized circle that meets C at right angles at R and S. It follows that C_1 is mapped to itself; for the fact that the image passes through R and S means that it must be a circle, and the radii of this circle through R and S must be along the tangents to C at R and S.

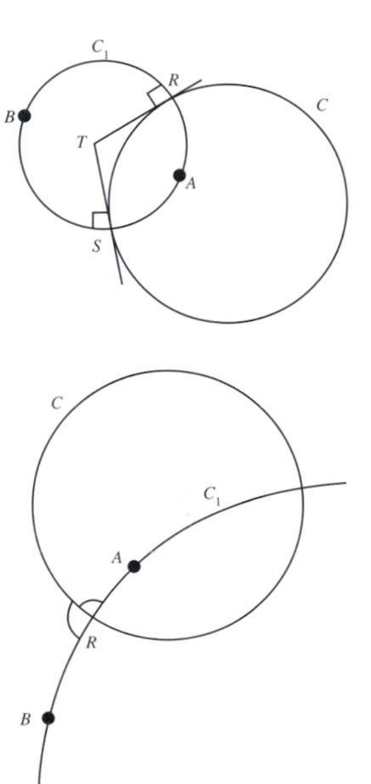

In fact, by a similar argument, every generalized circle through A and B that meets C at right angles maps to itself. The only way this can happen is if A inverts to B, and vice versa. Hence A and B are inverse points.

Next we show that if A and B are inverse points, then every generalized circle through A and B meets C at right angles.

Let A and B be inverse points, and let C_1 be a generalized circle through A and B that meets C at R. Under inversion in C, the points A and B swap over and R remains fixed, so C_1 maps to itself. It follows that both the angles that C_1 makes with the circle C must be equal. These angles must therefore be right angles, which is what we want to prove. ■

Theorem 4 has two important corollaries. The first follows directly from the Coaxal Circles Theorem.

Corollary 1 If C is an Apollonian circle defined by the point circles A and B, then A and B are inverse points with respect to C.

Proof Let C be an Apollonian circle with respect to the point circles A and B. Then by the Coaxal Circles Theorem, every generalized circle through A and B meets C at right angles. By Theorem 4, A and B are inverse points with respect to C. ■

The second corollary asserts that inverse points are preserved by inversive transformations.

Corollary 2 Let A and B be inverse points with respect to a generalized circle C, and let t be an inversive transformation. Then $t(A)$ and $t(B)$ are inverse points with respect to the generalized circle $t(C)$.

Proof

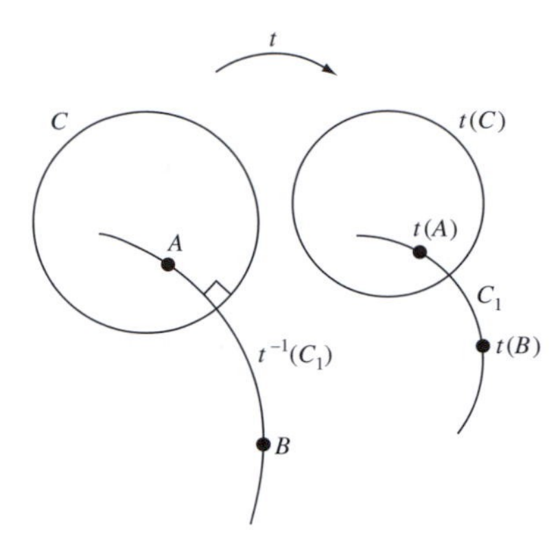

Let C_1 be an arbitrary generalized circle through $t(A)$ and $t(B)$. Then $t^{-l}(C_1)$ is a generalized circle that passes through A and B. By Theorem 4, $t^{-l}(C_1)$ meets C at right angles. But t preserves the magnitude of angles, so C_1 meets $t(C)$ at right angles. Since C_1 is an arbitrary generalized circle through $t(A)$ and $t(B)$, it follows from Theorem 4 that $t(A)$ and $t(B)$ are inverse points with respect to $t(C)$. ■

An immediate consequence of Corollary 2 is that if $\mathscr{F}$ is the family of *all* generalized circles that have two given points A and B as inverse points, then under an inversive transformation t the family $\mathscr{F}$ maps onto the family of *all* generalized circles that have $t(A)$ and $t(B)$ as inverse points.

At first sight this observation appears to be little more than a restatement of Corollary 2; however the restatement has a particular significance, as the following theorem shows.

Theorem 5 Let $\mathscr{F}$ be the family of all generalized circles that have A and B as inverse points. Then $\mathscr{F}$ is either a concentric family of circles with centres A or B, or the Apollonian family of circles with point circles A and B.

Proof First, suppose that either A or B is the point ∞; to be definite, assume $A = \infty$. Then each circle in $\mathscr{F}$ has ∞ and B as inverse points. But this can happen only if B is the centre of each circle in $\mathscr{F}$; in other words $\mathscr{F}$ is the family of concentric circles with centre B. A similar argument applies if $B = \infty$.

Next suppose that neither A nor B is the point ∞, and let t be the inversion in the circle of unit radius with centre A. By Theorem 2, t maps the family $\mathscr{F}$ to the family of all generalized circles with inverse points at $t(A) = \infty$ and $t(B)$, namely the family of concentric circles with centre $t(B)$. By Theorem 5, $\mathscr{F}$ is the Apollonian family of circles defined by the point circles A and B. ■

Roughly speaking, you can think of the family of concentric circles with centre A as an Apollonian family of circles defined by point circles A and ∞. Then an Apollonian family of circles defined by the point circles A and B maps to an Apollonian family of circles defined by the point circles $t(A)$ and $t(B)$.

In light of the remarks that precede this theorem, it follows that under an inversive transformation t an Apollonian family of circles defined by point circles A and B

EITHER maps to an Apollonian family of circles defined by the point circles $t(A)$ and $t(B)$,
OR maps to a concentric family of circles with centre $t(A)$ or $t(B)$.

Similarly a concentric family of circles with centre A either maps to the Apollonian family of circles defined by the point circles $t(A)$ and $t(\infty)$, or it maps to a concentric family of circles with centre $t(A)$ or $t(\infty)$.
So far, we have used inversive geometry to prove results about Apollonian families of circles, but similar methods can be used to prove results about other plane figures. The technique is to use an inversive transformation (often an inversion) to map one or more circles in the figure to extended straight lines. The additional symmetry that results from this transformation is then used to establish the required result.

For future reference, we also note here a fact that we met in the proof of Theorem 4 above.

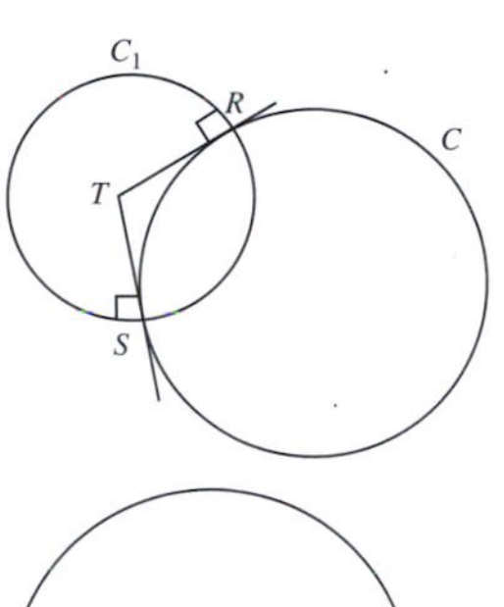

Corollary 3 Let C and C_1 be circles that meet at right angles at two points R and S. Then the centre of C_1 is the point T of intersection of the tangents to C at R and S; and C_1 maps to itself under inversion in C.

Example 2 Let C_1, C_2 and C_3 be circles in the plane such that C_2 and C_3 touch at a point P, C_1 and C_3 touch at a point Q, and C_1 and C_2 touch at a point R. Let C be the circle that passes through P, Q and R. Prove that C cuts C_1, C_2 and C_3 at right angles.

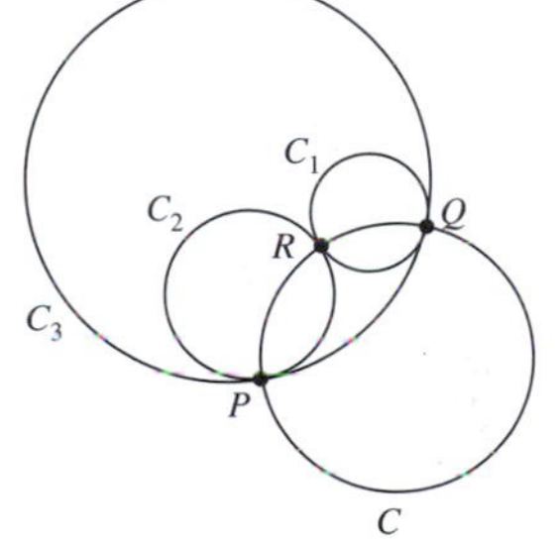

Solution Let t be an inversion in a circle with centre R. Then $t(C_1)$ and $t(C_2)$ are straight lines. Moreover, since C_1 and C_2 do not meet at any point other than P, it follows that $t(C_1)$ and $t(C_2)$ cannot meet in $\mathbb{C}$ and must therefore be parallel.

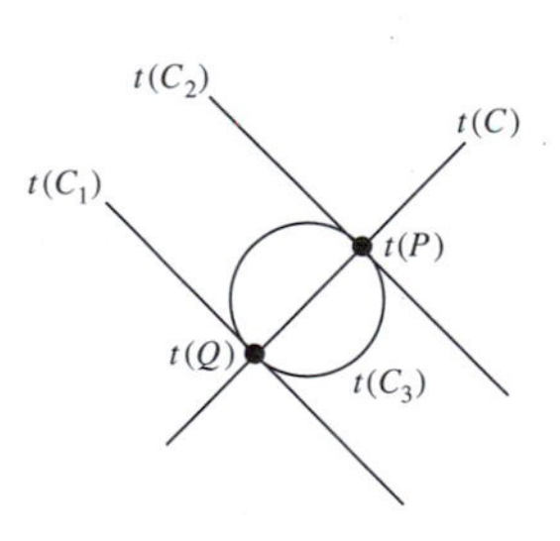

Now C_3 is tangential to C_1 and C_2, and does not pass through R, so $t(C_3)$ must be a circle that touches $t(C_1)$ and $t(C_2)$ tangentially at $t(Q)$ and $t(P)$, respectively. Next observe that C passes through R, so $t(C)$ is the extended line through $t(Q)$ and $t(P)$. Since the transformed figure has reflectional symmetry about $t(C)$, it follows that $t(C)$ cuts $t(C_1)$, $t(C_2)$ and $t(C_3)$ at right angles at $t(P)$ and $t(Q)$. But t preserves angles, so C cuts C_1, C_2 and C_3 at right angles at P and Q.

A similar argument with t replaced by inversion in a circle with centre P (or Q) shows that C cuts C_1 and C_2 at right angles at R. □

5.5.3 Two Circles Determine a Coaxal Family

It is clear that if we are given two circles in a coaxal family of mutually tangential circles, this determines the whole family.

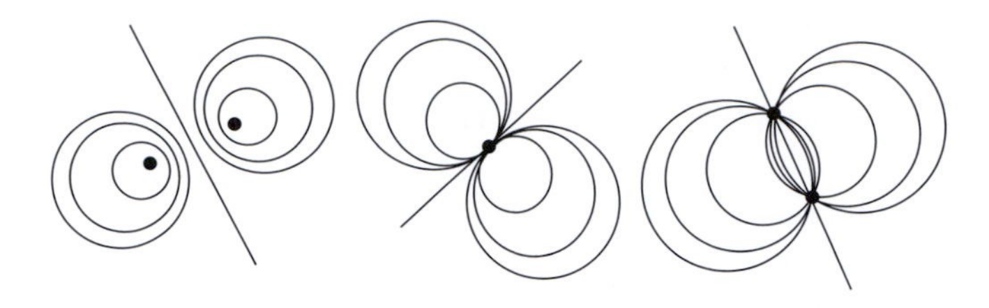

Similarly if we are given two circles in a coaxal family of circles through two fixed points, this determines the whole family. This raises the question as to whether two circles in an Apollonian family are sufficient to determine the whole family. We now prove that this is indeed the case. The crucial tool that we need is the following result.

Theorem 6 The Concentricity Theorem
Let C_1 and C_2 be any two non-intersecting circles in the plane. Then there is a inversive transformation that maps C_1 and C_2 onto a pair of concentric circles.

Proof If the circles are concentric then there is nothing to prove, so we may assume they are not concentric.

Step 1 Pick a point O on the circle C_1, and invert both circles in the circle of unit radius with centre O. Under this inversion C_1 maps to a line, C_1', and C_2 maps to a circle, C_2'.

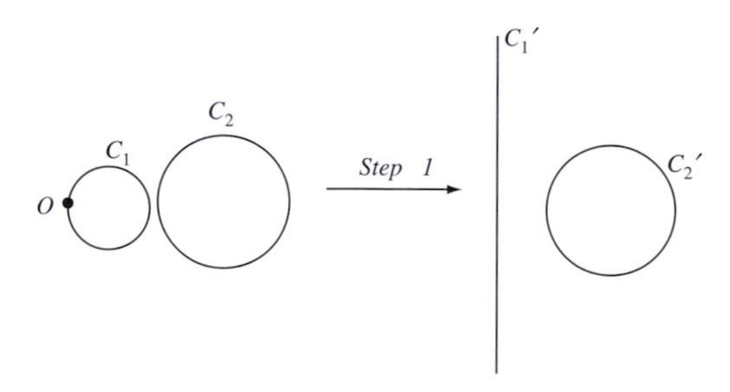

Step 2 Pick a point T on C_2' that does not lie on the perpendicular from the centre of C_2' to the line C_1', and let the tangent to C_2' at T intersect C_1' at the point U, say. Draw the circle C_3' with centre U and radius UT. This circle is perpendicular to C_2' at T, because a tangent to a circle is perpendicular to the radius at the point of contact. The circle C_3' is perpendicular to the line C_1' because its centre lies on the line C_1'.

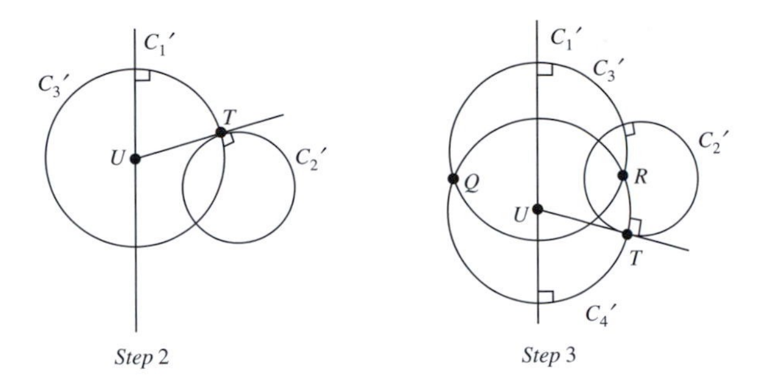

Step 3 Repeat the construct in Step 2 starting with another position for the point T, to obtain a circle C_4' perpendicular to both C_1' and C_2' at the respective points T. Let C_3' and C_4' meet at Q and R, say.

Step 4 Invert the figure again, this time in the unit circle with centre Q. Then the line C_1' maps to a circle C_1'', and the circle C_2' maps to a circle C_2''; the circles C_3' and C_4' pass through Q, so they invert to straight lines at right angles to C_1'' and C_2''. These lines are therefore diameters of the circles C_1'' and C_2'', and so the point where they meet (the image R'' of R) must be the centre of both circles.

As in the (second) proof of Theorem 1, the choice of 1 for the radius is not significant.

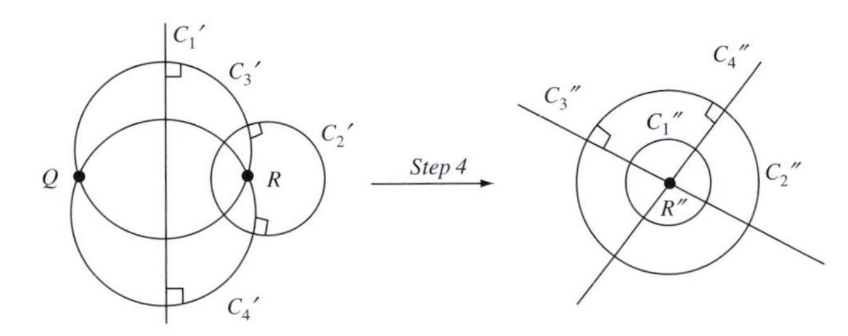

Step 5 Composing the inversion in Step 1 with the inversion in Step 4, we obtain an inversive transformation that maps C_1 and C_2 onto the pair of concentric circles C_1'' and C_2''. ∎

We can now prove the following beautiful result.

Theorem 7 Two Apollonian Circles Theorem
Let C_1 and C_2 be two non-concentric circles that do not intersect. Then there is a unique Apollonian family of circles that contains C_1 and C_2.

Proof By the Concentricity Theorem there is an inversive transformation t that maps C_1 and C_2 onto a pair of concentric circles $t(C_1)$ and $t(C_2)$. Let O be the common centre of these circles, and let $\mathscr{G}$ be the family of concentric circles with centre O. Under the inversive transformation t^{-1} the family $\mathscr{G}$ maps to the family of all generalized circles that have $t^{-1}(O)$ and $t^{-1}(\infty)$ as inverse points; we denote this family by $\mathscr{F}$. Now $\mathscr{F}$ cannot be a concentric family of circles, since it contains C_1 and C_2; so, by Theorem 5, it must be an Apollonian family of circles.

Theorem 6 above

To complete the proof we must show that $\mathcal{F}$ is the only Apollonian family of circles that contains C_1 and C_2. To do this suppose that $\mathcal{F}'$ is any Apollonian family of circles that contains C_1 and C_2. Then $\mathcal{F}'$ is mapped by t either to an Apollonian family of circles, or to a concentric family of circles. In fact the image of $\mathcal{F}'$ under t contains the concentric circles $t(C_1)$ and $t(C_2)$ and must therefore be the family $\mathcal{G}$ of concentric circles with centre O. It follows that $\mathcal{F}$ and $\mathcal{F}'$ are the same family of circles, for they are both mapped to $\mathcal{G}$ by t. ■

5.5.4 Some Applications of Inversion

As we mentioned earlier, inversion can often be used to prove results about plane figures, using a suitable inversion to map one or more circles in the figures to extended lines – the additional symmetry then being used to establish the desired result.

Our first application of inversion to prove a beautiful result is known as Steiner's Porism. A *porism* is a mathematical construction problem that has a surprising answer: either the construction cannot be carried out or it has infinitely many solutions.

Jakob Steiner (1796–1863) was a 19th-century Swiss geometer.

Theorem 8 Steiner's Porism
Let C_1 and C_2 be non-intersecting circles, with C_1 inside C_2. Then:

EITHER it is impossible to fit a chain of circles between C_1 and C_2, with each circle touching C_1 and C_2 and two other circles in the chain;

OR it is possible to construct such a chain, and the first circle in the chain can be placed in any convenient position.

C_2

C_1

chain of circles

Proof There are two possibilities.

EITHER It is impossible to fit a chain between C_1 and C_2 with the properties described, in which case the porism is established.

OR There is at least one chain F that fits between C_1 and C_2. To establish the porism in this case we must show that a chain also exists for each choice of a first circle C between C_1 and C_2.

By the Concentricity Theorem there is an inversive transformation t that maps C_1 and C_2 onto concentric circles C_1' and C_2'. Under t the chain F maps onto a chain of circles F' between C_1' and C_2', and C maps onto a circle C' which touches C_1' and C_2'. Since C_1' and C_2' have a common centre O, we can rotate the chain F' about O until its first circle is superimposed on C'. If we denote the rotated chain by G', then the inversive transformation t^{-1} maps G' back to a chain G between C_1 and C_2. Moreover, the first circle of G is C as required.

Recall that t^{-1} must be an inversive transformation, since t is.

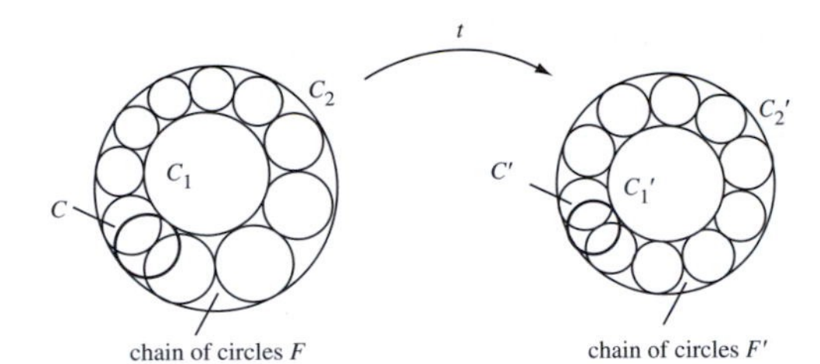

■

Another famous porism is **Poncelet's Porism**: If C_1 and C_2 are any two plane conics for which (for any given $n > 2$) it is possible to find one n-sided polygon which is simultaneously inscribed in C_1 and circumscribed around C_2, then it is possible to find infinitely many of them.

We omit a proof, as it is beyond the scope of this book.

Next, we prove a classical result known as Ptolemy's Theorem – but using inversion rather than using Ptolemy's Pythagorean geometry methods! A key tool in our proof is the following useful result.

Claudius Ptolemaus (c. 85–165 AD) was a geometer, astronomer and geographer in Alexandria, Egypt.

Lemma 1 Let a and b be points in $\mathbb{C} - \{O\}$. Then, under inversion in a circle C with centre O and radius r the distance between the images of a and b is $\frac{r^2}{|a|\cdot|b|}|a-b|$.

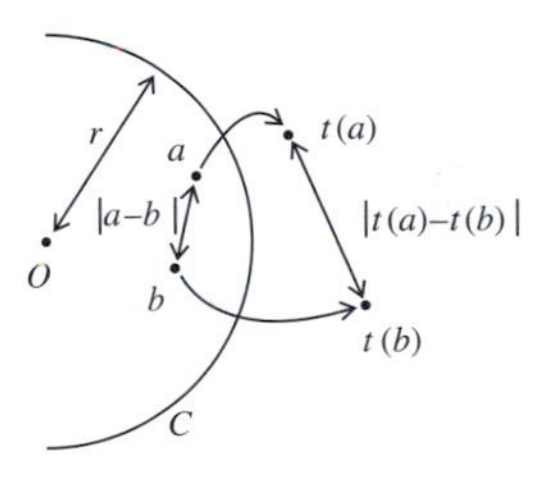

Proof Inversion in C may be represented in the complex plane by the transformation

$$t(z) = \frac{r^2}{\overline{z}}, \quad \text{where } z \in \mathbb{C} - \{O\}.$$

This is a special case of a general formula for inversion in C that you met in Theorem 3, Subsection 5.2.1.

It follows from this formula that

$$\begin{aligned} t(a) - t(b) &= \frac{r^2}{\overline{a}} - \frac{r^2}{\overline{b}} \\ &= \frac{r^2}{\overline{a} \cdot \overline{b}}(\overline{b} - \overline{a}), \end{aligned}$$

from which we may deduce that

$$|t(a) - t(b)| = \frac{r^2}{|a| \cdot |b|}|b - a|,$$

which is equivalent to the desired result. ■

The formula for distances in Lemma 1 plays a crucial role in our proof of Ptolemy's Theorem that the sum of the product of the lengths of opposite sides of a cyclic quadrilateral equals the product of the lengths of the diagonals.

Theorem 9 Ptolemy's Theorem
Let A, B, C, D be the vertices (in order round a circle) of a quadrilateral $ABCD$ inscribed in a circle. Then

$$AD \cdot BC + AB \cdot CD = AC \cdot BD.$$

Proof Let the transformation t be inversion of the figure $ABCD$ in the circle $\mathscr{C}$ with centre A and radius 1. This inversion maps A to ∞, the circle $\mathscr{C}$ to an extended line (which we will denote by ℓ), and the points B, C, D on $\mathscr{C}$ to points B', C', D' on ℓ.

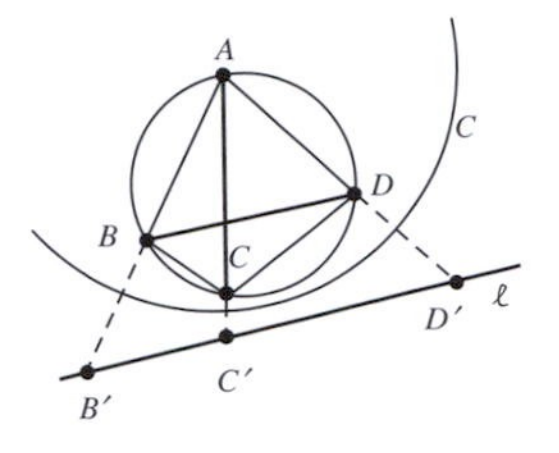

Since B, C, D are in order on $\mathscr{C}$, it follows that B', C', D' are in order on ℓ. Consequently,

$$B'C' + C'D' = B'D'. \tag{3}$$

Since t is self-inverse, the images of B', C', D' under inversion in $\mathscr{C}$ are B, C, D. Hence, if we apply the result of Lemma 1 in turn to the distances $B'C'$, $C'D'$, $B'D'$ under the inversion t, it follows from equation (3) that

$$\frac{BC}{AB \cdot AC} + \frac{CD}{AC \cdot AD} = \frac{BD}{AB \cdot AD}.$$

Multiplying both sides of this equation by the product $AB \cdot AC \cdot AD$, we obtain the desired result

$$AD \cdot BC + AB \cdot CD = AC \cdot BD. \quad \blacksquare$$

Our final application of inversion is particularly attractive visually.

Theorem 10 Shoemaker's Knife
Let D be a region in the upper half-plane, whose boundary consists of three semicircles S_1, S_2 and S_3 that meet at three points A, B and C on the x-axis, as shown. Then it is possible to fit a chain of circles into D between S_1 and S_2, with the first circle touching S_i, S_2 and S_3, and with each successive circle touching S_1, S_2 and the previous and following circles in the chain.

The name '*Shoemaker's Knife*' (or $\alpha\rho\beta\epsilon\lambda o\sigma$, *Arbelos*) is due to Archimedes.

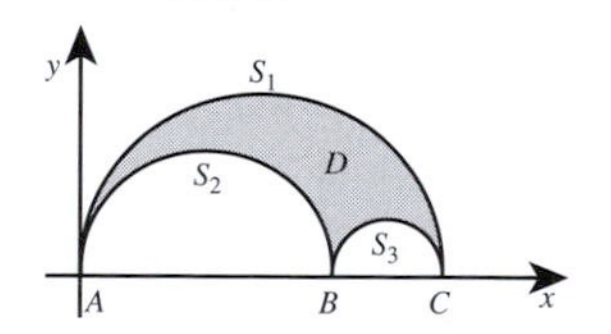

Proof Let t denote inversion in the circle $\mathscr{C}$ with centre A and radius 1.

Under t, the positive x-axis maps to itself, and points in the upper half-plane (the set of points above the x-axis) map to points in the upper half-plane.

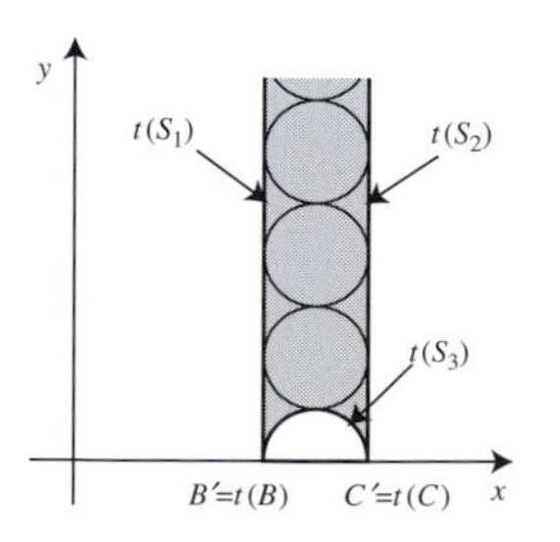

The circle of which S_3 is a portion is mapped by t to another circle, which crosses the x-axis at right angles, at points B' and C' say. The image of S_3 is therefore another semicircle in the upper half-plane, with endpoints B' and C'.

Under t, the origin A maps to ∞, so the semicircles S_1 and S_2 map to portions of (generalized) circles in the upper half-plane that join ∞ to B' and C', respectively. These portions do not intersect in the (ordinary) plane $\mathbb{C}$, and they meet the x-axis at right angles. Hence the images of S_1 and S_2 are two vertical half-lines with endpoints B' and C'.

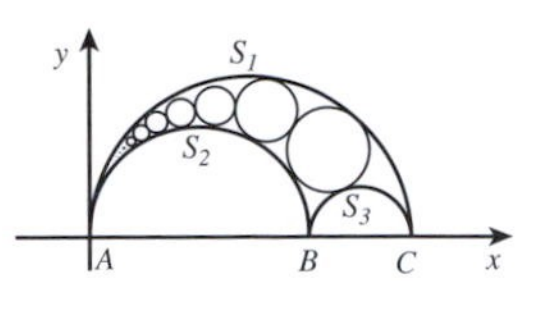

Now, it is obviously possible to fit a chain of circles into the vertical strip bounded by the semicircle $t(S_3)$ and the vertical lines $t(S_1)$ and $t(S_2)$, with the first circle touching $t(S_3)$ and with each successive circle touching $t(S_1)$, $t(S_2)$ and the previous and following circles in the chain.

We can transform this new configuration back into the original configuration by applying the inverse transformation t^{-1}. Since t^{-1} preserves circles and their tangencies, the required result now follows. ∎

Problem 3 Prove that the height above the x-axis of the centre of the n-th circle, C_n say, in the chain of circles constructed in Theorem 10 is n times the diameter of that circle.

Hint: Invert the figure in a circle with centre A that intersects C_n at right angles.

Problem 4 In the figure for Theorem 10, let A, B and C have coordinates (0,0), $(b, 0)$ and $(c, 0)$, respectively. Show that the centres of the circles in the chain constructed in Theorem 10 all lie on an ellipse with foci $F\left(\frac{1}{2}b, 0\right)$ and $F'\left(\frac{1}{2}c, 0\right)$, and major axis $\left[(0,0), \left(\frac{1}{2}(b+c), 0\right)\right]$.

Hint: Use the Sum of Focal Distances of the Ellipse theorem.

Theorem 5, Subsection 1.1.4

5.6 Exercises

Section 5.1

1. Determine the image under inversion in the unit circle $\mathscr{C}$ of each of the following points.
   (a) $(3, -4)$ (b) $(-1, 1)$ (c) $(9, 0)$ (d) $\left(\frac{1}{2}, -\frac{\sqrt{3}}{2}\right)$
2. Determine the image under inversion in the unit circle $\mathscr{C}$ of the following circles C (with the origin removed if it belongs to C):
   (a) the circle with centre $(3, -4)$ and radius 5;
   (b) the circle with centre $(1, 2)$ and radius 3.
3. Determine the image under inversion in the unit circle $\mathscr{C}$ of the following lines ℓ (with the origin removed if it belongs to ℓ):
   (a) the line $y + 3x = 5$;
   (b) the line $y + 2x = 0$.
4. Three circles C_1, C_2, C_3 pass through the origin O, and meet at three other distinct points D, E and F, as shown. The following sets of points are collinear: A, O and D; B, O and E; C, O and F. OA and OB are diameters of C_1 and C_2, respectively. Prove that OC is a diameter of C_3.
   Hint: Invert the figure in a unit circle with centre O, and then use the fact that the altitudes of a triangle are concurrent.

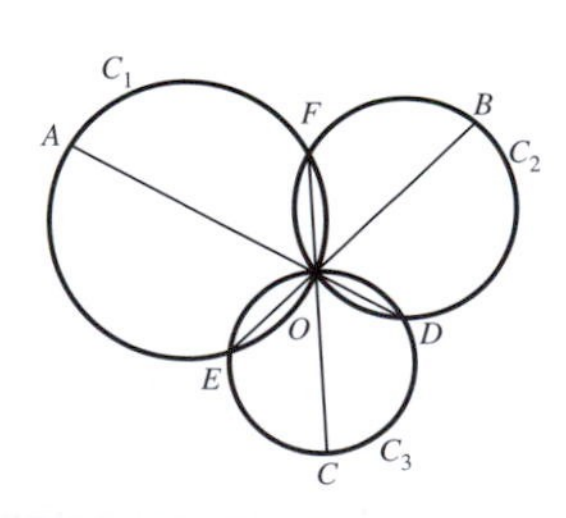

5. By means of a specific example, show that the centre of the image circle $t(C_2)$ of a circle C_2 under inversion t in another circle C_1 may not be the image of the centre of C_2 under t.
6. Let the origin O, P and Q be distinct points; and let P' and Q', respectively, be the images of P and Q under inversion in the circle with centre O and radius r. Prove that

$$P'Q' = \frac{r^2}{OP \cdot OQ} PQ.$$

(For simplicity, you may assume that O, Q, P are points inside the circle of inversion and occur in this order along the radius outwards from O. The result holds in all other cases too, by a similar argument.)
7. Let P and Q be distinct points in $\mathbb{C} - \{O\}$ with the origin O, P and Q not being collinear; and let P' and Q', respectively, be their images under inversion in the circle with centre O and radius r. Let ON and ON' be the perpendiculars from O to PQ and $P'Q'$ (extended, if necessary), respectively.
   (a) Prove that

$$P'Q' = \frac{r^2}{OP \cdot OQ} PQ.$$

   (b) Prove that

$$\frac{ON}{PQ} = \frac{ON'}{P'Q'}.$$

8. Let AB be a diameter of a circle C, and let chords AD and BE (extended, if necessary) of C intersect at a point F. Prove that the circle DEF is orthogonal to the circle C.

Section 5.2

1. Let t be the transformation defined by

$$t(z) = \tfrac{1}{2}(1 + \sqrt{3}i)z + 2i \quad (z \in \mathbb{C}).$$

   (a) Show that t represents an isometry.
   (b) Interpret t as the composite of a rotation and a translation.
   (c) Interpret t as a composite of reflections.
2. Determine the image under inversion in the unit circle $\mathscr{C}$ of each of the following points.
   (a) $-3 + 4i$ (b) $5 - 12i$
3. Let C be the circle of radius 2 with centre $1 + i$.
   (a) Write down the inversion in C of the extended complex plane, as a transformation of $\hat{\mathbb{C}}$.
   (b) Determine the image of i under the inversion in C.

4. Let t be the extended linear function defined by

$$t(z) = \begin{cases} -5iz + (2+6i), & \text{if } z \in \mathbb{C}, \\ \infty, & \text{if } z = \infty. \end{cases}$$

Express t as a composite of inversions of the extended complex plane.

5. Prove that two points z and z^* in the complex plane correspond to two diametrically-opposite points on the Riemann sphere if and only if $z \cdot \overline{(z^*)} = -1$.

Section 5.3

1. Which of the following formulas define a Möbius transformation? For each formula that does define a Möbius transformation, state the image of ∞ under M.
(a) $M(z) = -2\frac{z+1}{z-2}$ (b) $M(z) = -2z + \frac{1}{z-2}$ (c) $M(z) = \frac{2iz-2}{z+i}$
2. Decide which, if any, of the following matrices

$$\mathbf{A}_1 = \begin{pmatrix} -1 & -i \\ i & 1 \end{pmatrix}, \quad \mathbf{A}_2 = \begin{pmatrix} 1 & i \\ i & 1 \end{pmatrix}, \quad \mathbf{A}_3 = \begin{pmatrix} -i & i \\ 1 & -i \end{pmatrix},$$

are associated with each of the following Möbius transformations.
(a) $M_1(z) = \frac{z-1}{iz+1}$ (b) $M_2(z) = \frac{iz-1}{z-i}$ (c) $M_3(z) = \frac{-z-1}{z+i}$
3. Determine a formula for each of the following Möbius transformations, where M_1 and M_2 are the Möbius transformations defined in Exercise 2.
(a) $M_1 \circ M_1(z)$ (b) $M_1 \circ M_2(z)$ (c) $M_2 \circ M_1^{-1}(z)$
4. State which of the following transformations of $\hat{\mathbb{C}}$ onto itself are inversive transformations, giving a brief reason in each case.
(a) $t(z) = \frac{4\bar{z}-2}{3\bar{z}-2}$ (b) $t(z) = 5$ (c) $t(z) = \frac{i\bar{z}+2}{z-2i}$
5. Let t be the inversive transformation defined by

$$t(z) = \frac{z+i}{z-i}.$$

Determine the image of each of the following generalized circles under t:
(a) the extended line $\ell \cup \{\infty\}$, where ℓ is the line with equation $y = x$;
(b) the unit circle $\mathscr{C}$.

Section 5.4

1. For each set of three points below, determine the Möbius transformation that maps the three points to 0, 1, ∞, respectively.
(a) $i, -i, \infty$ (b) $\infty, 1, i$ (c) $2i, \infty, 3$ (d) $1, 2, 3$
2. Using the results of Exercise 1, determine the Möbius transformations that map:
(a) $1, 2, 3$ to $2i, \infty, 3$; (b) $\infty, 1, i$ to $i, -i, \infty$;
(c) $2i, \infty, 3$ to $\infty, 1, i$; (d) $i, -i, \infty$ to $1, 2, 3$.
3. Determine whether each of the following sets of points are collinear:
(a) $\frac{3}{2} + i, 2i, -6 + 6i$ (b) $1 + 2i, 4 - 5i, 10 - 20i$

4. Determine whether each of the following sets of points lies on a circle:
   (a) $i, -3, 1+2i, 2+4i$ (b) $1-2i, -4+i, 4+i, 1+6i$

Section 5.5

1. An Apollonian family of circles $\mathcal{F}$ includes (6,0) as a point circle and also the circle C with equation $x^2+y^2=4$. Determine the other point circle in the family.
2. The circles C_1 and C_2 in an Apollonian family of circles have the segments [0,8] and $\left[1, \frac{7}{2}\right]$, respectively, of the x-axis as diameters.
   (a) Determine the point circles of the family.
   (b) Hence determine the equation of the Apollonian circle in the family that passes through the point (1,1).
3. Determine the images under inversion in a unit circle with centre A of the following families $\mathcal{F}$ of coaxal circles.
   (a) $\mathcal{F}$ is the family of all generalized circles tangential to the y-axis at the origin, and $A=(0,0)$.
   (b) $\mathcal{F}$ is the family of all generalized circles through the points $A=(-1,0)$ and $B=(1,0)$.
   (c) $\mathcal{F}$ is the Apollonian family of all generalized circles with point circles $A=(-1,0)$ and $B=(1,0)$.
4. Let C_1 be the (extended) y-axis in the plane, and C_2 the circle with centre $(2,0)$ and radius 1. Find a sequence of inversions whose composite inversive transformation maps C_1 and C_2 onto concentric circles.
5. Let C_1 and C_2 be circles that touch at a point P, apart from which C_2 lies inside C_1. Prove that it is possible to fit a chain of circles between C_1 and C_2, with each circle touching C_1 and C_2 and two other circles in the chain, and with the first circle in the chain placed in any convenient position.
6. Prove that, if the Möbius transformation M maps $\mathscr{C}$, the unit circle with centre the origin, to itself, then M is necessarily of the form

$$M(z)=K\frac{z-\alpha}{\bar{\alpha}z-1}, \quad \text{where } |K|=1 \quad \text{and} \quad |\alpha|\neq 1.$$

7. Let $\mathcal{F}$ be an Apollonian family of circles with point circles A and B and radical axis ℓ. Prove that, for any given point P on the radical axis, all the tangents from P to circles in $\mathcal{F}$ are of equal length.
8. Let D be the region in the upper half-plane whose boundary consists of three semicircles S_1, S_2 and S_3 with diameter the segments $[-1, 0]$, $[0, 1]$ and $[-1, 1]$, respectively, of the x-axis, as shown.
   (a) Prove that it is possible to fit a chain of circles into D between S_1 and S_2, with the first circle touching S_1, S_2 and S_3, and with each successive circle touching S_1, S_2 and the previous and following circles in the chain.

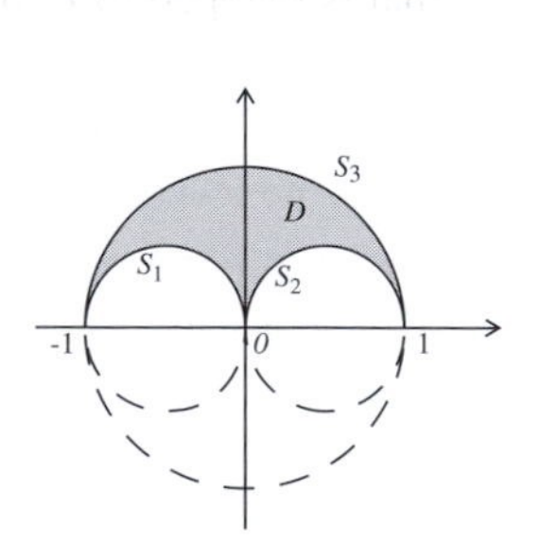

(b) Prove that the nth circle in the chain has its centre at a height $\frac{2n}{4n^2-1}$ above the x-axis, and has radius $\frac{1}{4n^2-1}$.

9. Let C_1, C_2, C_3 be three circles, external to each other, where the pairs C_1 and C_2, C_2 and C_3, and C_3 and C_1 each touch at a point. Prove that there are exactly two circles that touch all of C_1, C_2, C_3.
These circles are sometimes called *kissing circles* or *Soddy circles*, and were discovered by René Descartes (1643), then rediscovered by Jakob Steiner (1826) and Frederick Soddy (1936). Soddy published his discovery as a poem titled *The Kiss Precise*, in the scientific journal *Nature*; it starts:

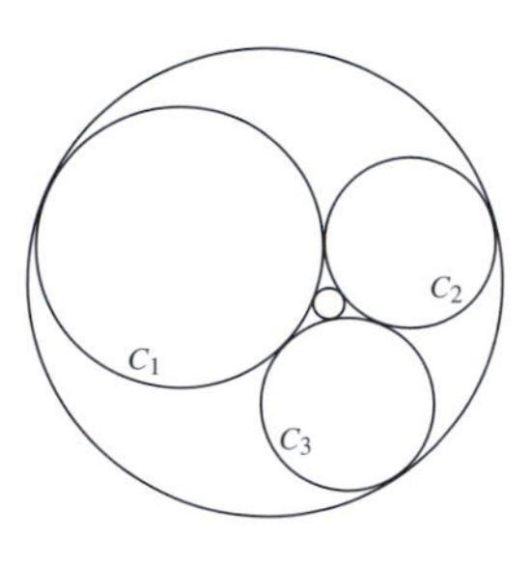

Frederick Soddy (1877–1956) won the 1921 Nobel Prize in chemistry for research into radioactive decay and the theory of isotopes.

For pairs of lips to kiss maybe
Involves no trigonometry.
'Tis not so when four circles kiss
Each one the other three.
To bring this off the four must be
As three in one or one in three.
If one in three, beyond a doubt
Each gets three kisses from without.
If three in one, then is that one
Thrice kissed internally.

Summary of Chapter 5

Section 5.1: Inversion

1. Under **reflection in a line** ℓ in the plane, a point A is mapped to an image point A' that lies an equal distance from ℓ, but on the opposite side of ℓ.
2. Let C be a circle in the plane, with centre O and radius r, and let A be any point other than O. If A' is the point on the line OA that lies on the same side of O as A and satisfies the equation $OA \cdot OA' = r^2$, then we call A' the **inverse** of A with respect to (or 'in') the circle C. The point O is called the **centre of inversion**, and C the **circle of inversion**. The transformation t defined by $t(A) = A'$ ($A \in \mathbb{R}^2 - \{O\}$) is known as **inversion in** C.
   There is no point to which O is mapped by the inversion, and no point that is mapped to O by the inversion.
   Inversion in a circle is a generalization of reflection in a line.
   We often use the term **inversion** to mean EITHER reflection in a line OR inversion in a circle.
3. If A is a point outside a circle C with centre O, AB and AC are the two tangents from A to C, and A' is the point of intersection of OA and BC, then A and A' are inverse points with respect to C.
4. Inversion in a circle C maps points outside C to points inside C, and vice-versa; points on C map to themselves.

Inversion in a circle is a one-one transformation of the plane minus the centre of inversion onto itself.

Inversion is a self-inverse transformation.

5. Inversion is not a Euclidean transformation; for example, it does not preserve lengths.

   Inversion is not an affine transformation; for example, it does not preserve straight lines.

6. Inversion in the unit circle $\mathscr{C}$ is the function $t : (x, y) \mapsto \left(\frac{x}{x^2+y^2}, \frac{y}{x^2+y^2}\right)$, where $(x, y) \in \mathbb{R}^2 - \{O\}$.

   Strategy To determine an equation for the image of a curve under inversion in the unit circle $\mathscr{C}$:

   1. write down an equation that relates the x- and y-coordinates of the points on the curve;
   2. replace x by $x/(x^2 + y^2)$ and y by $y/(x^2 + y^2)$, and simplify the resulting equation.

   If the curve passes through the origin we must first remove the origin from the curve. If we remove a point A from a curve, the curve is said to be **punctured** at A.

7. **Images of lines under inversion** Under inversion in a circle with centre O:

   (a) a line that does not pass through O maps onto a circle punctured at O;
   (b) a line punctured at O maps onto itself.

   Images of circles under inversion Under inversion in a circle with centre O:

   (a) a circle that does not pass through O maps onto a circle;
   (b) a circle punctured at O maps onto a line that does not pass through O.

   Even if inversion maps one circle onto another, it may not map centre to centre.

8. Let c_1 and c_2 be two curves that intersect at the point A, and let the tangents to the curves at A be ℓ_1 and ℓ_2, respectively. Then the **anti-clockwise angle** from c_1 to c_2 at A is the anti-clockwise angle from ℓ_1 to ℓ_2, and the **clockwise angle** from c_1 to c_2 at A is the clockwise angle from ℓ_1 to ℓ_2.

9. **Symmetry Lemma** Let ℓ be a line that does not pass through the point O. Then under inversion in a circle with centre O, ℓ maps to a circle C (punctured at O), and the tangent to C at O is parallel to ℓ.

10. **Angle Theorem** An inversion in any circle preserves the magnitude of angles between curves but reverses their orientation.

Section 5.2: Extending the Plane

1. The transformation $t(z) = z+c$ ($z \in \mathbb{C}$), where $c = a+ib$, is a **translation** through the vector (a, b).

   The transformation $t(z) = \bar{z}$ ($z \in \mathbb{C}$) is a **reflection** in the x-axis.

   The transformation $t(z) = az$ ($z \in \mathbb{C}$), where $|a| = 1$, $a = \cos\theta_0 + i\sin\theta_0$, is a **rotation** through an anti-clockwise angle θ_0 about the origin.

Each isometry t **of the plane** can be represented in the complex plane by one of the functions $t(z) = az + b$ or $t(z) = a\bar{z} + b$, where $a, b \in \mathbb{C}$, $|a| = 1$. Conversely, all such functions represent isometries.

Every isometry can be expressed as a composite of reflections.

2. The transformation $t(z) = kz$ $(z \in \mathbb{C})$, where k is real and positive, is a **scaling** by a **factor** k.
3. An inversion in a circle C of radius r with centre (a, b) may be represented in the complex plane by the transformation $t(z) = \frac{r^2}{\overline{z-c}} + c$ $(z \in \mathbb{C} - \{c\})$, where $c = a + ib$.

   In particular, inversion in the unit circle $\mathscr{C}$ may be represented by the transformation $t(z) = \frac{1}{\bar{z}}$ $(z \in \mathbb{C} - \{O\})$.
4. A **linear function** is a function of the form $t(z) = az + b$ $(z \in \mathbb{C})$, where $a, b \in \mathbb{C}$ and $a \neq 0$.

   It may be decomposed into a composite $t_2 \circ t_1$, where t_1 is the scaling $t_1(z) = |a|z$ and t_2 is the isometry $t_2(z) = (a/|a|)z + b$. This can be described geometrically as a scaling by the factor $|a|$, followed by a rotation through the angle $\operatorname{Arg}(a/|a|)$, followed by a translation through the vector $(\operatorname{Re} b, \operatorname{Im} b)$.

   Linear functions preserve angles, and map circles and lines to circles and lines.
5. The **reciprocal function** is the function of the form $t(z) = \frac{1}{z}$ $(z \in \mathbb{C} - \{O\})$.

   It may be decomposed into a composite $t_2 \circ t_1$, where t_1 is the inversion $t_1(z) = 1/\bar{z}$ and t_2 is the conjugation $t_2(z) = \bar{z}$.

   The reciprocal function preserves angles.
6. The **extended (complex) plane** $\hat{\mathbb{C}}$ is the union of the complex plane $\mathbb{C}$ and one extra point, the **point at infinity**, denoted by the symbol ∞.
7. An **extended line** is any line ℓ in the plane together with the point ∞. An extended line can be thought of as a circle of infinite radius.

   A **generalized circle** in the extended plane is a set that is either a circle or an extended line.
8. Let C be a generalized circle in the extended complex plane. Then an **inversion of the extended plane with respect to** C is a function t defined by one of the following rules.

   (a) If C is a circle of radius r with centre O, then

$$t(A) = \begin{cases} \text{the inverse of } A \text{ with respect to } C, & \\ & \text{if } A \in \mathbb{C} - \{O\}, \\ \infty, & \text{if } A = O, \\ O, & \text{if } A = \infty. \end{cases}$$

   (b) If C is an extended line $\ell \cup \{\infty\}$, then

$$t(A) = \begin{cases} \text{the reflection of } A \text{ in } \ell, & \text{if } A \in \mathbb{C}, \\ \infty, & \text{if } A = \infty. \end{cases}$$

Inversions of the extended plane map generalized circles onto generalized circles.

9. The **extended conjugation function**, **extended reciprocal function** and **extended linear functions** may be defined in the natural way as mappings of $\hat{\mathbb{C}}$ to itself.
10. The extended reciprocal function and the extended linear functions can be decomposed into a composite of inversions, and map generalized circles onto generalized circles.
11. **Strategy** To scale $\hat{\mathbb{C}}$ by a factor k:
    1. invert in the circle $\{z : |z| = 1\}$, then
    2. invert in the circle $\{z : |z| = \sqrt{k}\}$.

    Strategy To rotate $\hat{\mathbb{C}}$ through an angle θ:
    1. invert in the line $\{z : \text{Arg}\, z = 0\} \cup \{\infty\}$, then
    2. invert in the line $\left\{z : \text{Arg}\, z = \frac{1}{2}\theta\right\} \cup \{\infty\}$.

    Strategy To translate $\hat{\mathbb{C}}$ through a vector (a, b):
    1. invert in the line $\{(x, y) : ax + by = 0\} \cup \{\infty\}$, then
    2. invert in the line $\{(x, y) : ax + by = \frac{1}{2}(a^2 + b^2)\} \cup \{\infty\}$.
12. The **Riemann sphere** $\mathbb{S}$ is the sphere in $\mathbb{R}^3$ with centre the origin and radius 1. The points $N(0, 0, 1)$ and $S = (0, 0, -1)$ are called the **North Pole** and the **South Pole** of $\mathbb{S}$, respectively.

    Lines through the North Pole intersecting $\mathbb{S}$ at points $P'$and the (x, y)-plane at points P give a one–one correspondence between points on $\mathbb{S}$ and points in the plane. (We associate the North Pole with the point ∞ in the extended complex plane.)

    The one-one onto function $\pi : \mathbb{S} \to \hat{\mathbb{C}}$ which maps points on $\mathbb{S}$ to the associated points in the extended complex plane is called **stereographic projection**. If $(X, Y, Z) \in \mathbb{S}$ and $z = x + iy \in \hat{\mathbb{C}}$, then

    $$\pi(X, Y, Z) = \frac{X}{1 - Z} + i\frac{Y}{1 - Z}$$

    and

    $$\pi^{-1}(x + iy) = \left(\frac{2x}{x^2 + y^2 + 1}, \frac{2y}{x^2 + y^2 + 1}, \frac{x^2 + y^2 - 1}{x^2 + y^2 + 1}\right).$$
13. Under stereographic projection, circles on the Riemann sphere map onto generalized circles in $\hat{\mathbb{C}}$. In particular, circles on the sphere that pass through N map onto extended lines in $\hat{\mathbb{C}}$, and circles on the sphere that do not pass through N map onto ordinary circles in $\hat{\mathbb{C}}$.

    Stereographic projection preserves the magnitude of angles.

Section 5.3: Inversive Geometry

1. A transformation $t : \hat{\mathbb{C}} \to \hat{\mathbb{C}}$ is an **inversive transformation** if it can be expressed as a composite of inversions.

    Inversive geometry is the study of those properties of figures in $\hat{\mathbb{C}}$ that are preserved by inversive transformations.

The extended reciprocal function and the extended linear functions are inversive transformations.

2. Inversive transformations preserve the magnitude of angles, map generalized circles to generalized circles, and preserve tangency.

   The set of inversive transformations forms a group under the operation of composition of functions, called the **inversive group**.

3. Every Euclidean transformation is also an inversive transformation, and every Euclidean property is also an inversive property.

   The 'doubling map' $t(z) = 2z$ ($z \in \mathbb{C}$) is an affine transformation and an inversive transformation, but is not a Euclidean transformation.

   The transformation of $\mathbb{R}^2$ to itself given by $t(\mathbf{x}) = \begin{pmatrix} 2 & 0 \\ 0 & 1 \end{pmatrix} \mathbf{x}$ represents a horizontal shear; it is an affine transformation, but is not an inversive transformation.

4. A **Möbius transformation** is a function $t : \hat{\mathbb{C}} \to \hat{\mathbb{C}}$ of the form $M(z) = \frac{az+b}{cz+d}$, where $a, b, c, d \in \mathbb{C}$, and $ad - bc \neq 0$.

   If $c = 0$, we adopt the convention that $M(\infty) = \infty$; otherwise we adopt the convention that $M(-d/c) = \infty$ and $M(\infty) = a/c$.

   The extended linear functions and the extended reciprocal function are Möbius transformations.

   Every Möbius transformation is an inversive transformation.

5. Möbius transformations preserve the magnitude and orientation of angles, and map generalized circles onto generalized circles.

6. Let M be a Möbius transformation defined by $M(z) = \frac{az+b}{cz+d}$, where a, b, c, $d \in \mathbb{C}$, and $ad - bc \neq 0$; then $\mathbf{A} = \begin{pmatrix} a & b \\ c & d \end{pmatrix}$ is a matrix **associated** with M.

   A matrix associated with a Möbius transformation is invertible.

   If $\mathbf{A}$ is a matrix associated with a Möbius transformation M, then so is $c\mathbf{A}$ for any non-zero $c \in \mathbb{C}$. Every matrix associated with M has the form $c\mathbf{A}$ for some $c \in \mathbb{C} - \{O\}$.

7. Let M_1 and M_2 be Möbius transformations with associated matrices $\mathbf{A}_1$ and $\mathbf{A}_2$, respectively. Then $M_1 \circ M_2$ is a Möbius transformation with associated matrix $\mathbf{A}_1\mathbf{A}_2$.

   Strategy To compose two Möbius transformations M_1 and M_2:
   1. write down matrices $\mathbf{A}_1$ and $\mathbf{A}_2$ associated with M_1 and M_2;
   2. calculate $\mathbf{A}_1\mathbf{A}_2$;
   3. write down the Möbius transformation $M_1 \circ M_2$ with which $\mathbf{A}_1\mathbf{A}_2$ is associated.

8. The inverse of the Möbius transformation $M(z) = \frac{az+b}{cz+d}$ is also a Möbius transformation, and it may be written in the form $M^{-1}(z) = \frac{dz-b}{-cz+a}$.

9. The set of all Möbius transformations forms a group under composition of functions.

10. Every inversion t has the form $t(z) = M(\bar{z})$, where M is a Möbius transformation.

Every inversive transformation t can be represented in $\hat{\mathbb{C}}$ by one of the formulas $t(z) = \frac{az+b}{cz+d}$ or $t(z) = \frac{a\bar{z}+b}{c\bar{z}+d}$, where $a, b, c, d \in \mathbb{C}$, and $ad - bc \neq 0$.

The composite of an even number of inversions is a Möbius transformation and is called a **direct** inversive transformation (since it preserves the orientation of angles); the composite of an odd number of inversions is called an **indirect** inversive transformation (since it reverses the orientation of angles).

11. **Strategy** To determine the image of a generalized circle C under an inversive transformation t:
    1. write down three points z_1, z_2, z_3 on C;
    2. determine the images $t(z_1), t(z_2), t(z_3)$;
    3. the image $t(C)$ is the (unique) generalized circle through $t(z_1)$, $t(z_2)$, $t(z_3)$.

Section 5.4: Fundamental Theorem of Inversive Geometry

1. **Fundamental Theorem of Inversive Geometry**
   Let z_1, z_2, z_3 and w_1, w_2, w_3 be two sets of three points in the extended complex plane $\hat{\mathbb{C}}$. Then there is a unique Möbius transformation M which maps z_1 to w_1, z_2 to w_2, and z_3 to w_3.
2. **Strategy** To determine the Möbius transformation M which maps three given points z_1, z_2, z_3 onto the points 0, 1 and ∞, respectively:
    1. choose the appropriate form of mapping from the following formulas for M:

mapping	form of M
$z_1, z_2, z_3 \mapsto 0, 1, \infty$	$K\frac{z-z_1}{z-z_3}$
$\infty, z_2, z_3 \mapsto 0, 1, \infty$	$\frac{K}{z-z_3}$
$z_1, \infty, z_3 \mapsto 0, 1, \infty$	$\frac{z-z_1}{z-z_3}$
$z_1, z_2, \infty \mapsto 0, 1, \infty$	$K(z-z_1)$

    2. find the complex number K for which $M(z_2) = 1$.
3. **Strategy** To determine the Möbius transformation M which maps the points z_1, z_2, z_3 to the points w_1, w_2, w_3, respectively:
    1. find the Möbius transformation M_1 which maps the points z_1, z_2, z_3 to the points $0, 1, \infty$, respectively;
    2. find the Möbius transformation M_2 which maps the points w_1, w_2, w_3 to the points $0, 1, \infty$, respectively;
    3. calculate $M = M_2^{-1} \circ M_1$.
4. **Strategy** To determine whether z_1, z_2, z_3 and z_4 lie on a circle:
    1. find the Möbius transformation M which maps z_1, z_2, z_3 to $0, 1, \infty$, respectively;
    2. the points z_1, z_2, z_3, z_4 lie on a generalized circle if and only if $M(z_4)$ is real;

3. the generalized circle in Step 2 is a circle provided that $M(\infty)$ is not real.

If $M(z_4)$ and $M(\infty)$ are both real, then z_1, z_2, z_3, z_4 lie on a line.

5. Let C_1 and C_2 be generalized circles in the extended complex plane. Then there is a Möbius transformation that maps C_1 onto C_2.

   All generalized circles are inversive-congruent.

Section 5.5: Coaxal Families of Circles

1. **Apollonian Circles Theorem** Let A and B be two distinct points in the plane, and let k be a positive real number other than 1. Then the locus of points P that satisfy $PA : PB = k : 1$ is a circle whose centre lies on the line through A and B.

   The centre of the circle is not at either A or B. If $A = (-a, 0)$ and $B = (a, 0)$, then the circle has centre c and radius r where $c = \left(-a\left(\frac{1+k^2}{1-k^2}\right), 0\right)$ and $r = \frac{2ak}{|1-k^2|}$.

   If $k = 1$, then P lies on the perpendicular bisector of AB.

   Every positive value of k gives rise to a generalized circle, known as a **circle of Apollonius**; the family of all such circles is known as the **Apollonian family of circles** defined by the points A and B.

   If $0 < k < 1$, the Apollonian circle surrounds A; if $k > 1$, the Apollonian circle surrounds B. The points A and B are called **point circles**, corresponding to the cases $k = 0$ and '$k = \infty$', respectively. Every point in the plane lies on precisely one circle (or generalized circle) in each Apollonian family.

2. Let A and B be distinct points in the plane, and let t be the inversion in the circle with centre A and radius 1. Then the Apollonian family of circles defined by the point circles A and B is mapped by t to the family of all concentric circles with centre $t(B)$; and the family of all concentric circles with centre $t(B)$ is mapped by t to the Apollonian family of circles defined by the point circles A and B.

3. A **coaxal family of circles** in the plane is a family of (generalized) circles of one of the following types:

   1. an Apollonian family, with particular point circles;
   2. a family that intersect at one particular point;
   3. a family that intersect at two particular points.

   The extended line in each family is called the **radical axis** of the family.

   Coaxal Circles Theorem Let A and B be distinct points in the plane. Let $\mathscr{F}$ be the Apollonian family defined by the point circles A and B, and let $\mathscr{G}$ be the family of all generalized circles through A and B. Then every member of $\mathscr{F}$ is orthogonal to every member of $\mathscr{G}$.

4. Two points A and B in the extended complex plane are inverse points with respect to a generalized circle C if and only if every generalized circle through A and B meets C at right angles.

If C is an Apollonian circle defined by the point circles A and B, then A and B are inverse points with respect to C.

Let A and B be inverse points with respect to a generalized circle C, and let t be an inversive transformation. Then $t(A)$ and $t(B)$ are inverse points with respect to the generalized circle $t(C)$.

Let $\mathscr{F}$ be the family of all generalized circles that have A and B as inverse points. Then $\mathscr{F}$ is either a concentric family of circles with centre A or B, or an Apollonian family of circles with point circles A and B.

Let C and C_1 be circles that meet at right angles at two points R and S. Then the centre of C_1 is the point T of intersection of the tangents to C at R and S; and C_1 maps to itself under inversion in C.

5. **Concentricity Theorem** Let C_1 and C_2 be any two non-intersecting circles in the plane. Then there is an inversive transformation that maps C_1 and C_2 onto a pair of concentric circles.
6. **Two Apollonian Circles Theorem** Let C_1 and C_2 be two non-concentric circles that do not intersect. Then there is a unique Apollonian family of circles that contains C_1 and C_2.
7. A **porism** is a mathematical construction problem that has a surprising answer: either the construction cannot be carried out or it has infinitely many solutions.

   Steiner's Porism Let C_1 and C_2 be non-intersecting circles, with C_1 inside C_2. Then EITHER it is impossible to fit a chain of circles between C_1 and C_2, with each circle touching C_1 and C_2 and two other circles in the chain OR it is possible to construct such a chain, and the first circle in the chain can be placed in any convenient position.

   Poncelet's Porism If C_1 and C_2 are any two plane conics for which (for any given $n > 2$) it is possible to find one $n-$sided polygon which is simultaneously inscribed in C_1 and circumscribed around C_2, then it is possible to find infinitely many of them.
7. Let a and b be points in $\mathbb{C} - \{O\}$. Then, under inversion in a circle C with centre O and radius r the distance between the images of a and b is $\frac{r^2}{|a|\cdot|b|}|a-b|$.

   Ptolemy's Theorem Let A,B,C,D be the vertices (in order) of a quadrilateral inscribed in a circle. Then $AD \cdot BC + AB \cdot CD = AC \cdot BD$.
8. **Shoemaker's Knife/Arbelos** Let D be a region in the upper half-plane, whose boundary consists of three semicircles S_1, S_2 and S_3 that meet at three points A, B and C on the x-axis. Then it is possible to fit a chain of circles into D between S_1 and S_2, with the first circle touching S_1, S_2 and S_3, and with each successive circle touching S_1, S_2 and the previous and following circles in the chain.

6 Hyperbolic Geometry: the Poincaré Model

The book, Euclid's *Elements*, which underlay most geometrical teaching in Western Europe for over 2000 years, gave definitions of the basic terms in geometry and rules (called postulates) for their use. Many of Euclid's assumptions seem entirely uncontroversial, such as the assertions that through any two distinct points in a plane or in space there passes a unique line, and that a line can be extended indefinitely in both directions. On these foundations Euclid gave rigorous proofs of theorems in elementary geometry, which could be accepted as true because of the way they had been established.

Among the postulates for Euclidean geometry is one about parallel lines which is equivalent to the following statement.

The (Euclidean) Parallel Postulate Given any line ℓ and a point P not on ℓ, there is a unique line m in the same plane as P and ℓ which passes through P and does not meet ℓ.

Note that the parallel postulate makes two assertions: first that the parallel line exists, and second that it is unique.

Historically it was felt that the Parallel Postulate as it stands is not obvious. It implies that every line through P other than m eventually meets ℓ, but plainly this meeting place can be a long way away. How much confidence would you place in an assumption about lines meeting somewhere in the Virgo star cluster, some 100 million light years away?

Many Greek, Arab and later Western geometers had felt that the answer was 'not much', and had tried to delete the Euclidean Parallel Postulate from the list and instead derive it as a theorem. They all failed, and in the end the reason was laid bare: the Parallel Postulate cannot be made into a theorem in this way because there are internally consistent models of geometries which obey all the Euclidean postulates except the Parallel Postulate.

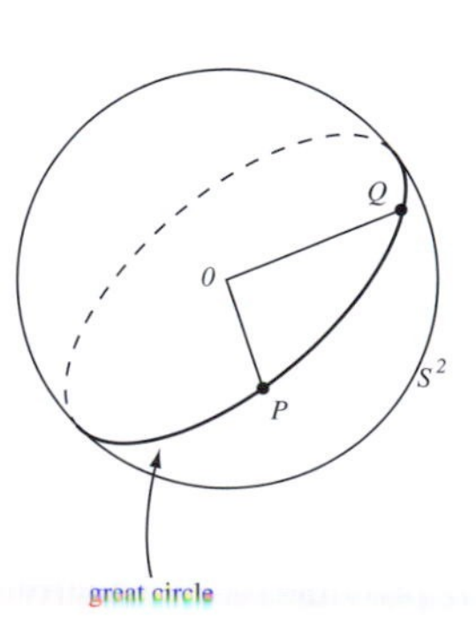

But how do we obtain geometries other than Euclidean geometry with its Parallel Postulate? One way to do this might be to insist that any two lines intersect. The geometry to which this gives rise is called *Elliptic Geometry*. One model of elliptic geometry is *Spherical Geometry* (for example, on the surface of the Earth), and we shall investigate this model in Chapter 7.

A 'line' PQ on the surface of a sphere (which we call S^2), such as on the Earth's surface, can be defined to be the *great circle* (that is, the intersection of the plane through the centre O of the sphere with the surface of the sphere) through the two points P and Q. Any two distinct great circles meet in two (diametrically opposite) points, such as P and P' shown in the margin, so any two lines in spherical geometry meet in two points.

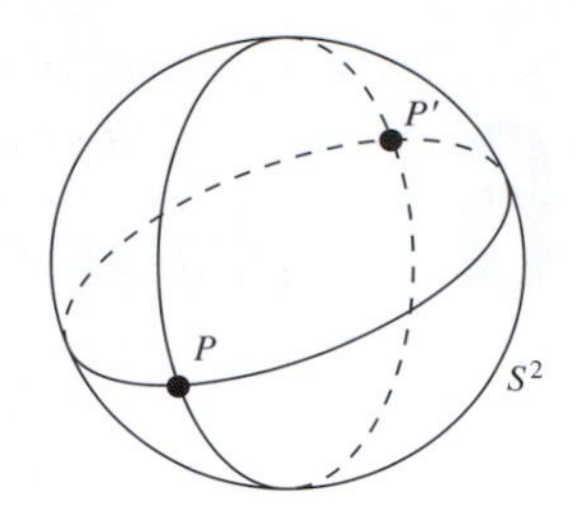

Generalizing this fact, in *Elliptic Geometry* we keep all the other postulates that collectively describe Euclidean geometry, remove the Parallel Postulate, and replace it with the following analogue.

The (Elliptic) Parallel Postulate Given any line ℓ and a point P not on ℓ, all lines through P meet ℓ. (That is, there are no lines through P that are parallel to ℓ.)

Another way to obtain a non-Euclidean geometry is to insist that the line through P that does not meet ℓ is not unique. We could again keep all the other postulates that collectively describe Euclidean geometry, remove the Parallel Postulate, and replace it with the following.

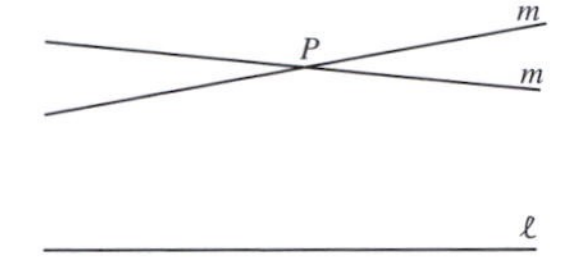

The (Hyperbolic) Parallel Postulate Given any line ℓ and a point P not on ℓ, there are at least two lines m through P that do not meet ℓ. (That is, there are at least two lines through P that are parallel to ℓ.)

The geometry whose postulates are those of Euclidean geometry but with this variant of the Euclidean parallel postulate is called *Hyperbolic Geometry*. The term *non-Euclidean Geometry* is often used to describe elliptic and hyperbolic geometries together.

In fact there are still other geometries (and other models of both hyperbolic and elliptic geometries), that lie beyond the scope of this book.

Non-Euclidean geometry was discovered independently in the late 1820s by the distinguished Russian mathematician Nicolai Ivanovich Lobachevskii (1792–1856), working at the University of Kazan, and the Hungarian János Bolyai (1802–1860), an Army officer whose father, Farkas (or Wolfgang) (1775–1856), was also a mathematician. Many of the same ideas were also known to Carl Friedrich Gauss (1777–1855), but he did not publish them. However, non-Euclidean geometry was not accepted until after these three were dead, when the German Bernhard Riemann (1826–1866) and the Italian Eugenio Beltrami (1835–1900) published their ideas about geometry in the late 1860s. Riemann gave the appropriate general setting, Beltrami a specific account of non-Euclidean geometry. Only then did the new geometry become rapidly accepted. Amongst other things, it spurred Felix Klein (1849–1925) to propose his view of geometry.

Non-Euclidean geometry was one of the most momentous mathematical discoveries of the 19th century. It had several revolutionary implications, because it provided a physically plausible description of space that differed markedly from Euclid's. It became possible to imagine that the universe is not Euclidean, and in the 19th century some mathematicians and astronomers, such as Gauss and his friend Bessel), entertained the idea quite seriously.

Friedrich Wilhelm Bessel (1784–1846) was a German mathematician and astronomer, after whom Bessel functions are named.

The theorems in non-Euclidean geometries often differ markedly from their Euclidean equivalents. For example, in Euclidean geometry the angles of any triangle sum to π; in elliptic geometry the angle sum of any triangle turns out to be strictly greater than π, whereas in hyperbolic geometry the angle sum of any triangle turns out to be strictly less than π.

In this chapter we shall concentrate on a model of hyperbolic geometry due to the French mathematician Henri Poincaré. In this model (which we will generally call 'hyperbolic geometry' for simplicity), the space of points is the interior of the unit disc $\mathscr{D} = \{z : |z| < 1\}$. All figures in this geometry will be drawn as they appear in this disc. In Section 6.1 we define *hyperbolic lines* and *hyperbolic angles*. We then introduce *hyperbolic reflections* in these lines, and obtain the *group of hyperbolic transformations*: a hyperbolic transformation is a composite of hyperbolic reflections. In Section 6.2 we see that every hyperbolic transformation can be described as either a Möbius transformation (of a certain form) or a Möbius transformation composed with complex conjugation.

You met Möbius transformations in Subsection 5.3.3.

The properties of hyperbolic geometry include angle and length, and so we can define *hyperbolic circles*, which we do in Section 6.3. We also obtain a formula relating Euclidean lengths and *hyperbolic lengths* (at least in special cases), and in Section 6.4 we give useful formulas for the study of hyperbolic triangles. In Section 6.5 we look at *area* in hyperbolic geometry, and at hyperbolic tilings, or *tessellations*, such as the one displayed on the cover of this book. Finally, in Section 6.6 we briefly discuss the half-plane model of hyperbolic geometry.

6.1 Hyperbolic Geometry: the Disc Model

6.1.1 What is Hyperbolic Geometry?

Note that we shall use the alternative notations z and (x, y) freely to describe points in $\mathscr{D}$, according to convenience.

The **points** or ***d*-points** (where '*d*' stands for 'disc') in (Poincaré's version of) hyperbolic geometry consist of the points in the unit disc

$$\mathscr{D} = \{z : |z| < 1\} = \left\{(x, y) : x^2 + y^2 < 1\right\}.$$

In sketches we usually illustrate $\mathscr{D}$ by drawing in its boundary

$$\mathscr{C} = \{z : |z| = 1\} = \left\{(x, y) : x^2 + y^2 = 1\right\},$$

but you should note that points on $\mathscr{C}$ are *not* points which belong to the geometry.

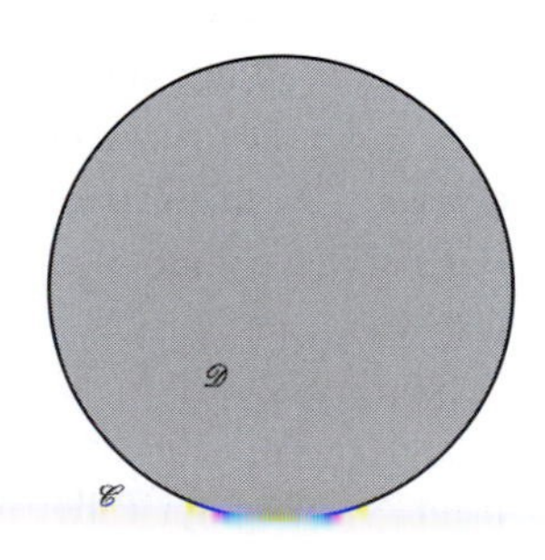

A crucial feature of this geometry is the concept of a *d-line,* or a *line in hyperbolic geometry.*

Definition A ***d*-line** is that part of a (Euclidean) generalized circle which meets $\mathscr{C}$ at right angles and which lies in $\mathscr{D}$.

A generalized circle is either a circle or a (Euclidean) line; see Subsection 5.2.3.

Every d-line is part of a generalized (Euclidean) circle which meets the boundary circle in two points. We call these two points the **boundary points** of the d-line. Notice that the boundary points of a d-line are not d-points – for they are not points in $\mathscr{D}$.

The sketch below illustrates various d-lines. Some of these d-lines are arcs of (Euclidean) circles; some are (Euclidean) line segments, in fact, diameters of $\mathscr{D}$.

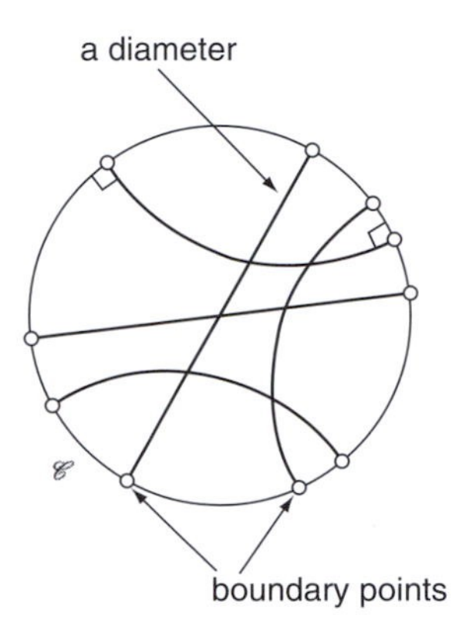

You may find the following figure a helpful reminder for the shapes of d-lines:

In fact, any d-line that is part of a Euclidean line must be a diameter of $\mathscr{D}$, since otherwise it cannot meet $\mathscr{C}$ at right angles.

Conversely, any d-line ℓ that is an arc of a Euclidean circle cannot pass through the origin. For if ℓ meets $\mathscr{C}$ at a point P, then the tangent ℓ' to ℓ at P is a radius of $\mathscr{C}$ and so passes through the origin O, which is the centre of $\mathscr{C}$. But then ℓ' is a line touching the Euclidean circle ℓ at P, and because circles lie entirely on one side of their tangents, ℓ' cannot meet ℓ again. Thus ℓ cannot pass through O.

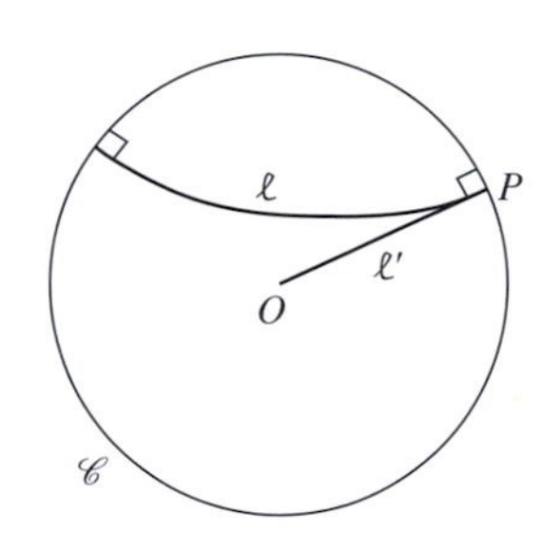

Indeed, if ℓ is a d-line that is an arc of a Euclidean circle, then there is a diameter of $\mathscr{D}$ that does not meet it. For, if ℓ meets $\mathscr{C}$ at the points P and Q, then the tangents to ℓ at P and Q are radii of $\mathscr{C}$ and so pass through the origin. They divide the disc $\mathscr{D}$ into four regions, one of which contains ℓ. Clearly, a diameter can be drawn that lies entirely in two of the other regions – in fact, you can see that there are infinitely many such diameters. Each of these divides $\mathscr{D}$ into two regions, one which contains ℓ and one which does not.

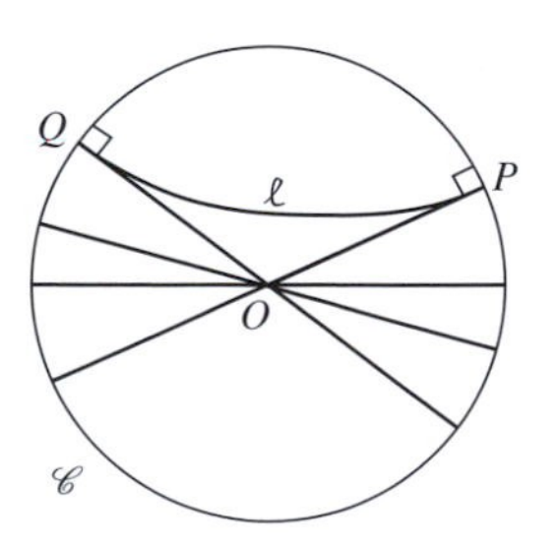

Notice also that if a d-line ℓ is part of a Euclidean circle meeting $\mathscr{C}$ at points P and Q, then its centre is the point R where the tangents to $\mathscr{C}$ at P and Q meet. So the point R lies outside the circle $\mathscr{C}$.

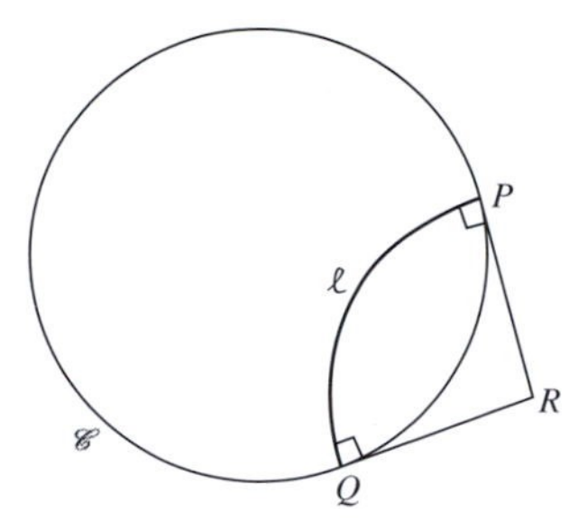

Sometimes it is clear from a (reasonably accurate) sketch whether or not part of a generalized circle is a d-line. For example, consider

$$\ell = \left\{(x, y) : x^2 + y^2 - 4y = 0\right\} \cap \mathscr{D}.$$

Since

$$x^2 + y^2 - 4y = 0 \Leftrightarrow x^2 + (y-2)^2 = 4,$$

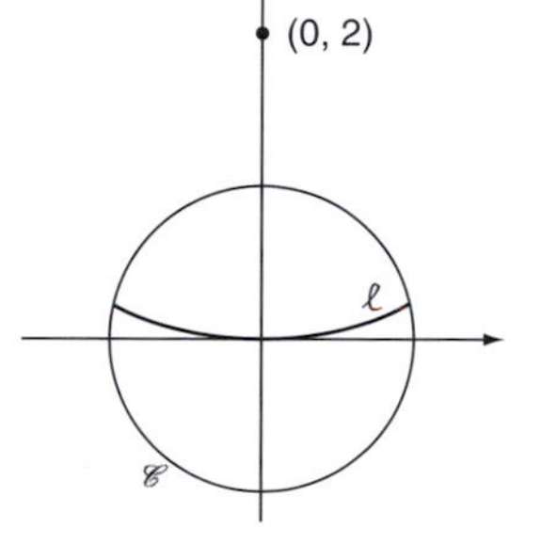

we can sketch ℓ, and deduce that it is not a d-line since it clearly does not meet $\mathscr{C}$ right angles. (The tangents to ℓ at its boundary points on $\mathscr{C}$ do not pass through the origin.)

Problem 1 Sketch the following parts of generalized circles, and determine which of them are d-lines.

$$\ell_1 = \{(x, y)\colon y = 3x\} \cap \mathscr{D}$$

$$\ell_2 = \{(x, y)\colon 3x + y = 1\} \cap \mathscr{D}$$

$$\ell_3 = \left\{(x, y)\colon x^2 + y^2 + 2x + 2y + 1 = 0\right\} \cap \mathscr{D}$$

Now, it is rather easy to be mislead by a sketch into believing that something is true in general whereas it is not! So, we now give an analytic criterion for identifying d-lines.

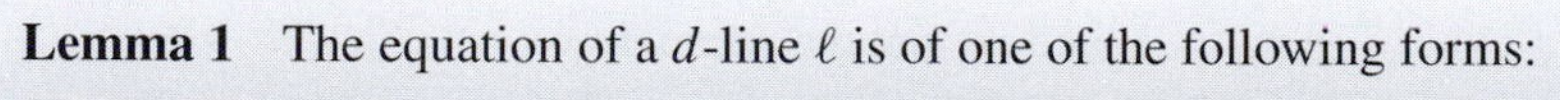

Lemma 1 The equation of a d-line ℓ is of one of the following forms:

$ax + by = 0$, where a and b are not both zero;

$x^2 + y^2 + fx + gy + 1 = 0$, where $f^2 + g^2 > 4$.

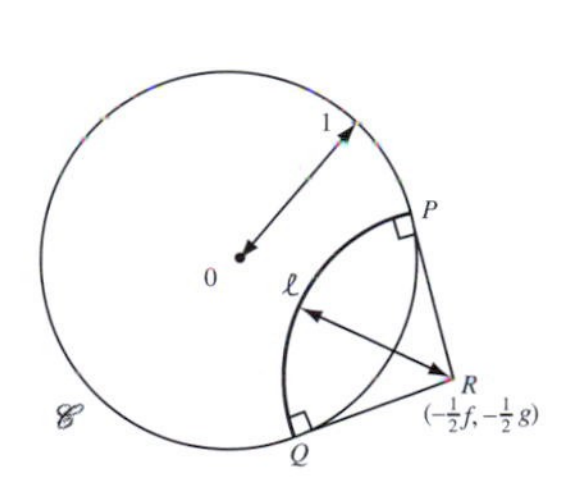

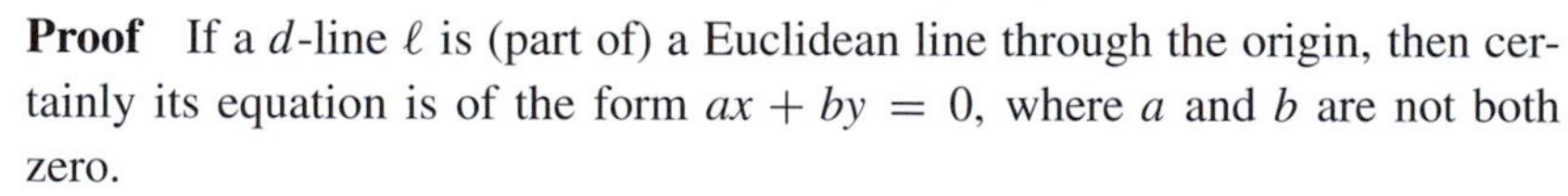

Proof If a d-line ℓ is (part of) a Euclidean line through the origin, then certainly its equation is of the form $ax + by = 0$, where a and b are not both zero.

The other possibility is that the d-line ℓ is (part of) a Euclidean circle C that intersects the boundary $\mathscr{C}$ of $\mathscr{D}$ at right angles. So, let C have equation

$$x^2 + y^2 + fx + gy + h = 0, \text{ for some real numbers } f,\ g \text{ and } h.$$

First, it follows from the Orthogonality Test that any circle with this equation intersects the unit circle $\mathscr{C}$ with equation

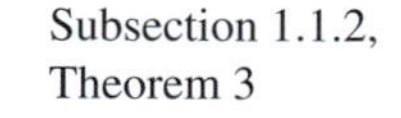

Subsection 1.1.2, Theorem 3

$$x^2 + y^2 - 1 = 0$$

if and only if

$$f \cdot 0 + g \cdot 0 = 2 \cdot (h + (-1));$$

that is, if and only if $h = 1$. So the equation of C must be of the form

$$x^2 + y^2 + fx + gy + 1 = 0, \text{ for some } f \text{ and } g.$$

Next, the centre $\left(-\frac{1}{2}f, -\frac{1}{2}g\right)$ of C must be outside $\mathscr{C}$, so that $\left(-\frac{1}{2}f\right)^2 + \left(-\frac{1}{2}g\right)^2 > 1$, which we can equivalently write in the form $f^2 + g^2 > 4$.

Finally, we need the circle C to meet $\mathscr{D}$. This happens if and only if

$$1 + \text{radius of } C > \text{distance of the centre of } C \text{ from } O;$$

since $h = 1$ we can write this requirement in the form

$$1 + \sqrt{\frac{1}{4}f^2 + \frac{1}{4}g^2 - 1} > \sqrt{\frac{1}{4}f^2 + \frac{1}{4}g^2}. \qquad (1)$$

Since we already know that $\sqrt{\frac{1}{4}f^2 + \frac{1}{4}g^2} > 1$, the inequality (1) follows at once if we can prove that

$$1 + \sqrt{t - 1} > \sqrt{t}, \quad \text{whenever } t > 1. \qquad (2)$$

Here we are writing t in place of $\frac{1}{4}f^2 + \frac{1}{4}g^2$.

So, let $F(t) = 1 + \sqrt{t-1} - \sqrt{t}$. Then

$$\begin{aligned} F(1) &= 1 + \sqrt{0} - \sqrt{1} \\ &= 0, \end{aligned}$$

and

$$\begin{aligned} F'(t) &= \frac{1}{2\sqrt{t-1}} - \frac{1}{2\sqrt{t}} \\ &= \frac{\sqrt{t} - \sqrt{t-1}}{2\sqrt{t-1}\sqrt{t}} > 0. \end{aligned}$$

It follows at once that $F(t) > F(1) = 0$ for $t > 1$, and so that the inequality (2) holds, as required. ∎

Thus, for example, let C be the circle in $\mathbb{R}^2$ with equation

$$x^2 + y^2 + 3x - 2y + 1 = 0,$$

In fact, C is the circle with centre $\left(-\frac{3}{2}, 1\right)$ and radius $\sqrt{\left(-\frac{3}{2}\right)^2 + (-1)^2 - 1} = \frac{3}{2}$.

and let ℓ be the part of C that lies in $\mathscr{D}$. Then ℓ is a d-line since its equation is of the form required by Lemma 1, with $f = 3$ and $g = -2$, and $f^2 + g^2 = 3^2 + (-2)^2 = 13 > 4$.

Note the following useful result that we obtain from putting $a = -\frac{1}{2}f$, $b = -\frac{1}{2}g$ into Lemma 1.

Corollary 1 Let $\alpha = a + ib$ be a point outside the unit circle $\mathscr{C}$. Then the d-line that is part of a Euclidean circle with centre α has equation

$$x^2 + y^2 - 2ax - 2by + 1 = 0.$$

We can now turn to the question of parallelism, the notion that gave rise to the discovery of non-Euclidean geometries in the first place.

Definition Two d-lines that do not meet in $\mathscr{D}$ are:
parallel if the generalized Euclidean circles of which they are parts meet at a point on $\mathscr{C}$;
ultra-parallel if the generalized Euclidean circles of which they are parts do not meet on $\mathscr{C}$.

Recall that the boundary $\mathscr{C}$ is *not* part of the unit disc $\mathscr{D}$.

Remarks

1. It follows that given any d-line ℓ and any point P in $\mathscr{D}$ which is not on ℓ, there exist exactly two d-lines through P which are parallel to ℓ, as the next figure shows. A rigorous proof that such d-lines exist will be found below (Problem 5).

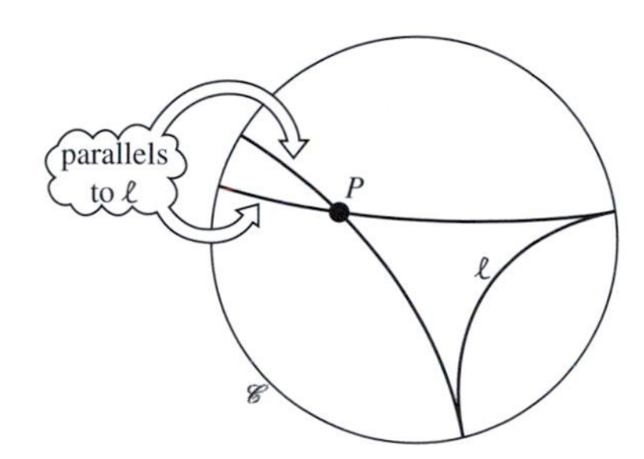

2. Similarly, corresponding to any given d-line ℓ and any point P in $\mathscr{D}$ which is not on ℓ, there exist infinitely many d-lines through P which are ultra-parallel to ℓ.

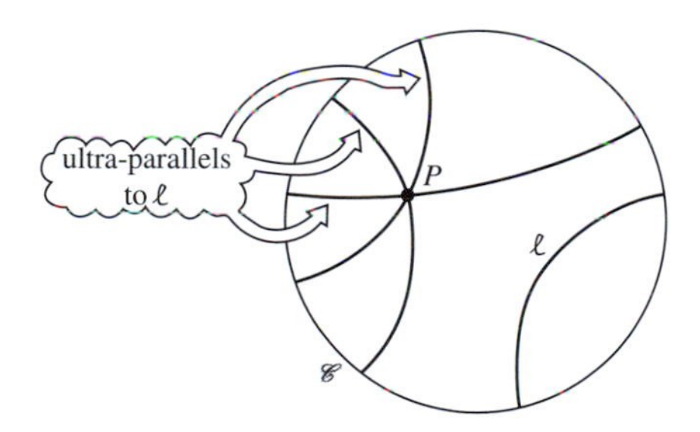

Problem 2 The following sets represent d-lines:

$$\ell_1 = \{(x, y) : y = x\} \cap \mathscr{D};$$
$$\ell_2 = \left\{(x, y) : x^2 + y^2 - 4x + 1 = 0\right\} \cap \mathscr{D};$$
$$\ell_3 = \left\{(x, y) : x^2 + y^2 - 2\sqrt{2}x + 1 = 0\right\} \cap \mathscr{D};$$
$$\ell_4 = \left\{(x, y) : x^2 + y^2 + 2x + 2y + 1 = 0\right\} \cap \mathscr{D}.$$

Sketch them, and decide from your figure which of the d-lines intersect each other, which are parallel, and which are ultra-parallel.

Problem 3 Sketch three d-lines ℓ_1, ℓ_2 and ℓ_3 with the property that ℓ_1 is parallel to ℓ_2, and ℓ_2 is parallel to ℓ_3, but ℓ_1 is not parallel to ℓ_3.

It follows that the relation 'is parallel to' is not an equivalence relation in hyperbolic geometry.

In Euclidean geometry and in inversive geometry, reflection and inversion played important roles; this is also the case in hyperbolic geometry. Obviously, reflection of the unit disc $\mathscr{D}$ in a diameter maps $\mathscr{D}$ onto itself; in fact, so does inversion of $\mathscr{D}$ in a d-line that does not pass through the origin.

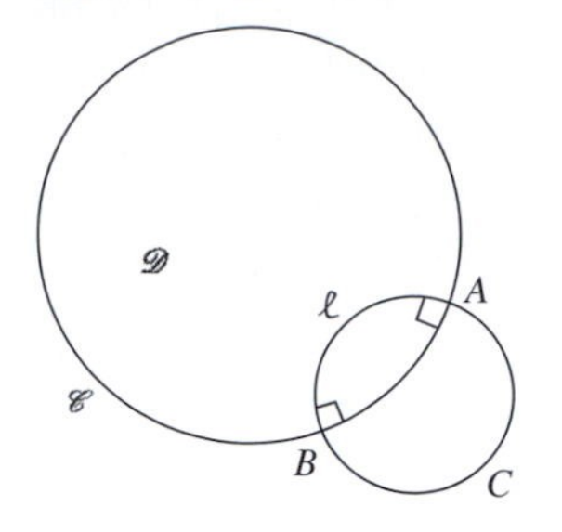

Theorem 1 Let ℓ be a d-line that is part of a Euclidean circle C. Then inversion in C maps $\mathscr{C}$ onto $\mathscr{C}$, and $\mathscr{D}$ onto $\mathscr{D}$.

Proof Let C meet $\mathscr{C}$ at the points A and B.

Under inversion in C, A and B map to themselves. Since inversion preserves angles, the circle $\mathscr{C}$ maps onto some circle that meets C at right angles at the points A and B. There is only one such circle, $\mathscr{C}$ itself; it follows that inversion in C maps $\mathscr{C}$ onto itself.

Subsection 5.1.3, Theorem 5

Inversion in C maps the shorter arc AB of $\mathscr{C}$ onto the longer arc AB of $\mathscr{C}$, and vice versa.

Hence the inside $\mathscr{D}$ of $\mathscr{C}$ must map either to the inside or to the outside of the image of $\mathscr{C}$, namely $\mathscr{C}$ itself. But the points of ℓ map to themselves under this inversion, so the image of $\mathscr{D}$ must be $\mathscr{D}$ itself. ■

Inversion in C maps the region bounded by ℓ and the shorter arc AB of $\mathscr{C}$ onto the region bounded by ℓ and the longer arc AB of $\mathscr{C}$, and vice versa.

We now know that inversion in a d-line maps the disc $\mathscr{D}$ onto itself. We also know that the composition of an inversion with itself is the identity map. Because the analogous properties are true of Euclidean reflections, we make the following definition.

Definition A **hyperbolic reflection** in a d-line ℓ is the restriction to $\mathscr{D}$ of the inversion in the generalized circle of which ℓ is part.

Recall that we use the term 'inversion' to mean *either* reflection in a line *or* inversion in a circle (see Subsection 5.1.1, 'Convention').

Remarks

1. If the d-line ℓ is part of a Euclidean circle C meeting the boundary circle $\mathscr{C}$ at points P and Q, then the point R where the tangents to $\mathscr{C}$ at P and Q meet is the centre of the inversion.
2. Notice that Theorem 1 may be reformulated as follows: hyperbolic reflection in any d-line maps the unit disc $\mathscr{D}$ onto itself.

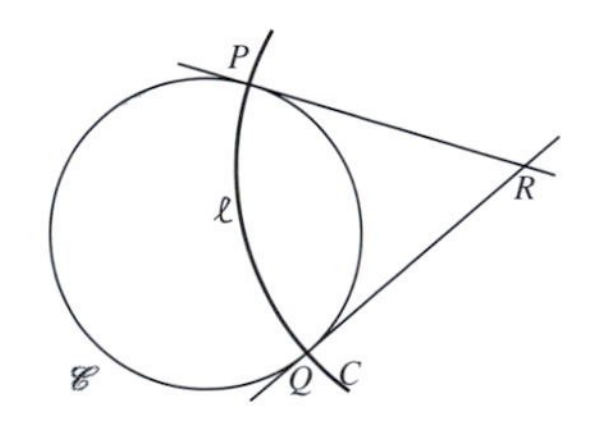

Hyperbolic reflections are the building blocks of transformations in hyperbolic geometry. Composites of a finite number of hyperbolic reflections are called **hyperbolic transformations**. The set of all such transformations under the operation of composition of functions forms a group, called the **hyperbolic group**, $G_{\mathscr{D}}$.

A proof of the fact that these form a group will be found as an Exercise in Section 6.7; it consists simply of checking the axioms for a group.

Definition **Hyperbolic geometry** consists of the unit disc, $\mathscr{D}$, together with the group $G_{\mathscr{D}}$ of hyperbolic transformations.

Remarks

1. Notice, in particular, that the identity mapping of $\mathscr{D}$ to itself belongs to $G_{\mathscr{D}}$, since it may be expressed as the finite composition $r \circ r^{-1}$ for any hyperbolic reflection r. Also, if $t_1 \circ t_2 \circ \ldots \circ t_n$ is any hyperbolic transformation, then its inverse is another hyperbolic transformation $t_n^{-1} \circ \ldots \circ t_2^{-1} \circ t_1^{-1}$.
2. Let the generalized circle C of which a d-line ℓ is part meet the boundary circle $\mathscr{C}$ in points P and Q, and let r be the hyperbolic reflection in ℓ. Then although r has domain the unit disc $\mathscr{D}$, and $P, Q \notin \mathscr{D}$, we sometimes find it convenient to call P and Q the boundary points of ℓ and to write the images of P and Q under inversion in C (of which r is a restriction) as $r(P)$ and $r(Q)$. The same convention is extended to the case where r is a hyperbolic transformation.

Next, we define the idea of angle in hyperbolic geometry.

Definition The **(hyperbolic) angle between two curves** (for example, two d-lines) through a given point A in $\mathscr{D}$ is the Euclidean angle between their (Euclidean) tangents at A.

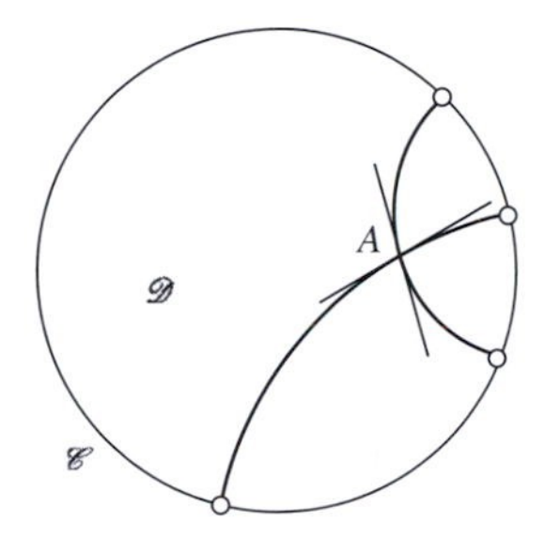

Now, Euclidean reflections and inversions both preserve the magnitudes of angles. It follows that hyperbolic transformations also preserve magnitudes of angles. Also, Euclidean reflections and inversions map generalized circles onto generalized circles. Combining this with the angle-preservation property, we deduce that hyperbolic reflections and inversions map d-lines onto d-lines – and so compositions of a finite number of such transformations also have this property. We summarize these facts in the following result.

Theorem 2 Hyperbolic transformations map d-lines onto d-lines, and preserve the magnitudes of angles.

In the remainder of Chapter 6 we study the properties of various figures under hyperbolic transformations, and obtain surprising (and sometimes beautiful) results!

6.1.2 Existence of *d*-lines

Through a typical point A of $\mathscr{D}$ there is at least one d-line: namely, the diameter through the origin O and A. But through the origin there are infinitely many d-lines: the diameters of $\mathscr{D}$. Is there more than one d-line through an arbitrary point A of $\mathscr{D}$?

The first step towards answering this question is the following useful result, which shows that there is a hyperbolic transformation which maps A to O.

Lemma 2 Origin Lemma
Let A be a point of $\mathscr{D}$ other than the origin O. Then there exists a d-line ℓ such that hyperbolic reflection in ℓ maps A to O.

Proof We seek a d-line ℓ which is part of a Euclidean circle with centre R, say, such that inversion in this circle maps A to O. Suppose that this circle meets $\mathscr{C}$ at the point P.

The condition that this inversion maps A to O is

$$RO \cdot RA = RP^2, \tag{3}$$

since RP is a radius of the circle we seek.

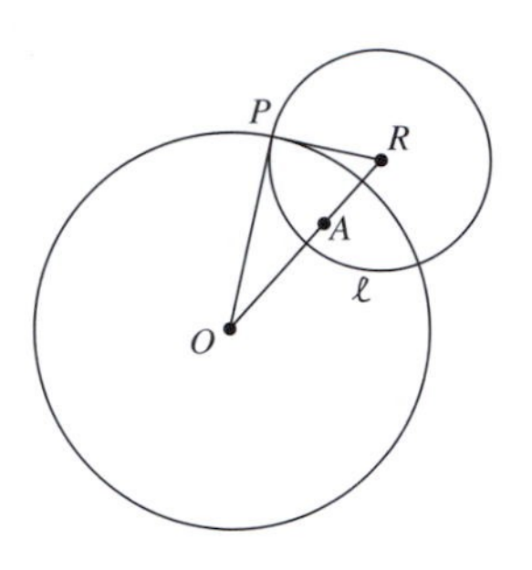

The condition that part of this circle is a d-line is that triangle $\triangle RPO$ is right-angled at P. By Pythagoras' Theorem, this implies that

$$RP^2 + PO^2 = RO^2,$$

which, since $OP = 1$, is equivalent to

$$RP^2 = RO^2 - 1. \tag{4}$$

Eliminating the radius RP^2 from equations (3) and (4), we deduce that

$$RO \cdot RA = RO^2 - 1.$$

This is equivalent to

$$RO^2 - RO \cdot RA = 1$$

or

$$RO \cdot (RO - RA) = 1.$$

But $RO - RA = AO$, so we deduce that

$$RO \cdot AO = 1$$

which is equivalent to

$$OA \cdot OR = 1.$$

This tells us that the circle we seek has for its centre the point R which is found by inverting the given point A in the boundary circle $\mathscr{C}$ – an unexpectedly memorable result! ■

The great value of the Origin Lemma is that it enables us to study any problem in hyperbolic geometry by mapping a suitably chosen point to the origin, thereby yielding a (frequently) simpler picture than before – yet without losing any generality. We shall use this method often in this chapter.

We can, for example, use this approach to answer the question of how many d-lines pass through a given point of $\mathscr{D}$.

Theorem 3 Let A be a point of $\mathscr{D}$. Then there exist infinitely many d-lines through A.

Proof Let A be the origin. As we said above, each of the infinitely many diameters of $\mathscr{D}$ passes through the origin, and each of these diameters is a d-line.

If A is not the origin, then by the Origin Lemma, there is a hyperbolic transformation, r say, that maps A to the origin O, through which pass infinitely many d-lines – all the diameters of $\mathscr{D}$.

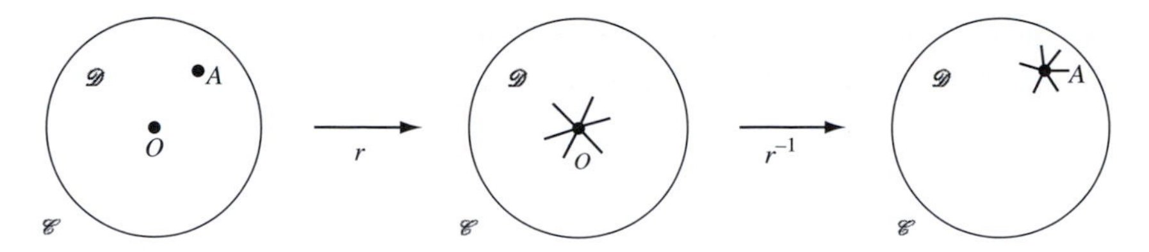

Since r^{-1} is also a hyperbolic transformation, it follows that the images of these diameters are also d-lines – and they pass through A. ■

For, hyperbolic transformations map d-lines to d-lines, by Theorem 2.

An important result in Euclidean geometry is that there is exactly one line through any two given points. There is an analogous result in hyperbolic geometry.

Theorem 4 Let A and B be any two distinct points of $\mathscr{D}$. Then there exists a unique d-line ℓ through A and B.

Proof (Existence) By the Origin Lemma, there is a hyperbolic transformation r that maps A to the origin O; let the image of B under r be the point B' of $\mathscr{D}$.

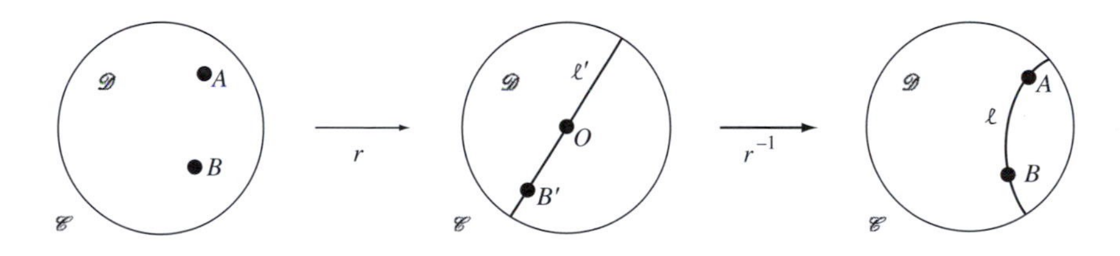

Then there is a unique d-line ℓ' (a diameter of $\mathscr{D}$) that passes through O and B'. Since r^{-1} is also a hyperbolic transformation, it follows that $\ell = r^{-1}(\ell')$ is also a d-line (by Theorem 2) – and it passes through A and B.

(Uniqueness) Suppose that ℓ_1 is another d-line through A and B. Then $r(\ell_1)$ is a d-line that passes through O and B'. It follows that $r(\ell_1) = \ell'$, so ℓ_1 must be the same as $r^{-1}(\ell') = \ell$. This proves the uniqueness of the d-line through A and B. ■

Problem 4 Let A_1 and A_2 be any two points of $\mathscr{D}$. Use the Origin Lemma to prove that there is a hyperbolic transformation that maps A_1 to A_2.

In fact, we can establish a result stronger than that of Problem 4: we can determine a hyperbolic transformation which maps A_1 to A_2 *and* maps any given d-line through A_1 to any given d-line through A_2.

Theorem 5 Let A_1 and A_2 be any two points of $\mathscr{D}$, and let ℓ_1 and ℓ_2 be d-lines through A_1 and A_2, respectively. Then there is a hyperbolic transformation which maps A_1 to A_2 and ℓ_1 to ℓ_2.

Proof By the Origin Lemma, there are hyperbolic transformations r_1 and r_2 which map A_1 and A_2, respectively, to the origin O. Let the images of ℓ_1 and ℓ_2 under r_1 and r_2 be the d-lines ℓ_1' and ℓ_2', respectively. Let r_3 be the hyperbolic transformation which rotates $\mathscr{D}$ about O so that ℓ_1' maps onto ℓ_2'.

We shall verify in Subsection 6.2.1 that any rotation of $\mathscr{D}$ about O is a hyperbolic transformation.

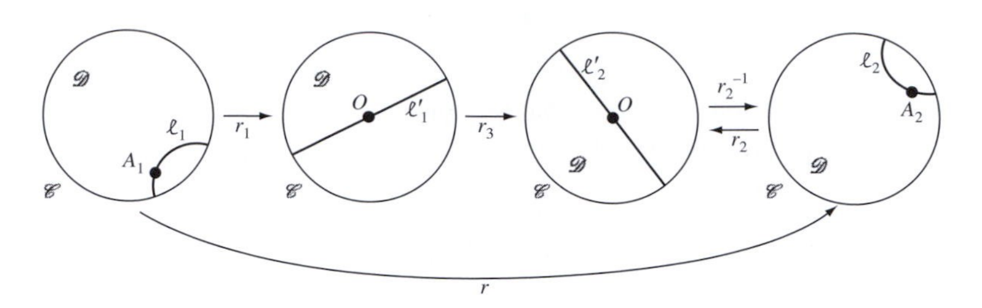

Now let r be the hyperbolic transformation given by

$$r = r_2^{-1} \circ r_3 \circ r_1.$$

Then r maps A_1 to A_2 and ℓ_1 to ℓ_2, as required. ■

Remark

We can choose the rotation r_3 to map the two parts of $\ell_1' - O$ onto the two parts of $\ell_2' - O$ in whichever way we please. Hence we can arrange for r to map the two parts of $\ell_1 - A_1$ onto the two parts of $\ell_2 - A_2$ in whichever way we please.

In the following example we show that parallel d-lines map to parallel d-lines under a hyperbolic transformation.

Example 1 Let A, B and C be points of $\mathscr{C}$ such that the d-lines AB and BC are parallel. Let r be a hyperbolic transformation under which the images of AB and BC are the d-lines $A'B'$ and $B'C'$, respectively, where A', B' and C' are points of $\mathscr{C}$. Show that $A'B'$ and $B'C'$ are parallel d-lines.

Note the convenient use of the label AB for a d-line even though A and B do not lie on that d-line (they lie on $\mathscr{C}$). There is no ambiguity since there is only one circle through A and B with centre the point where the tangents to $\mathscr{C}$ at A and B meet. Analogously, we shall say that a d-line passes through its boundary points.

Solution First, we show that $A'B'$ and $B'C'$ do not meet in $\mathscr{D}$. If they do, let them meet at the point P', say. Then the point P' corresponds to a point P, say, on the d-lines AB and BC; that is, $r^{-1}(P') = P$. But there is no such point, since AB and BC are parallel; it follows that $A'B'$ and $B'C'$ do not meet in $\mathscr{D}$.

Next, the generalized circles of which $A'B'$ and $B'C'$ are parts meet at a point B' on $\mathscr{C}$. So we conclude that the images of $A'B'$ and $B'C'$ are parallel d-lines. □

Problem 5 Show that, given a d-line ℓ and a point P not on ℓ, there are exactly two d-lines through P which are parallel to ℓ.
Hint: Use the Origin Lemma.

6.1.3 Inversion Preserves Inverse Points

Earlier, we proved the following result.

Subsection 5.5.2, Corollary 2

> Let A and B be inverse points with respect to a generalized circle C, and let t be an inversive transformation. Then $t(A)$ and $t(B)$ are inverse points with respect to the generalized circle $t(C)$.

We now consider a special case of this result, which we can interpret in terms of hyperbolic geometry.

We take C to be a generalized circle of which part (a d-line ℓ) lies inside the unit disc $\mathscr{D}$. Let A be a point in $\mathscr{D}$; then its inverse with respect to ℓ, B say, is also in $\mathscr{D}$. We take t to be an inversion for which the generalized circle of inversion C^* is such that part (a d-line ℓ^*) lies inside $\mathscr{D}$. Then, by the above result, $A' = t(A)$ and $B' = t(B)$ are inverse points in $\mathscr{D}$ with respect to the generalized circle $C' = t(C)$, part of which lies in $\mathscr{D}$, namely the d-line $\ell' = t(\ell)$.

See the diagram below.

Every inversion is an inversive transformation.

We now interpret this special case in terms of hyperbolic geometry.

Theorem 6 Let $A, B \in \mathscr{D}$ be inverse points with respect to the d-line ℓ, and let A', B' and ℓ' be the images of A, B and ℓ under inversion (hyperbolic reflection) in another d-line ℓ^*. Then A' and B' are inverse points with respect to inversion in ℓ'.

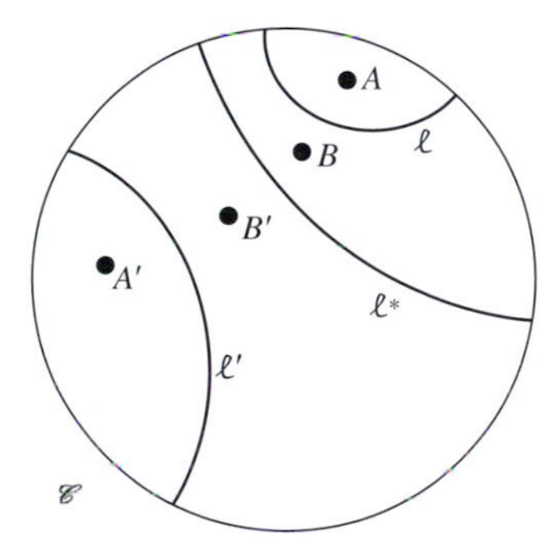

Remark

Note the convenient use of the term 'inversion' to mean 'hyperbolic reflection'. This is consistent with the use of this term in Chapter 5.

We shall make use of this theorem in Section 6.5, when we study the design on the cover of the book.

We finish this section with an illustration of the use of Theorem 6.

Example 2 Let ℓ be a d-line and P be a point in $\mathscr{D}$. Let Q be the image of P under inversion in the circle C of which ℓ is part. Let ℓ' be the d-line through P and Q, meeting ℓ at the point S.

(a) Show that inversion in C maps ℓ' to itself.

(b) Deduce that the d-lines ℓ and ℓ' meet at right angles.

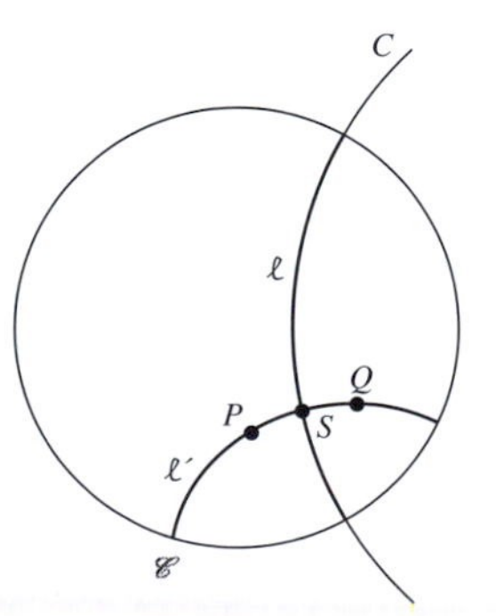

(c) Use Theorems 5 and 6 to show that there is a hyperbolic transformation mapping the figure to one in which the d-line ℓ is the x-axis and P and Q are complex conjugates.

Solution

(a) Inversion in C exchanges P and Q. So it maps ℓ', the unique d-line through P and Q, to itself.
(b) Inversion in C exchanges the angles ℓ' makes with ℓ, and maps angles to angles of equal magnitude. So the angles must be equal, and since their sum is π, each must be $\pi/2$. Thus ℓ and ℓ' meet at right angles.
(c) We use Theorem 5 to map the point S to O and the d-line ℓ to the x-axis. This makes the d-line ℓ' part of the y-axis. By Theorem 6, P and Q are mapped to points that are inverse with respect to the x-axis, and so are complex conjugates of each other. □

6.2 Hyperbolic Transformations

6.2.1 Hyperbolic Transformations and Möbius Transformations

Each element of the group $G_{\mathscr{D}}$ of hyperbolic transformations is the composite of a finite number of reflections in d-lines. In this subsection we shall establish an explicit formula for a hyperbolic transformation in terms of Möbius transformations.

Subsection 5.3.3

We start by considering reflection in a d-line ℓ which is part of a circle C with centre the point R and radius r. Suppose that the d-line has boundary points P and Q. We let the coordinates of the point R be (a, b), and write the complex number $a + ib$ as α, so that $|\alpha| > 1$ (which follows because ℓ is a d-line). We shall need the following observation: because the d-line ℓ is part of the circle with centre α and radius r, the triangle $\triangle RPO$ is right angled at P. So, by Pythagoras' Theorem, $RP^2 + PO^2 = RO^2$. This implies that

We shall call this d-line the 'd-line obtained from α'.

$$r^2 + 1 = a^2 + b^2.$$

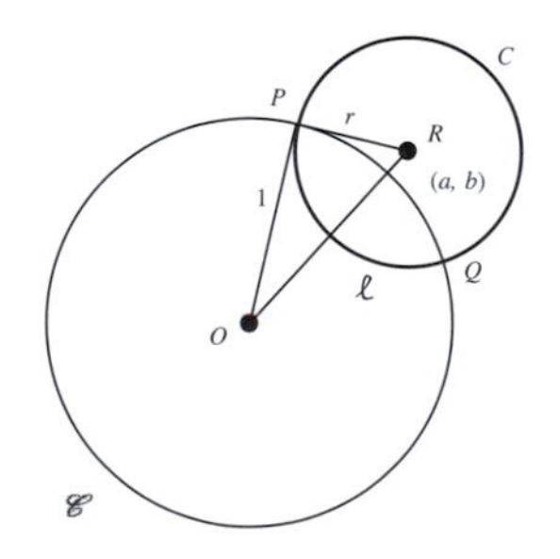

Using the fact that $a^2 + b^2 = \alpha\bar{\alpha}$, we may rewrite this equation as

$$r^2 - \alpha\bar{\alpha} = -1. \qquad (1)$$

We now use equation (1), and the fact that inversion t in the circle C has the form

$$t(z) = \frac{r^2}{\bar{z} - \bar{\alpha}} + \alpha \quad (z \in \mathbb{C} - \{\alpha\}),$$

This follows from Subsection 5.2.1, Theorem 3.

to obtain the form of the hyperbolic reflection ρ in the d-line ℓ.

Since ℓ is part of C, ρ is given by

$$\begin{aligned}\rho(z) &= \frac{r^2}{\overline{z - \alpha}} + \alpha \\ &= \frac{r^2 + \alpha\bar{z} - \alpha\bar{\alpha}}{\bar{z} - \bar{\alpha}} \\ &= \frac{\alpha\bar{z} - 1}{\bar{z} - \bar{\alpha}} \qquad (z \in \mathscr{D}),\end{aligned}$$

by equation (1). Thus we have proved the following lemma.

Lemma 1 The hyperbolic reflection ρ in the d-line ℓ that is part of a Euclidean circle with centre α is given by the hyperbolic transformation

$$\rho(z) = \frac{\alpha\bar{z} - 1}{\bar{z} - \bar{\alpha}} \qquad (z \in \mathscr{D}).$$

Recall that $|\alpha| > 1$.

Notice that we may write

$$\rho(z) = (M \circ B)(z), \qquad \text{for } z \in \mathscr{D},$$

where $M(z) = \frac{\alpha z - 1}{z - \bar{\alpha}}$ is a Möbius transformation, and $B(z) = \bar{z}$ is complex conjugation. You know from Chapter 5 that Möbius transformations are important in inversive geometry: they preserve the magnitudes of angles, and they map generalized circles to generalized circles. Since angles and generalized circles are important in hyperbolic geometry, it is not surprising that particular sorts of Möbius transformations turn up here too.

Subsection 5.3.3, Theorem 5

Problem 1 Find the point which has image 0 under reflection in the d-line obtained from α. Hence obtain a second proof of the Origin Lemma.

Reflection in a d-line which is a diameter of $\mathscr{D}$ is simply (Euclidean) reflection in that line. Recall that

Subsection 5.2.1

$$t_1(z) = \bar{z}$$

is reflection in the x-axis, and

$$t_2(z) = \alpha z,$$

where $\alpha = \cos\theta + i\sin\theta$ with $\theta = \operatorname{Arg}\alpha$, is rotation about the origin through the angle θ. It follows that the composite

$$\left(t_2 \circ t_1 \circ t_2^{-1}\right)(z) = \alpha(\alpha\bar{z}) = \alpha^2\bar{z}$$

For $t_2^{-1}(z) = \bar{\alpha}z$

is reflection in the line with equation $y = x\tan\theta$.

Thus hyperbolic reflection σ in the diameter on which $y = x\tan\theta$ is given by

$$\sigma(z) = \alpha^2\bar{z} \qquad (z \in \mathscr{D}),$$

where $\alpha = \cos\theta + i\sin\theta$.

Example 1 Find the composite $\sigma_2 \circ \sigma_1$ of the hyperbolic reflections

$$\sigma_1(z) = \alpha^2 \bar{z} \quad \text{and} \quad \sigma_2(z) = \beta^2 \bar{z},$$

where $\alpha = \cos\theta_1 + i\sin\theta_1$ and $\beta = \cos\theta_2 + i\sin\theta_2$, and interpret the transformation $\sigma_2 \circ \sigma_1$ geometrically.

Solution We have

$$(\sigma_2 \circ \sigma_1)(z) = \beta^2 \overline{(\alpha^2 \bar{z})} = \beta^2 \bar{\alpha}^2 z,$$

which is a Euclidean rotation about the origin of $\mathscr{D}$. □

The composite of reflections ρ and σ in two d-lines that are obtained from α and β, respectively, is found as follows. By Lemma 1, we have

$$\rho(z) = \frac{\alpha\bar{z} - 1}{\bar{z} - \bar{\alpha}} = (M \circ B)(z)$$

Recall that $M(z) = \frac{\alpha z - 1}{z - \bar{\alpha}}$ and $B(z) = \bar{z}$.

and

$$\sigma(z) = \frac{\beta\bar{z} - 1}{\bar{z} - \bar{\beta}} = (M' \circ B)(z),$$

where $M'(z) = \frac{\beta z - 1}{z - \bar{\beta}}$ is a Möbius transformation. So

$$(\sigma \circ \rho)(z) = (M' \circ B \circ M \circ B)(z).$$

But

$$(B \circ M)(z) = \frac{\overline{\alpha z} - 1}{\bar{z} - \alpha} = (\tilde{M} \circ B)(z),$$

where $\tilde{M}(z)$ is the Möbius transformation $\tilde{M}(z) = \frac{\bar{\alpha} z - 1}{z - \alpha}$. So

$$\begin{aligned}(\sigma \circ \rho)(z) &= (M' \circ B \circ M \circ B)(z) \\ &= (M' \circ \tilde{M} \circ B \circ B)(z) = (M' \circ \tilde{M})(z),\end{aligned}$$

because $(B \circ B)(z) = z$.

Now matrices associated with M' and $\tilde{M}$ are $\begin{pmatrix} \beta & -1 \\ 1 & -\bar{\beta} \end{pmatrix}$ and $\begin{pmatrix} \bar{\alpha} & -1 \\ 1 & -\alpha \end{pmatrix}$, respectively. Hence a matrix associated with the composite $M' \circ \tilde{M}$ is

Here we are using the strategy in Subsection 5.3.5 for finding composites.

$$\begin{pmatrix} \beta & -1 \\ 1 & -\bar{\beta} \end{pmatrix} \begin{pmatrix} \bar{\alpha} & -1 \\ 1 & -\alpha \end{pmatrix} = \begin{pmatrix} \bar{\alpha}\beta - 1 & \alpha - \beta \\ \bar{\alpha} - \bar{\beta} & \alpha\bar{\beta} - 1 \end{pmatrix},$$

so that the composite $\sigma \circ \rho$ of the hyperbolic reflections σ and ρ is of the form

$$(\sigma \circ \rho)(z) = \frac{az + b}{\bar{b}z + \bar{a}},$$

where $a = \bar{\alpha}\beta - 1, b = \alpha - \beta$.

For ease of reference, we record this last result as a theorem.

Theorem 1 The composite of the hyperbolic reflections

$$\rho(z) = \frac{\alpha\bar{z} - 1}{\bar{z} - \bar{\alpha}} \quad \text{and} \quad \sigma(z) = \frac{\beta\bar{z} - 1}{\bar{z} - \bar{\beta}} \qquad (z \in \mathscr{D})$$

is the hyperbolic transformation

$$(\sigma \circ \rho)(z) = \frac{(\bar{\alpha}\beta - 1)z + \alpha - \beta}{(\bar{\alpha} - \bar{\beta})z + \alpha\bar{\beta} - 1} \qquad (z \in \mathscr{D}).$$

Example 2 Show that the hyperbolic reflection ρ is its own inverse, by showing that $\rho(\rho(z)) = z$.

Solution From Theorem 1, we know that the composite of the reflection ρ with itself is given by

$$(\rho \circ \rho)(z) = \frac{(\bar{\alpha}\alpha - 1)z + \alpha - \alpha}{(\bar{\alpha} - \bar{\alpha})z + \alpha\bar{\alpha} - 1} = \frac{(\bar{\alpha}\alpha - 1)z}{\alpha\bar{\alpha} - 1} = z,$$

$\alpha\bar{\alpha} \neq 1$ since $|\alpha| > 1$.

as required. □

Example 3 Show that the composite $M_2 \circ M_1$ of the Möbius transformations

$$M_1(z) = \frac{az + b}{\bar{b}z + \bar{a}} \quad \text{and} \quad M_2(z) = \frac{cz + d}{\bar{d}z + \bar{c}}$$

is a Möbius transformation of the same form.

That is, of the form $\frac{ez+f}{\bar{f}z+\bar{e}}$, for some $e, f \in \mathbb{C}$.

Solution By the strategy for finding composites of Möbius transformations, a matrix associated with the composite $M_2 \circ M_1$ is the matrix product

$$\begin{pmatrix} c & d \\ \bar{d} & \bar{c} \end{pmatrix} \begin{pmatrix} a & b \\ \bar{b} & \bar{a} \end{pmatrix} = \begin{pmatrix} ca + d\bar{b} & cb + d\bar{a} \\ \bar{d}a + \bar{c}\bar{b} & \bar{d}b + \bar{c}\bar{a} \end{pmatrix}.$$

This product is of the required form, because

$$\overline{(ca + d\bar{b})} = \bar{d}b + \bar{c}\bar{a} \quad \text{and} \quad \overline{(cb + d\bar{a})} = \bar{d}a + \bar{c}\bar{b}. \qquad \square$$

It follows from this example that every composite of an even number of hyperbolic reflections in d-lines can be represented as a Möbius transformation, restricted to $\mathscr{D}$, of the form

$$M(z) = \frac{az + b}{\bar{b}z + \bar{a}}, \qquad z \in \mathscr{D}.$$

Similarly, it can be shown that every composite of an odd number of reflections can be represented as a Möbius transformation of this form composed with complex conjugation.

We omit the details of a proof of this.

It follows from the above discussion that any composite of a finite number of hyperbolic reflections, that is, any hyperbolic transformation, can be expressed in one of the forms

$$z \mapsto M(z) \quad \text{or} \quad z \mapsto M(\bar{z}) \quad (z \in \mathscr{D}),$$

where M is the Möbius transformation

$$M(z) = \frac{az+b}{\bar{b}z+\bar{a}}, \quad z \in \mathscr{D}.$$

Moreover, since $M(0) = b/\bar{a}$ and the image of 0 under a hyperbolic transformation must be in $\mathscr{D}$, we require that the above Möbius transformation must be such that

$$|b| < |a|.$$

The remaining question is: 'Do all such Möbius transformations represent hyperbolic transformations?' In fact the answer is YES, as the following theorem shows.

Theorem 2 The restriction to $\mathscr{D}$ of every Möbius transformation of the form $M(z) = \frac{az+b}{\bar{b}z+\bar{a}}$ with $|b| < |a|$ is a composite of two hyperbolic reflections, and is therefore a hyperbolic transformation.

When the context is clear, we often omit to say that the Möbius transformation we are considering is restricted to $\mathscr{D}$.

The proof is a little devious (and no less elegant for that). We first prove it for the special case when the Möbius transformation maps the origin to itself. Using this special case, we then prove the general case.

Proof

Case 1 *The Möbius transformation* $M(z) = \frac{az+b}{\bar{b}z+\bar{a}}$ *maps the origin to itself.*

We shall show that M is the composite of two reflections.

The condition $M(0) = 0$ implies that $b = 0$, so that the Möbius transformation is simply $M(z) = \frac{a}{\bar{a}}z$ — a rotation about the origin.

Now, if we let $a = re^{i\theta}$, then

$$\frac{a}{\bar{a}} = e^{2i\theta}.$$

So, if we let σ_1 and σ_2 be the reflections in diameters of $\mathscr{D}$ given by

$$\sigma_1(z) = \bar{z}, \quad z \in \mathscr{D}, \quad \text{and} \quad \sigma_2(z) = e^{2i\theta}\bar{z}, \quad z \in \mathscr{D},$$

then

$$\sigma_2 \circ \sigma_1(z) = e^{2i\theta}\overline{(\bar{z})} = e^{2i\theta}z,$$

so that $M = \sigma_2 \circ \sigma_1$, as required.

Recall that reflection in the diameter $y = x\tan\theta$ of $\mathscr{D}$ is a mapping of the form $z \mapsto \alpha^2\bar{z}$, where $\alpha = e^{i\theta} = \cos\theta + i\sin\theta$ and $z \in \mathscr{D}$.

Case 2 *The Möbius transformation* $M(z) = \frac{az+b}{\bar{b}z+\bar{a}}$ *does not map the origin to itself.*

We shall again show that M is the composite of two hyperbolic reflections.

Consider a hyperbolic transformation ρ given by $\rho(z) = (M' \circ B)(z)$, where $M'(z) = \frac{\alpha z - 1}{z - \bar{\alpha}}$ is a Möbius transformation, and $B(z) = \bar{z}$ is complex conjugation. We shall choose α so as to make ρ and $M \circ \rho$ hyperbolic reflections. This will show that $M = (M \circ \rho) \circ \rho^{-1}$ is a composite of two hyperbolic reflections.

Since M and M' are associated with the matrices

$$\begin{pmatrix} a & b \\ \bar{b} & \bar{a} \end{pmatrix} \quad \text{and} \quad \begin{pmatrix} \alpha & -1 \\ 1 & -\bar{\alpha} \end{pmatrix},$$

respectively, the composite $M \circ M'$ is associated with the matrix

$$\begin{pmatrix} a & b \\ \bar{b} & \bar{a} \end{pmatrix} \begin{pmatrix} \alpha & -1 \\ 1 & -\bar{\alpha} \end{pmatrix} = \begin{pmatrix} \alpha a + b & -(\bar{\alpha} b + a) \\ \alpha \bar{b} + \bar{a} & -\bar{b} - \bar{a}\bar{\alpha} \end{pmatrix},$$

so we deduce that

$$(M \circ M' \circ B)(z) = \frac{(a\alpha + b)\bar{z} - (\bar{\alpha} b + a)}{(\alpha \bar{b} + \bar{a})\bar{z} - \bar{b} - \bar{a}\bar{\alpha}}.$$

Now since $M(0) \neq 0$, it follows that $b \neq 0$. Hence we may choose $\alpha = -\overline{(a/b)}$, so that

$$\bar{\alpha} b + a = -(a/b)b + a = 0.$$

Next, since $|b| < |a|$, we deduce that $|\alpha| = \frac{|a|}{|b|} > 1$, and so the transformation $\rho = M' \circ B$ is a hyperbolic reflection as required. Moreover,

For, ρ is of the form required by Lemma 1.

$$(M \circ \rho)(z) = (M \circ M' \circ B)(z) = \frac{(a\alpha + b)\bar{z}}{-\bar{b} - \bar{a}\bar{\alpha}}$$

is of the form $-\frac{\gamma}{\bar{\gamma}}\bar{z}$, where $\gamma = a\alpha + b$, and so the transformation $M \circ \rho$ is a hyperbolic reflection. It follows that the transformation M is a composite of two hyperbolic reflections, $M \circ \rho$ and ρ^{-1}. ■

A similar argument shows that the more general transformation

$$z \mapsto \frac{a\bar{z} + b}{\bar{b}\bar{z} + \bar{a}}$$

is a composite of three reflections. For it is the composite $(M \circ B)(z)$ of the Möbius transformation

$$M(z) = \frac{az + b}{\bar{b}z + \bar{a}}$$

with complex conjugation $B(z) = \bar{z}$. Now, the Möbius transformation M is a composite of two reflections, and complex conjugation B is reflection in the x-axis. So $M \circ B$ is a composite of three reflections, as claimed.

Putting together all the above results, we have shown the following.

Theorem 3 Every hyperbolic transformation can be written as a composite of at most three hyperbolic reflections.

Example 4 Show that the composite $\sigma_2 \circ \sigma_1$ of the two reflections in lines through the origin given by $\sigma_1(z) = \alpha^2\bar{z}$ and $\sigma_2(z) = \beta^2\bar{z}$ can be written in the form $M(z) = \frac{az+b}{\bar{b}z+\bar{a}}$.

Solution From Example 1, we know that

$$(\sigma_2 \circ \sigma_1)(z) = \beta^2 \bar{\alpha}^2 z.$$

Since $|\alpha| = 1$ and $|\beta| = 1$,

$$\alpha^{-1} = \bar{\alpha} \quad \text{and} \quad \beta^{-1} = \bar{\beta},$$

and so

$$(\sigma_2 \circ \sigma_1)(z) = M(z),$$

where

$$M(z) = \frac{az+b}{\bar{b}z+\bar{a}} \quad \text{with } a = \beta\bar{\alpha} \text{ and } b = 0. \qquad \square$$

Problem 2 Let $M(z) = \frac{az+b}{\bar{b}z+\bar{a}}$ be a hyperbolic transformation mapping the origin to itself. Show that $b = 0$.

Hyperbolic Rotations and Translations

It follows from the Angle Theorem of inversive geometry that a single hyperbolic reflection reverses the orientation of angles between d-lines. So a composite of two such transformations leaves the orientation unchanged, while a composite of three reverses it again. We call a hyperbolic transformation that leaves orientation unchanged a **direct** hyperbolic transformation, and one that reverses orientation **indirect.** So, by Theorem 3 above, we deduce that

Subsection 5.1.3, Theorem 5

> a direct hyperbolic transformation can be written as a composite of at most two hyperbolic reflections

Theorem 1 gives the form of a composite of two reflections.

and

> an indirect hyperbolic transformation can be written as a composite of at most three hyperbolic reflections.

It is possible to say more about the direct transformations. Let r_1 and r_2 be reflections in the d-lines ℓ_1 and ℓ_2, respectively.

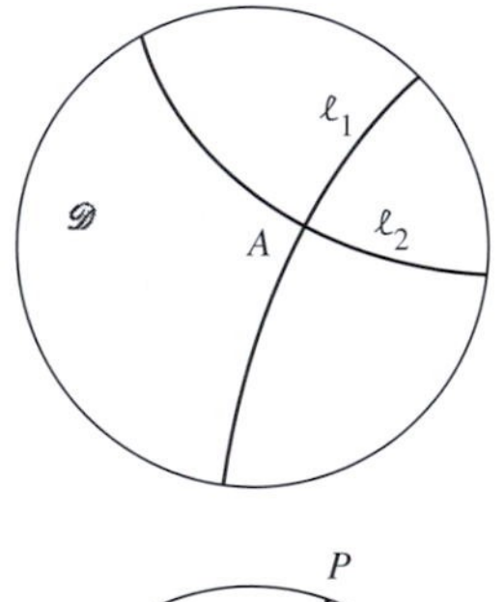

First, suppose that ℓ_1 and ℓ_2 intersect at some point A. (Certainly they cannot intersect at more than one point if they are distinct d-lines, since by Theorem 4 of Subsection 6.1.2 there is a unique d-line through any two points of $\mathscr{D}$.) Then the composition $r_2 \circ r_1$ leaves the point A fixed, moves the points of $\mathscr{D}$ 'around A', and does not alter the orientation of $\mathscr{D}$. Such a hyperbolic transformation is called a **(hyperbolic) rotation,** and has exactly one fixed point in $\mathscr{D}$.

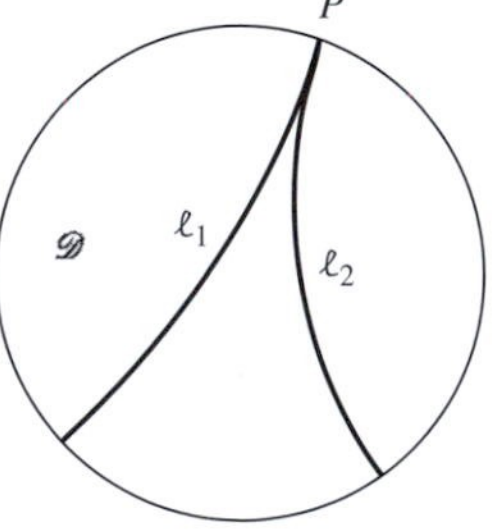

Next, suppose that the d-lines ℓ_1 and ℓ_2 are parallel, that is, the generalized circles of which they are parts do not meet in $\mathscr{D}$ but do meet at some point P on $\mathscr{C}$. Then the composite $r_2 \circ r_1$ moves the points of $\mathscr{D}$ 'around P', and does not alter the orientation of $\mathscr{D}$; it has no fixed point in $\mathscr{D}$, but it leaves all the parallel lines with P as their common boundary point as parallel lines ending at P. We can regard this as the limiting case of a rotation (about a point of $\mathscr{C}$), and so call it a **(hyperbolic) limit rotation.**

Finally, suppose that the d-lines ℓ_1 and ℓ_2 are ultra-parallel, that is, the generalized circles of which they are parts do not meet in $\mathscr{D}$ or on $\mathscr{C}$. Then the composite $r_2 \circ r_1$ moves all the points of $\mathscr{D}$ in one general direction, and does not alter the orientation of $\mathscr{D}$; but no point of $\mathscr{D}$ (or $\mathscr{C}$) remains fixed. Such a hyperbolic transformation is called a **(hyperbolic) translation.**

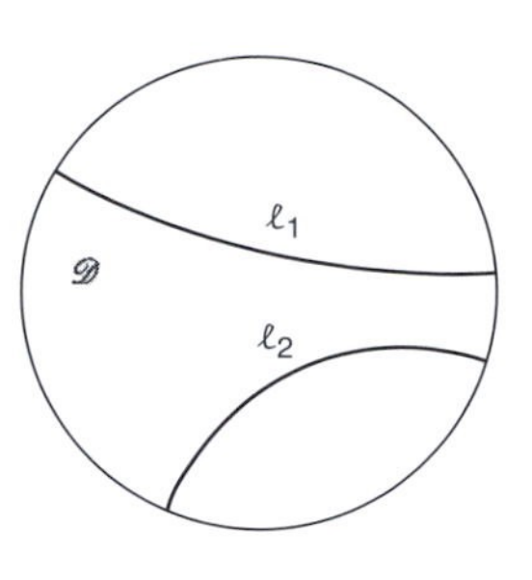

The analogy between Euclidean geometry and hyperbolic geometry is not exact, however. For instance, the composite of two Euclidean translations is independent of the order in which they are applied, whereas the composite of two hyperbolic translations may not be independent of the order in which they are applied.

6.2.2 The Canonical Form of a Hyperbolic Transformation

In this subsection we show how to write a hyperbolic transformation in the most suitable form for applications. We shall do this only for a direct transformation of the form $M(z) = \frac{az+b}{\bar{b}z+\bar{a}}$, where $|b| < |a|$. Our result is the following theorem.

Theorem 4 A direct hyperbolic transformation M can be written in the form

$$M(z) = K\frac{z-m}{1-\bar{m}z},$$

where K and m are complex numbers with $|K| = 1$ and $m \in \mathscr{D}$.

Remark

We call this form the **canonical form** of a direct hyperbolic transformation. It has the great advantage of showing that the transformation M maps the point m to the origin.

Proof We know from the previous subsection that a direct hyperbolic transformation can always be written in the form

$$M(z) = \frac{az+b}{\bar{b}z+\bar{a}}, \quad \text{with } |b| < |a|.$$

See the discussion before Theorem 2.

Indeed, on dividing the expression for $M(z)$ above and below by $\bar{a}$, we can write it as

$$M(z) = \frac{\frac{a}{\bar{a}}z + \frac{b}{\bar{a}}}{1 - \frac{-\bar{b}}{\bar{a}}z} = \frac{a}{\bar{a}}\left(\frac{z - \frac{-b}{a}}{1 - \frac{-\bar{b}}{\bar{a}}z}\right),$$

which is of the required form, with $K = a/\bar{a}$ and $m = -b/a$. Since $|a| = |\bar{a}|$, it follows that $|K| = 1$, as required. Since $|b| < |a|$, it follows that $|m| < 1$, as required. ∎

We can now find the form of every direct hyperbolic transformation that maps a point m of $\mathscr{D}$ to the origin. We know from Theorem 4 that it must be of the form

$$M(z) = K\frac{z - m'}{1 - \overline{m'}z}, \qquad \text{where } |K| = 1 \text{ and } |m'| < 1.$$

The condition that $M(m) = 0$ implies that $m = m'$, so we deduce that in fact *every* direct hyperbolic transformation mapping the point m to the origin is of the form

$$M(z) = K\frac{z - m}{1 - \overline{m}z}.$$

The direct hyperbolic transformations that map the origin to itself are therefore those for which $m = 0$, so they are of the form $M(z) = Kz$ with $|K| = 1$. These are (Euclidean) rotations of the disc $\mathscr{D}$ about the origin through an angle θ, where $K = \cos\theta + i\sin\theta$. In general, if all that is required is just one direct hyperbolic transformation sending a given point m to the origin, then we may set $K = 1$, and use the transformation $M(z) = \frac{z-m}{1-\overline{m}z}$.

$K = 1$ is the most convenient value to choose.

Example 5

(a) Find the general form of a direct hyperbolic transformation that maps the point $\frac{1}{2}i$ to the origin.
(b) Find one direct hyperbolic transformation that maps $\frac{3}{4}$ to the origin.

Solution

(a) The general form is

$$M(z) = K\frac{z - \frac{1}{2}i}{1 - \left(-\frac{1}{2}i\right)z} = K\frac{2z - i}{iz + 2}, \qquad \text{where } |K| = 1.$$

(b) Taking $K = 1$ in the formula of Theorem 4, as suggested above, one such transformation is

$$M(z) = \frac{z - \frac{3}{4}}{1 - \frac{3}{4}z}.$$

□

Problem 3 Determine all the direct hyperbolic transformations that map the origin to the origin and the line $y = x/\sqrt{3}$ to the line $y = \sqrt{3}x$.

Problem 4 For each of the following points, determine all the direct hyperbolic transformations that map it to the origin.
(a) $\frac{1}{4}i$ (b) $-\frac{1}{3} + \frac{2}{3}i$

The general form of the inverse of a direct hyperbolic transformation M mapping the point m to the origin is a direct hyperbolic transformation sending the origin to m. To find this inverse explicitly, we write

$$M(z) = K\frac{z-m}{1-\overline{m}z} \quad \text{(by Theorem 4)}$$
$$= \frac{Kz - Km}{-\overline{m}z+1},$$

and make use of the fact that M is the restriction to $\mathscr{D}$ of a Möbius transformation to write

By 'restriction to $\mathscr{D}$' we mean that we only consider those z in $\mathscr{D}$.

$$M^{-1}(z) = \frac{z+Km}{\overline{m}z+K}, \qquad (2)$$

where $|K| = 1$.

This form for M^{-1} follows immediately from Subsection 5.3.6, Theorem 7.

Problem 5 Determine the general form of the inverse of the direct hyperbolic transformation M which maps $\frac{3}{4}$ to the origin.

Problem 6 Prove that any direct hyperbolic transformation M that maps the diameter $(-1, 1)$ onto itself must be of the form

$$M(z) = \pm\frac{z-m}{1-mz}, \qquad \text{where } m \in (-1,1).$$

Since an *indirect* hyperbolic transformation can be written as the composite of at most three reflections, it follows that the general form of such a transformation is

$$z \mapsto \frac{a\bar{z}+b}{\bar{b}\bar{z}+\bar{a}}.$$

See the discussion preceding Theorem 2.

An argument similar to that for direct transformations shows that this transformation can always be written in the form

$$z \mapsto K\frac{\bar{z}-m}{1-\overline{m}\bar{z}}, \qquad \text{where } |k| = 1 \text{ and } |m| < 1.$$

We call this form the **canonical form** of an indirect hyperbolic transformation.

Theorem 4 gives the general form of a direct hyperbolic transformation that maps a given point m in $\mathscr{D}$ to the origin. By two applications of the theorem, therefore, we can obtain the general form of a direct hyperbolic transformation that maps any given point in $\mathscr{D}$ to any other given point in $\mathscr{D}$.

If we require a particular direct transformation that maps p to q, then we may choose a particular M_1 and a particular $\mathbf{A}_1$.

Strategy To determine the general form of the direct hyperbolic transformation that maps one point p in $\mathscr{D}$ to another point q in $\mathscr{D}$:

1. write down the general form of the direct transformation M_1 that maps p to 0, and a matrix $\mathbf{A}_1$ associated with M_1;
2. write down a direct transformation M_2 that maps q to 0, and a matrix $\mathbf{A}_2$ associated with M_2;

The constant arising in Step 1 means that it is sufficient to use any particular transformation in Step 2, since the final general transformation requires only one constant.

3. form the matrix product $\mathbf{A}_2^{-1}\mathbf{A}_1$ associated with the direct transformation $M_2^{-1} \circ M_1$, and hence write down the general form of the required direct transformation $M_2^{-1} \circ M_1$.

The following figure illustrates this strategy.

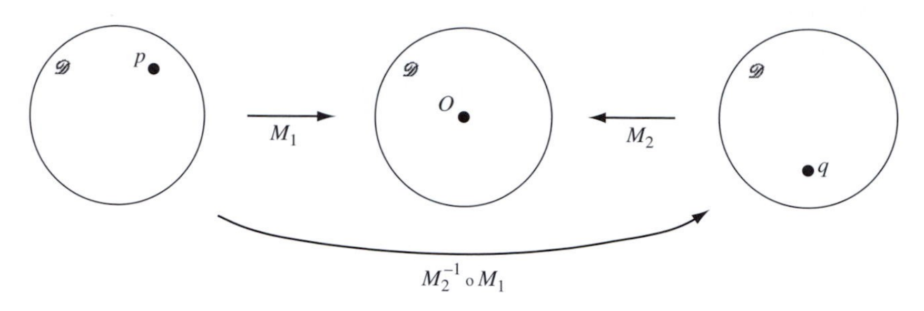

Example 6 Determine the general form of the direct hyperbolic transformation that maps $\frac{1}{2}i$ to $\frac{3}{4}$.

Solution We have already seen that the general form of the direct hyperbolic transformation M_1 that maps $\frac{1}{2}i$ to 0 is

Example 5, part (a).

$$M_1(z) = K\frac{2z - i}{iz + 2}, \qquad \text{where } |K| = 1;$$

a matrix associated with M_1 is

$$\mathbf{A}_1 = \begin{pmatrix} 2K & -iK \\ i & 2 \end{pmatrix}.$$

Also, the direct transformation

$$M_2(z) = \frac{z - \frac{3}{4}}{1 - \frac{3}{4}z}$$

Example 5, part (b).

maps $\frac{3}{4}$ to 0; a matrix associated with M_2 is

$$\mathbf{A}_2 = \begin{pmatrix} 1 & -\frac{3}{4} \\ -\frac{3}{4} & 1 \end{pmatrix}.$$

The inverse of $\mathbf{A}_2$ is

$$\mathbf{A}_2^{-1} = \tfrac{16}{7}\begin{pmatrix} 1 & \frac{3}{4} \\ \frac{3}{4} & 1 \end{pmatrix}.$$

So a matrix associated with the required direct transformation is

$$\begin{aligned}\mathbf{A}_2^{-1}\mathbf{A}_1 &= \tfrac{16}{7}\begin{pmatrix} 1 & \frac{3}{4} \\ \frac{3}{4} & 1 \end{pmatrix}\begin{pmatrix} 2K & -iK \\ i & 2 \end{pmatrix} \\ &= \tfrac{4}{7}\begin{pmatrix} 8K + 3i & -4iK + 6 \\ 6K + 4i & -3iK + 8 \end{pmatrix}.\end{aligned}$$

Hence any direct transformation that maps $\frac{1}{2}i$ to $\frac{3}{4}$ may be written in the form

$$M(z) = \frac{(8K + 3i)z + (-4iK + 6)}{(6K + 4i)z + (-3iK + 8)}, \qquad \text{where } |K| = 1. \qquad \square$$

The fraction $\frac{4}{7}$ disappears at this point, as it is simply a multiple of the whole matrix.

Problem 7

(a) Determine the general form of the direct hyperbolic transformation that maps $-\frac{1}{3}i$ to $\frac{2}{3}$.
(b) Determine the direct hyperbolic transformation that also maps i to 1.

Problem 8 Determine the general form of the direct hyperbolic transformation that maps $\frac{1}{2}$ to $\frac{1}{2}$.

These functions *fix* the point $\frac{1}{2}$.

Problem 9 For each of the following pairs of points, either find a direct hyperbolic transformation mapping the first point to 0 and the second point to $\frac{1}{2}$, or prove that no such transformation exists.

(a) $\frac{1}{2}i$ and 0 (b) $\frac{1}{2}$ and $\frac{2}{3}$ (c) $\frac{1}{3}(1+i)$ and $\frac{1}{3}(1-i)$.

6.3 Distance in Hyperbolic Geometry

In the Euclidean geometry of the plane, corresponding to any two points there exists a non-negative number called the distance between the points. This is given by the formula

$$d(z_1, z_2) = |z_1 - z_2|, \quad z_1, z_2 \in \mathbb{C}.$$

So $d(z_1, z_2) = d(z_2, z_1)$.

In this section, we introduce an analogous formula for the distance in hyperbolic geometry between any two points of the unit disc $\mathscr{D}$.

6.3.1 The Distance Formula

We begin by looking at various properties that we would expect any distance function d to have in any geometry whose points lie in the complex plane $\mathbb{C}$. 'Ordinary' Euclidean distance, for example, clearly possesses the following four properties.

Properties of a Distance Function d

1. $d(z_1, z_2) \geq 0$ for all z_1 and z_2;
   $d(z_1, z_2) = 0$ if and only if $z_1 = z_2$.
2. $d(z_1, z_2) = d(z_2, z_1)$ for all z_1 and z_2.
3. $d(z_1, z_3) + d(z_3, z_2) \geq d(z_1, z_2)$ for all z_1, z_2 and z_3.
4. $d(z_1, z_3) + d(z_3, z_2) = d(z_1, z_2)$ if and only if z_1, z_3 and z_2 lie in this order on a line.

Property 3 is known as the *Triangle Inequality.*

Property 1 asserts that the distance between any two points in the geometry is always positive, unless the two points coincide – in which case the distance between them must be zero.

Property 2 asserts that the distance from z_1 to z_2 is the same as the distance from z_2 to z_1.

Property 3 asserts that the distance from z_1 to z_2 is always less than (or equal to) the distance from z_1 to another point z_3 plus the distance from z_3 to z_2. In Euclidean geometry this property may be rewritten in the form

$$|z_1 - z_3| + |z_3 - z_2| \geq |z_1 - z_2|.$$

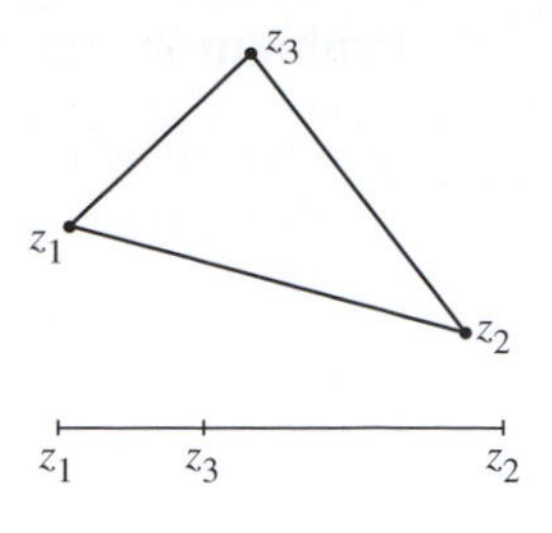

Property 4 asserts that distance along a line is additive.

However, there are some additional properties that we wish the distance function in hyperbolic geometry to possess.

Additional Properties of the Distance Function d in Hyperbolic Geometry

5. $d(z_1, z_2) = d(\bar{z}_1, \bar{z}_2)$ for all z_1 and z_2 in $\mathcal{D}$.
6. $d(z_1, z_2) = d(M(z_1), M(z_2))$ for all z_1 and z_2 in $\mathcal{D}$ and all direct hyperbolic transformations M in $G_{\mathcal{D}}$.

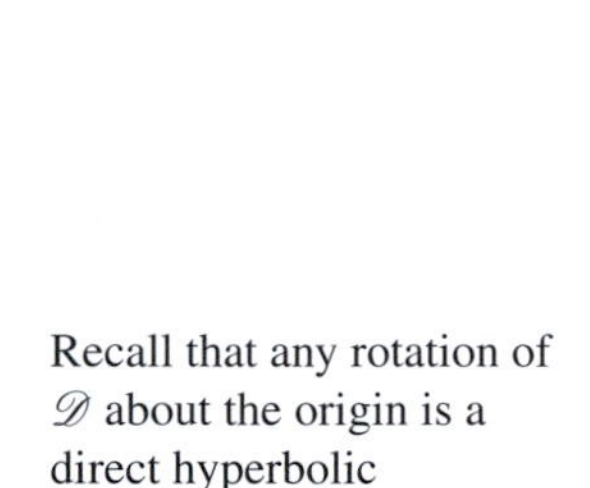

Properties 5 and 6 together assert that hyperbolic transformations of the unit disc $\mathcal{D}$ do not alter distances between points, since a hyperbolic transformation is either a direct transformation of the unit disc to itself (as in Property 6) or the composite of such a function with the conjugation function $z \mapsto \bar{z}$ (which, by Property 5, preserves distance).

These additional properties enable us to make some useful observations about the form of the distance function. First, the direct hyperbolic transformation

$$M : z \mapsto \frac{z - z_1}{1 - \bar{z}_1 z}, \quad \text{where } z_1 \in \mathcal{D},$$

maps z_1 to 0 and z_2 to $\frac{z_2 - z_1}{1 - \bar{z}_1 z_2}$. If R is the rotation of the unit disc $\mathcal{D}$ about the origin which sends $\frac{z_2 - z_1}{1 - \bar{z}_1 z_2}$ to the point $\left|\frac{z_2 - z_1}{1 - \bar{z}_1 z_2}\right|$, then overall the composite $R \circ M$ is a direct hyperbolic transformation which sends z_1 to 0 and z_2 to $\left|\frac{z_2 - z_1}{1 - \bar{z}_1 z_2}\right|$.

Recall that any rotation of $\mathcal{D}$ about the origin is a direct hyperbolic transformation (Subsection 6.2.1).

By Property 6 it follows that

$$d(z_1, z_2) = d\left(0, \left|\frac{z_2 - z_1}{1 - \bar{z}_1 z_2}\right|\right), \quad \text{for all } z_1, z_2 \in \mathcal{D}.$$

This shows that the distance $d(z_1, z_2)$ must be some function of the quantity $\left|\frac{z_2 - z_1}{1 - \bar{z}_1 z_2}\right|$ alone. But what function ensures that d has the Properties 1-6 above? Indeed, does such a function exist? In fact, there is essentially only one 'well-behaved' function which yields these properties, and it turns out to be the function $\tanh^{-1}$.

We verify this assertion in Subsection 6.3.5.

Definition The **hyperbolic distance** $d(z_1, z_2)$ between the points z_1 and z_2 in the unit disc $\mathcal{D}$ is defined by

$$d(z_1, z_2) = \tanh^{-1}\left(\left|\frac{z_2 - z_1}{1 - \bar{z}_1 z_2}\right|\right).$$

Of course before we can be sure that this is a reasonable definition of distance we need to check that d satisfies all the Properties 1–6. Clearly d satisfies Properties 1, 2, 5, and we shall prove the other properties later in this section; for the moment we assume that all the properties hold, and we use them to explore some of the practical consequences of the definition.

First observe that the formula for the distance between two given points is particularly simple in the case where one of the points lies at the origin. Indeed we have

Definition The **hyperbolic distance** $d(0,z)$ between the points 0 and z in the unit disc $\mathcal{D}$ is

$$d(0,z) = \tanh^{-1}(|z|).$$

Since this formula is simpler to remember, we shall tend to use it more often than the formula for $d(z_1, z_2)$!

Since $|z|$ is just the Euclidean distance of z from the origin, this equation tells us that we can obtain the hyperbolic distance of a point z from the origin by applying the inverse tanh function to the Euclidean distance of z from the origin. The graph of the inverse tanh function in the margin reveals two important characteristics of hyperbolic distance.

1. Near the origin the graph is nearly a straight line with slope 1, so for a point z near the origin the hyperbolic distance of z from 0 is approximately equal to the Euclidean distance of z from 0. (This is analogous to the situation on the surface of the Earth; small portions of the Earth's surface *look* flat and distances between its points are *approximately* Euclidean.)
2. As the point z approaches the boundary of the disc $\mathcal{D}$, the hyperbolic distance of z from the origin increases without bound. Indeed, as the Euclidean distance $|z|$ tends to 1, the hyperbolic distance $\tanh^{-1}(|z|)$ tends to ∞. From the point of view of someone living in the hyperbolic geometry, the boundary points appear to be 'infinitely far away' – an observation that is consistent with the idea that parallel lines 'meet' on the boundary.

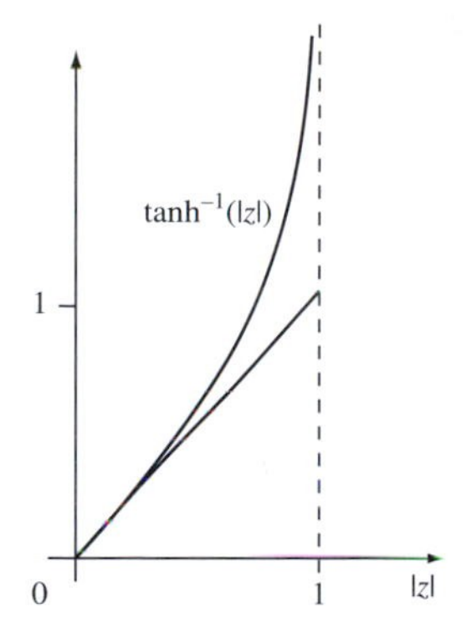

With the help of a calculator we can easily use the formula

$$d(0,z) = \tanh^{-1}(|z|) \tag{1}$$

to calculate the hyperbolic distance of a point from the origin.

At the end of this subsection we prove the alternative formula

$$d(0,z) = \tfrac{1}{2}\log_e\left(\tfrac{1+|z|}{1-|z|}\right)$$

which you can use if your calculator does not have a tanh function.

For example, the hyperbolic distance of $\frac{1}{2}$ from the origin is given by

$$d\left(0, \tfrac{1}{2}\right) = \tanh^{-1}\left(\frac{1}{2}\right) = \tanh^{-1}(0.5) \simeq 0.549.$$

But we can also use equation (1) to calculate the hyperbolic distance between any two points which lie on a diameter of the disc $\mathcal{D}$.

Example 1 Find the hyperbolic distance between the points 0.1 and 0.2.

Solution Here we use Property 4, which states that distances along a (hyperbolic) line are additive. In particular, this implies that

$$d(0.1, 0.2) = d(0, 0.2) - d(0, 0.1),$$

so we can write

$$\begin{aligned} d(0.1, 0.2) &= d(0, 0.2) - d(0, 0.1) \\ &= \tanh^{-1}(0.2) - \tanh^{-1}(0.1) \\ &\simeq 0.203 - 0.100 \\ &\simeq 0.102. \end{aligned}$$

□

Throughout Section 6.3 we shall work to the full accuracy of our calculator, but we shall record our results only to 3 or 4 decimal places.

Problem 1 Determine the hyperbolic distances $d\left(0, \frac{1}{3}i\right)$ and $d(0.8, 0.9)$.

By rearranging equation (1) we obtain the formula

$$|z| = \tanh d(0, z). \tag{2}$$

This can be used to locate a point, given its hyperbolic distance along a radius of the disc $\mathscr{D}$. For example, the point a that is at a hyperbolic distance 0.1 from the origin along the positive real axis is given by $a = \tanh 0.1 \simeq 0.0997$.

Problem 2 The following table (plotted in the margin) illustrates how points bunch up towards the boundary of $\mathscr{D}$ as their hyperbolic distance from 0 doubles.

Find the two missing entries in the table.

$d(0, a), a > 0$	0.2	0.4	0.8	1.6	3.2
a	0.197...	0.380...	0.664...		

If we have to calculate the distance between two points that do not lie on a diameter of $\mathscr{D}$, then we can use the distance formula given in the definition.

Example 2 Find the hyperbolic distance between the points $\frac{1}{2}$ and $\frac{1}{3}i$.

Solution From the definition of hyperbolic distance we have

$$\begin{aligned} d\left(\tfrac{1}{2}, \tfrac{1}{3}i\right) &= \tanh^{-1}\left(\left|\frac{\frac{1}{2} - \frac{1}{3}i}{1 - \overline{\frac{1}{3}i} \cdot \frac{1}{2}}\right|\right) \\ &= \tanh^{-1}\left(\left|\frac{3 - 2i}{6 + i}\right|\right) \\ &= \tanh^{-1}\left(\sqrt{\frac{13}{37}}\right) \\ &\simeq 0.6819. \end{aligned}$$

□

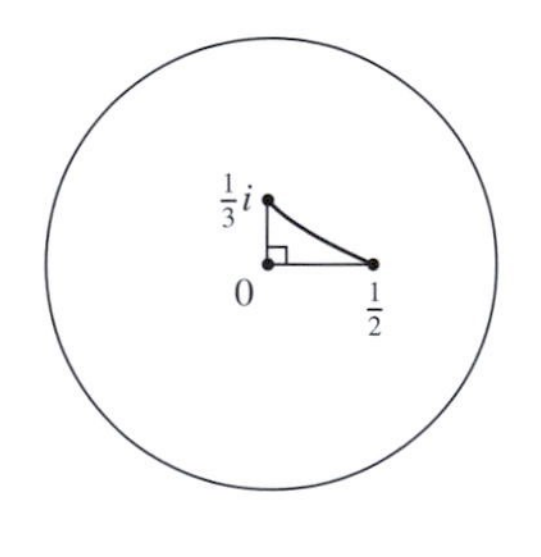

We can use the distances that we have calculated above to demonstrate that Pythagoras' Theorem fails to hold in hyperbolic geometry, at least without some rewording. For, if we consider the hyperbolic right-angled triangle with vertices at 0, $\frac{1}{3}i$ and $\frac{1}{2}$, then the square of the hypotenuse is

$$d\left(\tfrac{1}{2}, \tfrac{1}{3}i\right)^2 \simeq 0.6819^2 \simeq 0.4650,$$

whereas the sum of the squares of the other two sides is

$$d\left(0, \tfrac{1}{2}\right)^2 + d\left(0, \tfrac{1}{3}i\right)^2 \simeq 0.5493^2 + 0.3466^2 \simeq 0.4219.$$

We evaluated $d\left(0, \frac{1}{2}\right)$ before Example 1, and $d\left(0, \frac{1}{3}i\right)$ in Problem 1.

We shall see later on how Pythagoras' Theorem can be reformulated in hyperbolic geometry.

Subsection 6.4.3, Theorem 8

We end this subsection by giving an alternative formula for the hyperbolic distance of a point z from the origin. First, observe that

$$\begin{aligned} |z| &= \tanh d(0, z) \\ &= \frac{\sinh d(0, z)}{\cosh d(0, z)} \\ &= \frac{e^{d(0,z)} - e^{-d(0,z)}}{e^{d(0,z)} + e^{-d(0,z)}} \\ &= \frac{e^{2d(0,z)} - 1}{e^{2d(0,z)} + 1}. \end{aligned}$$

We can solve this equation to obtain an expression for $d(0, z)$ in terms of $|z|$. By cross-multiplication, we obtain

$$e^{2d(0,z)} - 1 = |z|\left(e^{2d(0,z)} + 1\right),$$

which implies that

$$e^{2d(0,z)}(1 - |z|) = 1 + |z|,$$

which is equivalent to

$$e^{2d(0,z)} = \frac{1 + |z|}{1 - |z|}.$$

Taking natural logarithms of both sides, we deduce that

$$2d(0, z) = \log_e\left(\frac{1 + |z|}{1 - |z|}\right),$$

and so

$$d(0, z) = \frac{1}{2}\log_e\left(\frac{1 + |z|}{1 - |z|}\right). \qquad (3)$$

This formula is useful if your calculator does not have a tanh function.

The formula for $d(0, z)$ is sometimes quoted in this form.

6.3.2 Hyperbolic Midpoints

Having described how to calculate hyperbolic distances we now introduce the idea of the hyperbolic midpoint of a hyperbolic line segment.

Definition A point m is the **hyperbolic midpoint** of the hyperbolic line segment joining a and b if m lies on this segment and

$$d(a, m) = d(m, b) = \frac{1}{2}d(a, b).$$

$d(a,m)$ $d(m,b)$
a m b

For simplicity, we confine our attention to midpoints of line segments that lie along a diameter of $\mathscr{D}$. The method depends on whether the endpoints of the segment lie on the same side of the origin O, or on opposite sides of O.

Example 3 Find the hyperbolic midpoint m of the line segment which joins each of the following pairs of points.

(a) $\frac{1}{4}i$ and $\frac{3}{4}i$ (b) $\frac{1}{4}i$ and $-\frac{3}{4}i$

Solution First observe that

$$d\left(0, \tfrac{1}{4}i\right) = \tanh^{-1}\left(\left|\tfrac{1}{4}i\right|\right) = \tanh^{-1}(0.25) = 0.255\ldots,$$

and

$$d\left(0, \tfrac{3}{4}i\right) = \tanh^{-1}\left(\left|\tfrac{3}{4}i\right|\right) = \tanh^{-1}(0.75) = 0.973\ldots.$$

(a) Here $\frac{3}{4}i$ and $\frac{1}{4}i$ lie on the same side of 0, so the midpoint must lie at a hyperbolic distance

$$\tfrac{1}{2}\left(d\left(0, \tfrac{3}{4}i\right) + d\left(0, \tfrac{1}{4}i\right)\right) \simeq \tfrac{1}{2}(0.973 + 0.255) = 0.614$$

from 0 on the radius through $\frac{3}{4}i$. That is, $m \simeq (\tanh 0.614)i \simeq 0.547i$. (Notice that this is further from 0 than the Euclidean midpoint $0.5i$.)

(b) Here $-\frac{3}{4}i$ and $\frac{1}{4}i$ lie on opposite sides of 0, so the midpoint lies at a hyperbolic distance

$$\tfrac{1}{2}\left(d\left(0, -\tfrac{3}{4}i\right) - d\left(0, \tfrac{1}{4}i\right)\right) \simeq \tfrac{1}{2}(0.973 - 0.255) = 0.359$$

from 0 on the radius through $-\frac{3}{4}i$. That is, $m \simeq -(\tanh 0.359)i \simeq -0.344i$. □

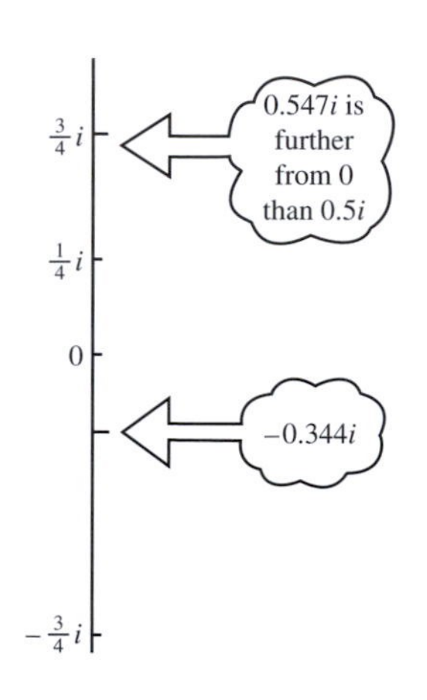

The following strategy generalizes the method used in Example 3.

Strategy To find the hyperbolic midpoint of a line segment joining two points p, q on a diameter of $\mathscr{D}$, where $|p| > |q|$:

1. calculate $d(0, p)$ and $d(0, q)$;
2. if p and q lie on opposite sides of 0, then calculate

$$d = \tfrac{1}{2}(d(0, p) - d(0, q)),$$

otherwise calculate

$$d = \tfrac{1}{2}(d(0, p) + d(0, q));$$

3. then the hyperbolic midpoint is the point m on the radius through 0 and p at a Euclidean distance $\tanh(d)$ from 0.

Problem 3 Find the hyperbolic midpoint m of the line segment which joins each of the following pairs of points:

$$0.5 \text{ and } 0.8; \qquad -0.2 \text{ and } 0.8.$$

Remark

To calculate the hyperbolic midpoint of a line segment which joins two *arbitrary* points p, q in $\mathcal{D}$ we would use a Möbius transformation M to map p to 0 and q to $M(q)$. After calculating the hyperbolic midpoint m' of the segment from 0 to $M(q)$, we would then obtain the midpoint of the original segment by calculating $M^{-1}(m')$.

6.3.3 Hyperbolic Circles

By analogy with Euclidean circles, we define a hyperbolic circle to be the locus of points which are a fixed hyperbolic distance from a fixed point.

Definition The **hyperbolic circle** with **hyperbolic radius** r and **hyperbolic centre** at c is the set defined by

$$\{z : d(c, z) = r, \quad z \in \mathcal{D}\}.$$

It is natural to ask what a hyperbolic circle looks like to the Euclidean eye. In the case of a hyperbolic circle with radius r centred at the origin, it is just the set of points given by

$$\{z : d(0, z) = r, \quad z \in \mathcal{D}\};$$

that is, the set

$$\{z : \tanh^{-1}(|z|) = r, \quad z \in \mathcal{D}\},$$

or

$$\{z : |z| = \tanh r, \quad z \in \mathcal{D}\}.$$

This is a Euclidean circle with radius $\tanh r$ centred at the origin! Even more remarkable is the fact that *every* hyperbolic circle is a Euclidean circle.

Theorem 1 Every hyperbolic circle is a Euclidean circle in $\mathscr{D}$, and vice versa.

Proof We have already established the result for circles centred at 0, so let C be any hyperbolic circle with hyperbolic centre $m \neq 0$. Let the diameter of $\mathscr{D}$ through m meet C at the points a and b, and let K be the Euclidean circle which has ab as a diameter and (Euclidean) centre p.

For the steps in the argument before Theorem 1 can be reversed.

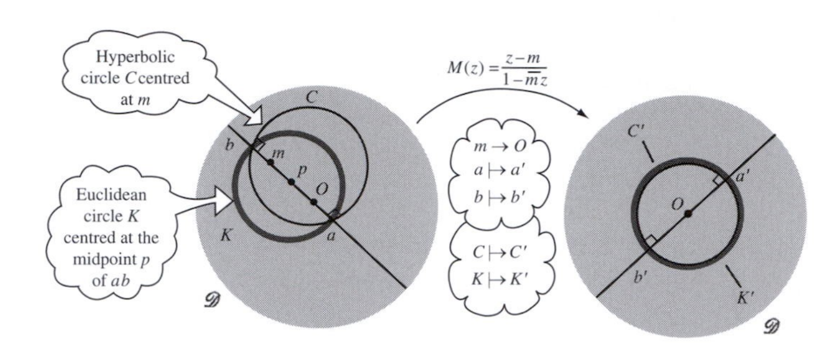

We will show that C is a Euclidean circle by showing that it coincides with K. To do this let M be the hyperbolic transformation defined by

$$M(z) = \frac{z - m}{1 - \overline{m}z}, \quad z \in \mathscr{D},$$

and let a' and b' be the images of a and b under M.

Since M preserves hyperbolic distances, and maps m to O, it must map C to a hyperbolic circle C' with centre O. But we already know that such a circle is also a Euclidean circle. Moreover, since the line segment ab passes through m, its image $a'b'$ passes through O and is therefore a diameter of C'.

Also, since M is angle-preserving, and since ab meets K at right angles, it follows that M maps K to a circle K' which meets $a'b'$ at right angles at the points a' and b'. This implies that $a'b'$ is also a diameter of K'. It follows that C' coincides with K', and hence that C coincides with K.

Theorem 2, Subsection 6.1.1

Conversely, let K be any Euclidean circle in $\mathscr{D}$ with Euclidean centre $p \neq 0$, and let the diameter of $\mathscr{D}$ through p meet K at the points a and b. If C is the hyperbolic circle through a and b with hyperbolic centre the hyperbolic midpoint m of ab, then we can use the same argument as before to deduce that K coincides with C, and so we conclude that K is a hyperbolic circle. ∎

The above proof shows that the Euclidean and hyperbolic centres of a circle lie on the same line through O. This observation enables us to write down the following strategy for finding the Euclidean centre and radius of a hyperbolic circle.

Strategy To find the Euclidean centre and radius of a hyperbolic circle C with hyperbolic centre m:

1. find the points a, b where Om meets C;
2. the Euclidean centre of C is the Euclidean midpoint of ab;
3. the Euclidean radius of C is $\frac{1}{2}|a - b|$.

Example 4 Find the Euclidean centre and radius of the hyperbolic circle

$$C = \left\{z : d\left(z, \tfrac{1}{2}\right) = \tfrac{1}{2}\right\}.$$

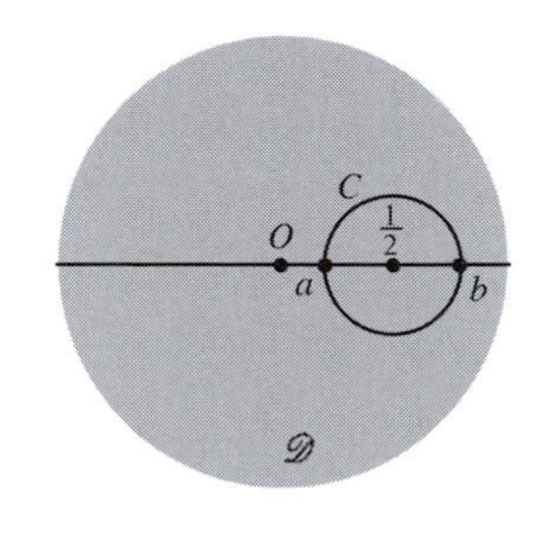

Solution Here the hyperbolic centre m is the point $\frac{1}{2}$, and so

$$d(0, m) = \tanh^{-1}\left(\tfrac{1}{2}\right) \simeq 0.549.$$

Since the hyperbolic radius of C is equal to $\frac{1}{2}$, it follows that Om meets C at the points a, b, where

$$d(0, a) \simeq 0.549 - 0.5 = 0.049,$$

$$d(0, b) \simeq 0.549 + 0.5 = 1.049.$$

So

$$a \simeq \tanh 0.049 \simeq 0.049,$$

$$b \simeq \tanh 1.049 \simeq 0.782.$$

Since a and b both lie on the same side of 0, we have

$$\text{Euclidean centre} = \tfrac{1}{2}(a + b) \simeq 0.415,$$

$$\text{Euclidean radius} = \tfrac{1}{2}|a - b| \simeq 0.366. \qquad \square$$

Problem 4 Determine the Euclidean centre and radius of the hyperbolic circle

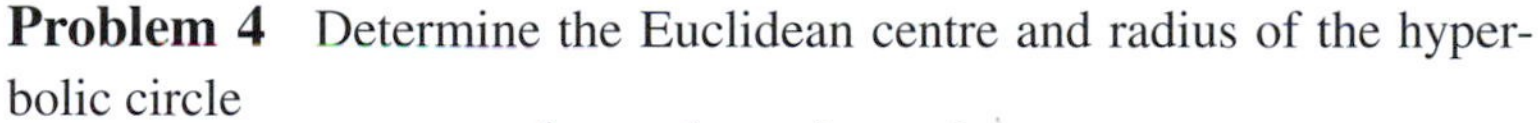

$$C = \left\{z : d\left(z, -\tfrac{1}{4}\right) = \tfrac{1}{2}\right\}.$$

A similar strategy can be used to find the hyperbolic centre and radius of a Euclidean circle.

Strategy To find the hyperbolic centre and radius of a Euclidean circle K with Euclidean centre p:

1. find the points a, b where Op meets K;
2. the hyperbolic centre of K is the hyperbolic midpoint of ab;
3. the hyperbolic radius of K is $\frac{1}{2}d(a, b)$.

Example 5 Find the hyperbolic centre and radius of the Euclidean circle

$$K = \left\{z : \left|z + \tfrac{1}{2}i\right| = \tfrac{1}{4}\right\}.$$

Solution Here the Euclidean centre p is the point $-\frac{1}{2}i$, and the Euclidean radius is $\frac{1}{4}$, so Op meets K at the points $a = -\frac{1}{4}i$ and $b = -\frac{3}{4}i$. Thus

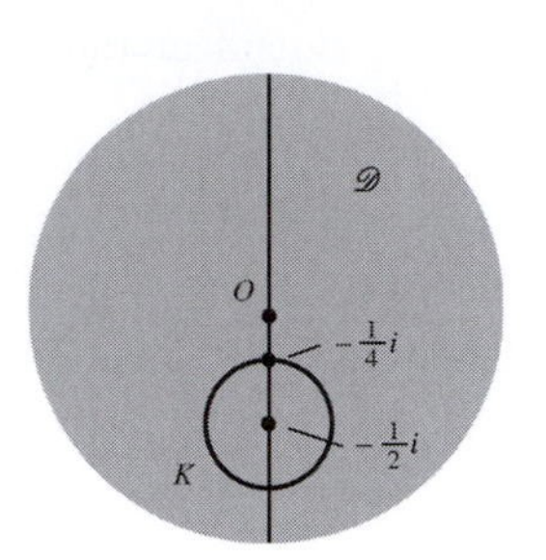

$$d(0, a) = \tanh^{-1}\left(\left|-\tfrac{1}{4}i\right|\right) = \tanh^{-1}(0.25) \simeq 0.255\ldots;$$

$$d(0, a) = \tanh^{-1}\left(\left|-\tfrac{3}{4}i\right|\right) = \tanh^{-1}(0.75) \simeq 0.973\ldots.$$

Since a and b both lie on the same side of O, the hyperbolic centre m of K is given by

$$d(0, m) = \tfrac{1}{2}(d(0, a) + d(0, b)) \simeq 0.614,$$

so that

$$\text{hyperbolic centre } m \simeq -i \cdot \tanh 0.614 \simeq -0.547i;$$

$$\text{hyperbolic radius} = \tfrac{1}{2}|d(0, a) - d(0, b)| \simeq 0.359. \qquad \square$$

Problem 5 Determine the hyperbolic centre and radius of the Euclidean circle

$$K = \left\{z : \left|z - \tfrac{1}{4}\right| = \tfrac{1}{2}\right\}.$$

We now use the fact that hyperbolic circles are also Euclidean circles to prove the Triangle Inequality property of hyperbolic distance.

Property 3 of a distance function, Subsection 6.3.1.

Theorem 2 For all z_1, z_2, z_3 in $\mathscr{D}$:

$$d(z_1, z_3) + d(z_3, z_2) \geq d(z_1, z_2).$$

Proof First let M be a hyperbolic transformation that maps z_1 to 0, and let $M(z_2) = b$ and $M(z_3) = c$. Then Ob is a straight line.

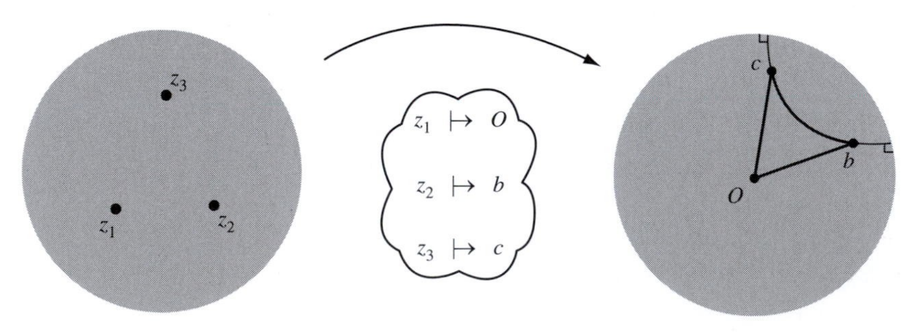

Taking O as centre we draw a hyperbolic circle C_2 through c to meet Ob at the point q; then $d(O,q) = d(O,c)$. Similarly, taking b as centre we draw a hyperbolic circle C_1 through c to meet Ob at the point p, where $d(b,p) = d(b,c)$.

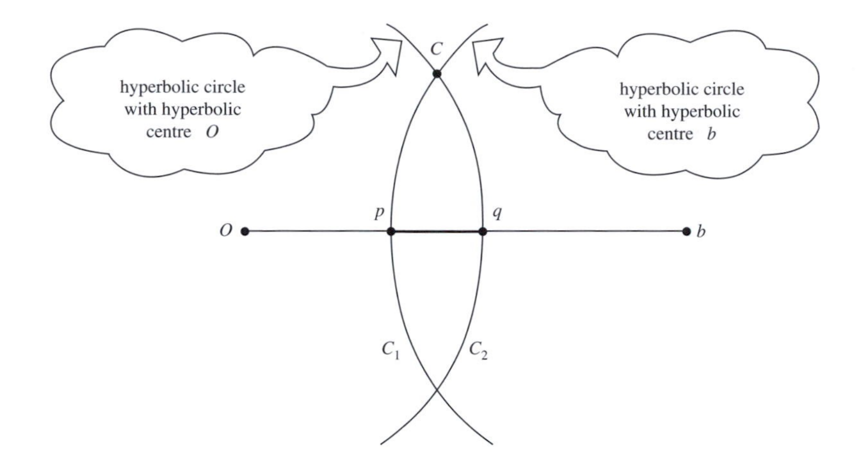

We have drawn this figure in the case that p and q lie between O and b. The argument also holds if q lies beyond b or p beyond O.

Since both circles are also Euclidean circles with centres on Ob, and since the circles intersect, it follows that

Recall that hyperbolic circles are Euclidean circles, by Theorem 1.

$$\begin{aligned} d(0,c) + d(c,b) &= d(0,q) + d(p,b) \\ &\geq d(0,q) + d(q,b) \\ &= d(0,b). \end{aligned}$$

Since M preserves hyperbolic distances, we deduce that

Property 6, Subsection 6.3.1

$$d(z_1, z_3) + d(z_3, z_2) \geq d(z_1, z_2),$$

as required. ■

We can deduce from the Triangle Inequality (Property 3) that in hyperbolic geometry the curve of shortest length or 'geodesic' between two points of $\mathscr{D}$ is the segment of the d-line that joins them. Rather loosely, we can express this fact as follows: 'Distances are measured along d-lines in hyperbolic geometry.'

We omit a formal definition of 'length of a curve', and so a proof of this claim.

6.3.4 Reflected Points in Hyperbolic Geometry

In Euclidean geometry the image A' of a point A under reflection in a line ℓ is an equal distance from the line and on the opposite side of the line, so that ℓ is the perpendicular bisector of the line segment AA'. We now prove that the same result also holds in hyperbolic geometry, where the distances and lines are hyperbolic ones.

To state the result, we need the concept of *hyperbolic line segment,* or *d-line segment*; this is just that part of the d-line through two points which lies between those points.

Theorem 3 Let A and A' be points in the unit disc $\mathcal{D}$ that are images of each other under reflection in a d-line ℓ. Then ℓ is the hyperbolic perpendicular bisector of the hyperbolic line segment AA'.

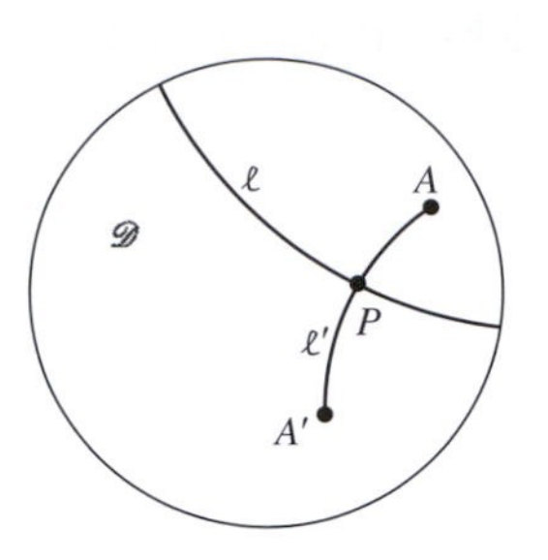

Proof Let the d-line ℓ' through A and A' meet ℓ at P.

Now A and A' map to each other under hyperbolic reflection in ℓ, and P remains invariant. But hyperbolic reflection maps d-lines to d-lines, and there is exactly one d-line through A and A'. Thus ℓ' must map onto itself under the hyperbolic reflection in ℓ, and the hyperbolic line segment PA must map onto the hyperbolic line segment PA'.

But hyperbolic reflection preserves angles and lengths. Since lengths are preserved, it follows that PA and PA' must be of equal hyperbolic length. Also, since the angles that PA and PA' make with the same part of ℓ at P map onto each other, they must be equal too; since they must add up to a total angle π at P (because APA' is a hyperbolic line segment through P), it follows that PA and PA' both meet ℓ at right angles. ■

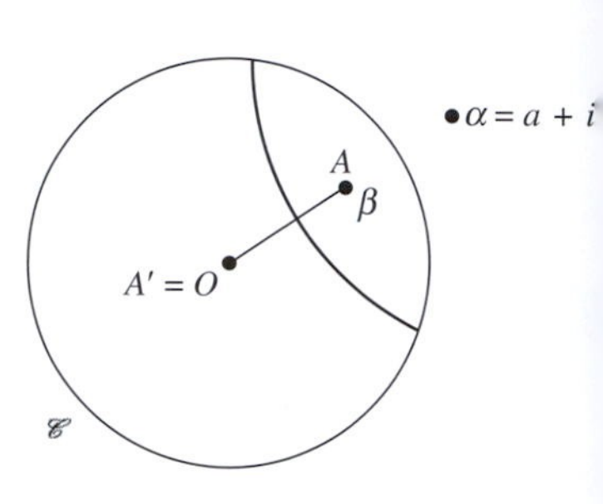

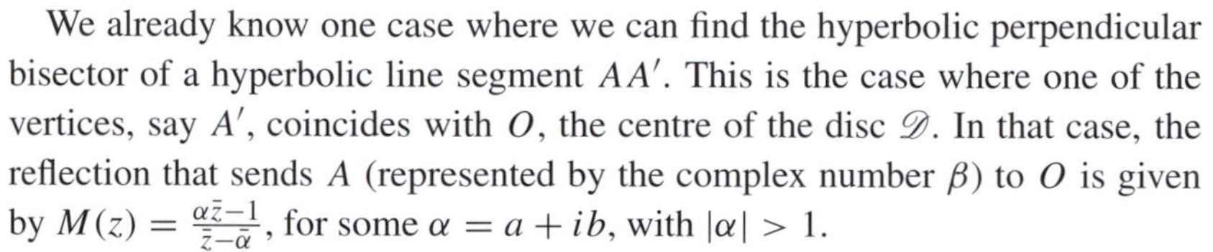

We already know one case where we can find the hyperbolic perpendicular bisector of a hyperbolic line segment AA'. This is the case where one of the vertices, say A', coincides with O, the centre of the disc $\mathcal{D}$. In that case, the reflection that sends A (represented by the complex number β) to O is given by $M(z) = \frac{\alpha\bar{z}-1}{\bar{z}-\bar{\alpha}}$, for some $\alpha = a + ib$, with $|\alpha| > 1$.

The equation of the d-line obtained from α in which the reflection takes place is, as we saw earlier,

By Corollary 1, Subsection 6.1.1

$$x^2 + y^2 - 2ax - 2by + 1 = 0.$$

Since $M(\beta) = 0$, it follows from the formula for M that $\alpha\bar{\beta} - 1 = 0$.

More generally, suppose we want to find a hyperbolic reflection that exchanges the points p and q in $\mathcal{D}$. Again, the hyperbolic reflection that does this must be of the form

$$M(z) = \frac{\alpha\bar{z} - 1}{\bar{z} - \bar{\alpha}}.$$

Subsection 6.2.1, Lemma 1

The condition $M(p) = q$ implies

$$q = \frac{\alpha\bar{p} - 1}{\bar{p} - \bar{\alpha}},$$

which is equivalent to

$$1 + \bar{p}q = \bar{\alpha}q + \alpha\bar{p}. \tag{4}$$

The condition $M(q) = p$ implies

$$p = \frac{\alpha\bar{q} - 1}{\bar{q} - \bar{\alpha}},$$

which is equivalent to

$$1 + \bar{q}p = \bar{\alpha}p + \alpha\bar{q}. \tag{5}$$

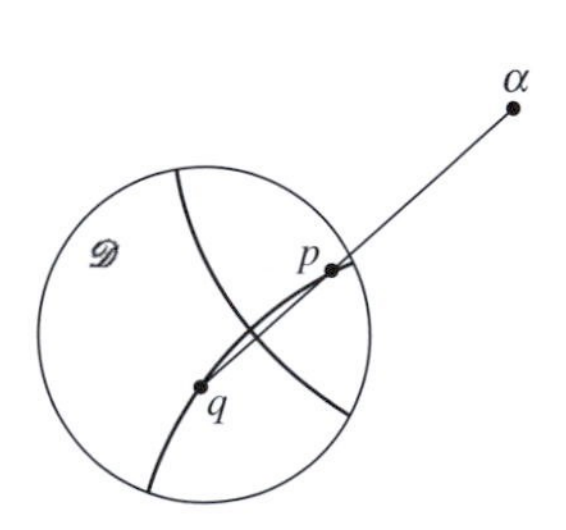

Equations (4) and (5) can be solved for α provided that $p\bar{p} - q\bar{q} \neq 0$. Indeed, if we subtract q times equation (5) from p times equation (4), and then divide by $p\bar{p} - q\bar{q}$, we obtain

$$\alpha = \frac{p - q + pq(\bar{p} - \bar{q})}{p\bar{p} - q\bar{q}}.$$

Notice that in the special case when $q = 0$, this formula gives $\alpha = 1/\bar{p}$, as it should.

If $p\bar{p} - q\bar{q} = 0$, then $|p| = |q|$, in which case the sought-for reflection is simply Euclidean reflection in the diameter of $\mathscr{D}$ which bisects the angle $\angle pOq$.

We summarize this discussion in the following lemma.

Lemma 1 Reflection Lemma

Let p and q be points in the unit disc $\mathscr{D}$. If $|p| \neq |q|$, then the hyperbolic reflection that maps p and q onto each other is given by

$$M(z) = \frac{\alpha\bar{z} - 1}{\bar{z} - \bar{\alpha}}, \quad \text{where } \alpha = \frac{p - q + pq(\bar{p} - \bar{q})}{p\bar{p} - q\bar{q}}.$$

The d-line in which this reflection takes place has equation

$$x^2 + y^2 - 2ax - 2by + 1 = 0, \quad \text{where } \alpha = a + ib.$$

By Lemma 1, Subsection 6.1.1

Example 6 Determine the equation of the hyperbolic perpendicular bisector of [0.2, 0.9], the line segment from 0.2 to 0.9.

Solution Using the Reflection Lemma, with $p = 0.9$ and $q = 0.2$, we find that

$$\alpha = \frac{0.7 + 0.9 \cdot 0.2 \cdot 0.7}{0.81 - 0.04} \simeq 1.0727.$$

So the equation of the d-line which is the (hyperbolic) perpendicular bisector of [0.2, 0.9] is

$$x^2 + y^2 - 2ax + 1 = 0, \quad \text{where } a \simeq 1.0727. \qquad \square$$

Problem 6

(a) Determine the equation of the hyperbolic perpendicular bisector ℓ of the line segment $[0.5, 0.8]$.

(b) Deduce the equations of the hyperbolic perpendicular bisectors of the line segments $[-0.8, -0.5]$ and $[0.5i, 0.8i]$.

We often write $[z_1, z_2]$ to mean the Euclidean line segment in $\mathbb{C}$ with endpoints z_1, z_2.

Problem 7

(a) Let A be a point in $\mathscr{D}$, ℓ a d-line through A, and $0 < \theta < \pi$. Prove that there are exactly two d-lines through A that make an angle θ with ℓ.

(b) Prove that, if ΔABC is a d-triangle, then there is a unique d-line ℓ that bisects $\angle BAC$, and that reflection in ℓ maps the d-lines containing BA and CA onto each other.

6.3.5 Proofs

In Subsection 6.3.1 we asserted that the hyperbolic distance between two points is invariant under hyperbolic transformations. We now supply a proof of this fact.

Theorem 4 The formula for hyperbolic distance

$$d(z_1, z_2) = \tanh^{-1}\left(\left|\frac{z_2 - z_1}{1 - \bar{z}_1 z_2}\right|\right), \quad \text{where } z_1, z_2 \in \mathscr{D},$$

satisfies Property 6; that is,

See Subsection 6.3.1.

$$d(z_1, z_2) = d(M(z_1), M(z_2))$$

for all z_1 and z_2 in $\mathscr{D}$ and all direct hyperbolic transformations M in $G_{\mathscr{D}}$.

Proof First, we define the expression $R(z_1, z_2)$ to be

$$R(z_1, z_2) = \left|\frac{z_2 - z_1}{1 - \bar{z}_1 z_2}\right|, \quad \text{where } z_1, z_2 \in \mathscr{D}. \tag{6}$$

It follows from the definition of $d(z_1, z_2)$ above that

We omit the details of this, for brevity.

$$d(z_1, z_2) = \tanh^{-1}(R(z_1, z_2));$$

hence d possesses Property 6 if we can prove that

$$R(z_1, z_2) = R(M(z_1), M(z_2)) \tag{7}$$

for all $z_1, z_2 \in \mathscr{D}$ and all $M \in G_{\mathscr{D}}$.

So, let $z_1, z_2 \in \mathscr{D}$ and $M \in G_{\mathscr{D}}$. Then the direct hyperbolic transformation M_1 in $G_{\mathscr{D}}$ given by

$$M_1 : z \mapsto \frac{z - z_1}{1 - \overline{z_1} z}, \quad z \in \mathscr{D},$$

maps z_1 to 0, and z_2 to $\dfrac{z_2 - z_1}{1 - \bar{z}_1 z_2}$.

Also, the direct hyperbolic transformation M_2 in $G_{\mathscr{D}}$ given by

$$M_2 : z \mapsto \frac{z - M(z_1)}{1 - \overline{M(z_1)} z}, \quad z \in \mathscr{D},$$

where $M \in G_{\mathscr{D}}$ so that $|M(z_1)| < 1$, maps $M(z_1)$ to 0, and $M(z_2)$ to $\frac{M(z_2) - M(z_1)}{1 - \overline{M(z_1)} M(z_2)}$.

Now, the composite mapping $M_2 \circ M \circ M_1^{-1}$ is also a transformation in $G_{\mathcal{D}}$. It maps $\mathcal{D}$ to $\mathcal{D}$, 0 to 0, and

Note that M_1^{-1} maps 0 to z_1, and $\frac{z_2 - z_1}{1 - \bar{z}_1 z_2}$ to z_2.

$$\frac{z_2 - z_1}{1 - \bar{z}_1 z_2} \quad \text{to} \quad \frac{M(z_2) - M(z_1)}{1 - \overline{M(z_1)}M(z_2)}. \tag{8}$$

But any direct transformation in $G_{\mathcal{D}}$ that maps 0 to 0 is simply a (Euclidean) rotation; so it follows from equation (8) that

See Case 1 of Theorem 2, Subsection 6.2.1.

$$\left|\frac{z_2 - z_1}{1 - \bar{z}_1 z_2}\right| = \left|\frac{M(z_2) - M(z_1)}{1 - \overline{M(z_1)}M(z_2)}\right|. \tag{9}$$

It follows from the definition of R that we can rewrite equation (9) in the desired form

We defined R in equation (6).

$$R(z_1, z_2) = R(M(z_1), M(z_2)). \qquad \blacksquare$$

Next, we stated earlier that the function d that we used to define the hyperbolic distance function in $\mathcal{D}$ was 'essentially' the only 'well-behaved' function with Properties 1-6 of Subsection 6.3.1. We now explain why this is so.

Theorem 5 Let $d(z_1, z_2)$ be any 'well-behaved' function defined for all $z_1, z_2 \in \mathcal{D}$ that satisfies Properties 1-6 of Subsection 6.3.1. Then

$$d(z_1, z_2) = K \tanh^{-1}\left(\left|\frac{z_2 - z_1}{1 - \overline{z_1} z_2}\right|\right), \text{ for some } K > 0. \tag{10}$$

Proof Let $z_1, z_2 \in \mathcal{D}$. Since hyperbolic distances are invariant under Möbius transformations, $d(z_1, z_2)$ is unchanged if we map z_1 to the origin by the direct hyperbolic transformation (a Möbius transformation)

By Property 6

$$z \mapsto \frac{z - z_1}{1 - \overline{z_1} z}.$$

Next, hyperbolic distances are invariant under rotations of $\mathcal{D}$, since such rotations are hyperbolic transformations. So, we may also assume that z_2 lies on the positive real axis.

Hence, in order to prove the formula (10) it is sufficient to prove that

$$d(0, z) = K \tanh^{-1}(z), \text{ for some } K > 0 \text{ and all } z \in (0, 1). \tag{11}$$

For simplicity, we will now simply write $d(z)$ in place of $d(0, z)$ whenever the context means that no confusion will arise.

Let $0 < a, c < 1$. Then the direct hyperbolic transformation (a Möbius transformation)

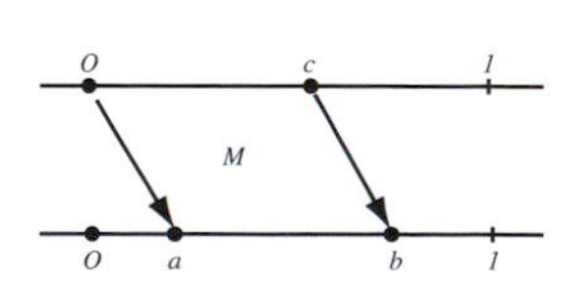

$$M(z) = \frac{z + a}{1 + az}, \quad z \in \mathcal{D}$$

maps 0 to a, and c onto $b = \frac{c+a}{1+ac}$. Since M maps $[0, 1)$ one-one to $[a, 1)$, we must have $a < b < 1$.

Now, hyperbolic distances are additive along a line, so that

By Property 4

$$d(b) = d(a) + d(a, b). \tag{12}$$

But, since hyperbolic distances are invariant under hyperbolic transformations, we have

By Property 6

$$d(a,b) = d(0,c) \\ = d(c),$$

so that it follows from equation (12) that

In this equation, a and c are independent variables.

$$d\left(\frac{c+a}{1+ac}\right) = d(a) + d(c). \tag{13}$$

Now we will assume that the function d is differentiable on $[0, 1)$. If we then differentiate both sides of equation (13) with respect to the variable c, we get

This is where we use the hypothesis that the distance function is 'well-behaved'.

$$d'\left(\frac{c+a}{1+ac}\right) \cdot \frac{(1+ac)\cdot 1 - (c+a)\cdot a}{(1+ac)^2} = 0 + d'(c);$$

and we may rewrite this equation in the form

$$d'\left(\frac{c+a}{1+ac}\right) = \frac{(1+ac)^2}{1-a^2} \cdot d'(c).$$

If we substitute 0 for c into this equation, we get

Here $d'_R(0)$ is the right derivative of d at 0.

$$d'(a) = \frac{d'_R(0)}{1-a^2} \\ = \frac{K}{1-a^2}, \text{ for some real } K.$$

Since d must be an increasing function on $[0, 1)$, we must have $K \geq 0$. But we cannot have $K = 0$, as it would then follow from the above equation that $d'(a) = 0$ for all $a \in [0, 1)$ - so that $d(0, a)$ takes the same value (which would have to be 0) for all $a \in [0, 1)$. Hence

This follows from Property 1.

$$d'(a) = \frac{K}{1-a^2}, \text{ for some } K > 0, \text{ and all } a \in (0, 1).$$

If we then integrate both sides of this formula from 0 to z, we get

$$d(z) = K \tanh^{-1}(z), \text{ for some } K > 0, \text{ and all } z \in [0, 1).$$ ■

Remarks

1. For any (complex) point z near the origin, equation (10) gives

$$d(0, z) = K \tanh^{-1}(|z|) \\ \simeq K|z|,$$

so it is natural to make the choice $K = 1$ for the definition of distance in hyperbolic geometry. Then $d(0, z) = \tanh^{-1}(|z|)$, for all $z \in \mathscr{D}$.

2. When $d(0, z) = \tanh^{-1}(|z|)$, $z \in \mathscr{D}$, we can reformulate the equation $d'(z) = \frac{1}{1-|z|^2}$ in the form

In advanced texts, the formula for hyperbolic distance is often used in this infinitesimal form.

$$\mathrm{d}s = \frac{|\mathrm{d}z|}{1-|z|^2},$$

relating infinitesimal hyperbolic distances $\mathrm{d}s$ to infinitesimal Euclidean distances $|\mathrm{d}z|$ in $\mathscr{D}$ near any point $z \in \mathscr{D}$.

6.4 Geometrical Theorems

6.4.1 Triangles

A **triangle** in hyperbolic geometry, or a ***d*-triangle,** consists of three points in the unit disc $\mathscr{D}$ that do not lie on a single d-line, together with the segments of the three d-lines joining them. One or two of the sides may be segments of diameters of $\mathscr{D}$; but, in general, the sides are parts of Euclidean circles.

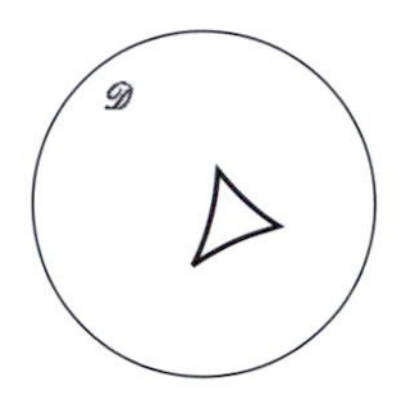

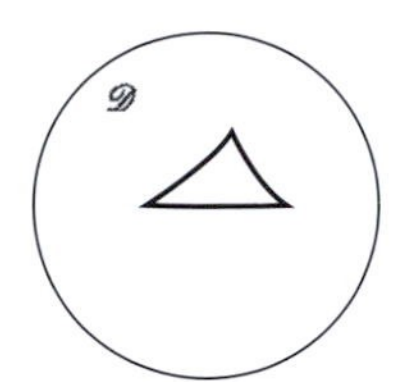

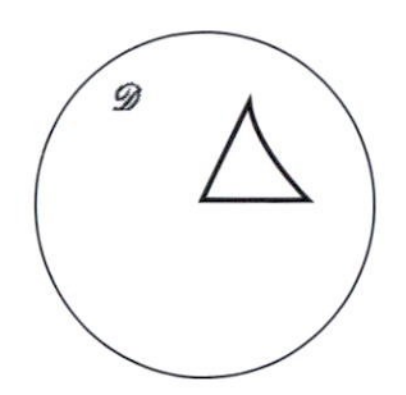

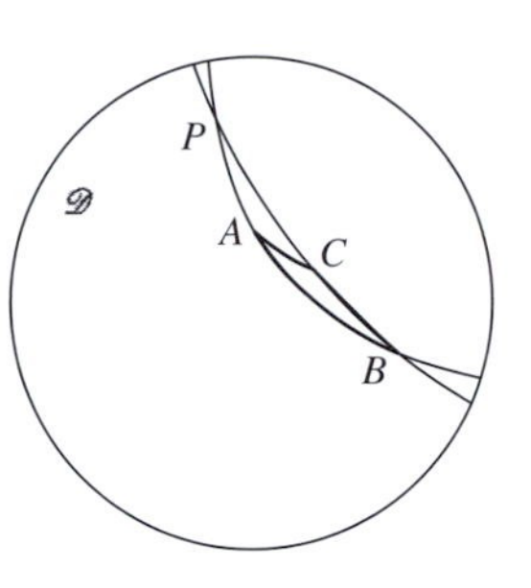

Notice that if ΔABC is a d-triangle, then the extended sides can meet only at A, B and C. For if the d-lines through A and B and through B and C meet at a point P distinct from B, then the two d-lines meet at B and P. But we saw earlier that there is a unique d-line through any two points of $\mathscr{D}$; so it would follow that AB and BC are part of a single d-line, a possibility that we have already excluded. So the sides of a d-triangle cannot 'overlap'.

Subsection 6.1.2, Theorem 4

One of the basic results in Euclidean geometry is that the sum of the angles of a (Euclidean) triangle is π. This is *not* true for d-triangles in hyperbolic geometry!

Theorem 1 The sum of the angles of a d-triangle is less than π.

Proof By the Origin Lemma, we can map the d-triangle ΔABC onto a d-triangle $\Delta OB'C'$ by any hyperbolic transformation that sends A to the origin O. Since hyperbolic transformations preserve angles, the sums of the angles of the two d-triangles are the same.

Subsection 6.1.2, Lemma 2

Then OB' and OC' are parts of Euclidean lines, and $B'C'$ is part of a Euclidean circle that 'bends towards' the origin. Thus the angles at B' and C' of the d-triangle $\Delta OB'C'$ are less than the corresponding angles of the Euclidean triangle $\Delta OB'C'$, and the angles of both triangles at O are the same.

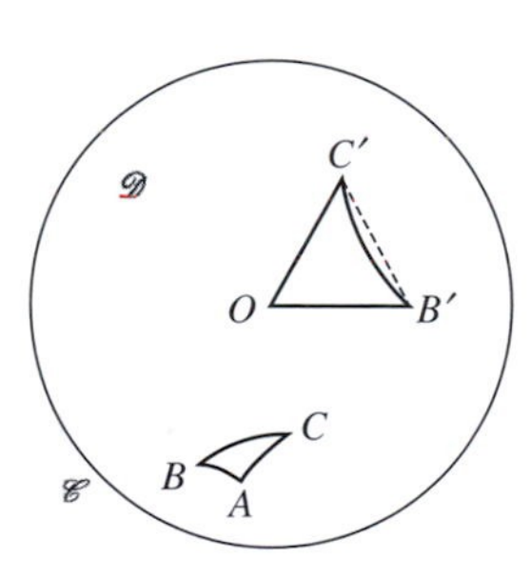

Since the sum of the angles of the Euclidean triangle $\Delta OB'C'$ is π, it follows that the sum of the angles of the d-triangle $\Delta OB'C'$ is (strictly) less than π. The result then follows. ■

Problem 1 Prove that each external angle of a d-triangle is greater than the sum of the opposite two internal angles. (See the diagram on the left below.)

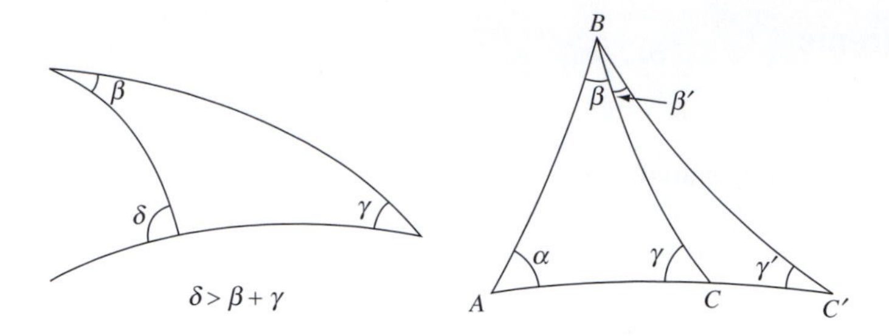

Problem 2 Let ΔABC and $\Delta ABC'$ be two hyperbolic triangles such that A, C and C' lie on a d-line in this order. Let the angles at A, B, C in the first triangle and at A, B, C' in the second triangle be α, β, γ and $\alpha, \beta + \beta', \gamma'$, respectively. Show that $\alpha + \beta + \gamma > \alpha + \beta + \beta' + \gamma'$. (See the diagram on the right, above.)

It will be seen from Problem 2 that the larger triangle, $\Delta ABC'$, has the smaller angle sum. This is true in general, not only when one triangle fits neatly inside another. In fact, it can be shown that the area of a hyperbolic triangle with angles α, β and γ is proportional to $\pi - (\alpha + \beta + \gamma)$, and we shall prove this in Subsection 6.5.1. You may for the moment use the following result as a useful memory aid:

small triangles have angle sums close to (but less than) π, and triangles with large areas have angle sums close to zero.

In Euclidean geometry, we can also prove that the sum of the angles of a (Euclidean) quadrilateral equals 2π. Now, a **quadrilateral** *ABCD* in hyperbolic geometry, or a ***d*-quadrilateral,** consists of four points A, B, C, D in $\mathscr{D}$ (no three of which lie on a single d-line), together with the segments *AB, BC, CD* and *DA* of the four d-lines joining them. We require also that no two of these segments meet except at one of the points *A, B, C* or *D*.

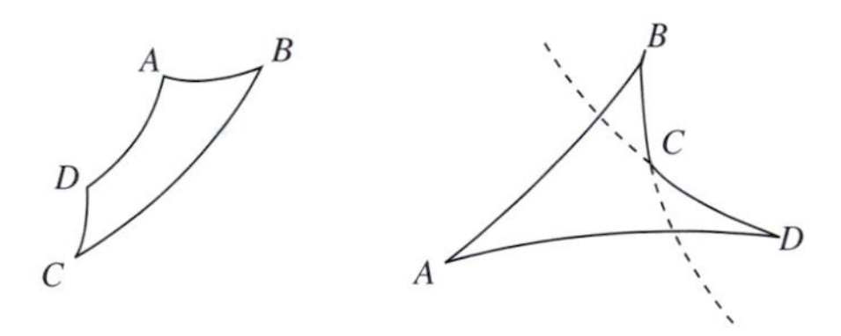

How large can the sum of the angles of a d-quadrilateral be?

Theorem 2 The sum of the angles of a d-quadrilateral is less than 2π.

Proof Any d-quadrilateral can be divided into two (non-overlapping) d-triangles by one or other of the d-lines joining alternate vertices.

We shall assume this fact without proof; see the figures below.

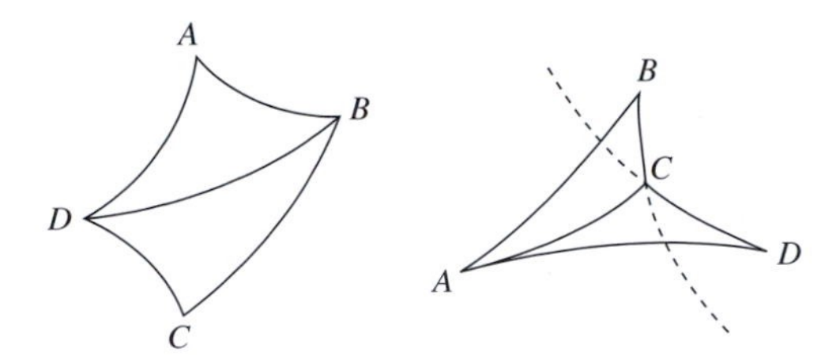

The angles of each d-triangle sum to less than π. The angles of the d-quadrilateral belong to one or other d-triangle, or partly to one and partly to the other. It follows that the angles of the d-quadrilateral sum to less than 2π. ∎

We shall use this fact later.

In hyperbolic geometry there are many theorems about d-triangles that are analogues of the corresponding theorems about Euclidean triangles, such as the following.

Theorem 3 Let ΔABC be a d-triangle in which $\angle ABC = \angle ACB$. Then the sides AB and AC are of equal length.

This is the hyperbolic analogue of the Euclidean result in Example 1, Subsection 2.1.1.

Proof Let D be the midpoint of the d-line segment BC. By applying the Origin Lemma, if necessary, we may assume that D coincides with O, the centre of the disc $\mathscr{D}$. (Although this is not strictly necessary for the proof, it simplifies the picture.) Then BC is part of a diameter of $\mathscr{D}$.

Let the d-line ℓ be the perpendicular bisector of BC; it is the diameter of $\mathscr{D}$ perpendicular to BC. Reflect (in both the Euclidean and hyperbolic senses) the triangle ΔABC in the d-line ℓ. Because reflections preserve length, and $DB = DC$, it follows that B and C change places. Suppose that A moves to some point A'. Since reflection preserves angles, it follows that $\angle A'BC = \angle ACB$. Also, recall that we are given that $\angle ACB = \angle ABC$, so $\angle A'BC = \angle ABC$. But this can happen only if A' lies on the d-line through A and B. Similarly, $\angle A'CB = \angle ABC = \angle ACB$, so A' must also lie on the d-line through A and C. This means that A and A' must coincide. Hence the d-line segment BA reflects to the d-line segment CA, and vice versa; so these d-line segments have the same length. ∎

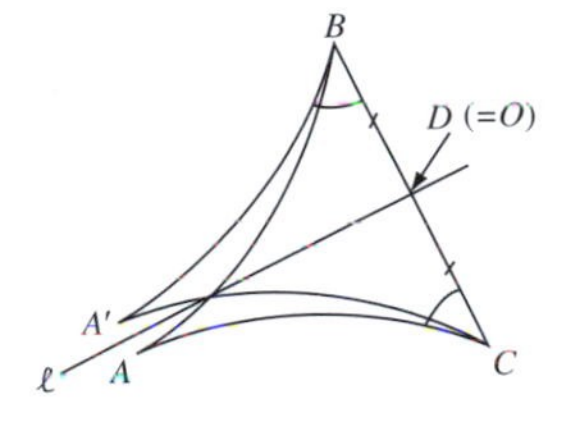

Problem 3 Let ΔABC be a d-triangle in which the sides AB and AC have equal hyperbolic length. Prove that $\angle ABC = \angle ACB$.
Hint: Consider reflection in the d-line that bisects angle $\angle BAC$.

The notion of mapping one figure onto another *exactly* by a transformation in the geometry (for example, the reflection in the d-line ℓ in the proof of Theorem 3) is one that we met previously: namely, congruence.

Definition Two figures in the unit disc $\mathscr{D}$ are ***d*-congruent** if there is a hyperbolic transformation that maps one onto the other.

For example, if A, B, C and D are four points in $\mathscr{D}$ such that the hyperbolic distances $d(A, B)$ and $d(C, D)$ are equal, then the d-line segment AB is congruent to the d-line segment CD. For, as we saw earlier, there is a hyperbolic transformation t that maps A to C and the d-line through A and B onto the d-line through C and D, in such a way that D and $t(B)$ lie on the same side of C along $t(\ell)$. Then, since B and D are the same hyperbolic distances from A and C, respectively, it follows that t must map B onto D. Thus AB and CD are congruent.

Subsection 6.1.2, Theorem 5, and the Remark following that theorem.

Now, since hyperbolic transformations preserve angles, it follows that if two d-triangles do not have corresponding angles equal (that is, they are not **similar**), then they certainly cannot be d-congruent.

Recall that two Euclidean figures are *similar* if one is a scale copy of the other; in particular, two Euclidean triangles are similar if they have corresponding angles equal.

However, the following result is still very surprising, because the analogous result is *false* in Euclidean geometry.

Theorem 4 Similar d-triangles are d-congruent.

Proof We have to prove that if the d-triangles ΔABC and ΔPQR have the angles at A, B and C and the angles at P, Q and R equal, respectively, then the two d-triangles are d-congruent.

We may apply a hyperbolic transformation to map A to the origin O; this does not change the angles of the triangle. To avoid complicated notation, we shall still denote this image d-triangle by ΔABC.

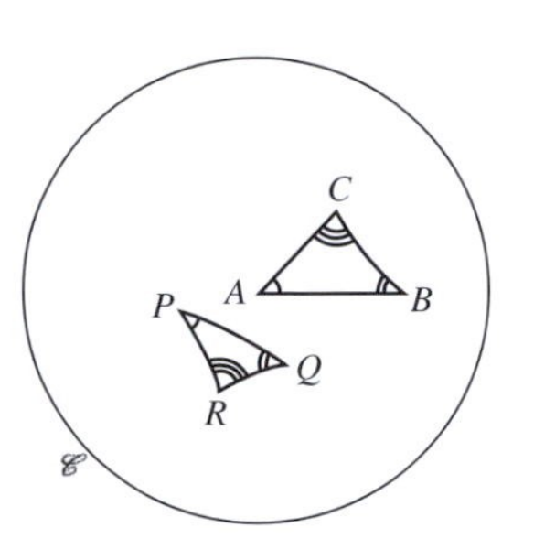

We can also apply a hyperbolic transformation to map P to the origin and the radius on which Q lies to the radius on which B lies. By reflecting in the d-line through O and B, if necessary, we can deduce from the fact that the angles at A and P are equal that the image of R lies on the same radius as (the image of) C. Again, we shall denote the image d-triangle still by the notation ΔPQR, for simplicity.

The following figure shows the result of the transformations described above.

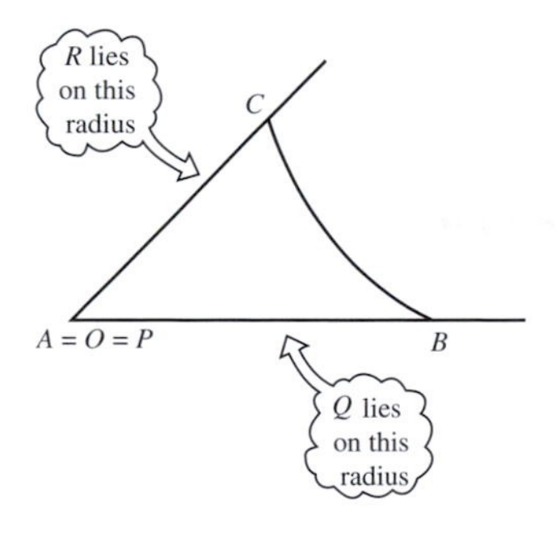

To prove the result, we have to show (in terms of the points obtained by the above preliminary mappings) that $B = Q$ and $C = R$. We proceed by considering the various possible situations that would arise if the result were false.

(a) *The d-line segment BC might lie between the origin and the d-line segment QR.* (The argument is similar if QR lies between O and BC.)

In this case,

$$\angle RCB = \pi - \angle OCB \quad \text{and} \quad \angle CBQ = \pi - \angle OBC,$$

so the angle sum of the d-quadrilateral $CBQR$ is

$$(\pi - \angle OCB) + (\pi - \angle OBC) + \angle OQR + \angle ORQ = 2\pi,$$

which is impossible, by Theorem 2.

(b) *B and Q may coincide, but not C and R.* (The argument is similar if R lies between O and C, or if C and R coincide, but not B and Q.)

In this case, if C lies between O and R, the external angle at C of the d-triangle ΔBCR is less than or equal to the sum of the opposite two internal angles, which is impossible by the result of Problem 1.

(c) *The d-line segments BC and QR may cross.*

Let the point of intersection of BC and QR be X. In this case, the external angle at B of the d-triangle ΔXBQ is less than or equal to the sum of the opposite two internal angles, which is impossible by the result of Problem 1.

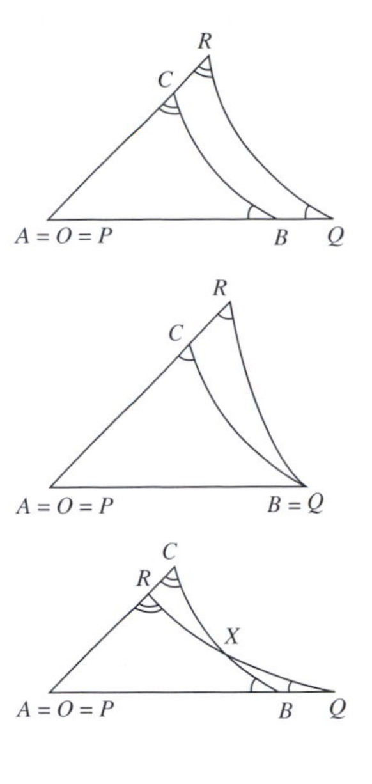

It follows that the only possibility is that $B = Q$ and $C = R$, as required. ■

In the following diagram, all the d-triangles have the same angles at their vertices, and so are similar. It follows from Theorem 4 that they are d-congruent to each other.

This looks very unlikely, since the triangles 'seem' to be getting smaller as they move away from the origin towards $\mathscr{C}$; however, it is a consequence of the way in which we defined both hyperbolic transformations and hyperbolic distance that all the triangles have sides of the same (hyperbolic) lengths as well as angles of the same sizes. This results from the fact that equal Euclidean distances on a ruler correspond to increasing hyperbolic distances as the ruler is moved outwards.

Asymptotic Triangles

The vertices of d-triangles lie in the unit disc $\mathscr{D}$. However, it is often useful to talk about figures in $\mathscr{D}$ with three sides that are d-lines but where one or more of the Euclidean circles or lines of which they are part meet on $\mathscr{C}$ rather than in $\mathscr{D}$. In this sense, we say that they are triangles 'with one or more vertices on $\mathscr{C}$'; such triangles are called **asymptotic triangles**.

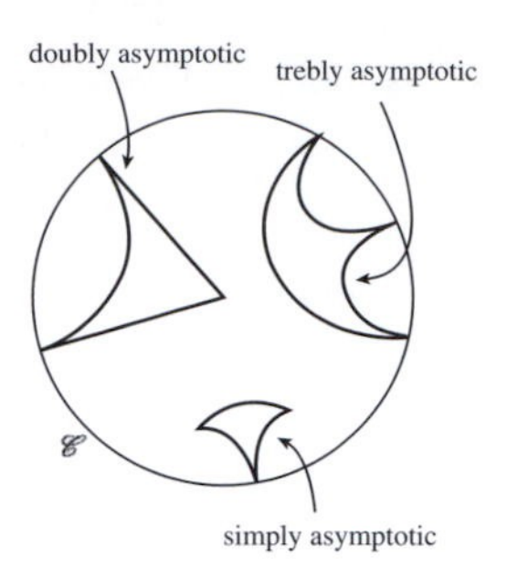

If an asymptotic triangle has one vertex on $\mathscr{C}$ is said to be **simply asymptotic**, if it has two vertices on $\mathscr{C}$ **doubly asymptotic**, and three vertices on $\mathscr{C}$ **trebly asymptotic**.

There is no hyperbolic transformation that maps points of $\mathscr{C}$ to points of $\mathscr{D}$, or vice-versa; so asymptotic triangles are *essentially* different from 'ordinary' d-triangles. However, it turns out that many of the results that hold for d-triangles hold also for asymptotic triangles, and that their proofs are similar.

Theorem 5 The angle sum of an asymptotic triangle is less than π. The angle sum of a trebly asymptotic triangle is zero.

We omit the proof of this result. The proof of the first assertion is similar to that of Theorem 1.

Problem 4 Prove that two doubly asymptotic triangles are d-congruent if and only if they have the same angle at their vertex in $\mathscr{D}$.

In hyperbolic geometry you should be careful not to assume that results are valid simply because they hold in Euclidean geometry. Sometimes asymptotic triangles are useful in constructing counter-examples.

We shall do exactly this at the end of Subsection 6.4.2.

6.4.2 Perpendicular Lines

In Euclidean geometry, two given lines ℓ and ℓ' have a common perpendicular if and only if they are parallel to each other. In hyperbolic geometry, the situation is somewhat different, as you will see. First, we must define the term *common perpendicular* in hyperbolic geometry.

Definition Let ℓ and ℓ' be two d-lines, and suppose that there exist points A on ℓ and A' on ℓ' such that the d-line segment AA' meets ℓ and ℓ' at right angles. Then AA' is a **common perpendicular** to ℓ and ℓ'.

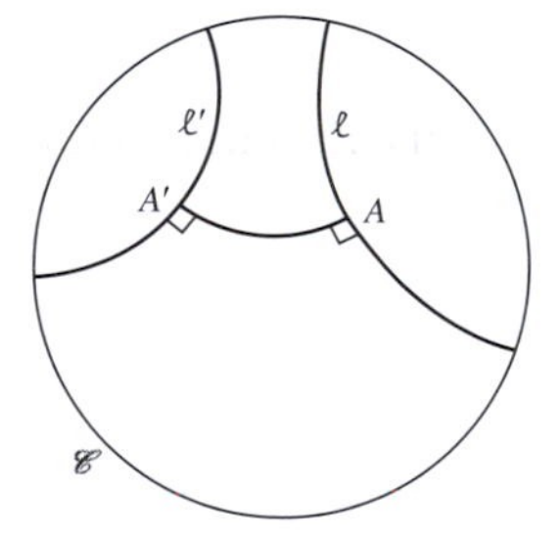

Theorem 6 Common Perpendiculars Theorem
Two d-lines have a common perpendicular if and only they are ultra-parallel. This common perpendicular is unique.

Proof First we show that if two d-lines ℓ and ℓ' have a common perpendicular, then they are ultra-parallel. Let the common perpendicular be the d-line segment AA', where A is on ℓ and A' is on ℓ'. By the Origin Lemma, we can find a transformation $r \in G_{\mathscr{D}}$ which maps the point A to the origin, O. Then the d-line $r(\ell)$ is a diameter of the disc $\mathscr{D}$, as shown on the left below. The

d-line $r(\ell')$ is part of a Euclidean circle whose centre R lies somewhere outside the disc $\mathscr{D}$ on the continuation of the diameter joining O to $r(A')$. The radius of this Euclidean circle is less than RO, and RO is perpendicular to $r(\ell)$, so this circle cannot meet $r(\ell)$, and hence $r(\ell)$ and $r(\ell')$ are ultra-parallel. It follows that the d-lines ℓ and ℓ' are ultra-parallel.

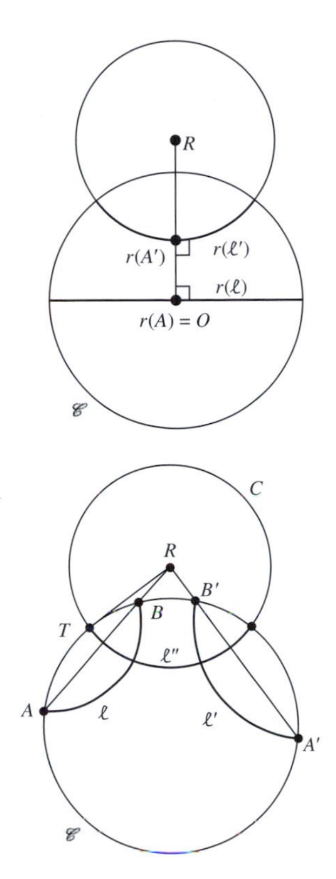

Secondly, we show that if two d-lines ℓ and ℓ' are ultra-parallel, then they have a common perpendicular. As shown in the figure on the right above, let the boundary points of ℓ be A and B, and the boundary points of ℓ' be A' and B'. Consider the *Euclidean* lines AB and $A'B'$.

If the Euclidean lines AB and $A'B'$ are not parallel, then they meet at a point R outside $\mathscr{D}$, as shown on the right above. Let RT be a tangent from R to the boundary circle $\mathscr{C}$. Consider the circle C with centre R and radius RT. This circle meets $\mathscr{C}$ at right angles, so the part of it in $\mathscr{D}$ is a d-line, ℓ'' say.

The Euclidean triangles ΔRTB and ΔRAT are similar, since $\angle TRB = \angle ART$ (being the same angle) and $\angle RTB = \angle RAT$ (since the exterior angle equals the interior opposite angle for a tangent and chord of a circle). Hence, in particular, $RB/RT = RT/RA$ so that $RA \cdot RB = RT^2$. In other words, A and B are inverse points with respect to the circle with centre R and radius RT.

Inversion in C exchanges A and B, and exchanges A' and B'. So it maps the d-line ℓ to itself, and the d-line ℓ' to itself. It follows that ℓ'' intersects ℓ and ℓ' at right angles, and so a common perpendicular to the d-lines ℓ and ℓ' exists.

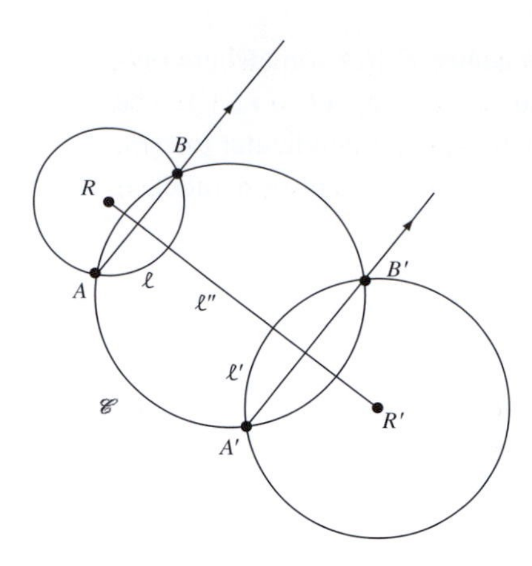

If the Euclidean lines AB and $A'B'$ are parallel, then there is a diameter ℓ'' of $\mathscr{D}$ that is the perpendicular bisector of AB and $A'B'$. It therefore passes through the centres R and R' of the Euclidean circles of which ℓ and ℓ' are parts, and so it is perpendicular to both ℓ and ℓ'. Since ℓ'' is a diameter of $\mathscr{D}$ it is a d-line, and the common perpendicular we seek exists.

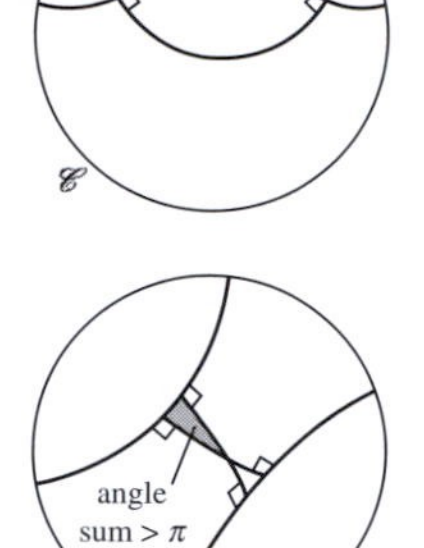

Finally, the common perpendicular to two ultra-parallel lines is unique. For, if there were two, and they were disjoint (as shown on the top illustration alongside), they and the two ultra-parallel lines would form four sides of a hyperbolic quadrilateral all of whose angles were right angles. But then the angle sum of such a quadrilateral would be 2π, which is impossible. Similarly, if there were two common perpendiculars that intersected each other (as shown in the lower illustration alongside), we should have a d-triangle whose angle sum was $\geq \pi$, which is impossible.

It follows that the common perpendicular to two ultra-parallel lines is unique. ■

In Euclidean geometry, the altitudes of a triangle play an interesting role; this is also the case in hyperbolic geometry.

Before we can talk about altitudes, we need to prove that through any point P not on a given d-line ℓ there is necessarily another (unique) d-line p such that p meets ℓ at right angles.

This fact is needed in order to prove that altitudes of a given d-triangle *exist*.

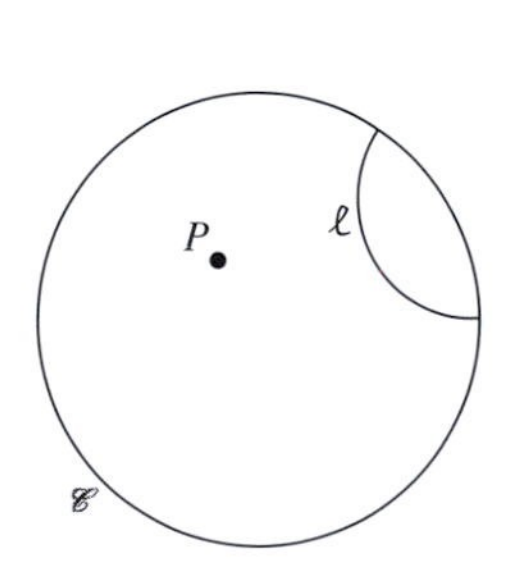

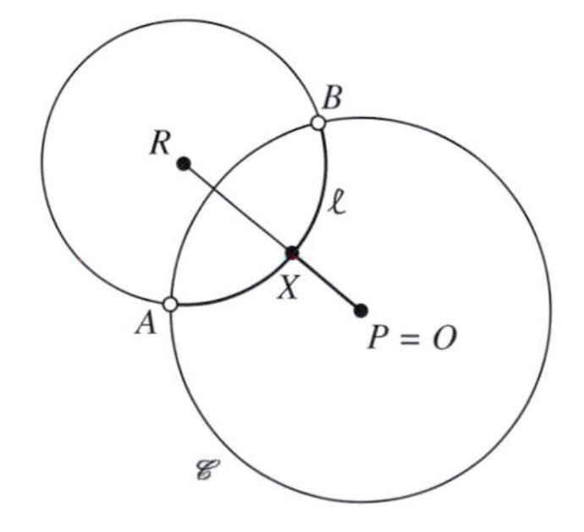

By a preliminary hyperbolic transformation, if necessary, we may assume that P is the origin. Let ℓ have boundary points A and B, and let R be the centre of the Euclidean circle of which ℓ is a part. Let the (Euclidean) line OR meet ℓ at X; OR and ℓ intersect at right angles at X, so that the d-line segment OX is a perpendicular from O to ℓ. The diameter which includes OX meets ℓ at right angles, as required. It is unique because no other line through O can meet ℓ at right angles.

Thus it is possible to 'drop' a perpendicular from a point to a d-line, and so it makes sense to talk about *the* altitude through a vertex of a d-triangle.

Definition Let ℓ be a d-line which passes through one vertex A of a d-triangle ΔABC and which is perpendicular to the side BC at the point D. The d-line segment AD is an **altitude** of the triangle ΔABC.

Theorem 7 Altitude Theorem
Let the sides AB and AC of a d-triangle ΔABC be of equal hyperbolic length, and let the angle at A be θ. Then the hyperbolic length of the altitude through A of the triangle is less than some number that depends only on θ.

Proof Map the d-triangle ΔABC onto a d-triangle $\Delta OB'C'$ by any hyperbolic transformation that sends A to the origin O. Since hyperbolic transformations preserve (hyperbolic) lengths and angles, OB' and OC' are of equal length and $\angle B'OC' = \theta$.

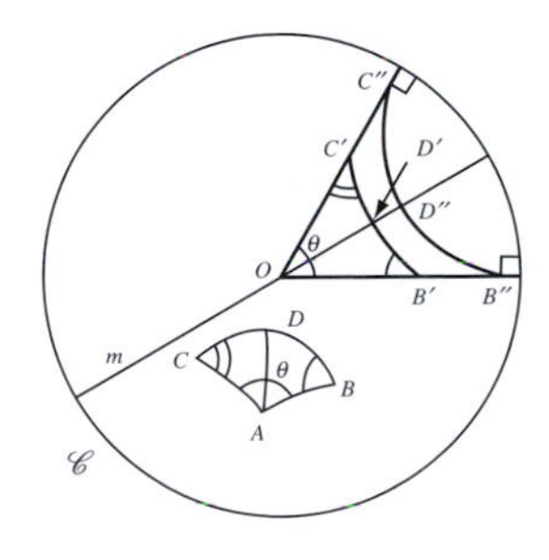

Let the diameter m of $\mathscr{D}$ that bisects the angle $\angle B'OC'$ meet the d-line segment $B'C'$ at D'. It is clear, by symmetry, that the d-line segment OD' must be perpendicular to the d-line segment $B'C'$, so that OD' (which is part of a Euclidean line) is the altitude of the d-triangle $\Delta OB'C'$ through O. Since d-triangles have a unique altitude through each vertex, the altitude AD of the d-triangle ΔABC must map onto the altitude OD', and so they are of equal hyperbolic length.

Then the length of OD' is less than the length of OD'', where D'' lies on m and the d-line joining the boundary points B'' and C'' of the lines OB' and OC' (as shown). It follows that the hyperbolic length of the original altitude AD is less than the hyperbolic length of OD'', which clearly depends only on θ. ∎

Problem 5 Prove that the d-triangles $\Delta OB'D'$ and $\Delta OC'D'$ in the proof of Theorem 7 are d-congruent.

Problem 6 Determine a (numerical) upper bound for the hyperbolic length of the altitude AD of an isosceles d-triangle ΔABC in which AB and AC are of equal hyperbolic length and the angle at A is a right angle.
Hint: This situation is a special case of that in Theorem 7: $\theta = \pi/2$, so B'' and C'' in the proof of that theorem are 1 and i, respectively.

In Euclidean geometry the altitudes of a triangle are concurrent; but this is not true in general in hyperbolic geometry. For example, let A, B and C be the points $-1, 0$ and $\frac{1+i}{\sqrt{2}}$. Then the altitudes through A and C of the d-triangle ΔABC are the d-lines with endpoints $-1, -i$ and $\frac{1+i}{\sqrt{2}}, \frac{1-i}{\sqrt{2}}$, respectively. These d-lines do not meet anywhere in $\mathcal{D}$ or on $\mathcal{C}$. So certainly the altitudes of a doubly asymptotic triangle are not concurrent.

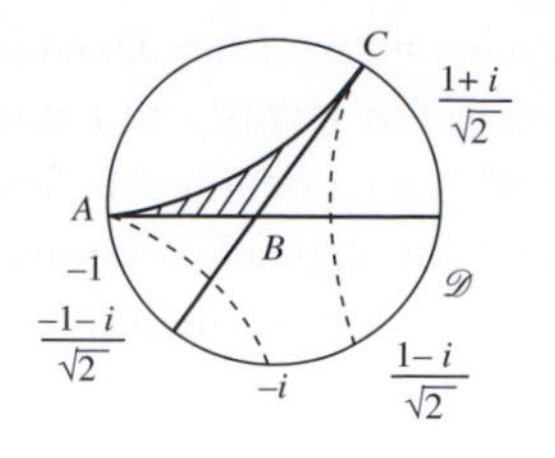

Finally, if we choose a (non-asymptotic) d-triangle with one vertex at O and the other two vertices in $\mathcal{D}$ but very close to A and C, it is clear that the altitudes of *this* d-triangle cannot be concurrent.

6.4.3 Right-Angled *d*-Triangles

In Euclidean geometry, Pythagoras' Theorem plays a central role in many calculations. As we have seen, however, the theorem does not hold in hyperbolic geometry if we simply replace Euclidean lines by d-lines and Euclidean distances by hyperbolic distances in its statement.

Paragraph following Example 2 of Subsection 6.3.1.

However it seems reasonable to imagine that hyperbolic functions might play some role in a version of Pythagoras' Theorem in hyperbolic geometry, and indeed this is the case.

Theorem 8 Pythagoras' Theorem
Let ΔABC be a d-triangle in which the angle at C is a right angle. If a, b and c are the hyperbolic lengths of BC, CA and AB, then

$$\cosh 2c = \cosh 2a \times \cosh 2b.$$

We prove Theorem 8 later in this subsection.

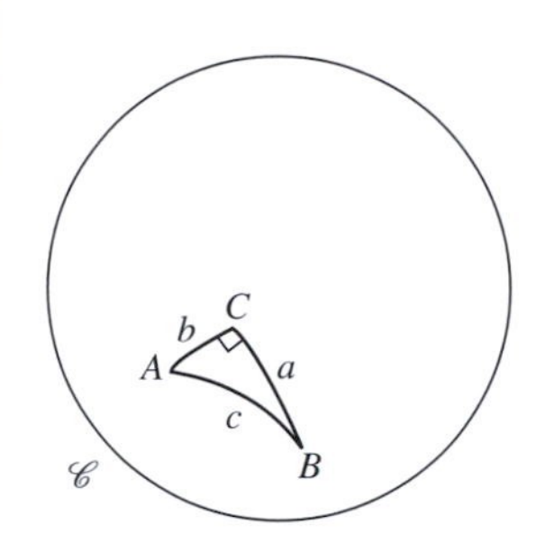

Problem 7 Use Pythagoras' Theorem to determine the hyperbolic lengths of the sides of the d-triangle with vertices $0, (1+i)/2\sqrt{2}$ and $(-1+i)/5\sqrt{2}$, which has a right angle at O.

Problem 8 Use Pythagoras' Theorem to determine the hyperbolic lengths of the sides of the d-triangles with vertices at the following points:

(a) $0, \frac{3}{4}$ and $\frac{3}{4}i$; (b) $0, r$ and ir (where $0 < r < 1$).

We can use these ideas to provide another solution to Problem 6 above.

Example 1 Prove that, if ΔABC is an isosceles d-triangle in which AB and AC are of equal hyperbolic lengths and the angle at A is a right angle, then the hyperbolic length of the altitude AD is less than $\frac{1}{2}\cosh^{-1}\left(\sqrt{2}\right) \simeq 0.4407$ $\left(\text{which is } \frac{1}{2}\log_e\left(1+\sqrt{2}\right)\right)$.

In Problem 6 we found that this upper bound $\frac{1}{2}\cosh^{-1}\left(\sqrt{2}\right)$ was equal to $\tanh^{-1}\left(\sqrt{2}-1\right)$; in fact the two numbers are equal. Such unexpected identities involving hyperbolic functions occur surprisingly often!

Solution The given triangle is d-congruent to the d-triangle $\Delta OB'C'$ with vertices at $0, r$ and ir, for some r with $0 < r < 1$. Let OD' be an altitude of this triangle. Let the hyperbolic lengths of the sides OB' and OC' be denoted by b, of side $B'C'$ by a, and of the altitude OD' by d.

It follows from Problem 5 (or by symmetry) that D' is the Euclidean midpoint and the hyperbolic midpoint of the d-line segment $B'C'$, so that the d-line segments $B'D'$ and $C'D'$ both have hyperbolic length $\frac{1}{2}a$.

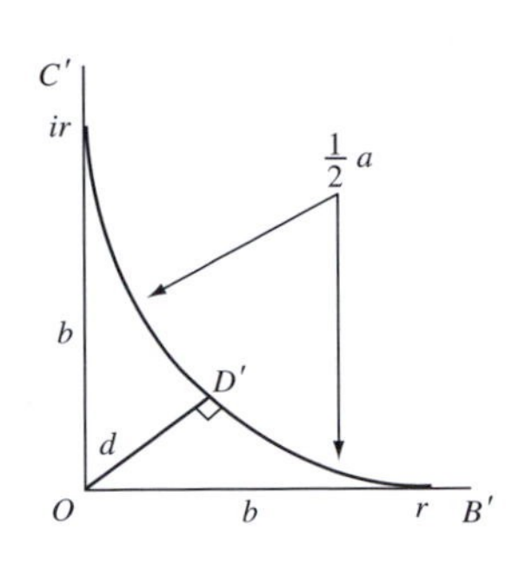

By applying Pythagoras' Theorem to the d-triangle $\Delta OB'D'$, we find that

$$\cosh 2b = (\cosh 2d) \times \left(\cosh 2\left(\tfrac{1}{2}a\right)\right)$$

$$= (\cosh 2d) \times \cosh a. \qquad (1)$$

Next, by applying Pythagoras' Theorem to the d-triangle $\Delta OB'C'$, we find that

$$\cosh 2a = \cosh^2 2b. \qquad (2)$$

Now, we may rewrite equation (2) in the form

$$2\cosh^2 a - 1 = \cosh^2 2b; \qquad (3)$$

Here we use the identity $\cosh^2 2x = 2\cosh^2 x - 1$, for $x \in \mathbb{R}$.

and if we then substitute for $\cosh a$ from equation (1) into equation (3), we obtain

$$2\left(\frac{\cosh 2b}{\cosh 2d}\right)^2 - 1 = \cosh^2 2b.$$

We may rearrange this equation in the form

$$\left(\frac{\cosh 2d}{\cosh 2b}\right)^2 = \frac{2}{1 + \cosh^2 2b},$$

so that

$$\cosh 2d = \frac{\sqrt{2}\cosh 2b}{\sqrt{(1 + \cosh^2 2b)}}$$

$$= \frac{\sqrt{2}}{\sqrt{\left(1 + \frac{1}{\cosh^2 2b}\right)}}$$

$$< \sqrt{2}.$$

It follows that $d < \frac{1}{2}\cosh^{-1}\left(\sqrt{2}\right) \simeq 0.4407$, as claimed. □

Pythagoras' Theorem takes a simple, if unexpected, form in hyperbolic geometry. Moreover, the relationship closely approximates the Euclidean one if the sides of the triangle are very small, which is what you would expect because small hyperbolic triangles have angle sums nearly equal to π and so are nearly Euclidean themselves. To see the relationship, we expand the cosh terms as power series, using the formula

Recall that

$$\cosh x = 1 + \frac{x^2}{2!} + \frac{x^4}{4!} + \ldots = \sum_{n=0}^{\infty} \frac{x^{2n}}{(2n)!}$$

for all $x \in \mathbb{R}$.

$$\cosh 2c = 1 + \frac{(2c)^2}{2!} + \text{higher powers of } 2c.$$

If we ignore the higher powers of $2c$, as we may when $2c$ is very small, we find

$$\cosh 2c \simeq 1 + 2c^2$$

and

$$\cosh 2a \times \cosh 2b \simeq \left(1 + 2a^2\right)\left(1 + 2b^2\right)$$
$$= 1 + 2a^2 + 2b^2 + 4a^2b^2.$$

If we ignore the term $4a^2b^2$ on the grounds that it is as small as other terms we have already dropped, then we are left with

$$1 + 2c^2 \simeq 1 + 2a^2 + 2b^2,$$

which reduces to

$$c^2 \simeq a^2 + b^2.$$

So for 'small' d-triangles the hyperbolic version of Pythagoras' Theorem is essentially the same as the Euclidean version.

In Euclidean geometry there is a simple connection between the lengths of the sides of a right-angled triangle and its angles. An analogous result exists in hyperbolic geometry also.

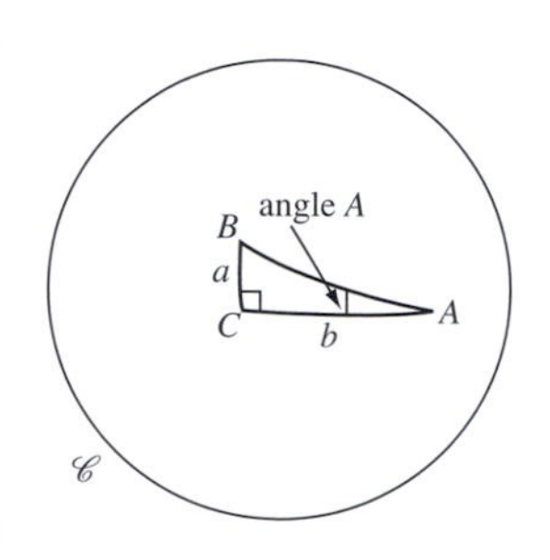

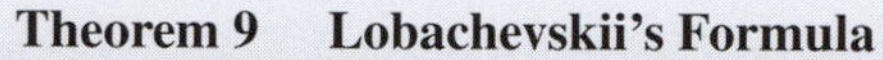

Theorem 9 Lobachevskii's Formula
Let the d-triangle ΔABC have a right angle at C, and let the hyperbolic lengths of sides AC and BC be b and a, respectively. Then

$$\tan A = \frac{\tanh 2a}{\sinh 2b}.$$

We prove this result later in the subsection.

For example, let the points A, B, C be $\frac{1}{2}$, $\frac{1}{2}i$ and 0, respectively. Then it follows from Lobachevskii's Formula that

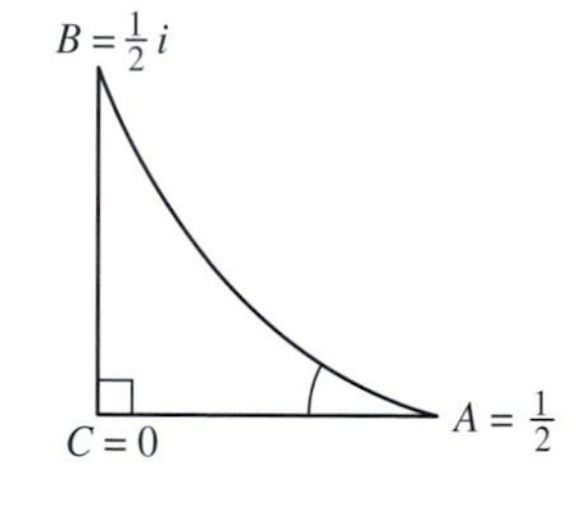

$$\tan A = \frac{\tanh 2d\left(0, \frac{1}{2}i\right)}{\sinh 2d\left(0, \frac{1}{2}\right)}$$
$$= \frac{\tanh 2d\left(0, \frac{1}{2}\right)}{\sinh 2d\left(0, \frac{1}{2}\right)}$$
$$= \frac{1}{\cosh 2d\left(0, \frac{1}{2}\right)}$$
$$\simeq \frac{1}{1.6667} \simeq 0.6,$$

so that $A \simeq \tan^{-1} 0.6 \simeq 0.5404$ radians.

Remark

It can be shown that when the triangle is very small, and so nearly Euclidean, $\tanh 2a$ is approximately $2a$ and $\sinh 2b$ is approximately $2b$, so that $\tan A \simeq a/b$, which is the Euclidean value.

Problem 9 Determine the angle at A in the d-triangle ΔABC with vertices $-\frac{3}{5}i, -\frac{4}{5}$ and 0, respectively.

We can also use these ideas to answer another question that loomed very large in the discovery of hyperbolic geometry.

Let ℓ' be the d-line from a point P that is perpendicular to a d-line ℓ not through P, and let p be the length of the segment of ℓ' from P to ℓ. What is the angle φ between ℓ' and a d-line through P which is parallel to ℓ?

The angle φ is called 'the angle of parallelism'.

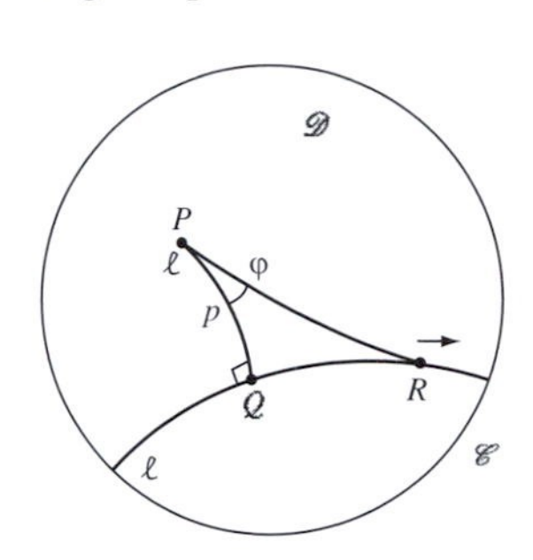

Let Q be the foot of the perpendicular from P to ℓ, and R a point on ℓ different from Q. As the point R moves away from Q towards $\mathscr{C}$, the hyperbolic length of $QR, d(Q, R)$, tends to infinity; it follows that $\tanh 2d(Q, R)$ tends to 1 as R approaches $\mathscr{C}$.

From Lobachevskii's Formula,

$$\tan \angle QPR = \frac{\tanh 2d(Q, R)}{\sinh 2p},$$

and so, if we let R approach $\mathscr{C}$ along ℓ, it follows that

$$\tan \varphi = \frac{1}{\sinh 2p}.$$

We have proved the following result.

Corollary 1 Angle of Parallelism
Let ℓ be a d-line, and P a point of $\mathscr{D}$ that does not lie on ℓ. Then the angle φ between the perpendicular from P to ℓ (of hyperbolic length p) and either d-line through P that is parallel to ℓ is given by

$$\tan \varphi = \frac{1}{\sinh 2p}.$$

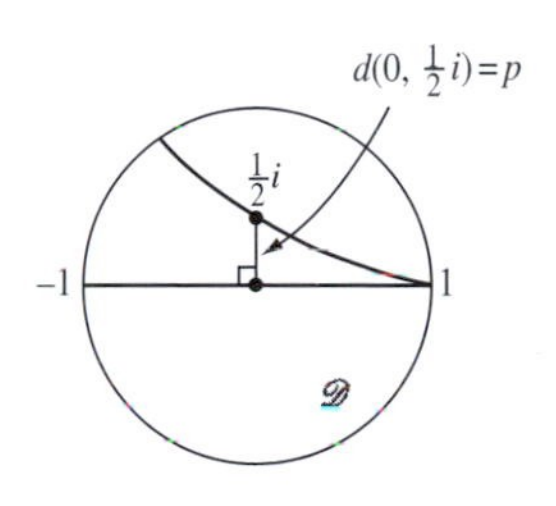

For example, the hyperbolic distance of the point $\frac{1}{2}i$ from the diameter $\ell = (-1, 1)$ is $\tanh^{-1} 0.5 \simeq 0.5493$. It follows that the angle φ between the perpendicular from $\frac{1}{2}i$ to ℓ and either parallel to ℓ through $\frac{1}{2}i$ is given by

$$\tan \varphi = \frac{1}{\sinh\left(2 \tanh^{-1} 0.5\right)} \simeq \frac{1}{\sinh 1.0986};$$

hence, $\varphi \simeq 0.6435$ radians.

In fact, the remarkable formula

$$\sinh\left(2 \tanh^{-1} x\right) = \frac{2x}{1 - x^2}$$

gives $\tan \varphi = \frac{3}{4}$ exactly.

Problem 10 Determine the angle between the perpendicular from the point $\frac{3}{4}$ to the d-line ℓ with equation $x = 0$ and either parallel to ℓ through $\frac{3}{4}$.

Finally we give, without proof, another useful formula.

Theorem 10 Sine Formula
Let ΔABC be a d-triangle right-angled at C. Let a and c be the hyperbolic lengths of BC and AB. Then

$$\sin A = \frac{\sinh 2a}{\sinh 2c}.$$

A proof of this result will be found as an Exercise in Section 6.7.

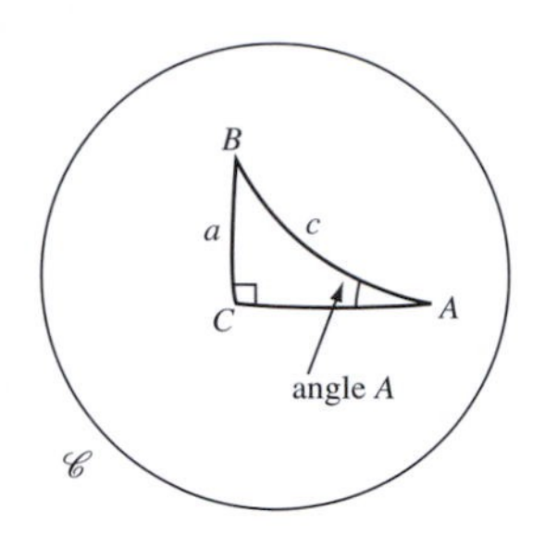

Remark

This result agrees with the expected one when the triangle is small, for in such cases $\sinh 2a \simeq 2a$ and $\sinh 2c \simeq 2c$, so $\sin A \simeq a/c$, which is the Euclidean result.

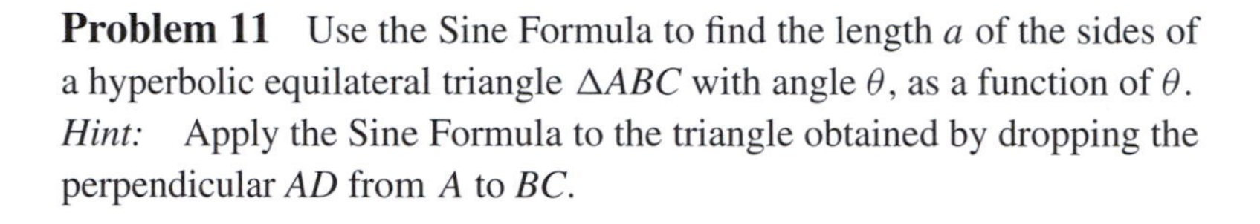

Problem 11 Use the Sine Formula to find the length a of the sides of a hyperbolic equilateral triangle ΔABC with angle θ, as a function of θ.
Hint: Apply the Sine Formula to the triangle obtained by dropping the perpendicular AD from A to BC.

Proofs

We now supply proofs of two results that you met earlier in this subsection.

The formulas in Theorems 8 and 9 depend on the following remarkable formula:

$$\cosh\left(2\tanh^{-1} x\right) = \frac{1+x^2}{1-x^2}. \tag{4}$$

The substitution $x = \tanh t$ shows that equation (4) holds if and only if

$$\cosh 2t = \frac{1+\tanh^2 t}{1-\tanh^2 t},$$

and this formula follows from the definitions of cosh and tanh, as you can check.

For future convenience, we now list various expressions that we have now met for the inverse tanh function.

For $x \in (-1, 1)$, the following are all equal to $\tanh^{-1} x$:

$$\frac{1}{2}\log_e \frac{1+x}{1-x}, \ \frac{1}{2}\sinh^{-1}\frac{2x}{1-x^2}, \ \frac{1}{2}\cosh^{-1}\frac{1+x^2}{1-x^2}.$$

Theorem 8 Pythagoras' Theorem
Let ΔABC be a d-triangle in which the angle at C is a right angle. If a, b and c are the hyperbolic lengths of BC, CA and AB, then

$$\cosh 2c = \cosh 2a \times \cosh 2b.$$

Proof To prove Pythagoras' Theorem in hyperbolic geometry, by the Origin Lemma we may assume that C is at the centre of $\mathscr{D}$, the point A is at the point a' on the horizontal diameter and the point B is at ib' on the vertical diameter, where a' and b' are real and positive.

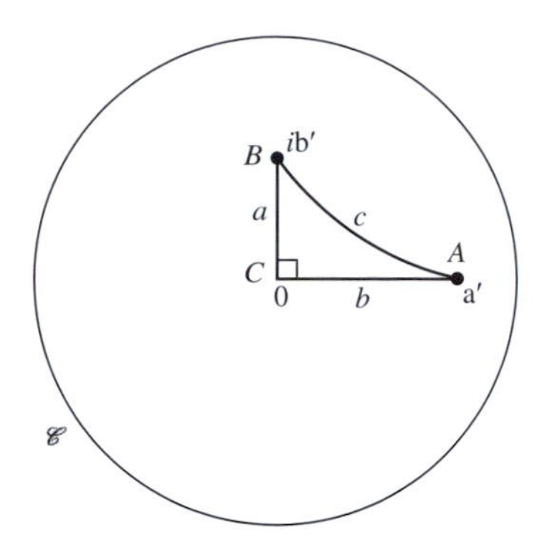

Here a, b, c are real because they are the hyperbolic lengths of the sides of the d-triangle ΔABC.

From the Distance Formula, it follows that

$$a = \tanh^{-1} b' \quad \text{and} \quad b = \tanh^{-1} a'. \tag{5}$$

We now map $\mathscr{D}$ to itself by the hyperbolic transformation

$$M(z) = \frac{z - a'}{1 - a'z}.$$

Under M, A goes to the origin (which we shall also call A', for clarity), B goes to the point B', with complex coordinates

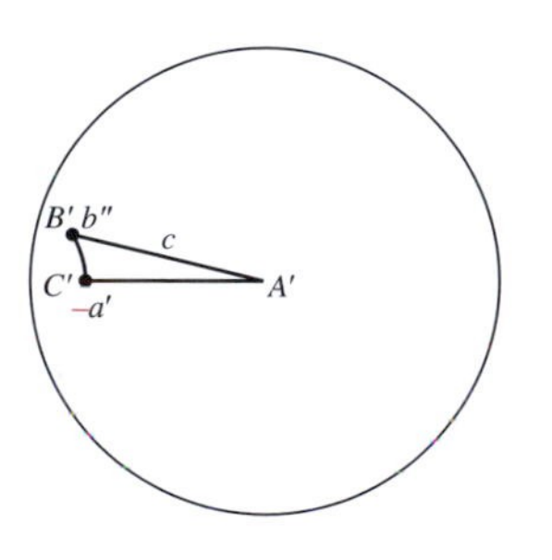

$$b'' = \frac{ib' - a'}{1 - ia'b'},$$

and C goes to the point C', with coordinates $-a'$.

Because the hyperbolic transformation preserves lengths, the hyperbolic length c of AB is equal to the hyperbolic length of $A'B'$.

To find c, the hyperbolic length of AB, we first find the modulus of b''. We calculate its square:

$$|b''|^2 = \frac{|ib' - a'|^2}{|1 - ia'b'|^2} = \frac{a'^2 + b'^2}{1 + a'^2 b'^2}. \tag{6}$$

Now we observe from formula (4) and equation (5) above that

$$\cosh 2a = \cosh\left(2 \tanh^{-1} b'\right) = \frac{1 + b'^2}{1 - b'^2},$$

$$\cosh 2b = \cosh\left(2 \tanh^{-1} a'\right) = \frac{1 + a'^2}{1 - a'^2};$$

also, since $c = \tanh^{-1} |b''|$, we get from the formula (4) that

$$\cosh 2c = \cosh\left(2 \tanh^{-1} |b''|\right) = \frac{1 + |b''|^2}{1 - |b''|^2}.$$

But, it then follows from equation (6) that

$$\begin{aligned}\cosh 2c = \frac{1+|b''|^2}{1-|b''|^2} &= \frac{1+\frac{a'^2+b'^2}{1+a'^2b'^2}}{1-\frac{a'^2+b'^2}{1+a'^2b'^2}} \\ &= \frac{1+a'^2+b'^2+a'^2b'^2}{1-a'^2-b'^2+a'^2b'^2} \\ &= \frac{1+b'^2}{1-b'^2} \times \frac{1+a'^2}{1-a'^2}, \\ &= \cosh\left(2\tanh^{-1} b'\right) \times \cosh\left(2\tanh^{-1} a'\right) \\ &= \cosh 2a \times \cosh 2b.\end{aligned}$$

This proves the analogue of Pythagoras' Theorem in hyperbolic geometry. ∎

Theorem 9 Lobachevskii's Formula
Let the d-triangle ΔABC have a right angle at C, and let the hyperbolic lengths of sides AC and BC be b and a, respectively. Then

$$\tan A = \frac{\tanh 2a}{\sinh 2b}.$$

Proof We proceed as in the proof of Theorem 8, by mapping the d-triangle ΔABC with vertices at a', ib' and 0, respectively, by the hyperbolic transformation $M(z) = \frac{z-a'}{1-a'z}$ onto the d-triangle with vertices at 0,

$$b'' = \frac{ib'-a'}{1-ia'b'} = \frac{-a'\left(1+b'^2\right)+ib'\left(1-a'^2\right)}{1+a'^2b'^2} \tag{7}$$

We obtain the second fraction in equation (7) by multiplying the numerator and denominator of the first fraction by $1+ia'b'$.

and $-a'$, respectively, where $b = \tanh^{-1} a'$ and $a = \tanh^{-1} b'$, so that

$$a' = \tanh b \quad \text{and} \quad b' = \tanh a.$$

The mapping also sends the angle A in ΔABC onto an angle of equal size (which we also denote by A by a slight abuse of notation) at the origin. Since the triangle with vertices at $0, b''$ and $-a'$ is right-angled at $-a'$, it follows from equation (7) that

$$\tan A = \frac{b'\left(1-a'^2\right)}{a'\left(1+b'^2\right)}. \tag{8}$$

Now, since $a' = \tanh b$, we can write

$$\frac{1-a'^2}{a'} = \frac{1-\tanh^2 b}{\tanh b}$$
$$= \frac{\cosh^2 b - \sinh^2 b}{\sinh b \cdot \cosh b} = \frac{2}{\sinh 2b}.$$

Here we use the identities

$$\cosh^2 x - \sinh^2 x = 1$$

and

$$\sinh 2x = 2 \sinh x \cosh x$$

for $x \in \mathbb{R}$.

Similarly, we can verify that

$$\frac{b'}{1+b'^2} = \tfrac{1}{2} \tanh 2a.$$

Hence, if we substitute these expressions for $\frac{1-a'^2}{a'}$ and $\frac{b'}{1+b'^2}$ into equation (8), we find that

$$\tan A = \frac{\tanh 2a}{\sinh 2b}. \qquad \blacksquare$$

6.4.4 Equidistant Curves to *d*-Lines

In the Euclidean plane $\mathbb{R}^2$, the set of points at a fixed distance from a given line ℓ is a pair of straight lines parallel to ℓ. In the hyperbolic disc $\mathscr{D}$ there are two curves equidistant from a given d-line, though neither is itself a d-line.

Theorem 11 Let ℓ be a d-line that ends at points A and B on $\mathscr{C}$, and let ℓ' be that part in $\mathscr{D}$ of a (Euclidean) circle through A and B, and let α be the angle (in $\mathscr{D}$) between ℓ and ℓ'. Then all points on ℓ' are the same hyperbolic distance d from ℓ, where

$$d = \tfrac{1}{2} \log_e \left\{\tan\left(\tfrac{\alpha}{2} + \tfrac{\pi}{4}\right)\right\}.$$

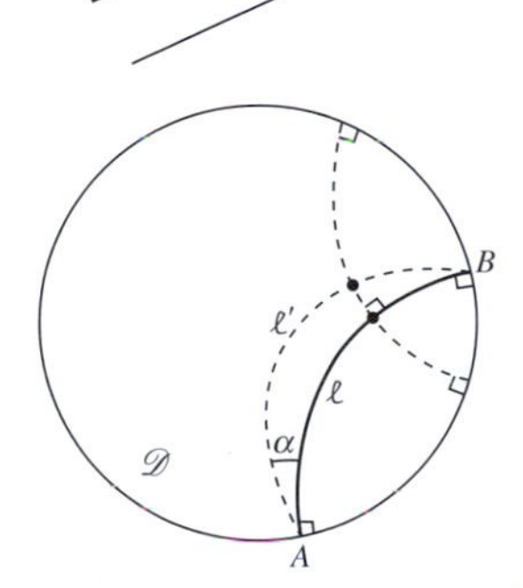

Proof First, map $\mathscr{C}$ and $\mathscr{D}$ to themselves by a Möbius transformation M that maps A and B to -1 and 1 (not necessarily respectively) and maps ℓ' to a curve in the upper half of $\mathscr{D}$. Since Möbius transformations preserve angles and generalized circles, it follows that the image of the d-line AB is the diameter $[-1, 1]$, and the image of ℓ' is an arc of a circle in the upper half of $\mathscr{D}$ through ± 1 making an angle α (in $\mathscr{D}$) with $[-1, 1]$.

It is not hard to verify that such an M exists.

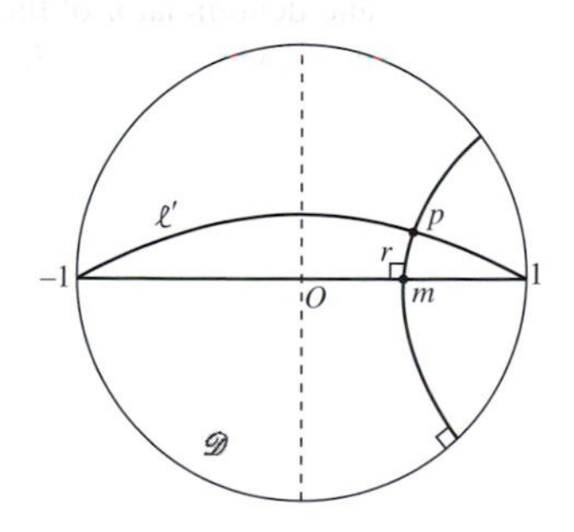

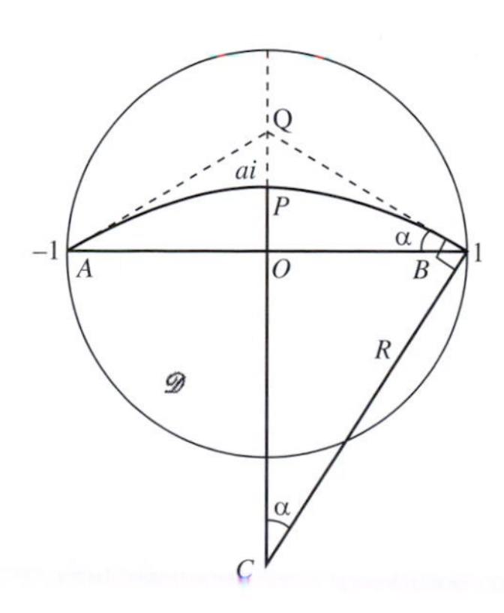

In view of the above remarks, it is sufficient to assume that the d-line ℓ is the diameter $(-1, 1)$ of $\mathscr{D}$. Let p be a point on ℓ', and m the foot of the perpendicular r from p to $(-1, 1)$. Let M_1 be the Möbius transformation defined by

Similarly, we shall refer below to ℓ' rather than to $M(\ell')$.

$$M_1(z) = \frac{z - m}{1 - mz}.$$

Then M_1 maps $\mathscr{D}$ to itself, and $(-1, 1)$ to itself. It maps m to 0, ℓ' to an arc of a (Euclidean) circle, with radius R and centre C (say), through ± 1; and it maps r to a portion of a d-line through O that is perpendicular to $(-1, 1)$– and so to a portion of the imaginary axis. It follows that the point $P = M_1(p)$ is of the form ai, for some real number $a \in (-1, 1)$. Since M_1 is a hyperbolic transformation it does not alter hyperbolic distances, so that

Property 6, Subsection 6.3.1

$$\begin{aligned} d(p, m) &= d(0, ai) = d(0, a) \\ &= d, \text{ say.} \end{aligned}$$

Now, if Q is the point where the perpendicular to CB at B meets OC, the triangles ΔCBQ and ΔBOQ are similar, since $\angle CBQ = \frac{\pi}{2} = \angle BOQ$ and $\angle BCQ = \angle OQB$. It follows that $\angle BCQ = \angle OBQ = \alpha$. Also, from ΔOBC we see that $R \sin \alpha = OB = 1$. Hence

$$\begin{aligned} a &= R - R\cos\alpha \\ &= \frac{1 - \cos\alpha}{\sin\alpha} \\ &= \frac{2\sin^2\left(\frac{\alpha}{2}\right)}{2\sin\left(\frac{\alpha}{2}\right)\cos\left(\frac{\alpha}{2}\right)} = \tan\left(\frac{\alpha}{2}\right). \end{aligned}$$

Here we use the identities

$$\cos 2x = 1 - 2\sin^2 x$$

and

$$\sin 2x = 2\sin x \cos x$$

for $x \in \mathbb{R}$.

It follows that

$$\begin{aligned} d &= \tanh^{-1} a = \tanh^{-1}\left(\tan\left(\frac{\alpha}{2}\right)\right) \\ &= \tfrac{1}{2}\log_e\left(\frac{1 + \tan\left(\frac{\alpha}{2}\right)}{1 - \tan\left(\frac{\alpha}{2}\right)}\right) \\ &= \tfrac{1}{2}\log_e\left(\tan\left(\tfrac{\alpha}{2} + \tfrac{\pi}{4}\right)\right). \end{aligned}$$

Recall from equation (3) of Subsection 6.3.1 that

$$\tanh^{-1} k = \tfrac{1}{2}\log_e\left(\frac{1+k}{1-k}\right)$$

for any $k \geq 0$.

This completes the proof of the theorem. ■

For example, the portion in the upper half of $\mathscr{D}$ of the Euclidean circle through ± 1 that makes an angle of $\pi/4$ with the segment $[-1, 1]$ is a set of points equidistant in $\mathscr{D}$ from $[-1, 1]$, with the distance of separation being

$$d = \frac{1}{2}\log_e\left(\tan\left(\frac{\pi}{8} + \frac{\pi}{4}\right)\right) = \frac{1}{2}\log_e\left(\tan\left(\frac{3\pi}{8}\right)\right) \simeq 0.4407.$$

6.5 Area

6.5.1 Area of a *d*-triangle

In the Euclidean plane $\mathbb{R}^2$, the area of a set is defined in terms of the union of areas of small rectangles. In the hyperbolic disc $\mathscr{D}$ the situation is somewhat different; d-quadrilaterals cannot have four right angles and so do not conveniently exhaust areas in an analogous way.

Theorem 2, Subsection 6.4.1

We want our definition of area in hyperbolic geometry to have such properties as the following:

- The area of a d-triangle is non-negative, and is zero only if its Euclidean area is zero;
- d-congruent d-triangles (and d-congruent asymptotic d-triangles) should have the same area;
- if one d-triangle can be fitted inside another, it should have smaller area;
- area should be additive.

Then we can determine the area of any figure in $\mathscr{D}$ that can be divided up into d-triangles.

Throughout this subsection, all 'lines' and 'line segments' will be d-lines and d-line segments, but to avoid constant repetition we will simply drop the 'd'.

Notice first that as d-triangles grow in size their angle sum decreases. For example, let ΔABC be contained in $\Delta AB'C$, where A, B and B' are collinear, as shown. Denote by α, β, γ and α, β', γ' the angles of the two d-triangles, as shown. Consider the d-triangle $\Delta BCB'$. The angles at B and C in this are $\pi - \beta$ and $\gamma' - \gamma$. Then, since the sum of the angles of a d-triangle is less than π, it follows that

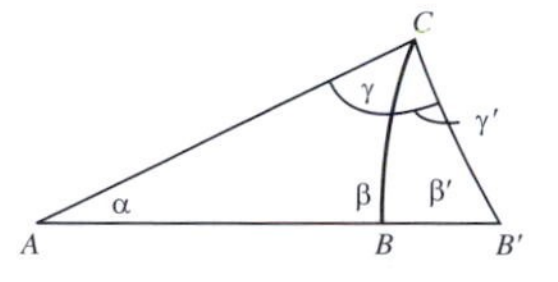

Theorem 1, Subsection 6.4.1

$$(\pi - \beta) + (\gamma' - \gamma) + \beta' < \pi,$$

so that $\beta + \gamma > \beta' + \gamma'$; hence $\alpha + \beta + \gamma > \alpha + \beta' + \gamma'$, as asserted.

> **Problem 1** Prove that a trebly asymptotic d-triangle can be divided into two doubly asymptotic right-angled d-triangles.
> *Hint:* Consider a new vertex sliding along one edge; how do the angles vary?

Now, since 'very small' d-triangles are d-congruent to small d-triangles near the origin, where d-lines are 'approximately Euclidean' lines, the sum of the angles of 'very small' d-triangles is close to π.

At the opposite end of the size scale, we can use the result of Problem 1 above to prove the surprising fact that all trebly asymptotic d-triangles are d-congruent to each other.

Theorem 1 below

Theorem 1 All trebly asymptotic d-triangles are d-congruent to each other.

Proof It is sufficient to prove that trebly asymptotic d-triangles are d-congruent to the trebly asymptotic d-triangle with vertices -1, 1 and i.

So, let ΔABC be any trebly asymptotic d-triangle, where (for convenience) we assume that the vertices A, B, C occur in clockwise order on $\mathscr{C}$. Let ℓ denote the line AB, and let D be any point of ℓ in $\mathscr{D}$.

Then, by the Origin Lemma, we can map $\mathscr{D}$ to itself and D onto the origin O by a Möbius transformation, M_1 say. It follows that M_1 maps AB onto a diameter of $\mathscr{D}$.

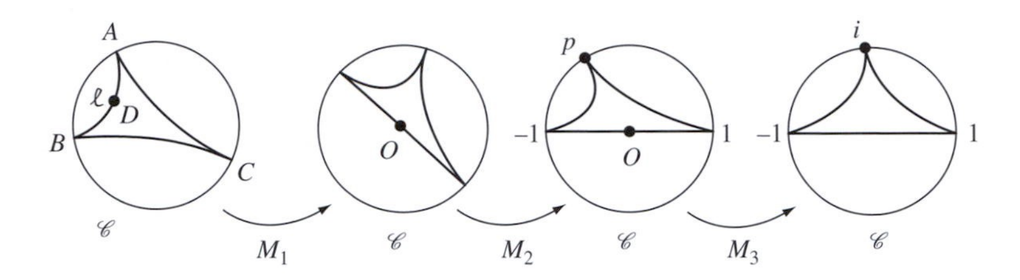

We can then map $\mathscr{D}$ to itself by another Möbius transformation, M_2 say, that rotates the disc about the origin and maps A and B onto -1 and 1, respectively. Since Möbius transformations are direct hyperbolic transformations, they preserve the orientation of points; it follows that the image of C under $M_2 \circ M_1$ is some point p, say, on the upper half of $\mathscr{C}$.

Now, for any real number $m \in (-1, 1)$, the Möbius transformation

$$M_3(z) = \frac{z - m}{1 - mz}$$

maps $\mathscr{D}$ to $\mathscr{D}$, $\mathscr{C}$ to $\mathscr{C}$, -1 to -1, 1 to 1, and the upper halves of $\mathscr{D}$ and $\mathscr{C}$ to themselves. It also maps p to i if m is chosen so that

$$i = \frac{p - m}{1 - mp},$$

which we can rewrite in the form

$$m = \frac{p - i}{1 - pi}. \tag{1}$$

Then, using the facts that $|p| = 1$ and $\text{Im}\, p > 0$, we can check that equation (1) indeed provides a value of m that is real and strictly between -1 and 1.

We omit the details.

Composing the three Möbius transformations that we have met, the composite Möbius transformation $M_3 \circ M_2 \circ M_1$ maps the trebly asymptotic d-triangle ΔABC onto the trebly asymptotic d-triangle with vertices -1, 1 and i, as required. ■

Trebly asymptotic d-triangles will play a crucial role in our discussion of hyperbolic area. First, we see that every d-triangle can be fitted inside some trebly asymptotic d-triangle.

Theorem 2 Let ΔABC be a d-triangle. Then there exists a trebly asymptotic d-triangle ΔDEF that contains ΔABC.

Proof First, we construct a trebly asymptotic d-triangle ΔDEF as follows: extend the segment AB beyond B to meet $\mathscr{C}$ at D, the segment BC beyond C to meet $\mathscr{C}$ at E, and the segment CA beyond A to meet $\mathscr{C}$ at F. Then draw the trebly asymptotic d-triangle ΔDEF.

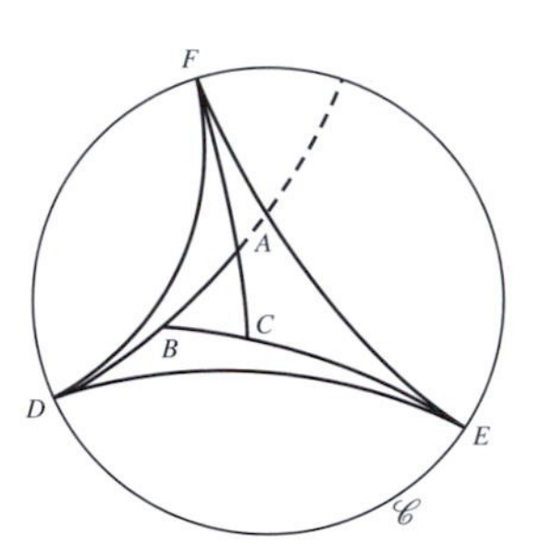

Now, the d-line containing the points A, B, D cuts off a minor arc of $\mathscr{C}$ that contains F. It follows that the d-line through D and F lies on that side of the d-line through A, B, D, and (in particular) that the d-triangles ΔDAF and ΔABC do not overlap.

Similarly, the d-triangles ΔEBD and ΔABC do not overlap, and also the d-triangles ΔFCE and ΔABC do not overlap. Also, none of the d-triangles ΔDAF, ΔEBD, ΔFCE overlap.

It follows that ΔABC is contained within ΔDEF, as required. ■

Then, since every d-triangle can be fitted inside some trebly asymptotic d-triangle, the following completely surprising result (which holds whatever our definition of hyperbolic area is) proves the existence of an upper bound for the area of all d-triangles.

Theorem 3 All trebly asymptotic d-triangles have the same (hyperbolic) area, and this is a finite number.

Note that the area must be finite, whatever definition of hyperbolic area we choose to use.

Proof Using the result of Problem 1, we can divide a given trebly asymptotic d-triangle into six singly asymptotic d-triangles, as shown in the margin diagram. Therefore it is sufficient to prove that the area of a singly asymptotic d-triangle is finite.

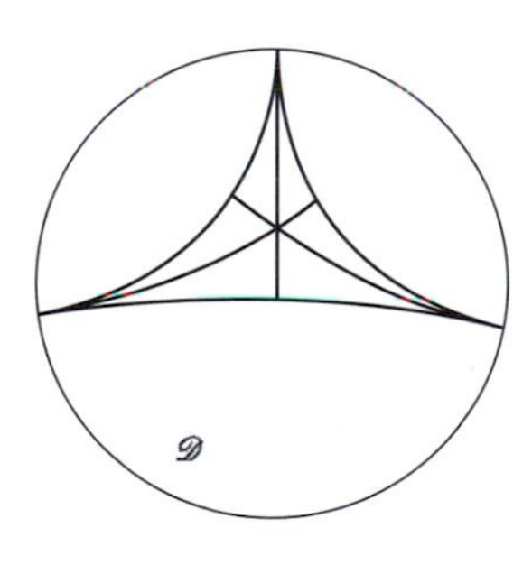

Let ΔABC be a singly-asymptotic d-triangle, with C on $\mathscr{C}$ (as in the left figure below). First, extend AB to meet $\mathscr{C}$ at a point C', and bisect $\angle BAC$ by AD, where D is a point on CC'. Next, (hyperbolically) reflect $\angle BAC$ in AD; since AB has finite hyperbolic length and AC has infinite hyperbolic length (since C lies on $\mathscr{C}$), this reflection maps B to some point A_1, say, on AC, as shown in the right diagram below. Also, the reflection must map BC onto A_1C'; and, by symmetry, BC and A_1C' must meet at some point, B_1 say, on AD.

Next, bisect $\angle C'BC$ by BD_0, where D_0 is on $C'C$, and bisect $\angle C'A_1C$ by A_1D_1, where D_1 is on $C'C$, as shown below.

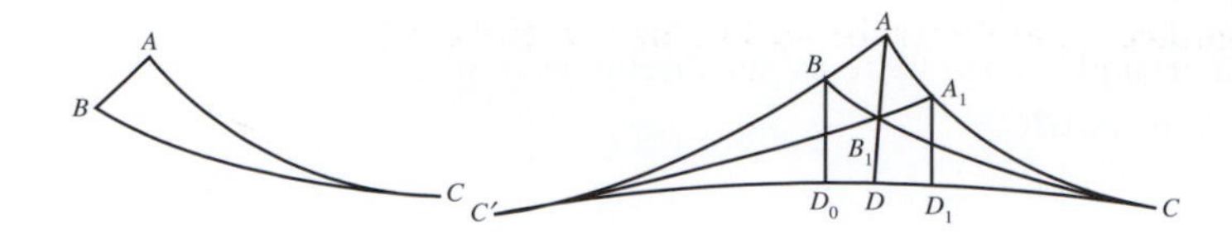

Our strategy of proof is now as follows. The d-pentagon $AA_1D_1D_0B$ and its boundary all lie in $\mathscr{D}$, and so at a minimum positive distance from $\mathscr{C}$, the boundary of $\mathscr{D}$. Also, by adding additional line segments B_1D_0 and B_1D_1 we can divide the d-pentagon into eight d-triangles; each d-triangle has finite area, and so the d-pentagon must also have finite area. We will show that ΔBAC can be cut up and reassembled (iteratively) inside this d-pentagon, and so has an area less than or equal to the area of the d-pentagon; in particular, it must have finite area.

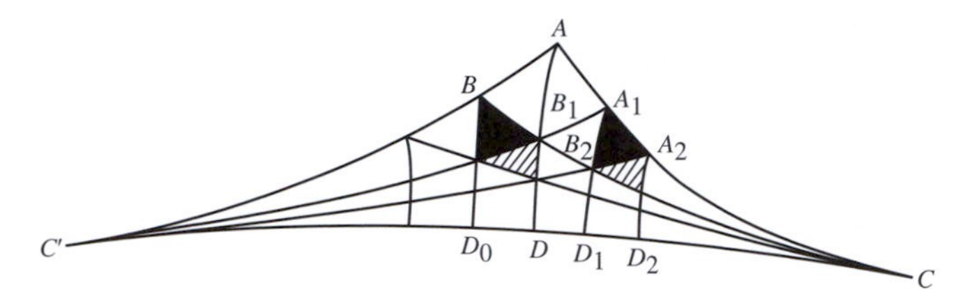

We now do to ΔB_1A_1C what we did previously to ΔBAC. Reflect (in a hyperbolic way) ΔB_1A_1C in A_1D_1 : B_1 maps to some point A_2 on A_1C and B_1C maps to A_2C', which meets B_1C at some point B_2 on A_1D_1.

$\Delta B_2A_1A_2$ is congruent to several d-triangles in the d-pentagon; for convenience we choose the black one on the left in the figure above. We then repeat the process for ΔB_2A_2C. We bisect ΔB_2A_2C with A_2D_2, where D_2 is some point on $C'C$. The cross-hatched d-triangles in the above figure are d-congruent – one is in ΔBAC and one is in the d-pentagon.

Continuing indefinitely, we do for $\Delta B_{n+1}A_{n+1}C$ what we have just done for ΔB_nA_nC. With each pair of hyperbolic reflections we get two new d-triangles in ΔBAC and two new d-triangles in the d-pentagon. These d-triangles are d-congruent by means of a sequence of hyperbolic reflections in the d-lines $A_{n+1}D_{n+1}, A_nD_n, \ldots, A_1D_1$ and AD.

In this way, we reassemble the asymptotic d-triangle *inside* the d-pentagon; hence the asymptotic d-triangle must have finite area.

It follows that all trebly asymptotic d-triangles have finite area, and so (as we noted before the theorem that all trebly asymptotic d-triangles are d-congruent) they have the same area. ■

Problem 2

(a) Prove that a doubly asymptotic d-triangle is a subset of a suitably chosen trebly asymptotic d-triangle, and so is finite.

(b) Prove that the area of a doubly asymptotic d-triangle depends only on the angle at its vertex in $\mathscr{D}$.

We can now give Gauss' magnificent proof that the area of each d-triangle is proportional to the quantity

$$\pi - \text{(angle sum)}.$$

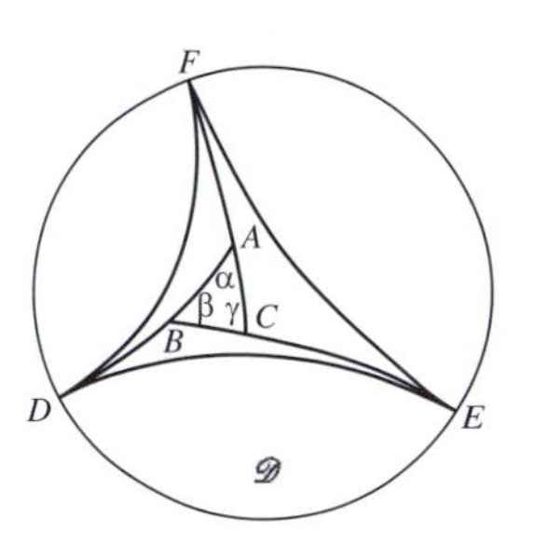

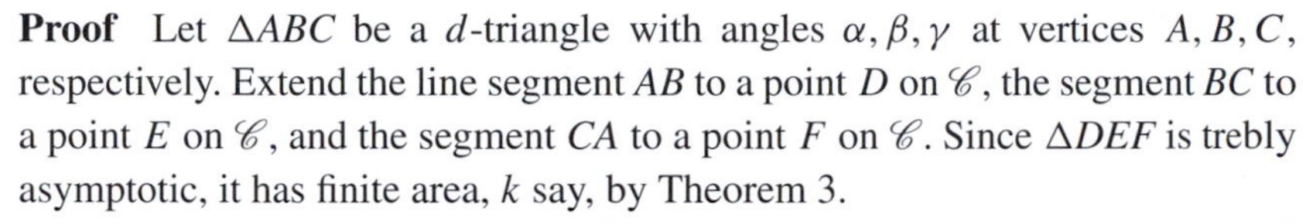

Theorem 4 The area of a d-triangle with angles α, β, γ is given by $K(\pi - (\alpha + \beta + \gamma))$, where K is the same constant for all d-triangles.

Proof Let ΔABC be a d-triangle with angles α, β, γ at vertices A, B, C, respectively. Extend the line segment AB to a point D on $\mathscr{C}$, the segment BC to a point E on $\mathscr{C}$, and the segment CA to a point F on $\mathscr{C}$. Since ΔDEF is trebly asymptotic, it has finite area, k say, by Theorem 3.

Now, we have seen that any doubly asymptotic d-triangle has an area which depends only on the angle at its vertex in $\mathscr{D}$, and so only on the 'exterior angle' of the triangle at its vertex in $\mathscr{D}$. So, we define a function f such that the area of any doubly asymptotic d-triangle with exterior angle θ, say, in $\mathscr{D}$ is $f(\theta)$.

Problem 2, part (b)

Then, it is clear from the margin figure above that in ΔFAD this exterior angle equals α; so, the (hyperbolic) area of ΔFAD is $f(\alpha)$.

If the segment FAC extended meets $\mathscr{C}$ at X (as in the left figure below), then ΔDXF is trebly asymptotic and is composed of the two doubly asymptotic d-triangles ΔFAD and ΔDAX. Then, if we denote the area of each trebly-asymptotic d-triangle (such as ΔDXF) by k, say, and notice that (from our definition of f) the area of ΔDAX is $f(\pi - \alpha)$ since its exterior angle at A is $\pi - \alpha$, it follows that

$$k = f(\alpha) + f(\pi - \alpha). \tag{2}$$

Here we use the fact that (hyperbolic) area is additive.

Our next objective is to use equation (2) to determine an explicit formula for f.

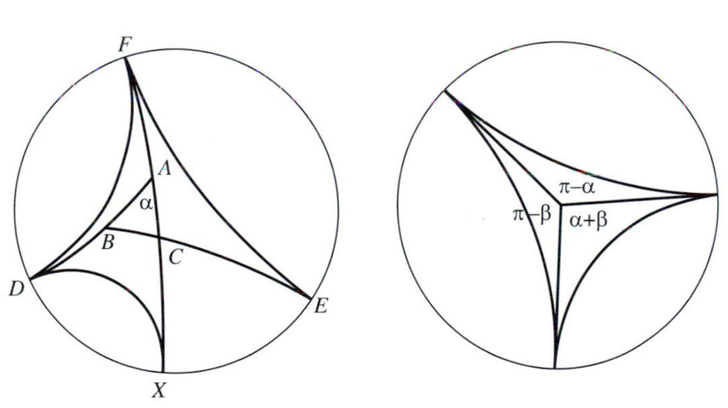

So, consider any trebly asymptotic d-triangle, as shown in the figure on the right above. By drawing radii from the center of $\mathscr{D}$ to its vertices on $\mathscr{C}$, we can divide this d-triangle into three doubly asymptotic d-triangles. If we denote the angles at O of these three d-triangles by $\pi - \alpha$, $\pi - \beta$ and $\alpha + \beta$, as shown above, we have

$$k = f(\alpha) + f(\beta) + f(\pi - (\alpha + \beta))$$

By equation (2), $f(\pi - (\alpha + \beta)) = k - f(\alpha + \beta)$, so that

$$f(\alpha) + f(\beta) + k - f(\alpha + \beta) = k,$$

that is,

$$f(\alpha) + f(\beta) = f(\alpha + \beta). \qquad (3)$$

It follows from equation (3) that f is linear, so that $f(\alpha) = \lambda\alpha$ for some constant λ.

We omit the details of this fact, which uses a continuity argument.

Using this formula for f, we deduce from equation (2) that

$$\lambda(\alpha) + \lambda(\pi - \alpha) = k,$$

so that $\lambda = \frac{k}{\pi}$. Hence $f(\alpha) = \frac{k}{\pi}\alpha$.

Returning to the original d-triangle ΔABC in the margin figure above, the area of ΔABC is equal to the area of ΔDEF minus the sum of the areas of the three doubly asymptotic d-triangles ΔFAD, ΔDBE and ΔCAF. Since the latter four areas are k, $f(\alpha)$, $f(\beta)$ and $f(\gamma)$, respectively, it follows that

We again use the fact that (hyperbolic) area is additive.

$$\text{area of } \Delta ABC = k - (f(\alpha) + f(\beta) + f(\gamma)).$$

Using the formula for f that we have just found, we can write this equation as

$$\text{area of } \Delta ABC = k - \left(\frac{k}{\pi}\alpha + \frac{k}{\pi}\beta + \frac{k}{\pi}\gamma\right),$$

and so as

$$\text{area of } \Delta ABC = K(\pi - (\alpha + \beta + \gamma)),$$

where $K = k/\pi$. ■

Note that here K is arbitrary.

It is usual to take $K = 1$ (which is equivalent to taking the area of a trebly asymptotic d-triangle to be π). Then

$$\text{area of } \Delta ABC = \pi - (\alpha + \beta + \gamma).$$

The expression $\pi - (\alpha + \beta + \gamma)$ is called the *angular defect* of ΔABC.

Now this result is all very well, but how do we know whether, given any positive real numbers α, β, γ with $\alpha + \beta + \gamma < \pi$, there actually is a d-triangle whose angles are α, β, γ?

Recall that, if α, β, γ are positive real numbers with $\alpha + \beta + \gamma = \pi$, then such a Euclidean triangle exists.

A key tool in showing that such a d-triangle exists, is the following result.

Lemma 1 Let P be a point on a d-line ℓ of $\mathscr{D}$, and let $0 < \beta < \frac{\pi}{2}$. Then there are exactly two d-lines through P that make an acute angle β with ℓ, and each is a (hyperbolic) reflection of the other in ℓ.

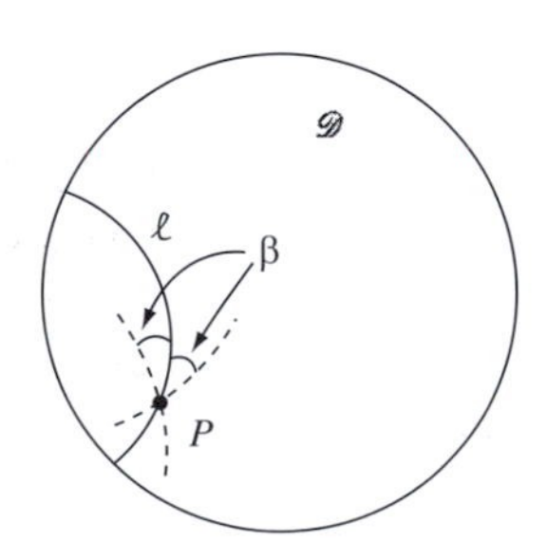

Proof Since all the properties in the statement of the Lemma are hyperbolic properties, and there is a hyperbolic transformation (for example, a Möbius transformation) that maps the point P to the origin O, it is sufficient to prove the result when P is the origin.

Then there are (by Euclidean geometry) exactly two diameters of $\mathscr{D}$ (and diameters are d-lines) through the origin that make an acute angle β with the given diameter ℓ, and each is a reflection of the other in ℓ. This completes the proof. ■

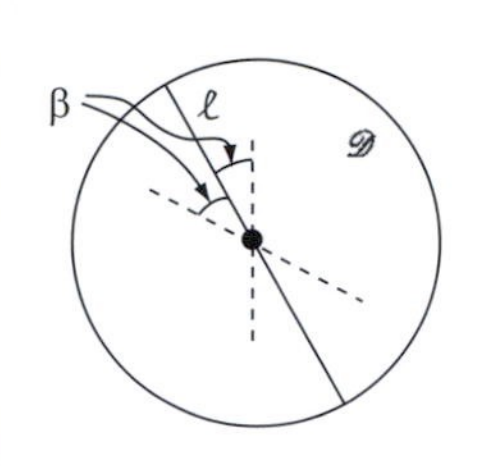

We are now in a position to prove the final result in this subsection.

Theorem 5 Let α, β, γ be positive real numbers with $0 < \alpha+\beta+\gamma < \pi$. Then there is a d-triangle in $\mathscr{D}$ with angles α, β, γ.

Proof Since $0 < \alpha + \beta + \gamma < \pi$, at least one of the numbers lies strictly between 0 and $\frac{\pi}{2}$; let us suppose that this number is β. Next, let A, P, Q be the points in $\mathscr{D}$ with coordinates 0, 1, $\cos\alpha + i \sin\alpha$, and let ℓ be a variable d-line in $\mathscr{D}$ whose initial position is as the d-line with endpoints P and Q. Then the angles of the doubly-asymptotic d-triangle ΔAPQ are α, 0 and 0.

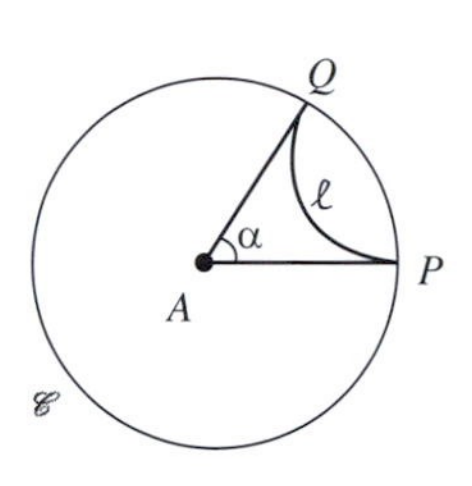

The first step is to keep fixed the endpoint of ℓ that is originally at Q, and let the endpoint that is originally at P move round $\mathscr{C}$ away from Q. As this endpoint moves away from P, the angle θ between ℓ and the radius AP varies continuously, starting at 0. By the time that the moving endpoint reaches Q', the image of Q under Euclidean reflection in the radius AP, the angle θ has reached the value $\frac{\pi}{2}$. It follows, by continuity and the intermediate value property, that there is some point, R say, on AP such that for the d-line through P and R the corresponding value of the angle θ is β.

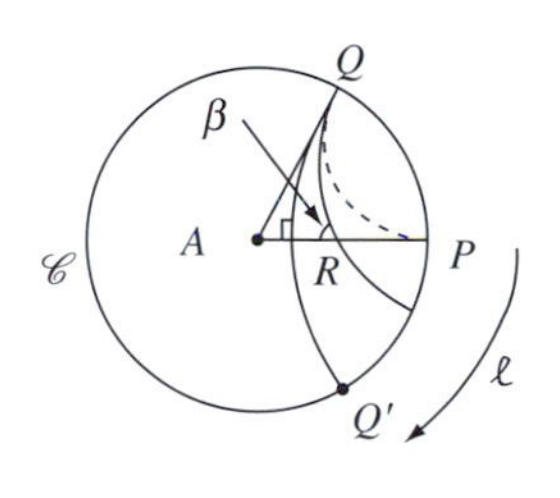

Now, it follows from Lemma 1 that through every point S between A and R there is a d-line that makes an angle β with the radius AP; also, if we assume that this angle β lies above the radius AP and contains part of the second quadrant of $\mathscr{D}$ in its interior, this d-line is unique.

We now let S move along AP from R towards A, taking with it the corresponding d-line ℓ. Denote by T the point where the variable d-line ℓ intersects the radius AQ. Then, as S moves along AP from R towards A, T moves along AQ from Q towards A.

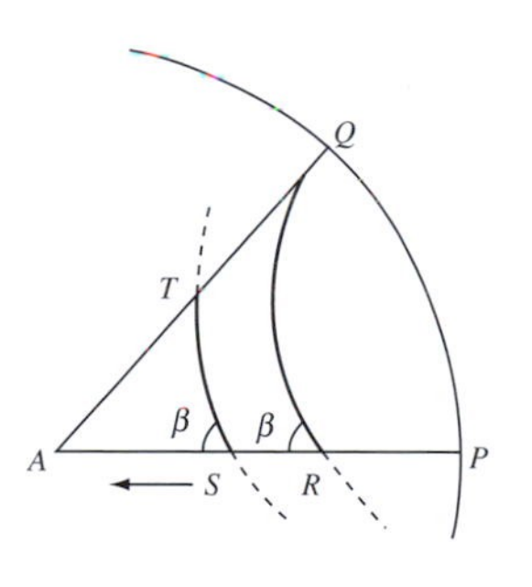

The d-lines ℓ_1 and ℓ_2 corresponding to successive positions S_1 and S_2 of S do not meet in $\mathscr{D}$. For, if they met at a point X in $\mathscr{D}$, then the external angle at S_2 of the d-triangle ΔXS_1S_2 would be less than or equal to the sum of the opposite two internal angles, which is impossible by the result of Problem 1 in Subsection 6.4.1. (The following figure shows the situation when X is in the upper/lower half of $\mathscr{D}$, respectively.)

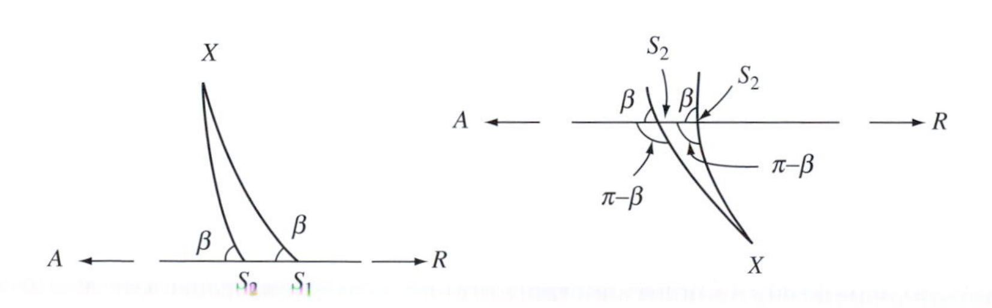

Hence as S moves monotonically along AP from R towards A, T moves monotonically along AQ from Q towards A. As S approaches A, the curvature of the d-line ℓ decreases, and ℓ approaches the diameter d of $\mathscr{D}$ that makes an angle β with the diameter containing AP.

As T moves monotonically along AQ from Q towards A, the angle at T between the radial line AT and ℓ starts at 0, and approaches the value $\pi - (\alpha + \beta)$ as T approaches A. It follows, by continuity and the intermediate value property, that there is some point, C say, on AQ such that at C the d-line ℓ makes an angle γ with the radius through C.

If we denote by B the point where the d-line ℓ through C meets AP, then the triangle ΔABC has the desired angles α, β, γ. ■

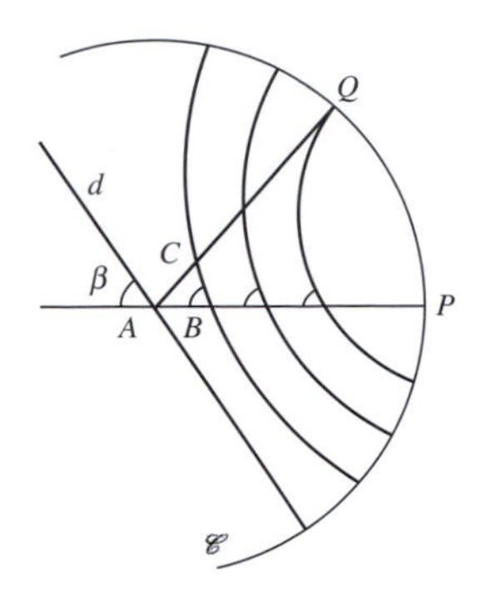

6.5.2 Tessellations

The term *tessellate* means 'fit together exactly a number of identical shapes, leaving no spaces', as in the making of a mosaic. For example, we may tessellate the Euclidean plane using rectangles of a given size in several ways, and we may tessellate it in other ways as well – as illustrated below. In these examples, we may move a given occurrence of any shape onto any other occurrence by a Euclidean transformation, and so by a combination of Euclidean reflections.

You saw in Theorem 2 of Subsection 5.2.1 that every Euclidean transformation is a composite of reflections.

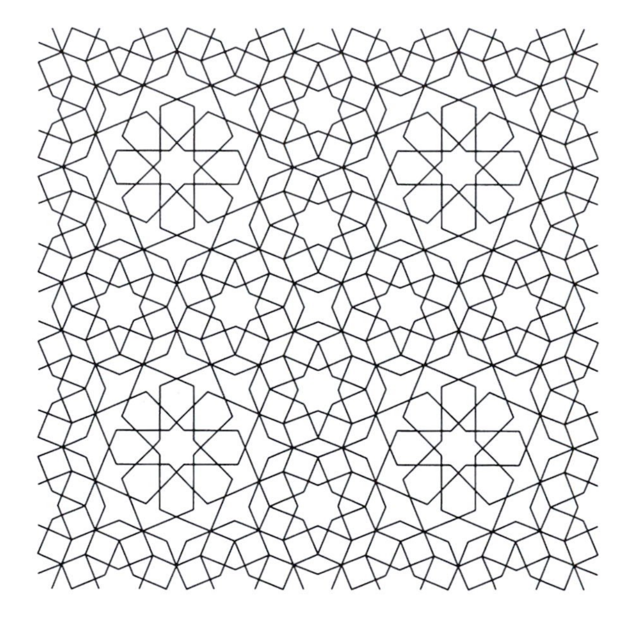

Euclidean tessellations using a single figure are sometimes called 'wallpaper patterns', for obvious reasons! With each such figure we can associate its group of symmetries; it is an interesting piece of mathematics that there are only 17 different wallpaper patterns, in the sense that the symmetry group of any wallpaper pattern is isomorphic to one of only 17 non-isomorphic groups.

Patterns like those in the figure above are found among the tilings of the Alhambra Palace in Granada (Spain), and elsewhere in Moorish art.

Tessellations occur in hyperbolic geometry also. For example, we may start with a symmetric figure, a symmetric d-triangle ΔABC, in $\mathscr{D}$ bounded by three d-lines (as in the following figure), reflect ΔABC in each d-line in turn to obtain 3 copies each d-congruent to ΔABC, and so on. This produces a hyperbolic tessellation of $\mathscr{D}$, as shown in the figure below; for clarity we have shaded all the copies that can be obtained from ΔABC by direct hyperbolic transformations. All the copies are d-congruent to each other, even though to a Euclidean eye they seem to be of various different shapes and sizes!

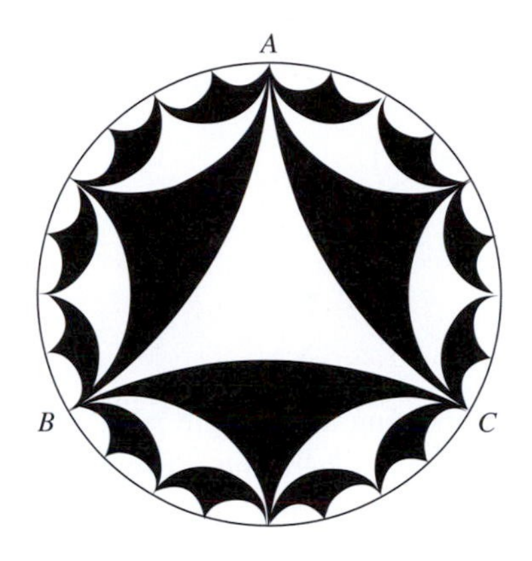

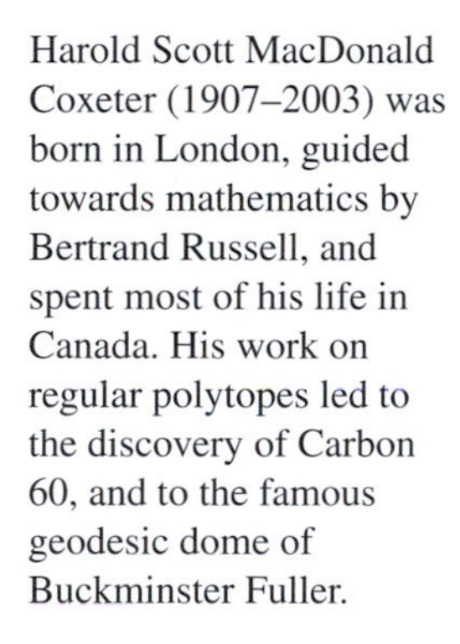
Harold Scott MacDonald Coxeter (1907–2003) was born in London, guided towards mathematics by Bertrand Russell, and spent most of his life in Canada. His work on regular polytopes led to the discovery of Carbon 60, and to the famous geodesic dome of Buckminster Fuller.

Another tessellation of $\mathscr{D}$ was suggested to the graphic artist M. C. Escher by the geometer H. S. M. Coxeter, and resulted in the picture below, *Circle Limit IV*. The angels and devils rest on a tessellation of $\mathscr{D}$ by d-triangles.

Maurits Cornelis Escher (1898–1972) received his graphic art training in Arnhem and Haarlem, and worked in Italy, Switzerland, Belgium and Holland.

6.5.3 Kaleidoscopes

We now explore the ideas underlying the (Euclidean) kaleidoscope, and explain how hyperbolic tessellations can be constructed.

A kaleidoscope is a machine that produces wallpaper patterns using reflections in mirrors.

A Euclidean Kaleidoscope

If two mirrors meet at right angles, an angle of $\pi/2$, any point between the mirrors has an image in each mirror; then, each image itself has an image in the mirrors – giving a total of four points in all, lying on a circle centred at the vertex where the mirrors meet.

If the mirrors meet at an angle of $\pi/3$, there are 6 points in all; and, in general, if the mirrors meet at an angle π/k there are $2k$ points in all.

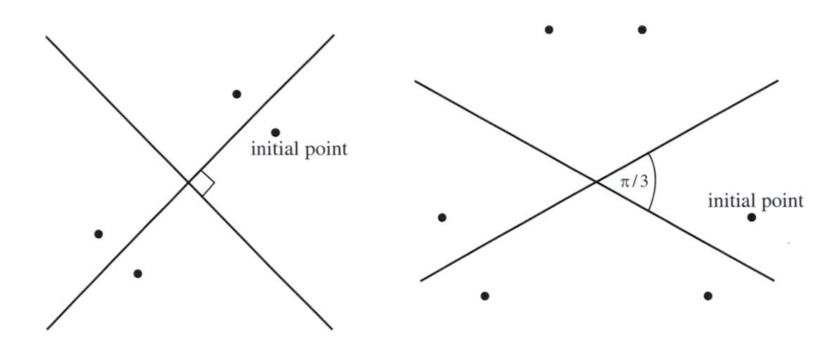

A real kaleidoscope has three mirrors, each pair meeting at an angle π/k, where k is some integer. For example, consider the case where the mirrors meet at angles of $\pi/2, \pi/3$ and $\pi/6$. Corresponding to these vertices there are 4 points, 6 points and 12 points – making a total (in the first instance) of 17 points. But each of these points in turn has images in the mirrors, and so on . . .. The whole pattern of images spreads through the plane!

The total number is 17 rather than 22, since some of them are repeated.

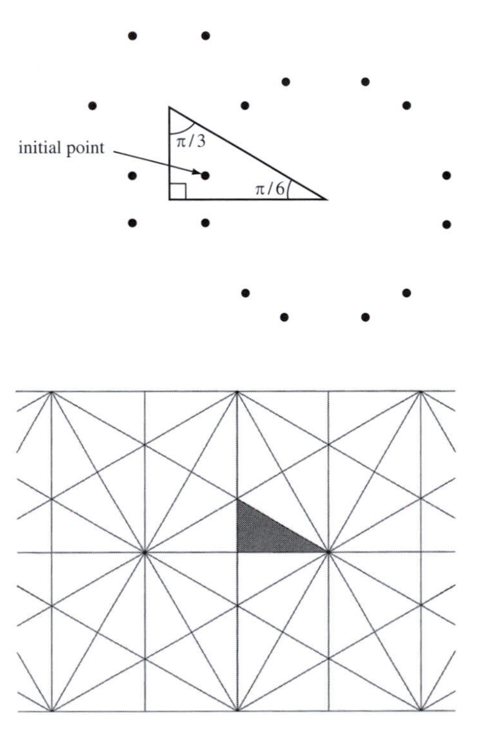

Since reflection is a Euclidean isometry, the process of reflections fills the whole plane with congruent shapes.

A 'Hyperbolic Kaleidoscope'

Hyperbolic reflections and transformations are hyperbolic isometries; that is, they map figures without altering the hyperbolic distances between their points even though the images may look (to a Euclidean eye) very different in size and shape from the originals. In particular, they map d-triangles to d-congruent d-triangles.

If two mirrors along d-lines meet at an angle of $\pi/2$, any point between the mirrors has an image under hyperbolic reflection in each mirror; then, each image itself has an image in the mirrors. Again, we obtain a total of four points in all, lying on a hyperbolic circle centred at the vertex.

Similarly, if the mirrors along d-lines meet at an angle of $\pi/3$, there are 6 points in all; and, in general, if the mirrors meet at an angle of π/k, there are $2k$ points in all.

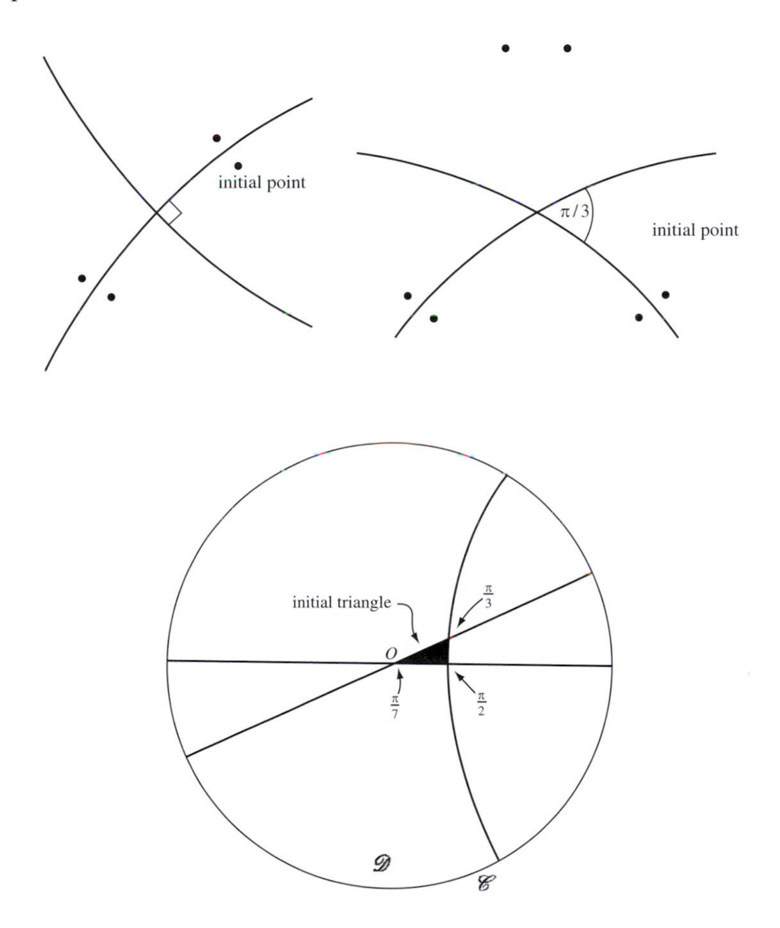

Now construct a hyperbolic analogue of the Euclidean kaleidoscope, with three mirrors along d-lines in $\mathscr{D}$: two of them lying along d-lines through the origin that meet at O at an angle of $\pi/7$, and the third lying along a d-line that meets the first two at angles of $\pi/2$ and $\pi/3$. We reflect the triangle in each side in turn (using hyperbolic reflections), and continue to reflect indefinitely in the three d-lines.

It is not hard to verify that such a d-line must exist.

Note that $\frac{\pi}{7} + \frac{\pi}{2} + \frac{\pi}{3} < \pi$.

Since hyperbolic reflection is a hyperbolic isometry, the process of reflections fills the whole unit disc with d-congruent d-triangles. The figure that we end up with is just the design on the front of the book and in Subsection 6.4.1.

6.6 Hyperbolic Geometry: the Half-Plane Model

In previous sections we described the disc model of hyperbolic geometry, in which the unit disc $\mathcal{D}$ is the space of points, and the transformations of this space as being of the form

$$z \mapsto M(z) = \frac{az+b}{\bar{b}z+\bar{a}} \quad \text{or} \quad z \mapsto M(\bar{z}) = \frac{a\bar{z}+b}{\bar{b}\bar{z}+\bar{a}}, \quad \text{where } |b| < |a|.$$

However there is a completely *equivalent* hyperbolic geometry, in which the space (the set of points) is the upper half-plane

That is, there is a one–one correspondence between points of the two spaces and the transformations in the two geometries.

$$H = \{z : z \in \mathbb{C}, \quad \operatorname{Im} z > 0\},$$

and the h-lines in H (the analogues of d-lines in $\mathcal{D}$) are the restriction to H of vertical lines in the plane together with the restriction to H of circles with centre on the real axis in $\mathbb{C}$. Notice that the h-lines are all arcs of generalized circles that meet the boundary of H at right angles, just as d-lines are all arcs of generalized circles that meet the boundary $\mathcal{C}$ of $\mathcal{D}$ at right angles.

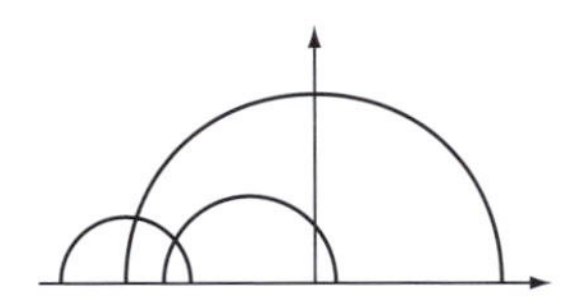

The Möbius transformation $z \mapsto i\frac{1+z}{1-z}$ maps $\mathcal{D}$ one–one onto H, and provides a convenient method for 'transferring' the geometric properties in $\mathcal{D}$ to corresponding properties in H. For example, it maps the d-lines in $\mathcal{D}$ to the h-lines in H.

The hyperbolic transformations in H are the mappings of H to itself given by

$$z \mapsto M(z) = \frac{az+b}{cz+d} \quad \text{or} \quad z \mapsto M(-\bar{z}) = \frac{a(-\bar{z})+b}{c(-\bar{z})+d},$$

where a, b, c and d are real, and $ad - bc > 0$. These are the composites of reflections in h-lines, which are just inversions in the Euclidean circles of which the h-lines form a part.

The geometry in H has many properties similar to those in $\mathcal{D}$:

- the sum of the angles of an h-triangle is less than π;
- two h-triangles are h-congruent if their angles are equal in pairs;
- the area of an h-triangle with angles α, β, γ is $K(\pi - (\alpha+\beta+\gamma))$, for some $K > 0$;
- h-circles are Euclidean circles in H, and vice-versa.

Generally we take $K = 1$, as in Subsection 6.5.1.

Infinitesimal hyperbolic distances ds in H are related to their Euclidean lengths by the formula $ds = \frac{|dz|}{y}$, from which it can be shown that geodesics in H are the h-lines.

Recall that a *geodesic* is the curve of shortest length joining two points in the space.

Finally, we comment that the hyperbolic geometry of H turns out to be of great importance in many unexpected branches of mathematics – such as Number Theory, in which tessellations of H such as the following occur naturally.

6.7 Exercises

Section 6.1

1. Sketch the following d-lines:

$$\ell_1 = \{(x, y) : x + y = 0\} \cap \mathscr{D};$$
$$\ell_2 = \left\{(x, y) : x^2 + y^2 + 2\sqrt{2}y + 1 = 0\right\} \cap \mathscr{D};$$
$$\ell_3 = \left\{(x, y) : x^2 + y^2 - \sqrt{5}y + 1 = 0\right\} \cap \mathscr{D};$$
$$\ell_4 = \left\{(x, y) : x^2 + y^2 - 3x + 1 = 0\right\} \cap \mathscr{D}.$$

   Hence decide which of the d-lines intersect, which are parallel and which are ultra-parallel.
2. Sketch three d-lines ℓ_1, ℓ_2 and ℓ_3 with the following properties: ℓ_1 and ℓ_3 are parallel, ℓ_2 intersects ℓ_1 and ℓ_3, and the angles at which ℓ_2 crosses ℓ_1 and ℓ_3 are not equal.

   (This shows that the *Corresponding Angles Theorem* in Euclidean geometry does not hold in hyperbolic geometry.)
3. Verify that the points $\left(-\frac{1}{\sqrt{2}}, \frac{1}{\sqrt{2}}\right)$ and $\left(-\frac{1}{\sqrt{2}}, -\frac{1}{\sqrt{2}}\right)$ lie on the circle C with equation

$$x^2 + y^2 + 2\sqrt{2}x + 1 = 0.$$

Hence write down the equations of the diameters of $\mathscr{D}$ which are parallel to the d-line ℓ which is part of C.

4. Use the Origin Lemma to show that, given two d-lines meeting at right angles, there is a hyperbolic reflection mapping them to two perpendicular diameters of $\mathscr{D}$.
5. Let ℓ_1, ℓ_2 and ℓ_3 be any three d-lines such that hyperbolic reflection in any one exchanges the other two.
   (a) Prove that there is some point P of $\mathscr{D}$ which lies on each of the three d-lines.
   (b) Use the Origin Lemma to prove that the acute angle between any two of the d-lines is $\frac{\pi}{3}$.
6. Verify that the following equations are the equations of generalized circles of which d-lines are a part, and sketch the corresponding d-lines.
   (a) $x^2 + y^2 + 4x + 1 = 0$
   (b) $x^2 + y^2 + 2x - 2y + 1 = 0$
7. Use the fact that the circle with radius a and centre $(1, a)$, where $a > 0$, touches the x-axis to write down the equation of a d-line through the point $\frac{1}{2}i$ that is parallel to the d-line with equation $y = 0$.

Section 6.2

1. Determine the general form of the direct hyperbolic transformations that map 0 to 0 and the diameter $y = x$ of $\mathscr{D}$ to the diameter $y = 0$ of $\mathscr{D}$.
2. (a) Determine the general form of the direct hyperbolic transformations that map each of the following points to 0:

$$-\tfrac{1}{2}, \quad \tfrac{1}{2} - \tfrac{1}{2}i.$$

   (b) Determine a direct hyperbolic transformation that maps $-\frac{1}{2}$ to $\frac{1}{2} - \frac{1}{2}i$.
   (c) Determine the general form of the direct hyperbolic transformation that maps $\frac{1}{2} - \frac{1}{2}i$ to $-\frac{1}{2}$.
3. By drawing a sketch, or in some other way, decide whether the effect of successive hyperbolic reflections in d-lines ℓ_1 and then ℓ_2 given by the following equations is a (hyperbolic) rotation, a limit rotation or a translation.
   (a) $\ell_1 : x^2 + y^2 - 2x - 2y + 1 = 0$; $\ell_2 : x^2 + y^2 + 2x + 2y + 1 = 0$.
   (b) $\ell_1 : x^2 + y^2 - 2x - 2y + 1 = 0$; $\ell_2 : x^2 + y^2 - 2x + 2y + 1 = 0$.
4. Let M be a direct hyperbolic transformation mapping the point p in $\mathscr{D}$ to the point q. Let M_q be a direct hyperbolic transformation mapping the point q to 0.
   (a) Show that the composite transformation $M' = M_q \circ M$ maps p to 0.
   (b) Deduce that M is the composite of a hyperbolic transformation mapping p to 0 followed by a hyperbolic transformation mapping 0 to q.

5. (a) Prove that the set of all direct hyperbolic transformations forms a group under the operation of composition of functions.
   (b) Prove that the set of all hyperbolic transformations forms a group under the operation of composition of functions.

Section 6.3

1. Determine $d\left(-\frac{1}{5}i, 0\right)$ and $d(0.9, 0.99)$.
2. Determine the positive number a for which $d(0,a) = 2$, and the negative number b for which $d(0,b) = 10$.
3. Determined $d\left(-\frac{1}{2}, \frac{1}{2}i\right)$.
4. (a) Determine the hyperbolic midpoints of the d-line segments $[0.1, 0.9]$ and $[-0.1, 0.9]$.
   (b) Determine the hyperbolic centre and radius of the circle $C = \{z : |z - 0.5| = 0.4\}$.
   (c) Determine the equation of the hyperbolic perpendicular bisector ℓ of $[0.1, 0.9]$.
5. Determine the Euclidean centre and radius of the hyperbolic circle $C = \left\{z : d\left(\frac{1}{2}, z\right) = 1\right\}$.
6. Let $0 < a < b < 1$, and let the d-line ℓ with equation $x^2+y^2-2cx+1 = 0$, where $c > 1$, intersect the radius $[0, 1)$ between $A(a, 0)$ and $B(b, 0)$.
   (a) Prove that ℓ intersects AB at the point $\left(c - \sqrt{c^2 - 1}, 0\right)$.
   Now assume that ℓ is the hyperbolic perpendicular bisector of AB.
   (b) Use the Reflection Lemma to prove that
   $$c = \frac{1 + ab}{a + b}.$$
   (c) Deduce from parts (a) and (b) that the hyperbolic midpoint of AB is
   $$\left(\frac{1 + ab - \sqrt{(1 - a^2)(1 - b^2)}}{a + b}, 0\right).$$
7. Let ℓ be a d-line that is not a diameter of $\mathscr{D}$ and whose endpoints on $\mathscr{C}$ subtend an angle $\theta, 0 < \theta < \pi$, at the centre of $\mathscr{D}$. Prove that the hyperbolic length of the d-line segment from the origin to ℓ that is perpendicular to ℓ is $\tanh^{-1}\left(\cot \frac{1}{4}(\theta + \pi)\right)$.

Section 6.4

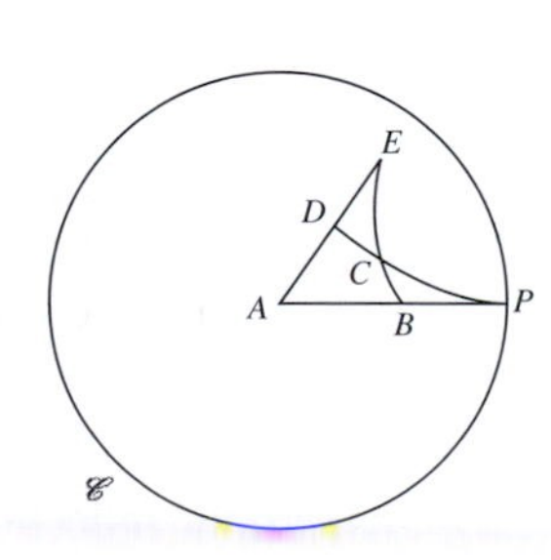

1. $ABCD$ is a d-quadrilateral. The d-lines through A and B and through D and C are parallel, with a common boundary point P on $\mathscr{C}$; the d-lines through D and A and through C and B meet at a point E in $\mathscr{D}$, as in the figure in the margin (in which A is at the origin).
   (a) Prove that it is impossible that $\angle EAB = \angle EDC$.
   Hint: Consider the Euclidean triangle ΔADP.

(b) Prove that it is impossible that $\angle EBA = \angle ECD$.
Hint: Consider the Euclidean triangle ΔPBC, with B at the centre of $\mathscr{D}$.

(c) Prove that the following combination is impossible:

$$\angle EAB = \angle ECD \quad \text{and} \quad \angle EBA = \angle EDC.$$

(d) Prove that the following combination is impossible:

$$\angle EAB = \angle EBA \quad \text{and} \quad \angle EDC = \angle ECD.$$

2. Let ℓ_1 and ℓ_2 be a pair of perpendicular diameters of $\mathscr{D}$. Sketch ℓ_1 and ℓ_2, and some of the ultra-parallels to ℓ_1 for which ℓ_2 is their common perpendicular with ℓ_1.
3. $ABCD$ is a d-quadrilateral such that the hyperbolic length of each side is 1. The d-line segments AC and BD are of equal hyperbolic length and intersect at right angles at the origin O, their common midpoint, and OE is an altitude of the d-triangle ΔOAB. Also, P is the hyperbolic midpoint of OE.
Determine the hyperbolic length of OE, the Euclidean length of OP, the angle $\angle ABC$, the hyperbolic length of OB, the Euclidean length of BD, and the angle φ between PE and either d-line through P that is parallel to the d-line through A and B.

ABCD is a hyperbolic square of side-length 1.

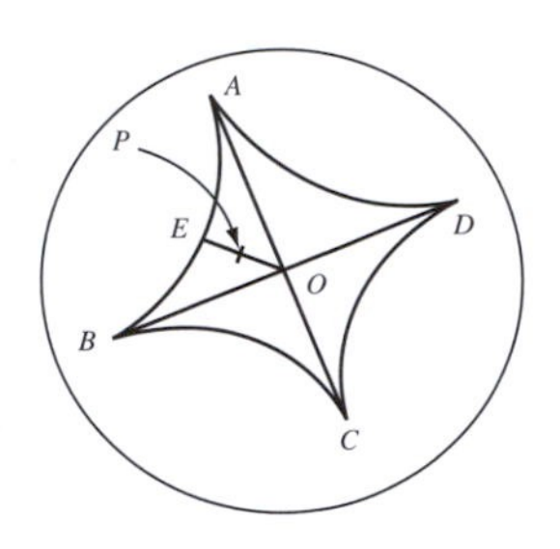

4. Let ΔABC be a d-triangle with $\angle ABC = \pi/3$, and let the sides BA and BC have the same hyperbolic length c. Let BD be the altitude from B to the side AC. Show that the hyperbolic length b of AC is greater than c.
Hint: Apply the Sine Formula to triangle ΔBDA.

Proof of the Sine Formula (Subsection 6.4.3, Theorem 10).

5. Show that if ΔABC is a d-triangle right-angled at C, then

$$\sin A = \frac{\sinh 2a}{\sinh 2c},$$

where a and c are the hyperbolic lengths of the sides BC and AB.
Hint: If $\tan\phi = t$, then $\sin^2\phi = \frac{t^2}{1+t^2}$.
6. Determine the hyperbolic lengths of the sides of the d-triangle with angles $\frac{\pi}{2}$, $\frac{\pi}{6}$ and $\frac{\pi}{6}$.

Section 6.5

1. Determine the hyperbolic area of the d-triangle ΔABC with vertices at the points $A(0,0)$, $B\left(\frac{1}{2},0\right)$ and $C\left(0,\frac{1}{2}\right)$.
2. Let A, B, C and D be the points (0,0), (1,0), (0,1) and $(-1,0)$, respectively, and let ℓ be the d-line BC.
(a) Determine the vertices of the d-triangle that is the image of the d-triangle ΔABC under hyperbolic reflection in ℓ.
(b) Determine the vertices of the d-triangle that is the image of the d-triangle ΔACD under hyperbolic reflection in ℓ.
(c) Sketch the d-triangles ΔABC and ΔACD, and their images under hyperbolic reflection in ℓ.

(d) Determine the areas of the d-triangles ΔABC and ΔACD, and of their images under hyperbolic reflection in ℓ.

Hint: You may assume that Theorem 4 of Subsection 6.5.1 holds for asymptotic triangles, with $K = 1$.

Summary of Chapter 6

Introduction

1. **The (Euclidean) Parallel Postulate** Given any line ℓ and a point P not on ℓ, there is a unique line m in the same plane as P and ℓ which passes through P and does not meet ℓ.
2. **The (Elliptic) Parallel Postulate** Given any line ℓ and a point P not on ℓ, all lines through P meet ℓ. (That is, there are no lines through P that are parallel to ℓ.)
3. **The (Hyperbolic) Parallel Postulate** Given any line ℓ and a point P not on ℓ, there are at least two lines m through P that do not meet ℓ. (That is, there are at least two lines through P that are parallel to ℓ.)

Section 6.1: Hyperbolic Geometry: the Disc Model

1. In (Poincaré's model of) hyperbolic geometry, **points** (or ***d*-points**, where 'd' stands for 'disc') consists of the points in the unit disc $\mathscr{D} = \{z : |z| < 1\} = \{(x, y) : x^2 + y^2 < 1\}$.

   The points on the boundary $\mathscr{C} = \{z : |z| = 1\} = \{(x, y) : x^2 + y^2 = 1\}$ of $\mathscr{D}$ are not points that belong to the geometry.
2. A ***d*-line** (or **line in hyperbolic geometry**) is that part of a (Euclidean) generalized circle which meets $\mathscr{C}$ at right angles and which lies in $\mathscr{D}$. A d-line may be an arc of a (Euclidean) circle or a Euclidean line segment that is a diameter of $\mathscr{D}$. The points where the corresponding generalised circle meets $\mathscr{C}$ are called the **boundary points** of the d-line.

   d-lines that are arcs of Euclidean circles all lie in one-half of the disc $\mathscr{D}$, and all curve away from the diameter bounding the half-disc.

   The equation of a d-line ℓ is of the form $ax + by = 0$, where a and b are not both zero, or $x^2 + y^2 + fx + gy + 1 = 0$, where $f^2 + g^2 > 4$.
3. Two d-lines that do not meet in $\mathscr{D}$ are **parallel** if the generalized Euclidean circles of which they are parts meet at a point on $\mathscr{C}$, or **ultra-parallel** if the generalized Euclidean circles of which they are parts do not meet on $\mathscr{C}$.

   Corresponding to any given d-line ℓ and any point P in $\mathscr{D}$ that is not on ℓ, there exist exactly two d-lines through P that are parallel to ℓ, and infinitely many d-lines through P that are ultra-parallel to ℓ.
4. Let ℓ be a d-line that is part of a Euclidean circle C. Then inversion in C maps $\mathscr{C}$ onto $\mathscr{C}$, and $\mathscr{D}$ onto $\mathscr{D}$.

5. A **hyperbolic reflection** in a d-line ℓ is the restriction to $\mathscr{D}$ of the inversion in the generalized circle C of which ℓ is part.
   The centre of such a circle C is the point of intersection of the tangents to $\mathscr{C}$ where $\mathscr{C}$ and C intersect.
6. **Hyperbolic transformations** are the composites of a finite number of hyperbolic reflections, and form a group called the **hyperbolic group**, $G_{\mathscr{D}}$, under the operation of composition of functions.
   Hyperbolic geometry consists of the unit disc $\mathscr{D}$ together with the group $G_{\mathscr{D}}$ of hyperbolic transformations.
7. The (**hyperbolic**) **angle between two curves** through a given point A in $\mathscr{D}$ is the Euclidean angle between their (Euclidean) tangents at A.
8. Hyperbolic transformations map d-lines onto d-lines, and preserve the magnitudes of angles.
9. **Origin Lemma** Let A be a point of $\mathscr{D}$ other than the origin O. Then there exists a d-line ℓ such that hyperbolic reflection in ℓ maps A to O.
10. Let A be a point of $\mathscr{D}$. Then there exist infinitely many d-lines through A.
    Let A and B be two distinct points of $\mathscr{D}$. Then there exists a unique d-line ℓ through A and B.
11. Let A_1 and A_2 be any two points of $\mathscr{D}$, and let ℓ_1 and ℓ_2 be d-lines through A_1 and A_2, respectively. Then there is a hyperbolic transformation which maps A_1 to A_2 and ℓ_1 to ℓ_2.
12. Let $A, B \in \mathscr{D}$ be inverse points with respect to the d-line ℓ, and let A', B' and ℓ' be the images of A, B and ℓ under hyperbolic reflection in another d-line ℓ^*. Then A' and B' are inverse points with respect to inversion in ℓ'.

Section 6.2: Hyperbolic Transformations

1. Hyperbolic reflection ρ in the d-line ℓ which is part of a Euclidean circle C with centre α is given by the formula

$$\rho(z) = \frac{\alpha\bar{z} - 1}{\bar{z} - \bar{\alpha}}, \ z \in \mathscr{D}.$$

   Hyperbolic reflection is self-inverse.
   Hyperbolic reflection σ in the diameter of $\mathscr{D}$ on which $y = x \tan\theta$ is given by the formula $\sigma(z) = \alpha^2\bar{z}$, $z \in \mathscr{D}$, where $\alpha = \cos\theta + i\sin\theta$.
2. The composite of the hyperbolic reflections $\rho(z) = \frac{\alpha\bar{z}-1}{\bar{z}-\bar{\alpha}}$ and $\sigma(z) = \frac{\beta\bar{z}-1}{\bar{z}-\bar{\beta}}$ is $(\sigma \circ \rho)(z) = \frac{(\bar{\alpha}\beta-1)\bar{z}+\alpha-\beta}{(\bar{\alpha}-\bar{\beta})\bar{z}+\alpha\bar{\beta}-1}$.
3. The restriction to $\mathscr{D}$ of every Möbius transformation of the form $M(z) = \frac{az+b}{\bar{b}z+\bar{a}}$, with $|b| < |a|$, is a composite of two hyperbolic reflections, and is therefore a hyperbolic transformation.
4. Every hyperbolic transformation can be expressed as a composite of at most three hyperbolic reflections.
5. The composition of an even number of hyperbolic reflections does not change the orientation of the unit disc, and is called a **direct** hyperbolic

transformation. A direct hyperbolic transformation can be expressed as a composite of at most two hyperbolic reflections.

Any direct hyperbolic transformation is given by $z \mapsto M(z) = \frac{az+b}{\overline{b}z+\overline{a}}$, $z \in \mathscr{D}$, for some a, b with $|b| < |a|$.

The composition of an odd number of hyperbolic reflections changes the orientation of the unit disc, and is called an **indirect** hyperbolic transformation.

Any indirect hyperbolic transformation is given by $z \mapsto M(\overline{z})$, where $M(z) = \frac{az+b}{\overline{b}z+\overline{a}}$, $z \in \mathscr{D}$, for some a,b with $|b| < |a|$.

6. The composite of hyperbolic reflections in d-lines ℓ_1 and ℓ_2 that intersect at a point in $\mathscr{D}$ is called a (**hyperbolic**) **rotation** and has exactly one fixed point in $\mathscr{D}$.

   The composite of hyperbolic reflections in d-lines ℓ_1 and ℓ_2 that do not meet in $\mathscr{D}$ but do meet at a point on $\mathscr{C}$ is called a (**hyperbolic**) **limit rotation**, and has no fixed points in $\mathscr{D}$ but leaves one point of $\mathscr{C}$ fixed.

   The composite of hyperbolic reflections in d-lines ℓ_1 and ℓ_2 that do not meet in $\mathscr{D}$ or on $\mathscr{C}$ is called a (**hyperbolic**) **translation**, and has no fixed points in $\mathscr{D}$ or on $\mathscr{C}$.

   The composite of two hyperbolic translations may depend on the order in which they are applied.

7. A direct hyperbolic transformation M can be written in the (canonical) form $M(z) = K\frac{z-m}{1-\overline{m}z}$, where K and m are complex numbers with $|K| = 1$ and $m \in \mathscr{D}$.

   Conversely, every direct hyperbolic transformation M that maps a point m of $\mathscr{D}$ to the origin is of this form.

   A direct hyperbolic transformation M that maps the diameter $(-1, 1)$ onto itself is of the form $M(z) = \pm\frac{z-m}{1-mz}$, where $m \in (-1, 1)$.

   An indirect hyperbolic transformation M can be written in the (canonical) form $z \mapsto K\frac{\overline{z}-m}{1-\overline{m}\overline{z}}$, where $|K| = 1$ and $m \in \mathscr{D}$.

8. **Strategy** To determine the general form of the direct hyperbolic transformation that maps one point p in $\mathscr{D}$ to another point q in $\mathscr{D}$:
   1. write down the general form of the direct hyperbolic transformation M_1 that maps p to 0, and a matrix $\mathbf{A}_1$ associated with M_1;
   2. write down a direct hyperbolic transformation M_2 that maps q to 0, and a matrix $\mathbf{A}_2$ associated with M_2;
   3. form the matrix product $\mathbf{A}_2^{-1}\mathbf{A}_1$ associated with the direct transformation $M_2^{-1} \circ M_1$, and hence write down the general form of the transformation.

Section 6.3: Distance in Hyperbolic Geometry

1. **Properties of a Distance Function d**
   1. $d(z_1, z_2) \geq 0$ for all z_1 and z_2;
      $d(z_1, z_2) = 0$ if and only if $z_1 = z_2$.
   2. $d(z_1, z_2) = d(z_2, z_1)$ for all z_1 and z_2.

3. $d(z_1, z_3) + d(z_3, z_2) \geq d(z_1, z_2)$ for all z_1, z_2 and z_3. (Triangle Inequality)
4. $d(z_1, z_3) + d(z_3, z_2) = d(z_1, z_2)$ if and only if z_1, z_2 and z_3 lie in this order on a line.

Additional Properties of the Distance Function *d* in Hyperbolic Geometry

5. $d(z_1, z_2) = d(\overline{z_1}, \overline{z_2})$ for all z_1 and z_2 in $\mathscr{D}$.
6. $d(z_1, z_2) = d(M(z_1), M(z_2))$ for all z_1 and z_2 in $\mathscr{D}$ and all direct hyperbolic transformations M in $G_{\mathscr{D}}$.

2. The **hyperbolic distance** $d(0, z)$ between the origin and any point $z \in \mathscr{D}$ is given by $d(0, z) = \tanh^{-1}(|z|) = \frac{1}{2}\log_e\left(\frac{1+|z|}{1-|z|}\right)$. Also, $|z| = (e^{2d(0,z)} - 1)/(e^{2d(0,z)} + 1)$.

   In general, the hyperbolic distance between two arbitrary points z_1 and z_2 in $\mathscr{D}$ is given by $d(z_1, z_2) = \tanh^{-1}\left(\left|\frac{z_2 - z_1}{1 - \overline{z_1}z_2}\right|\right)$.
3. For small z, $d(0, z) \simeq |z|$; as $|z| \to 1$, $d(0, z) \to +\infty$; for $z \neq 0$, $d(0, z) > |z|$.
4. A point m in $\mathscr{D}$ is the **hyperbolic midpoint** of the hyperbolic line segment joining a and b in $\mathscr{D}$ if m lies on this segment and $d(a, m) = d(m, b) = \frac{1}{2}d(a, b)$.
5. **Strategy** To find the hyperbolic midpoint of a hyperbolic line segment joining two points p, q on a diameter of $\mathscr{D}$, where $|p| > |q|$:
   1. calculate $d(0, p)$ and $d(0, q)$;
   2. if p and q lie on opposite sides of O, then calculate $d = \frac{1}{2}(d(0, p) - d(0, q))$; otherwise calculate $d = \frac{1}{2}(d(0, p) + d(0, q))$;
   3. then the hyperbolic midpoint is the point m on the radius through O and p at a Euclidean distance $\tanh(d)$ from O.
6. The **hyperbolic circle** with **hyperbolic radius** r and **hyperbolic centre** c is the set $\{z : d(c, z) = r,\ z \in \mathscr{D}\}$.
7. Every hyperbolic circle in $\mathscr{D}$ is a Euclidean circle, and vice versa.

   Strategy To find the Euclidean centre and radius of a hyperbolic circle C with hyperbolic centre m:
   1. find the points a, b where Om meets C;
   2. the Euclidean centre of C is the Euclidean mid-point of ab;
   3. the Euclidean radius of C is $\frac{1}{2}|a - b|$.

   Strategy To find the hyperbolic centre and radius of a Euclidean circle K with Euclidean centre p:
   1. find the points a, b where Op meets K;
   2. the hyperbolic centre of K is the hyperbolic mid-point of ab;
   3. the hyperbolic radius of K is $\frac{1}{2}d(a, b)$.
8. Let A and A' be points in the unit disc $\mathscr{D}$ that are images of each other under reflection in a d-line ℓ. Then ℓ is the (hyperbolic) perpendicular bisector of the hyperbolic line segment AA'.
9. **Reflection Lemma** Let p and q be points of $\mathscr{D}$. If $|p| \neq |q|$, then the hyperbolic reflection that maps p and q onto each other is given by

$M(z) = (\alpha\bar{z} - 1)/(\bar{z} - \bar{\alpha})$, where

$$\alpha = (p - q + pq(\bar{p} - \bar{q}))/(p\bar{p} - q\bar{q}).$$

The d-line in which this reflection takes place has equation

$$x^2 + y^2 - 2ax - 2by + 1 = 0, \text{ where } \alpha = a + ib.$$

If $|p| = |q|$, then the reflection that maps p and q onto each other is reflection in the diameter of $\mathscr{D}$ that bisects the angle $\angle pOq$.

Section 6.4: Geometrical theorems

1. A ***d*-triangle** consists of three points in the unit disc $\mathscr{D}$ that do not lie on a single d-line, together with the segments of the three d-lines joining them.
2. The sum of the angles of a d-triangle is less than π.

   Each external angle of a d-triangle is greater than the sum of the opposite two internal angles.

   Small triangles have angle sums close to (but less than) π, and triangles with large areas have angle sums close to zero.
3. A ***d*-quadrilateral** *ABCD* consists of four points A, B, C, D in $\mathscr{D}$ (no three of which lie on a single d-line), together with the segments *AB*, *BC*, *CD* and *DA* of the four d-lines joining them. We also require that no two of these segments meet except at one of the points A, B, C or D.

   The sum of the angles of a d-quadrilateral is less than 2π.
4. In a d-triangle ΔABC the angles $\angle ABC$ and $\angle ACB$ are equal if and only if the sides *AB* and *AC* are of equal (hyperbolic) length.
5. Two figures in the unit disc $\mathscr{D}$ are ***d*-congruent** if there is a hyperbolic transformation that maps one onto the other.

   Similar d-triangles (that is, d-triangles with corresponding angles equal) are d-congruent.
6. An **asymptotic triangle** is a d-triangle, except that one or more of its vertices lie on $\mathscr{C}$ rather than in $\mathscr{D}$. Simply asymptotic, doubly asymptotic and trebly asymptotic triangles have 1, 2 and 3 vertices on $\mathscr{C}$, respectively.

   The angle sum of an asymptotic triangle is less than π. The angle sum of a trebly asymptotic triangle is zero.
7. Let ℓ and ℓ' be two d-lines, and suppose that there exist points A on ℓ and A' on ℓ' such that the d-line segment AA' meets ℓ and ℓ' at right angles. Then AA' is a **common perpendicular** to ℓ and ℓ'.

   Common Perpendiculars Theorem Two d-lines have a common perpendicular if and only if they are ultra-parallel. This common perpendicular is unique.
8. Through any point P of $\mathscr{D}$ not on a d-line ℓ, there is a unique d-line that meets ℓ at right angles.

   Let ℓ be a d-line which passes through one vertex A of a d-triangle ΔABC and which is perpendicular to the side BC at the point D. The d-line segment AD is an **altitude** of the triangle ΔABC.

   Altitude Theorem Let the sides *AB* and *AC* of a d-triangle ΔABC be of equal hyperbolic length, and let the angle at A be θ. Then the hyperbolic length of the altitude through A of the triangle is less than some number that depends only on θ.

9. **Pythagoras' Theorem** Let ΔABC be a d-triangle in which the angle at C is a right angle. If a, b and c are the hyperbolic lengths of BC, CA and AB, then $\cosh 2c = \cosh 2a \times \cosh 2b$.
10. **Lobachevskii's Formula** Let the d-triangle ΔABC have a right angle at C, and let the hyperbolic lengths of AC and BC be b and a, respectively. Then $\tan A = \tanh 2a / \sinh 2b$.
11. **Angle of Parallelism** Let ℓ be a d-line, and P a point of $\mathscr{D}$ that does not lie on ℓ. Then the angle φ between the perpendicular from P to ℓ (of hyperbolic length p) and either d-line through P that is parallel to ℓ is given by $\tan \varphi = 1 / \sinh 2p$.
12. **Sine Formula** Let ΔABC be a d-triangle right-angled at C. Let a and c be the hyperbolic lengths of BC and AB, respectively. Then $\sin A = \sinh 2a / \sinh 2c$.
13. Let ℓ be a d-line that ends at points A and B on $\mathscr{C}$, and let ℓ' be that part in $\mathscr{D}$ of a (Euclidean) circle through A and B, and let α be the angle (in $\mathscr{D}$) between ℓ and ℓ'. Then all points on ℓ' are the same distance $\frac{1}{2}\log_e\left\{\tan\left(\frac{\alpha}{2}+\frac{\pi}{4}\right)\right\}$ from ℓ.
14. For $x \in (-1, 1)$, the following are all equal to $\tanh^{-1} x$: $\frac{1}{2}\log_e \frac{1+x}{1-x}$, $\frac{1}{2}\sinh^{-1}\frac{2x}{1-x^2}$, $\frac{1}{2}\cosh^{-1}\frac{1+x^2}{1-x^2}$.

Section 6.5: Area

1. **Area** in hyperbolic geometry has the following properties:
   - the area of a d-triangle is non-negative, and is zero only if its Euclidean area is zero;
   - d-congruent d-triangles (and d-congruent asymptotic d-triangles) have the same area;
   - if one d-triangle can be fitted inside another, it has a smaller area;
   - area is additive.
2. All trebly asymptotic d-triangles are d-congruent to each other.
3. Let ΔABC be a d-triangle. Then there exists a trebly asymptotic d-triangle ΔDEF that contains ΔABC.
4. All trebly asymptotic d-triangles have the same (hyperbolic) area, and this is a finite number.
5. The area of a d-triangle with angles α, β, γ is given by $K(\pi - (\alpha + \beta + \gamma))$, where K is the same constant for all d-triangles. It is usual to take $K = 1$, so that the area of a d-triangle is its **angular defect**.

   Let α, β, γ be positive real numbers with $0 < \alpha + \beta + \gamma < \pi$. Then there is a d-triangle in $\mathscr{D}$ with angles α, β, γ.
6. Let P be a point on a d-line ℓ of $\mathscr{D}$, and let $0 < \beta < \frac{\pi}{2}$. Then there are exactly two d-lines through P that make an acute angle β with ℓ, and each is a (hyperbolic) reflection of the other in ℓ.

7. **Tessellate** means 'fit together exactly a number of identical shapes, leaving no spaces'.

   Tessellations of the Euclidean plane using a single figure are sometimes called **wallpaper patterns**. There are only 17 different wallpaper patterns, in the sense that the symmetry group of any wallpaper pattern is isomorphic to one of only 17 non-isomorphic groups.
8. A **kaleidoscope** is a machine that produces wallpaper patterns using reflections in mirrors.

   Successive reflections of a point in two plane mirrors that meet at an angle π/k give $2k$ points in all.

   Successive (hyperbolic) reflections of a point in two mirrors along d-lines in $\mathcal{D}$ that meet at an angle π/k give $2k$ points in all. We may obtain a **hyperbolic tessellation** of $\mathcal{D}$ (with d-congruent d-triangles) by successive hyperbolic reflections starting from a d-triangle in $\mathcal{D}$ formed by three mirrors along d-lines at angles of $\pi/7$ (at the origin), $\pi/2$ and $\pi/3$.

Section 6.6: Hyperbolic Geometry: the Half-Plane Model

1. In the half-plane model, the set of points is the upper half-plane $H = \{z : z \in \mathbb{C}, \operatorname{Im} z > 0\}$; h-lines are the restriction to H of vertical lines in the plane together with the restriction to H of circles with centre on the real axis in $\mathbb{C}$; and the hyperbolic transformations are the mappings of H to itself given by $z \mapsto (az+b)/(cz+d)$ or $z \mapsto (-a\bar{z}+b)/(-c\bar{z}+d)$, where a, b, c, d are real and $ad-bc>0$.

7 Elliptic Geometry: the Spherical Model

In the introduction to Chapter 6, we saw that in Elliptic Geometry all the postulates of Euclidean Geometry hold except that the Euclidean Parallel Postulate is replaced by the following.

We usually take the sphere to have radius 1.

The (Elliptic) Parallel Postulate Given any line ℓ and a point P not on ℓ, all lines through P meet ℓ. (That is, there are no lines through P that are parallel to ℓ.)

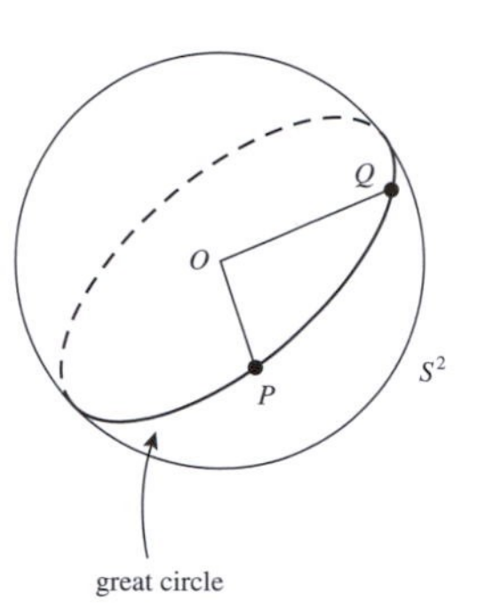

In this chapter we study the geometry of the sphere (which we call S^2), such as the surface of the Earth, as a model of elliptic geometry.

A 'line' joining two points P and Q on the surface of a sphere is the curve of shortest length (that is, a *geodesic*) on the sphere between the points: it is actually the *great circle* (that is, the intersection of the plane through the centre O of the sphere with the surface of the sphere) through the two points. Any two distinct great circles meet in two (diametrically opposite) points, so any two lines in spherical geometry meet in two points.

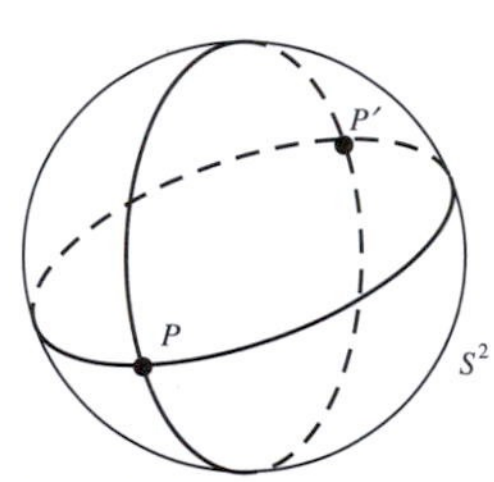

The group of transformations in spherical geometry is the group of isometries of the sphere. Clearly a rotation of the sphere is an isometry, and we shall see that the group of isometries is composed of rotations of the sphere together with reflections in great circles. It turns out that any orientation-preserving isometry of the sphere is in fact a single rotation.

Recall that an *isometry* is a one–one mapping that preserves distances.

In Section 7.1, we establish a system of coordinates for the sphere; these are essentially the familiar coordinates of *latitude* and *longitude*. Then, in Section 7.2, we obtain matrix representations for the isometries of the sphere: every isometry can be written as a 3×3 matrix.

Next, in Section 7.3, we define a *triangle* on the sphere, find a formula expressing its area in terms of the sum of its angles, and show that every spherical triangle has a *dual triangle* whose sides and angles have the magnitudes of the angles and sides, respectively, of the original triangle.

There is an attractive formula for the distance between two points on the sphere in terms of their coordinates. There is also an attractive generalization of Pythagoras' Theorem in spherical geometry, and in proving it we are led to the trigonometric formulas connecting the sides and angles of a spherical triangle. These establish that of the six measurements that describe a spherical triangle (three side lengths, three angles) any set of three determines the other three. This is not the case in Euclidean geometry (where similar, non-congruent figures exist), but it is also true in hyperbolic geometry. Indeed, there are closer analogies between spherical geometry and hyperbolic geometry than there are between either geometry and Euclidean geometry.

Many aspects of spherical geometry may be studied more easily if we first map the sphere onto a plane. This can be done in many ways, although, none of the mappings (or *projections*) is an isometry of the sphere onto the plane. In Section 7.4, we revisit *stereographic projection*, which maps the sphere onto the plane in a way that preserves angles and generalized circles. We also show how to obtain the group of isometries of the stereographic image of the sphere on the plane; and revisit coaxal circles. Finally, in Section 7.5, we look briefly at the problem of mapping the spherical Earth onto a flat map.

7.1 Spherical Space

7.1.1 Spherical Geometry

Spherical geometry is the study of geometry on the surface of a sphere. It is simplest to choose this sphere to have unit radius; we therefore concentrate on what we call the *unit sphere* in Euclidean 3-dimensional space $\mathbb{R}^3$, namely the sphere whose centre is at the origin and which has radius 1. In terms of the usual x-, y-, z-Cartesian coordinates in $\mathbb{R}^3$ with origin O at the centre of the sphere, the unit sphere has the equation $x^2 + y^2 + z^2 = 1$. We shall denote it by the symbol S^2.

It is usually straightforward to modify our results to the case of a sphere of arbitrary radius, R say.

The positive x-, y-, z-axes meet the sphere S^2 at the points $A(1,0,0)$, $B(0, 1, 0)$, $C(0, 0, 1)$, respectively. For obvious reasons the point C is sometimes called the *North Pole* of the sphere, and denoted by N. Every point (x, y, z) on the sphere S^2 has a diametrically opposite or *antipodal* point $(-x, -y, -z)$; for example, the antipodal point to the North Pole is the *South Pole* $S(0,0,-1)$. The plane through O perpendicular to the line ON cuts the sphere in a circle called the *equator*.

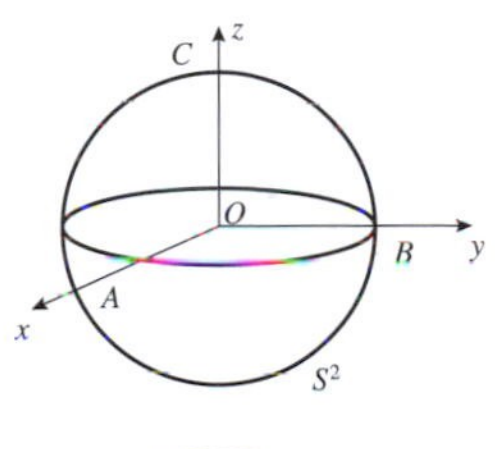

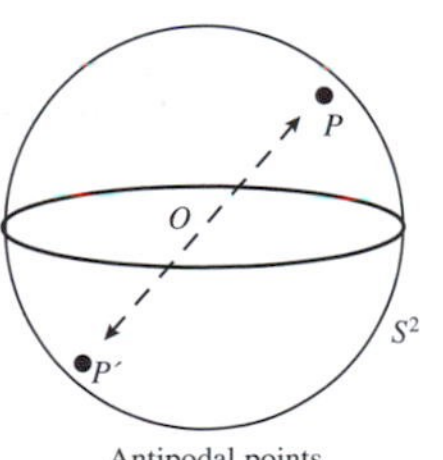

Antipodal points

Example 1 Prove that any plane that cuts the sphere S^2 in more than one point cuts it in a circle.

Solution Let OT be a radius of the sphere perpendicular to the given plane. A rotation of the sphere about this axis maps the plane to itself, and the sphere to itself; so it maps their intersection to itself. Each point on the intersection traces out a circle (of the same radius and centre) under the rotation, so the intersection is a circle. □

The analogue in spherical geometry of a straight line in the plane is a curve on S^2 called a great circle.

Definition A **great circle** is the circle cut out on the sphere S^2 by a plane through the centre of the sphere.

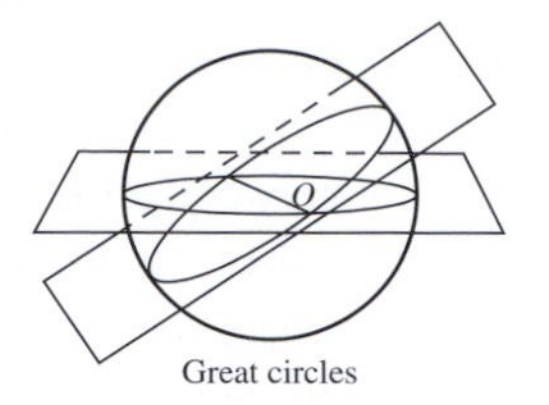

Great circles

Note that, by Example 1, a great circle is indeed a circle. Moreover, it is a circle of maximal radius 1 on the sphere, which gives rise to the name 'great' circle.

Definitions Any great circle through N is called a **circle of longitude**. The part of the circle of longitude through $A(1, 0, 0)$ which has positive x-coordinates we call the **Greenwich Meridian**. The curve of intersection of S^2 with a plane parallel to the equator is a circle called a **circle of latitude**. The curve of intersection of S^2 with a plane that is not a great circle is called a **little circle**.

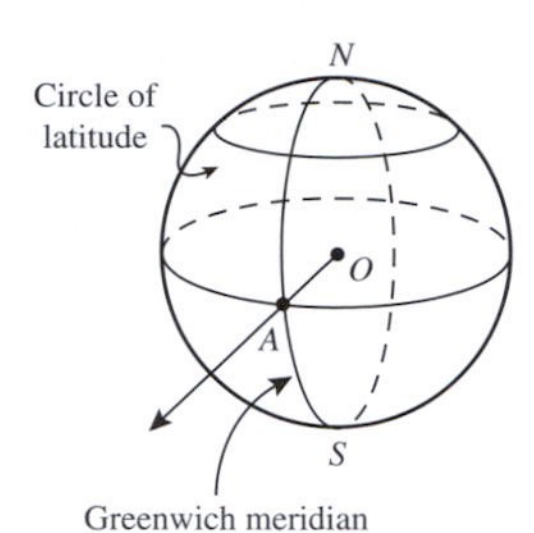

The equator is a circle of latitude; circles of latitude other than the equator are little circles.

We define distance in spherical geometry (on S^2) in the natural way.

Definitions Let P and Q be any two points on S^2, and C a great circle through P and Q. Then either arc with endpoints P and Q is said to be a **line** on S^2 joining P and Q, or a **line** PQ. The **length** of a given line on S^2 is defined to be the Euclidean length of the corresponding circular arc; the **distance** (or **spherical distance**) between two points P and Q on S^2 is the length of the shorter of the two lines PQ, and is often denoted simply as PQ.

Let P, Q and R be any three points on S^2 that do not lie on a single great circle. Then a line PQ, a line QR and a line RP form a **triangle** $\triangle PQR$.

Thus 'the line PQ' is not unique.

Thus there are several triangles $\triangle PQR$.

A property of great circles that we state without proof is that *great circles are geodesics on* S^2; that is, curves of shortest length between any of their points. This is intuitively clear for the circles of longitude from the North Pole, for instance. Note that little circles are not geodesics on S^2.

The distance between any two points on S^2 is a portion of a great circle through the points. This circle has radius 1, and so has circumference 2π – the same as the angle that the great circle subtends at its center (the centre of the sphere). It follows, by proportion, that the distance between any two points on S^2 is equal to the angle that they subtend at the centre of S^2.

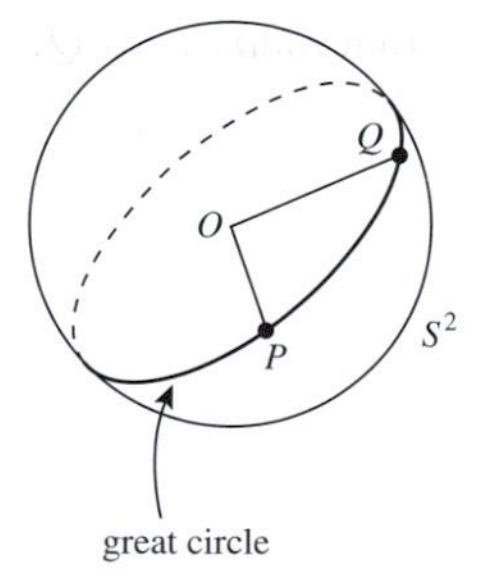

Theorem 1 The distance between any two points P and Q on S^2 is equal to the angle that they subtend at the centre of S^2.

Thus, for example, the distance from the North Pole to any point on the equator is $\pi/2$, and the distance from the North Pole to the South Pole is π.

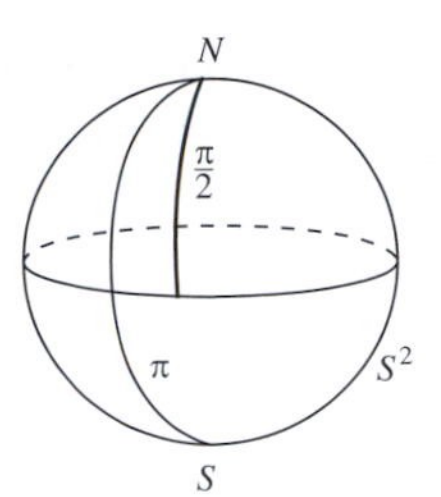

With the above definition, spherical distance possesses all the 'natural' properties that a distance function should possess. (We listed these in Subsection 6.3.1.)

For example, it satisfies the Triangle Inequality; that is, if P, Q and R are any three points on S^2, then the side lengths of the triangle ΔPQR satisfy the inequality $PQ + QR \geq PR$. This can be proved in a similar way to the Triangle Inequality in hyperbolic geometry (Theorem 2, Subsection 6.3.3).

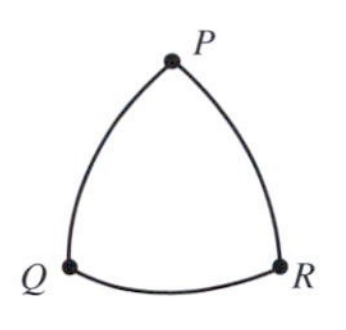

Great circles in spherical geometry play many of the roles that lines play in Euclidean plane geometry. There is always a great circle joining any two points on S^2, because there is always a plane passing through them and the origin. The plane and the corresponding great circle are unique unless the points are diametrically opposite, in which case there are infinitely many great circles joining the points.

However, one way in which spherical geometry differs markedly from Euclidean plane geometry comes from observing that any two great circles meet not in one, but in two (diametrically opposite) points. They do so because the planes that define them have a common point, the origin, and therefore a common line in $\mathbb{R}^3$, which cuts S^2 in the given points. So although great circles are analogous to the straight lines of the plane, the analogy breaks down almost at once. There are certainly no 'parallel lines' on the sphere, because any two great circles meet in two points.

It is sometimes helpful to observe that corresponding to each point P on S^2 there is a unique great circle of points $\pi/2$ distant from P, which plays the role of the equator with P as a 'pole'. We shall call this great circle the **great circle associated with the point** P. Conversely, given any great circle there are two points of S^2 which play the role of its 'poles'.

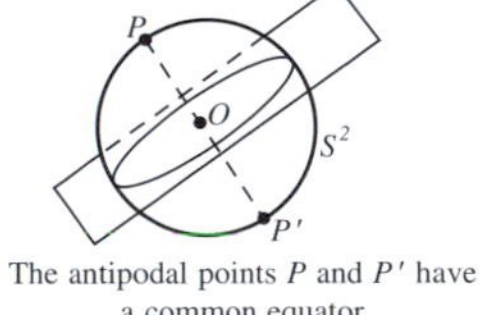

The antipodal points P and P' have a common equator

Example 2 Let P and Q be any two points on S^2. Show that there is a rotation of S^2 that maps P to Q.

Solution Consider the great circle through P and Q, and let T be one of its poles. The rotation of S^2 about the axis OT and through the angle $\angle POQ$ (in the appropriate direction) sends P to Q, as required. □

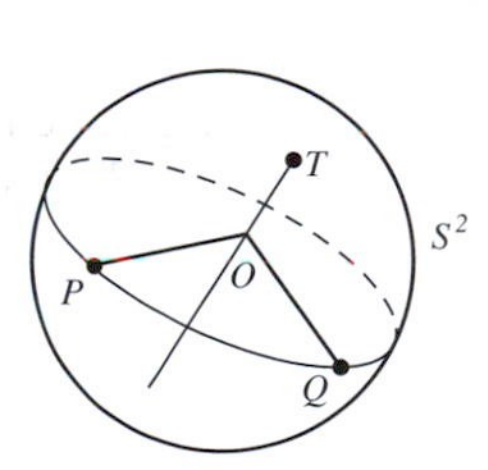

Problem 1 Determine whether rotations of S^2 exist with the following properties.

(a) $(1,0,0) \mapsto (0,-1,0)$
(b) $(1,0,0) \mapsto (0,1,0)$ and $(0,1,0) \mapsto (0,0,1)$
(c) $(1,0,0) \mapsto (0,-1,0)$ and $(0,0,1) \mapsto (0,1,0)$

7.1.2 Coordinates on S^2

We now set up a system of *spherical polar coordinates* on S^2. We start by defining the direction along the equator from $A(1,0,0)$ to $B(0,1,0)$ to be the positive direction on the equator. Then for any point P_1 on the equator let ϕ be the angle $\angle AOP_1$ measured in the positive direction along the equator; then we assign the coordinates $(\cos\phi, \sin\phi, 0)$ to P_1, where $0 \le \phi < 2\pi$.

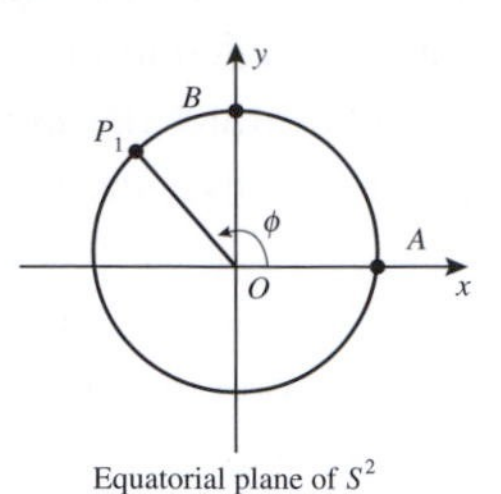

Equatorial plane of S^2

Next, let P be any point on S^2 (other than the North and South Poles) that does not lie on the equator. Let P_1 be the point on the equator that lies on the arc NS of the great circle through N and P.

Then let ϕ be the angle $\angle AOP_1$ measured in the positive direction along the equator ($0 \le \phi < 2\pi$), and let θ be the angle $\angle NOP$ ($0 < \theta < \pi$); we assign the coordinates $(\cos\phi\sin\theta, \sin\phi\sin\theta, \cos\theta)$ to P.

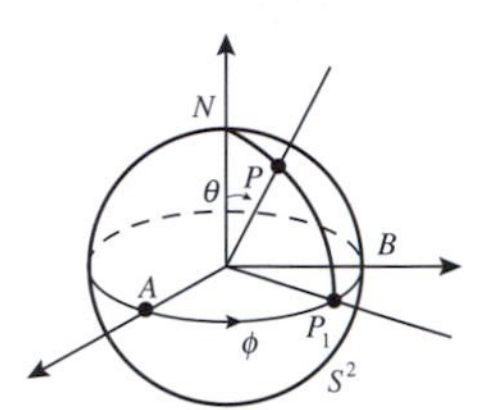

P has spherical coordinates $(\cos\phi \sin\theta, \sin\phi, \sin\theta, \cos\theta)$.

For example, the point of S^2 with spherical polar coordinates $\left(\cos\frac{\pi}{3}\sin\frac{\pi}{4}, \sin\frac{\pi}{3}\sin\frac{\pi}{4}, \cos\frac{\pi}{4}\right)$ is the point $\left(\frac{1}{2\sqrt{2}}, \frac{\sqrt{3}}{2\sqrt{2}}, \frac{1}{\sqrt{2}}\right)$.

Problem 2 Determine the angles ϕ and θ in the spherical polar representation $(\cos\phi\sin\theta, \sin\phi\sin\theta, \cos\theta)$ of the points $\left(0, -\frac{\sqrt{3}}{2}, \frac{1}{2}\right)$ and $\left(\frac{1}{\sqrt{14}}, -\frac{\sqrt{2}}{\sqrt{14}}, \frac{3}{\sqrt{14}}\right)$.

In this way every point on S^2 other than the North and South Poles has been given spherical polar coordinates involving appropriate values of ϕ and θ. At the North Pole we take θ to be zero and let ϕ take any value in the interval $[0, 2\pi)$, and at the South Pole we take θ to be π and let ϕ take any value in the interval $[0, 2\pi)$; the non-uniqueness of coordinates at the Poles will cause us no problems.

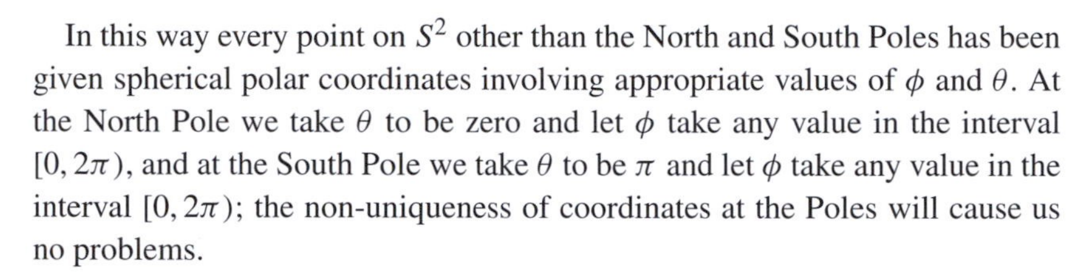

Definitions Let P be a point on S^2 with spherical polar coordinates $(\cos\phi\sin\theta, \sin\phi\sin\theta, \cos\theta)$, where $0 \le \phi < 2\pi$ and $0 \le \theta \le \pi$. The angle θ of P is called its **colatitude**; the quantity $\theta' = \frac{\pi}{2} - \theta$ is called its **latitude** (note that $-\frac{\pi}{2} \le \theta' \le \frac{\pi}{2}$). The angle ϕ of P is called its **longitude**.

Sometimes we shall find it convenient to regard ϕ as being defined on $\mathbb{R}$ rather than on $[0, 2\pi)$, as a periodic function with period 2π.

Notice, for example, that the distance of a point P with colatitude θ on S^2 from the North Pole is simply θ.

Problem 3 Determine the latitude, colatitude and longitude of the point $\left(\frac{1}{2\sqrt{2}}, \frac{\sqrt{3}}{2\sqrt{2}}, \frac{1}{\sqrt{2}}\right)$ on S^2.
Hint: You met this point just before Problem 2 above.

Problem 4 Determine the spherical polar coordinates of the point 45° W of the Greenwich meridian on S^2 and 30° S.

Note that 30° S means '30° below the equatorial plane'. Also, sometimes we take the phrase 'of the Greenwich meridian' as understood when describing longitude.

Notice also that the set of points on the sphere with a common colatitude θ forms a circle, C_θ say; this is a little circle if $\theta \ne \pi/2$, and a great circle (the equator) if $\theta = \pi/2$. Since θ measures the distance of each point of the circle C_θ from the North Pole N, we shall say that C_θ is a *circle on the sphere with centre N*. Since any point on S^2 may be chosen as the North Pole (with a suitable choice of coordinate axes), it follows that a circle on S^2, defined as the set of points equidistant from a given point, is either a great circle or a little circle.

Problem 5 Show that, given any two great circles on S^2, there are two great circles bisecting the angles between them, each at right angles to the other.
Hint: The angle between two curves on S^2 at one of their points of intersection is the angle between their tangents at that point.

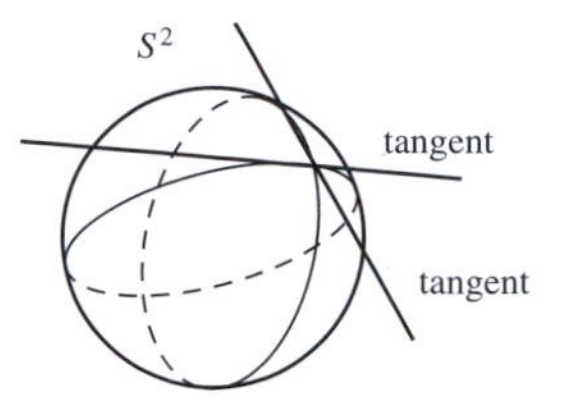

7.2 Spherical Transformations

7.2.1 Isometries of S^2

An isometry of S^2 is a mapping of the unit sphere to itself that preserves distances between points. In fact the isometries of a sphere form a group, since

- the composition of two isometries is also an isometry;
- the identity mapping is an isometry, and it is the identity element of the group;
- the inverse of any isometry is an isometry;
- the composition of isometries is associative, since the composition of functions is always associative.

We have therefore established the following theorem.

Theorem 1 The isometries of S^2 form a group.

We call this group the *group of spherical isometries*, and denote it by the symbol $S(2)$.

We shall see at the end of this section that the rotations of S^2 form a group, which is identical to the group of direct isometries of the sphere. A reflection of S^2 in a plane through the centre of the unit sphere is not a direct isometry of S^2, but every composition of an even number of such reflections is a direct isometry of S^2.

A transformation is *direct* if it preserves the *orientation* of angles, and *indirect* if it reverses them.

We now consider the three *elementary rotations* of S^2 (that is, its rotations about the coordinates axes) and their composites.

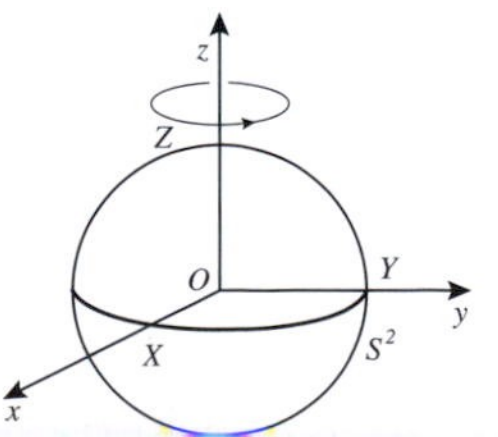

The elementary rotation about OZ leaves the z-axis fixed, and rotates the (x, y)-plane. The positive direction of this rotation is anticlockwise as seen from above (that is, as viewed from a point of the positive z-axis), from the positive x-axis towards the positive y-axis. So a rotation of S^2 about OZ through an angle γ, which we denote by $R(Z,\gamma)$, is given by

$$R(Z,\gamma)\colon S^2 \to S^2$$

$$\mathbf{x} \mapsto \mathbf{Ax}$$

where $\mathbf{x} = \begin{pmatrix} x \\ y \\ z \end{pmatrix}$ and $\mathbf{A}$ is the matrix $\begin{pmatrix} \cos\gamma & -\sin\gamma & 0 \\ \sin\gamma & \cos\gamma & 0 \\ 0 & 0 & 1 \end{pmatrix}$.

We say that $\mathbf{A}$ is the matrix *associated with* (or *of*) the elementary rotation $R(Z,\gamma)$.

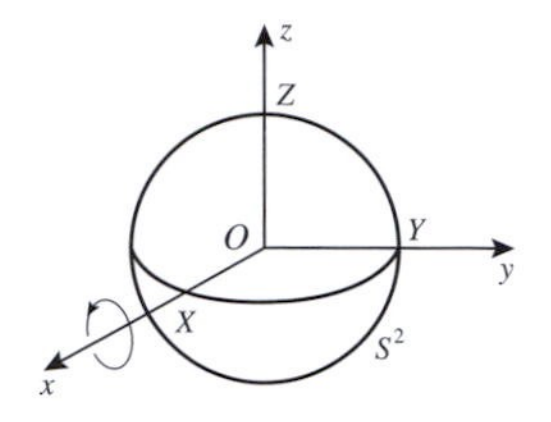

Next, the elementary rotation about OX leaves the x-axis fixed, and rotates the (y,z)-plane. The positive direction of this rotation is taken to be anticlockwise as seen from above (that is, as viewed from a point of the positive x-axis), from the positive y-axis towards the positive z-axis. So a rotation of S^2 about OX through an angle α, which we denote by $R(X,\alpha)$, is given by

$$R(x,\alpha)\colon S^2 \to S^2$$

$$\mathbf{x} \mapsto \mathbf{Ax}$$

where $\mathbf{x} = \begin{pmatrix} x \\ y \\ z \end{pmatrix}$ and $\mathbf{A}$ is the matrix $\begin{pmatrix} 1 & 0 & 0 \\ 0 & \cos\alpha & -\sin\alpha \\ 0 & \sin\alpha & \cos\alpha \end{pmatrix}$.

Here $\mathbf{A}$ is the matrix *associated with* (or *of*) the elementary rotation $R(X,\alpha)$.

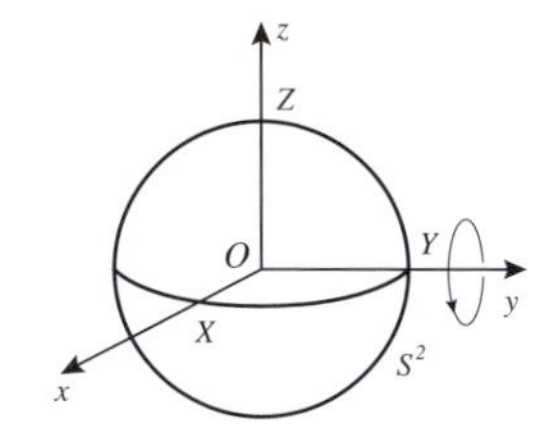

Finally, the elementary rotation about OY leaves the y-axis fixed, and rotates the (x,z)-plane. The positive direction of this rotation is taken to be anticlockwise as seen from above (that is, as viewed from a point of the positive y-axis), from the positive z-axis towards the positive x-axis. So a rotation of S^2 about OY through an angle β, which we denote by $R(Y,\beta)$, is given by

$$R(Y,\beta)\colon S^2 \to S^2$$

$$\mathbf{x} \mapsto \mathbf{Ax}$$

where $\mathbf{x} = \begin{pmatrix} x \\ y \\ z \end{pmatrix}$ and $\mathbf{A}$ is the matrix $\begin{pmatrix} \cos\beta & 0 & \sin\beta \\ 0 & 1 & 0 \\ -\sin\beta & 0 & \cos\beta \end{pmatrix}$.

Here $\mathbf{A}$ is the matrix *associated with* (or *of*) the elementary rotation $R(y,\beta)$.

It is easy to verify that the determinants of the elementary rotations $R(X,\alpha)$, $R(Y,\beta)$ and $R(Z,\gamma)$ are all equal to 1. Since any rotation can be expressed as a finite composition of these rotations, it follows that the determinant of the matrix of any rotation of S^2 also has determinant 1.

We will prove this fact in Theorem 7, Subsection 7.2.3.

Problem 1 Determine the images of the points $(1,0,0), (0,1,0)$, $(0,0,1)$ and $\left(\frac{1}{\sqrt{14}}, -\frac{2}{\sqrt{14}}, \frac{3}{\sqrt{14}}\right)$ under the transformation $R\left(Y, \frac{\pi}{4}\right)$.

Notice that the simple cases of these rotations when α, β or $\gamma = \pi/2$ correspond to rotations of one axis onto another. For example,

$$R\left(Y, \tfrac{\pi}{2}\right) = \begin{pmatrix} \cos\frac{\pi}{2} & 0 & \sin\frac{\pi}{2} \\ 0 & 1 & 0 \\ -\sin\frac{\pi}{2} & 0 & \cos\frac{\pi}{2} \end{pmatrix} = \begin{pmatrix} 0 & 0 & 1 \\ 0 & 1 & 0 \\ -1 & 0 & 0 \end{pmatrix};$$

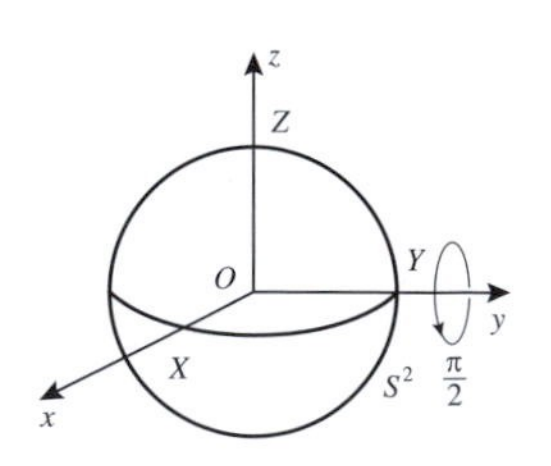

under this mapping, $(1,0,0) \mapsto (0,0,-1), (0,1,0) \mapsto (0,1,0)$ and $(0,0,1) \mapsto (1,0,0)$.

Example 1 Show that $R\left(X, \frac{\pi}{2}\right) \cdot R\left(Y, \frac{\pi}{2}\right) \neq R\left(Y, \frac{\pi}{2}\right) \cdot R\left(X, \frac{\pi}{2}\right)$.

This shows that multiplication of the matrices of elementary rotations (and hence composition of elementary rotations) is not commutative.

Solution Here

$$R\left(X, \tfrac{\pi}{2}\right) \cdot R\left(Y, \tfrac{\pi}{2}\right) = \begin{pmatrix} 1 & 0 & 0 \\ 0 & \cos\frac{\pi}{2} & -\sin\frac{\pi}{2} \\ 0 & \sin\frac{\pi}{2} & \cos\frac{\pi}{2} \end{pmatrix} \times \begin{pmatrix} \cos\frac{\pi}{2} & 0 & \sin\frac{\pi}{2} \\ 0 & 1 & 0 \\ -\sin\frac{\pi}{2} & 0 & \cos\frac{\pi}{2} \end{pmatrix}$$

$$= \begin{pmatrix} 1 & 0 & 0 \\ 0 & 0 & -1 \\ 0 & 1 & 0 \end{pmatrix} \begin{pmatrix} 0 & 0 & 1 \\ 0 & 1 & 0 \\ -1 & 0 & 0 \end{pmatrix}$$

$$= \begin{pmatrix} 0 & 0 & 1 \\ 1 & 0 & 0 \\ 0 & 1 & 0 \end{pmatrix},$$

and

$$R\left(Y, \tfrac{\pi}{2}\right) \cdot R\left(X, \tfrac{\pi}{2}\right) = \begin{pmatrix} \cos\frac{\pi}{2} & 0 & \sin\frac{\pi}{2} \\ 0 & 1 & 0 \\ -\sin\frac{\pi}{2} & 0 & \cos\frac{\pi}{2} \end{pmatrix} \times \begin{pmatrix} 1 & 0 & 0 \\ 0 & \cos\frac{\pi}{2} & -\sin\frac{\pi}{2} \\ 0 & \sin\frac{\pi}{2} & \cos\frac{\pi}{2} \end{pmatrix}$$

$$= \begin{pmatrix} 0 & 0 & 1 \\ 0 & 1 & 0 \\ -1 & 0 & 0 \end{pmatrix} \begin{pmatrix} 1 & 0 & 0 \\ 0 & 0 & -1 \\ 0 & 1 & 0 \end{pmatrix}$$

$$= \begin{pmatrix} 0 & 1 & 0 \\ 0 & 0 & -1 \\ -1 & 0 & 0 \end{pmatrix}.$$

Thus $R\left(X, \frac{\pi}{2}\right) \cdot R\left(Y, \frac{\pi}{2}\right) \neq R\left(Y, \frac{\pi}{2}\right) \cdot R\left(X, \frac{\pi}{2}\right)$, as required. □

Problem 2 Calculate $R(Z, \gamma) \cdot R(Y, \beta)$ and $R(Y, \beta) \cdot R(Z, \gamma)$.

Problem 3 For the elementary rotation $R(X, \alpha)$, verify that the transpose of its matrix is equal to the matrix of the elementary rotation $R(X, -\alpha)$. State and prove the corresponding results for the elementary rotations $R(Y, \beta)$ and $R(Z, \gamma)$.

Notice that the inverse of an elementary rotation through a given angle is an elementary rotation through the same angle but in the opposite direction;

that is,

$$R(X,\alpha)^{-1} = R(X,-\alpha),$$
$$R(Y,\beta)^{-1} = R(Y,-\beta)$$
and $$R(Z,\gamma)^{-1} = R(Z,-\gamma).$$

Combining this observation with the result of Problem 3, we obtain the agreeable consequence that the inverse of a matrix representing an elementary rotation is equal to the transpose of that matrix. For example the matrix $\mathbf{A}$ of the elementary rotation $R(Y,\beta)$ is

$$\mathbf{A} = \begin{pmatrix} \cos\beta & 0 & \sin\beta \\ 0 & 1 & 0 \\ -\sin\beta & 0 & \cos\beta \end{pmatrix};$$

then $\mathbf{A}^{-1} = \mathbf{A}^T$.

Recall that any square matrix $\mathbf{A}$ with the property that $\mathbf{A}^{-1} = \mathbf{A}^T$ or $\mathbf{A}\mathbf{A}^T = \mathbf{I}$ is an *orthogonal* matrix.

We saw earlier that we can rotate S^2 in such a way as to map any particular point onto any other. We now see how we can determine the matrix of such a rotation explicitly.

Subsection 7.1.1, Example 2

First, we see how to rotate S^2 so as to send the point $A(1,0,0)$ to any point $P(\cos\phi\sin\theta, \sin\phi\sin\theta, \cos\theta)$ of S^2. First we rotate S^2 to send A to the point $P_1(\cos\phi, \sin\phi, 0)$; then we rotate S^2 to send P_1 to P.

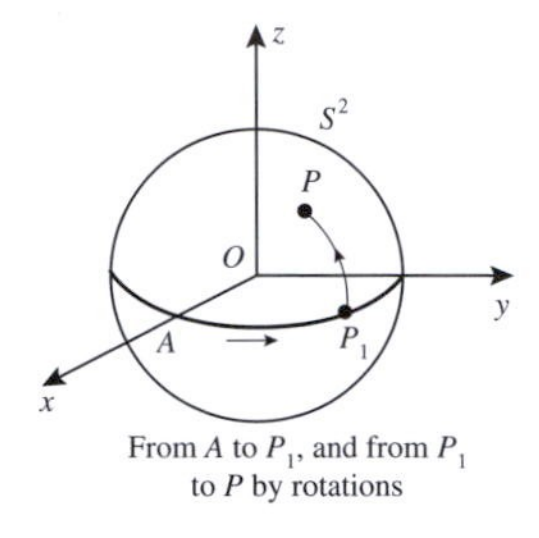

From A to P_1, and from P_1 to P by rotations

The first rotation is the elementary rotation $R(Z,\phi)$; but what is the second rotation? It is a rotation of S^2 'towards the North Pole' through an angle $\theta' = \frac{\pi}{2} - \theta$, so it is something like $R(Y,-\theta')$ or $R\left(Y,\theta - \frac{\pi}{2}\right)$. But it is not a rotation about the y-axis; instead, it is a rotation of S^2 about the direction where the first rotation sent the y-axis. We may perform such a rotation in several stages: rotate the sphere first by the reverse transformation $R(Z,-\phi)$, then by the elementary rotation $R\left(Y,\theta - \frac{\pi}{2}\right)$, and then finally by the transformation $R(Z,\phi)$.

For $\theta' = \frac{\pi}{2} - \theta$, and $R(Z,\phi)^{-1} = R(Z,-\phi)$.

Thus the second rotation is the composition mapping

$$R(Z,\phi)R(Y,-\theta')R(Z,-\phi) = R(Z,\phi)R(Y,-\theta')R(Z,\phi)^{-1}.$$

It follows that the final transformation that sends A to P is given by

$$\left(R(Z,\phi)R(Y,-\theta')R(Z,\phi)^{-1}\right)R(Z,\phi) = R(Z,\phi)R(Y,-\theta').$$

Theorem 2 A rotation of S^2 that maps $A(1,0,0)$ to $P(\cos\phi\sin\theta, \sin\phi\sin\theta, \cos\theta)$ is given by the composition $R(Z,\phi)R(Y,-\theta')$, where $\theta' = \frac{\pi}{2} - \theta$.

Note that there are many such rotations. *Any* composition of this particular rotation with a subsequent rotation about OP has the required properties.

Problem 4 Determine the matrix $\mathbf{A}$ of a rotation $\mathbf{x} \mapsto \mathbf{A}\mathbf{x}$ of S^2 that maps $A(1,0,0)$ to $P\left(\frac{1}{2\sqrt{2}}, \frac{\sqrt{3}}{2\sqrt{2}}, \frac{1}{\sqrt{2}}\right)$. Verify your answer by direct calculation.

Hint: You met this point P in Problem 3 of Subsection 7.1.2.

Example 2 Determine a rotation that maps the point $P(\cos\phi\sin\theta, \sin\phi\sin\theta, \cos\theta)$ of S^2 to $N(0,0,1)$, in terms of elementary rotations.

Solution By Theorem 2, one rotation that sends A to P is given by the composition $R(Z,\phi)R(Y,-\theta')$, where $\theta' = \frac{\pi}{2} - \theta$. It follows that the mapping

$$\begin{aligned}\left(R(Z,\phi)R(Y,-\theta')\right)^{-1} &= R(Y,-\theta')^{-1}R(Z,\phi)^{-1}\\ &= R(Y,\theta')R(Z,-\phi)\end{aligned}$$

sends P to A.
Now at the North Pole $\phi = 0$ and $\theta = 0$, so that $\theta' = \frac{\pi}{2}$. Hence one rotation that sends A to N is $R\left(Y, -\frac{\pi}{2}\right)$. It follows that one rotation that sends P to N is

$$\begin{aligned}&R\left(Y,-\frac{\pi}{2}\right)\circ R\left(Y,\theta'\right)R\left(Z,-\phi\right)\\ &= R\left(Y,\theta'-\frac{\pi}{2}\right)R\left(Z,-\phi\right)\\ &= R\left(Y,-\theta\right)R\left(Z,-\phi\right).\end{aligned}$$

□

For $R(Y,\beta_1)R(Y,\beta_2) = R(Y,\beta_1+\beta_2)$, for any angles β_1, β_2.

Problem 5 Determine the matrix $\mathbf{A}$ of a rotation $\mathbf{x} \mapsto \mathbf{Ax}$ of S^2 that maps $P\left(\frac{1}{2\sqrt{2}}, \frac{\sqrt{3}}{2\sqrt{2}}, \frac{1}{\sqrt{2}}\right)$ to N. Verify your answer by direct calculation.
Hint: You met this point P in Problem 4.

Problem 6 Determine the matrix $\mathbf{A}$ of a rotation $\mathbf{x} \mapsto \mathbf{Ax}$ of S^2 that maps $Q\left(\frac{1}{2}, -\frac{1}{2}, -\frac{1}{\sqrt{2}}\right)$ to $P\left(\frac{1}{2\sqrt{2}}, \frac{\sqrt{3}}{2\sqrt{2}}, \frac{1}{\sqrt{2}}\right)$. Verify your answer by direct calculation.

7.2.2 Reflections in Great Circles

Reflection in the *equatorial plane* of S^2 (that is, in the plane $z = 0$ that contains the equator of S^2) is the mapping $\mathbf{x} \mapsto \mathbf{Ax}$ that sends the point $\mathbf{x} = (x, y, z)$ to the point $(x, y, -z)$, so $\mathbf{A} = \begin{pmatrix} 1 & 0 & 0\\ 0 & 1 & 0\\ 0 & 0 & -1\end{pmatrix}$. Each point lies vertically above or below its image, unless it lies in the equatorial plane – in which case it coincides with its image.

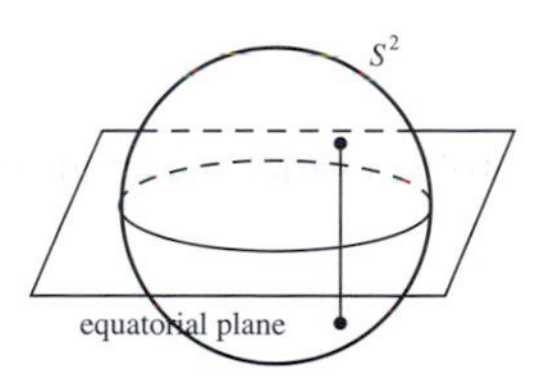

Problem 7 Evaluate the determinant of the above matrix $\mathbf{A}$.

Next, let π be any plane that passes through the centre O of the sphere S^2. The equation of π must be of the form $ax+by+cz = 0$, since it passes through O; if we define the vectors $\mathbf{a}$ and $\mathbf{x}$ to be (a,b,c) and (x,y,z), respectively, then we can express the equation of π in the form

$$\mathbf{a} \cdot \mathbf{x} = 0.$$

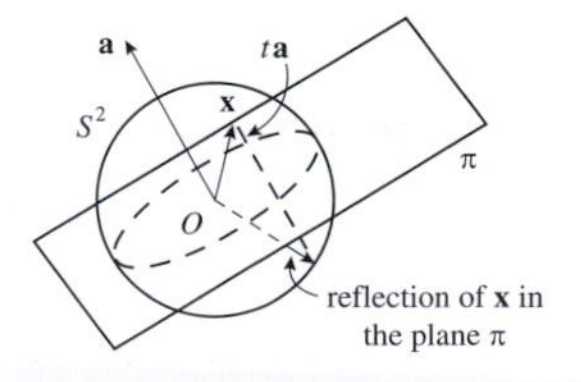

This defines a plane with the vector $\mathbf{a}$ as a normal to it; without loss of generality, we shall assume that a, b and c have been chosen so that $\mathbf{a}$ has length 1.

Let r be the reflection of S^2 in the plane π; we shall find a matrix for this mapping r of S^2 to itself.

We shall interchangeably call this *reflection in* π and *reflection in the great circle* in which π intersects S^2.

The image of a point $\mathbf{x}$ of S^2 is the point $r(\mathbf{x})$. Now the line joining the points $\mathbf{x}$ and $r(\mathbf{x})$ is a line parallel to the normal vector $\mathbf{a}$, so we must have that

$$r(\mathbf{x}) = \mathbf{x} + t\mathbf{a}, \qquad \text{for some value of the parameter } t.$$

For $r(\mathbf{x}) - \mathbf{x} = t\mathbf{a}$, for some t.

We can find this value of t by observing that since $\mathbf{x}$ lies on S^2 so does its image $r(\mathbf{x})$, and so

$$(\mathbf{x} + t\mathbf{a}) \cdot (\mathbf{x} + t\mathbf{a}) = 1.$$

This implies that

$$\mathbf{x} \cdot \mathbf{x} + 2t\, \mathbf{x} \cdot \mathbf{a} + t^2 \mathbf{a} \cdot \mathbf{a} = 1,$$

so that

$$2t\, \mathbf{x} \cdot \mathbf{a} + t^2 = 0.$$

For $\mathbf{x} \cdot \mathbf{x} = 1$ and $\mathbf{a} \cdot \mathbf{a} = 1$.

Now, if $\mathbf{x} \cdot \mathbf{a} = 0$, then the point $\mathbf{x}$ lies in the plane π in which we are reflecting, and then as you would expect $t = 0$ and $r(\mathbf{x}) = \mathbf{x}$.

If $\mathbf{x} \cdot \mathbf{a} \neq 0$, then we may exclude the possibility that $t = 0$. It follows from the equation $2t\mathbf{x} \cdot \mathbf{a} + t^2 = 0$ that $t = -2\mathbf{x} \cdot \mathbf{a}$, and so

For then clearly $r(\mathbf{x}) \neq \mathbf{x}$.

$$r(\mathbf{x}) = \mathbf{x} - 2(\mathbf{x} \cdot \mathbf{a})\mathbf{a}.$$

To represent this mapping via a matrix, we express this equation in terms of coordinates and group the terms carefully:

$$\begin{aligned} r(x, y, z) &= (x, y, z) - 2(ax + by + cz)(a, b, c) \\ &= \left(\left(1 - 2a^2\right)x - 2aby - 2acz, -2abx + \left(1 - 2b^2\right)y \right. \\ &\qquad \left. -2bcz, -2acx - 2bcy + \left(1 - 2c^2\right)z\right), \end{aligned}$$

so that it can be expressed in matrix form as

$$\begin{pmatrix} x \\ y \\ z \end{pmatrix} \mapsto \begin{pmatrix} 1 - 2a^2 & -2ab & -2ac \\ -2ab & 1 - 2b^2 & -2bc \\ -2ac & -2bc & 1 - 2c^2 \end{pmatrix} \begin{pmatrix} x \\ y \\ z \end{pmatrix}.$$

Theorem 3 Reflection of S^2 in the plane π with equation $ax+by+cz = 0$, where $a^2 + b^2 + c^2 = 1$, is given by the mapping $\mathbf{x} \mapsto \mathbf{Ax}$ where

$$\mathbf{A} = \begin{pmatrix} 1 - 2a^2 & -2ab & -2ac \\ -2ab & 1 - 2b^2 & -2bc \\ -2ac & -2bc & 1 - 2c^2 \end{pmatrix}.$$

Notice that in the particular case when the plane is the equatorial plane, with equation $z = 0$ (so that $a = b = 0$ and $c = 1$), Theorem 3 gives the matrix for the mapping $\mathbf{x} \mapsto \mathbf{Ax}$ to be

$$\mathbf{A} = \begin{pmatrix} 1 & 0 & 0 \\ 0 & 1 & 0 \\ 0 & 0 & -1 \end{pmatrix}$$

that you met above.

Problem 8 Determine the matrix $\mathbf{A}$ such that the mapping $\mathbf{x} \mapsto \mathbf{Ax}$ represents reflection of S^2 in the plane π with equation $3x+4y-5z=0$.

Remark

There is another way to find the matrix representing a reflection in a great circle. Let the point P be one of the polar points of the plane π through O that meets S^2 in the given great circle. Let R be a rotation taking P to the North Pole N, and let r be reflection in the equatorial plane of S^2. Then the composition mapping $R^{-1}rR$ represents reflection of S^2 in π.

We omit the details of this calculation.

We now establish a result relating products of reflections of S^2 to rotations of S^2. It is the exact analogue of a result in Euclidean geometry. The result holds also for general reflections and rotations of $\mathbb{R}^3$, so we prove it in that more general setting.

Subsection 5.2.1, margin note alongside Problem 4

Theorem 4 The product of any two reflections of $\mathbb{R}^3$ in planes through O that meet in a common line is a rotation about that common line.

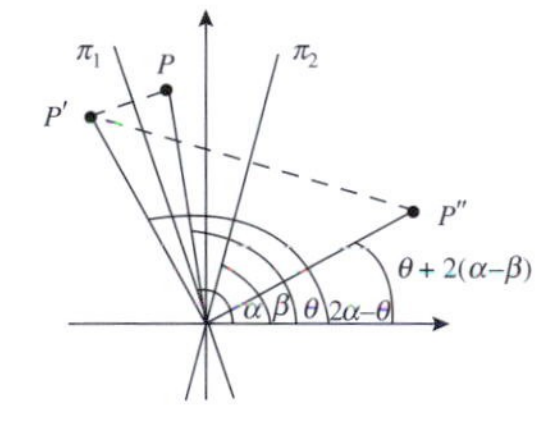

Proof By choosing coordinate axes suitably, we may arrange that the common line is the z-axis. The planes are therefore vertical, and it follows that each reflection leaves the z-coordinate of each point unaltered. It is therefore sufficient to look at the effect of each reflection and composite on planes perpendicular to the z-axis; we shall discuss the (x, y)-plane, but a similar argument applies to all planes parallel to this one.

Let the first plane π_1 make an angle α with the positive x-axis and the second plane π_2 make an angle β with the positive x-axis. Let a point P in the (x, y)-plane have plane polar coordinates (r, θ). Then the image of P under reflection in π_1 is the point P' with polar coordinates $(r, 2\alpha - \theta)$. The image of P' under the reflection in π_2 is the point P'' with polar coordinates $(r, 2\beta - (2\alpha - \theta)) = (r, \theta + 2(\alpha - \beta))$. This is the image of P under rotation about the common line of π_1 and π_2, through twice the angle between the planes. ∎

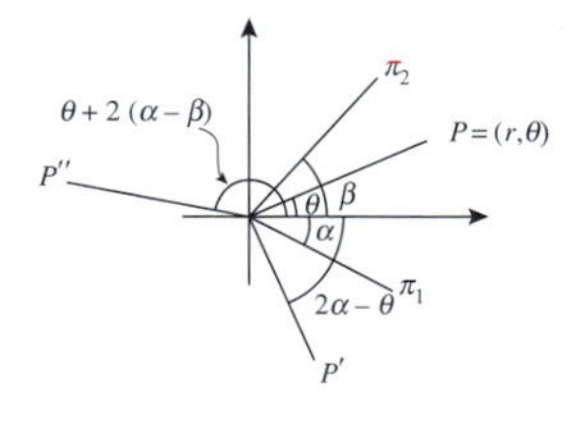

We deduce from this theorem that the composite of reflections in two great circles on S^2 is a rotation of S^2 about the line common to the planes defining the great circles, and through twice the angle between them. We can also deduce that every rotation is a product of two reflections; these reflections are in planes that have the axis of rotation as their common line, and are separated by half the angle of the rotation (but they are otherwise arbitrary).

Example 3 Prove that the image of a circle C on S^2 under an isometry t of S^2 to itself is a circle.

Solution Let the circle C be cut out by a plane π, and let ℓ be the line through the origin perpendicular to π. Let ℓ meet S^2 at the points P and P'. Then all points of C are the same distance from P.

Since t is an isometry, the points of $t(C)$ are all the same distance from $t(P)$. It follows that they lie on a circle, which must be the circle cut out on S^2 by the plane $t(\pi)$. □

We now establish the following important result.

Theorem 5 The Product Theorem
Every isometry of S^2 is a product of at most three reflections in great circles.

Proof The key idea in this proof is the observation that *an isometry of S^2 is known completely when its effect on three points (which do not lie on a common great circle) is known*. We shall prove this observation in the course of proving the theorem.

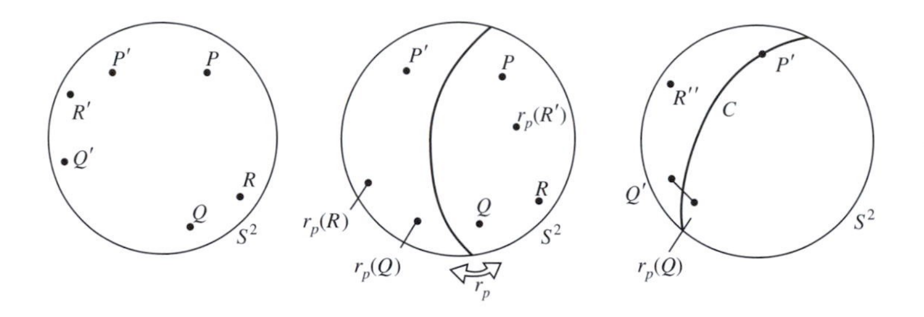

Let P, Q and R be three points not lying on the same great circle, and let P', Q' and R' be their images under an isometry t of S^2 to itself. Let r_P be a reflection of S^2 in a great circle that sends P to P'. Consider the angle in $\mathbb{R}^3$ with arms $P'r_P(Q)$ and $P'Q'$, and let C be the great circle that bisects this angle and the segment joining $r_P(Q)$ and Q'. Reflection in C fixes P' and sends $r_p(Q)$ to Q'. Suppose that this reflection sends $r_P(R')$ to R''.

See left figure above.
See centre figure above.
See right figure above.

The distance of R'' from P' is the same as the distance RP (call it d_P), and the distance of R'' from Q' is the same as the distance QR (call it d_Q), since both reflections are isometries. So the point R'' lies on the intersection of the circle with centre P' and radius d_P and the circle with centre Q' and radius d_Q; so too does the point R'.

Then it may be that R' and R'' coincide, in which case we have found a product of two reflections sending R to R'. Or it may be that R' and R'' do not coincide but a reflection in the great circle through P' and Q' maps R'' to R', and so three reflections have been used.

We now claim that any isometry fixing three points on S^2 fixes every point of S^2. Let P, Q and R be the points, let M be any other point on S^2, and let t be the isometry fixing the points P, Q and R. Then $t(M)$ lies on the circle with centre P and radius PM and on the circle with centre Q and radius QM. Of the two points satisfying this condition, only one also lies on the circle with centre R and radius RM, and that is the point M itself; so the image of M under t is M, and therefore t fixes every point of S^2. It follows that if two isometries t_1 and t_2 agree on three points, then they agree on every point, since the composite $t_2^{-1} \circ t_1$ satisfies the conditions of our last remark.

We deduce that the isometry mapping the points P, Q and R to the points P', Q' and R' must be the product of the two or three reflections that we described above. This completes the proof. ■

It does not follow that these reflections are uniquely determined – recall the case of the rotation discussed above.

Finally in this subsection we show that the Isosceles Triangle Theorem in spherical geometry can be established using a reflection in a great circle.

Theorem 6 Isosceles Triangle Theorem
Let a triangle ΔPQR on S^2 have sides PQ of length r, QR of length p, and RP of length q, and let $p = q$. Also, let the angles at the vertices P, Q and R be α, β and γ, respectively. Then $\alpha = \beta$.

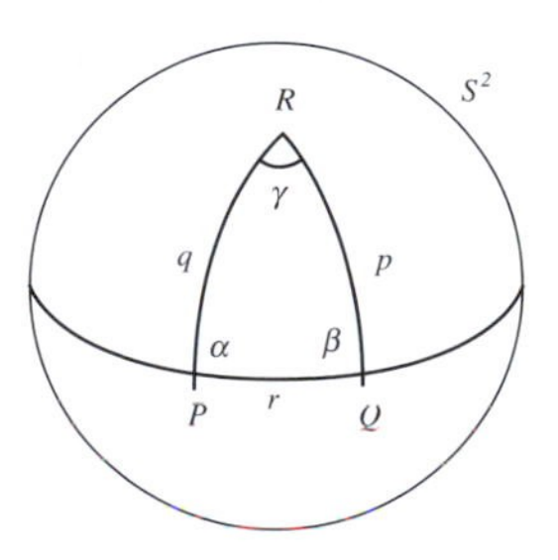

Proof Consider the reflection in the great circle that is the internal bisector of the angle $\angle PRQ$.

It maps the great circle through R and P to the great circle through R and Q; and, since it is an isometry and $p = q$, it therefore maps P to Q, and Q to P. It maps the great circle through P and Q to itself, and so it interchanges the angles α and β, which shows that $\alpha = \beta$. ■

7.2.3 Direct Isometries

The Product Theorem allows us to deduce that every isometry is either a reflection, a rotation or the composite of a rotation and a reflection. For if it is a composite of two reflections, then it is a rotation; and if it is a composite of three reflections, then the composite of the first two is a rotation.

Theorem 5

It is clear that every rotation of S^2 is an direct isometry of S^2.

We shall now see that every direct isometry of S^2 is a product of rotations. For each reflection reverses the orientation of angles, and so the product of two reflections preserves the orientation of angles, while the product of three reflections reverses it again. So a direct isometry must be the product of an *even* number of reflections and therefore be a rotation.

For the product of any pair of reflections is a rotation.

It is instructive to give a direct proof of this result that does not use the Product Theorem, because reflections are not easy to handle geometrically or algebraically.

First, any direct isometry of S^2 that fixes the North Pole N must map each circle of latitude to itself, and so must be a rotation of that circle. Let D and E be points on different circles of latitude; they remain the same distance apart, so the rotation of their two circles of latitude must be through the same angle. It follows that every circle of latitude on S^2 is rotated equally, and so the isometry fixing N must be a rotation about ON.

Next, let M be a direct isometry that sends a point P to P' and a point Q to Q', and let R be a rotation that sends P to P'. Then $M \circ R^{-1}$ is an isometry that fixes P'. We may assume that $P' = N$ (if need be by additionally composing with a rotation that sends $P'$to N). Then the isometry $M \circ R^{-1} = R(Z, \gamma)$, by the remarks in the previous paragraph; so the direct isometry $M = R(Z, \gamma) \circ R$ is a product of two rotations. We deduce that *every* direct isometry is a product of rotations.

We shall now see that every direct isometry of S^2 is a single rotation.

Since every 3×3 matrix with real entries (such as a product of rotations must be) must have a real eigenspace, then any such 3×3 matrix describing an isometry of S^2 to itself must leave the points where its eigenspace meets S^2 fixed. Since it is also a direct symmetry, it must therefore be a rotation about those points. It follows that every direct isometry of S^2 to itself is simply a rotation.

See Appendix 2.

We can deduce from this also the fact that the composite of two rotations of S^2 is also a rotation of S^2, since the composite is itself a direct isometry.

We summarize these facts as follows.

Theorem 7

(a) Every isometry of S^2 is a reflection, a rotation, or the composite of a reflection and a rotation.
(b) Every direct isometry of S^2 is a rotation, and vice versa.

Problem 9 Prove that the composite of an even number of reflections of S^2 in great circles is a rotation of S^2.

7.3 Spherical Trigonometry

7.3.1 Spherical Triangles

Recall that a triangle on S^2, or a *spherical triangle*, consists of the arcs of three great circles joining three points on S^2. We first establish that spherical triangles exist with any given angles; more precisely, we establish the following result.

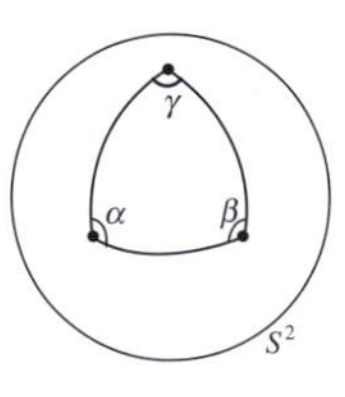

Theorem 1 Given any three angles α, β and γ with $\alpha+\beta+\gamma > \pi$, there exists a spherical triangle with those angles.

In fact we also require that $\alpha+\beta+\gamma < 5\pi$ in view of Theorem 3 below – but we shall not go into this here.

Proof We use the idea of stereographic projection, under which circles on S^2 correspond to generalized circles (that is, lines or circles) in the extended plane.

Theorem 7 of Subsection 5.2.4 (in which stereographic projection was introduced).

By a rotation of S^2, if necessary, we may assume that one of the vertices of a triangle on the unit sphere is at the South Pole, S.

Under stereographic projection a triangle on S^2 with one vertex at the South Pole S corresponds to a circular-arc triangle in the plane (that is, a 'triangle' in $\mathbb{C}$ whose sides are arcs of generalized circles) with one vertex at O, the centre of the unit circle $\mathscr{C}$, and two straight sides. It follows that to prove the theorem it is enough to construct a triangle of this form in $\mathbb{C}$ that has the specified angles. This we now do; in our construction we assume that α and β lie between 0 and $\frac{\pi}{2}$, for simplicity.

Recall that stereographic projection is angle-preserving (Theorem 8 of Subsection 5.2.4).

Consider the circular-arc triangle ΔABP in $\mathbb{C}$ that has angles $\frac{\pi}{2}$ at A, $\frac{\pi}{2}$ at B, and θ at P, where PA and PB are straight lines (so that P is the centre of the circle through A and B). Let D be the point on PB such that the angle $\angle BAD$ equals α, let C be the point on PA such that the angle $\angle ABC$ equals β, let AD and BC meet at O, and let the angle $\angle AOB$ equal x (as shown). Our first task is to evaluate x.

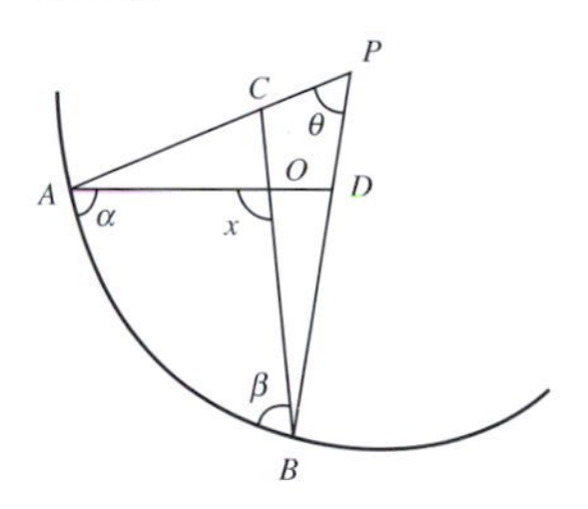

Both the angles $\angle AOC$ and $\angle BOD$ are equal to $\pi - x$. By considering the triangles ΔAOC and ΔBOD in turn, we deduce that

$$\angle PCO = (\pi - x) + \left(\tfrac{\pi}{2} - \alpha\right) = \tfrac{3\pi}{2} - x - \alpha,$$

and

$$\angle PDO = (\pi - x) + \left(\tfrac{\pi}{2} - \beta\right) = \tfrac{3\pi}{2} - x - \beta.$$

Then the sum of the angles of the quadrilateral $PCOD$ is

$$2\pi = \theta + \left(\tfrac{3\pi}{2} - x - \alpha\right) + x + \left(\tfrac{3\pi}{2} - x - \beta\right),$$

so that

$$x = \pi + \theta - \alpha - \beta.$$

It follows that $x = \gamma$ if and only if

$$\theta = \alpha + \beta + \gamma - \pi.$$

This is possible, since by hypothesis $\alpha+\beta+\gamma > \pi$.

Hence if we repeat the construction with the value $\alpha+\beta+\gamma-\pi$ for θ, the circular-arc triangle ΔABO has the desired angles α, β and γ. It follows from the earlier discussion that by stereographic projection we can then construct the required spherical triangle on S^2 with these angles. ■

A modification of the above argument gives the following important result, which we state without proof.

Theorem 2 The sum of the angles of any spherical triangle is greater than π.

In view of Theorems 1 and 2, it is useful to assign a name to the difference between the sum of the angles of a spherical triangle and π: this measures how far the triangle departs from being Euclidean.

Definition The **angular excess** of a spherical triangle is the difference between the sum of the angles of the triangle and π.

Problem 1 Construct spherical triangles with angular excess of

(a) 0.01 radians, (b) 3.14 radians.

Subsection 7.1.1

Recall that earlier when we defined a distance PQ between two given points P and Q on S^2, we noted that there were two 'line segments' PQ. Now, for definiteness, we call the shorter of these two the **strict line segment** PQ. (If both are of length π, we choose either in some explicit way.)

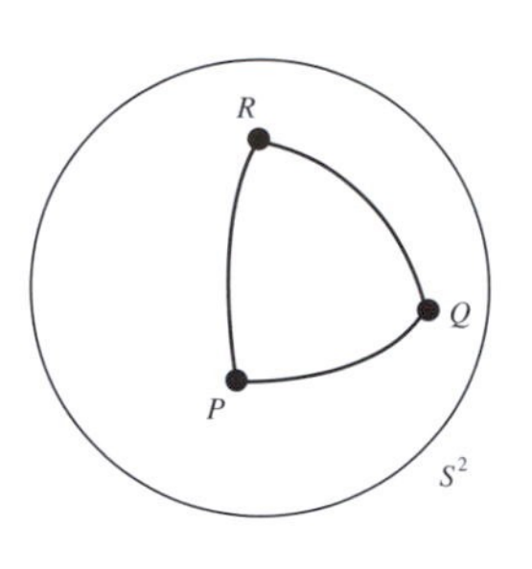

Similarly, when we defined a triangle ΔPQR with three given vertices P, Q and R on S^2, we noted that there were several such triangles. For definiteness, we now call the triangle whose sides are strict line segments the **strict triangle** ΔPQR. Then it is possible to rotate the sphere in such a way that the whole of the strict triangle can be seen at one time. (You should convince yourself that this is the case.) The **inside** of the strict triangle ΔPQR is then defined to be the region of S^2 bounded by the three strict line segments. The **outside** is the complement of the inside.

We now investigate the regions into which the corresponding great circles defining the strict triangle ΔPQR divide S^2. Notice first that the equatorial plane defining each of the relevant great circles cuts S^2 into two regions. A second great circle (and equatorial plane) divides each of these regions into two, giving four regions. Finally a third great circle (and equatorial plane) divides each of these regions into two, giving eight regions: so there are eight spherical triangles with given vertices P, Q and R on S^2.

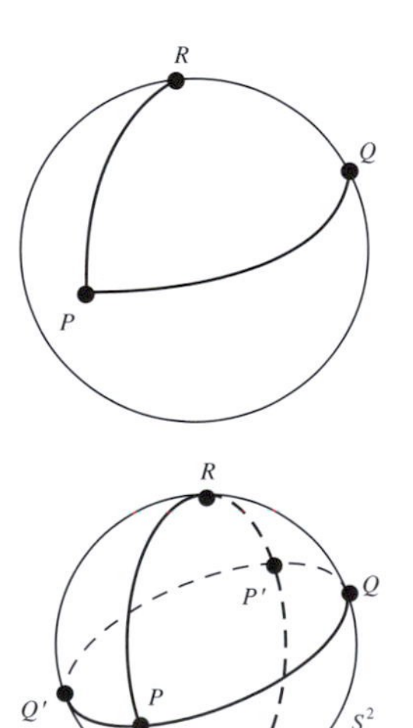

Now let us look at the strict triangle ΔPQR more closely. For definiteness, we will assume that the vertices P, Q and R are chosen in clockwise order as viewed by an observer at the centre of the sphere. Rotate the sphere to move R to the North Pole, and Q to some point in the (y, z)-plane. By our choice of the orientation of ΔPQR and the fact that the line segments PR and PQ are strict line segments, it follows that this rotation leaves P on 'the front' of the sphere, where we can see it.

The great circle through R and Q is the 'rim' of S^2 that we can see. So the great circle RP meets it again at R', say, the diametrically opposite point to R; and the great circle QP meets it again at Q', say, the diametrically opposite point to Q. Similarly the great circles PR and PQ meet again at P', say, the diametrically opposite point to P.

So, associated with the strict triangle ΔPQR in this particular position on S^2 there are three great circles, each meeting in two diametrically opposite points; a total of six points in all. The great circles define eight triangles:

- the strict triangle ΔPQR and the triangles $\Delta PQ'R$, $\Delta PQ'R'$ and $\Delta PR'Q$ (all visible on the 'front' of S^2), and
- four triangles $\Delta P'QR$, $\Delta P'Q'R$, $\Delta P'Q'R'$ and $\Delta P'R'Q$ 'round the back'.

In fact, each of the eight triangles has a diametrically opposite triangle to which it is congruent (under the map $(x, y, z) \mapsto (-x, -y, -z)$). For example, the triangle ΔPQR is diametrically opposite to the triangle $\Delta P'Q'R'$.

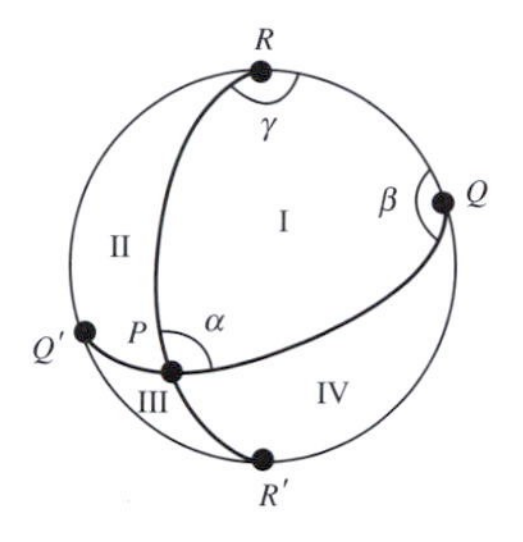

In the figure in the margin we have marked the four 'visible' triangles ΔPQR, $\Delta PQ'R$, $\Delta PQ'R'$ and $\Delta PR'Q$ with Roman numerals I, II, III and IV, respectively. Similarly, we denote their diametrically opposite triangles $\Delta P'Q'R'$, $\Delta P'QR'$, $\Delta P'QR$ and $\Delta P'RQ'$ with Roman numerals I′, II′, III′ and IV′, respectively.

In order to label equal angles conveniently, let us denote by α the angle $\angle QPR$ in the strict triangle ΔPQR, by β the angle $\angle PQR$, and by γ the angle $\angle QRP$. A useful observation is that two great circles meet at the same angles at each of their points of intersection; so, for example, the angles $\angle QPR$ and $\angle Q'P'R'$ are equal.

Example 1 Show that the triangles ΔPQR and $\Delta P'Q'R'$ have their angles equal in pairs.

Solution The angle $\angle QPR$ at P in ΔPQR is formed by two great circles that meet again at P', forming the angle $\angle Q'P'R'$ at P' in $\Delta P'Q'R'$, so they must be equal.

A similar argument shows that the angles at vertices Q and Q', and the angles at vertices R and R', are also equal. □

Problem 2 Determine the angles in each of triangles II, III, and IV, in terms of α, β and γ.

Area of a Triangle

We next find the area of a spherical triangle in terms of its angles. In this discussion, we introduce the term **lune** to denote a two-sided polygon defined by two great circles, choosing arbitrarily which of the four such regions on the sphere we mean.

The area of a lune is clearly proportional to the corresponding angle between the two great circles, the *angle of the lune*. Now when the lune has angle $\frac{\pi}{2}$, its area is π (for the lune is then one-quarter of S^2, which has area 4π); so by proportion the area of a lune of angle μ, say, is 2μ.

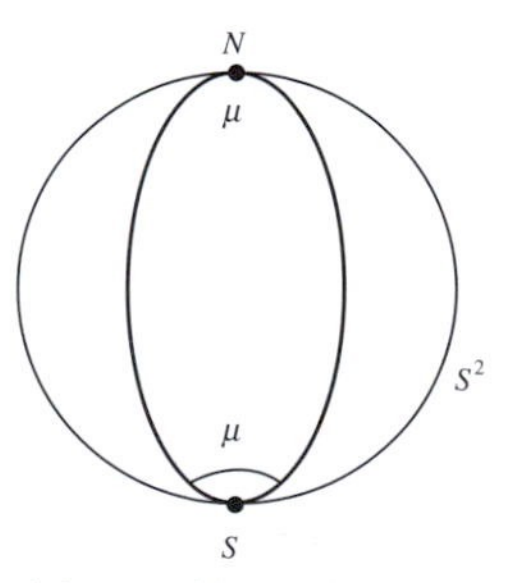

A lune with vertices at N and S, and angle μ.

The 'front' of S^2 as shown in the margin consists of two lunes with vertices R and R', each formed of two strict triangles, I + IV and II + III (using the above notation for triangles). The lune I + IV has area 2γ , since it has angle γ. With a slight abuse of notation we express this fact about areas in the form

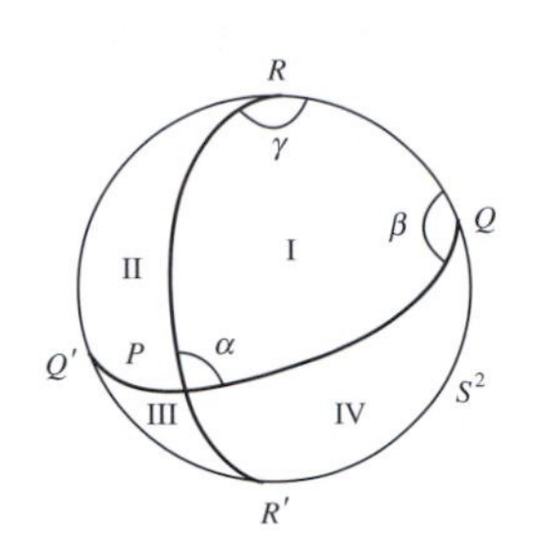

$$\mathrm{I} + \mathrm{IV} = 2\gamma.$$

A similar argument involving the lune with vertices Q and Q' yields the equation

$$\mathrm{I} + \mathrm{II} = 2\beta,$$

and then with the lune with vertices P and P' yields the equation

$$\mathrm{I} + \mathrm{III}' = 2\alpha.$$

Since triangles III and III$'$ are congruent, we may rewrite this last equation as

$$\mathrm{I} + \mathrm{III} = 2\alpha.$$

Adding the three equations, we obtain

$$3 \cdot \mathrm{I} + \mathrm{II} + \mathrm{III} + \mathrm{IV} = 2(\alpha + \beta + \gamma).$$

Also, the area of the front of S^2 is

$$\mathrm{I} + \mathrm{II} + \mathrm{III} + \mathrm{IV} = 2\pi.$$

It follows that

$$2 \cdot \mathrm{I} + 2\pi = 2(\alpha + \beta + \gamma)$$

or

$$\mathrm{I} = (\alpha + \beta + \gamma) - \pi.$$

This completes the proof of the following theorem for triangles, in the special case that they are strict triangles.

The general case is proved by dividing an arbitrary triangle into strict triangles, and applying the special case to each; we omit the details.

Theorem 3 The area of a spherical triangle is equal to its angular excess.

Problem 3 Construct a spherical triangle of area $\frac{3\pi}{4}$.

7.3.2 Dual Triangles

The somewhat fiddly nature of the definition of a strict triangle pays off handsomely in a feature of spherical geometry that was often exploited in the 18th century by Euler and Lagrange, two of the pioneer figures in spherical geometry.

We required that the sides of a strict triangle be less than or equal to π in length; for simplicity in what follows, we shall ignore the case of equality.

Let A, B and C be the vertices of a strict triangle, with angles α, β and γ at the vertices, respectively, and opposite sides of length a, b and c, respectively. Each side and angle of the triangle $\triangle ABC$ has magnitude between 0 and π.

We now construct a new spherical triangle $\triangle A'B'C'$, called the **dual** of the original triangle $\triangle ABC$, with vertices A', B' and C', as follows. Its angles at A', B' and C' are $\pi - a$, $\pi - b$ and $\pi - c$, respectively; and its sides are $A'B'$ of length $\pi - \gamma$, $B'C'$ of length $\pi - \alpha$, and $C'A'$ of length $\pi - \beta$.

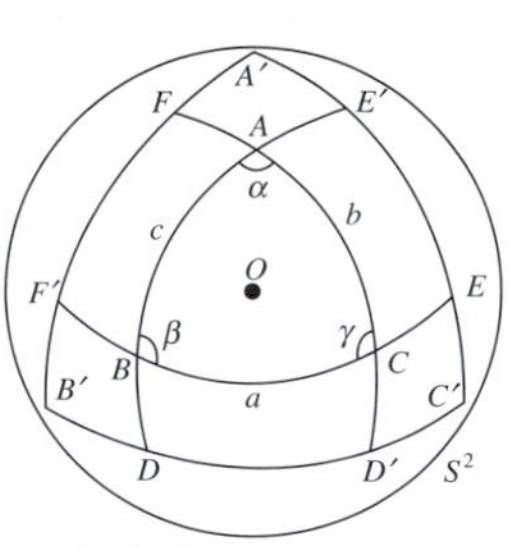

$\triangle A'B'C'$ is dual to $\triangle ABC$.

One way of constructing such a triangle is as follows. Extend the segments AB and AC to segments AD and AD', each of length $\frac{\pi}{2}$, and draw the great circle through D and D' (this is the equator for A). Similarly, extend the segments BC and BA to segments BE and BE', each of length $\frac{\pi}{2}$, and draw the great circle through E and E' (this is the equator for B); and extend the segments CA and CB to segments CF and CF', each of length $\frac{\pi}{2}$, and draw the great circle through F and F' (this is the equator for C). Then either of the strict triangles defined by the great circles DD', EE' and FF' is a spherical triangle with the desired properties.

For example, choose the one that overlaps $\triangle ABC$, and denote its vertex on EE' and FF' by A', its vertex on FF' and DD' by B', and its vertex on DD' and EE' by C'. We now check that the triangle $\triangle A'B'C'$ is indeed dual to the original triangle $\triangle ABC$.

The lengths of CF' and CB' are both $\frac{\pi}{2}$, since $A'FF'B'$ is the equator for C. But B' is on the equator $B'DD'C'$ for A, so $B'A = \frac{\pi}{2}$. Thus A and C are on the equator for B', so that the great circle $FACD'$ defined by A and C must be the equator for B'. This means that D' lies on the equator for B', so $B'D' = \frac{\pi}{2}$. Similarly $DC' = \frac{\pi}{2}$.

Now consider the equator $B'DD'C'$ for A, and denote by O the centre of the sphere. The angle $\angle DOD'$ is equal to $\angle BAC = \alpha$ (since both $\angle DOD'$ and $\angle BAC$ are just the angle between the planes OAB and OAC); thus DD' lies on a (great) circle of radius 1, and so the length of DD' is also α (since DD' is equal to $\angle DOD'$). It follows that the length of $B'C'$ is

$$B'C' = B'D' + DC' - DD' = \tfrac{\pi}{2} + \tfrac{\pi}{2} - \alpha = \pi - \alpha.$$

Similarly the length of $C'A'$ is $\pi - \beta$, and the length of $A'B'$ is $\pi - \gamma$.

Next, by construction $CF' = \frac{\pi}{2}$ and $BE = \frac{\pi}{2}$. It follows that the length of $F'E$ is

$$F'E = F'C + BE - BC = \tfrac{\pi}{2} + \tfrac{\pi}{2} - a = \pi - a.$$

Hence the angle $\angle F'OE$ equals $\pi - a$ (since $F'BCE$ is a circle of radius 1 with centre O and hence $\angle F'OE$ equals the length of $F'E$), so that the angle $\angle F'A'E$ also equals $\pi - a$ (since both are just the angle between the planes $OA'F'$ and $OA'E$). It follows that the angle $\angle B'A'C'$ equals $\pi - a$, as required. Similarly, $\angle A'B'C' = \pi - b$ and $\angle B'C'A' = \pi - c$.

We have therefore established the following theorem.

Theorem 4 Dual Triangles Theorem
With every strict spherical triangle Δ there is associated another triangle Δ', called its *dual*, whose angles are the complements of the sides of Δ and whose sides are the complements of the angles of Δ.

Here by the 'complement' of a quantity x we mean the quantity $\pi - x$.

Here is an example of how the theorem proves useful. Let ΔABC be a strict triangle with $\angle ABC = \angle ACB$. Then, if $\Delta A'B'C'$ is its dual (which exists by Theorem 4), we have $A'C' = A'B'$; it then follows from the Isosceles Triangle Theorem that $\angle A'B'C' = \angle A'C'B'$. It then follows by duality that $AC = AB$. This is the converse of the Isosceles Triangle Theorem, in the case of strict triangles.

Section 7.2.2, Theorem 6

We ask you to prove the general case of the converse in Section 7.6.

Problem 4 Let A, B and N be the points (1, 0, 0), (0, 1, 0) and (0, 0, 1) on S^2. Prove that the triangle ΔABN is congruent to any of its duals.

7.3.3 Distance between Two Points

In real life we are often interested in the (spherical) distance between points on the surface E of the Earth, whose radius is about 4000 miles. The comparable problem for S^2 is to find a formula for the distance between two points of S^2. Then a scaling map

In metric units, the radius of the Earth is about 6378 km.

$$t : S^2 \to E$$
$$\mathbf{x} \mapsto 4000\mathbf{x}$$

Here we take the unit of measurement for the radius S^2 to be 1 mile.

provides an answer to the original question concerning E.

Let P and Q be points on S^2 with Cartesian coordinates (p_1, p_2, p_3) and (q_1, q_2, q_3), respectively, whose distance apart on S^2 is d. It follows that, if O denotes the centre of S^2, the angle $\angle POQ$ also equals d.

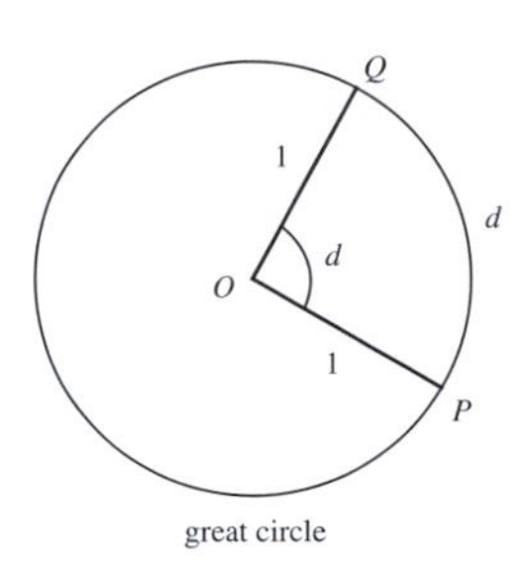

great circle

Then the dot product of the vectors $\overrightarrow{OP}$ and $\overrightarrow{OQ}$ is

$$\overrightarrow{OP} \cdot \overrightarrow{OQ} = p_1q_1 + p_2q_2 + p_3q_3 = 1 \cdot 1 \cdot \cos d,$$

so that $\cos d = p_1q_1 + p_2q_2 + p_3q_3$.

Theorem 5 The distance d on S^2 between the points (p_1, p_2, p_3) and (q_1, q_2, q_3) is given by

$$\cos d = p_1q_1 + p_2q_2 + p_3q_3.$$

In the expression $\cos d$, d must be taken in radians, as usual.

In applications to E, we must remember that if a point has latitude θ' and longitude ϕ then it has colatitude $\theta = \frac{\pi}{2} - \theta'$ and Cartesian coordinates $(\cos\phi \sin\theta, \sin\phi \sin\theta, \cos\theta)$.

The latitudes and longitudes of various cities are approximately as follows.

City	Latitude	Longitude
New York	41°	−74°
Rio de Janeiro	−23°	−43°
Sydney	−34°	151°
Tokyo	36°	140°

Remember that (geographers') longitude West of the Greenwich meridian is *negative*.

Example 2 Estimate the distance between New York and Sydney, taking the radius of the Earth as 4000 miles.

Solution The colatitudes of New York and Sydney are $90° - 41° = 49°$ and $90° + 34° = 124°$, so that the coordinates of the corresponding points P and Q on S^2 are

$$\begin{aligned}&(\cos(-74°)\sin(49°), \sin(-74°)\sin(49°), \cos(49°))\\ &\quad\simeq (0.2756 \cdot 0.7547, -0.9613 \cdot 0.7547, 0.6561)\\ &\quad\simeq (0.208, -0.725, 0.656),\end{aligned}$$

and

$$\begin{aligned}&(\cos(151°)\sin(124°), \sin(151°)\sin(124°), \cos(124°))\\ &\quad\simeq (-0.8746 \cdot 0.8290, 0.4848 \cdot 0.8290, -0.5592)\\ &\quad\simeq (-0.725, 0.402, -0.559).\end{aligned}$$

Hence the distance between New York and Sydney is approximately $4000d$ miles, where

$$\begin{aligned}\cos d &\simeq -0.208 \cdot 0.725 - 0.725 \cdot 0.402 - 0.656 \cdot 0.559\\ &\simeq -0.151 - 0.291 - 0.367 = -0.809.\end{aligned}$$

Thus $d \simeq \cos^{-1}(-0.809) \simeq 2.513$ radians, and the required distance $4000d \simeq 4000 \cdot 2.51 \simeq 10\,050$ miles. □

Problem 5 Estimate the distance between Tokyo and Rio de Janeiro, taking the radius of the Earth as 4000 miles.

Right-Angled Triangles

It only makes sense to discuss Pythagoras' Theorem in those geometries with a *distance*. We have already seen Pythagoras' Theorem in hyperbolic geometry, so what is the analogue in spherical geometry?

Theorem 8 of Subsection 6.4.3

Theorem 6 Pythagoras' Theorem
Let $\triangle ABC$ be a triangle on S^2 in which the angle at C is a right angle. If a, b and c are the lengths of BC, CA and AB, then

$$\cos c = \cos a \times \cos b.$$

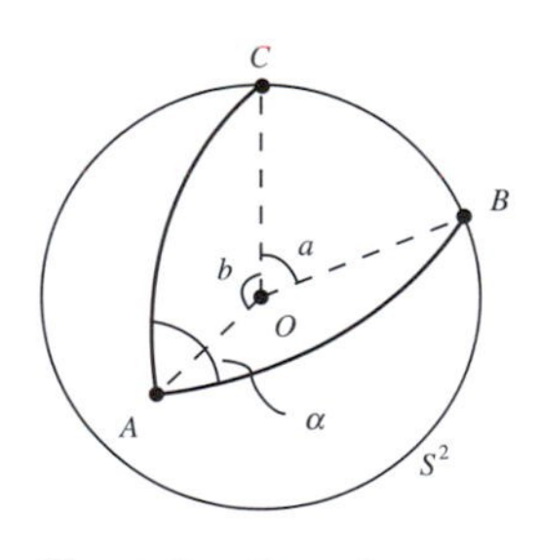

Proof We may rotate S^2 so that C is the North Pole N $(0, 0, 1)$ of S^2, A lies on the Greenwich meridian and has coordinates $(\sin b, 0, \cos b)$, and B has coordinates $(0, \sin a, \cos a)$.

Since the sphere has radius 1, at $A, \theta = b$ and $\phi = 0$; and at $B, \phi = \frac{\pi}{2}$ and $\theta = a$.

We then apply the rotation $R(Y, -b)$ to S^2. This maps A to C, C to some point C', and B to some point B'.

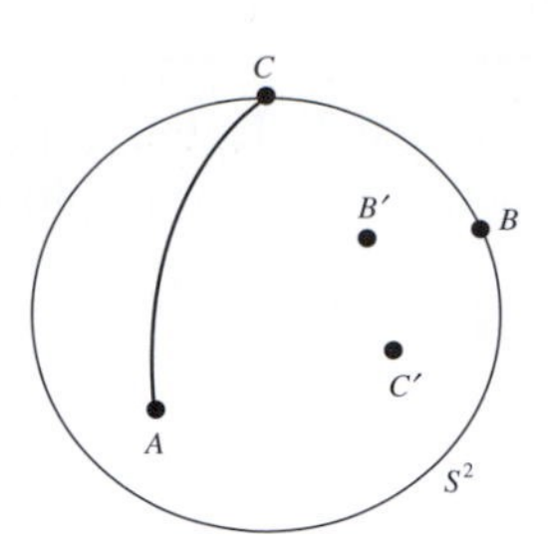

Let α denote the angle $\angle CAB$. Then the rotation $R(Y, -b)$ maps the angle $\angle CAB$ to the angle $\angle C'CB'$, so that the angle $\angle C'CB' = \alpha$; it follows that the angle $\angle ACB' = \pi - \alpha$ since $\angle ACB' + \angle B'CC' = \pi$. Hence the longitude of B' is $\pi - \alpha$. Also, the colatitude of B' is

$$\text{length of CB}' = \text{length of AB} = c.$$

So the coordinates of B' are

$$\begin{aligned}(\cos(\pi - \alpha)\sin c, \sin(\pi - \alpha)\sin c, \cos c)\\ = (-\cos\alpha\ \sin c,\ \sin\alpha\ \sin c, \cos c).\end{aligned} \qquad (1)$$

But we also have as coordinates of B'

$$\begin{aligned}R(Y, -b)(B) &= \begin{pmatrix}\cos b & 0 & -\sin b\\ 0 & 1 & 0\\ \sin b & 0 & \cos b\end{pmatrix}\begin{pmatrix}0\\ \sin a\\ \cos a\end{pmatrix}\\ &= \begin{pmatrix}-\sin b\cos a\\ \sin a\\ \cos b\cos a\end{pmatrix}.\end{aligned} \qquad (2)$$

Here we use the formulas for coordinates under isometries given in Subsection 7.2.1.

Equating the third coordinates in (1) and (2), we get the formula

$$\cos c = \cos a \times \cos b,$$

as required. ■

For example, if $a = 1, b = 2$ and $\angle ACB = \frac{\pi}{2}$, then

$$\cos c = \cos 1 \times \cos 2 \simeq 0.5403 \cdot (-0.4161) \simeq -0.2248,$$

so that $c \simeq 1.798$.

Problem 6 $\triangle PQR$ is a triangle on S^2 in which the angle at R is a right angle and the lengths of PQ and PR are 1.7 and 1.9, respectively. Estimate the length of QR.

Remark

For all values of x, $\cos x = 1 - \frac{x^2}{2!} +$ higher powers of x; so, when x is small, $\cos x \simeq 1 - \frac{1}{2}x^2$. It follows that, if we are dealing with small triangles on S^2, then the conclusion of Theorem 6 may be interpreted in the form

$$1 - \tfrac{1}{2}c^2 \simeq \left(1 - \tfrac{1}{2}a^2\right) \cdot \left(1 - \tfrac{1}{2}b^2\right) \simeq 1 - \tfrac{1}{2}a^2 - \tfrac{1}{2}b^2,$$

Here we drop the term in a^2b^2, since it is smaller than the remaining terms.

so that $c^2 \simeq a^2 + b^2$. Thus Theorem 6 is a close approximation to the Euclidean version of Pythagoras' Theorem in $\mathbb{R}^2$.

The area formula for S^2 suggests that the angles of a spherical triangle determine its size as well as its shape. In fact, any three pieces of the six possible

Theorem 3

pieces of information about a triangle (its three angles and the lengths of its three sides) determine the remaining three.

We prove this fact here for right angled triangles; a proof for general triangles can then be constructed by dividing a general triangle into two right angled triangles using an altitude.

Returning to the formula in the proof of Theorem 6 for the coordinates of B' and equating the first and second coordinates, we obtain

$$\cos\alpha \sin c = \sin b \cos a \quad \text{and} \quad \sin\alpha \sin c = \sin a.$$

Dividing the second of these by the first, we get $\tan\alpha = \tan a / \sin b$.

Re-arranging the above formulas to make them easier to remember, we have the following result.

Theorem 7 Let $\triangle ABC$ be a triangle on S^2 in which the angle at C is a right angle. If a, b and c are the lengths of BC, CA and AB, and α denotes the angle $\angle CAB$, then

$$\cos c = \cos a \times \cos b,$$

$$\sin\alpha = \frac{\sin a}{\sin c}, \quad \text{and} \quad \tan\alpha = \frac{\tan a}{\sin b}.$$

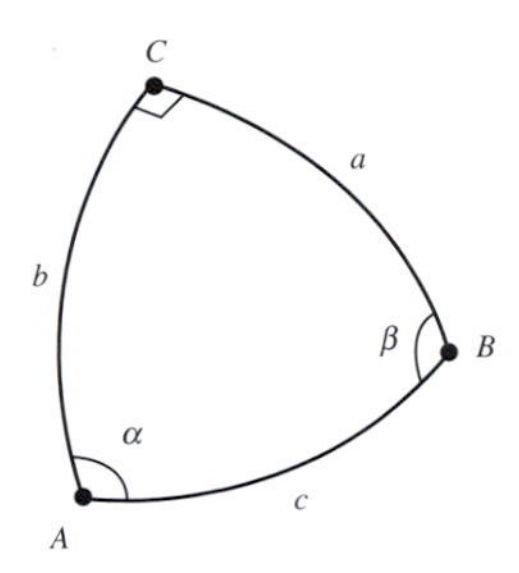

In the proof of Theorem 6 and before Theorem 7

A similar discussion to that above gives that

$$\cos\beta \sin c = \sin a \cos b \quad \text{and} \quad \sin\beta \sin c = \sin b,$$

so that

$$\sin\beta = \frac{\sin b}{\sin c} \quad \text{and} \quad \tan\beta = \frac{\tan b}{\sin a}.$$

From these formulas and those in Theorem 7 above, it follows that in a right-angled triangle two further pieces of information about the triangle enable us to determine all its angles and sides.

Problem 7 $\triangle PQR$ is a triangle on S^2 in which the angle at R is a right angle and the lengths of PQ and PR are 1.7 and 1.9, respectively. Estimate the angle $\angle QPR$.

7.3.4 Spherical Trigonometry

In this subsection we deduce various formulas of spherical trigonometry for strict triangles by dropping perpendiculars from a vertex of a spherical triangle to the opposite side, giving two right-angled ones for which we already have formulas. First, we verify that we can always drop a perpendicular from a point to a line. It follows that a spherical triangle has 3 altitudes.

Given a strict segment AB and a point C on S^2 that form a triangle $\triangle ABC$, there is always a great circle through C perpendicular to AB. This great circle meets the great circle of which AB is part in two points, which we call D and D'. Since D and D' are antipodal points, they are a distance π apart, so one of three things must happen.

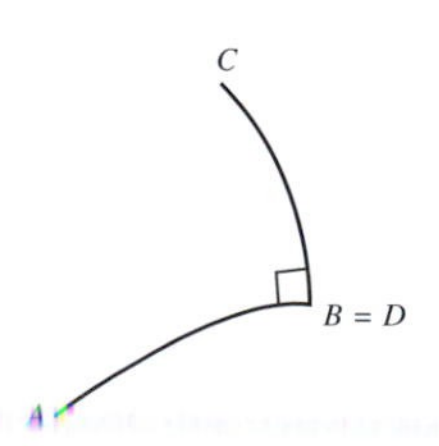

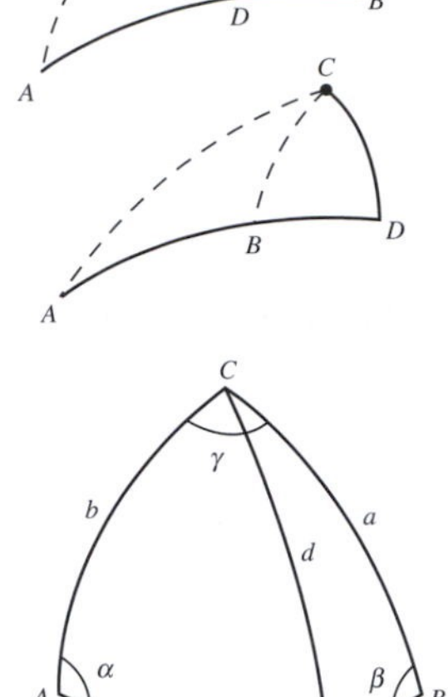

1. One of D and D' coincides with A or B. The corresponding strict line segment is the **perpendicular** from C to AB. This case will not lead us to any new results, so we do not consider it further.
2. One of D and D', say D, lies between A and B. Then we call CD the **perpendicular** from C to AB; it lies inside $\triangle ABC$.
3. One of D and D', say D, is such that the distance AD is less than π and B lies between A and D. Then we again call CD the **perpendicular** from C to AB; in this case it lies outside $\triangle ABC$.

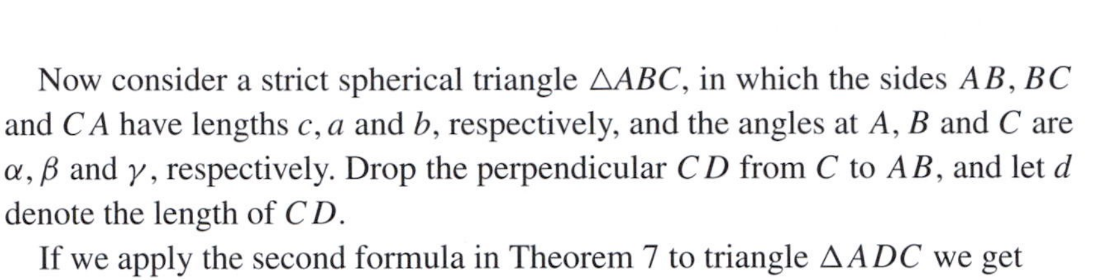

Now consider a strict spherical triangle $\triangle ABC$, in which the sides AB, BC and CA have lengths c, a and b, respectively, and the angles at A, B and C are α, β and γ, respectively. Drop the perpendicular CD from C to AB, and let d denote the length of CD.

If we apply the second formula in Theorem 7 to triangle $\triangle ADC$ we get

$$\sin \alpha = \frac{\sin d}{\sin b},$$

so that $\sin d = \sin \alpha \times \sin b$. Applying a similar argument to triangle $\triangle BDC$ we get $\sin d = \sin \beta \times \sin a$.

Thus, $\sin \alpha \times \sin b = \sin \beta \times \sin a$, which is more memorable in the form

$$\frac{\sin \alpha}{\sin a} = \frac{\sin \beta}{\sin b}.$$

Then, by a similar argument involving the perpendicular from A to BC, we obtain the following result.

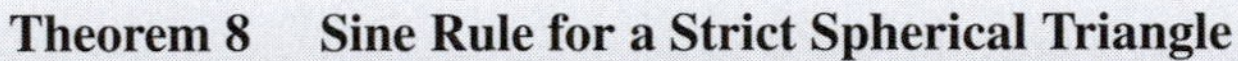

Theorem 8 Sine Rule for a Strict Spherical Triangle
Let $\triangle ABC$ be a strict triangle on S^2, in which the sides AB, BC and CA have lengths c, a and b, respectively, and the angles at A, B and C are α, β and γ respectively. Then

$$\frac{\sin \alpha}{\sin a} = \frac{\sin \beta}{\sin b} = \frac{\sin \gamma}{\sin c}.$$

We can use an argument similar to that in the proof of Pythagoras' Theorem to find further results.

Theorem 6

Let $\triangle ABC$ be a strict triangle on S^2. We may rotate S^2 so that C is the North Pole $N(0, 0, 1)$, A lies on the Greenwich meridian and has coordinates $(\sin b, 0, \cos b)$, and B has coordinates $(\cos \gamma \sin a, \sin \gamma \sin a, \cos a)$.

We then apply the rotation $R(Y, -b)$ to S^2. This maps A to C, C to some point C', and B to some point B'. As we saw earlier, the coordinates of B' are $(-\cos \alpha \sin c, \sin \alpha \sin c, \cos c)$.

In the proof of Theorem 6

But we also have as coordinates of B'

$$R(Y,-b)(B) = \begin{pmatrix} \cos b & 0 & -\sin b \\ 0 & 1 & 0 \\ \sin b & 0 & \cos b \end{pmatrix} \begin{pmatrix} \cos\gamma \sin a \\ \sin\gamma \sin a \\ \cos a \end{pmatrix}$$

$$= \begin{pmatrix} \cos b \cos\gamma \sin a - \sin b \cos a \\ \sin\gamma \sin a \\ \sin b \cos\gamma \sin a + \cos b \cos a \end{pmatrix}.$$

Equating the third coordinates in the two expressions for B' we get

$$\cos c = \sin b \cos\gamma \sin a + \cos b \cos a.$$

Theorem 9 Cosine Rule for Sides of a Strict Spherical Triangle
Let $\triangle ABC$ be a strict triangle on S^2, in which the sides AB, BC and CA have lengths c, a and b, respectively, and the angles at A, B and C are α, β and γ, respectively. Then

$$\cos c = \cos a \cos b + \sin a \sin b \cos\gamma.$$

Notice that this formula reduces to Pythagoras' Theorem when γ is a right-angle.

Example 3 Let $\triangle ABC$ be a strict isosceles triangle on S^2, in which the sides AB, BC and CA have lengths c, a and a, respectively, and the angles at A, B and C are α, α and γ, respectively. Prove that

(a) $\cos c = \cos^2 a + \sin^2 a \cos\gamma$; (b) $\cos\alpha = \dfrac{\tan\left(\frac{1}{2}c\right)}{\tan a}$.

Solution Putting $b = a$ in the conclusion of Theorem 9, we obtain the formula (a).

Next, by applying Theorem 9 to find $\cos a$, we get

$$\cos a = \cos a \cos c + \sin a \sin c \cos\alpha.$$

We may then rearrange this formula to get

$$\cos\alpha = \frac{\cos a(1-\cos c)}{\sin a \sin c}$$

$$= \frac{\cos a \cdot 2\sin^2\left(\frac{1}{2}c\right)}{\sin a \cdot 2\sin\left(\frac{1}{2}c\right)\cos\left(\frac{1}{2}c\right)} = \frac{\tan\left(\frac{1}{2}c\right)}{\tan a},$$

as required. □

Here we use the identities

$$\cos 2x = 1 - 2\sin^2 x$$

and

$$\sin 2x = 2\sin x \cos x$$

for $x \in \mathbb{R}$.

Problem 8 Let $\triangle ABC$ be a strict equilateral triangle on S^2, in which the sides have length a and the angles are equal to α. Prove that

$$\cos\alpha = \frac{\cos a}{1+\cos a},$$

We can now use any dual triangle $\Delta A'B'C'$ of ΔABC to prove a new trigonometric result. The dual triangle has sides of length $\pi - \alpha, \pi - \beta$ and $\pi - \gamma$, and has angles of magnitude $\pi - a, \pi - b$ and $\pi - c$. By applying Theorem 9 to the dual triangle and using the formulas $\sin(\pi - x) = \sin x$ and $\cos(\pi - x) = -\cos x$, we get

You met dual triangles in Subsection 7.3.2.

$$-\cos\gamma = \cos\alpha\cos\beta - \sin\alpha\sin\beta\cos c;$$

and we can reformulate this formula as follows.

Theorem 10 Cosine Rule for Angles of a Strict Spherical Triangle
Let ΔABC be a strict triangle on S^2, in which the sides AB, BC and CA have lengths c, a and b, respectively, and the angles at A, B and C are α, β and γ, respectively. Then

$$\cos\gamma = \sin\alpha\sin\beta\cos c - \cos\alpha\cos\beta.$$

7.4 Spherical Geometry and the Extended Complex Plane

Earlier we studied the Riemann sphere and stereographic projection of the sphere onto the extended complex plane $\hat{\mathbb{C}} = \mathbb{C} \cup \{\infty\}$.

Subsection 5.2.4

Here we denote the Riemann sphere by S^2 rather than $\mathbb{S}$.

7.4.1 Stereographic Projection

In this section we regard stereographic projection as a mapping π of the sphere S^2 onto an extended plane $\mathbb{R}^2 \cup \{\infty\}$, under which the point (X, Y, Z) maps onto the point $\pi(X, Y, Z) = \left(\frac{X}{1-Z}, \frac{Y}{1-Z}\right)$; the inverse mapping is given by

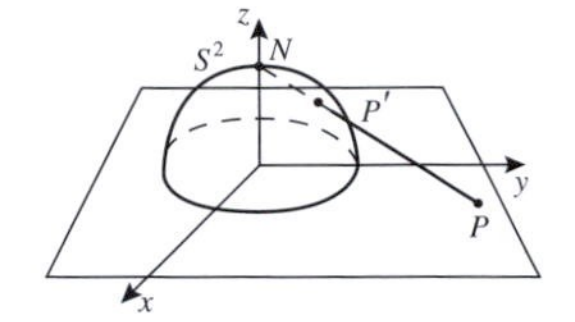

$$\pi^{-1}(x, y) = \left(\frac{2x}{x^2 + y^2 + 1}, \frac{2y}{x^2 + y^2 + 1}, \frac{x^2 + y^2 - 1}{x^2 + y^2 + 1}\right).$$

Example 1 Let C denote a great circle on S^2 that is the intersection of S^2 with the plane $aX + bY + cZ = 0$, where $a^2 + b^2 + c^2 = 1, c \neq 0$.

Since $c \neq 0$, C does not pass through the North Pole N of S^2.

(a) Determine the point N' that is the image of N under reflection in C.
(b) Determine the images of N' and C in $\mathbb{R}^2$ under stereographic projection π.
(c) Verify that $\pi(N')$ is the centre of the circle $\pi(C)$.

Solution

(a) The point N' has coordinates

We follow the approach of Subsection 7.2.2.

$$\begin{aligned}&(0, 0, 1) - 2\{(0, 0, 1) \cdot (a, b, c)\}(a, b, c)\\ &\quad = (0, 0, 1) - 2c(a, b, c) = \left(-2ac, -2bc, 1 - 2c^2\right).\end{aligned}$$

(b) From the formula for π, the image of N' under π is

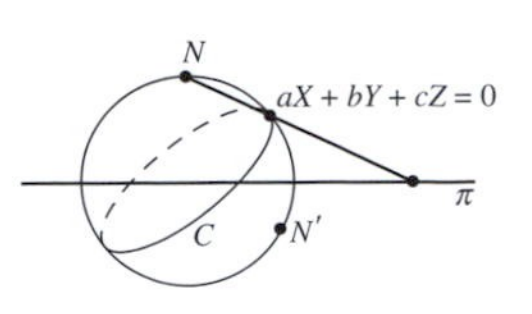

$$\pi(-2ac, -2bc, 1-2c^2) = \left(\frac{-2ac}{1-(1-2c^2)}, \frac{-2bc}{1-(1-2c^2)}\right)$$
$$= \left(-\frac{a}{c}, -\frac{b}{c}\right).$$

Here we use the fact $c \neq 0$.

From the formula for π^{-1} the image $\pi(C)$ of C has equation

$$a\frac{2x}{x^2+y^2+1} + b\frac{2y}{x^2+y^2+1} + c\frac{x^2+y^2-1}{x^2+y^2+1} = 0;$$

we may rewrite this in the form

$$2ax + 2by + c\left(x^2+y^2-1\right) = 0,$$

or

$$x^2 + y^2 + 2\frac{a}{c}x + 2\frac{b}{c}y - 1 = 0.$$

Again we use the fact $c \neq 0$.

(c) Since the equation of the circle $\pi(C)$ may be written in the form

$$\left(x+\frac{a}{c}\right)^2 + \left(y+\frac{b}{c}\right)^2 = 1 + \frac{a^2+b^2}{c^2} = \frac{1}{c^2},$$

the centre of $\pi(C)$ is the point $\left(-\frac{a}{c}, -\frac{b}{c}\right)$, that is, the point $\pi(N')$. □

Problem 1 Let $P = (x, y)$ be a point in $\mathbb{R}^2$ other than the origin. Let P' be the point on S^2 that corresponds to P under stereographic projection, P'' the reflection of P' in the (X, Y)-plane, and Q the point in $\mathbb{R}^2$ that corresponds to P'' under stereographic projection.

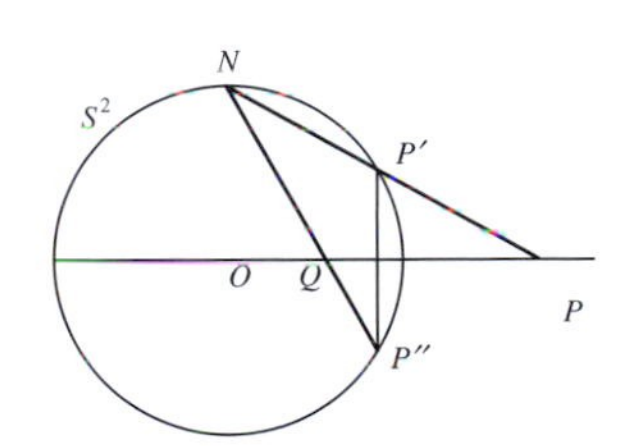

(a) Determine the coordinates of P', P'' and Q.
(b) Prove that P and Q are inverse points with respect to the unit circle $x^2 + y^2 = 1$ in $\mathbb{R}^2$.

There is a simple and useful expression for the image under stereographic projection of a point on S^2 with given colatitude and longitude.

Example 2 The point P' on S^2 with colatitude θ and longitude ϕ has coordinates $(\cos\phi\sin\theta, \sin\phi\sin\theta, \cos\theta)$. Prove that the image P of P' under stereographic projection has coordinates $\left(\frac{\cos\phi}{\tan\left(\frac{1}{2}\theta\right)}, \frac{\sin\phi}{\tan\left(\frac{1}{2}\theta\right)}\right)$.

Recall that the latitude of P is $\theta' = \frac{\pi}{2} - \theta$.

Solution From the formula for π, the image P of P' under π is the point

$$\left(\frac{\cos\phi\sin\theta}{1-\cos\theta}, \frac{\sin\phi\sin\theta}{1-\cos\theta}\right)$$

$$= \left(\frac{\cos\phi\cdot 2\sin\left(\frac{1}{2}\theta\right)\cos\left(\frac{1}{2}\theta\right)}{2\sin^2\left(\frac{1}{2}\theta\right)}, \frac{\sin\phi\cdot 2\sin\left(\frac{1}{2}\theta\right)\cos\left(\frac{1}{2}\theta\right)}{2\sin^2\left(\frac{1}{2}\theta\right)}\right)$$

$$= \left(\frac{\cos\phi}{\tan\left(\frac{1}{2}\theta\right)}, \frac{\sin\phi}{\tan\left(\frac{1}{2}\theta\right)}\right),$$

as required. □

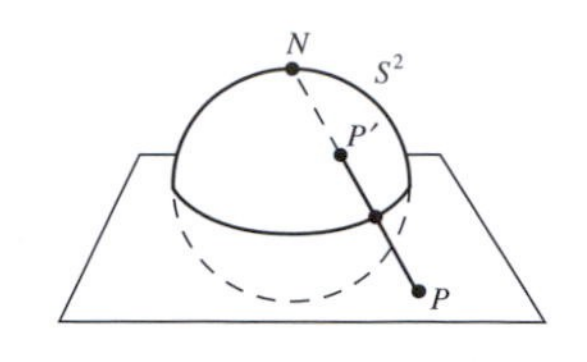

Next, we obtain a neat formula relating the distance on S^2 between the South Pole S and a point P' on S^2 to the distance of the corresponding point $P = \pi(P')$ from O in $\mathbb{R}^2$.

Let P be a point $(x, 0)$ in $\mathbb{R}^2, x > 0$. The corresponding point $P' = \pi^{-1}(P)$ on S^2 has coordinates $\left(\frac{2x}{1+x^2}, 0, \frac{x^2-1}{x^2+1}\right)$; it follows from the definition of spherical polar coordinates on S^2 as $(\cos\phi\sin\theta, \sin\phi\sin\theta, \cos\theta)$, where θ is colatitude, that the colatitude θ of P' satisfies the equation

$$\tan\theta = \frac{2x}{x^2-1}. \tag{1}$$

Then if the spherical distance of P' from S is d, the colatitude θ of P' also satisfies the equation $\theta + d = \pi$, so that

$$\tan\theta = -\tan d. \tag{2}$$

Then, if we denote by T the quantity $\tan\left(\frac{1}{2}d\right)$, it follows from equations (1) and (2) that $\frac{2x}{x^2-1} = -\frac{2T}{1-T^2}$. We can write this equation in the form $(T-x)(1+xT) = 0$; and, since both x and T are positive it follows that $T = x$. In other words $\tan^{-1}x = \tan^{-1}T = \frac{1}{2}d$, so that $d = 2\tan^{-1}x$.

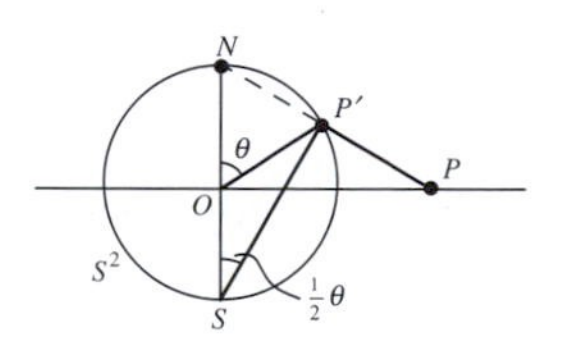

Since a rotation of S^2 and $\mathbb{R}^2$ about the Z-axis does not change any of the distances in the above argument, we deduce the following result.

Theorem 1 Let a point P' on S^2 map under stereographic projection to a point P in $\mathbb{R}^2$ with coordinates $(r\cos\theta, r\sin\theta)$. Then the spherical distance of P' from the South Pole S is $2\tan^{-1}(r)$.

Here Q does not represent colatitude.

Problem 2 Let P and Q be the points $(2, 2)$ and $(3, 3)$ in $\mathbb{R}^2$, and P' and Q' the corresponding points on S^2 under the mapping π^{-1}. Use Theorem 1 to determine the spherical distance $P'Q'$.

7.4.2 Conjugate Transformations

We have seen a correspondence between points of the sphere S^2 and points of the extended plane. We now investigate the correspondence between spherical transformations and Möbius transformations.

Subsection 7.4.1

Earlier we saw that a spherical transformation is the composition of at most three reflections in great circles, and that the composition of any two reflections is a rotation. Here we shall see that reflection in a great circle corresponds to inversion in the image of that circle under stereographic projection, and that a rotation of S^2 corresponds to a particular kind of Möbius transformation. Thus we have two equivalent descriptions of spherical transformations, one on the sphere using 3×3 matrices, and one in the extended plane using Möbius transformations.

Theorems 5 and 4 of Subsection 7.2.2

Firstly, we establish the following pleasant result.

Theorem 2 Let P' and Q' be points on S^2 that are mirror images under reflection in a great circle C', and let the images of P', Q' and C' under stereographic projection π be P, Q and C, respectively. Then P and Q are inverse points with respect to C.

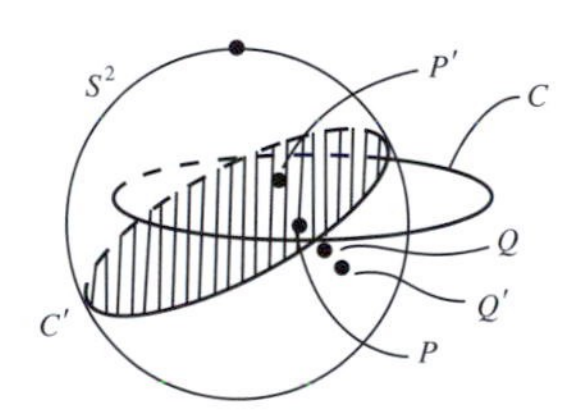

Proof The great circle lies in a plane, Π say, through the origin. Π consists of those points in $\mathbb{R}^3$ that are equidistant from the points P' and Q'. So any sphere through P' and Q' has its centre somewhere in Π, and so any circle on S^2 through P' and Q' also has its centre in Π. Thus any circle C'' on S^2 through P' and Q' meets the great circle C' at right angles.

Under stereographic projection π, the circles C' and C'' map to generalized circles; since stereographic projection is angle-preserving, these images meet at right angles.

Now let C^* be any generalized circle through P and Q. The image of C^* under π^{-1} is a circle on S^2 through P' and Q'; so $\pi^{-1}(C^*)$ is some circle C'' on S^2 that meets C at right angles. Then, since stereographic projection is angle-preserving, it follows that C^* meets C at right angles.

But C^* is any circle through P and Q, so any circle through P and Q meets C at right angles. It follows that P and Q are inverse points with respect to C, as required. ■

Here we are using Theorem 4 of Subsection 5.5.2.

Notice that, if C_1' and C_2' are a great circle and a little circle on S^2, and C_1 and C_2 are their images in the extended plane under π, then there exists a

Möbius transformation that maps C_1 onto C_2, but there cannot be a rotation of S^2 that maps C_1' onto C_2'. Hence there are Möbius transformations of the extended plane that do not correspond to rotations of the sphere.

Now, inversion in a generalized circle in the extended plane is necessarily a Möbius transformation composed with a complex conjugation. It follows from Theorem 2 that any spherical transformation corresponds to a Möbius transformation, composed (if need be) with a complex conjugation.

By Theorem 10, Subsection 5.3.7

So each rotation R of S^2 corresponds to some Möbius transformation of the extended plane by means of stereographic projection π. If P is a point in the extended plane, it follows that $\pi^{-1}(P)$ is a point on S^2; $R\pi^{-1}(P)$ is the image of that point after the rotation, and $\pi R\pi^{-1}(P)$ is the projection of that image onto the extended plane.

For both are orientation-preserving mappings.

So, $\pi R\pi^{-1}$ is a mapping of the extended plane to itself. We therefore make the following definition.

Definition Let R denote a rotation of S^2, and π stereographic projection. Then the transformation $\pi R\pi^{-1}$ of the extended plane is called the **conjugate transformation** of the rotation R.

It follows from the above discussion that the image of the group of rotations of S^2 under π is a group of Möbius transformations.

Lemma The conjugate transformation of a rotation of S^2 is a Möbius transformation of the extended plane.

This was proved in the earlier discussion.

Indeed it can be proved that the mapping $R \mapsto \pi R\pi^{-1}$ maps the group of rotations of S^2 isomorphically onto a group of Möbius transformations of the extended plane, but that it is not the whole group of Möbius transformations. We now characterize the Möbius transformations that are conjugate to rotations of S^2.

We omit a proof of this fact.

We saw this in the discussion above.

Theorem 3 A Möbius transformation M that is conjugate to a rotation of S^2 is of the form $M(z) = \frac{az+b}{-\bar{b}z+\bar{a}}$, where $a, b \in \mathbb{C}$.

Proof There is one family of rotations of S^2 that we can easily relate to Möbius transformations, namely rotations around the Z-axis. Their conjugate transformations are simply rotations about the origin through the same angle. Our strategy will be to use reflections of S^2 to reduce a given rotation of S^2 to one about the Z-axis.

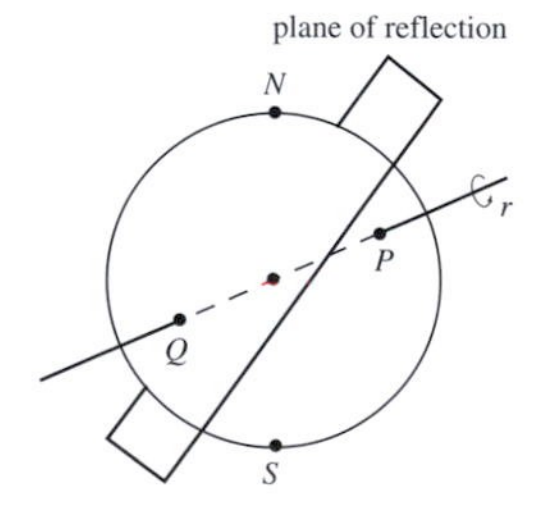

Let R denote a rotation of S^2 about the (diametrically opposite) points P and Q of S^2, through an angle θ, and let r denote the reflection in a suitable great circle of S^2 that maps P to the North Pole N; so that $r(P) = N$. Then since r is a reflection, it is its own inverse, so that $r(N) = P$. Similarly $r(Q) = S$, the South Pole, and $r(S) = Q$.

So we shall often use r and r^{-1} interchangeably below.

The spherical transformation $r^{-1}Rr$ is a direct isometry of S^2, and so is a rotation of S^2. Since

$$\begin{aligned}\left(r^{-1}Rr\right)(N) &= \left(r^{-1}R\right)(r(N))\\ &= \left(r^{-1}R\right)(P)\\ &= r^{-1}R(P) = r^{-1}(P) = N,\end{aligned}$$

the point N is a fixed point of the rotation; similarly S is also a fixed point. So $r^{-1}Rr$ is a rotation of S^2 about the Z-axis.

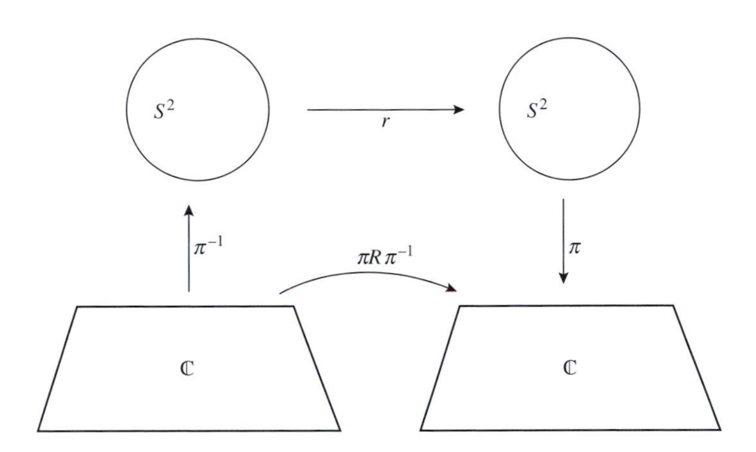

It follows that the transformation $\pi(r^{-1}Rr)\pi^{-1}$ is a rotation of the (x, y)-plane about the origin. Denoting this by M, it follows that $M(z) = e^{i\theta}z$, for some real number θ.

Now

$$\begin{aligned}M &= \pi\left(r^{-1}Rr\right)\pi^{-1}\\ &= \pi r^{-1}\left(\pi^{-1}\pi\right)R\left(\pi^{-1}\pi\right)r\pi^{-1}\\ &= \left(\pi r\pi^{-1}\right)\left(\pi R\pi^{-1}\right)\left(\pi r\pi^{-1}\right);\end{aligned}$$

For $\pi^{-1}\pi$ is the identity map of S^2 to itself.

so that, since we know M, in order to know $\pi R\pi^{-1}$ it is now sufficient to find $\pi r\pi^{-1}$.

Since N and P map to each other under the reflection r in a suitable great circle of S^2, the mapping $\pi r\pi^{-1}$ must be inversion in a circle in the plane.

By Theorem 2 and $r^{-1} = r$.

Now the great circle of S^2 is the intersection of the sphere $X^2+Y^2+Z^2 = 1$ with some plane $aX + bY + cZ = 0$; we may assume that $a^2 + b^2 + c^2 = 1$. Since the plane cannot be vertical, $c \neq 0$.

This assumption will simplify our calculations.

By applying the formula for π^{-1} to the equation of the plane, we see that the image of the great circle is a circle C in the (x, y)-plane with equation

$$2ax + 2by + c\left(x^2 + y^2 - 1\right) = 0,$$

or

$$x^2 + y^2 + 2\frac{a}{c}x + 2\frac{b}{c}y - 1 = 0.$$

The circle C has centre $\left(-\frac{a}{c}, -\frac{b}{c}\right)$, and the square of its radius is

For, $a^2 + b^2 + c^2 = 1$

$$1 + \left(\frac{a}{c}\right)^2 + \left(\frac{b}{c}\right)^2 = \frac{1}{c^2}.$$

Next we express things in terms of complex numbers. Let $\alpha = -\frac{a}{c} - i\frac{b}{c}$, so that $\alpha\bar{\alpha} = |\alpha|^2 = \frac{a^2+b^2}{c^2}$. Now inversion in C is a mapping of the form

α is the centre of the circle in the plane.

$$\begin{aligned} z \mapsto \; & \frac{(1/c)^2}{\overline{z-\alpha}} + \alpha \\ = \; & \frac{\alpha\bar{z} - |\alpha|^2 + (1/c)^2}{\bar{z} - \bar{\alpha}} \\ = \; & \frac{\alpha\bar{z} + 1}{\bar{z} - \bar{\alpha}}, \end{aligned}$$

You met this formula in Theorem 3, Subsection 5.2.1.

For, $-|\alpha|^2 + (1/c)^2 = 1$

so that

$$\pi r \pi^{-1}(z) = \frac{\alpha\bar{z} + 1}{\bar{z} - \bar{\alpha}}.$$

Then

$$M(\pi r \pi^{-1})(z) = e^{i\theta} \cdot \frac{\alpha\bar{z} + 1}{\bar{z} - \bar{\alpha}},$$

so that

$$\begin{aligned} \pi R \pi^{-1}(z) &= \frac{\alpha\left(e^{-i\theta} \cdot \frac{\bar{\alpha}z+1}{z-\alpha}\right) + 1}{e^{-i\theta} \cdot \frac{\bar{\alpha}z+1}{z-\alpha} - \bar{\alpha}} \\ &= \frac{\alpha e^{-i\theta}(\bar{\alpha}z + 1) + (z - \alpha)}{e^{-i\theta}(\bar{\alpha}z + 1) - \bar{\alpha}(z - \alpha)} \\ &= \frac{z\left(|\alpha|^2 e^{-i\theta} + 1\right) + (\alpha e^{-i\theta} - \alpha)}{z(\bar{\alpha}e^{-i\theta} - \bar{\alpha}) + (e^{-i\theta} + |\alpha|^2)} \\ &= \frac{z\left(|\alpha|^2 e^{-i\theta/2} + e^{i\theta/2}\right) + \left(\alpha e^{-i\theta/2} - \alpha e^{i\theta/2}\right)}{z\left(\bar{\alpha}e^{-i\theta/2} - \bar{\alpha}e^{i\theta/2}\right) + \left(e^{-i\theta/2} + |\alpha|^2 e^{i\theta/2}\right)}; \end{aligned}$$

thus $\pi R \pi^{-1}(z) = \frac{az+b}{-\bar{b}z+\bar{a}}$, where $a = |\alpha|^2 e^{-i\theta/2} + e^{i\theta/2}$ and $b = \alpha e^{-i\theta/2} - \alpha e^{i\theta/2}$, as required. ■

Remark

Notice that the Möbius transformations $M(z) = \frac{az+b}{\bar{b}z+\bar{a}}$ conjugate to a rotation of S^2 are rather like the Möbius transformations representing direct hyperbolic transformations $M(z) = \frac{az+b}{\bar{b}z+\bar{a}}$, with $|b| < |a|$.

You met these Möbius transformations in Theorem 2, Subsection 6.2.1.

Problem 3 The two fixed points of a rotation of S^2 are distinct from N and S. Show that they map under stereographic projection to points in $\mathbb{C}$ of the form s and $-1/\bar{s}$.

Example 3 Prove that a Möbius transformation M that is conjugate to a rotation $R(Y,\beta)$ of S^2 is of the form $M(z) = \frac{az-b}{bz+a}$ where a, b are real.

Solution A rotation $R(Y,\beta)$ of S^2 fixes the points $(0,1,0)$ and $(0,-1,0)$, which map under stereographic projection to the points i and $-i$, respectively.

By Theorem 3, a conjugate Möbius transformation must be of the form

$$M(z) = \frac{cz+d}{-\bar{d}z+\bar{c}};$$

We use c and d to avoid using the letters a and b for several different things.

this fixes i and $-i$ if

$$i = \frac{ci+d}{-\bar{d}i+\bar{c}} \quad \text{and} \quad -i = \frac{-ci+d}{\bar{d}i+\bar{c}},$$

which we can rewrite in the form

$$\bar{d}+\bar{c}i = ci+d \quad \text{and} \quad \bar{d}-\bar{c}i = -ci+d.$$

Adding these equations gives $2\bar{d} = 2d$, so that d is real, and subtracting the equations gives $2\bar{c}i = 2ci$, so that c is real.

It follows that M has the desired form $M(z) = \frac{az-b}{bz+a}$, with a in place of c and $-b$ in place of d. □

Problem 4 Prove that a Möbius transformation M that is conjugate to a rotation $R(X,\alpha)$ of S^2 is of the form $M(z) = \frac{az+ib}{ibz+a}$, where a, b are real.

We end with some interesting observations concerning Möbius transformations. A *fixed point* of a Möbius transformation $M(z) = \frac{az+b}{cz+d}$, $ad-bc \neq 0$, is a point z for which $M(z) = z$, that is, $\frac{az+b}{cz+d} = z$ or $cz^2 + (d-a)z - b = 0$. Since a quadratic equation can have at most two roots, it follows that a Möbius transformation can have at most two fixed points.

Problem 5 Determine the fixed points in $\mathbb{C}$ and in $\hat{\mathbb{C}}$ of the following Möbius transformations:

$$z \mapsto z+1; \quad z \mapsto -\frac{4}{z+4}; \quad z \mapsto -\frac{4}{z+5}.$$

Example 4 Prove that a Möbius transformation M with distinct fixed points a and b in $\mathbb{C}$ maps the family $\mathscr{A}$ of Apollonian circles defined by the point circles a and b to itself.

Solution Let $\mathcal{B}$ be the family of generalized circles through a and b.

Since M has fixed points a and b, it maps the family $\mathcal{B}$ onto itself. Also, since a Möbius transformation preserves angles, M thus maps the family of generalized circles that is orthogonal to those in $\mathcal{B}$ to itself. This family is the family $\mathcal{A}$ of Apollonian circles defined by the point circles a and b. □

A Möbius transformation maps generalized circles onto generalized circles.

By Theorem 3, Subsection 5.5.1

Example 5 Prove that a Möbius transformation M with distinct fixed points a and b in $\mathbb{C}$ *either* maps each circle in the family $\mathcal{A}$ of Apollonian circles defined by the point circles a and b to itself *or* it maps no circle in $\mathcal{A}$ to itself.

Solution By a suitable Möbius transformation f, we can map the point a to 0 and b to ∞. Then f maps the family $\mathcal{A}$ of Apollonian circles defined by the point circles a and b to the family $\mathcal{B}$ of concentric circles with centre 0.

Then $f \circ M \circ f^{-1}$ is a Möbius transformation, and it maps 0 to 0 and ∞ to ∞. It follows that $f \circ M \circ f^{-1}$ is of the form $z \mapsto Kz$, for some non-zero complex number K.

It is easy to verify that a Möbius transformation of $\hat{\mathbb{C}}$ that has 0 and ∞ as its fixed points is necessarily of the form $z \mapsto Kz$.

If $|K| = 1$, then $f \circ M \circ f^{-1}$ maps each concentric circle in $\mathcal{B}$ to itself; but if $|K| \neq 1$ then $f \circ M \circ f^{-1}$ maps no circle in $\mathcal{B}$ to itself.

If we then apply the mapping f^{-1} to return to the original situation, the required result follows at once. □

7.4.3 Coaxal Circles

Stereographic projection enables us to interpret the theory of coaxal circles, including all its special cases, very elegantly on the sphere S^2.

You met coaxal circles in Subsection 5.5.2.

Earlier you saw that, if C is the circle cut out on S^2 by the plane with equation $aX + bY + cZ + d = 0$, then stereographic projection maps C to:

In the proof of Theorem 7 of Subsection 5.2.4

(a) the circle with equation

$$x^2 + y^2 + 2\frac{a}{c+d}x + 2\frac{b}{c+d}y - \frac{c-d}{c+d} = 0, \quad \text{if } c \neq -d;$$

(b) the straight line with equation

$$ax + by - \frac{c-d}{2} = 0, \quad \text{if } c = -d.$$

If $c = -d$, the circle C passes through N.

Problem 6 Prove that a circle in $\mathbb{R}^2$ with equation

$$x^2 + y^2 + 2\alpha x + 2\beta y + \gamma = 0$$

corresponds under stereographic projection to the circle cut out on S^2 by the plane with equation

$$2\alpha X + 2\beta Y + (1-\gamma)Z + (1+\gamma) = 0.$$

Definition The circle in $\mathbb{R}^2$ with equation

$$x^2 + y^2 + 2\alpha x + 2\beta y + \gamma = 0$$

and the plane with equation

$$2\alpha X + 2\beta Y + (1 - \gamma)Z + (1 + \gamma) = 0$$

are said to be **associated**.

That is, they correspond to each other under stereographic mapping.

We now use the idea of associated planes and circles to revisit the theory of coaxal circles.

Let C_1 and C_2 be distinct circles in $\mathbb{R}^2$ with equations

We allow C_1 and C_2 to be point circles.

$$x^2 + y^2 + 2\alpha_1 x + 2\beta_1 y + \gamma_1 = 0$$

and

$$x^2 + y^2 + 2\alpha_2 x + 2\beta_2 y + \gamma_2 = 0,$$

respectively; then their corresponding associated planes π_1 and π_2 have equations

$$2\alpha_1 X + 2\beta_1 Y + (1 - \gamma_1)Z + (1 + \gamma_1) = 0$$

and

$$2\alpha_2 X + 2\beta_2 Y + (1 - \gamma_2)Z + (1 + \gamma_2) = 0.$$

Now, either π_1 and π_2 are parallel or they are not parallel; they cannot be the same plane, since the associated circles are distinct.

First, suppose that π_1 and π_2 are not parallel, so that they have a common line, ℓ say.

For arbitrary p and q, $p \neq -q$, let C_{pq} be the generalized circle with equation

$$\frac{p}{p+q}\left(x^2 + y^2 + 2\alpha_1 x + 2\beta_1 y + \gamma_1\right) + \frac{q}{p+q}\left(x^2 + y^2 + 2\alpha_2 x + 2\beta_2 y + \gamma_2\right) = 0,$$

and let π_{pq} be its associated plane, with equation

$$\frac{p}{p+q}\left(2\alpha_1 X + 2\beta_1 Y + (1 - \gamma_1) Z + (1 + \gamma_1)\right) + \frac{q}{p+q}\left(2\alpha_2 X + 2\beta_2 Y + (1 - \gamma_2) Z + (1 + \gamma_2)\right) = 0.$$

The associated plane π_{pq} passes through the line ℓ; so, as p and q vary, π_{pq} gives a family of planes through ℓ. There are three cases to consider.

Case 1: The line ℓ does not meet S^2.

Here the family of planes cuts out a family of circles in S^2 consisting of two point circles P and Q and a set of disjoint circles in between these.

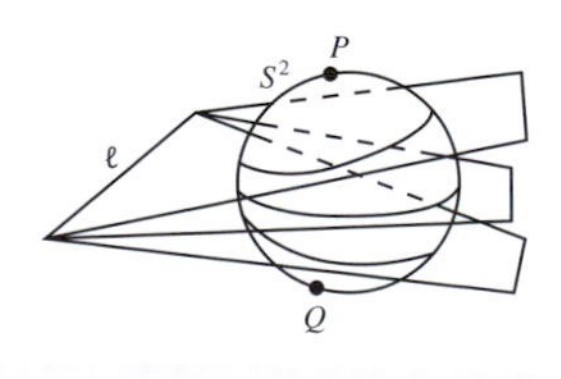

Under stereographic projection from N, the associated circles in the plane form an Apollonian family: a family of coaxal circles with two point circles,

Case 2: The line ℓ touches S^2 (at the point P, say).
Here, every plane of the family meets S^2. The plane tangent to the sphere at the point P meets it in a single point, P. All the other planes cut S^2 in a circle. All these circles have a common tangent at P, which is the line ℓ.

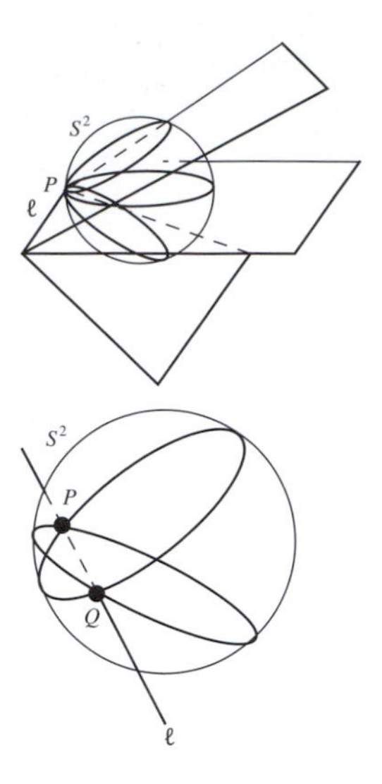

Under stereographic projection from N, the associated circles in the plane form a family of coaxal circles having a common tangent. If N and P are distinct, the image circles are 'true' circles except for one straight line; if N coincides with P, the images are parallel lines.

Case 3: The line ℓ cuts S^2 in two points, say P and Q.
Here, every plane of the family meets S^2. Each plane cuts the sphere in a circle passing through the points P and Q.

Under stereographic projection from N, the associated circles in the plane form a family of coaxal circles with two common points. If N, P and Q are distinct, the image circles are 'true' circles except for one straight line; if N coincides with either P or Q, say P, the images are straight lines passing through the image of Q.

On the other hand, if the associated planes π_1 and π_2 are parallel then the family of planes π_{pq} consists of the planes parallel to π_1 and π_2, and the circles they cut out on the sphere consist of two point circles and a set of disjoint circles in between these.

This is similar to the situation in Case 1 above.

7.5 Planar Maps

We use maps of the Earth on a daily basis: for example. street maps of towns, road maps of countries, geological maps of continents, and for land, sea and air travel. Most of our maps are on flat sheets of paper, rather than globes; and we have to reduce the actual dimensions by a convenient *scale factor*, such as 1:12,000,000 for a continent, 1:100,000 for a map of a large city, or 1:25,000 for a detailed neighbourhood map.

Here, scale factor = length on map divided by length on Earth's surface.

We will assume in this section that the Earth is a sphere. In fact, the shape of the Earth is closer to that of an *oblate spheroid*, a sphere flattened near each pole and with a bulge around the equator. This bulge results from the rotation of the Earth, and causes its equatorial diameter to be over 26 miles (43 km) greater than its pole-to-pole diameter

Now an ideal planar map would represent shapes on the Earth's surface in such a way that relative distances, relative areas, shape, and angles are preserved under the operation of mapping. Unfortunately, no ideal map is possible!

For example, let ΔABC be a triangle on S^2, and let P, Q, R be the points on the planar map that correspond to A, B, C, respectively. Then the sides of ΔABC are line segments on S^2, that is, parts of great circles; and the sum of its angles exceeds π. The sides of ΔPQR are line segments in the plane, and the sum of its angles is π. So we cannot simultaneously have geodesics on S^2

Subsection 7.3.1, Theorem 2

mapping onto geodesics in the plane AND preserve angles. This shows that an ideal planar map cannot exist.

Recall that 'geodesics' are curves of shortest length between two points.

Furthermore, Euler showed in 1775 that it is not possible to map a portion of the Earth's surface onto a planar map in such a way that relative distances (that is, distances up to a scaling factor) are preserved.

Leonhard Euler (1707–1783) was a prolific Swiss analyst, who worked for Frederick the Great (Prussia) and Catherine the Great (Russia) even after becoming blind.

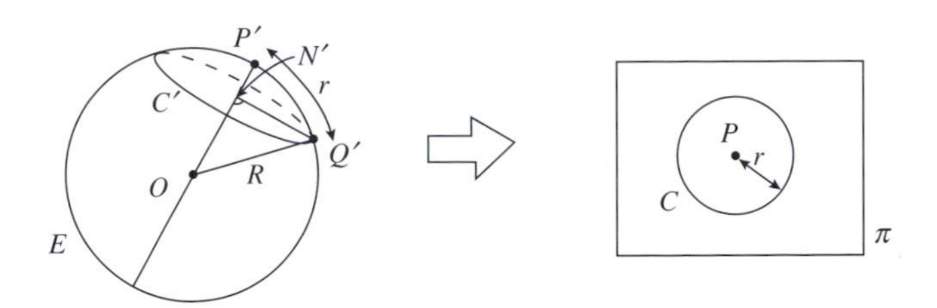

To see this, let E denote the surface of the Earth, a sphere with centre O and radius R; and let $P' \in E$ be the centre of a circle C' on E of radius r (we will suppose that $r < \frac{1}{2}\pi R$, so that C' lies in a hemisphere on E with P' as pole). Suppose that some ideal mapping represents the hemisphere onto a flat sheet, π say, and let P, C on π represent P', C' on E.

Here r is a distance measured along the surface.

Since the mapping scales all distances by the same scale factor, c say, all points on E at a distance r from P' map to points at a distance cr from P; hence C' is represented on π by a circle C of radius cr – so that the length of C is $2\pi cr$.

Next, let Q' be any point on C', and N' the foot of the perpendicular from Q' to OP'. Then $\angle P'OQ' = \frac{r}{R}$, so that $Q'N' = R\sin(\angle P'OQ') = R\sin\left(\frac{r}{R}\right)$. But the circle C' is a circle in $\mathbb{R}^3$ with radius $Q'N'$, so that its length is $2\pi R\sin\left(\frac{r}{R}\right)$. Hence its representation C on π must have length $c \times 2\pi R\sin\left(\frac{r}{R}\right)$.

We deduce from the last two paragraphs that

$$2\pi cr = 2\pi cR\sin\left(\frac{r}{R}\right),$$

so that

$$\frac{r}{R} = \sin\left(\frac{r}{R}\right).$$

This equality is impossible, since $\frac{\sin x}{x} < 1$ for all x with $0 < x < \frac{\pi}{2}$.

This is a standard result in Calculus.

So we have proved that it is not possible to represent a portion of the Earth's surface on a planar map in such a way that relative distances are preserved. It follows from this, too, that no (precisely accurate) planar map of a portion of E can have a fixed scale factor for the whole map.

In fact, there are three principal types of *map projections* for representing the spherical Earth, E, on a flat sheet of paper, π. In their work, cartographers often use the term *meridians* to denote lines of latitude, and *parallels* to denote lines of longitude: we will follow this convention here.

In addition there are various purely mathematical approaches, for special purposes.

Azimuthal Projections

Azimuthal projections of E onto π are ones where we project E onto a sheet π that is tangent to E at some point, often the South Pole S. We can think of the projection mapping in the following way: a light source placed at some point (the *centre* of the projection) on the perpendicular to π at the point of contact projects each point P' on E onto a corresponding point P on π. Wherever we place the source, meridians map to radial lines from the point of contact S, parallels to concentric circles with centre S, and directions from the point of contact are preserved.

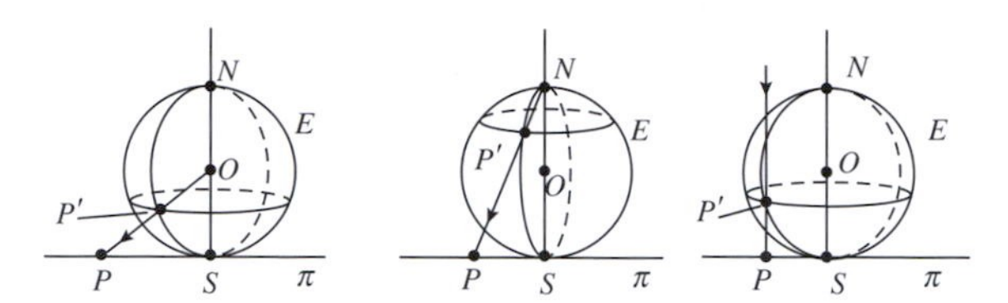

If the source is placed at the centre of E, the projection is called the *gnomic projection*, and we obtain a map of the lower hemisphere of E. Since any plane through O meets E in a great circle and π in a line, each line in π comes from a great circle on E. Hence geodesics on E are mapped onto straight lines on π.

By making the tangent point N, we could obtain a map of the upper hemisphere.

If the light source is placed at the North Pole N of E, we obtain a map of the whole of $E-\{N\}$ on π. This projection is called the *stereographic projection* of E onto π, and is the stereographic projection that you met earlier. It preserves angles, and so is called a *conformal* projection.

Subsection 5.2.4

If the light source is placed 'at infinity' along the direction *SON*, and we consider only the upper hemisphere or the lower hemisphere of E, then we obtain a map of one hemisphere of E onto π that is actually circular. This projection is called the *orthogonal* or *orthographic projection*. It preserves distances along parallels.

Cylindrical projections

Cylindrical projections of E onto π are ones where we project E onto a sheet π that is rolled in the form of a cylinder round E, touching it at its Equator; then we slit the sheet and flatten it.

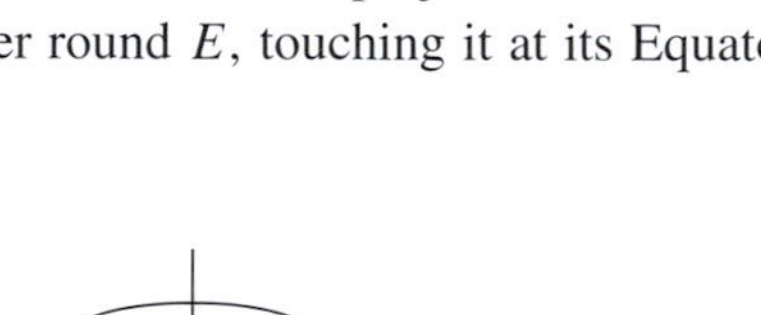

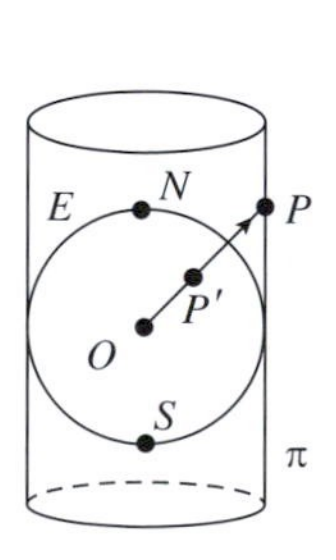

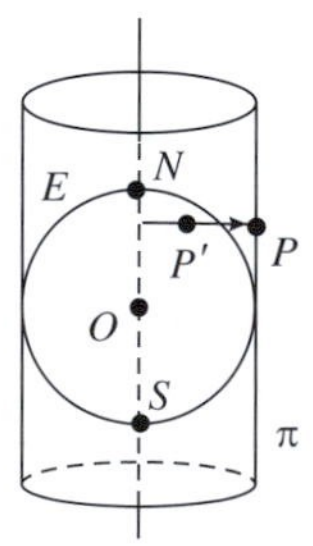

In one cylindrical projection, a light source placed at the centre O of E, the *centre* of the projection, projects each point P' on E onto a corresponding point P on π. Parallels map to horizontal circles that become horizontal lines when we unwrap the sheet, and meridians to vertical lines. The equator is mapped to itself, and areas near the North and South Poles become greatly enlarged.

Another cylindrical projection projects each point P' on E horizontally onto the nearest point P on π. Again, parallels map to horizontal circles that become horizontal lines when we unwrap the sheet, and meridians to vertical lines. This projection preserves the relative sizes of areas, so it is called an *equal area projection*.

The well-known *Mercator projection* is a cylindrical map where the parallels are horizontal and the meridians vertical, but the spacing of the parallels is chose cunningly so that *paths of constant compass bearing* (that is, paths making the same angle with every meridian they cross) are shown as straight lines. This projection is invaluable in navigation, since sailors and aircrew using it do not need to constantly change the compass bearing of the course they follow.

Gerardus Mercator (Latinized form of Gerard Kremer) (1512–1594) was a Flemish map maker who devised this projection in 1569, and introduced the term 'atlas'.

Conical projections

Conical projections of E onto π are ones where we project E onto a sheet π that is rolled in the form of a cone round E, touching it at some parallel; then we slit the sheet along a meridian and flatten it, taking on the shape of a sector of a circle.

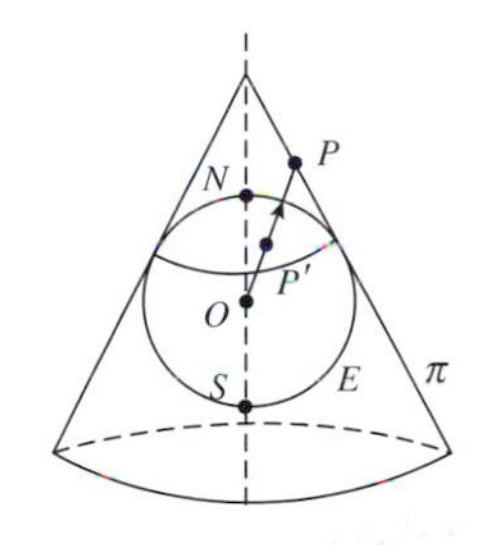

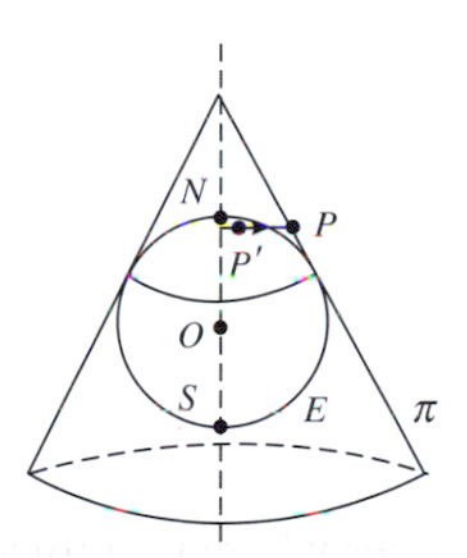

We may, for example, project each point P' on E to the corresponding point P on π radially out from the centre of E, or horizontally out from the axis of the cone to the nearest point on π. Parallels map to arcs of circles whose centres are at the vertex V of the flattened cone, and meridians to straight lines radially out from V. Conic projections are particularly useful for mapping large areas in the mid-latitudes that are wide from east to west (such as Canada).

There are many variations of these fundamental types, all designed for special purposes; but a full discussion is beyond the scope of this book. Map-making is a fascinating subject!

You should look at a good Atlas now to see the various types of projections that it uses.

7.6 Exercises

Section 7.1

1. (a) Determine the equation of the great circle on S^2 that passes through the points $\left(\frac{1}{\sqrt{6}}, -\frac{1}{\sqrt{6}}, \frac{2}{\sqrt{6}}\right)$ and $\left(-\frac{1}{\sqrt{6}}, \frac{2}{\sqrt{6}}, \frac{1}{\sqrt{6}}\right)$.
   (b) Determine the equation of the little circle on S^2 that passes through the points $\left(\frac{1}{\sqrt{6}}, \frac{-1}{\sqrt{6}}, \frac{2}{\sqrt{6}}\right)$, $(0, 0, 1)$ and $\left(\frac{-1}{\sqrt{6}}, \frac{2}{\sqrt{6}}, \frac{1}{\sqrt{6}}\right)$.
2. Determine the spherical polar representation of the point $\left(\frac{\sqrt{3}}{2\sqrt{2}}, -\frac{\sqrt{3}}{2\sqrt{2}}, -\frac{1}{2}\right)$ on S^2. Hence state the latitude, colatitude, and longitude of this point.
3. Determine the coordinates on S^2 of San Francisco (38° N, 123° W) and Brisbane (27° S, 153° E).

Section 7.2

1. Determine the images of the point $\left(\frac{3}{5\sqrt{2}}, \frac{4}{5\sqrt{2}}, -\frac{1}{\sqrt{2}}\right)$ under the mappings $R(X, \frac{\pi}{3})$, $R(Y, \frac{3\pi}{4})$ and $R(Z, \frac{5\pi}{4})$.
2. Determine the matrix $\mathbf{A}$ of a rotation $\mathbf{x} \mapsto \mathbf{Ax}$ of S^2 that maps $\left(\frac{3}{5\sqrt{2}}, \frac{4}{5\sqrt{2}}, -\frac{1}{\sqrt{2}}\right)$ to $\left(-\frac{4}{5\sqrt{2}}, -\frac{3}{5\sqrt{2}}, \frac{1}{\sqrt{2}}\right)$. Verify your answer by direct calculation.
3. (a) Determine the matrix $\mathbf{A}$ such that the mapping $r : \mathbf{x} \mapsto \mathbf{Ax}$ represents reflection of S^2 in the plane π with equation $x - 2y + z = 0$.
   (b) Hence determine the image under r of the point $P\left(\frac{1}{\sqrt{14}}, -\frac{2}{\sqrt{14}}, \frac{3}{\sqrt{14}}\right)$.
   (c) Prove that the midpoint of P and $r(P)$ lies in the plane π.
4. Let a triangle ΔPQR on S^2 have sides PQ of length r, QR of length p, and RP of length q, and angles α, β and γ at the vertices P, Q and R, respectively; let $\alpha = \beta$. Prove that $p = q$.

   This is the converse of Theorem 6 of Subsection 7.2.2.
5. Express the rotation $R(X, \frac{\pi}{4})$ of S^2 as the composition of two reflections of S^2 in great circles.

Section 7.3

1. Construct the following figures on S^2:
   (a) a spherical triangle of area $(2\pi - 0.1)$;
   (b) a lune of area $\pi/3$.
2. An equilateral strict spherical triangle has area $\frac{\pi}{12}$. Determine:
   (a) the magnitudes of its angles;
   (b) the lengths of its sides;
   (c) the area of a dual triangle.
3. Estimate the distance between Beijing (40° N, 117° E) and Lima (12° S, 77° W), taking the radius of the Earth as 4000 miles.
4. Let ΔABC be a triangle on S^2 in which the angle at C is a right angle. If a, b and c denote the lengths of BC, CA and AB, and α and β denote the angles $\angle CAB$ and $\angle CBA$, prove that:

(a) $\cos c = \cot\alpha \times \cot\beta$;
(b) $\cos\beta = \tan a \times \cot c = \cos b \times \sin\alpha$;
(c) $\sin a = \tan b \times \cot\beta = \sin\alpha \times \sin c$.

5. Let the spherical triangle ΔABC have sides AB, BC and CA of lengths 1, 0.6 and 1.3, respectively. Estimate the length of the perpendicular from B to CA.
6. Let the figure $ABCD$ on S^2 be a strict spherical 'square' – that is, have four sides of the same length r and four angles of the same magnitude α. Determine r in terms of α.

A 'strict' spherical square is one whose whole can be seen at one time after a suitable rotation of S^2.

7. Prove that the distance d on S^2 between the points with spherical polar coordinates $(\cos\phi_1 \sin\theta_1, \sin\phi_1 \sin\theta_1, \cos\theta_1)$ and $(\cos\phi_2 \sin\theta_2, \sin\phi_2 \sin\theta_2, \cos\theta_2)$ is given by.

$$\cos d = \sin\theta_1 \sin\theta_2 \cos(\phi_1 - \phi_2) + \cos\theta_1 \cos\theta_2.$$

Sometimes this formula can save you a bit of arithmetic.

Section 7.4

1. Let P and Q be the points $\left(\frac{1}{2}, \frac{1}{2}\right)$ and $\left(-\frac{1}{4}, -\frac{1}{4}\right)$ in $\mathbb{R}^2$, and P' and Q' be the corresponding points on S^2 under the mapping π^{-1}. Use Theorem 1 of Subsection 7.4.1 to determine the spherical distance $P'Q'$.
2. Determine the Möbius transformation that is conjugate to the rotation $R\left(Y, \frac{\pi}{4}\right)$ of S^2.
3. (a) Prove that a Möbius transformation that maps 0 to ∞ and ∞ to 0 is of the form $z \mapsto \frac{K}{z}$, for some non-zero complex number K.
   (b) Prove that a Möbius transformation M that maps a to b and b to a (where $a \neq b$) maps the family $\mathscr{A}$ of Apollonian circles defined by the point circles a and b to itself, and maps exactly one circle in $\mathscr{A}$ to itself.

Summary of Chapter 7

Introduction

The (Elliptic) Parallel Postulate Given any line ℓ and a point P not on ℓ, all lines through P meet ℓ. (That is, there are no lines through P that are parallel to ℓ.)

Section 7.1: Spherical Space

1. The **unit sphere** in $\mathbb{R}^3$ is $S^2 = \{(x, y, z) : x^2 + y^2 + z^2 = 1\}$. Each point (x, y, z) on S^2 has an **antipodal point** $(-x, -y, -z)$. The **North Pole** of S^2 is $N(0, 0, 1)$, and the **South Pole** is $S(0, 0, -1)$. The **equator** is the intersection of S^2 with the horizontal plane through the origin.
2. A **great circle** is the circle cut out on S^2 by a plane through the centre of S^2.

A **circle of longitude** is a great circle through N. The **Greenwich Meridian** is the part of the circle of longitude through $A\,(1, 0, 0)$ which has positive x-coordinates.

A **circle of latitude** is the curve of intersection of S^2 with a plane parallel to the equator. A **little circle** is the curve of intersection of S^2 with a plane that is not a great circle.

3. A **line** on S^2 joining two points P and Q, or a **line** PQ, is either arc with endpoints P and Q of a great circle through P and Q.
4. The **length** of a line on S^2 is the Euclidean length of the corresponding circular arc; the **distance** (or **spherical distance**) between two points P and Q on S^2 is the length of the shorter of the two lines PQ, and is often denoted simply as PQ.

   For any three points P, Q and R on S^2 that do not lie on a single great circle, the lines PQ, QR and RP form a **triangle** ΔPQR.
5. Great circles are **geodesics** on S^2; that is, curves of shortest distance between points.
6. The distance between any two points P and Q on S^2 is equal to the angle that they subtend at the centre of S^2. There is always a great circle joining any two points on S^2.

   Spherical distance possesses all the natural properties that a distance function should possess.
7. Any two lines on S^2 meet in two points.
8. Corresponding to each point P on S^2 the **great circle associated with P** is the great circle of points $\pi/2$ distant from P. Conversely, for each great circle there are two points of S^2 which play the role of its 'poles'.
9. For any two points P and Q on S^2, there is a rotation of S^2 that maps P to Q.
10. Each point P on S^2 has **spherical polar coordinates** $(\cos\phi \sin\theta, \sin\phi \sin\theta, \cos\theta)$, where $0 \leq \phi < 2\pi$ and $0 \leq \theta \leq \pi$.

    The **colatitude** of P is the angle θ; the **latitude** of P is $\theta' = \frac{\pi}{2} - \theta$. The **longitude** of P is the angle ϕ.

    The distance of P from the North Pole is its colatitude θ.
11. A **circle** on S^2 is the set of points on S^2 that are equidistant from a given point of S^2; it is either a little circle or a great circle.

Section 7.2: Sperical transformations

1. An **isometry** of S^2 is a mapping of S^2 to itself that preserves distances between points. The isometries of S^2 form a group, the **group of spherical isometries**, denoted by the symbol $S\,(2)$.

   A direct isometry of S^2 preserves the orientation of angles on S^2; an **indirect isometry** of S^2 reverses them. Reflection of S^2 in a plane through the centre of S^2 is an **indirect isometry** of S^2.
2. The three elementary rotations of S^2 are its rotations about the coordinate axes.

The transformation

$$R(Z,\gamma): \quad S^2 \to S^2$$
$$\mathbf{x} \mapsto \mathbf{Ax}$$

where $\mathbf{x} = \begin{pmatrix} x \\ y \\ z \end{pmatrix}$ and $\mathbf{A} = \begin{pmatrix} \cos\gamma & -\sin\gamma & 0 \\ \sin\gamma & \cos\gamma & 0 \\ 0 & 0 & 1 \end{pmatrix}$ is a rotation of S^2 about the positive z-axis through an angle γ.

The transformation

$$R(X,\alpha): \quad S^2 \to S^2$$
$$\mathbf{x} \mapsto \mathbf{Ax}$$

where $\mathbf{x} = \begin{pmatrix} x \\ y \\ z \end{pmatrix}$ and $\mathbf{A} = \begin{pmatrix} 1 & 0 & 0 \\ 0 & \cos\alpha & -\sin\alpha \\ 0 & \sin\alpha & \cos\alpha \end{pmatrix}$ is a rotation of S^2 about the positive x-axis through an angle α.

The transformation

$$R(Y,\beta): \quad S^2 \to S^2$$
$$\mathbf{x} \mapsto \mathbf{Ax}$$

where $\mathbf{x} = \begin{pmatrix} x \\ y \\ z \end{pmatrix}$ and $\mathbf{A} = \begin{pmatrix} \cos\beta & 0 & \sin\beta \\ 0 & 1 & 0 \\ -\sin\beta & 0 & \cos\beta \end{pmatrix}$ is a rotation of S^2 about the positive y-axis through an angle β.

3. The **inverse of an elementary rotation** through a given angle is an elementary rotation through the same angle but in the opposite direction. Its matrix is the transpose of the matrix of the original elementary rotation.
4. A rotation of S^2 that maps $A(1,0,0)$ to $P(\cos\phi\sin\theta, \sin\phi\sin\theta, \cos\theta)$ is given by the composition $R(Z,\phi)\,R(Y,-\theta')$, where $\theta' = \frac{\pi}{2} - \theta$.
5. The **equatorial plane** of S^2 is the plane $z = 0$ that contains the equator of S^2. **Reflection in the equatorial plane** of S^2 is the mapping $\mathbf{x} \mapsto \mathbf{Ax}$ that sends the point $\mathbf{x} = (x,y,z)$ to the point $(x,y,-z)$; it has matrix $\mathbf{A} = \begin{pmatrix} 1 & 0 & 0 \\ 0 & 1 & 0 \\ 0 & 0 & -1 \end{pmatrix}$.
6. Reflection of S^2 in the plane with equation $ax + by + cz = 0$, where $a^2 + b^2 + c^2 = 1$, is given by the mapping $\mathbf{x} \mapsto \mathbf{Ax}$ where $\mathbf{A} = \begin{pmatrix} 1-2a^2 & -2ab & -2ac \\ -2ab & 1-2b^2 & -2bc \\ -2ac & -2bc & 1-2c^2 \end{pmatrix}$.
7. The product of any two reflections of $\mathbb{R}^3$ in planes through O that meet in a common line is a rotation about that common line.
8. Every rotation is a product of two reflections; these reflections are in planes that have the axis of rotation as their common line, and are separated by half the angle of the rotation.
9. The image of a circle on S^2 under an isometry of S^2 to itself is a circle.
10. **The Product Theorem** Every isometry of S^2 is a product of at most three reflections in great circles.

    An isometry of S^2 is known completely when its effect on three points (which do not lie on a common great circle) is known.

11. **Isosceles Triangle Theorem** Let a triangle ΔPQR on S^2 have sides PQ of length r, QR of length p, and RP of length q, and let $p = q$. Also, let the angles at the vertices P, Q and R be α, β and γ, respectively. Then $\alpha = \beta$.
12. Every isometry of S^2 is a reflection, a rotation, or the composite of a reflection and a rotation.

    Every direct isometry of S^2 is a rotation, and vice versa.

Section 7.3: Spherical Trigonometry

1. Given any three angles α, β and γ with $(5\pi >)\,\alpha + \beta + \gamma > \pi$, there exists a spherical triangle with those angles.
2. The sum of the angles of any spherical triangle is greater than π. The **angular excess** of a spherical triangle is the difference between the sum of the angles of the triangle and π.
3. The **strict line segment** PQ is the shorter of the two line segments PQ. (If both are of length π, we choose either in some explicit way.)

    The **strict triangle** ΔPQR is that triangle ΔPQR whose sides are strict line segments, and its inside is the region of S^2 that they bound; we can rotate S^2 so that the whole of the strict triangle can be seen at one time.
4. A **lune** is a 2-sided polygon bounded by two great circles
5. The area of a spherical triangle is equal to its angular excess.
6. **Dual Triangles Theorem** With every strict spherical triangle Δ there is associated another triangle Δ', called its *dual*, whose angles are the complements of the sides of Δ and whose sides are the complements of the angles of Δ. (The 'complement' of a quantity x is the quantity $\pi - x$.)
7. The distance d on S^2 between the points (p_1, p_2, p_3) and (q_1, q_2, q_3) is given by $\cos d = p_1q_1 + p_2q_2 + p_3q_3$. The distance between corresponding points on the surface of the Earth is then given (in miles) by the scaling map $\mathbf{x} \mapsto 4000\mathbf{x}$.
8. **Pythagoras' Theorem** Let ΔABC be a triangle on S^2 in which the angle at C is a right angle. If a, b and c are the lengths of BC, CA and AB, then $\cos c = \cos a \times \cos b$.

    Also $\sin\alpha = \sin a \big/ \sin c$ and $\tan\alpha = \tan a \big/ \sin b$.
9. We can always drop a perpendicular from a point on S^2 to a line on S^2.
10. Let ΔABC be a **strict triangle** on S^2, in which the sides AB, BC and CA have lengths c, a and b, respectively, and the angles at A, B and C are α, β and γ, respectively. Then

    Sine Rule $\frac{\sin\alpha}{\sin a} = \frac{\sin\beta}{\sin b} = \frac{\sin\gamma}{\sin c}$;

    Cosine Rule for Sides $\cos c = \cos a \cos b + \sin a \sin b \cos\gamma$;

    Cosine Rule for Angles $\cos\gamma = \sin\alpha \sin\beta \cos c - \cos\alpha \cos\beta$.

Section 7.4: Spherical Geometry and the Extended Complex Plane

1. The image of the point P $(\cos\phi \sin\theta, \sin\phi \sin\theta, \cos\theta)$ under stereographic projection π has coordinates $\left(\cos\phi \big/ \tan\left(\frac{1}{2}\theta\right), \sin\phi \big/ \tan\left(\frac{1}{2}\theta\right)\right)$.

2. Let a point P' on S^2 map under stereographic projection to a point P in $\mathbb{R}^2$ with coordinates $(r\cos\theta, r\sin\theta)$. Then the spherical distance of P' from the South Pole S is $2\tan^{-1}(r)$.
3. Let P' and Q' be points on S^2 that are mirror images under reflection in a great circle C', and let the images of P', Q' and C' under stereographic projection π be P, Q and C, respectively. Then P and Q are inverse points with respect to C.
4. Let R denote a rotation of S^2, and π stereographic projection. Then the transformation $\pi R\pi^{-1}$ of the extended plane is the **conjugate transformation** of the rotation R.

   The conjugate transformation of a rotation of S^2 is a Möbius transformation of the extended plane.
5. A Möbius transformation M that is conjugate to a rotation of S^2 is of the form $M(z) = (az+b)/\left(-\overline{b}z+\overline{a}\right)$, where $a, b \in \mathbb{C}$.

   A Möbius transformation can have at most two fixed points.
6. The circle in $\mathbb{R}^2$ with equation $x^2 + y^2 + 2\alpha x + 2\beta y + \gamma = 0$ and the plane with equation $2\alpha X + 2\beta Y + (1-\gamma)Z + (1+\gamma) = 0$ are said to be **associated**.

Section 7.5: Planar Maps

1. It is not possible to map a portion of the Earth's surface onto a planar map in such a way that relative distances are preserved. There is no ideal planar map.
2. **Azimuthal projections** of the Earth E onto a flat sheet of paper π are ones where we project E onto π placed tangent to E at some point, often the South Pole S.

   The **gnomic projection** uses the centre of E as the centre of projection, stereographic projection uses the North Pole of E as the centre of projection, and *orthogonal* or *orthographic projection* uses rays from infinity parallel to the polar axis.
3. **Cylindrical projections** of E onto π are ones where we project E onto a sheet π that is rolled in the form of a cylinder round E, touching it at its Equator; then we slit the sheet along a meridian and flatten it.
4. **Conical projections** of E onto π are ones where we project E onto a sheet π that is rolled in the form of a cone round E, touching it at some parallel; then we slit the sheet along a meridian and flatten it, taking on the shape of a sector of a circle.

8 The Kleinian View of Geometry

We now describe in detail what we have called the Kleinian view of geometry. As we said earlier, this is a powerful way in which the geometries we have presented can be unified and seen as different aspects of the same idea.

Chapter 0

8.1 Affine Geometry

Klein argued that in any geometry there are properties of figures that one discusses, but these properties vary from one geometry to another. In Euclidean geometry, segments have lengths, lines cross at angles one can measure; it makes sense to ask if a given triangle is equilateral, and so on. But none of these properties makes sense in affine geometry. There one may speak only of ratios of lengths along parallel lines, and all triangles are (affine) congruent. Klein interpreted this as follows: in each geometry we are studying the same space of points, the familiar plane $\mathbb{R}^2$. In Euclidean geometry we admit only the transformations that preserve length. These are transformations of the form

$$t(\mathbf{x}) = \mathbf{U}\mathbf{x} + \mathbf{a},$$

where $\mathbf{U}$ is an orthogonal 2×2 matrix and $\mathbf{a} \in \mathbb{R}^2$.

Because every transformation in Euclidean geometry preserves lengths, and therefore angles, it makes sense to speak of distance in this geometry.

In affine geometry, however, we admit all transformations of the form

$$t(\mathbf{x}) = \mathbf{A}\mathbf{x} + \mathbf{b},$$

where $\mathbf{A}$ is an invertible 2×2 matrix and $\mathbf{b} \in \mathbb{R}^2$. Because every transformation in affine geometry preserves ratios of lengths along parallel lines, it makes sense to speak of such ratios in this geometry. But because there are affine transformations that do not preserve length, it does not make sense to speak of distance in this geometry.

Klein observed that rather more is true. In both geometries, the allowable transformations form a group. This was a novel observation at the time, for Group Theory as such was only just being discovered. Moreover, the group of Euclidean transformations is a subgroup of the group of affine transformations.

This observation enabled Klein to unify the two geometries. He said that both geometries concern the same space $\mathbb{R}^2$, that each geometry involves a group of transformations which act on this space, and that the Euclidean group is a subgroup of the affine group. In this sense Euclidean geometry is a *subgeometry* of affine geometry.

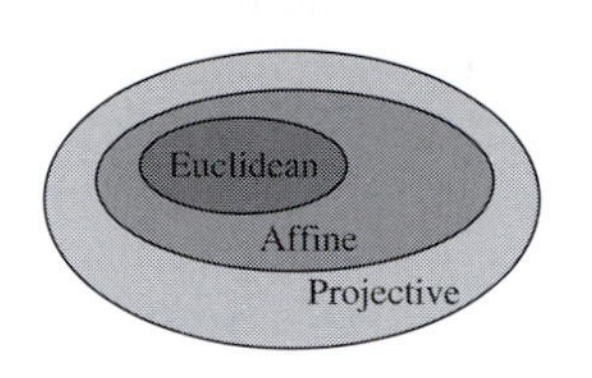

The hard work comes with the remaining geometries. How can affine geometry and projective geometry be unified? Projective geometry is not about the plane $\mathbb{R}^2$ but about projective space $\mathbb{RP}^2$. Its transformations are written as 3×3 homogeneous matrices, not as 2×2 matrices. To accomplish the unification, we must first establish that the space $\mathbb{R}^2$ can be regarded as a subset of $\mathbb{RP}^2$.

Here, *homogeneous* means that the matrix can be multiplied by any non-zero real number without affecting the transformation.

To do this, we consider the subset of Points in $\mathbb{RP}^2$ that pierce the standard embedding plane $z = 1$. We shall temporarily denote this subset by $\mathbb{RP}_0^2$, so that

$$\mathbb{RP}_0^2 = \{[x, y, z] : z \neq 0\}.$$

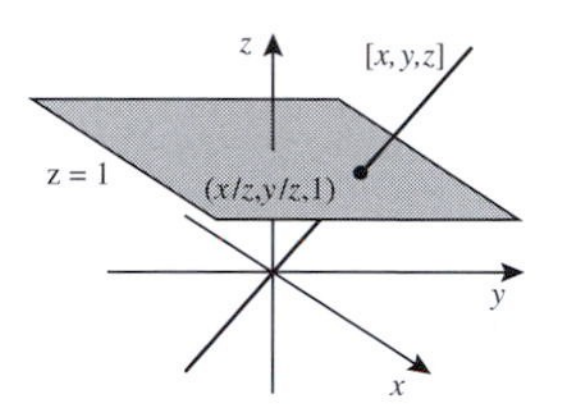

Now each point (x, y) in $\mathbb{R}^2$ can be identified with the Point $[x, y, 1]$ in $\mathbb{RP}_0^2$; and, conversely, each Point $[x, y, z] = [x/z, y/z, 1]$ in $\mathbb{RP}_0^2$ can be identified with the point $(x/z, y/z)$ in $\mathbb{R}^2$. This provides us with a one–one correspondence between $\mathbb{R}^2$ and $\mathbb{RP}_0^2$ which enables us to regard $\mathbb{R}^2$ as a subset of $\mathbb{RP}^2$.

Next (following Klein) we prove a result which shows that the affine group can be considered as a subgroup of the projective group.

Theorem 1 The group of affine transformations can be identified with the subgroup of projective transformations that map $\mathbb{RP}_0^2$ onto itself.

Proof A projective transformation which maps $\mathbb{RP}_0^2$ onto itself is also one that maps the ideal Line $z=0$ onto itself. It is therefore sufficient to characterize the projective transformations which map ideal Points to ideal Points.

It is clear that the set of projective transformations which map the Line $z = 0$ onto itself is a group. For, if t is a transformation which maps ideal Points to ideal Points, then so is its inverse t^{-1}; and, if t and s are projective transformations which map ideal Points to ideal Points, then so is $t \circ s$. We denote this group by $P_0(2)$.

Recall that the group of all projective transformations is denoted by $P(2)$.

Let $\mathbf{A} = \begin{pmatrix} a & b & c \\ d & e & f \\ g & h & k \end{pmatrix}$ be a matrix associated with a projective transformation t. Then t maps ideal Points to ideal Points if and only if

$$\begin{pmatrix} a & b & c \\ d & e & f \\ g & h & k \end{pmatrix} \begin{pmatrix} x \\ y \\ 0 \end{pmatrix} = \begin{pmatrix} ? \\ ? \\ 0 \end{pmatrix},$$

where we do not care what numbers the ?'s stand for, so long as they are not both zero. It follows that t maps ideal Points to ideal Points if and only if

$$gx + hy + 0 = 0.$$

This must hold for all $x \neq 0$ and all $y \neq 0$, so $g = h = 0$. Since $\mathbf{A}$ is invertible, it follows that $\det \mathbf{A} = k(ae-db) \neq 0$, so $k \neq 0$ and $(ae-db) \neq 0$. Moreover, since the matrices associated with a projective transformation are determined only up to a non-zero real multiple, we can divide $\mathbf{A}$ by k. If we then rewrite a/k as a, b/k as b, and so on, we see that the transformations in $P_0(2)$ are precisely those projective transformations which have an associated matrix of the form

$$\begin{pmatrix} a & b & c \\ d & e & f \\ 0 & 0 & 1 \end{pmatrix}, \quad \text{where } ae - db \neq 0.$$

Let us now consider the effect that such a transformation has on Points in $\mathbb{RP}_0^2$. If $[x', y', 1]$ is the image of $[x, y, 1]$ under the transformation, then

$$\begin{pmatrix} x' \\ y' \\ 1 \end{pmatrix} = \begin{pmatrix} a & b & c \\ d & e & f \\ 0 & 0 & 1 \end{pmatrix} \begin{pmatrix} x \\ y \\ 1 \end{pmatrix} = \begin{pmatrix} ax + by + c \\ dx + ey + f \\ 1 \end{pmatrix}.$$

This equation provides us with a relationship between x, y and x', y' which we can rewrite in the form

$$\begin{pmatrix} x' \\ y' \end{pmatrix} = \begin{pmatrix} a & b \\ d & e \end{pmatrix} \begin{pmatrix} x \\ y \end{pmatrix} + \begin{pmatrix} c \\ f \end{pmatrix}.$$

Since $ae-db \neq 0$ it follows that we can map the point (x, y) associated with the Point $[x, y, 1]$ to the point (x', y') associated with the Point $[x', y', 1]$ by an affine transformation. We conclude that the group $P_0(2)$ may be identified with the affine group $A(2)$. ∎

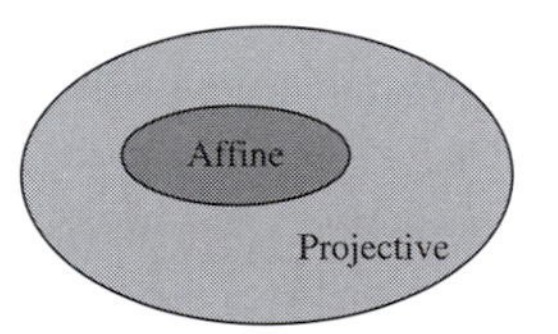

It follows that the space $\mathbb{R}^2$ of affine geometry can be considered as a subset of projective space $\mathbb{RP}^2$, and that the group $A(2)$ of affine geometry can be considered as a subgroup of the projective group $P(2)$. In other words, affine geometry is a *subgeometry* of projective geometry.

It is important to realize that in the process of extracting affine space from projective space we lose all the ideal Points on the ideal Line, and that this omission can have a significant effect on figures. For example, in projective geometry any two Lines meet at a Point. But if the Point of intersection is an ideal Point, then the corresponding affine figure consists of two lines that do not meet. This gives rise to the affine concept of parallel lines, which is meaningless in projective geometry. The fact that 'parallelism' is an affine property can be checked easily by recalling that an affine transformation corresponds to a projective transformation which maps ideal Points to ideal Points. Under such a transformation, Lines which intersect at an ideal Point map to Lines which intersect at an ideal Point. This is an alternative proof of the following theorem that we first proved in Chapter 2.

Theorem 2 An affine transformation maps parallel straight lines to parallel straight lines.

Subsection 2.3.3, Theorem 3

Parallelism is not the only new property to arise from the omission of ideal Points. In affine geometry, three distinct types of (non-degenerate) conic arise because there are three ways in which a projective conic can meet the ideal Line. In the first figure below, the projective conic and the Line cross at two Points; the corresponding picture in $\mathbb{R}^2$ is that of a hyperbola. In the second figure, the projective conic and the Line touch at one Point; the corresponding picture in $\mathbb{R}^2$ is that of a parabola. In the third figure, the projective conic and the Line do not meet at any Point; the corresponding picture in $\mathbb{R}^2$ is that of an ellipse.

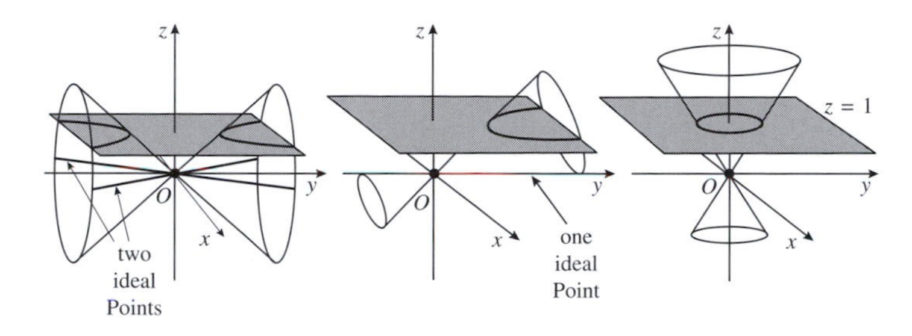

So hyperbolas, parabolas and ellipses are plane conics which correspond to projective conics with two, one and no ideal Points missing, respectively. Since affine transformations correspond to projective transformations which map ideal Points to ideal Points, they cannot change the number of ideal Points on a projective conic, and hence cannot change the type of the plane conic. This is an alternative proof of the following theorem first proved in Chapter 2.

Subsection 2.5.1, Theorem 4

Theorem 3 Affine transformations map ellipses to ellipses, parabolas to parabolas, and hyperbolas to hyperbolas.

So far, we have discussed some affine properties that do not make sense in projective geometry. By contrast, any property in projective geometry must also be a property in affine geometry. Indeed, *any property that is preserved by the group of all projective transformations must also be preserved by the subgroup of all affine transformations*.

For example, tangency is a property of projective conics, so every projective transformation maps tangents to tangents. In particular, a projective transformation which corresponds to an affine transformation must map a tangent whose Point of contact is not an ideal Point to a tangent whose Point of contact is not an ideal Point. Interpreting this observation as an affine result, we obtain the following theorem which we first met in Chapter 2.

Theorem 4 Let t be an affine transformation, and let ℓ be a tangent to a conic C. Then $t(\ell)$ is a tangent to the conic $t(C)$.

Subsection 2.5.1, Theorem 7

Notice that in this proof of the theorem we did not have to consider tangents whose Point of contact is an ideal Point. But to what does such a tangent correspond in $\mathbb{R}^2$? The answer depends on how many ideal Points lie on the projective conic.

A projective conic which passes through one ideal Point corresponds to a parabola in $\mathbb{R}^2$. Since such a projective conic meets the ideal Line at precisely one Point, the tangent at that Point must be the ideal Line $z = 0$. This tangent does not correspond to anything in $\mathbb{R}^2$!

Recall that a Line is a *tangent* to a projective conic if it meets the conic at precisely one Point.

More interesting are those projective conics which pass through two ideal Points. Such a projective conic corresponds to a hyperbola in $\mathbb{R}^2$. The tangents which touch the projective conic at its two ideal Points correspond to lines in $\mathbb{R}^2$, and these lines are defined to be the asymptotes of the hyperbola.

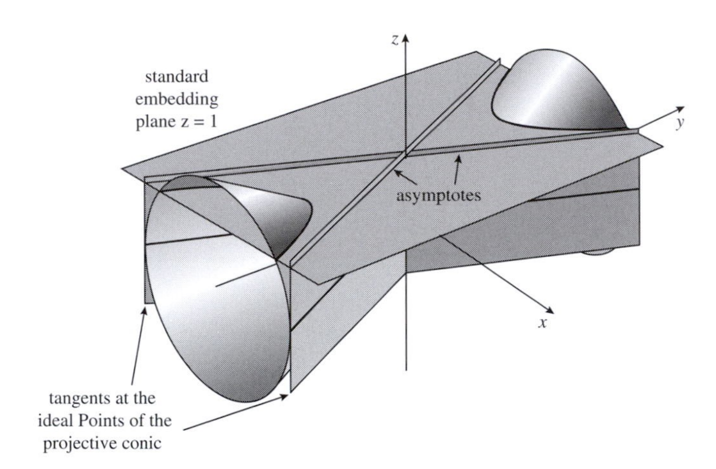

Since affine transformations correspond to those projective transformations which map ideal Points to ideal Points, it follows that tangents at ideal Points map to tangents at ideal Points, and hence that asymptotes map to asymptotes. This proves the following theorem which you met in Chapter 2.

Subsection 2.5.1, Theorem 6

Theorem 5 Let t be an affine transformation, and let H be a hyperbola with asymptotes ℓ_1 and ℓ_2. Then $t(H)$ has asymptotes $t(\ell_1)$ and $t(\ell_2)$.

We now summarize what we have done so far.

Geometry	Space	Transformations	Invariants
Projective	$\mathbb{RP}^2 = \mathbb{R}^2$ + ideal Line	$t : [\mathbf{x}] \mapsto [\mathbf{Ax}]$, where $\mathbf{A}$ is a 3×3 invertible matrix	incidence, collinearity, cross-ratio
Affine	$\mathbb{R}^2$	$t : \mathbf{x} \mapsto \mathbf{Ax} + \mathbf{b}$, where $\mathbf{A}$ is a 2×2 invertible matrix	ratios along parallel lines
Euclidean	$\mathbb{R}^2$	$t : \mathbf{x} \mapsto \mathbf{Ux} + \mathbf{a}$, where $\mathbf{U}$ is a 2×2 orthogonal matrix	length, angle

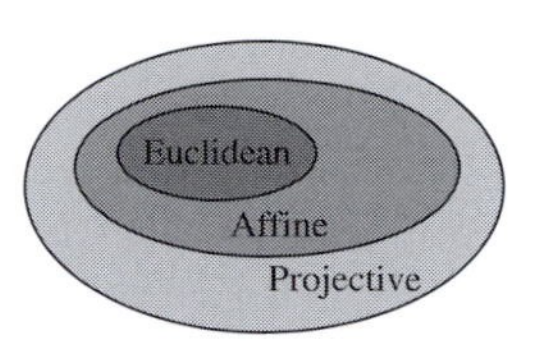

We can think of the geometries in this table as forming a hierarchy of geometries with projective geometry at the top. Any property of a geometry in the hierarchy is also a property of the geometries further down the hierarchy. Also, theorems that are valid in one geometry in the hierarchy also hold in geometries further down.

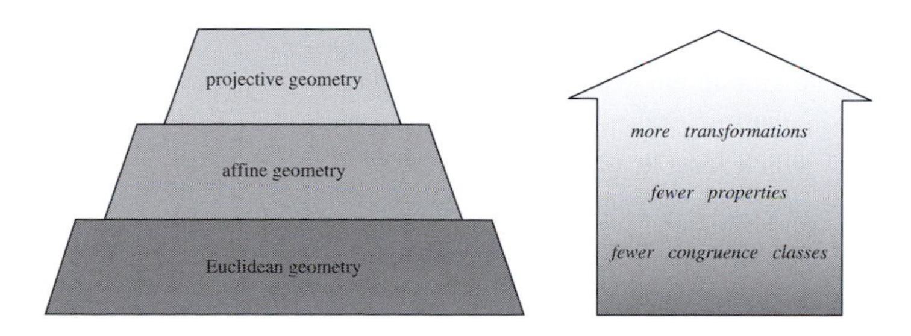

As we move up the hierarchy, the group of transformations becomes larger and so the congruence classes of figures merge together as more and more figures become congruent to each other. In Euclidean geometry, two triangles are congruent if and only if they have the same side lengths, and two conics are congruent if and only if they are of the same type (hyperbola, parabola or ellipse) and have the same shape and size. In affine geometry, any two triangles are congruent, and conics are congruent if and only if they are of the same type. In projective geometry, any two triangles are congruent and any two projective conics are congruent.

8.2 Projective Reflections

The cross-ratio enables us to introduce a type of projective transformation that generalizes reflection in Euclidean geometry.

Definitions Let F be a Point that does not lie on a Line ℓ. Then a **projective reflection** in ℓ with **centre** F and non-zero **parameter** k is a mapping

$$r : \mathbb{RP}^2 \to \mathbb{RP}^2$$
$$P \mapsto Q$$

where

(a) $Q = P$, if P lies on ℓ;
(b) the cross-ratio $(FRPQ) = k$, where R is the Point of intersection of FP and ℓ.

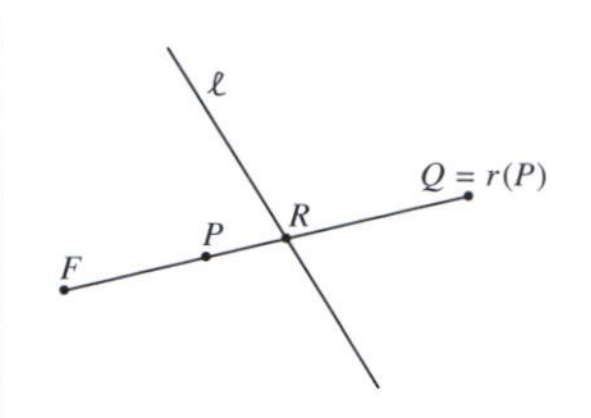

Remark

Provided that P does not lie on ℓ, then by varying the parameter k we can arrange that the Point Q is any Point on the Line FP other than the Points F, R and P.

We now indicate why this map is called a projective reflection.

Let ℓ' be the representation of a Line ℓ in an embedding plane π, and consider all the lines in π that are perpendicular to ℓ'. Since these lines are parallel, they must correspond to Lines in $\mathbb{RP}^2$ which meet at an ideal Point, F say. Let r denote the projective reflection in ℓ with centre F and parameter $k = -1$. Then, in the embedding plane π the effect of r is to map each (non-ideal) point P' to its Euclidean reflection in ℓ'.

r fixes the Points of ℓ so that its effect in π is to fix the points of ℓ'.

Next consider a Point P that does not lie on ℓ, and let $Q = r(P)$. Then PQ passes through F and meets ℓ at the Point R, where $(FRPQ) = -1$. For the embedding plane π, F is an ideal Point, so that

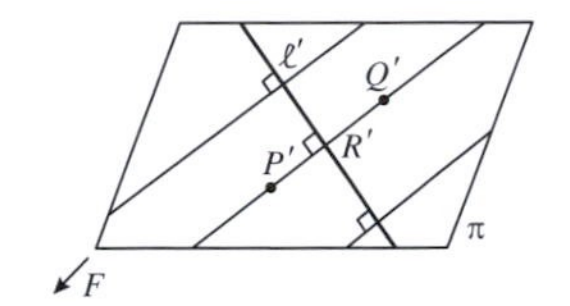

$$(FRPQ) = \frac{Q'R'}{P'R'} = -1$$

(where R', P' and Q' are the representations of R, P and Q in π). The point R' is therefore the midpoint of $P'Q'$. Moreover, since PQ passes through F, $P'Q'$ must be perpendicular to ℓ', so Q' is the Euclidean reflection of P' in ℓ', as suggested.

We now relate projective reflections to projective transformations.

Theorem 1 A projective reflection is a projective transformation.

Proof Consider the projective reflection r in the Line ℓ with centre F and parameter k, and let $Q = r(P)$ where P is a Point of $\mathbb{RP}^2$ that does not belong to $\ell \cup \{F\}$. Then $(FRPQ) = k$, where R is the Point of intersection of PQ and ℓ.

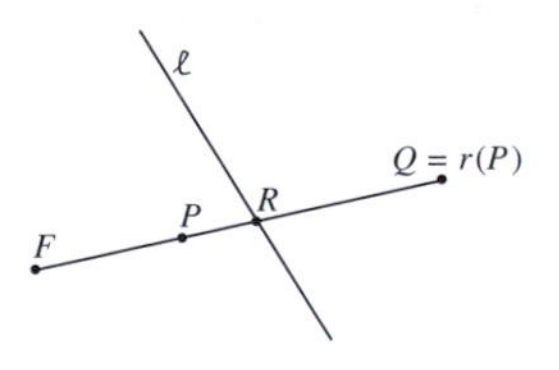

By the Fundamental Theorem of Projective Geometry there is a projective transformation t which maps F to $F' = [0, 0, 1]$ and which maps the Line ℓ to the Line ℓ' with equation $z = 0$. (This can be achieved by mapping a pair of Points on ℓ to $[1,0,0]$ and $[0,1,0]$.) If $P' = t(P)$, $Q' = t(Q)$ and $R' = t(R)$, then, since t preserves cross-ratio, it follows that $(F'R'P'Q') = (FRPQ) = k$.

Now, if $P' = [a,b,c]$, then the Line through P' and F' has equation $bx = ay$, and so it meets the Line $z = 0$ at $R' = [a,b,0]$. As Q' also lies on the Line $bx = ay$, we can write $Q' = [a,b,d]$, for some d; here $d \neq 0$, since Q' does not lie on ℓ.

F' P' R' Q'

Since

$$(a,b,c) = 1 \cdot (a,b,0) + c \cdot (0,0,1)$$

and

$$(a,b,d) = 1 \cdot (a,b,0) + d \cdot (0,0,1),$$

if follows that $k = (F'R'P'Q') = c/d$, so that $d = c/k$. Hence the mapping $s : \mathbb{RP}^2 \mapsto \mathbb{RP}^2$ which sends P' to Q' is the projective transformation defined by

$$s([a,b,c]) = [a,b,c/k].$$

Also, s fixes the Points of $\ell' \cup \{F'\}$. Thus the transformation r which maps P to Q must be equal to the composite $t^{-1} \circ s \circ t$ and so must be a projective transformation, as required. ■

Remarks

1. If the parameter $k = 1$, then s and hence $r = t \circ s \circ t^{-1}$ are the identity map.
2. If $k \neq 1$, then the only Points fixed by r are F and the Points of ℓ; for, if

$$s([a,b,c]) = [a,b,c/k] = [a,b,c],$$

then either $c = 0$ (in which case $[a,b,c]$ lies on ℓ') or $a = b = 0$ (in which case $[a,b,c] = F'$).
3. If $k = 0$, then s, and hence r, is not a projective transformation.

We showed earlier that every projective transformation can be expressed as a composite of perspectivity transformations. In fact, a similar result holds for projective reflections.

Theorem 4, Subsection 3.3.4

Theorem 2 Every projective transformation can be expressed as a composite of a finite number of projective reflections.

We omit a proof, since we shall not use this result.

8.3 Hyperbolic Geometry and Projective Geometry

First we introduce a new way of representing lines in hyperbolic geometry that will be useful in explaining how hyperbolic geometry is related to projective geometry. In this representation lines appear straight, but angles are no longer represented accurately. The realization that this could be done was one of the crucial discoveries made by Klein around the time he was preparing his Erlangen Program.

To obtain the new representation we use two mappings, stereographic projection and orthogonal projection. Stereographic projection from the North Pole of S^2 maps the Southern hemisphere onto the interior of the unit disc $\mathcal{D}$, and the equator to the boundary of $\mathcal{D}$. Its inverse therefore maps $\mathcal{D}$ onto the Southern hemisphere. It is an angle-preserving map, so it sends an arc of a generalized circle in $\mathcal{D}$, which is perpendicular to the equator, to an arc of a circle on the sphere perpendicular to the equator – such an image may be thought of as a circular arc 'hanging vertically downwards'.

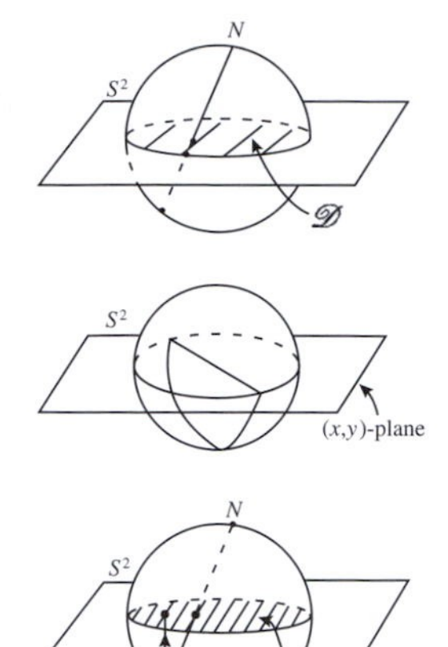

Orthogonal projection maps each point (x, y, z) in $\mathbb{R}^3$ along a line parallel to the z-axis to the point $(x, y, 0)$ in the (x, y)-plane. It maps the Southern hemisphere and the equator back onto $\mathcal{D}$. It maps a circular arc on the sphere perpendicular to the equator onto a segment of a straight line.

Definition The **linearization map** L of the unit disc $\mathcal{D}$ onto itself consists of the inverse of the stereographic projection map followed by the orthogonal projection map.

Then L fixes every point of the boundary of the unit disc $\mathcal{D}$, and maps arcs of generalized circles perpendicular to the boundary of $\mathcal{D}$ to segments of straight lines.

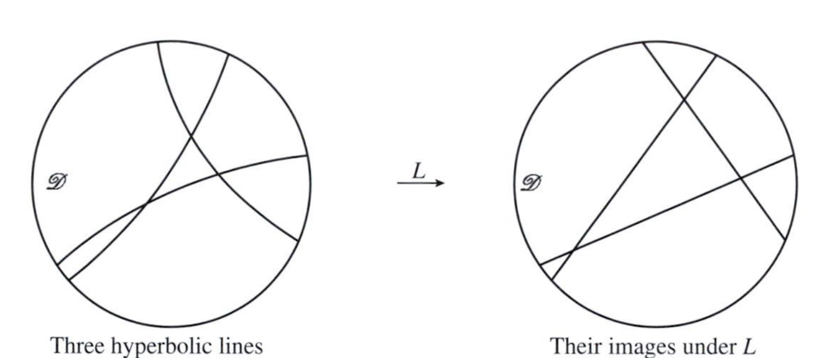

Three hyperbolic lines in the Poincaré model

Their images under L

The linearization map transforms the Poincaré model of hyperbolic geometry into a new model of hyperbolic geometry in which d-lines are represented by straight line segments. We call this new model of hyperbolic geometry the *Beltrami–Klein model*, after the Italian mathematician Eugenio Beltrami (1833–1900) who discovered it, and the German mathematician Felix Klein, who connected it to projective geometry.

It is important to realize that although d-lines look straight in this model, angles are not represented accurately.

By considering the definition of L as the composite of the inverse of stereographic projection followed by orthogonal projection, it is clear that the image of the d-line joining two points on the boundary of $\mathcal{D}$ is the straight line segment joining the points. To find the image of a point P of $\mathcal{D}$ under L, consider

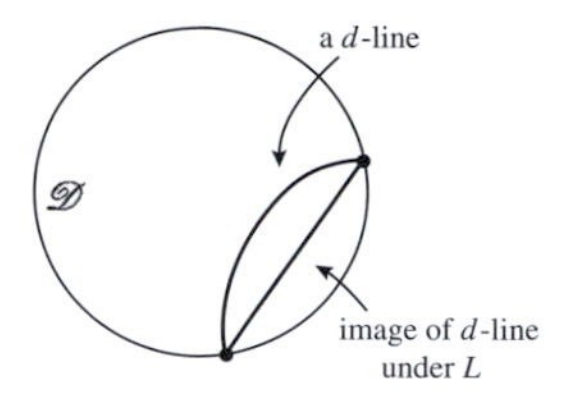

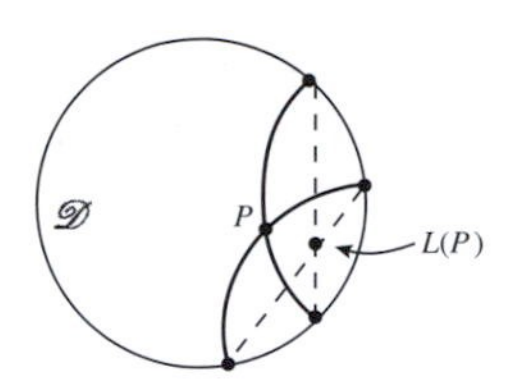

two d-lines through P and locate their end-points on the boundary of $\mathscr{D}$. Then $L(P)$ is the point where the corresponding chords intersect. Notice that, if it helps, we can always assume that one of the d-lines is a diameter of $\mathscr{D}$.

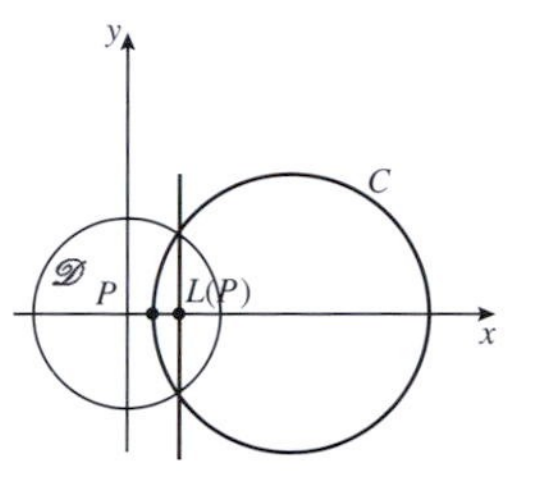

Example 1 Find the image under L of the point $P(r, 0)$ in $\mathscr{D}$.

Solution Notice first that it is sufficient to consider the image under L of the x-axis and the d-line through P at right angles to the x-axis.

The transformation L maps the x-axis to itself (as a line, not pointwise). The d-line through P perpendicular to the x-axis is part of the circle C with equation

$$x^2 + y^2 - 2ax + 1 = 0. \qquad (1)$$

The centre of C is on the x-axis, so there is no y-term in equation (1).

Since $P(r, 0)$ lies on C, it follows from equation (1) that $r^2 - 2ar + 1 = 0$, so that $a = \frac{r^2+1}{2r}$.

The circle C with equation (1) meets the boundary of $\mathscr{D}$ where

$$x^2 + y^2 - 2ax + 1 = 0 \quad \text{and} \quad x^2 + y^2 = 1.$$

The chord of $\mathscr{D}$ through the two points of intersection of C and the boundary of $\mathscr{D}$ is found by subtracting these two equations, and so it has equation

$$2 - 2ax = 0, \quad \text{or} \quad ax = 1.$$

It follows that the image $L(P)$ of P is the point $\left(\frac{1}{a}, 0\right) = \left(\frac{2r}{1+r^2}, 0\right)$. □

According to the Kleinian view of geometry, to define a geometry we must specify a space of points and a group of transformations acting on that space. Our new Beltrami–Klein model of hyperbolic geometry agrees with the (earlier) Poincaré model as far as the space of points is concerned: the space is the unit disc $\mathscr{D}$. The group of transformations is different, although isomorphic to the original group; the new group consists of all transformations of $\mathscr{D}$ to itself of the form LtL^{-1}, where t is a hyperbolic transformation. Clearly, if t is a hyperbolic transformation mapping $\mathscr{D}$ to itself and d-lines to d-lines, then LtL^{-1} is a transformation mapping $\mathscr{D}$ to itself and chords to chords.

To connect transformations of the form LtL^{-1} with projective geometry, we shall show that every such transformation LtL^{-1} is a projective transformation which maps $\mathscr{C}$ to itself and $\mathscr{D}$ to itself. It is sufficient to prove this in the case

$\mathscr{C} = \{(x, y) : x^2 + y^2 = 1\}$

that t is a hyperbolic reflection; for, if $t = t_1 \circ t_2 \circ \ldots \circ t_n$ is a decomposition of t into hyperbolic reflections then it follows that

$$LtL^{-1} = \left(L \circ t_1 \circ L^{-1}\right) \circ \left(L \circ t_2 \circ L^{-1}\right) \circ \ldots \circ \left(L \circ t_n \circ L^{-1}\right)$$

is a decomposition of LtL^{-1} into projective transformations with the required properties; then it follows that LtL^{-1} is a projective transformation with the required properties.

Theorem 1 Let t be the hyperbolic reflection in the d-line d joining the boundary points A and B of $\mathscr{D}$, and let F be the pole of the (Euclidean) line ℓ through A and B. Then LtL^{-1} is the projective reflection r in ℓ with centre F and parameter -1.

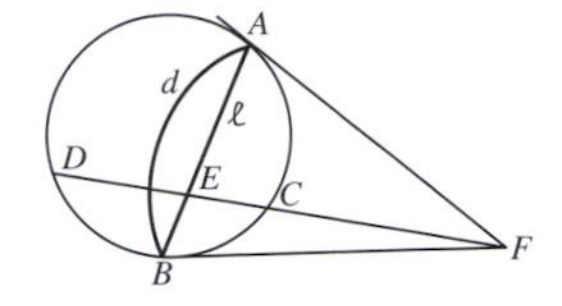

Proof First we show that LtL^{-1} agrees with r on $\mathscr{C}$. To do this, observe that lines through F are perpendicular to d, so the hyperbolic reflection r maps a given point C of $\mathscr{C}$ to the other point D of $\mathscr{C}$ that lies on FC. Since L maps $\mathscr{C}$ to itself, it follows that $LtL^{-1}(C) = D$.

Now let E be the point of intersection of FC with ℓ. Then, since F is the polar of ℓ with respect to $\mathscr{C}$, it follows that the cross-ratio $(FECD) = -1$. Hence

By the result of Exercise 7 on Subsection 4.3, in Subsection 4.6.

$$r(C) = D = LtL^{-1}(C),$$

as required.

Next we show that LtL^{-1} agrees with r on $\mathscr{D}$. To do this, let P be an arbitrary point of $\mathscr{D}$. Then let AP meet $\mathscr{C}$ at D, and let AC meet FP at Q. We will verify that $r(P) = Q$.

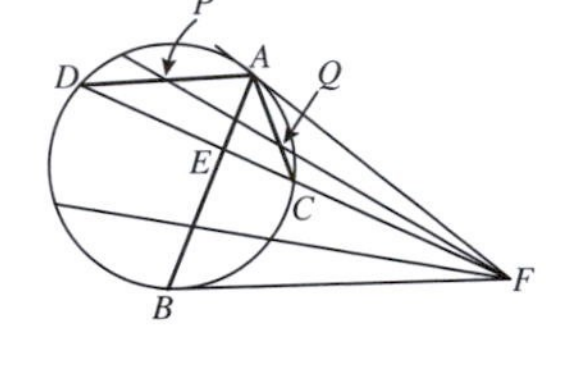

We know that $r(A) = A$ and that $r(D) = C$, so that r must map AD to AC. We also know that r maps the line FP to itself, so that it must map P (the point of intersection of AD and FP) to the point of intersection of AC and FP), namely the point Q. Thus $r(P) = Q$.

Now we consider the transformation LtL^{-1}. It certainly maps chords of $\mathscr{D}$ to chords; so, to find $LtL^{-1}(P)$ we may take AD and FP as two Lines through P, and find their images. The images of AD and FP are AC and FP, so that the image of P is Q. That is, $LtL^{-1}(P) = Q$.

Thus LtL^{-1} and r agree on both $\mathscr{C}$ and $\mathscr{D}$. ■

We have thus shown that every hyperbolic transformation t corresponds to a projective transformation LtL^{-1} which fixes $\mathscr{C}$ and maps its interior $\mathscr{D}$ to itself. In a similar way we can verify that every projective transformation that fixes $\mathscr{C}$ and maps $\mathscr{D}$ to itself corresponds to a hyperbolic transformation.

Here we think of $\mathscr{C}$ as being embedded in a suitable embedding plane.

We omit the details of this.

This shows that hyperbolic geometry can be considered as the sub-geometry of projective geometry in which the space of points is the set of points inside the fixed projective conic $\mathscr{C}$, and its group of transformations corresponds to those projective transformations mapping $\mathscr{C}$ to itself and the inside $\mathscr{D}$ of $\mathscr{C}$ to itself.

We defined the *inside* of a projective conic in Subsection 4.1.2.

In fact, since all non-degenerate conics are projectively equivalent, we can replace $\mathscr{C}$ in the above discussion by any non-degenerate conic E. This may be achieved by replacing the linearization map L by the composite map $s \circ L \circ s^{-1}$, where s is a projective transformation that maps $\mathscr{C}$ onto E.

We omit the details.

Defining Hyperbolic Distance in the Projective Setting

The fact that hyperbolic geometry may be considered a sub-geometry of projective geometry enables us to give a new method for defining distance in hyperbolic geometry. Projective transformations preserve cross-ratio, which involves four Points; distance involves only two points. So we shall have a ready-made sense of distance in the Beltrami–Klein model of projective geometry as soon as we can connect two interior Points of a projective conic (for simplicity, we shall take $\mathscr{C}$ as our projective conic) with a set of four collinear Points. This is easy – we join the two given Points P and Q by a Line that meets the projective conic in the Points A and B.

Historically, it was this discovery that excited Klein.

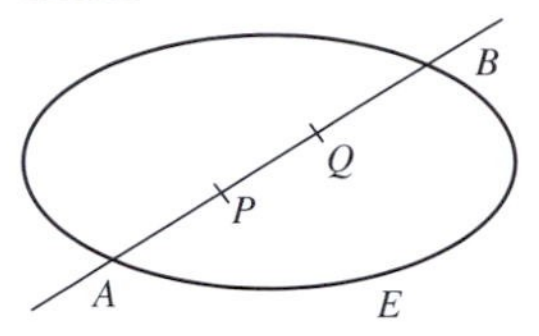

We know that the cross-ratio $(ABPQ)$ is unaltered by any projective transformation, and so in particular by the projective transformations that map the interior of the projective conic E to itself.

We now investigate whether we can sensibly define the hyperbolic distance d from P to Q by the formula $d(P,Q) = (ABPQ)$. The first thing we should want is that distance along a line is additive: that is, if R also lies on the d-line through P and Q, and R lies between P and Q, then is it true that

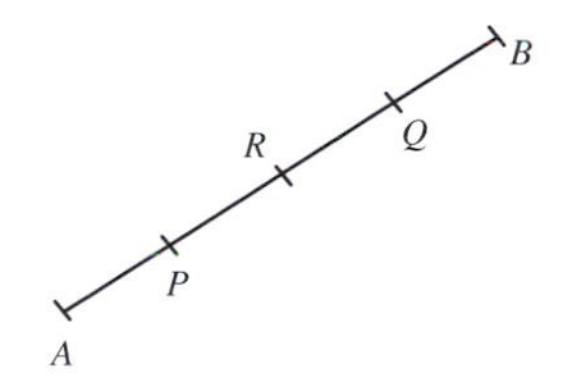

$$d(P,Q) = d(P,R) + d(R,Q)?$$

It follows from the definition of cross-ratio that

$$d(P,Q) = (ABPQ) = \frac{AP}{PB}\Big/\frac{AQ}{QB},$$

$$d(P,R) = (ABPR) = \frac{AP}{PB}\Big/\frac{AR}{RB},$$

and

$$d(R,Q) = (ABRQ) = \frac{AR}{RB}\Big/\frac{AQ}{QB}.$$

In this discussion we shall not bother to distinguish between Points and the corresponding points in a suitable embedding plane, for simplicity.

Hence, if we use the definition $d(P,Q) = (ABPQ)$, then $d(P,Q) \neq d(P,R) + d(R,Q)$ but instead $d(P,Q) = d(P,R) \cdot d(R,Q)$. This suggests that in the Beltrami–Klein model we should instead define distance between arbitrary points P and Q of $\mathscr{D}$ by the formula

$$d(P,Q) = k \log_e(ABPQ), \quad \text{for some number } k,$$

Recall that $\log_e(xy) = \log_e x + \log_e y$, for any positive numbers x and y.

since then distance d is additive. (In fact, the arbitrariness in the choice of the constant k is analogous to the freedom we have to choose the radius of the sphere in spherical geometry. However, we choose the sign of k so that distances are non-negative.)

To define a distance function d for the Poincaré model, let the conic E be the unit circle $\mathscr{C}$ and define the distance d between two points P and Q in $\mathscr{D}$ to be the distance d' between the corresponding points $P' = L(P)$ and $Q' = L(Q)$ in the Beltrami–Klein model. That is, we define

Notice that we use d' here, temporarily, to avoid confusion.

$$d(P, Q) = d'(P', Q').$$

To verify that this definition agrees with that given when we introduced hyperbolic geometry earlier, let P be the origin and let Q be an arbitrary point $(r, 0)$ on the positive x-axis. Then, by Example 1, $P' = L(P)$ is also at the origin and $Q' = L(Q) = (s, 0)$, where $s = \frac{2r}{1+r^2}$. Since the x-axis meets $\mathscr{C}$ at the points $A(-1, 0)$ and $B(1, 0)$, it follows that

Subsection 6.3.1

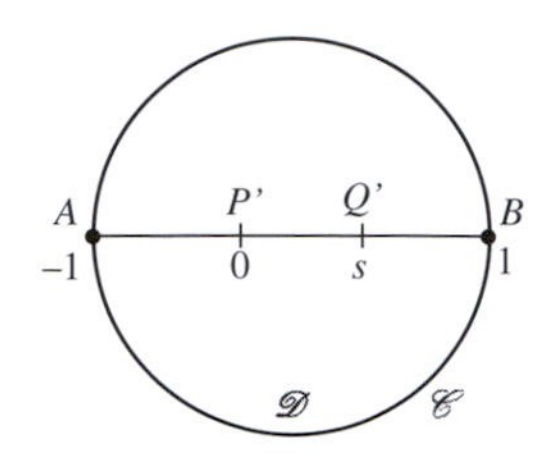

$$\begin{aligned}
d(P, Q) &= d'(P', Q') \\
&= k \log_e(ABP'Q') \\
&= k \log_e\left(\frac{1}{1} \Big/ \frac{1+s}{1-s}\right) \\
&= k \log_e\left(\frac{1 - \frac{2r}{1+r^2}}{1 + \frac{2r}{1+r^2}}\right) \\
&= k \log_e\left(\frac{1 - 2r + r^2}{1 + 2r + r^2}\right) \\
&= -k \log_e\left(\left(\frac{1+r}{1-r}\right)^2\right) \\
&= -2k \log_e\left(\frac{1+r}{1-r}\right).
\end{aligned}$$

Here we use that

$$\log_e\left(\frac{1}{x}\right) = -\log_e(x)$$

and

$$\log_e\left(x^2\right) = 2\log_e(x).$$

By choosing $k = -\frac{1}{4}$ and representing P and Q by the complex numbers 0 and $z = r + 0i$, respectively, we obtain

$$d(0, z) = \frac{1}{2} \log_e\left(\frac{1+|z|}{1-|z|}\right).$$

This is the logarithm formula for hyperbolic distance in $\mathscr{D}$ that you met earlier.

Subsection 6.3.1

8.4 Elliptic Geometry: the Spherical Model

To bring (the spherical model of) elliptic geometry into the Kleinian fold, we need to connect its space S^2 to the space $\mathbb{RP}^2$ of projective points and its group to the group of projective transformations.

With each point of S^2 we can associate a line through the origin, the line that joins the given point to the origin; this map assigns the same line to a pair of antipodal points. Conversely, given a projective point we can associate it with the pair of diametrically opposite points that it cuts out on S^2. In this way, we see that there is a 2–1 map of S^2 to $\mathbb{RP}^2$ that sends each pair of antipodal points on S^2 to the line through the origin (the projective point) that they define.

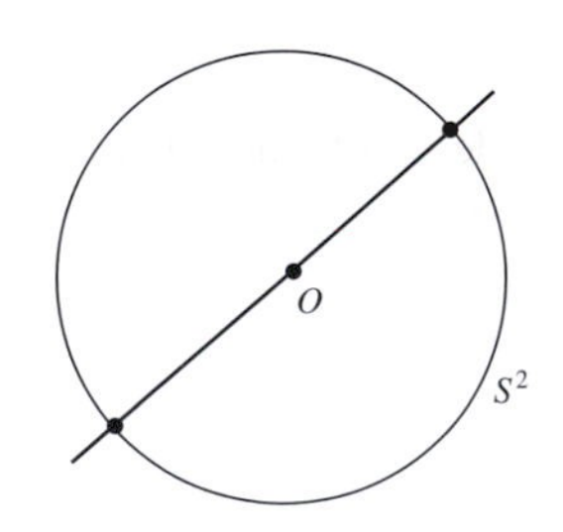

Since a spherical transformation sends antipodal points to antipodal points, it maps projective points to projective points. Also, every spherical transformation maps great circles to great circles, and so maps projective lines to projective lines. So it is clear that spherical transformations are projective transformations, and the group of spherical transformations is a subgroup of the projective group.

Projective Geometry and Spherical Geometry

Spherical geometry is a subgeometry of projective geometry which preserves the Euclidean angle between two lines through the origin in $\mathbb{R}^3$; the exact description of the subgroup is quite subtle, however, because the map from points of S^2 to Points in $\mathbb{RP}^2$ is 2–1.

The description lies beyond the scope of this book.

One method of representing a projective point by a single point on S^2 is to keep exactly half of S^2, in some sensible way. For example, let E^2 be the subset of S^2 defined by

Here we give an informal description of the representation.

$$E^2 = \text{northern hemisphere}, \{(x, y, z) : z > 0\}$$
$$\cup \text{ half the equator}, \{(x, y, z) : z = 0, x > 0\} \cup \{B(0, 1, 0)\}.$$

Then every line through the origin meets E^2 in exactly one point, and every projective line meets E^2 in half a great circle.

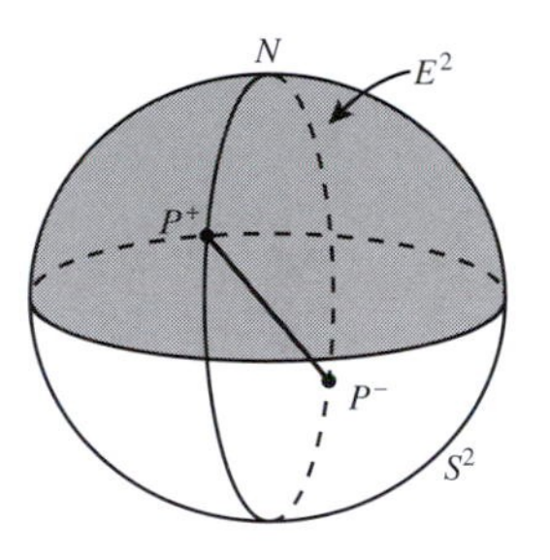

The 2–1 map from S^2 to $\mathbb{RP}^2$ therefore gives a 1–1 map from S^2 to E^2: every point of $\mathbb{RP}^2$ corresponds to a unique point of E^2, and every point of E^2 defines a unique Point of $\mathbb{RP}^2$. So E^2 provides 'a complete picture' of projective space $\mathbb{RP}^2$.

To see what's odd about E^2, consider the Points corresponding to the points $(\sin\theta, 0, \cos\theta)$ on S^2 as they trace out the Line $y = 0$ in $\mathbb{RP}^2$. Each such Point gives rise to two points on S^2:

$$P^+ = (\sin\theta, 0, \cos\theta) \quad \text{and} \quad P^- = (-\sin\theta, 0, -\cos\theta).$$

As θ goes from 0 to $\pi/2$, the point P^+ lies in E^2. But when $\theta > \pi/2$, the point P^- lies in E^2, rather than the point P^+, and this continues as θ goes from $\pi/2$ to $3\pi/2$ (including $\theta = 3\pi/2$). Then as θ goes from $3\pi/2$ to 2π, it is again the point P^+ that lies in E^2.

Here we discuss the points of S^2 that correspond to the complete Line $y = 0$.

So in E^2 the image of the Line is one-half of a great circle, described twice. There is an awkward switch in E^2 from the Euclidean point corresponding to the projective point being P^+ to being P^-. We cope with this by imagining the pairs of antipodal points to be 'joined up'; that is, we agree to regard antipodal points in the equator of S^2 as 'being the same point'.

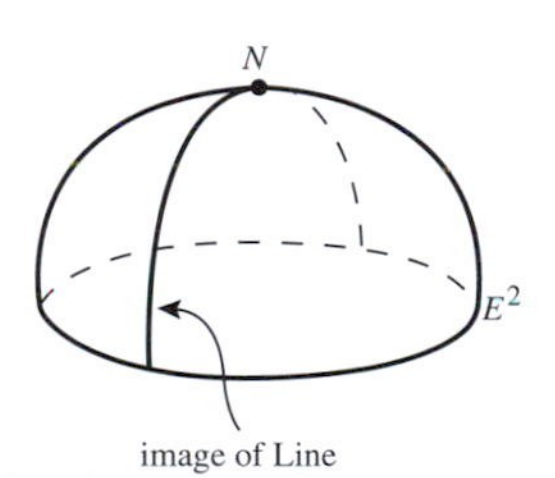

Conics appear equally odd. For example, let E denote a projective conic made up of Euclidean lines through the origin of $\mathbb{R}^3$ that form a right circular (Euclidean) cone in $\mathbb{R}^3$. This cone meets S^2 in two *little circles*.

Suppose first that the points P^+ all lie in E^2, so that the cone does not meet the equatorial plane in S^2. Then the points in E^2 that correspond to E form a little circle on S^2.

Suppose next that the cone cuts the equatorial plane of S^2. Then we must switch from P^+ to P^- as the Point describes the Line, and the points in E^2 that correspond to E form two curves (parts of little circles) on S^2.

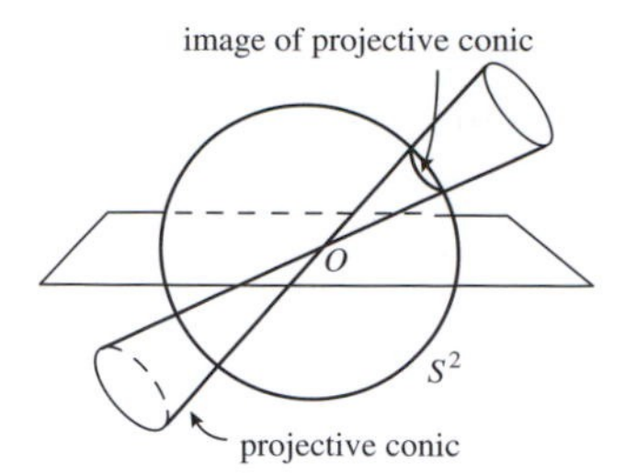

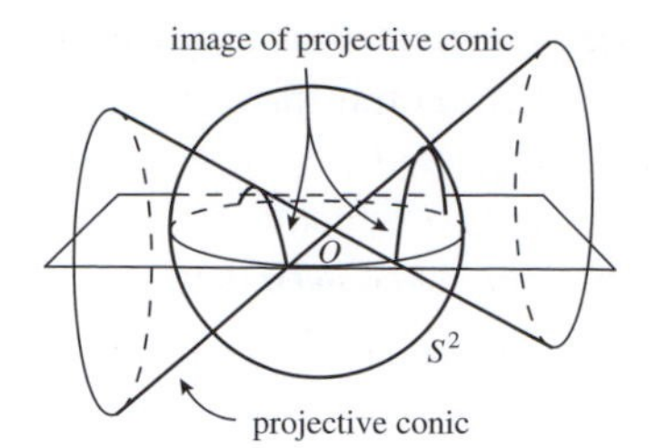

To see what's really odd about this, 'thicken up' the Line until it is a strip bounded by two little circles in S^2. The boundaries of the strip can be taken to be two projective conics, one on either side of the Line. Since the image of the Line meets the equator in S^2, we know that the images of the (associated) projective conics also meet the equator of S^2.

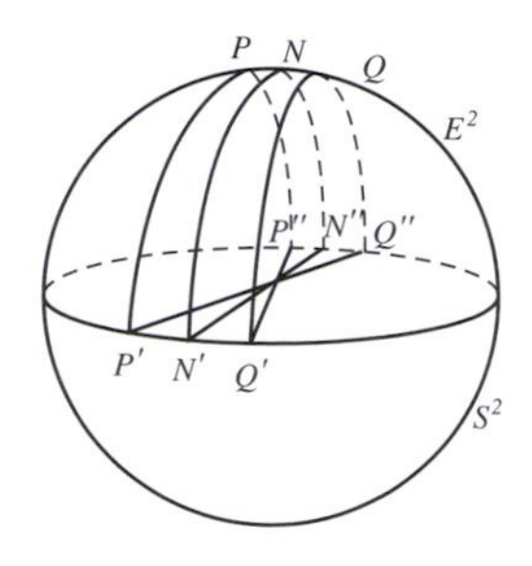

Now trace the image of the Line on S^2. As you trace the Line from N to N', the curve PP' (which is part of a projective conic $PP'P''$) is to your right, and the curve QQ' (which is part of a projective conic $QQ'Q''$) is to your left.

When you reach the equator, the equatorial points N' and N'' are identified, and you can complete your journey round S^2 to return to N. But the point P' is identified with Q'', so that the moving point that started from P returns to Q; similarly, the point Q' is identified with P'', so that the moving point that started from Q returns to P.

The curve $P''PP'$ that was on your right is now on your left, and the curve $Q''QQ'$ that was on your left is now on your right. The image of the strip is a strip in which the ends have been joined after receiving a half-twist. The image is a *Möbius band*, which illustrates a significant mathematical point: 2-dimensional real projective space $\mathbb{RP}^2$ is not *orientable* – that is, it cannot be given an everywhere-consistent sense of left and right.

Möbius bands are beloved of conjurers!

That is, of *orientation*.

Inversive Geometry and Spherical Geometry

In inversive geometry, we study transformations which map generalized circles to generalized circles and preserve the magnitude of angles. Every transformation in inversive geometry is composed of a sequence of inversions.

In spherical geometry, we study transformations which map great circles to great circles and preserve the magnitude of angles. Every transformation in spherical geometry is composed of a sequence of reflections in great circles, which are therefore inversions. It follows that spherical geometry is a subgeometry of inversive geometry, in exact analogy with the way that Euclidean geometry is a subgeometry of affine geometry.

We omit the full details.

8.5 Euclidean Geometry

Spherical geometry appears easier to understand than hyperbolic geometry because we can produce spheres S^2 as a model of the geometry, but there seems to be no surface in Euclidean 3-dimensional space which is a good model of hyperbolic geometry. One may exhibit 'pieces' of hyperbolic geometry, but it was shown by David Hilbert around 1900 that *no* surface embedded in 3-dimensional space is a complete model of hyperbolic 2-dimensional space.

David Hilbert (1862–1943), a professor at Göttingen University, was a mathematical universalist, influential in the foundations of mathematics; he is best known for his 1900 '23 Problems' collection that profoundly influenced the course of 20th-century mathematics.

But it can be argued that space might be hyperbolic (indeed, some would say that, according to Einstein's theory of General Relativity, it is *more nearly* hyperbolic than Euclidean). This raises the question: Is Euclidean geometry a subgeometry of hyperbolic geometry? We sketch a discussion of the answer to this interesting question.

Remarkably, the answer to the question is YES: Euclidean geometry *is* a subgeometry of hyperbolic geometry

To see that this is so, we need first to 'go up a dimension'. Hyperbolic 3-dimensional space can be modelled on the interior of the *unit ball*: a hyperbolic point is a point (x, y, z) such that $x^2 + y^2 + z^2 < 1$, and a d-line is an arc of a generalized circle perpendicular to S^2. All the other definitions in hyperbolic geometry generalize in the obvious way, including that of inversion and hyperbolic reflection. In particular, the transformations of hyperbolic geometry correspond to composites of inversions in spheres perpendicular to S^2.

It lies beyond our scope to deal with the matter fully.

$S^2 = \{(x, y, z): x^2 + y^2 + z^2 = 1\}$.

In the unit ball *horospheres* are spheres that touch S^2 at a single point. For simplicity, we now fix our attention on the horosphere H_N that touches S^2 at N, and restrict our attention to the subgroup G_N, say, of hyperbolic transformations that map H_N to itself. Then we have the following remarkable result.

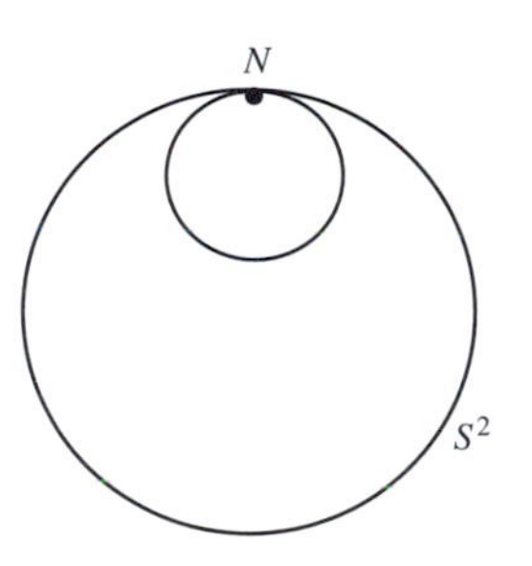

Theorem 1 The geometry on H_N is Euclidean geometry, and the group G_N acts on H_N as Euclidean isometries.

Proof Invert the whole figure in a sphere with centre N. The point N is mapped to 'infinity'; S^2 maps to a plane $\pi : z = a$, say, and H_N maps to a plane parallel to π, the plane $\pi' : z = b$, say. An element of G_N is an inversion in a generalized sphere through N and perpendicular to H_N; following composition with the first inversion, the element of G_N becomes an inversion in an unbounded generalized sphere perpendicular to the plane π'. These generalized spheres must in fact be planes perpendicular to the plane π'.

Now, inversion in a plane is Euclidean reflection in that plane. So the subgroup mapping the image of H_N to itself is the group generated by all Euclidean reflections, which is the group of Euclidean isometries. It follows that the geometry on the plane π' is Euclidean geometry. The desired result then follows by applying the inverse of the original inversion. ■

If indeed we had grown up believing that we lived in a hyperbolic 3-dimensional space, we should therefore be very familiar with spheres and horospheres!

Summary of Chapter 8

Section 8.1: Affine Geometry

1. The group of affine transformations can be identified with the subgroup of projective transformations that map $\mathbb{RP}_0^2$ onto itself.
2. Euclidean geometry is a subgeometry of affine geometry, which is itself a subgeometry of projective geometry; the three form a hierarchy. The higher up the geometry in the hierarchy, it has more transformations, fewer properties and fewer congruence classes.

Section 8.2: Projective Reflections

1. Let F be a Point that does not lie on a Line ℓ. Then a **projective reflection** r in ℓ with centre F and non-zero parameter k is a mapping of $\mathbb{RP}^2$ to itself, with P mapping to Q, where (i) $Q = P$ if P lies on ℓ and (ii) the cross-ratio $(FRPQ) = k$, where R is the Point of intersection of FP and ℓ.
2. A projective reflection is a projective transformation. Every projective transformation can be expressed as a composite of a finite number of projective reflections.

Section 8.3: Hyperbolic Geometry and Projective Geometry

1. **Orthogonal projection** maps $(x, y, z) \in \mathbb{R}^3$ to $(x, y, 0) \in \mathbb{R}^3$ or equivalently, to $(x, y) \in \mathbb{R}^2$.
2. The **linearization map** L of the unit disc $\mathscr{D}$ onto itself consists of the inverse of the stereographic projection map followed by the orthogonal projection map.
3. In the **Beltrami–Klein model** of hyperbolic geometry the space is $\mathscr{D}$ and the d-lines are straight-line segments.
4. Let t be the hyperbolic reflection in the d-line d joining the boundary points A and B of $\mathscr{D}$, and let F be the pole of the (Euclidean) line ℓ through A and B. Then LtL^{-1} is the projective reflection r in ℓ with centre F and parameter -1.
5. Every projective transformation that fixes $\mathscr{C}$ and maps $\mathscr{D}$ to itself corresponds to a hyperbolic transformation.

   Hyperbolic geometry can be considered as a subgeometry of projective geometry, where the points are points inside a fixed projective conic $\mathscr{C}$ and the group of transformations corresponds to those projective transformations mapping $\mathscr{C}$ onto itself and the inside of $\mathscr{C}$ to itself.
6. In the Beltrami–Klein model, we can define the **distance** between two points P and Q of $\mathscr{D}$ (where PQ meets $\mathscr{C}$ at A and B) by $d(P, Q) = k \log_e (ABPQ)$, where k is chosen so that distances are non-negative. This is equivalent to the definition of distance in the Poincaré model.

Section 8.4: Elliptic Geometry: the Spherical Model

1. There is a 2–1 map of S^2 to $\mathbb{RP}^2$ that sends each pair of antipodal points on S^2 to the line through the origin that they define. There is a 1-1 map of $E^2 = \{(x, y, z) : z > 0\} \cup \{(x, y, z) : z = 0, x > 0\} \cup \{(0, 1, 0)\}$ to $\mathbb{RP}^2$.
2. Spherical transformations are projective transformations; the group of spherical transformations is a subgroup of the projective group.
3. Spherical geometry is a subgeometry of inversive geometry.

Section 8.4: Euclidean Geometry

1. Euclidean geometry is a subgeometry of hyperbolic geometry
2. A point in 3-dimensional hyperbolic space is a point (x, y, z) with $x^2+y^2+z^2 < 1$, and a d-line is an arc of a generalized circle perpendicular to S^2.
3. A **horosphere** is a sphere in $\mathbb{R}^3$ that touches S^2 at a single point.

   Let H_N be the horosphere that touches S^2 at N, and G_N the subgroup of hyperbolic transformations that map H_N to itself. Then the geometry on H_N is Euclidean geometry, and the group G_N acts on H_N as Euclidean isometries.

Special Symbols

$\{a, b, ..., t\}$	set consisting of the elements $a, b, \ldots, t$
$\{x : x \text{ satisfies some condition}\}$	set consisting of the elements x that satisfy some condition
(a, b)	$\{x : a < x < b\}$
$[a, b]$	$\{x : a \leq x \leq b\}$
$[a, b)$	$\{x : a \leq x < b\}$
$(a, b]$	$\{x : a < x \leq b\}$
$\circ$	composition of functions (from right to left)
$\in$	belongs to
$\notin$	does not belong to
$\subset$	is a subset of
$\not\subset$	is not a subset of
$\subseteq$	equals or is a subset of
$\cup$	set-theoretic union
$\cap$	set-theoretic intersection
$\emptyset$	empty set
$\equiv$	is equivalent to, is congruent to
$\mapsto$	maps to
$f : A \to B$ $x \mapsto y$	the function f maps the set A to the set B, with the element x mapping to the element y
$\det \mathbf{A}$	determinant of a matrix $\mathbf{A}$
$\operatorname{adj} \mathbf{A}$	adjoint of a matrix $\mathbf{A}$
$\mathbf{A}^T$	transpose of a matrix $\mathbf{A}$
$\mathbf{A}^{-1}$	inverse of a matrix $\mathbf{A}$
$\mathbf{x} \cdot \mathbf{y}$	dot or scalar product of vectors $\mathbf{x}$ and $\mathbf{y}$
$\Vert\mathbf{v}\Vert$	length of a vector $\mathbf{v}$
t^{-1}	inverse of the transformation t
$\mathbb{R}$	set of real numbers
$\mathbb{R}^+$	set of positive real numbers
$\mathbb{R}^n$	n-dimensional Euclidean space
$\mathbb{C}$	set of complex numbers
$\mathbb{N}$	set of natural numbers

Δ	triangle
Σ	sum
$\angle$	angle
$A(2)$	set of all affine transformations of $\mathbb{R}^2$
$\mathbb{RP}^2$	real projective space of dimension 2
$P(2)$	group of projective transformations of $\mathbb{RP}^2$
$[\mathbf{x}]$	homogeneous coordinates of a Point in $\mathbb{RP}^2$
$(ABCD)$	cross-ratio of the ordered set of points A, B, C, D
$s,\ s_i,\ s_{ij}$	Joachimsthal's notation for conics
$\mathscr{D}$	unit disc, $\{z : z \in \mathbb{C} : \lvert z\rvert < 1\}$ or $\left\{(x, y) : (x, y) \in \mathbb{R}^2,\ x^2 + y^2 < 1\right\}$
$\mathscr{C}$	unit circle, $\{z : z \in \mathbb{C} : \lvert z\rvert = 1\}$ or $\left\{(x, y) : (x, y) \in \mathbb{R}^2,\ x^2 + y^2 = 1\right\}$
$\operatorname{Re} z$	real part of a complex number z
$\operatorname{Im} z$	imaginary part of a complex number z
$\overline{z}$	conjugate of a complex number z
$\lvert z\rvert$	modulus of a real or complex number z
$\hat{\mathbb{C}}$	extended complex plane $\mathbb{C} \cup \{\infty\}$
$A - B$	set of points in A that are not in B
$\mathbb{S}$	Riemann sphere
$G_{\mathscr{D}}$	group of hyperbolic transformations
$d(z_1, z_2)$	hyperbolic distance between z_1 and z_2
H	upper half-plane, $\{z : z \in \mathbb{C},\ \operatorname{Im} z > 0\}$
S^2	unit sphere in $\mathbb{R}^3$, $\left\{(x, y, z) : (x, y, z) \in \mathbb{R}^3,\ x^2 + y^2 + z^2 = 1\right\}$
$S(2)$	group of spherical isometries
$R(X, \alpha)$	rotation about the x-axis in $\mathbb{R}^3$, through an angle α
$R(Y, \beta)$	rotation about the y-axis in $\mathbb{R}^3$, through an angle β
$R(Z, \gamma)$	rotation about the z-axis in $\mathbb{R}^3$, through an angle γ
$\pi(X, Y, Z)$	stereographic projection of (X, Y, Z)
$\mathscr{A}$	family of Apollonian circles
$\mathbb{RP}_0^2$	Points in $\mathbb{RP}^2$ that pierce the standard embedding plane $z = 1$, $\left\{[x, y, z] \in \mathbb{RP}^2 : z \neq 0\right\}$
$P_0(2)$	group of projective transformations mapping the Line $z = 0$ to itself
L	linearization map of unit disc $\mathscr{D}$ onto itself
E^2	union of $\{(x, y, z) : z > 0\}$ and $\{(x, y, z) : z = 0,\ x > 0\} \cup \{(0, 1, 0)\}$
H_N	horosphere that touches S^2 at N
G_N	group of hyperbolic transformations that map H_N to itself

Further Reading

[1] Akopyan, A. V. and Zaslavsky, A. A., *Geometry of Conics*, Mathematical World 26, American Mathematical Society, Providence, 2007. ISBN 978-08218-4323-9

[2] Anderson, James W., *Hyperbolic Geometry*, Springer Undergraduate Mathematics Series, Springer-Verlag, London, 1999. ISBN 1-85233-156-9

[3] Ayers, Frank, *Theory and Problems of Projective Geometry,* Schaum's Outline Series, Schaum Publishing Co., McGraw Hill, New York, 1967. ISBN-13: 978-0070026575

[4] Beardon, Alan, *The Geometry of Discrete Groups,* Graduate Texts in Mathematics 91, Springer-Verlag, New York, corrected printing, 1985. ISBN 0-387-90788-2

[5] Cederberg, Judith N., *A Course in Modern Geometries,* Springer-Verlag, New York, 1989. ISBN 0-387-96922-5

[6] Cohn-Vossen, S. and Hilbert, David, *Geometry and the Imagination,* Chelsea, New York, 1990. ISBN 0-8218-1998-4

[7] Courant, Richard and Robbins, Herbert (revised by Ian Stewart), *What is Mathematics?,* Oxford University Press, Oxford, 2nd edition, 1996. ISBN 0-19-510519-2

[8] Coxeter, H. S. M., *Introduction to Geometry,* Wiley Classics Library, Wiley, New York, 2nd edition, 1989. ISBN 0-471-50458-0

[9] Coxeter, H. S. M., *Projective Geometry,* Springer-Verlag, New York, 2nd edition, 1987. ISBN 0-387-96532-7

[10] Coxeter, H. S. M., *The Real Projective Plane,* Springer-Verlag, New York, 3rd edition, 1992. ISBN 0-387-97889-5

[11] Coxeter, H. S. M. and Greitzer, S. L., *Geometry Revisited*, New Mathematical Library 19, Mathematical Association of America, Washington, 1967. ISBN 0-88385-619-0

[12] Eves, Howard, *Fundamentals of Modern Elementary Geometry,* Jones and Bartlett, Boston, 1992. ISBN 0-86720-247-5

[13] Feeman, Timothy G., *Portraits of the Earth: a Mathematician Looks at Maps,* Mathematical World 18, American Mathematical Society, Providence, 2002. ISBN 0-8218-3255-7

[14] Fenn, Roger, *Geometry,* Springer Undergraduate Mathematics Series, Springer-Verlag, London, 2001. ISBN 1-85233-058-9

[15] Gray, Jeremy J., *Ideas of Space: Euclidean, Non-Euclidean and Relativistic,* Clarendon Press, Oxford, 2nd edition, 1989. ISBN 0-19-853934-7

[16] Greenberg, Marvin Jay, *Euclidean and Non-Euclidean Geometry,* W. H. Freeman and Co., New York, 3rd edition, 1993. ISBN 0-7167-2446-4

[17] Iversen, B., *Hyperbolic Geometry,* LMS Student Texts 25, Cambridge University Press, Cambridge, 1987. ISBN 0-521-43508-0

[18] Pedoe, Dan, *A Course of Geometry for Colleges and Universities*, Cambridge University Press, Cambridge, 1970. ISBN 0-521-07638-2

[19] Rees, Elmer G., *Notes on Geometry,* Universitext, Springer-Verlag, Berlin, New York, 1983. ISBN 3-540-12053-X

[20] Silvester, John R., *Geometry, Ancient and Modern,* Oxford University Press, Oxford, 2001. ISBN 0-19-850825-5

[21] Stahl, Saul, *The Poincaré Half-Plane*, Jones and Bartlett, Boston, 1993. ISBN 0-86720-298-X

[22] Steinhaus, Hugo, *Mathematical Snapshots,* Oxford University Press, New York, 3rd American edition, 1983. ISBN 0-19-503267-5

[23] Wells, David G., *The Penguin Dictionary of Curious and Interesting Geometry,* Penguin Books, London, 1998. ISBN-13: 9780140118131

In addition, the World Wide Web contains a massive amount of interesting and relevant information that can be accessed using a Search Engine (such as GOOGLE). A little care is needed, however, as not all such information is accurate or clearly presented. However the 'MacTutor History of Mathematics archive' at http://www-history.mcs.st-andrews.ac.uk/ is a reliable and fascinating source of much historical information.

Appendix 1: A Primer of Group Theory

A **binary operation** $\circ$ defined on a set G is a map from $G \times G$ to G; that is, if g_1 and g_2 are elements of G then $g_1 \circ g_2 \in G$. We call the binary operation $\circ$ **composition** and $g_1 \circ g_2$ the **product** of g_1 and g_2, and we say that the binary operation is **closed**. An **isometry** of a figure F in $\mathbb{R}^2$ or $\mathbb{R}^3$ is a mapping of F to $\mathbb{R}^2$ or $\mathbb{R}^3$ that preserves distances; it is a **direct** isometry if it can be carried out by a 'continuous' transformation within the space concerned, and **indirect** otherwise.

Definition Let G be a set and $\circ$ a binary operation defined on G. Then $(G, \circ)$, or G for short, is a **group** if the following four axioms hold.

G1 CLOSURE For all $g_1, g_2 \in G$, $g_1 \circ g_2 \in G$.

G2 IDENTITY There exists an *identity element* $e \in G$ such that for all $g \in G$,

$$g \circ e = g = e \circ g.$$

G3 INVERSES For each $g \in G$, there exists an *inverse element* $g^{-1} \in G$ such that

$$g \circ g^{-1} = e = g^{-1} \circ g.$$

G4 ASSOCIATIVITY For all $g_1, g_2, g_3 \in G$,

$$g_1 \circ (g_2 \circ g_3) = (g_1 \circ g_2) \circ g_3.$$

In any group, the identity element is unique and each element has a unique inverse. If g belongs to a group, then $(g^{-1})^{-1} = g$.

Examples of groups include:

- The integers Z with the binary operation of addition, $+$;
- The real numbers $\mathbb{R}$ with the binary operation of addition, $+$;
- The positive real numbers $\mathbb{R}^+$ with the binary operation of multiplication, $\times$;
- The set of all translations of the plane $\mathbb{R}^2$;
- The set of all rotations and translations of the plane and their composites;

- The set of all composites of all reflections of the plane (this is the **isometry group** of the plane);
- The set of all 2×2 matrices with non-zero determinant.

If the binary operation $\circ$ of a group G is **commutative** (that is, $g_1 \circ g_2 = g_2 \circ g_1$ for all $g_1, g_2 \in G$), we say that the group is **commutative** or **Abelian**.

A subset H of a group G is a **subgroup** of G if it is a group with respect to the same binary operation. This can be reformulated as follows:

Definition Let $(G, \circ)$ be a group with identity e, and let H be a subset of G. Then (H, o) is a **subgroup** of $(G, \circ)$ if the following axioms hold:

SG1	CLOSURE	For all h_1 and $h_2 \in H$, the composite $h_1 \circ h_2 \in H$.
SG2	IDENTITY	The identity element $e \in H$.
SG3	INVERSES	For each $h \in H$, the inverse element $h^{-1} \in H$.

The **order** of an element g of a group G is the least positive integer n such that $g^n = e$. If there is no such integer, we say that g has **infinite order**. A **cyclic** group G is one in which every element is of the form g^n for some fixed element $g \in G$; g is called a **generator** of G.

The **order** of a group is the number of elements in it. If this number is infinite, we say that the group is of **infinite order**. For a group G of **finite order**, **Lagrange's Theorem** asserts that the order of any subgroup of G divides the order of G.

Two groups that have the same algebraic structure are called isomorphic.

Definition Two groups $(G, \circ)$ and $(H, \star)$ are **isomorphic** if there exists a mapping $\phi : G \to H$ such that the following properties hold:

1. ϕ is one-one and onto;
2. for all $g_1, g_2 \in G$, $\phi(g_1 \circ g_2) = \phi(g_1) \star \phi(g_2)$.

Such a mapping ϕ is called an **isomorphism**.

Two elements x, y of a group $(G, \circ)$ are **conjugates** of each other in G if there exists an element $g \in G$ such that $y = g \circ x \circ g^{-1}$ (which we often write simply as $y = gxg^{-1}$). We say that g **conjugates** x to y, and that g is a **conjugating element**.

Let H be a subgroup of a group $(G, \circ)$, and let $g \in G$; then the set $\boldsymbol{gH} = \{g \circ h : h \in H\}$ is a **left coset** of H in G. It is a subset of G, but in general not a subgroup of G. Two left cosets $g_1 H$ and $g_2 H$ are either equal or have no elements in common. A **right coset** is defined similarly as a set of the form $\boldsymbol{Hg} = \{h \circ g : h \in H\}$.

For any given group G and subgroup H and any two elements g_1 and g_2 of G, there is usually no particular relation between a left coset $g_1 H$ and a right coset $H g_2$; a right coset need not be a left coset, nor a left coset a right coset. If however every right coset is a left coset, and every left coset a right coset, then the subgroup H of G is called a normal subgroup of G.

> **Definition** A subgroup H is a **normal subgroup** of a group G if, for each element $g \in G$, $gH = Hg$.

Note that this does *not* require that $gh = hg$ for each h.

For example, the group of all translations of the plane is a normal subgroup of the group Γ of all isometries of the plane; but the group of all rotations about a fixed point P in the plane, although a subgroup of Γ, is not a normal subgroup of Γ.

If H is a normal subgroup of G, we can talk about the **cosets** of H in G rather than the left cosets and the right cosets. In this case, the cosets themselves form a group under the binary operation $\cdot$, called **set composition**, defined for subsets A, B of G as follows:

$$A \cdot B = \{a \circ b : a \in A, b \in B\}.$$

(In particular, $g_1H \cdot g_2H = (g_1 \circ g_2)H$, for all g_1, g_2 in G.) This group is called the **quotient group** of G by H, and is denoted by ***G/H***. The identity element of G/H is the coset $eH = H$, and the inverse of gH is $g^{-1}H$.

Every subgroup of an Abelian group G is a normal subgroup of G.

Appendix 2: A Primer of Vectors and Vector Spaces

Vectors

A **vector** is a quantity that is determined by its magnitude and direction. A **scalar** is a quantity that is determined by its magnitude. A vector in $\mathbb{R}^2$ or in $\mathbb{R}^3$ can be represented by a line whose length is a measure of the **magnitude** (or **length**) of the vector and whose direction is the same as that of the vector. A vector **represented** by a line segment from A to B is often written as $\overrightarrow{AB}$.

The **vector space** $\mathbb{R}^2$ is the set of ordered pairs of real numbers with the operations of addition and scalar multiplication defined as follows:

$$(x_1, y_1) + (x_2, y_2) = (x_1 + x_2, y_1 + y_2);$$
$$\alpha(x, y) = (\alpha x, \alpha y), \quad \text{for any real number } \alpha.$$

In $\mathbb{R}^2$ the vectors **i** and **j** are unit vectors in the positive direction of the x- and y-axes, respectively. Any **vector v** in $\mathbb{R}^2$ can be expressed as a sum of the form

$$\mathbf{v} = x\mathbf{i} + y\mathbf{j}, \quad \text{for some real numbers } x \text{ and } y;$$

often we simply write $\mathbf{v} = (x, y)$, for brevity. The numbers x, y are the **components** of **v** in the x- and y-directions. Similarly, in $\mathbb{R}^3$ the vectors **i**, **j** and **k** are unit vectors in the positive direction of the x-, y- and z-axes, respectively. Any **vector v** in $\mathbb{R}^3$ can be expressed as a sum of the form $\mathbf{v} = x\mathbf{i} + y\mathbf{j} + z\mathbf{k}$, for some real numbers x, y and z.

The **vector space** $\mathbb{R}^3$ is defined similarly.

The **dot** (or **scalar**) **product** of two vectors **u** and **v** is

$$\mathbf{u} \cdot \mathbf{v} = ||\mathbf{u}|| \times ||\mathbf{v}|| \times \cos\theta,$$

where $||\mathbf{u}||$ and $||\mathbf{v}||$ denote the lengths of the vectors **u** and **v** and θ is the **angle** between them. It follows that the angle θ between two vectors **u** and **v** is given by

$$\cos\theta = \frac{\mathbf{u} \cdot \mathbf{v}}{||\mathbf{u}|| \times ||\mathbf{v}||}.$$

If $\mathbf{u} = x_1\mathbf{i} + y_1\mathbf{j}$ and $\mathbf{v} = x_2\mathbf{i} + y_2\mathbf{j}$, then $\mathbf{u} \cdot \mathbf{v} = x_1x_2 + y_1y_2$.

Two non-zero vectors **u** and **v** are **orthogonal** if they are at right angles; then $\mathbf{u} \cdot \mathbf{v} = 0$.

The **length** of a vector **v** is given in terms of the dot product as $||\mathbf{v}|| = \sqrt{\mathbf{v} \cdot \mathbf{v}}$.

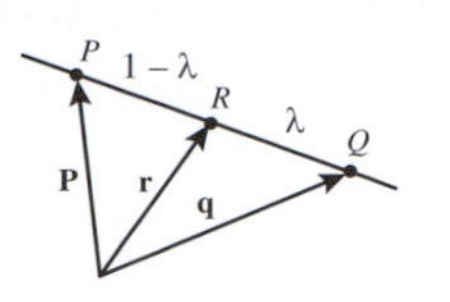

Position Vectors

The **position vector** $\mathbf{v} = x\mathbf{i} + y\mathbf{j}$ is the vector in $\mathbb{R}^2$ whose starting point is the origin and whose finishing point is the point with Cartesian coordinates (x, y). A position vector in $\mathbb{R}^3$ is defined similarly.

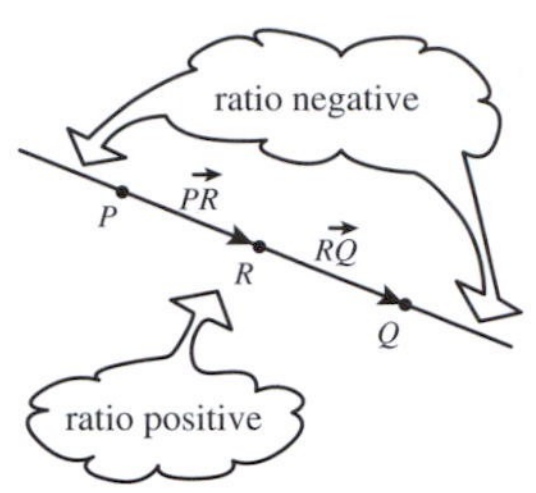

Let R be a point on the line which passes through two points P and Q in $\mathbb{R}^2$, and suppose that **p**, **q** and **r** are the position vectors of P, Q and R, respectively. Then there is a real number λ such that

$$\mathbf{r} = \lambda\mathbf{p} + (1 - \lambda)\mathbf{q}.$$

The coefficients of **p** and **q** in this expression can be used to calculate the ratio in which the point R divides the line segment PQ; namely,

$$\frac{PR}{RQ} = \frac{1 - \lambda}{\lambda}.$$

This ratio is actually a 'signed' ratio. If $0 < \lambda < 1$, then the ratio is positive and R lies between P and Q. But if $\lambda > 1$ or if $\lambda < 0$, then the ratio is negative and R lies beyond P or beyond Q, respectively.

This negative ratio can be interpreted as meaning that the vectors $\overrightarrow{PR}$ and $\overrightarrow{RQ}$ point in opposite directions.

Another way in which we can calculate ratios is to use the Cartesian coordinates of the points concerned. Indeed, provided that the line through P, Q and R is not parallel to the y-axis, then it is clear from the figure in the margin that

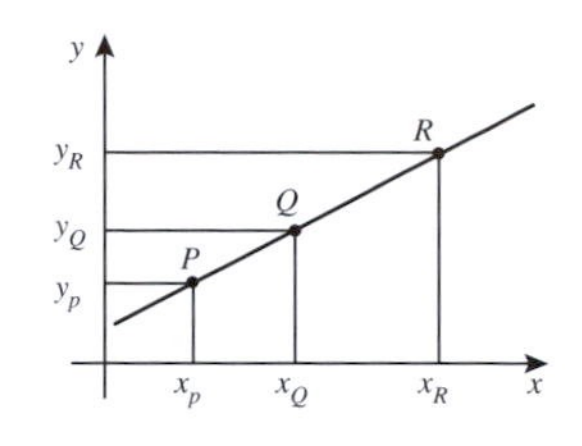

$$\frac{PR}{RQ} = \frac{x_R - x_P}{x_Q - x_R},$$

where x_P, x_Q and x_R are the x-coordinates of the points P, Q and R, respectively.

Similarly, if the line through P, Q and R is not parallel to the x-axis, then

$$\frac{PR}{RQ} = \frac{y_R - y_P}{y_Q - y_R},$$

where y_P, y_Q and y_R are the y-coordinates of the points P, Q and R, respectively.

Often the line through P, Q and R will be parallel to neither of the axes, in which case both formulas will work. In such cases we can pick the formula that makes the calculations simplest. One advantage of the coordinate approach is that we do not have to worry about whether the ratio is positive or negative, for the coordinates take care of the sign automatically.

Section Formula The position vector **r** of the point R that divides the line segment ℓ joining the points P and Q with position vectors **p** and **q** in the ratio $(1 - \lambda) : \lambda$ is

$$\mathbf{r} = \lambda\mathbf{p} + (1 - \lambda)\mathbf{q}.$$

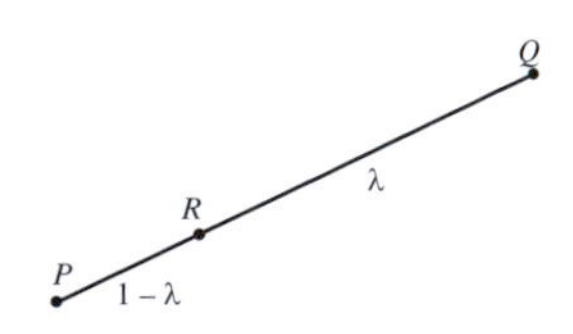

It follows that the **vector form of the equation of a line** through the points with position vectors $\mathbf{p}$ and $\mathbf{q}$ is

$$\mathbf{r} = \lambda\mathbf{p} + (1-\lambda)\mathbf{q}, \quad \text{where } \lambda \text{ varies over all real numbers.}$$

Vector Spaces

Definitions A **real vector space** consists of a set V of elements (called **vectors**) and a binary operation $+$ defined on V such that $(V,+)$ is an Abelian group, and such that for each $\mathbf{v} \in V$ and each real number $\alpha, \alpha\mathbf{v} \in V$. Moreover,

$$1\mathbf{v} = \mathbf{v},$$

$$\alpha(\beta\mathbf{v}) = (\alpha\beta)\mathbf{v},$$

and the two distributive laws hold:

$$\alpha(\mathbf{v}+\mathbf{w}) = \alpha\mathbf{v} + \alpha\mathbf{w},$$

$$(\alpha+\beta)\mathbf{v} = \alpha\mathbf{v} + \beta\mathbf{v}.$$

Examples of real vector spaces include:

- the set of real numbers with the operations of (ordinary) addition and of multiplication by real numbers;
- the set of ordered pairs of real numbers, $\mathbb{R}^2 = \{(x,y) : x, y \in \mathbb{R}\}$, where $(x_1,y_1)+(x_2,y_2) = (x_1+x_2, y_1+y_2)$ and $\alpha(x,y) = (\alpha x, \alpha y)$;
- the set of ordered triples of real numbers, $\mathbb{R}^3 = \{(x,y,z) : x,y,z \in \mathbb{R}\}$, where $(x_1,y_1,z_1)+(x_2,y_2,z_2) = (x_1+x_2, y_1+y_2, z_1+z_2)$ and $\alpha(x,y,z) = (\alpha x, \alpha y, \alpha z)$.

A set $E = \{\mathbf{e}_1, \mathbf{e}_2, \ldots, \mathbf{e}_n, \ldots\}$ of vectors is **linearly independent** if the only solution of the equation $\alpha_1\mathbf{e}_1 + \alpha_2\mathbf{e}_2 + \ldots + \alpha_n\mathbf{e}_n + \ldots = \mathbf{0}$ is $\alpha_1 = \alpha_2 = \ldots = 0$; if it has another solution, then E is said to be **linearly dependent.**

Given any set $E = \{\mathbf{e}_1, \mathbf{e}_2, \ldots, \mathbf{e}_n, \ldots\}$ of vectors in a vector space V, the set of all possible vectors of the form $\{\alpha_1\mathbf{e}_1 + \alpha_2\mathbf{e}_2 + \ldots + \alpha_n\mathbf{e}_n + \ldots : \alpha_1, \alpha_2, \ldots, \alpha_n, \ldots \text{ real}\}$ is called the **span** of E, denoted by $\langle E\rangle$. If E is any subset of V, then $\langle E\rangle$ is a vector space too, called a **vector subspace** of V.

A **basis** of a vector space V is a subset E of V such that E is a linearly independent set and E spans V. Then each vector $\mathbf{v}$ in V can be expressed uniquely in the form

$$\mathbf{v} = \alpha_1\mathbf{e}_1 + \alpha_2\mathbf{e}_2 + \ldots + \alpha_n\mathbf{e}_n + \ldots, \quad \text{where } \alpha_1, \alpha_2, \ldots, \alpha_n, \ldots \text{ are real.}$$

All bases of a given vector space V have the same number of elements. If this number is finite, n say, then V is said to be **finite-dimensional** or to have **dimension** n; if this number is infinite, then V is said to be **infinite-dimensional**.

The **standard basis** of $\mathbb{R}^n$ is the set of n vectors

$$\{(1,0,0,\ldots,0),(0,1,0,\ldots,0),\ldots,(0,0,\ldots,1)\}.$$

A basis E of a vector space V is an **orthogonal basis** if each pair of vectors in E are at right angles to each other; it is an **orthonormal basis** if, in addition, the length of each vector in E is 1.

Linear Transformations

Definition A **linear transformation** t is a mapping of one vector space V into another vector space W such that

(a) $t(\mathbf{u}+\mathbf{v}) = t(\mathbf{u}) + t(\mathbf{v})$, for all $\mathbf{u},\mathbf{v}\in V$; and
(b) $t(\alpha\mathbf{v}) = \alpha t(\mathbf{v})$, for all $\mathbf{v}\in V$, and all real α.

Let $t : V \to W$ be a linear transformation of a (**domain**) vector space V of dimension n and with a basis $E = \{\mathbf{e}_1, \mathbf{e}_2, \ldots, \mathbf{e}_n\}$ into a (**codomain**) vector space W of dimension m and with a basis $F = \{\mathbf{f}_1, \mathbf{f}_2, \ldots, \mathbf{f}_m\}$. Then, if

$$\begin{aligned}
t(\mathbf{e}_1) &= a_{11}\mathbf{f}_1 + a_{21}\mathbf{f}_2 + \ldots + a_{m1}\mathbf{f}_m,\\
t(\mathbf{e}_2) &= a_{12}\mathbf{f}_1 + a_{22}\mathbf{f}_2 + \ldots + a_{m2}\mathbf{f}_m,\\
&\ldots\\
t(\mathbf{e}_n) &= a_{1n}\mathbf{f}_1 + a_{2n}\mathbf{f}_2 + \ldots + a_{mn}\mathbf{f}_m,
\end{aligned}$$

the rectangular array or **matrix** denoted by

$$\mathbf{A} = (a_{ij}) = \begin{pmatrix} a_{11} & a_{12} & \ldots & a_{1n} \\ a_{21} & a_{22} & \ldots & a_{2n} \\ & & \vdots & \\ a_{m1} & a_{m2} & \ldots & a_{mn} \end{pmatrix}$$

with m rows and n columns **represents** the linear transformation t; we also say that t has **matrix A**. The matrix **A** is said to be of **size** or **dimension** $m \times n$. A 1-row matrix is called a **row matrix**, and a 1-column matrix is called a **column matrix**. An $m \times n$ matrix is **square** if $m = n$; that is, if it has the same number of rows as columns. A square matrix $\mathbf{A} = (a_{ij})$ is **diagonal** if the entries not on the **principal diagonal** from top left to bottom right of the array are zero; that is, $a_{ij} = 0$ for $i \neq j$. The **identity matrix** $\mathbf{I}_n$ (or **I**, for short) is the $n \times n$ diagonal matrix whose entries on the principal diagonal are all 1s.

An **anticlockwise rotation** of $\mathbb{R}^2$ through an angle θ is represented by the matrix

$$\begin{pmatrix} \cos\theta & -\sin\theta \\ \sin\theta & \cos\theta \end{pmatrix},$$

a **dilatation** of $\mathbb{R}^2$ by a factor k by

$$\begin{pmatrix} k & 0 \\ 0 & k \end{pmatrix},$$

and a **reflection** of $\mathbb{R}^2$ in the line $y = kx$ by

$$\begin{pmatrix} \dfrac{1-k^2}{1+k^2} & \dfrac{2k}{1+k^2} \\ \dfrac{2k}{1+k^2} & \dfrac{1-k^2}{1+k^2} \end{pmatrix}.$$

Two matrices **A** and **B** are **equal** if they are of the same size, and corresponding entries are equal. The sum, **A + B**, of matrices **A** and **B** is defined only if **A** and **B** are of the same size; the entries of **A + B** are obtained by adding corresponding entries of **A** and **B**. For any real number α and matrix **A**, the matrix $\alpha\mathbf{A}$ is a matrix of the same size as **A** but whose entries are α times the corresponding entries of **A**. The difference $\mathbf{A} - \mathbf{B}$ of matrices **A** and **B** is simply $\mathbf{A} + (-1)\mathbf{B}$.

The **transpose** $\mathbf{A}^T$ of a matrix **A** of size $m \times n$ is the matrix of size $n \times m$ obtained by interchanging the ith row for the ith column; that is, if $\mathbf{A} = (a_{ij})$ then $\mathbf{A}^T = (a_{ji})$.

The **product AB** of matrices **A** and **B** of sizes $m \times n$ and $k \times l$ is only defined if $n = k$; in this case, **AB** is the $m \times l$ matrix whose entry in the ith row and jth column is the dot product of the ith row of **A** and the jth column of **B**.

If $E = \{\mathbf{e}_1, \mathbf{e}_2, \ldots, \mathbf{e}_n\}$ is a basis for V and $F = \{\mathbf{f}_1, \mathbf{f}_2, \ldots, \mathbf{f}_m\}$ is a basis for W, and if $\mathbf{v} = \alpha_1\mathbf{e}_1 + \alpha_2\mathbf{e}_2 + \ldots + \alpha_n\mathbf{e}_n \in V$, then we can express **v** (in terms of its components with respect to E) in the matrix form

$$\mathbf{v} = \begin{pmatrix} \alpha_1 \\ \alpha_2 \\ \vdots \\ \alpha_n \end{pmatrix},$$

and the image of **v** under t (in terms of its components with respect to F) in the matrix form

$$t(\mathbf{v}) = \mathbf{A}\mathbf{v} = \begin{pmatrix} a_{11} & a_{12} & \ldots & a_{1n} \\ a_{21} & a_{22} & \ldots & a_{2n} \\ & & \vdots & \\ a_{m1} & a_{m2} & \ldots & a_{mn} \end{pmatrix} \begin{pmatrix} \alpha_1 \\ \alpha_2 \\ \vdots \\ \alpha_n \end{pmatrix}$$

$$= \begin{pmatrix} a_{11}\alpha_1 + a_{12}\alpha_2 + \ldots + a_{1n}\alpha_n \\ a_{21}\alpha_1 + a_{22}\alpha_2 + \ldots + a_{2n}\alpha_n \\ \vdots \\ a_{m1}\alpha_1 + a_{m2}\alpha_2 + \ldots + a_{mn}\alpha_n \end{pmatrix},$$

so that

$$t(\mathbf{v}) = (a_{11}\alpha_1 + a_{12}\alpha_2 + \ldots + a_{1n}\alpha_n)\mathbf{f}_1 + (a_{21}\alpha_1 + a_{22}\alpha_2 + \ldots + a_{2n}\alpha_n)\mathbf{f}_2 + \ldots + (a_{m1}\alpha_1 + a_{m2}\alpha_2 + \ldots + a_{mn}\alpha_n)\mathbf{f}_m.$$

Let s be a linear transformation mapping a vector space V (of dimension n and with a basis E) into a vector space W (of dimension m and with a basis F) with matrix $\mathbf{A}$; let t be a linear transformation mapping the vector space W (with the same basis F) into a vector space K (of dimension k and with a basis G) with matrix $\mathbf{B}$. Then the **composite** mapping $t \circ s$ is a linear transformation with matrix $\mathbf{BA}$. In general, $\mathbf{AB} \neq \mathbf{BA}$ even when both products are defined. If the product $\mathbf{AB}$ is defined, then $(\mathbf{AB})^T = \mathbf{B}^T\mathbf{A}^T$.

A linear transformation s of a vector space V into a vector space W is **invertible** if there is a linear transformation t of W into V such that

$$s \circ t = t \circ s = \text{the identity mapping of } V \text{ to itself.}$$

If V and W are of dimensions n and m, then s is invertible only if $m = n$; if s is invertible and has matrix $\mathbf{A}$, then the matrix of t is denoted by $\mathbf{A}^{-1}$, the **(matrix) inverse** of $\mathbf{A}$. If $\mathbf{A}$ is **invertible**, then $\mathbf{AA}^{-1} = \mathbf{A}^{-1}\mathbf{A} = \mathbf{I}_n$.

If $\mathbf{A}$ and $\mathbf{B}$ are invertible $n \times n$ matrices, then $\mathbf{AB}$ is invertible and $(\mathbf{AB})^{-1} = \mathbf{B}^{-1}\mathbf{A}^{-1}$. A square matrix $\mathbf{A}$ is **orthogonal** if $\mathbf{A}^{-1} = \mathbf{A}^T$, or, equivalently, if $\mathbf{AA}^T = \mathbf{I}$ or $\mathbf{A}^T\mathbf{A} = \mathbf{I}$.

Determinants

Associated with any square matrix $\mathbf{A}$ is a number called the **determinant** of $\mathbf{A}$, denoted by det $\mathbf{A}$, det($\mathbf{A}$) or $|\mathbf{A}|$.

If $\mathbf{A}$ is a 1×1 matrix (a), then det $\mathbf{A} = a$. If $\mathbf{A}$ is a 2×2 matrix $\begin{pmatrix} a & b \\ c & d \end{pmatrix}$, then

$$\det \mathbf{A} = \begin{vmatrix} a & b \\ c & d \end{vmatrix} = ad - bc.$$

If $\mathbf{A}$ is a 3×3 matrix

$$\begin{pmatrix} a & b & c \\ d & e & f \\ g & h & i \end{pmatrix},$$

then

$$\begin{aligned} \det \mathbf{A} &= \begin{vmatrix} a & b & c \\ d & e & f \\ g & h & i \end{vmatrix} \\ &= a\begin{vmatrix} e & f \\ h & i \end{vmatrix} - b\begin{vmatrix} d & f \\ g & i \end{vmatrix} + c\begin{vmatrix} d & e \\ g & h \end{vmatrix} \\ &= aei - afh - bdi + bfg + cdh - ceg. \end{aligned}$$

In each term of the second last line above, the coefficient is an entry in the top row of the determinant, with alternate + and − signs inserted, and the associated 2×2 determinant is obtained by deleting the row and column of $\mathbf{A}$ that contain that entry. ($n \times n$ determinants may be defined analogously.)

The determinant has the following properties:

$$\det \mathbf{I} = 1; \qquad \det \mathbf{A}^T = det\mathbf{A};$$

$$\det(\mathbf{AB}) = det\mathbf{A} \cdot det\mathbf{B}; \quad \det \mathbf{A}^{-1} = 1/(det\mathbf{A}).$$

A square matrix $\mathbf{A}$ is invertible if and only if $\det \mathbf{A} \neq 0$. If $\mathbf{A}$ is a 2×2 matrix $\begin{pmatrix} a & b \\ c & d \end{pmatrix}$, then

$$\mathbf{A}^{-1} = \frac{1}{\det \mathbf{A}} \begin{pmatrix} d & -b \\ -c & a \end{pmatrix}.$$

There are several ways of finding the inverse of a 3×3 matrix; we give only one here! The **adjoint matrix**, adj $\mathbf{A}$, associated with the 3×3 matrix

$$\mathbf{A} = \begin{pmatrix} a & b & c \\ d & e & f \\ g & h & i \end{pmatrix},$$

is the 3×3 matrix

$$\text{adj } \mathbf{A} = \begin{pmatrix} \begin{vmatrix} e & f \\ h & i \end{vmatrix} & -\begin{vmatrix} d & f \\ g & i \end{vmatrix} & \begin{vmatrix} d & e \\ g & h \end{vmatrix} \\ -\begin{vmatrix} b & c \\ h & i \end{vmatrix} & \begin{vmatrix} a & c \\ g & i \end{vmatrix} & -\begin{vmatrix} a & b \\ g & h \end{vmatrix} \\ \begin{vmatrix} b & c \\ e & f \end{vmatrix} & -\begin{vmatrix} a & c \\ d & f \end{vmatrix} & \begin{vmatrix} a & b \\ d & e \end{vmatrix} \end{pmatrix}^T$$

obtained by replacing each entry in $\mathbf{A}$ by the 2×2 determinant (called the **cofactor** of that entry) obtained by deleting the row and column of $\mathbf{A}$ that contain that entry, together with alternate + and − signs as indicated, and then taking the transpose. Then

$$\mathbf{A}^{-1} = \frac{1}{\det \mathbf{A}}(\text{adj } \mathbf{A}).$$

The equation of the line through two points $A = (a_1, a_2)$, $B = (b_1, b_2)$ in the plane $\mathbb{R}^2$ is

$$\begin{vmatrix} a_1 & b_1 & x \\ a_2 & b_2 & y \\ 1 & 1 & 1 \end{vmatrix} = 0 \quad \text{or} \quad \begin{vmatrix} a_1 & a_2 & 1 \\ b_1 & b_2 & 1 \\ x & y & 1 \end{vmatrix} = 0;$$

and three points $A = (a_1, a_2)$, $B = (b_1, b_2)$ and $C = (c_1, c_2)$ in the plane are collinear if and only if

$$\begin{vmatrix} a_1 & b_1 & c_1 \\ a_2 & b_2 & c_2 \\ 1 & 1 & 1 \end{vmatrix} = 0 \quad \text{or} \quad \begin{vmatrix} a_1 & a_2 & 1 \\ b_1 & b_2 & 1 \\ c_1 & c_2 & 1 \end{vmatrix} = 0.$$

Eigenvectors and Eigenvalues

An **eigenvector** of a linear transformation t of $\mathbb{R}^2$ (or $\mathbb{R}^3$) to itself is a non-zero vector $\mathbf{v}$ such that $t(\mathbf{v}) = \lambda\mathbf{v}$ for some real number λ called the corresponding **eigenvalue.** The **eigenspace** $S(\lambda)$ is the vector subspace of $\mathbb{R}^2$ (or $\mathbb{R}^3$) spanned by all eigenvectors of t with λ as the corresponding eigenvalue; an eigenspace may have dimension greater than 1.

If a linear transformation of $\mathbb{R}^2$(or $\mathbb{R}^3$) to itself has matrix $\mathbf{A}$, then a non-zero vector $\mathbf{v}$ such that $\mathbf{Av} = \lambda\mathbf{v}$ for some real number λ is called an **eigenvector** of $\mathbf{A}$; λ is called the corresponding **eigenvalue**.

The eigenvalues λ of an $n \times n$ matrix $\mathbf{A}$ satisfy the equation $\det(\mathbf{A} - \lambda\mathbf{I}) = 0$, called the **characteristic equation** of $\mathbf{A}$. A 3×3 matrix necessarily has at least one real eigenvalue. The **trace** of $\mathbf{A}$, denoted by tr $\mathbf{A}$, is the sum of the elements along the principal diagonal of $\mathbf{A}$, and equals the sum of the eigenvalues of $\mathbf{A}$.

To find the eigenvalues and eigenvectors of a matrix $\mathbf{A}$, first solve the characteristic equation to determine the eigenvalues λ; then solve the simultaneous **eigenvector equations** $(\mathbf{A} - \lambda\mathbf{I})\mathbf{v} = \mathbf{0}$ to find the corresponding eigenvectors and eigenspaces.

An $n \times n$ matrix $\mathbf{A}$ is **diagonalizable** if there is an $n \times n$ diagonal matrix $\mathbf{D}$ such that $\mathbf{D} = \mathbf{P}^{-1}\mathbf{AP}$ for some invertible $n \times n$ matrix $\mathbf{P}$; the entries on the principal diagonal of $\mathbf{D}$ are the eigenvalues of $\mathbf{A}$. An $n \times n$ matrix $\mathbf{A}$ is diagonalizable if and only if it possesses n linearly independent eigenvectors.

A (square) matrix $\mathbf{A}$ is **symmetric** if $\mathbf{A}^T = \mathbf{A}$; an $n \times n$ symmetric matrix has n real eigenvalues. A symmetric $n \times n$ matrix $\mathbf{A}$ is **orthogonally diagonalizable;** that is, there is an $n \times n$ diagonal matrix $\mathbf{D}$ such that $\mathbf{D} = \mathbf{P}^{-1}\mathbf{AP}$ for some orthogonal $n \times n$ matrix $\mathbf{P}$. The columns of $\mathbf{P}$ are eigenvectors of $\mathbf{A}$ of length 1, and are mutually orthogonal.

Appendix 3: Solutions to the Problems

Chapter 1

Section 1.1

1. Here we use the standard formula for the equation of a circle of given centre and radius given in Theorem 1.
   (a) This circle has equation
   $$(x-0)^2+(y-0)^2=1^2,$$
   which can be rewritten in the form
   $$x^2+y^2=1.$$
   (b) This circle has equation
   $$(x-0)^2+(y-0)^2=4^2,$$
   which can be rewritten in the form
   $$x^2+y^2=16.$$
   (c) This circle has equation
   $$(x-3)^2+(y-4)^2=2^2,$$
   which can be rewritten in the form
   $$x^2+y^2-6x-8y+21=0.$$
   (d) This circle has equation
   $$(x-3)^2+(y-4)^2=3^2,$$
   which can be rewritten in the form
   $$x^2+y^2-6x-8y+16=0.$$
2. Since the origin lies on the circle, its coordinates (0, 0) must satisfy the equation of the circle. Thus
   $$0^2+0^2+f\cdot 0+g\cdot 0+h=0,$$
   which reduces to the condition that $h=0$.
3. (a) We can use the general formula of Theorem 2 for centre and radius, with $f=-2, g=-6$ and $h=1$. It follows that the circle has centre $(1,3)$ and radius $\sqrt{9}=3$.
   (b) Here the coefficients of x^2 and y^2 are both 3, so we divide the equation by 3 to get it into standard form (that is, with the coefficients of x^2 and y^2 both 1); the equation becomes
   $$x^2+y^2-4x-16y=0.$$
   We can use the general formula of Theorem 2 for centre and radius, with $f=-4, g=-16$ and $h=0$. It follows that the circle has centre $(2,8)$ and radius $\sqrt{68}=2\sqrt{17}$.
4. (a) If we complete the square in the equation
   $$x^2+y^2+x+y+1=0, \qquad (1)$$
   we obtain the equation
   $$\left(x+\tfrac{1}{2}\right)^2+\left(y+\tfrac{1}{2}\right)^2=-\tfrac{1}{2}.$$
   This equation represents the empty set, since its left-hand side is always non-negative whereas its right-hand side is negative.

   (Alternatively, we can apply the Remark following Theorem 2 of Subsection 1.1.2 to the quantity $\frac{1}{4}f^2+\frac{1}{4}g^2-h$. In equation (1), we have $f=1, g=1$ and $h=1$. Thus
   $$\tfrac{1}{4}f^2+\tfrac{1}{4}g^2-h=\tfrac{1}{4}+\tfrac{1}{4}-1<0,$$
   so that the set must be the empty set.)

   (b) If we complete the square in the equation
   $$x^2+y^2-2x+4y+5=0, \qquad (2)$$

we obtain the equation

$$(x-1)^2 - 1 + (y+2)^2 - 4 + 5 = 0,$$

or

$$(x-1)^2 + (y+2)^2 = 0.$$

Thus the set in the plane represented by equation (2) is the single point $(1, -2)$.

(Alternatively, we can apply the Remark following Theorem 2 of Subsection 1.1.2 to the quantity $\frac{1}{4}f^2 + \frac{1}{4}g^2 - h$. In equation (2), we have $f = -2, g = 4$ and $h = 5$. Thus

$$\tfrac{1}{4}f^2 + \tfrac{1}{4}g^2 - h = 1 + 4 - 5 = 0,$$

so that the set must be a single point $\left(-\frac{1}{2}f, -\frac{1}{2}g\right) = (1, -2)$.)

(c) Here the coefficients of x^2 and y^2 are both 2; so we divide the equation by 2 to get it into standard form as

$$x^2 + y^2 + \tfrac{1}{2}x - \tfrac{3}{2}y - \tfrac{5}{2} = 0. \quad (3)$$

Then, if we complete the square in equation (3) we obtain the equation

$$\left(x + \tfrac{1}{4}\right)^2 - \tfrac{1}{16} + \left(y - \tfrac{3}{4}\right)^2 - \tfrac{9}{16} - \tfrac{5}{2} = 0,$$

or

$$\left(x + \tfrac{1}{4}\right)^2 + \left(y - \tfrac{3}{4}\right)^2 = \tfrac{25}{8}.$$

Thus the set in the plane represented by equation (3) is a circle with centre $\left(-\frac{1}{4}, \frac{3}{4}\right)$ and radius $\sqrt{\frac{25}{8}} = \frac{5}{4}\sqrt{2}$.

(Alternatively, we can apply Theorem 2 to the quantity $\frac{1}{4}f^2 + \frac{1}{4}g^2 - h$. In equation (3), we have $f = \frac{1}{2}, g = -\frac{3}{2}$ and $h = -\frac{5}{2}$. Thus

$$\tfrac{1}{4}f^2 + \tfrac{1}{4}g^2 - h = \tfrac{1}{16} + \tfrac{9}{16} + \tfrac{5}{2} = \tfrac{25}{8},$$

so that the set is the circle with centre $\left(-\frac{1}{4}, \frac{3}{4}\right)$ and radius $\sqrt{\frac{25}{8}} = \frac{5}{4}\sqrt{2}$.)

5. Here we use the result of Theorem 3.

(a) The condition that the intersecting circles

$$C_1 : x^2 + y^2 - 4x - 4y + 7 = 0$$
$$C_2 : x^2 + y^2 + 2x - 8y + 5 = 0$$

are orthogonal is that

$$(-4)2 + (-4)(-8) = 2(7 + 5);$$

that is, that $24 = 24$. This is the case; hence C_1 and C_2 are orthogonal.

(b) Here we must divide the given equation of C_2 by 3 in order to get it in the form $x^2 + y^2 + \cdots = 0$, before using Theorem 3. The condition that the intersecting circles

$$C_1 : x^2 + y^2 + 3x - 6y + 5 = 0$$
$$C_2 : x^2 + y^2 + \tfrac{4}{3}x + \tfrac{1}{3}y - 5 = 0$$

are orthogonal is that

$$(3)\left(\tfrac{4}{3}\right) + (-6)\left(\tfrac{1}{3}\right) = 2(5 - 5);$$

that is, that $4 - 2 = 0$. This is not the case; hence, C_1 and C_2 are not orthogonal.

6. First we write the equation of the second circle in the (normalized) form

$$x^2 + y^2 + \tfrac{5}{2}x - 3y + \tfrac{3}{2} = 0,$$

in which the cofficients of x^2 and y^2 are 1.

Then, by Theorem 4 (with $k = -1$) it follows that the equation of the line through the points of intersection of the two given circles is

$$x^2 + y^2 - 3x + 4y - 1 - 1\left(x^2 + y^2 + \tfrac{5}{2}x - 3y + \tfrac{3}{2}\right) = 0,$$

or

$$-\tfrac{11}{2}x + 7y - \tfrac{5}{2} = 0;$$

we can write this in the simpler form

$$11x - 14y + 5 = 0.$$

7. (a) The parabola E is the parabola in standard form where $4a = 1$, or $a = \frac{1}{4}$. It follows that the focus of E is $\left(\frac{1}{4}, 0\right)$, the vertex is $(0, 0)$, the axis is the x-axis, and the equation of the directrix is $x = -\frac{1}{4}$.

(b) The coordinates of P and Q are (t_1^2, t_1) and (t_2^2, t_2), respectively. So, if $t_1^2 \neq t_2^2$, the slope of PQ is given by

$$m = \frac{t_1 - t_2}{t_1^2 - t_2^2} = \frac{1}{t_1 + t_2}.$$

Since (t_1^2, t_1) lies on the line PQ, it follows that the equation of PQ is

$$y - t_1 = \frac{1}{t_1 + t_2}\left(x - t_1^2\right).$$

Multiplying both sides by $t_1 + t_2$, we get

$$(t_1 + t_2)(y - t_1) = x - t_1^2,$$

so that

$$(t_1 + t_2)y - t_1^2 - t_1 t_2 = x - t_1^2,$$

or

$$(t_1 + t_2)y = x + t_1 t_2. \quad (4)$$

If, however, $t_1^2 = t_2^2$, then since $t_1 \neq t_2$ we have $t_1 = -t_2$. Thus PQ is parallel to the y-axis, and so has equation $x = t_1^2$; so in this case too, PQ has equation (4).

(c) The midpoint of PQ is the point

$$\left(\tfrac{1}{2}\left(t_1^2 + t_2^2\right), \tfrac{1}{2}(t_1 + t_2)\right).$$

This is the focus $\left(\tfrac{1}{4}, 0\right)$ if

$$\left(\tfrac{1}{2}\left(t_1^2 + t_2^2\right), \tfrac{1}{2}(t_1 + t_2)\right) = \left(\tfrac{1}{4}, 0\right).$$

Comparing the second coordinates, we deduce that $t_2 = -t_1$. Comparing the first coordinates, we deduce that

$$\tfrac{1}{2}\left(t_1^2 + t_2^2\right) = \tfrac{1}{4}$$

so that $t_1^2 = \frac{1}{4}$. It follows that $t_1 = \pm\frac{1}{2}$, and so that $t_2 = \mp\frac{1}{2}$, respectively.

When $t = \frac{1}{2}$, the point $(t^2, t) = \left(\frac{1}{4}, \frac{1}{2}\right)$; and when $t = -\frac{1}{2}$, the point $(t^2, t) = \left(\frac{1}{4}, -\frac{1}{2}\right)$. It follows that the points P and Q must be $\left(\frac{1}{4}, \frac{1}{2}\right)$ and $\left(\frac{1}{4}, -\frac{1}{2}\right)$.

Remark

It follows that the chord PQ must be the vertical chord through the focus.

8.

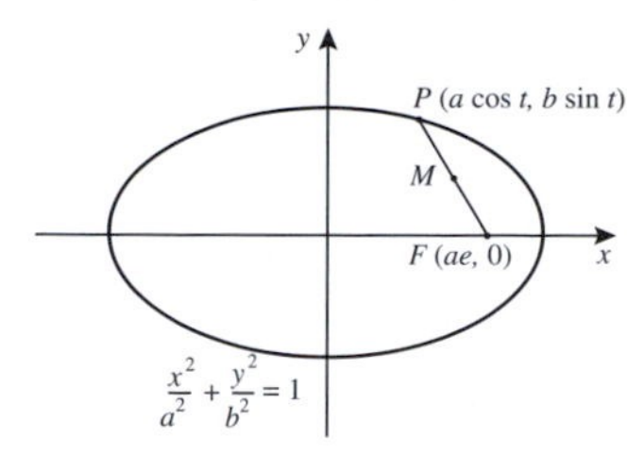

Let P have coordinates $(a\cos t, b\sin t)$. Since the coordinates of F are $(ae, 0)$, the coordinates of M, the midpoint of FP, are

$$\left(\tfrac{1}{2}(a\cos t + ae), \tfrac{1}{2}(a\sin t + 0)\right).$$

Thus M lies on the curve in $\mathbb{R}^2$ with parametric equations

$$x = \tfrac{1}{2}(a\cos t + ae), \quad y = \tfrac{1}{2}b\sin t.$$

We can rearrange these equations in the form

$$\cos t = \frac{2x - ae}{a}, \quad \sin t = \frac{2y}{b};$$

squaring and adding these, we get

$$\left(\frac{2x - ae}{a}\right)^2 + \left(\frac{2y}{b}\right)^2 = 1.$$

We can rearrange this equation in the form

$$\frac{\left(x - \frac{1}{2}ae\right)^2}{(a/2)^2} + \frac{y^2}{(b/2)^2} = 1;$$

thus M lies on an ellipse with centre $\left(\frac{1}{2}ae, 0\right)$, the point midway between the origin and F.

9.

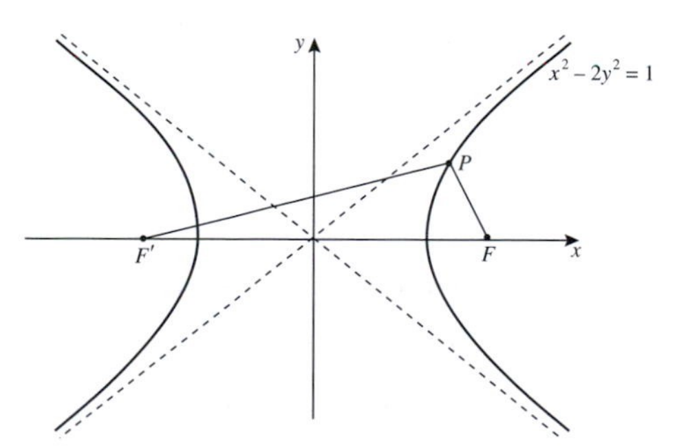

(a) This hyperbola is of the form $\frac{x^2}{a^2} - \frac{y^2}{b^2} = 1$ with $a = 1$ and $b^2 = \frac{1}{2}$, so that $b = 1/\sqrt{2}$.

If e denotes the eccentricity of the hyperbola E, so that $b^2 = a^2(e^2 - 1)$, we have

$$\tfrac{1}{2} = e^2 - 1;$$

it follows that $e^2 = \frac{3}{2}$ and so $e = \sqrt{\frac{3}{2}}$.

In general the foci are $(\pm ae, 0)$; it follows that here the foci are $\left(\pm\sqrt{\frac{3}{2}}, 0\right)$.

(b) Let F and F' be $\left(\sqrt{\frac{3}{2}}, 0\right)$ and $\left(-\sqrt{\frac{3}{2}}, 0\right)$, respectively. (It does not really matter which way round these are chosen.)

Then the slope of FP is

$$\frac{\frac{1}{\sqrt{2}}\tan t - 0}{\sec t - \sqrt{\frac{3}{2}}} = \frac{\tan t}{\sqrt{2}\sec t - \sqrt{3}},$$

where we may assume that $\sec t \neq \sqrt{\frac{3}{2}}$, since FP is not parallel to the y-axis; and the slope of $F'P$ is

$$\frac{\frac{1}{\sqrt{2}}\tan t - 0}{\sec t + \sqrt{\frac{3}{2}}} = \frac{\tan t}{\sqrt{2}\sec t + \sqrt{3}},$$

where we may assume that $\sec t \neq -\sqrt{\frac{3}{2}}$, since $F'P$ is not parallel to the y-axis.

(c) When FP is perpendicular to $F'P$, we have that

$$\frac{\tan t}{\sqrt{2}\sec t - \sqrt{3}} \cdot \frac{\tan t}{\sqrt{2}\sec t + \sqrt{3}} = -1.$$

We may rewrite this in the form

$$\frac{\tan^2 t}{2\sec^2 t - 3} = -1,$$

so that $2\sec^2 t - 3 + \tan^2 t = 0$; since $\sec^2 t = 1 + \tan^2 t$, it follows that we must have $3\tan^2 t = 1$. Since we are looking for a point P in the first quadrant, we choose $\tan t = 1/\sqrt{3}$.

When $\tan t = 1/\sqrt{3}$, we have $\sec^2 t = 1 + \frac{1}{3} = \frac{4}{3}$. Since we are looking for a point P in the first quadrant, we choose $\sec t = 2/\sqrt{3}$.

It follows that the required point P has coordinates

$$\left(\tfrac{2}{\sqrt{3}}, \tfrac{1}{\sqrt{2}} \cdot \tfrac{1}{\sqrt{3}}\right) = \left(\tfrac{2}{\sqrt{3}}, \tfrac{1}{\sqrt{6}}\right).$$

Section 1.2

1. We use the formula of Theorem 1 to find the slope of the tangent to the curve at the point with parameter t, where t is not a multiple of π.

   Since

$$x(t) = 2\cos t + \cos 2t + 1 \quad \text{and}$$
$$y(t) = 2\sin t + \sin 2t,$$

we have

$$x'(t) = -2\sin t - 2\sin 2t \quad \text{and}$$
$$y'(t) = 2\cos t + 2\cos 2t.$$

Hence the slope of the curve at this point is

$$\frac{y'(t)}{x'(t)} = \frac{2\cos t + 2\cos 2t}{-2\sin t - 2\sin 2t} = -\frac{\cos t + \cos 2t}{\sin t + \sin 2t}.$$

In particular, at the point with parameter $t = \pi/3$, this slope is

$$-\frac{\cos\pi/3 + \cos 2\pi/3}{\sin\pi/3 + \sin 2\pi/3} = -\frac{\frac{1}{2} - \frac{1}{2}}{\sqrt{3}/2 + \sqrt{3}/2} = 0.$$

Thus the tangent at this point is horizontal. Also,

$$y(\pi/3) = 2\sin(\pi/3) + \sin(2\pi/3) = 2 \cdot \frac{\sqrt{3}}{2} + \frac{\sqrt{3}}{2} = \frac{3}{2}\sqrt{3}.$$

It follows that the equation of the tangent at the point with parameter $t = \pi/3$ is $y = \frac{3}{2}\sqrt{3}$.

Next, at the point with parameter $t = \pi/2$, the slope of the curve is

$$-\frac{\cos\pi/2 + \cos\pi}{\sin\pi/2 + \sin\pi} = -\frac{0 - 1}{1 + 0} = 1.$$

Also,

$$x(\pi/2) = 2\cos(\pi/2) + \cos(\pi) + 1 = -1 + 1 = 0$$

and

$$y(\pi/2) = 2\sin(\pi/2) + \sin(\pi) = 2 + 0 = 2.$$

It follows that the equation of the tangent at the point with parameter $t = \pi/2$ is

$$y - 2 = 1(x - 0), \text{ or } y = x + 2.$$

2.

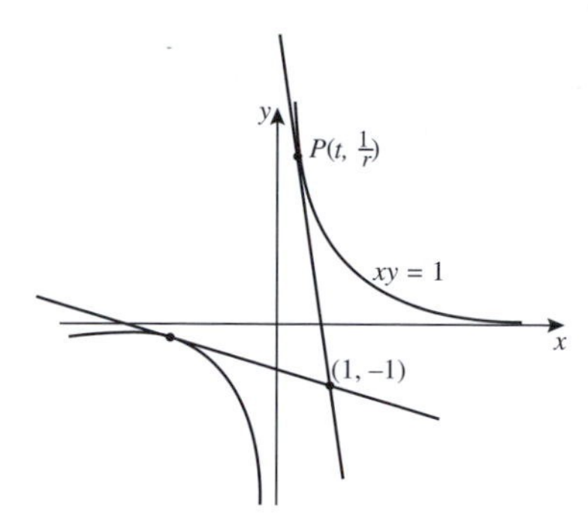

(a) Here $x'(t) = 1$ and $y'(t) = -1/t^2$; it follows that the slope of the tangent at the point with parameter t is

$$\frac{y'(t)}{x'(t)} = \frac{-1/t^2}{1} = -\frac{1}{t^2}.$$

It follows that the equation of the tangent at the point P is

$$y - \frac{1}{t} = -\frac{1}{t^2}(x - t),$$

or

$$y = -\frac{x}{t^2} + \frac{2}{t}.$$

(b) The line with equation $y = -\frac{x}{t^2} + \frac{2}{t}$ passes through the point $(1, -1)$ if

$$-1 = -\frac{1}{t^2} + \frac{2}{t}.$$

We can rewrite this equation in the form

$$t^2 + 2t - 1 = 0,$$

or

$$(t + 1)^2 = 2;$$

it follows that the values of t at the two points on the hyperbola for which the tangents pass through $(1, -1)$ are

$$t = -1 \pm \sqrt{2}.$$

When $t = -1 + \sqrt{2}$, the equation of the tangent is

$$y = -\frac{x}{\left(\sqrt{2} - 1\right)^2} + \frac{2}{\sqrt{2} - 1} = -\frac{x}{3 - 2\sqrt{2}} + \frac{2}{\sqrt{2} - 1}.$$

When $t = -1 - \sqrt{2}$, the equation of the tangent is

$$y = -\frac{x}{\left(-1 - \sqrt{2}\right)^2} + \frac{2}{-1 - \sqrt{2}} = -\frac{x}{3 + 2\sqrt{2}} - \frac{2}{1 + \sqrt{2}}.$$

3. The rectangular hyperbola $xy = 1$ has parametric equations $x = t, y = 1/t$ (where $t \neq 0$). You found in Problem 2(a) that the slope of the tangent at the point with parameter t is

$$\frac{y'(t)}{x'(t)} = -\frac{1}{t^2}.$$

Since $-\frac{1}{t^2} = -\frac{y_1}{x_1}$, the slope of the tangent at the point (x_1, y_1) may be written in a convenient form as $-\frac{y_1}{x_1}$. (The slope may be expressed in many other forms involving x_1 and y_1, but this particular form saves some algebra later in the calculation.)

Then the equation of the tangent to the hyperbola $xy = 1$ at the point (x_1, y_1) is

$$y - y_1 = -\frac{y_1}{x_1}(x - x_1).$$

Multiplying both sides by x_1, we may express this as

$$x_1 y - x_1 y_1 = -x y_1 + x_1 y_1,$$

so that

$$x_1 y + x y_1 = 2x_1 y_1 = 2;$$

dividing this by 2, we obtain the required equation.

4. The required equations may be obtained by simply substituting numbers into the appropriate equation in Theorem 2 or Problem 3.

(a) The equation of the tangent to the unit circle $x^2 + y^2 = 1$ at $\left(-\frac{1}{2}, \frac{1}{2}\sqrt{3}\right)$ is

$$x\left(-\tfrac{1}{2}\right) + y\left(\tfrac{1}{2}\sqrt{3}\right) = 1,$$

which may be written in the form

$$\sqrt{3}y = x + 2.$$

(b) The equation of the tangent to the rectangular hyperbola $xy = 1$ at $\left(-4, -\frac{1}{4}\right)$ is

$$\tfrac{1}{2}\left(x\left(-\tfrac{1}{4}\right) - 4y\right) = 1,$$

which may be written in the form

$$x + 16y = -8.$$

(c) The equation of the tangent to the parabola $y^2 = x$ at $(1, -1)$ is

$$y(-1) = \tfrac{1}{2}(x + 1),$$

which may be written in the form

$$x + 2y = -1.$$

5. Since $2^2 + 3^2 > 1$, the point $(2, 3)$ lies outside the unit circle. Hence, by Theorem 3, the polar of the point $(2, 3)$ with respect to the unit circle has the equation

$$2x + 3y = 1.$$

6.

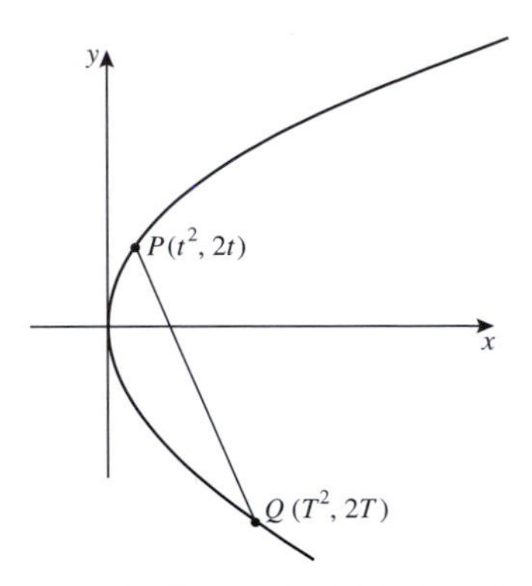

(a) We saw in Example 2(a) that the slope of the tangent at the point P with parameter t (where $t \neq 0$) is $1/t$. Since the normal and the tangent at P are perpendicular to each other, it follows that the slope m of the normal at P must satisfy the equation $m \cdot (1/t) = -1$. Hence $m = -t$.

(b) The normal at P is thus the line through the point $\left(t^2, 2t\right)$ with slope $-t$, and so has equation

$$y - 2t = -t\left(x - t^2\right),$$

or

$$y = -tx + 2t + t^3. \qquad (1)$$

(c) Let Q be the point on the parabola with parameter T, say; thus its coordinates are $\left(T^2, 2T\right)$. Since Q lies on the line with equation (1), it follows that

$$2T = -tT^2 + 2t + t^3;$$

we can rearrange this equation in the form

$$2(T - t) = -t\left(T^2 - t^2\right).$$

Since $T \neq t$, we may divide through by $T - t$, to get

$$\begin{aligned} 2 &= -t(T + t) \\ &= -tT - t^2, \end{aligned}$$

so that $tT = -2 - t^2$; it follows that $T = -\frac{2}{t} - t$.

7.

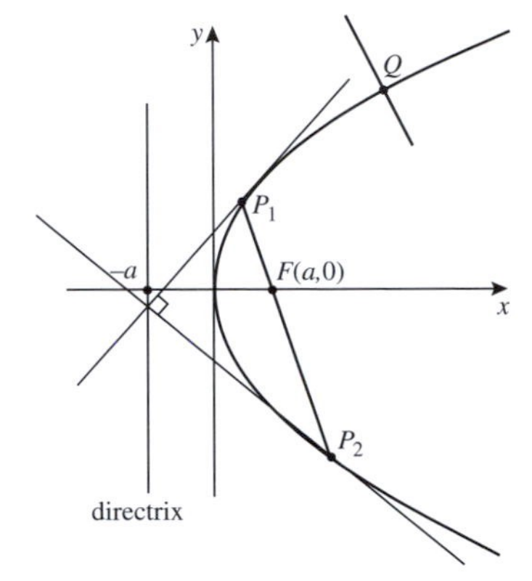

(a) P_1 has coordinates $\left(at_1^2, 2at_1\right)$ and P_2 has coordinates $\left(at_2^2, 2at_2\right)$. So, if $t_1^2 \neq t_2^2$ the slope of the chord P_1P_2 is

$$\begin{aligned} \frac{2at_2 - 2at_1}{at_2^2 - at_1^2} &= 2\frac{t_2 - t_1}{t_2^2 - t_1^2} \\ &= \frac{2}{t_2 + t_1}. \end{aligned}$$

It follows that the equation of P_1P_2 is

$$y - 2at_1 = \frac{2}{t_1 + t_2}\left(x - at_1^2\right)$$

or

$$(t_1 + t_2)(y - 2at_1) = 2\left(x - at_1^2\right). \qquad (2)$$

If, however, $t_1^2 = t_2^2$ we must have $t_1 = -t_2$ since P_1 and P_2 are distinct. The chord P_1P_2 is then parallel to the y-axis, so that we can write its equation as $x = at_1^2$. Thus the equation of the chord is given by equation (2) in this case too.

(b) If the chord P_1P_2 passes through the focus $(a, 0)$, the coordinates of $(a, 0)$ must satisfy equation (2); hence

$$(t_1 + t_2)(-2at_1) = 2\left(a - at_1^2\right),$$

so that

$$-t_1^2 - t_1t_2 = 1 - t_1^2.$$

It follows that $t_1t_2 = -1$.

(c) It follows from Example 2(a) that the equations of the tangents at P_1 and P_2 are

$$t_1y = x + at_1^2$$

and

$$t_2y = x + at_2^2,$$

respectively.

Now it follows also from part (a) of Example 2 that the slopes of the tangents at P_1 and P_2 are $1/t_1$ and $1/t_2$, respectively. These tangents are perpendicular if

$$\left(\frac{1}{t_1}\right) \cdot \left(\frac{1}{t_2}\right) = -1,$$

and we can rewrite this condition in the form $t_1t_2 = -1$.

We have already seen in part (b) that $t_1t_2 = -1$, and so we deduce that the tangents at P_1 and P_2 are indeed perpendicular.

(d) The equations of the tangents at P_1 and P_2 are

$$t_1y = x + at_1^2 \quad \text{and} \quad t_2y = x + at_2^2,$$

respectively. By subtracting these, we see that at the point (x, y) of intersection,

$$(t_1 - t_2)y = a\left(t_1^2 - t_2^2\right),$$

so that

$$y = a(t_1 + t_2).$$

It then follows from the equation $t_1y = x + at_1^2$ that

$$t_1a(t_1 + t_2) = x + at_1^2,$$

so that

$$\begin{aligned} x &= at_1t_2 \\ &= -a \quad (\text{since } t_1t_2 = -1). \end{aligned}$$

The point of intersection is therefore $(-a, a(t_1 + t_2))$.

Since the first coordinate of the point of intersection is $-a$, it follows that the point of intersection lies on the directrix of the parabola.

(e) Since (by the result of Example 2(a)) the tangent at Q has slope $1/t$, when $t \neq 0$, it follows that in this case the normal at Q has slope $-t$. When $t = 0$, the point Q is the origin, the vertex of the parabola; so in this case too the slope of the normal is $-t$.

Hence in general the equation of the normal at Q is

$$y - 2at = -t(x - at^2). \qquad (3)$$

If this normal passes through $F(a, 0)$, then the coordinates of F must satisfy equation (3); that is,

$$-2at = -t(a - at^2).$$

We can divide through by a and then rearrange the terms in this equation to get

$$0 = t(1 + t^2).$$

It follows that $t = 0$, and so Q must be the vertex of the parabola.

8.

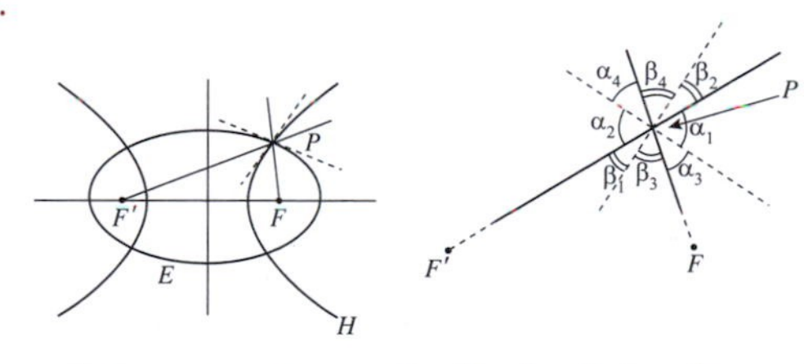

Let $\alpha_1, \alpha_2, \alpha_3, \alpha_4, \beta_1, \beta_2, \beta_3, \beta_4$ be the angles indicated in the above diagram.

Then

$$\begin{aligned} \alpha_1 &= \alpha_2 \quad \text{(vertically opposite angles)} \\ &= \alpha_3 \quad \text{(by the Reflection Property for the ellipse)} \\ &= \alpha_4 \quad \text{(vertically opposite angles),} \end{aligned}$$

and

$$\begin{aligned}\beta_1 = \beta_2 &\quad \text{(vertically opposite angles)}\\ = \beta_3 &\quad \text{(by the Reflection Property for the hyperbola)}\\ = \beta_4 &\quad \text{(vertically opposite angles).}\end{aligned}$$

Since $\alpha_1+\alpha_2+\alpha_3+\alpha_4+\beta_1+\beta_2+\beta_3+\beta_4 = 2\pi$, it follows that

$$\alpha_i + \beta_j = \tfrac{1}{2}\pi \quad \text{for any } i \text{ and any } j.$$

In particular, $\alpha_3 + \beta_3 = \frac{1}{2}\pi$, so that the tangents to E and H are perpendicular. In other words, E and H intersect at right angles.

9. Let the point $P(a\cos t, b\sin t)$ lie on the ellipse in standard form with equation $\frac{x^2}{a^2}+\frac{y^2}{b^2} = 1$, $a \geq b > 0$, and let the perpendicular from the focus $F(ae, 0)$ to the tangent at P meet that tangent at T.

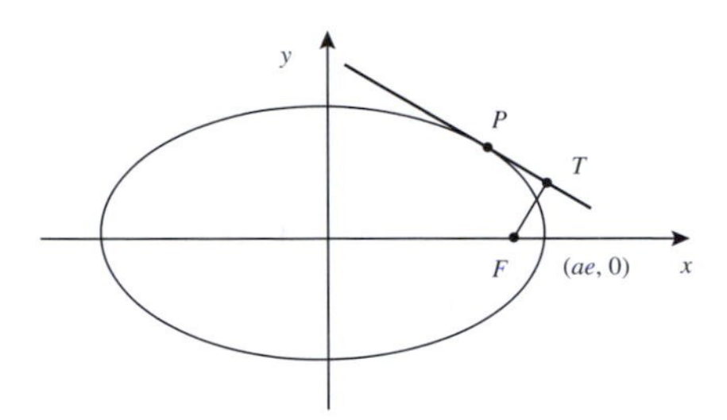

By Theorem 2 of Subsection 1.2.1, the tangent at P has equation

$$\frac{x \cdot a\cos t}{a^2} + \frac{y \cdot b\sin t}{b^2} = 1,$$

which we may rewrite in the form

$$bx\cos t + ay\sin t = ab. \qquad (*)$$

From this we see that, if $t \notin \{-\pi/2, 0, \pi/2, \pi\}$, the slope of the tangent PT is $-(b/a)\cot t$, so that the slope of the perpendicular FT must be $(a/b)\tan t$. Since FT also passes through $F(ae, 0)$, FT has equation

$$\begin{aligned} y - \frac{a}{b}\tan t \cdot x &= -\frac{a}{b}\tan t \cdot ae\\ &= -\frac{a^2 e}{b}\tan t,\end{aligned}$$

which we may rewrite in the form

$$ax\sin t - by\cos t = a^2 e\sin t. \qquad (**)$$

Then the coordinates of the point $T(x, y)$ of intersection of PT and FT must satisfy both equations (*) and (**). So, squaring each of (*) and (**) and adding, we find that the coordinates of T must satisfy the equation

$$\left(x^2+y^2\right)\left(b^2\cos^2 t + a^2\sin^2 t\right) = a^2\left(b^2 + a^2e^2\sin^2 t\right).$$

We then rewrite this equation in the form

$$\begin{aligned} x^2+y^2 &= a^2\frac{b^2+a^2e^2\sin^2 t}{b^2\cos^2 t + a^2\sin^2 t}\\ &= a^2\frac{(a^2-a^2e^2)+a^2e^2\sin^2 t}{(a^2-a^2e^2)(1-\sin^2 t)+a^2\sin^2 t}\\ &= a^2\frac{1-e^2+e^2\sin^2 t}{1-\sin^2 t - e^2 + e^2\sin^2 t + \sin^2 t}\\ &= a^2.\end{aligned}$$

It follows that the point T must lie on the auxiliary circle $x^2+y^2 = a^2$, as required.

If $t = 0$ or π, the tangent to the ellipse at P is a vertical line perpendicular to FP; so the tangent at P meets the perpendicular to it from F at P – which lies on the auxiliary circle.

Finally, if $t = \pm\pi/2$, the tangent to the ellipse at P is a horizontal line with equation $y = \pm b$. The point T where PT is perpendicular to FT must thus satisfy the equations $x = ae$ and $y = \pm b$; this lies on the auxiliary circle, since

$$\begin{aligned} x^2+y^2 &= a^2e^2 + (\pm b)^2\\ &= a^2e^2 + a^2\left(1-e^2\right) = a^2.\end{aligned}$$

Section 1.3

1. (a) The equation of the conic

$$11x^2 + 4xy + 14y^2 - 4x - 28y - 16 = 0$$

can be written in matrix form $\mathbf{x}^T\mathbf{A}\mathbf{x}+\mathbf{J}^T\mathbf{x}+H = 0$, where

$$\mathbf{A} = \begin{pmatrix}11 & 2\\ 2 & 14\end{pmatrix}, \quad \mathbf{J} = \begin{pmatrix}-4\\ -28\end{pmatrix},$$

$$H = -16 \quad \text{and} \quad \mathbf{x} = \begin{pmatrix}x\\ y\end{pmatrix}.$$

(b) The equation of the conic

$$x^2 - 4xy + 4y^2 - 6x - 8y + 5 = 0$$

can be written in matrix form $\mathbf{x}^T\mathbf{A}\mathbf{x}+\mathbf{J}^T\mathbf{x}+H=0$, where

$$\mathbf{A} = \begin{pmatrix} 1 & -2 \\ -2 & 4 \end{pmatrix}, \quad \mathbf{J} = \begin{pmatrix} -6 \\ -8 \end{pmatrix},$$

$$H = 5 \quad \text{and} \quad \mathbf{x} = \begin{pmatrix} x \\ y \end{pmatrix}.$$

2. (a) We saw in Problem 1(a) that the matrix form of the equation of this conic is $\mathbf{x}^T\mathbf{A}\mathbf{x}+\mathbf{J}^T\mathbf{x}+H=0$, where

$$\mathbf{A} = \begin{pmatrix} 11 & 2 \\ 2 & 14 \end{pmatrix}, \quad \mathbf{J} = \begin{pmatrix} -4 \\ -28 \end{pmatrix},$$

$$H = -16 \quad \text{and} \quad \mathbf{x} = \begin{pmatrix} x \\ y \end{pmatrix};$$

that is

$$(x\ y)\begin{pmatrix} 11 & 2 \\ 2 & 14 \end{pmatrix}\begin{pmatrix} x \\ y \end{pmatrix} + (-4\ -28)\begin{pmatrix} x \\ y \end{pmatrix} - 16 = 0.$$

First we diagonalize $\mathbf{A}$. Its characteristic equation is

$$\begin{aligned} 0 = \det(\mathbf{A} - \lambda\mathbf{I}) &= \begin{vmatrix} 11-\lambda & 2 \\ 2 & 14-\lambda \end{vmatrix} \\ &= \lambda^2 - 25\lambda + 150 \\ &= (\lambda - 15)(\lambda - 10), \end{aligned}$$

so that the eigenvalues of $\mathbf{A}$ are $\lambda = 15$ and $\lambda = 10$. The eigenvector equations of $\mathbf{A}$ are

$$(11-\lambda)x + 2y = 0,$$
$$2x + (14-\lambda)y = 0.$$

When $\lambda = 15$, these equations become

$$-4x + 2y = 0,$$
$$2x - y = 0,$$

so that we may take as a corresponding eigenvector $\begin{pmatrix} 1 \\ 2 \end{pmatrix}$, which we normalize to have unit length as $\begin{pmatrix} 1/\sqrt{5} \\ 2/\sqrt{5} \end{pmatrix}$.

When $\lambda = 10$, the eigenvector equations of $\mathbf{A}$ become

$$x + 2y = 0,$$
$$2x + 4y = 0,$$

so that we may take as a corresponding eigenvector $\begin{pmatrix} 2 \\ -1 \end{pmatrix}$ which we normalize to have unit length as $\begin{pmatrix} 2/\sqrt{5} \\ -1/\sqrt{5} \end{pmatrix}$.

Now $\begin{vmatrix} 1/\sqrt{5} & 2/\sqrt{5} \\ 2/\sqrt{5} & -1/\sqrt{5} \end{vmatrix} = -\frac{1}{5} - \frac{4}{5} = -1$, so interchanging the order of the eigenvectors as columns of $\mathbf{P}$ – in order that $\det \mathbf{P} = +1$, so that then $\mathbf{P}$ represents a rotation rather than a rotation composed with a reflection – we take as our rotation of the plane the transformation $\mathbf{x} = \mathbf{P}\mathbf{x}'$, where $\mathbf{P} = \begin{pmatrix} 2/\sqrt{5} & 1/\sqrt{5} \\ -1/\sqrt{5} & 2/\sqrt{5} \end{pmatrix}$.

This transformation changes the equation of the conic to the form

$$(\mathbf{P}\mathbf{x}')^T\mathbf{A}(\mathbf{P}\mathbf{x}') + \mathbf{J}^T(\mathbf{P}\mathbf{x}') + H = 0$$

or

$$(\mathbf{x}')^T(\mathbf{P}^T\mathbf{A}\mathbf{P})\mathbf{x}' + (\mathbf{J}^T\mathbf{P})\mathbf{x}' + H = 0.$$

Since $\mathbf{P}^T\mathbf{A}\mathbf{P} = \begin{pmatrix} 10 & 0 \\ 0 & 15 \end{pmatrix}$, this is the equation

$$(x'\ \ y')\begin{pmatrix} 10 & 0 \\ 0 & 15 \end{pmatrix}\begin{pmatrix} x' \\ y' \end{pmatrix}$$
$$+(-4\ -28)\begin{pmatrix} \frac{2}{\sqrt{5}} & \frac{1}{\sqrt{5}} \\ -\frac{1}{\sqrt{5}} & \frac{2}{\sqrt{5}} \end{pmatrix}\cdot\begin{pmatrix} x' \\ y' \end{pmatrix} - 16 = 0.$$

We may rewrite this equation in the form $10x'^2 + 15y'^2 + 4\sqrt{5}x' - 12\sqrt{5}y' - 16 = 0$, or

$$10\left(x'^2 + \tfrac{2}{\sqrt{5}}x'\right) + 15\left(y'^2 - \tfrac{4}{\sqrt{5}}y'\right) - 16 = 0;$$

so that, on completing the square, we have

$$10\left(x' + \tfrac{1}{\sqrt{5}}\right)^2 - 2 + 15\left(y' - \tfrac{2}{\sqrt{5}}\right)^2 - 12 - 16 = 0,$$

or

$$10\left(x' + \tfrac{1}{\sqrt{5}}\right)^2 + 15\left(y' - \tfrac{2}{\sqrt{5}}\right)^2 - 30 = 0$$

or

$$\frac{\left(x' + \frac{1}{\sqrt{5}}\right)^2}{3} + \frac{\left(y' - \frac{2}{\sqrt{5}}\right)^2}{2} = 1. \qquad (2)$$

This is the equation of an ellipse.

From equation (2) it follows that the centre of the ellipse is the point where $x' + \frac{1}{\sqrt{5}} = 0$ and $y' - \frac{2}{\sqrt{5}} = 0$, that is, where $x' = -\frac{1}{\sqrt{5}}$ and $y' = \frac{2}{\sqrt{5}}$. From the equation $\mathbf{x} = \mathbf{P}\mathbf{x}'$, it follows that in terms of the original coordinate system this is the point

$$\begin{pmatrix} x \\ y \end{pmatrix} = \begin{pmatrix} \frac{2}{\sqrt{5}} & \frac{1}{\sqrt{5}} \\ -\frac{1}{\sqrt{5}} & \frac{2}{\sqrt{5}} \end{pmatrix} \begin{pmatrix} -\frac{1}{\sqrt{5}} \\ \frac{2}{\sqrt{5}} \end{pmatrix} = \begin{pmatrix} 0 \\ 1 \end{pmatrix},$$

that is, the point $(0, 1)$.

Since $3 > 2$, it also follows from equation (2) that the major axis of the ellipse has equation $y' - \frac{2}{\sqrt{5}} = 0$, or $y' = \frac{2}{\sqrt{5}}$; and the minor axis has equation $x' + \frac{1}{\sqrt{5}} = 0$, or $x' = -\frac{1}{\sqrt{5}}$.

Finally, since the matrix $\mathbf{P}$ is orthogonal we can rewrite the equation $\mathbf{x} = \mathbf{P}\mathbf{x}'$ in the form $\mathbf{x}' = \mathbf{P}^{-1}\mathbf{x} = \mathbf{P}^T\mathbf{x}$, so that

$$\begin{pmatrix} x' \\ y' \end{pmatrix} = \begin{pmatrix} \frac{2}{\sqrt{5}} & -\frac{1}{\sqrt{5}} \\ \frac{1}{\sqrt{5}} & \frac{2}{\sqrt{5}} \end{pmatrix} \begin{pmatrix} x \\ y \end{pmatrix}$$

or as a pair of equations

$$x' = \tfrac{2}{\sqrt{5}}x - \tfrac{1}{\sqrt{5}}y,$$
$$y' = \tfrac{1}{\sqrt{5}}x + \tfrac{2}{\sqrt{5}}y.$$

It follows that the equation, $y' = \frac{2}{\sqrt{5}}$, of the major axis of the ellipse can be expressed in terms of the original coordinate system as

$$\tfrac{1}{\sqrt{5}}x + \tfrac{2}{\sqrt{5}}y = \tfrac{2}{\sqrt{5}} \quad \text{or} \quad x + 2y = 2.$$

Similarly, the equation, $x' = -\frac{1}{\sqrt{5}}$, of the minor axis of the ellipse can be expressed in terms of the original coordinate system as

$$\tfrac{2}{\sqrt{5}}x - \tfrac{1}{\sqrt{5}}y = -\tfrac{1}{\sqrt{5}} \quad \text{or} \quad 2x - y = -1.$$

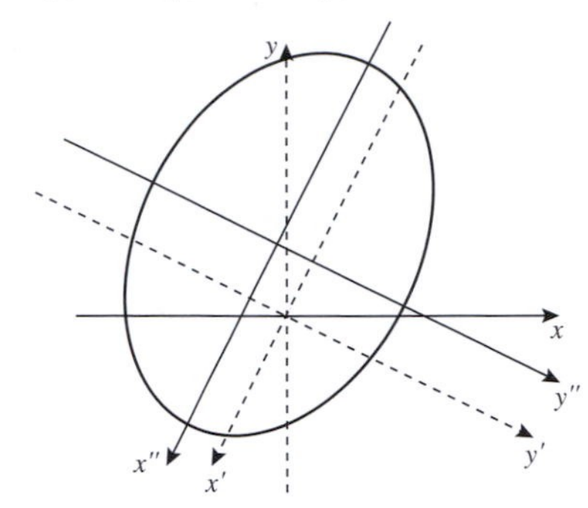

(b) We saw in Problem 1(b) that the matrix form of the equation of this conic is $\mathbf{x}^T\mathbf{A}\mathbf{x} + \mathbf{J}^T\mathbf{x} + H = 0$, where

$$\mathbf{A} = \begin{pmatrix} 1 & -2 \\ -2 & 4 \end{pmatrix} \quad \mathbf{J} = \begin{pmatrix} -6 \\ -8 \end{pmatrix},$$

$$H = 5 \quad \text{and} \quad \mathbf{x} = \begin{pmatrix} x \\ y \end{pmatrix};$$

that is

$$(x \quad y)\begin{pmatrix} 1 & -2 \\ -2 & 4 \end{pmatrix}\begin{pmatrix} x \\ y \end{pmatrix} + (-6 \quad -8)\begin{pmatrix} x \\ y \end{pmatrix} + 5 = 0.$$

First we diagonalize $\mathbf{A}$. Its characteristic equation is

$$0 = \det(\mathbf{A} - \lambda\mathbf{I}) = \begin{vmatrix} 1-\lambda & -2 \\ -2 & 4-\lambda \end{vmatrix} = \lambda^2 - 5\lambda = \lambda(\lambda - 5),$$

so that the eigenvalues of $\mathbf{A}$ are $\lambda = 0$ and $\lambda = 5$. The eigenvector equations of $\mathbf{A}$ are

$$(1-\lambda)x - 2y = 0,$$
$$-2x + (4-\lambda)y = 0.$$

When $\lambda = 0$, these equations become

$$x - 2y = 0,$$
$$-2x + 4y = 0,$$

so that we may take as a corresponding eigenvector $\begin{pmatrix} 2 \\ 1 \end{pmatrix}$, which we normalize to have unit length as $\begin{pmatrix} 2/\sqrt{5} \\ 1/\sqrt{5} \end{pmatrix}$.

When $\lambda = 5$, the eigenvector equations of **A** become

$$-4x - 2y = 0,$$
$$-2x - y = 0,$$

so that we may take as a corresponding eigenvector $\begin{pmatrix} 1 \\ -2 \end{pmatrix}$, which we normalize to have unit length as $\begin{pmatrix} 1/\sqrt{5} \\ -2/\sqrt{5} \end{pmatrix}$.

Now $\begin{vmatrix} 2/\sqrt{5} & 1/\sqrt{5} \\ 1/\sqrt{5} & -2/\sqrt{5} \end{vmatrix} = -\frac{4}{5} - \frac{1}{5} = -1$, so interchanging the order of the eigenvectors as columns of **P** – in order that $\det \mathbf{P} = +1$, so that then **P** represents a rotation rather than a rotation composed with a reflection – we take as our rotation of the plane the transformation $\mathbf{x} = \mathbf{P}\mathbf{x}'$, where $\mathbf{P} = \begin{pmatrix} 1/\sqrt{5} & 2/\sqrt{5} \\ -2/\sqrt{5} & 1/\sqrt{5} \end{pmatrix}$. This transformation changes the equation of the conic to the form

$$(\mathbf{P}\mathbf{x}')^T \mathbf{A}(\mathbf{P}\mathbf{x}') + \mathbf{J}^T(\mathbf{P}\mathbf{x}') + H = 0$$

or

$$(\mathbf{x}')^T(\mathbf{P}^T\mathbf{A}\mathbf{P})\mathbf{x}' + (\mathbf{J}^T\mathbf{P})\mathbf{x}' + H = 0.$$

Since $\mathbf{P}^T\mathbf{A}\mathbf{P} = \begin{pmatrix} 5 & 0 \\ 0 & 0 \end{pmatrix}$, this is the equation

$$(x' \quad y')\begin{pmatrix} 5 & 0 \\ 0 & 0 \end{pmatrix}\begin{pmatrix} x' \\ y' \end{pmatrix}$$
$$+(-6 \; -8)\begin{pmatrix} \frac{1}{\sqrt{5}} & \frac{2}{\sqrt{5}} \\ -\frac{2}{\sqrt{5}} & \frac{1}{\sqrt{5}} \end{pmatrix}\begin{pmatrix} x' \\ y' \end{pmatrix} + 5 = 0,$$

which we can rewrite in the form

$$5x'^2 + 2\sqrt{5}x' - 4\sqrt{5}y' + 5 = 0.$$

We may rewrite this equation in the form

$$5\left(x'^2 + \tfrac{2}{\sqrt{5}}x'\right) - 4\sqrt{5}y' + 5 = 0$$

so that, on completing the square, we have

$$5\left(x' + \tfrac{1}{\sqrt{5}}\right)^2 - 4\sqrt{5}y' + 4 = 0,$$

or

$$5\left(x' + \tfrac{1}{\sqrt{5}}\right)^2 - 4\sqrt{5}\left(y' - \tfrac{1}{\sqrt{5}}\right) = 0,$$

or

$$\left(x' + \tfrac{1}{\sqrt{5}}\right)^2 = \tfrac{4}{\sqrt{5}}\left(y' - \tfrac{1}{\sqrt{5}}\right). \quad (3)$$

This is the equation of a parabola. (It is not quite in standard form $(y'')^2 = 4ax''$, but in the similar form $(x'')^2 = 4ay''$; the argument will be similar.)

From equation (3) it follows that the vertex of the parabola is the point where $x' + \frac{1}{\sqrt{5}} = 0$ and $y' - \frac{1}{\sqrt{5}} = 0$, that is, where $x' = -\frac{1}{\sqrt{5}}$ and $y' = \frac{1}{\sqrt{5}}$. From the equation $\mathbf{x} = \mathbf{P}\mathbf{x}'$, it follows that in terms of the original coordinate system this is the point

$$\begin{pmatrix} x \\ y \end{pmatrix} = \begin{pmatrix} \frac{1}{\sqrt{5}} & \frac{2}{\sqrt{5}} \\ -\frac{2}{\sqrt{5}} & \frac{1}{\sqrt{5}} \end{pmatrix}\begin{pmatrix} -\frac{1}{\sqrt{5}} \\ \frac{1}{\sqrt{5}} \end{pmatrix}$$
$$= \begin{pmatrix} \frac{1}{5} \\ \frac{3}{5} \end{pmatrix},$$

that is, the point $\left(\frac{1}{5}, \frac{3}{5}\right)$.

It also follows from equation (3) that the axis of the parabola has equation $x' + \frac{1}{\sqrt{5}} = 0$, or $x' = -\frac{1}{\sqrt{5}}$.

Then, since the matrix **P** is orthogonal we can rewrite the equation $\mathbf{x} = \mathbf{P}\mathbf{x}'$ in the form $\mathbf{x}' = \mathbf{P}^{-1}\mathbf{x} = \mathbf{P}^T\mathbf{x}$, so that

$$\begin{pmatrix} x' \\ y' \end{pmatrix} = \begin{pmatrix} \frac{1}{\sqrt{5}} & -\frac{2}{\sqrt{5}} \\ \frac{2}{\sqrt{5}} & \frac{1}{\sqrt{5}} \end{pmatrix}\begin{pmatrix} x \\ y \end{pmatrix}$$

or as a pair of equations

$$x' = \tfrac{1}{\sqrt{5}}x - \tfrac{2}{\sqrt{5}}y,$$
$$y' = \tfrac{2}{\sqrt{5}}x + \tfrac{1}{\sqrt{5}}y.$$

It follows that the equation, $x' = -\frac{1}{\sqrt{5}}$, of the axis of the parabola can be expressed in terms of the original coordinate system as

$$\tfrac{1}{\sqrt{5}}x - \tfrac{2}{\sqrt{5}}y = -\tfrac{1}{\sqrt{5}} \quad \text{or} \quad x - 2y = -1.$$

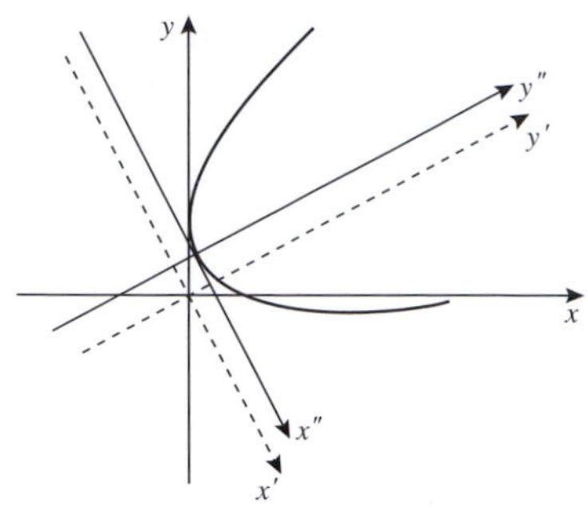

3. Here we use Theorem 3.
   (a) The matrix of the non-degenerate conic is $\mathbf{A} = \begin{pmatrix} 3 & -4 \\ -4 & 2 \end{pmatrix}$, so that
   $$\det \mathbf{A} = 6 - 16 = -10 < 0,$$
   so that the conic is a hyperbola.
   (b) The matrix of the non-degenerate conic is $\mathbf{A} = \begin{pmatrix} 1 & 4 \\ 4 & 16 \end{pmatrix}$, so that
   $$\det \mathbf{A} = 16 - 16 = 0,$$
   so that the conic is a parabola.
   (c) The matrix of the non-degenerate conic is $\mathbf{A} = \begin{pmatrix} 52 & -36 \\ -36 & 73 \end{pmatrix}$, so that
   $$\det \mathbf{A} = 52 \cdot 73 - 36^2 = 3796 - 1296 = 2500 > 0,$$
   so that the conic is an ellipse.

Section 1.4

1. The equation of the quadric surface given by
$$2x^2 + 5y^2 - z^2 + xy - 3yz - 2xz - 2x - 6y + 10z - 12 = 0$$
may be written in matrix form $\mathbf{x}^T\mathbf{A}\mathbf{x} + \mathbf{J}^T\mathbf{x} + M = 0$ where
$$\mathbf{A} = \begin{pmatrix} 2 & \frac{1}{2} & -1 \\ \frac{1}{2} & 5 & -\frac{3}{2} \\ -1 & -\frac{3}{2} & -1 \end{pmatrix}, \quad \mathbf{J} = \begin{pmatrix} -2 \\ -6 \\ 10 \end{pmatrix},$$
$$\mathbf{x} = \begin{pmatrix} x \\ y \\ z \end{pmatrix} \quad \text{and} \quad M = -12.$$

Similarly, the equation of the quadric surface given by
$$y - yz = xz$$
may be rewritten as
$$yz + xz - y = 0,$$
and so in matrix form $\mathbf{x}^T\mathbf{A}\mathbf{x} + \mathbf{J}^T\mathbf{x} + M = 0$ where
$$\mathbf{A} = \begin{pmatrix} 0 & 0 & \frac{1}{2} \\ 0 & 0 & \frac{1}{2} \\ \frac{1}{2} & \frac{1}{2} & 0 \end{pmatrix}, \quad \mathbf{J} = \begin{pmatrix} 0 \\ -1 \\ 0 \end{pmatrix},$$
$$\mathbf{x} = \begin{pmatrix} x \\ y \\ z \end{pmatrix} \quad \text{and} \quad M = 0.$$

2. We saw in Problem 1 that the equation of this quadric E can be written in matrix form as $\mathbf{x}^T\mathbf{A}\mathbf{x} + \mathbf{J}^T\mathbf{x} + M = 0$, where
$$\mathbf{A} = \begin{pmatrix} 0 & 0 & \frac{1}{2} \\ 0 & 0 & \frac{1}{2} \\ \frac{1}{2} & \frac{1}{2} & 0 \end{pmatrix}, \quad \mathbf{J} = \begin{pmatrix} 0 \\ -1 \\ 0 \end{pmatrix},$$
$$\mathbf{x} = \begin{pmatrix} x \\ y \\ z \end{pmatrix} \quad \text{and} \quad M = 0,$$
that is, as
$$\begin{pmatrix} x & y & z \end{pmatrix} \begin{pmatrix} 0 & 0 & \frac{1}{2} \\ 0 & 0 & \frac{1}{2} \\ \frac{1}{2} & \frac{1}{2} & 0 \end{pmatrix} \begin{pmatrix} x \\ y \\ z \end{pmatrix} + \begin{pmatrix} 0 & -1 & 0 \end{pmatrix} \begin{pmatrix} x \\ y \\ z \end{pmatrix} = 0.$$

We start by diagonalizing the matrix $\mathbf{A}$. Its characteristic equation is
$$\begin{aligned} 0 = \det(\mathbf{A} - \lambda\mathbf{I}) &= \begin{vmatrix} -\lambda & 0 & \frac{1}{2} \\ 0 & -\lambda & \frac{1}{2} \\ \frac{1}{2} & \frac{1}{2} & -\lambda \end{vmatrix} \\ &= -\lambda\left(\lambda^2 - \tfrac{1}{4}\right) + \tfrac{1}{2}\left(\tfrac{1}{2}\lambda\right) \\ &= -\lambda^3 + \tfrac{1}{2}\lambda \\ &= -\lambda\left(\lambda^2 - \tfrac{1}{2}\right), \end{aligned}$$

so that the eigenvalues of $\mathbf{A}$ are $\lambda = 0, \frac{1}{\sqrt{2}}$ and $-\frac{1}{\sqrt{2}}$. The eigenvector equations of $\mathbf{A}$ are

$$\begin{aligned} -\lambda x + \tfrac{1}{2}z &= 0, \\ -\lambda y + \tfrac{1}{2}z &= 0, \\ \tfrac{1}{2}x + \tfrac{1}{2}y - \lambda z &= 0. \end{aligned}$$

When $\lambda = 0$, these equations become

$$\begin{aligned} \tfrac{1}{2}z &= 0, \\ \tfrac{1}{2}z &= 0, \\ \tfrac{1}{2}x + \tfrac{1}{2}y &= 0. \end{aligned}$$

From the first two equations we get $z = 0$; it then follows from the third equation that $x + y = 0$. So we may take as a corresponding eigenvector $\begin{pmatrix} 1 \\ -1 \\ 0 \end{pmatrix}$, which we normalize to have unit length as $\begin{pmatrix} \frac{1}{\sqrt{2}} \\ \frac{-1}{\sqrt{2}} \\ 0 \end{pmatrix}$.

Similarly, when $\lambda = \frac{1}{\sqrt{2}}$, we may take as a corresponding eigenvector $\begin{pmatrix} 1 \\ 1 \\ \sqrt{2} \end{pmatrix}$, which we normalize to have unit length as $\begin{pmatrix} \frac{1}{2} \\ \frac{1}{2} \\ \frac{1}{\sqrt{2}} \end{pmatrix}$; and when $\lambda = \frac{-1}{\sqrt{2}}$, we may take as a corresponding eigenvector $\begin{pmatrix} 1 \\ 1 \\ -\sqrt{2} \end{pmatrix}$, which we normalize to have unit length as $\begin{pmatrix} 1/2 \\ 1/2 \\ -1/\sqrt{2} \end{pmatrix}$.

Now

$$\begin{vmatrix} \frac{1}{\sqrt{2}} & \frac{1}{2} & \frac{1}{2} \\ -\frac{1}{\sqrt{2}} & \frac{1}{2} & \frac{1}{2} \\ 0 & \frac{1}{\sqrt{2}} & -\frac{1}{\sqrt{2}} \end{vmatrix} = \begin{vmatrix} \frac{1}{\sqrt{2}} & \frac{1}{2} & \frac{1}{2} \\ 0 & 1 & 1 \\ 0 & \frac{1}{\sqrt{2}} & -\frac{1}{\sqrt{2}} \end{vmatrix}$$

$$= \frac{1}{\sqrt{2}} \begin{vmatrix} 1 & 1 \\ \frac{1}{\sqrt{2}} & -\frac{1}{\sqrt{2}} \end{vmatrix}$$

$$= \frac{1}{\sqrt{2}} \left(-\frac{2}{\sqrt{2}} \right) = -1;$$

Since interchanging the first and second eigenvectors interchanges the first two columns of the determinant, in order to have det $\mathbf{P} = 1$ we choose to take as a convenient rotation of $\mathbb{R}^3$ the transformation $\mathbf{x} = \mathbf{P}\mathbf{x}'$, where $\mathbf{P} = \begin{pmatrix} 1/2 & 1/\sqrt{2} & 1/2 \\ 1/2 & -1/\sqrt{2} & 1/2 \\ 1/\sqrt{2} & 0 & -1/\sqrt{2} \end{pmatrix}$. This transformation changes the equation of the quadric to the form

$$(\mathbf{P}\mathbf{x}')^T \mathbf{A} (\mathbf{P}\mathbf{x}') + \mathbf{J}^T (\mathbf{P}\mathbf{x}') + M = 0$$

or

$$(\mathbf{x}')^T (\mathbf{P}^T \mathbf{A} \mathbf{P}) \mathbf{x}' + (\mathbf{J}^T \mathbf{P}) \mathbf{x}' + M = 0.$$

Then since $\mathbf{P}^T \mathbf{A} \mathbf{P} = \begin{pmatrix} \frac{1}{\sqrt{2}} & 0 & 0 \\ 0 & 0 & 0 \\ 0 & 0 & -\frac{1}{\sqrt{2}} \end{pmatrix}$, the equation of the quadric becomes

$$(x' \quad y' \quad z') \begin{pmatrix} \frac{1}{\sqrt{2}} & 0 & 0 \\ 0 & 0 & 0 \\ 0 & 0 & -\frac{1}{\sqrt{2}} \end{pmatrix} \begin{pmatrix} x' \\ y' \\ z' \end{pmatrix}$$

$$+ (0 \quad -1 \quad 0) \begin{pmatrix} \frac{1}{2} & \frac{1}{\sqrt{2}} & \frac{1}{2} \\ \frac{1}{2} & -\frac{1}{\sqrt{2}} & \frac{1}{2} \\ \frac{1}{\sqrt{2}} & 0 & -\frac{1}{\sqrt{2}} \end{pmatrix} \begin{pmatrix} x' \\ y' \\ z' \end{pmatrix} = 0,$$

which we can rewrite in the form

$$\tfrac{1}{\sqrt{2}}(x')^2 - \tfrac{1}{\sqrt{2}}(z')^2 - \tfrac{1}{2}x' + \tfrac{1}{\sqrt{2}}y' - \tfrac{1}{2}z' = 0$$

or

$$(x')^2 - (z')^2 - \tfrac{1}{\sqrt{2}}x' + y' - \tfrac{1}{\sqrt{2}}z' = 0.$$

Completing the square in this equation, we get

$$\left(x' - \tfrac{1}{2\sqrt{2}}\right)^2 - \tfrac{1}{8} - \left(z' + \tfrac{1}{2\sqrt{2}}\right)^2 + \tfrac{1}{8} + y' = 0$$

or

$$\left(x' - \tfrac{1}{2\sqrt{2}}\right)^2 - \left(z' + \tfrac{1}{2\sqrt{2}}\right)^2 + y' = 0.$$

We now make the transformation

$$\mathbf{x}'' = \mathbf{x}' + \begin{pmatrix} -\frac{1}{2\sqrt{2}} \\ 0 \\ \frac{1}{2\sqrt{2}} \end{pmatrix}, \qquad (1)$$

so that we can rewrite the equation of E in the form

$$(x'')^2 - (z'')^2 + y'' = 0. \qquad (2)$$

It follows from equation (2) that E must be a hyperbolic paraboloid.

From equation (2) it follows that the centre of E is the point where $x'' = 0$, $y'' = 0$ and $z'' = 0$; that is, the point where $x' = 1/(2\sqrt{2})$, $y' = 0$ and $z' = -1/(2\sqrt{2})$. From the equation $\mathbf{x} = \mathbf{P}\mathbf{x}'$, it follows that in terms of the original coordinate system this is the point

$$\begin{pmatrix} x \\ y \\ z \end{pmatrix} = \begin{pmatrix} \frac{1}{2} & \frac{1}{\sqrt{2}} & \frac{1}{2} \\ \frac{1}{2} & -\frac{1}{\sqrt{2}} & \frac{1}{2} \\ \frac{1}{\sqrt{2}} & 0 & -\frac{1}{\sqrt{2}} \end{pmatrix} \begin{pmatrix} \frac{1}{2\sqrt{2}} \\ 0 \\ -\frac{1}{2\sqrt{2}} \end{pmatrix} = \begin{pmatrix} 0 \\ 0 \\ \frac{1}{2} \end{pmatrix},$$

that is, the point $\left(0, 0, \frac{1}{2}\right)$.

3. The line m through the points $(\sqrt{2}, 0, 1)$ and $(0, -\sqrt{2}, -1)$ can be parametrized as

$$\begin{aligned} &\mu(\sqrt{2}, 0, 1) + (1 - \mu)(0, -\sqrt{2}, -1) \\ &= (\mu\sqrt{2}, (\mu - 1)\sqrt{2}, 2\mu - 1), \end{aligned}$$

where $\mu \in \mathbb{R}$. Points on the line lie in E since

$$\begin{aligned} &(\mu\sqrt{2})^2 + \left((\mu - 1)\sqrt{2}\right)^2 - (2\mu - 1)^2 \\ &= 2\mu^2 + 2(\mu^2 - 2\mu + 1) - (4\mu^2 - 4\mu + 1) = 1. \end{aligned}$$

4. (a) Let the point $(x_0, y_0, z_0) \in A$, so that $x_0 + y_0 = 0$ and $z_0 = 0$. Then

$$\begin{aligned} x_0^2 - y_0^2 + z_0 &= (x_0 + y_0)(x_0 - y_0) \\ &\quad + z_0 = 0, \end{aligned}$$

so that (x_0, y_0, z_0) lies in E.

Similarly, the point $(x_0, y_0, z_0) \in B$ lies in E.

(b) The point $(\lambda - \mu, \lambda + \mu, 4\lambda\mu)$ lies in E, since

$$\begin{aligned} &(\lambda - \mu)^2 - (\lambda + \mu)^2 + (4\lambda\mu) \\ &= \left(\lambda^2 - 2\lambda\mu + \mu^2\right) \\ &\quad - (\lambda^2 + 2\lambda\mu + \mu^2) + 4\lambda\mu = 0. \end{aligned}$$

(c) Let λ be a fixed number. Then

$$\begin{aligned} \ell_\lambda &= \{(\lambda - \mu, \lambda + \mu, 4\lambda\mu) : \mu \in \mathbb{R}\} \\ &= \{(\lambda, \lambda, 0) + \mu(-1, 1, 4\lambda) : \mu \in \mathbb{R}\}, \end{aligned}$$

so that ℓ_λ is a line through the point $(\lambda, \lambda, 0)$ parallel to the vector $(-1, 1, 4\lambda)$. Clearly the point $(\lambda, \lambda, 0)$ lies on B.

This line passes through the point $(\lambda - 1, \lambda + 1, 4\lambda)$ since it is the point on ℓ_λ where $\mu = 1$.

When $\lambda = 0$,

$$\begin{aligned} \ell_0 &= \{(-\mu, \mu, 0) : \mu \in \mathbb{R}\} \\ &= \{\mu(-1, 1, 0) : \mu \in \mathbb{R}\}; \end{aligned}$$

so that this line is the line through the origin parallel to the vector $(-1, 1, 0)$. Thus, $\ell_0 = A$.

(d) Let μ be a fixed number. Then

$$\begin{aligned} m_\mu &= \{(\lambda - \mu, \lambda + \mu, 4\lambda\mu) : \lambda \in \mathbb{R}\} \\ &= \{(-\mu, \mu, 0) + \lambda(1, 1, 4\mu) : \lambda \in \mathbb{R}\}, \end{aligned}$$

so that m_μ is a line through the point $(-\mu, \mu, 0)$ parallel to the vector $(1, 1, 4\mu)$. Clearly the point $(-\mu, \mu, 0)$ lies on A.

This line passes through the point $(1 - \mu, 1 + \mu, 4\mu)$ since it is the point on m_μ where $\lambda = 1$.

When $\mu = 0$,

$$\begin{aligned} m_0 &= \{(\lambda, \lambda, 0) : \lambda \in \mathbb{R}\} \\ &= \{\lambda(1, 1, 0) : \lambda \in \mathbb{R}\}, \end{aligned}$$

so that this line is the line through the origin parallel to the vector $(1, 1, 0)$. Thus $m_0 = B$.

(e) Consider $\mathscr{L}$ first. Two distinct lines in $\mathscr{L}$ may be expressed in parametric form as

$$\{(\lambda_1 - \mu, \lambda_1 + \mu, 4\lambda_1\mu) : \mu \in \mathbb{R}\}$$

and

$$\left\{(\lambda_2 - \mu', \lambda_2 + \mu', 4\lambda_2\mu') : \mu' \in \mathbb{R}\right\},$$

for some real numbers μ and μ'.

These lines meet where

$$\lambda_1 - \mu = \lambda_2 - \mu',$$
$$\lambda_1 + \mu = \lambda_2 + \mu',$$

and

$$4\lambda_1\mu = 4\lambda_2\mu'.$$

Adding the first two equations gives $2\lambda_1 = 2\lambda_2$, so that $\lambda_1 = \lambda_2$. Thus the two lines can only meet if $\lambda_1 = \lambda_2$, so that all lines in the family $\mathscr{L}$ are disjoint.

A similar argument shows that all lines in the family $\mathscr{M}$ are disjoint.

Finally, if a line in $\mathscr{L}$ intersects a line in $\mathscr{M}$ they can have at most one point in common since the lines are non-parallel – for, one is parallel to $(-1, 1, 0)$ and one is parallel to $(1, 1, 0)$.

So, let the line

$$\ell_{\lambda_1} = \{(\lambda_1 - \mu_1, \lambda_1 + \mu_1, 4\lambda_1\mu_1) : \mu_1 \text{ varies over } \mathbb{R}\} \quad (3)$$

in $\mathscr{L}$ and the line

$$m_{\mu_2} = \{(\lambda_2 - \mu_2, \lambda_2 + \mu_2, 4\lambda_2\mu_2) : \lambda_2 \text{ varies over } \mathbb{R}\} \quad (4)$$

in $\mathscr{M}$ meet at a point (a, b, c).

For this to happen we must have

$$(\lambda_1 - \mu_1, \lambda_1 + \mu_1, 4\lambda_1\mu_1) = (\lambda_2 - \mu_2, \lambda_2 + \mu_2, 4\lambda_2\mu_2),$$

for some real numbers $\lambda_1, \mu_1, \lambda_2$ and μ_2. We deduce from this equation that

$$\left.\begin{aligned} \lambda_1 - \mu_1 &= \lambda_2 - \mu_2, \\ \lambda_1 + \mu_1 &= \lambda_2 + \mu_2, \\ 4\lambda_1\mu_1 &= 4\lambda_2\mu_2. \end{aligned}\right\} \quad (5)$$

By adding the first two equations in (5) we find that $2\lambda_1 = 2\lambda_2$, so that $\lambda_1 = \lambda_2$; and by subtracting the first two equations in (5) we find that $2\mu_1 = 2\mu_2$, so that $\mu_1 = \mu_2$.

For simplicity, we now assign the values λ and μ to the common values of λ_1, λ_2 and μ_1, μ_2, respectively.

It follows that the line in $\mathscr{L}$

$$\ell_\lambda = \{(\lambda - \mu, \lambda + \mu, 4\lambda\mu) : \mu \text{ is a specific real number}\}$$

and the line in $\mathscr{M}$

$$m_\mu = \{(\lambda - \mu, \lambda + \mu, 4\lambda\mu) : \lambda \text{ is a specific real number}\}$$

meet at one point (a, b, c) so long as (a, b, c) is that common value

$$\begin{aligned} &(\lambda_1 - \mu_1, \lambda_1 + \mu_1, 4\lambda_1\mu_1) \\ &= (\lambda_2 - \mu_2, \lambda_2 + \mu_2, 4\lambda_2\mu_2) \\ &= (\lambda - \mu, \lambda + \mu, 4\lambda\mu). \end{aligned}$$

This happens when

$$\left.\begin{aligned} a &= \lambda - \mu, \\ b &= \lambda + \mu, \\ c &= 4\lambda\mu; \end{aligned}\right\}$$

so that

$$\lambda = \tfrac{1}{2}(a + b) \text{ and } \mu = \tfrac{1}{2}(b - a).$$

Chapter 2

Section 2.1

1. First, reflect the figure in the line through O, the centre of the circle, and P. Under this reflection, P remains fixed, and the circle maps onto itself. In particular, the point A maps to a point A' on the circle, and so the tangent PA maps onto the line PA'.

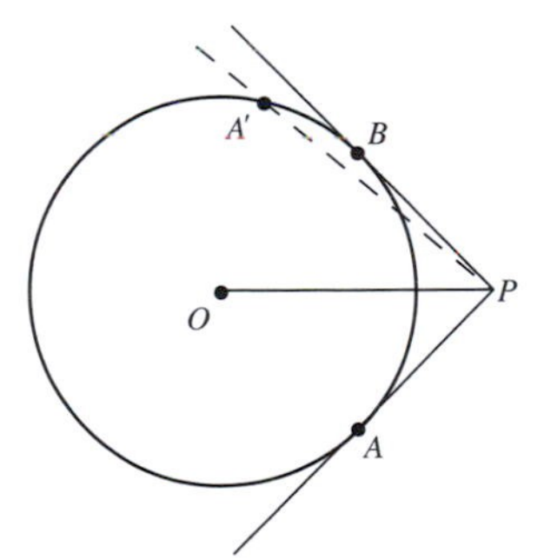

Now the tangent PA meets the circle at a single point A, so the image of the tangent must meet the circle at a single point. But the only way in which that can happen is if A' coincides with B.

Hence the line segment PA is reflected onto the line segment PB. Since reflection preserves lengths, it follows that $PA = PB$.

2. It is sufficient to show that there is an isometry which maps ΔABC onto ΔDEF. To construct this isometry, we start with the translation which maps A to D. This translation maps ΔABC onto $\Delta DB'C'$, where B' and C' are the images of B and C, respectively.

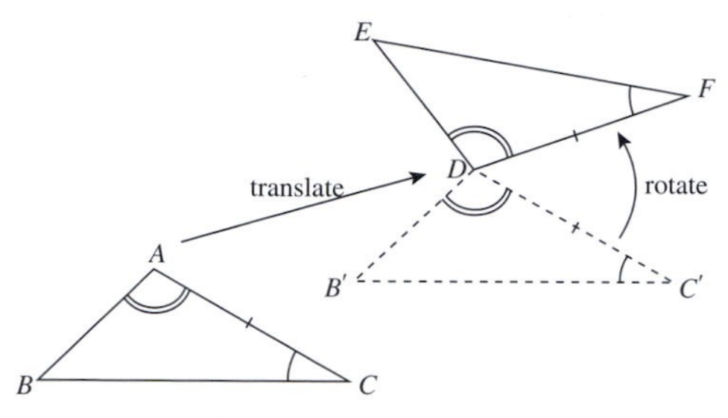

Since $DC' = AC = DF$, we can now rotate the point C' about D until it coincides with the point F. This rotation maps $\Delta DB'C'$ onto the triangle $\Delta DB''F$ shown below, where B'' is the image of B' under the rotation.

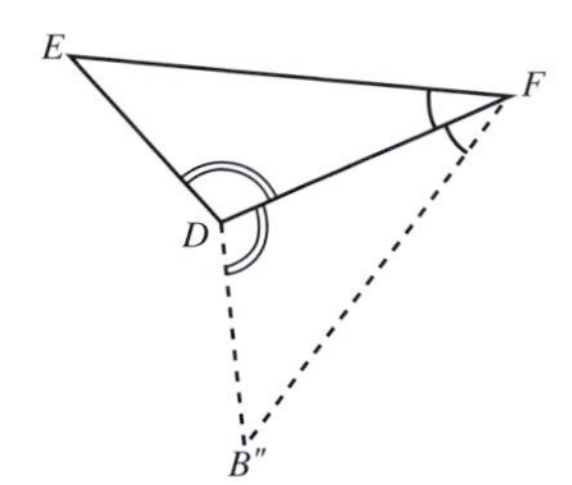

Now notice that

$$\begin{aligned}\angle EDF &= \angle BAC \quad \text{(given)}\\ &= \angle B'DC' \quad \text{(translation)}\\ &= \angle B''DF \quad \text{(rotation)},\end{aligned}$$

so either B'' lies on DE or the reflection of B'' in the line DF lies on DE. Also

$$\begin{aligned}\angle DFE &= \angle ACB \quad \text{(given)}\\ &= \angle DC'B' \quad \text{(translation)}\\ &= \angle DFB'' \quad \text{(rotation)},\end{aligned}$$

so either B'' lies on FE or the reflection of B'' in the line DF lies on FE. It follows that either B'' coincides with E or the reflection of B'' in the line DF coincides with E.

So, composing the translation, the rotation and (if necessary) a reflection, we obtain the required isometry which maps ΔABC onto ΔDEF. Since isometries preserve length and angle, it follows that $BC = EF, AB = DE$ and $\angle ABC = \angle DEF$.

3. Here we use the fact that a matrix $\mathbf{U}$ is orthogonal if $\mathbf{U}^T\mathbf{U} = \mathbf{I}$. We have

$$\begin{pmatrix}\cos\theta & \sin\theta\\ -\sin\theta & \cos\theta\end{pmatrix}\begin{pmatrix}\cos\theta & -\sin\theta\\ \sin\theta & \cos\theta\end{pmatrix} = \begin{pmatrix}1 & 0\\ 0 & 1\end{pmatrix}$$

and

$$\begin{pmatrix}\cos\theta & \sin\theta\\ \sin\theta & -\cos\theta\end{pmatrix}\begin{pmatrix}\cos\theta & \sin\theta\\ \sin\theta & -\cos\theta\end{pmatrix} = \begin{pmatrix}1 & 0\\ 0 & 1\end{pmatrix}.$$

So both matrices are orthogonal for all real θ.

4. First, $t_1 \circ t_2(\mathbf{x})$ is equal to

$$\begin{aligned}&t_1\left(\begin{pmatrix}-\frac{4}{5} & \frac{3}{5}\\ \frac{3}{5} & \frac{4}{5}\end{pmatrix}\mathbf{x} + \begin{pmatrix}-2\\ 1\end{pmatrix}\right)\\ &= \begin{pmatrix}\frac{3}{5} & -\frac{4}{5}\\ \frac{4}{5} & \frac{3}{5}\end{pmatrix}\left(\begin{pmatrix}-\frac{4}{5} & \frac{3}{5}\\ \frac{3}{5} & \frac{4}{5}\end{pmatrix}\mathbf{x} + \begin{pmatrix}-2\\ 1\end{pmatrix}\right) + \begin{pmatrix}1\\ -2\end{pmatrix}\\ &= \begin{pmatrix}\frac{3}{5} & -\frac{4}{5}\\ \frac{4}{5} & \frac{3}{5}\end{pmatrix}\begin{pmatrix}-\frac{4}{5} & \frac{3}{5}\\ \frac{3}{5} & \frac{4}{5}\end{pmatrix}\mathbf{x} + \begin{pmatrix}-2\\ -1\end{pmatrix} + \begin{pmatrix}1\\ -2\end{pmatrix}\\ &= \begin{pmatrix}-\frac{24}{25} & -\frac{7}{25}\\ -\frac{7}{25} & \frac{24}{25}\end{pmatrix}\mathbf{x} + \begin{pmatrix}-1\\ -3\end{pmatrix}.\end{aligned}$$

Next, $t_2 \circ t_1(\mathbf{x})$ is equal to

$$\begin{aligned}&t_2\left(\begin{pmatrix}\frac{3}{5} & -\frac{4}{5}\\ \frac{4}{5} & \frac{3}{5}\end{pmatrix}\mathbf{x} + \begin{pmatrix}1\\ -2\end{pmatrix}\right)\\ &= \begin{pmatrix}-\frac{4}{5} & \frac{3}{5}\\ \frac{3}{5} & \frac{4}{5}\end{pmatrix}\left(\begin{pmatrix}\frac{3}{5} & -\frac{4}{5}\\ \frac{4}{5} & \frac{3}{5}\end{pmatrix}\mathbf{x} + \begin{pmatrix}1\\ -2\end{pmatrix}\right) + \begin{pmatrix}-2\\ 1\end{pmatrix}\\ &= \begin{pmatrix}-\frac{4}{5} & \frac{3}{5}\\ \frac{3}{5} & \frac{4}{5}\end{pmatrix}\begin{pmatrix}\frac{3}{5} & -\frac{4}{5}\\ \frac{4}{5} & \frac{3}{5}\end{pmatrix}\mathbf{x} + \begin{pmatrix}-2\\ -1\end{pmatrix} + \begin{pmatrix}-2\\ 1\end{pmatrix}\\ &= \begin{pmatrix}0 & 1\\ 1 & 0\end{pmatrix}\mathbf{x} + \begin{pmatrix}-4\\ 0\end{pmatrix}.\end{aligned}$$

5. (a) Since $\mathbf{U}$ is an orthogonal matrix, it follows that $\mathbf{U^{-1}} = \mathbf{U^T}$. Taking the transpose of both sides, we have

$$\left(\mathbf{U}^{-1}\right)^T = \left(\mathbf{U}^T\right)^T = \mathbf{U} = \left(\mathbf{U}^{-1}\right)^{-1}.$$

Thus $\mathbf{U}^{-1}$ is an orthogonal matrix, and so t_2 is a Euclidean transformation.

(b) We have

$$\begin{aligned} t_1 \circ t_2(\mathbf{x}) &= t_1\left(\mathbf{U}^{-1}\mathbf{x} - \mathbf{U}^{-1}\mathbf{a}\right) \\ &= \mathbf{U}\left(\mathbf{U}^{-1}\mathbf{x} - \mathbf{U}^{-1}\mathbf{a}\right) + \mathbf{a} \\ &= (\mathbf{x} - \mathbf{a}) + \mathbf{a} \\ &= \mathbf{x} \end{aligned}$$

and

$$\begin{aligned} t_2 \circ t_1(\mathbf{x}) &= t_2(\mathbf{U}x + a) \\ &= \mathbf{U}^{-1}(\mathbf{Ux} + \mathbf{a}) - \mathbf{U}^{-1}\mathbf{a} \\ &= \left(\mathbf{x} + \mathbf{U}^{-1}\mathbf{a}\right) - \mathbf{U}^{-1}\mathbf{a} \\ &= \mathbf{x}, \end{aligned}$$

so t_2 is the inverse of t_1.

6. We have

$$\begin{pmatrix} \frac{3}{5} & -\frac{4}{5} \\ \frac{4}{5} & \frac{3}{5} \end{pmatrix}^{-1} = \begin{pmatrix} \frac{3}{5} & \frac{4}{5} \\ -\frac{4}{5} & \frac{3}{5} \end{pmatrix}$$

and

$$\begin{pmatrix} \frac{3}{5} & \frac{4}{5} \\ -\frac{4}{5} & \frac{3}{5} \end{pmatrix}\begin{pmatrix} 1 \\ -2 \end{pmatrix} = \begin{pmatrix} -1 \\ -2 \end{pmatrix},$$

so that

$$t^{-1}(\mathbf{x}) = \begin{pmatrix} \frac{3}{5} & \frac{4}{5} \\ -\frac{4}{5} & \frac{3}{5} \end{pmatrix}\mathbf{x} + \begin{pmatrix} 1 \\ 2 \end{pmatrix}.$$

7. (a) Not congruent
(b) Congruent
(c) Not congruent
(d) Congruent

8. Suppose that we are given three plane figures F_1, F_2 and F_3 such that

$$F_1 \text{ is congruent to } F_3 \qquad (1)$$

and

$$F_2 \text{ is congruent to } F_3. \qquad (2)$$

It follows from (2) and the symmetric property of congruence that

$$F_3 \text{ is congruent to } F_2. \qquad (3)$$

Hence from (1) and (3) and the transitive property of congruence, F_1 is congruent to F_2, as required.

Section 2.2

1. We use the fact that a 2 x 2 matrix is invertible if and only if its determinant is non-zero.

Each transformation is of the form

$$\mathbf{x} \mapsto \mathbf{Ax} + \mathbf{b},$$

where $\mathbf{A}$ is a 2 x 2 matrix, and so it is an affine transformation if and only if the determinant of the matrix $\mathbf{A}$ is non-zero.

(a) Here,

$$\begin{vmatrix} 1 & 3 \\ 1 & 2 \end{vmatrix} = 2 - 3 = -1,$$

which is non-zero; hence t_1 is an affine transformation.

(b) Here,

$$\begin{vmatrix} -6 & 5 \\ 3 & 2 \end{vmatrix} = -12 - 15 = -27,$$

which is non-zero; hence t_2 is an affine transformation.

(c) Here,

$$\begin{vmatrix} -2 & -1 \\ 8 & 4 \end{vmatrix} = -8 + 8 = 0;$$

hence t_3 is not an affine transformation.

(d) Here, $\mathbf{b} = \mathbf{0}$ and

$$\begin{vmatrix} 5 & -3 \\ -2 & 2 \end{vmatrix} = 10 - 6 = 4,$$

which is non-zero; hence t_4 is an affine transformation.

2. (a) Here, $t_1 \circ t_2(\mathbf{x})$ is equal to

$$\begin{aligned} & t_1\left(\begin{pmatrix} -6 & 5 \\ 3 & 2 \end{pmatrix}\mathbf{x} + \begin{pmatrix} 2 \\ 1 \end{pmatrix}\right) \\ &= \begin{pmatrix} 1 & 3 \\ 1 & 2 \end{pmatrix}\left(\begin{pmatrix} -6 & 5 \\ 3 & 2 \end{pmatrix}\mathbf{x} + \begin{pmatrix} 2 \\ 1 \end{pmatrix}\right) \\ &\quad + \begin{pmatrix} 4 \\ -2 \end{pmatrix} \\ &= \begin{pmatrix} 1 & 3 \\ 1 & 2 \end{pmatrix}\begin{pmatrix} -6 & 5 \\ 3 & 2 \end{pmatrix}\mathbf{x} + \begin{pmatrix} 5 \\ 4 \end{pmatrix} + \begin{pmatrix} 4 \\ -2 \end{pmatrix} \\ &= \begin{pmatrix} 3 & 11 \\ 0 & 9 \end{pmatrix}\mathbf{x} + \begin{pmatrix} 9 \\ 2 \end{pmatrix}. \end{aligned}$$

Since

$$\begin{vmatrix} 3 & 11 \\ 0 & 9 \end{vmatrix} = 27 - 0 = 27 \neq 0,$$

it follows that $t_1 \circ t_2$ is an affine transformation.

(b) Here, $t_2 \circ t_4(\mathbf{x})$ is equal to

$$t_2\left(\begin{pmatrix} 5 & -3 \\ -2 & 2 \end{pmatrix}\mathbf{x}\right)$$

$$= \begin{pmatrix} -6 & 5 \\ 3 & 2 \end{pmatrix}\left(\begin{pmatrix} 5 & -3 \\ -2 & 2 \end{pmatrix}\mathbf{x}\right) + \begin{pmatrix} 2 \\ 1 \end{pmatrix}$$

$$= \begin{pmatrix} -40 & 28 \\ 11 & -5 \end{pmatrix}\mathbf{x} + \begin{pmatrix} 2 \\ 1 \end{pmatrix}.$$

Since

$$\begin{vmatrix} -40 & 28 \\ 11 & -5 \end{vmatrix} = 200 - 308 = -108 \neq 0,$$

it follows that $t_2 \circ t_4$ is an affine transformation.

3. As described in Appendix 2, the inverse of a 2×2 matrix $\mathbf{A} = \begin{pmatrix} a & b \\ c & d \end{pmatrix}$ is $\mathbf{A}^{-1} = \frac{1}{ad-bc}\begin{pmatrix} d & -b \\ -c & a \end{pmatrix}$. Hence

$$\begin{pmatrix} 1 & 3 \\ 1 & 2 \end{pmatrix}^{-1} = \begin{pmatrix} -2 & 3 \\ 1 & -1 \end{pmatrix}$$

and

$$\begin{pmatrix} -2 & 3 \\ 1 & -1 \end{pmatrix}\begin{pmatrix} 4 \\ -2 \end{pmatrix} = \begin{pmatrix} -14 \\ 6 \end{pmatrix},$$

so that

$$t^{-1}(\mathbf{x}) = \begin{pmatrix} -2 & 3 \\ 1 & -1 \end{pmatrix}\mathbf{x} + \begin{pmatrix} 14 \\ -6 \end{pmatrix}.$$

Section 2.3

1. Let (x, y) be an arbitrary point on the line $3x - y + 1 = 0$, and let (x', y') be the image of (x, y) under t.Then

$$\begin{pmatrix} x' \\ y' \end{pmatrix} = \begin{pmatrix} \frac{1}{2} & -\frac{1}{2} \\ -1 & 2 \end{pmatrix}\begin{pmatrix} x \\ y \end{pmatrix} + \begin{pmatrix} -\frac{3}{2} \\ 4 \end{pmatrix}.$$

Since the inverse of the inverse of any invertible transformation is the original transformation, it follows from Example 1 that under t^{-1}, we have

$$\begin{pmatrix} x \\ y \end{pmatrix} = \begin{pmatrix} 4 & 1 \\ 2 & 1 \end{pmatrix}\begin{pmatrix} x' \\ y' \end{pmatrix} + \begin{pmatrix} 2 \\ -1 \end{pmatrix}.$$

Thus

$$x = 4x' + y' + 2 \quad \text{and} \quad y = 2x' + y' - 1.$$

Hence the image under t of the line $3x - y + 1 = 0$ has equation

$$3(4x' + y' + 2) - (2x' + y' - 1) + 1 = 0.$$

Dropping the dashes and simplifying, we obtain

$$5x + y + 4 = 0.$$

2. The argument here is similar to that of Problem 1. For, if (x, y) is an arbitrary point on the circle $x^2 + y^2 = 1$ and (x', y') is the image of (x, y) under t, then under t^{-1} we have

$$\begin{pmatrix} x \\ y \end{pmatrix} = \begin{pmatrix} 4 & 1 \\ 2 & 1 \end{pmatrix}\begin{pmatrix} x' \\ y' \end{pmatrix} + \begin{pmatrix} 2 \\ -1 \end{pmatrix}.$$

Thus

$$x = 4x' + y' + 2 \quad \text{and} \quad y = 2x' + y' - 1.$$

Hence the image under t of the circle $x^2 + y^2 = 1$ has equation

$$(4x' + y' + 2)^2 + (2x' + y' - 1)^2 = 1.$$

Dropping the dashes and simplifying. we obtain

$$10x^2 + 6xy + y^2 + 6x + y + 2 = 0.$$

3. First, we take $\mathbf{b} = \begin{pmatrix} 2 \\ 3 \end{pmatrix}$.

Next, we construct the matrix $\mathbf{A}$ whose first column is

$$\begin{pmatrix} 1 \\ 6 \end{pmatrix} - \begin{pmatrix} 2 \\ 3 \end{pmatrix} = \begin{pmatrix} -1 \\ 3 \end{pmatrix},$$

and whose second column is

$$\begin{pmatrix} 3 \\ -1 \end{pmatrix} - \begin{pmatrix} 2 \\ 3 \end{pmatrix} = \begin{pmatrix} 1 \\ -4 \end{pmatrix};$$

thus

$$\mathbf{A} = \begin{pmatrix} -1 & 1 \\ 3 & -4 \end{pmatrix}.$$

The required affine transformation t is therefore

$$t(\mathbf{x}) = \begin{pmatrix} -1 & 1 \\ 3 & -4 \end{pmatrix}\mathbf{x} + \begin{pmatrix} 2 \\ 3 \end{pmatrix} \quad (\mathbf{x} \in \mathbb{R}^2).$$

4. First, we take $\mathbf{b} = \begin{pmatrix} 1 \\ -2 \end{pmatrix}$. Next, we construct the matrix $\mathbf{A}$ whose first column is

$$\begin{pmatrix} 2 \\ 1 \end{pmatrix} - \begin{pmatrix} 1 \\ -2 \end{pmatrix} = \begin{pmatrix} 1 \\ 3 \end{pmatrix},$$

and whose second column is

$$\begin{pmatrix} -3 \\ 5 \end{pmatrix} - \begin{pmatrix} 1 \\ -2 \end{pmatrix} = \begin{pmatrix} -4 \\ 7 \end{pmatrix};$$

thus

$$\mathbf{A} = \begin{pmatrix} 1 & -4 \\ 3 & 7 \end{pmatrix}.$$

The required affine transformation t is therefore

$$t(\mathbf{x}) = \begin{pmatrix} 1 & -4 \\ 3 & 7 \end{pmatrix} \mathbf{x} + \begin{pmatrix} 1 \\ -2 \end{pmatrix} \quad (\mathbf{x} \in \mathbb{R}^2).$$

5. First, we find the affine transformation t_1 which maps $(0,0)$, $(1,0)$ and $(0,1)$ to $(1,-1)$, $(2,-2)$ and $(3,-4)$, respectively. This transformation has the form $t_1(\mathbf{x}) = \mathbf{A}\mathbf{x} + \mathbf{b}$, where $\mathbf{b} = \begin{pmatrix} 1 \\ -1 \end{pmatrix}$ and

$$\mathbf{A} = \begin{pmatrix} 2-1 & 3-1 \\ -2+1 & -4+1 \end{pmatrix} = \begin{pmatrix} 1 & 2 \\ -1 & -3 \end{pmatrix};$$

that is,

$$t_1(\mathbf{x}) = \begin{pmatrix} 1 & 2 \\ -1 & -3 \end{pmatrix} \mathbf{x} + \begin{pmatrix} 1 \\ -1 \end{pmatrix}.$$

Next, we find the affine transformation t_2 which maps $(0,0)$, $(1,0)$ and $(0,1)$ to $(8,13)$, $(3,4)$ and $(0,-1)$, respectively. This transformation has the form $t_2(\mathbf{x}) = \mathbf{A}\mathbf{x} + \mathbf{b}$, where $\mathbf{b} = \begin{pmatrix} 8 \\ 13 \end{pmatrix}$ and

$$\mathbf{A} = \begin{pmatrix} 3-8 & 0-8 \\ 4-13 & -1-13 \end{pmatrix} = \begin{pmatrix} -5 & -8 \\ -9 & -14 \end{pmatrix};$$

that is,

$$t_2(\mathbf{x}) = \begin{pmatrix} -5 & -8 \\ -9 & -14 \end{pmatrix} \mathbf{x} + \begin{pmatrix} 8 \\ 13 \end{pmatrix}.$$

We now require the formula for the inverse transformation t_1^{-1}. Since

$$\begin{pmatrix} 1 & 2 \\ -1 & -3 \end{pmatrix}^{-1} = -\begin{pmatrix} -3 & -2 \\ 1 & 1 \end{pmatrix} = \begin{pmatrix} 3 & 2 \\ -1 & -1 \end{pmatrix}$$

and

$$\begin{pmatrix} 3 & 2 \\ -1 & -1 \end{pmatrix} \begin{pmatrix} 1 \\ -1 \end{pmatrix} = \begin{pmatrix} 1 \\ 0 \end{pmatrix},$$

it follows that

$$t_1^{-1}(\mathbf{x}) = \begin{pmatrix} 3 & 2 \\ -1 & -1 \end{pmatrix} \mathbf{x} - \begin{pmatrix} 1 \\ 0 \end{pmatrix}.$$

The required affine transformation t is therefore $t(\mathbf{x}) = t_2 \circ t_1^{-1}(\mathbf{x})$, where $t_2 \circ t_1^{-1}(\mathbf{x})$ is equal to

$$\begin{aligned} & t_2\left(\begin{pmatrix} 3 & 2 \\ -1 & -1 \end{pmatrix} \mathbf{x} - \begin{pmatrix} 1 \\ 0 \end{pmatrix}\right) \\ &= \begin{pmatrix} -5 & -8 \\ -9 & -14 \end{pmatrix}\left(\begin{pmatrix} 3 & 2 \\ -1 & -1 \end{pmatrix} \mathbf{x} - \begin{pmatrix} 1 \\ 0 \end{pmatrix}\right) \\ &\quad + \begin{pmatrix} 8 \\ 13 \end{pmatrix} \\ &= \begin{pmatrix} -7 & -2 \\ -13 & -4 \end{pmatrix} \mathbf{x} - \begin{pmatrix} -5 \\ -9 \end{pmatrix} + \begin{pmatrix} 8 \\ 13 \end{pmatrix} \\ &= \begin{pmatrix} -7 & -2 \\ -13 & -4 \end{pmatrix} \mathbf{x} + \begin{pmatrix} 13 \\ 22 \end{pmatrix}. \end{aligned}$$

Section 2.4

1. (a)

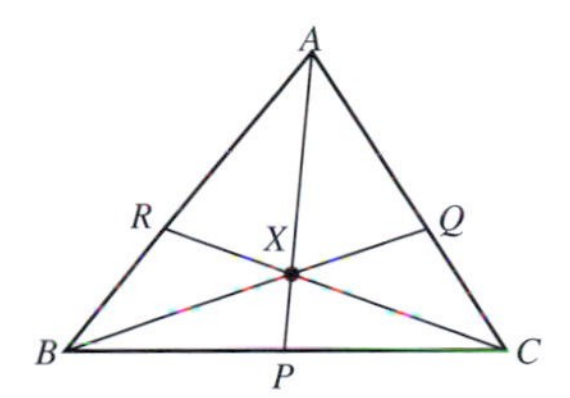

By Ceva's Theorem, we have

$$\frac{AR}{RB} \cdot \frac{BP}{PC} \cdot \frac{CQ}{QA} = 1.$$

First, $\frac{AR}{RB} = \frac{3}{2}$. Next, $\frac{AQ}{QC} = \frac{3}{2}$ and so $\frac{CQ}{QA} = \frac{2}{3}$.
It follows that

$$\frac{3}{2} \cdot \frac{BP}{PC} \cdot \frac{2}{3} = 1,$$

so $\frac{BP}{PC} = 1$.

Remark

This means that P is the midpoint of BC.

(b)

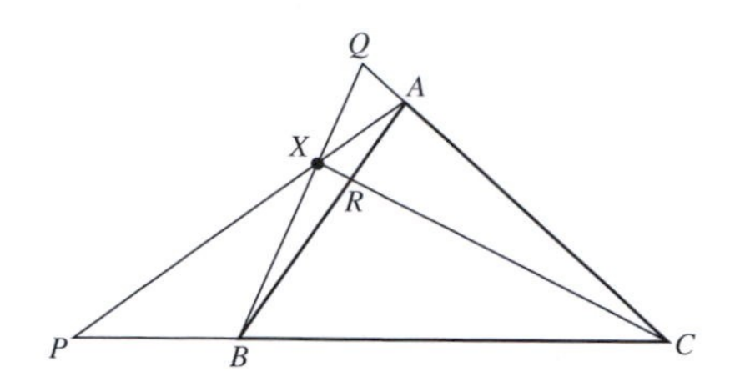

By Ceva's Theorem, we have

$$\frac{AR}{RB} \cdot \frac{BP}{PC} \cdot \frac{CQ}{QA} = 1.$$

Since $\frac{AR}{RB} = \frac{1}{2}$ and $\frac{BP}{PC} = -\frac{2}{7}$, it follows that

$$\frac{1}{2} \cdot \left(-\frac{2}{7}\right) \cdot \frac{CQ}{QA} = 1,$$

so $\frac{CQ}{QA} = -7$.

Remark

This means that Q lies on CA beyond A, and the length of CQ is seven times the length of AQ.

(c)

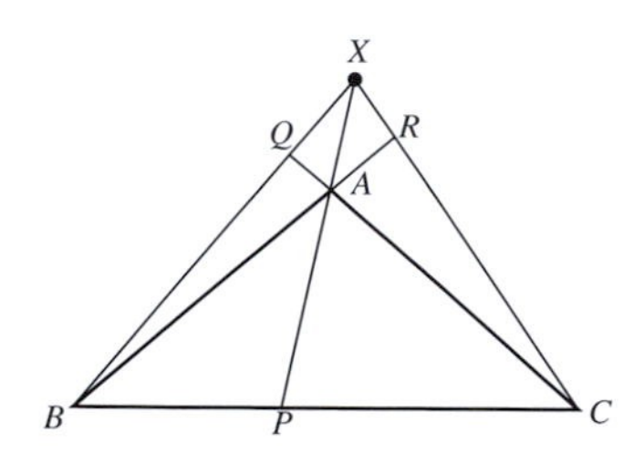

By Ceva's Theorem, we have

$$\frac{AR}{RB} \cdot \frac{BP}{PC} \cdot \frac{CQ}{QA} = 1.$$

We are given that $\frac{BP}{PC} = \frac{5}{7}$ and $\frac{CQ}{QA} = -7$, so

$$\frac{AR}{RB} \cdot \frac{5}{7} \cdot (-7) = 1.$$

Hence $\frac{AR}{RB} = -\frac{1}{5}$.

Remark

This means that R lies on BA beyond A, and the length of AR is one fifth of the length of RB.

2.

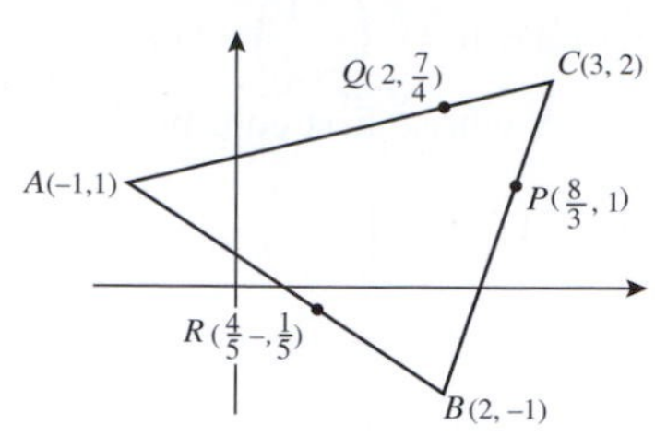

(a) Here we use the formula for calculating ratios given at the beginning of Appendix 2, just above the Section Formula. This gives

$$\frac{BP}{PC} = \frac{x_P - x_B}{x_C - x_P} = \frac{\frac{8}{3} - 2}{3 - \frac{8}{3}} = 2,$$

$$\frac{CQ}{QA} = \frac{x_Q - x_C}{x_A - x_Q} = \frac{2 - 3}{-1 - 2} = \frac{1}{3},$$

$$\frac{AR}{RB} = \frac{x_R - x_A}{x_B - x_R} = \frac{\frac{4}{5} + 1}{2 - \frac{4}{5}} = \frac{3}{2}.$$

Thus

P divides BC in the ratio 2 : 1,

Q divides CA in the ratio 1 : 3,

R divides AB in the ratio 3 : 2.

(b) It follows from part (a) that

$$\frac{AR}{RB} \cdot \frac{BP}{PC} \cdot \frac{CQ}{QA} = \frac{3}{2} \cdot 2 \cdot \frac{1}{3} = 1.$$

Thus by the converse to Ceva's Theorem, the lines AP, BQ and CR are concurrent.

3. (a)

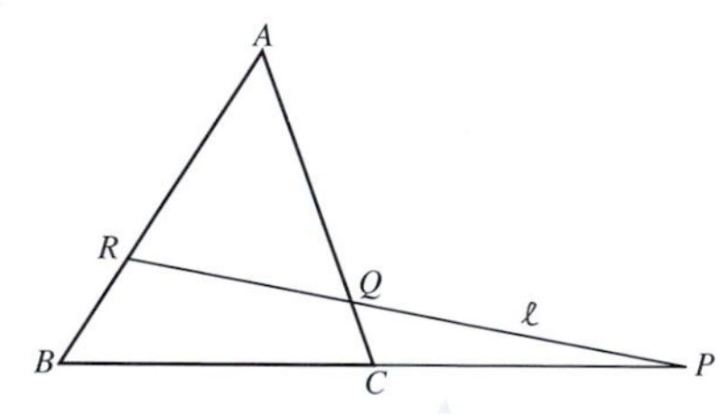

By Menelaus' Theorem, we have

$$\frac{AR}{RB} \cdot \frac{BP}{PC} \cdot \frac{CQ}{QA} = -1.$$

We are given that $\frac{AR}{RB} = 2$ and $\frac{BP}{PC} = -2$, so

$$2. (-2). \frac{CQ}{QA} = -1.$$

Hence $\frac{CQ}{QA} = \frac{1}{4}$.

Remark

This means that Q lies on CA between C and A, and the length of CQ is one quarter of the length of QA.

(b)

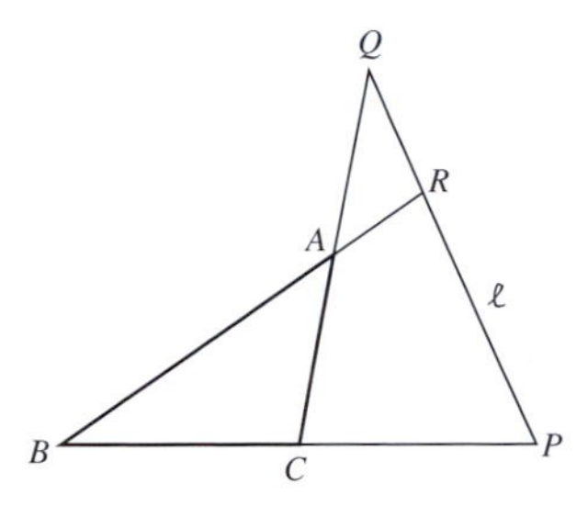

By Menelaus' Theorem, we have

$$\frac{AR}{RB} \cdot \frac{BP}{PC} \cdot \frac{CQ}{QA} = -1.$$

We are given that $\frac{AR}{RB} = -\frac{1}{4}$ and $\frac{BP}{PC} = -2$, so

$$\left(-\frac{1}{4}\right) \cdot (-2) \cdot \frac{CQ}{QA} = -1.$$

Hence $\frac{CQ}{QA} = -2$.

Remark

This means that Q lies on CA beyond A, and the length of CQ is twice the length of QA.

4.

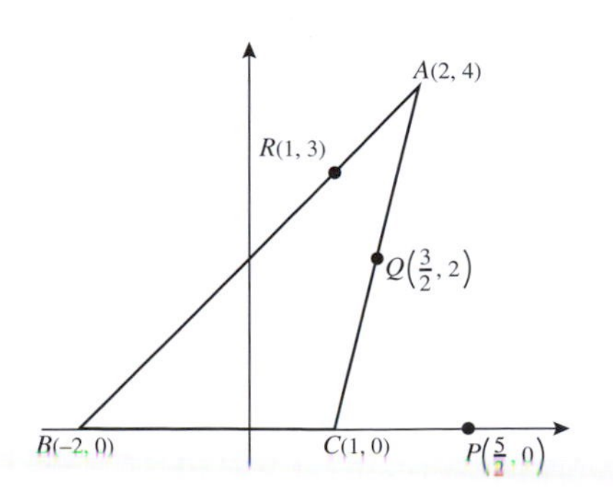

(a) Here we have

$$\frac{BP}{PC} = \frac{x_P - x_B}{x_C - x_P} = \frac{\frac{5}{2} + 2}{1 - \frac{5}{2}} = -3,$$

$$\frac{CQ}{QA} = \frac{x_Q - x_C}{x_A - x_Q} = \frac{\frac{3}{2} - 1}{2 - \frac{3}{2}} = 1,$$

$$\frac{AR}{RB} = \frac{x_R - x_A}{x_B - x_R} = \frac{1 - 2}{-2 - 1} = \frac{1}{3}.$$

Thus

P divides BC in the ratio $-3 : 1$,

Q divides CA in the ratio $1 : 1$,

R divides AB in the ratio $1 : 3$.

(b) It follows from part (a) that

$$\frac{AR}{RB} \cdot \frac{BP}{PC} \cdot \frac{CQ}{QA} = \frac{1}{3} \cdot (-3) \cdot 1 = -1.$$

Thus by the converse to Menelaus' Theorem, the points P, Q and R are collinear.

5.

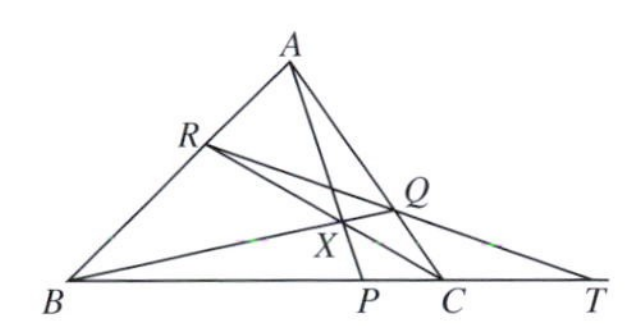

By Ceva's Theorem, we have

$$\frac{AR}{RB} \cdot \frac{BP}{PC} \cdot \frac{CQ}{QA} = 1. \quad (1)$$

Also, by Menelaus' Theorem, we have

$$\frac{AR}{RB} \cdot \frac{BT}{TC} \cdot \frac{CQ}{QA} = -1. \quad (2)$$

Comparing (1) and (2), we deduce that

$$\frac{BT}{TC} = -\frac{BP}{PC} = -k.$$

6.

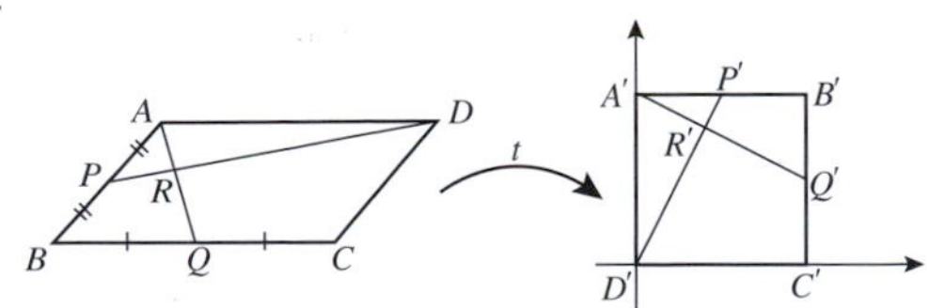

By the Fundamental Theorem of Affine Geometry (Theorem 1, Subsection 2.3.2), there exists a unique affine transformation t which maps A, D and C to $A'(0,1)$, $D'(0,0)$ and $C'(1,0)$, respectively.

(a) Since t maps AD onto the vertical line $A'D'$, and BC is parallel to AD, it follows that the image of BC under t must be a vertical line. Also, since t maps DC onto the horizontal line $D'C'$, and AB is parallel to DC, the image of AB under t must be a horizontal line. It follows that B', the image of B under t, must be the point with coordinates $(1,1)$.

(b) Since P is the midpoint of AB, its image $P'=t(P)$ must be the midpoint of $A'B'$ since ratios along a line are preserved by t. Hence $P' = \left(\frac{1}{2}, 1\right)$.

Since the slope of $D'P'$ is 2 and the line passes through the origin, the equation of the line $D'P'$ must be

$$y = 2x. \tag{3}$$

Next, since Q is the midpoint of BC, its image $Q' = t(Q)$ must be the midpoint of $B'C'$. Hence $Q' = \left(1, \frac{1}{2}\right)$.

Then the slope of the line $A'Q'$ must be

$$\frac{1-\frac{1}{2}}{0-1} = -\frac{1}{2}.$$

Since the line passes through the point $A'(0,1)$, the equation of the line $A'Q'$ must be

$$y - 1 = -\frac{1}{2}(x-0);$$

that is,

$$y = -\frac{1}{2}x + 1. \tag{4}$$

Now, R', the image of R under t, must lie on the lines $D'P'$ and $A'Q'$, so that its coordinates must satisfy both (3) and (4). Substituting for y from (3) into (4), we obtain

$$2x = -\frac{1}{2}x + 1,$$

so that $x = \frac{2}{5}$. It follows from (3) that $y = \frac{4}{5}$. Thus R' is the point $\left(\frac{2}{5}, \frac{4}{5}\right)$.

Comparing the y-coordinates 1, $\frac{4}{5}$ and 0 of P', R' and D', we obtain $P'R' : R'D' = 1:4$.

Since ratios along a line are preserved by the affine transformation t^{-1}, it follows that

$$PR : RD = 1:4.$$

Finally, comparing the x-coordinates 0, $\frac{2}{5}$ and 1 of A', R' and Q', we obtain $A'R' : R'Q' = 2:3$. Since ratios along a line are preserved under the affine transformation t^{-1}, it follows that

$$AR : RQ = 2:3.$$

7. The matrix $\mathbf{M}$ for the triangle of reference ΔABC is

$$\mathbf{M} = \begin{pmatrix} 1 & 2 & 1 \\ 1 & 2 & 2 \\ 1 & 1 & 1 \end{pmatrix},$$

whose inverse is

$$\begin{pmatrix} 0 & -1 & 2 \\ 1 & 0 & -1 \\ -1 & 1 & 0 \end{pmatrix}.$$

It follows from the representation (7) that the point $(-1,1)$ has barycentric coordinates with respect to the triangle of reference ΔABC given by

$$\begin{pmatrix} 0 & -1 & 2 \\ 1 & 0 & -1 \\ -1 & 1 & 0 \end{pmatrix}\begin{pmatrix} -1 \\ 1 \\ 1 \end{pmatrix} = \begin{pmatrix} 1 \\ -2 \\ 2 \end{pmatrix};$$

namely, barycentric coordinates $(1,-2,2)$.

8. (a) By Theorem 6, the points with barycentric coordinates $(1,1,-1)$, $(4,-2,-1)$ and $\left(\frac{1}{2}, 2, -\frac{3}{2}\right)$ are collinear if and only if

$$\begin{vmatrix} 1 & 4 & \frac{1}{2} \\ 1 & -2 & 2 \\ -1 & -1 & -\frac{3}{2} \end{vmatrix} = 0.$$

Now,

$$\begin{vmatrix} 1 & 4 & \frac{1}{2} \\ 1 & -2 & 2 \\ -1 & -1 & -\frac{3}{2} \end{vmatrix} = \begin{vmatrix} -2 & 2 \\ -1 & -\frac{3}{2} \end{vmatrix} - 4\begin{vmatrix} 1 & 2 \\ -1 & -\frac{3}{2} \end{vmatrix} + \frac{1}{2}\begin{vmatrix} 1 & -2 \\ -1 & -1 \end{vmatrix}$$

$$= (3+2) - 4\left(-\tfrac{3}{2}+2\right) + \tfrac{1}{2}(-1-2)$$

$$= 5 - 4\left(\tfrac{1}{2}\right) + \tfrac{1}{2}(-3)$$

$$= \tfrac{3}{2}$$

$$\neq 0.$$

It follows that the points are not collinear.

(b) By Theorem 6, the points with barycentric coordinates $(1,1,-1)$, $(2,-2,1)$, $(-1,7,-5)$ are collinear if and only if

$$\begin{vmatrix} 1 & 2 & -1 \\ 1 & -2 & 7 \\ -1 & 1 & -5 \end{vmatrix} = 0.$$

Now,

$$\begin{vmatrix} 1 & 2 & -1 \\ 1 & -2 & 7 \\ -1 & 1 & -5 \end{vmatrix} = \begin{vmatrix} -2 & 7 \\ 1 & -5 \end{vmatrix} - 2\begin{vmatrix} 1 & 7 \\ -1 & -5 \end{vmatrix} - \begin{vmatrix} 1 & -2 \\ -1 & 1 \end{vmatrix}$$

$$= (10-7) - 2(-5+7) - (1-2)$$

$$= 3 - 2(2) - (-1)$$

$$= 3 - 4 + 1$$

$$= 0.$$

It follows that the points are collinear.

By the Corollary to Theorem 6, the equation of the line through the points $(1,1,-1)$ and $(2,-2,1)$ is

$$\begin{vmatrix} 1 & 2 & \xi \\ 1 & -2 & \eta \\ -1 & 1 & \zeta \end{vmatrix} = 0.$$

Now,

$$\begin{vmatrix} 1 & 2 & \xi \\ 1 & -2 & \eta \\ -1 & 1 & \zeta \end{vmatrix} = \begin{vmatrix} -2 & \eta \\ 1 & \zeta \end{vmatrix} - 2\begin{vmatrix} 1 & \eta \\ -1 & \zeta \end{vmatrix} + \xi\begin{vmatrix} 1 & -2 \\ -1 & 1 \end{vmatrix}$$

$$= (-2\zeta - \eta) - 2(\zeta + \eta) + \xi(1-2)$$

$$= -\xi - 3\eta - 4\zeta.$$

It follows that the equation of the desired line is

$$\xi + 3\eta + 4\zeta = 0.$$

Section 2.5

1.

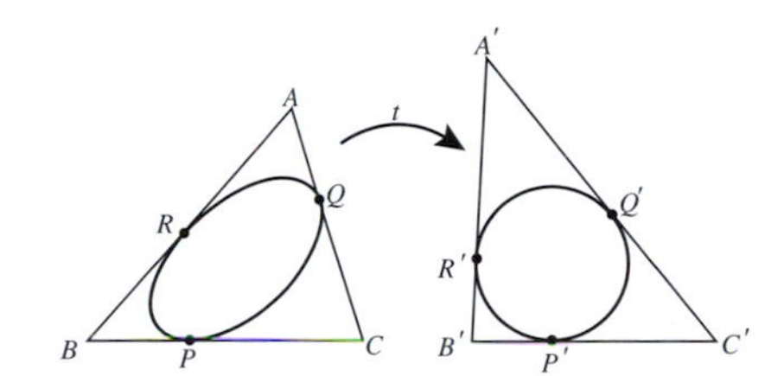

First, map the ellipse onto the unit circle, by some affine transformation t. Since tangency is preserved by affine transformations, the image under t of the triangle ΔABC is another triangle $\Delta A'B'C'$, whose sides are tangential to the unit circle.

These sides touch the unit circle at $P' = t(P)$, $Q' = t(Q)$ and $R' = t(R)$.

By Problem 1 of Section 2.1, the two tangents from a point to a circle are of equal length, and so (ignoring the directions of line segments)

$$A'Q' = A'R', \quad B'P' = B'R' \quad \text{and}$$
$$C'P' = C'Q'.$$

In terms of signed distances, it follows that

$$\frac{A'R'}{R'B'} \cdot \frac{B'P'}{P'C'} \cdot \frac{C'Q'}{Q'A'} = \pm 1;$$

in fact the product must equal 1 since P, Q and R are internal points of the sides of the triangle and therefore each of the above three fractions is positive.

Since ratios of lengths along a line are not changed by the inverse affine transformation t^{-1}, we deduce that

$$\frac{AR}{RB} \cdot \frac{BP}{PC} \cdot \frac{CQ}{QA} = 1.$$

It follows from the converse to Ceva's Theorem that the lines AP, BQ and CR are concurrent.

2. 

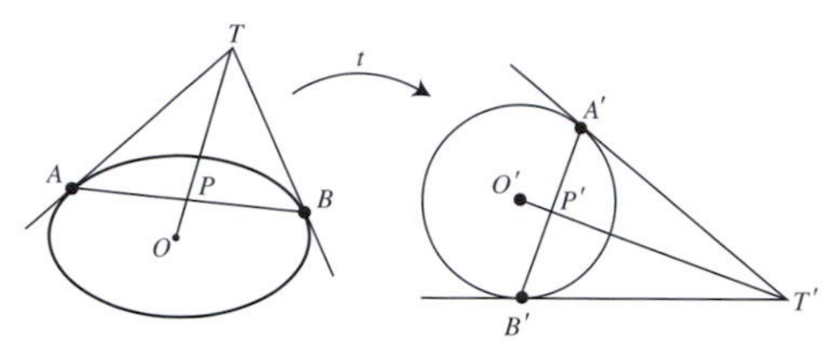

First, map the ellipse onto the unit circle, by some affine transformation t. Since tangency is preserved by affine transformations, the images under t of the tangents TA and TB are tangents $T'A'$ and $T'B'$ to the unit circle.

Let P' be the point of intersection of the chord $A'B'$ with the line joining T' to O', the centre of the unit circle. By symmetry, the triangles $\Delta\, T'A'P'$ and $\Delta\, T'B'P'$ are Euclidean-congruent and so $A'P' = B'P'$; in other words, P' is the midpoint of $A'B'$.

Let the line joining T to the centre O of the ellipse meet AB at P. Then, since P' is the midpoint of $A'B'$ and since midpoints of line segments and centres of ellipses are preserved by the inverse transformation t^{-1}, $P = t^{-1}(P')$ is the midpoint of $AB = t^{-1}(A'B')$. Hence OP bisects all chords of H that are parallel to ℓ.

3.

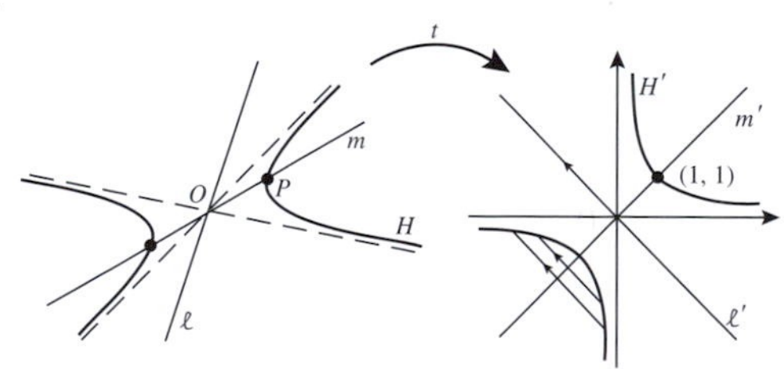

First, map the hyperbola H onto the rectangular hyperbola $H' = \{(x, y) : xy = 1\}$ by some affine transformation t, in such a way that t maps P to the point $(1, 1)$. Since the property of being the centre of the hyperbola is preserved under affine transformations, t maps the centre, O, of H to the centre of H', namely the origin.

Let m' be the image of OP under t. Then m' passes through the origin and the point $(1, 1)$, so its equation is $y = x$. Clearly, H' is symmetric with respect to m'. Now let ℓ' be the line with equation $y = -x$; this is perpendicular to m'. By symmetry, m' bisects all chords of the rectangular hyperbola which are parallel to ℓ'.

But the properties of parallelism and of ratios along a line are preserved by affine transformations, so if ℓ is the line $t^{-1}(\ell')$, then OP bisects all chords of H which are parallel to ℓ.

Chapter 3

Section 3.2

1. (a) This represents the same Point as [1, 2, 3], for if $\lambda = 2$, then

$$[2, 4, 6] = [\lambda, 2\lambda, 3\lambda] = [1, 2, 3].$$

(b) This does not represent the same Point as [1, 2, 3], for there is no λ that satisfies

$$1 = \lambda, \quad 2 = 2\lambda, \quad -3 = 3\lambda.$$

(c) This represents the same Point as [1, 2, 3], for if $\lambda = -1$, then

$$[-1, -2, -3] = [\lambda, 2\lambda, 3\lambda] = [1, 2, 3].$$

(d) This does not represent the same Point as [1, 2, 3], for there is no λ that satisfies

$$11 = \lambda, \quad 12 = 2\lambda, \quad 13 = 3\lambda.$$

2. In each case we multiply by the least common multiple of the denominators (or any integer multiple of the least common multiple) to obtain:

(a) $\left[\frac{3}{4}, \frac{1}{2}, -\frac{1}{8}\right] = [6, 4, -1]$ (multiply by 8);

(b) $\left[0, 4, \frac{2}{3}\right] = [0, 12, 2]$ (multiply by 3) $=[0, 6, 1]$;

(c) $\left[\frac{1}{6}, -\frac{1}{3}, -\frac{1}{2}\right] = [1, -2, -3]$ (multiply by 6).

3. Dividing by the first coordinate in each, we obtain:

$$[2,3,-5] = \left[1, \tfrac{3}{2}, -\tfrac{5}{2}\right];$$
$$[-8,-12,20] = \left[1, \tfrac{3}{2}, -\tfrac{5}{2}\right];$$
$$\left[\sqrt{2}, \sqrt{3}, -\sqrt{5}\right] = \left[1, \sqrt{\tfrac{3}{2}}, -\sqrt{\tfrac{5}{2}}\right];$$
$$[4,-6,10] = \left[1, -\tfrac{3}{2}, \tfrac{5}{2}\right];$$
$$[-20,-30,50] = \left[1, \tfrac{3}{2}, -\tfrac{5}{2}\right];$$
$$[74,148,0] = [1,2,0].$$

Hence the homogeneous coordinates

$$[2,3,-5], \quad [-8,-12,20], \quad [-20,-30,50]$$

all represent the same Point. The other homogeneous coordinates represent different Points.

4. In each case we seek an equation of the form $ax + by + cz = 0$ which is satisfied by the homogeneous coordinates of the given pair of Points.
   (a) An equation for the Line through $[0,1,0]$ and $[0,0,1]$ is $x = 0$.
   (b) An equation for the Line through $[2,2,3]$ and $[3,3,7]$ is $x = y$.

5. We use the strategy for determining an equation for the Line through two given Points given in Subsection 3.2.2.
   (a) An equation for the Line through the Points $[2,5,4]$ and $[3,1,7]$ is

$$\begin{vmatrix} x & y & z \\ 2 & 5 & 4 \\ 3 & 1 & 7 \end{vmatrix} = 0.$$

Now

$$\begin{vmatrix} x & y & z \\ 2 & 5 & 4 \\ 3 & 1 & 7 \end{vmatrix} = x\begin{vmatrix} 5 & 4 \\ 1 & 7 \end{vmatrix} - y\begin{vmatrix} 2 & 4 \\ 3 & 7 \end{vmatrix} + z\begin{vmatrix} 2 & 5 \\ 3 & 1 \end{vmatrix}$$
$$= 31x - 2y - 13z,$$

so an equation for the Line is

$$31x - 2y - 13z = 0.$$

   (b) An equation for the Line through the Points $[-2,-4,5]$ and $[3,-2,-4]$ is

$$\begin{vmatrix} x & y & z \\ -2 & -4 & 5 \\ 3 & -2 & -4 \end{vmatrix} = 0.$$

Now

$$\begin{vmatrix} x & y & z \\ -2 & -4 & 5 \\ 3 & -2 & -4 \end{vmatrix} = x\begin{vmatrix} -4 & 5 \\ -2 & -4 \end{vmatrix} - y\begin{vmatrix} -2 & 5 \\ 3 & -4 \end{vmatrix} + z\begin{vmatrix} -2 & -4 \\ 3 & -2 \end{vmatrix}$$
$$= 26x + 7y + 16z,$$

so an equation for the Line is

$$26x + 7y + 16z = 0.$$

6. We use the strategy for deciding whether three Points are collinear given in Subsection 3.2.2.
   (a) The Points $[1,2,3]$, $[1,1,-2]$ and $[2,1,-9]$ are collinear if and only if

$$\begin{vmatrix} 1 & 2 & 3 \\ 1 & 1 & -2 \\ 2 & 1 & -9 \end{vmatrix} = 0.$$

Now

$$\begin{vmatrix} 1 & 2 & 3 \\ 1 & 1 & -2 \\ 2 & 1 & -9 \end{vmatrix} = 1\begin{vmatrix} 1 & -2 \\ 1 & -9 \end{vmatrix} - 2\begin{vmatrix} 1 & -2 \\ 2 & -9 \end{vmatrix} + 3\begin{vmatrix} 1 & 1 \\ 2 & 1 \end{vmatrix}$$
$$= 1(-9+2) - 2(-9+4) + 3(1-2)$$
$$= -7 + 10 - 3$$
$$= 0.$$

It follows that the three given Points are collinear.

   (b) The Points $[1,2,-1]$, $[2,1,0]$ and $[0,-1,3]$ are collinear if and only if

$$\begin{vmatrix} 1 & 2 & -1 \\ 2 & 1 & 0 \\ 0 & -1 & 3 \end{vmatrix} = 0.$$

Now

$$\begin{vmatrix} 1 & 2 & -1 \\ 2 & 1 & 0 \\ 0 & -1 & 3 \end{vmatrix} = 1\begin{vmatrix} 1 & 0 \\ -1 & 3 \end{vmatrix} - 2\begin{vmatrix} 2 & 0 \\ 0 & 3 \end{vmatrix} - 1\begin{vmatrix} 2 & 1 \\ 0 & -1 \end{vmatrix}$$
$$= 1(3-0) - 2(6-0) - 1(-2-0)$$
$$= 3 - 12 + 2$$
$$= -7 \neq 0.$$

It follows that the three given Points are not collinear.

7. We have already shown that $[1,0,0]$, $[0,1,0]$, $[1,1,1]$ are not collinear, so this leaves three other cases to consider.

   First we check that $[1,0,0]$, $[0,0,1]$, $[1,1,1]$ are not collinear. This follows because $[1,0,0]$ and $[0,0,1]$ lie on the Line $y = 0$, whereas $[1,1,1]$ does not.

   Next we check that $[0,1,0]$, $[0,0,1]$, $[1,1,1]$ are not collinear. This follows because $[0,1,0]$ and $[0,0,1]$ lie on the Line $x = 0$, whereas $[1,1,1]$ does not.

   Finally we check that $[1,0,0]$, $[0,1,0]$, $[0,0,1]$ are not collinear. This follows because $[1,0,0]$, $[0,1,0]$ lie on the Line $z = 0$, whereas $[0,0,1]$ does not.

8. (a) At the Point of intersection $[x, y, z]$ of the two Lines, we have

$$x - y - z = 0, \text{ and} \qquad (1)$$
$$x + 5y + 2z = 0. \qquad (2)$$

   Subtracting equation (1) from equation (2), we obtain $6y + 3z = 0$, so $z = -2y$.

   Next, substituting $-2y$ in place of z in equation (1), we obtain $x - y + 2y = 0$, so $x = -y$.

   It follows that the homogeneous coordinates of the Point of intersection are $[-y, y, -2y]$ (where $y \neq 0$), which we may rewrite equivalently as $[-1, 1, -2]$.

   (b) At the Point of intersection $[x, y, z]$ of the two Lines, we have

$$x + 2y - z = 0, \text{ and} \qquad (3)$$
$$2x + y - 4z = 0. \qquad (4)$$

   Subtracting twice equation (3) from equation (4), we obtain $-3y - 2z = 0$, so $y = -\frac{2}{3}z$.

   Next, substituting $-\frac{2}{3}z$ in place of y in equation (3), we obtain $x - \frac{4}{3}z - z = 0$, so $x = \frac{7}{3}z$.

   It follows that the homogeneous coordinates of the Point of intersection are $\left[\frac{7}{3}z, -\frac{2}{3}z, z\right]$ (where $z \neq 0$), which we may rewrite equivalently as $[7, -2, 3]$.

9. First, we find equations for the two Lines, using the determinant formula.

   An equation for the Line through the Points $[1, 2, -3]$ and $[2, -1, 0]$ is

$$\begin{vmatrix} x & y & z \\ 1 & 2 & -3 \\ 2 & -1 & 0 \end{vmatrix} = 0.$$

Now

$$\begin{vmatrix} x & y & z \\ 1 & 2 & -3 \\ 2 & -1 & 0 \end{vmatrix} = x\begin{vmatrix} 2 & -3 \\ -1 & 0 \end{vmatrix} - y\begin{vmatrix} 1 & -3 \\ 2 & 0 \end{vmatrix} + z\begin{vmatrix} 1 & 2 \\ 2 & -1 \end{vmatrix}$$
$$= x(0-3) - y(0+6) + z(-1-4)$$
$$= -3x - 6y - 5z.$$

Hence an equation for the Line may be written as

$$3x + 6y + 5z = 0. \qquad (5)$$

Next, an equation for the Line through the Points $[1, 0, -1]$ and $[1, 1, 1]$ is

$$\begin{vmatrix} x & y & z \\ 1 & 0 & -1 \\ 1 & 1 & 1 \end{vmatrix} = 0.$$

Now

$$\begin{vmatrix} x & y & z \\ 1 & 0 & -1 \\ 1 & 1 & 1 \end{vmatrix} = x\begin{vmatrix} 0 & -1 \\ 1 & 1 \end{vmatrix} - y\begin{vmatrix} 1 & -1 \\ 1 & 1 \end{vmatrix}$$
$$+ z\begin{vmatrix} 1 & 0 \\ 1 & 1 \end{vmatrix}$$
$$= x(0+1) - y(1+1) + z(1-0)$$
$$= x - 2y + z.$$

Hence an equation for the Line may be written as

$$x - 2y + z = 0. \qquad (6)$$

At the Point of intersection $[x, y, z]$ of the two Lines, both equations (5) and (6) hold.

Adding three times equation (6) to equation (5), we obtain $6x + 8z = 0$, so $z = -\frac{3}{4}x$.

Next, substituting $z = -\frac{3}{4}x$ into equation (6), we obtain

$$x - 2y - \frac{3}{4}x = 0, \text{ so } y = \frac{1}{8}x.$$

Hence the homogeneous coordinates for the Point of intersection of the two Lines are $\left[x, \frac{1}{8}x, -\frac{3}{4}x\right]$ (where $x \neq 0$), or, equivalently, $[8, 1, -6]$.

10. In this particular case, the homogeneous coordinates of the Points are particularly simple, so we can write down equations for the two Lines without using determinants.

    An equation for the Line through the Points $[1, 0, 0]$ and $[0, 1, 0]$ is $z = 0$ (since this is of the right form, and passes through the two Points).

    An equation for the Line through the Points $[0, 0, 1]$ and $[1, 1, 1]$ is $x = y$ (since this is of the right form, and passes through the two Points).

    The two Lines meet where $z = 0$ and $x = y$, so the homogeneous coordinates for their Point of intersection are $[x, x, 0]$ (where $x \neq 0$), or, equivalently, $[1, 1, 0]$.

11. In Problem 10 we found that the Point of intersection of $z = 0$ and $x = y$ is $[1, 1, 0]$. Similarly, the Point of intersection of $y = 0$ and $z = x$ is $[x, 0, x]$ or $[1, 0, 1]$, and the Point of intersection of $x = 0$ and $y = z$ is $[0, z, z]$ or $[0, 1, 1]$.

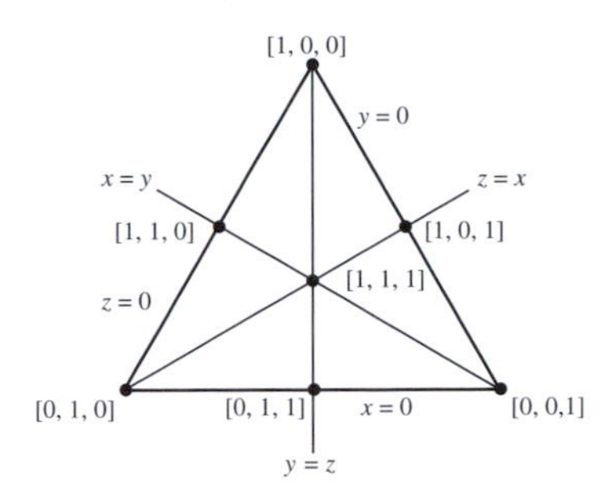

12. The ideal Points for π consist of lines through the origin of $\mathbb{R}^3$ that are parallel to π. These are the Points that lie on the ideal Line $y = 0$.

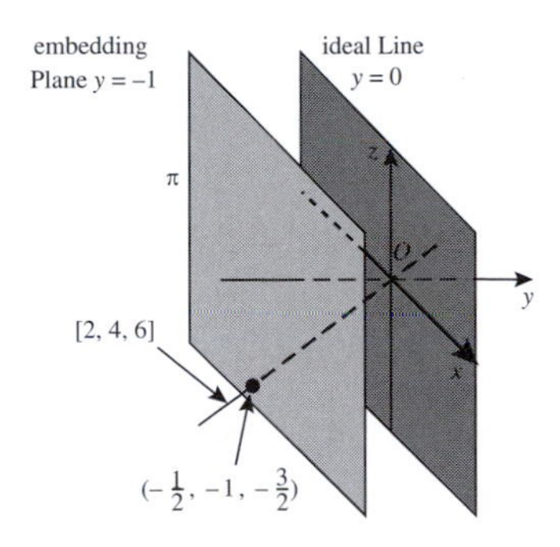

The Euclidean point of π which corresponds to the Point $[2, 4, 6]$ is that multiple of $(2, 4, 6)$ which lies on the plane $y = -1$. That is, $-\frac{1}{4}(2, 4, 6) = \left(-\frac{1}{2}, -1, -\frac{3}{2}\right)$.

Section 3.3

1. (a) The mapping

$$t : [x, y, z] \mapsto [-2y+3z, -x+5y-z, -3x]$$

can be expressed in the form $[\mathbf{x}] \mapsto [\mathbf{Ax}]$, where

$$\mathbf{A} = \begin{pmatrix} 0 & -2 & 3 \\ -1 & 5 & -1 \\ -3 & 0 & 0 \end{pmatrix}.$$

Now

$$\det \mathbf{A} = \begin{vmatrix} 0 & -2 & 3 \\ -1 & 5 & -1 \\ -3 & 0 & 0 \end{vmatrix}$$

$$= 0\begin{vmatrix} 5 & -1 \\ 0 & 0 \end{vmatrix} - (-2)\begin{vmatrix} -1 & -1 \\ -3 & 0 \end{vmatrix} + 3\begin{vmatrix} -1 & 5 \\ -3 & 0 \end{vmatrix}$$

$$= 0 + 2 \times (-3) + 3 \times 15$$

$$= 39 \neq 0,$$

so $\mathbf{A}$ is invertible. It follows that t is a projective transformation, and that $\mathbf{A}$ is a matrix associated with t.

(b) The mapping

$$t : [x, y, z] \mapsto [x - 7y + 4z, -x + 5y - z, x - 9y + 7z]$$

can be expressed in the form $[\mathbf{x}] \mapsto [\mathbf{Ax}]$, where

$$\mathbf{A} = \begin{pmatrix} 1 & -7 & 4 \\ -1 & 5 & -1 \\ 1 & -9 & 7 \end{pmatrix}.$$

Now

$$\det \mathbf{A} = \begin{vmatrix} 1 & -7 & 4 \\ -1 & 5 & -1 \\ 1 & -9 & 7 \end{vmatrix}$$

$$= 1\begin{vmatrix} 5 & -1 \\ -9 & 7 \end{vmatrix} + 7\begin{vmatrix} -1 & -1 \\ 1 & 7 \end{vmatrix} + 4\begin{vmatrix} -1 & 5 \\ 1 & -9 \end{vmatrix}$$

$$= 1 \times 26 + 7 \times (-6) + 4 \times 4$$

$$= 0,$$

so $\mathbf{A}$ is not invertible. It follows that t is not a projective transformation.

(c) The mapping

$$t : [x, y, z] \mapsto [x - 1 + z, 2y - 4z + 5, 2x]$$

cannot be expressed in the form $[\mathbf{x}] \mapsto [\mathbf{Ax}]$, where $\mathbf{A}$ is a 3×3 matrix whose entries are real numbers. Hence t cannot be a projective transformation.

2. (a) The image of the Point $[1, 2, -1]$ under t is given by

$$\left[\begin{pmatrix} 1 & 1 & -1 \\ -1 & -2 & 1 \\ 4 & -3 & 4 \end{pmatrix}\begin{pmatrix} 1 \\ 2 \\ -1 \end{pmatrix}\right] = \left[\begin{pmatrix} 4 \\ -6 \\ -6 \end{pmatrix}\right],$$

that is, the Point $[4, -6, -6] = [-2, 3, 3]$.

(b) The image of the Point $[1, 0, 0]$ under t is given by

$$\left[\begin{pmatrix} 1 & 1 & -1 \\ -1 & -2 & 1 \\ 4 & -3 & 4 \end{pmatrix}\begin{pmatrix} 1 \\ 0 \\ 0 \end{pmatrix}\right] = \left[\begin{pmatrix} 1 \\ -1 \\ 4 \end{pmatrix}\right],$$

that is, the Point $[1, -1, 4]$.

(c) The image of the Point $[0, 1, 0]$ under t is given by

$$\left[\begin{pmatrix} 1 & 1 & -1 \\ -1 & -2 & 1 \\ 4 & -3 & 4 \end{pmatrix}\begin{pmatrix} 0 \\ 1 \\ 0 \end{pmatrix}\right] = \left[\begin{pmatrix} 1 \\ -2 \\ -3 \end{pmatrix}\right],$$

that is, the Point $[1, -2, -3]$.

(d) The image of the Point $[0, 0, 1]$ under t is given by

$$\left[\begin{pmatrix} 1 & 1 & -1 \\ -1 & -2 & 1 \\ 4 & -3 & 4 \end{pmatrix}\begin{pmatrix} 0 \\ 0 \\ 1 \end{pmatrix}\right] = \left[\begin{pmatrix} -1 \\ 1 \\ 4 \end{pmatrix}\right],$$

that is, the Point $[-1, 1, 4]$.

(e) The image of the Point $[1, 1, 1]$ under t is given by

$$\left[\begin{pmatrix} 1 & 1 & -1 \\ -1 & -2 & 1 \\ 4 & -3 & 4 \end{pmatrix}\begin{pmatrix} 1 \\ 1 \\ 1 \end{pmatrix}\right] = \left[\begin{pmatrix} 1 \\ -2 \\ 5 \end{pmatrix}\right],$$

that is the Point $[1, -2, 5]$.

3. Since the matrix $\mathbf{A}$ which represents the transformation in Example 1 in Subsection 3.3.1 has 2 as its top left-hand entry, we obtain the required matrix by dividing each entry of $\mathbf{A}$ by 4. This gives the matrix

$$\begin{pmatrix} \frac{1}{2} & 0 & \frac{1}{4} \\ -\frac{1}{4} & \frac{1}{2} & -\frac{3}{4} \\ \frac{1}{4} & -\frac{1}{4} & \frac{5}{4} \end{pmatrix}.$$

4. Matrices associated with t_1 and t_2 are

$$\mathbf{A}_1 = \begin{pmatrix} 2 & 1 & 0 \\ -1 & 0 & 1 \\ 0 & 1 & 1 \end{pmatrix} \text{ and } \mathbf{A}_2 = \begin{pmatrix} 5 & 8 & 0 \\ 3 & 5 & 0 \\ 0 & 0 & 2 \end{pmatrix},$$

respectively. It follows that a matrix associated with the projective transformation $t_1 \circ t_2$ is $\mathbf{A}_1\ \mathbf{A}_2$. Now

$$\mathbf{A}_1\mathbf{A}_2 = \begin{pmatrix} 2 & 1 & 0 \\ -1 & 0 & 1 \\ 0 & 1 & 1 \end{pmatrix} \begin{pmatrix} 5 & 8 & 0 \\ 3 & 5 & 0 \\ 0 & 0 & 2 \end{pmatrix} = \begin{pmatrix} 13 & 21 & 0 \\ -5 & -8 & 2 \\ 3 & 5 & 2 \end{pmatrix},$$

so we conclude that $t_1 \circ t_2$ is the transformation

$$[x, y, z] \mapsto [13x + 21y, -5x - 8y + 2z, 3x + 5y + 2z].$$

Next, t_1^{-1} has an associated matrix $\mathbf{A}_1^{-1}$ given by

$$\mathbf{A}_1^{-1} = \begin{pmatrix} 1 & 1 & -1 \\ -1 & -2 & 2 \\ 1 & 2 & -1 \end{pmatrix}.$$

The projective transformation t_1^{-1} is therefore

$$[x, y, z] \longmapsto [x+y-z, -x-2y+2z, x+2y-z].$$

5. The equation of the Line can be written in the form $\mathbf{Lx} = 0$, where

$$\mathbf{L} = (1 \quad 2 \quad -1).$$

From Problem 4 we know that t_1^{-1} has an associated matrix

$$\mathbf{A}_1^{-1} = \begin{pmatrix} 1 & 1 & -1 \\ -1 & -2 & 2 \\ 1 & 2 & -1 \end{pmatrix},$$

so

$$\mathbf{LA}_1^{-1} = (1 \quad 2 \quad -1) \begin{pmatrix} 1 & 1 & -1 \\ -1 & -2 & 2 \\ 1 & 2 & -1 \end{pmatrix} = (-2 \quad -5 \quad 4).$$

The required image is therefore the Line

$$-2x - 5y + 4z = 0.$$

6. First we consider the images under t_1. The image of the Point $[1, -1, 1]$ under t_1 is

$$\left[\begin{pmatrix} -4 & -1 & 1 \\ -3 & -2 & 1 \\ 4 & 2 & -1 \end{pmatrix} \begin{pmatrix} 1 \\ -1 \\ 1 \end{pmatrix}\right] = \left[\begin{pmatrix} -2 \\ 0 \\ 1 \end{pmatrix}\right],$$

that is, the Point $[-2, 0, 1]$.

Similarly, the image of the Point $[1, -2, 2]$ under t_1 is

$$\left[\begin{pmatrix} -4 & -1 & 1 \\ -3 & -2 & 1 \\ 4 & 2 & -1 \end{pmatrix} \begin{pmatrix} 1 \\ -2 \\ 2 \end{pmatrix}\right] = \left[\begin{pmatrix} 0 \\ 3 \\ -2 \end{pmatrix}\right],$$

that is, the Point $[0, 3, -2]$.

Finally, the image of the Point $[-1, 2, -1]$ under t_1 is

$$\left[\begin{pmatrix} -4 & -1 & 1 \\ -3 & -2 & 1 \\ 4 & 2 & -1 \end{pmatrix} \begin{pmatrix} -1 \\ 2 \\ -1 \end{pmatrix}\right] = \left[\begin{pmatrix} 1 \\ -2 \\ 1 \end{pmatrix}\right],$$

that is, the Point $[1, -2, 1]$.

Next, we consider the images under t_2. The image of the Point $[1, -1, 1]$ under t_2 is

$$\left[\begin{pmatrix} -8 & -6 & -2 \\ -3 & 4 & 7 \\ 6 & 0 & -4 \end{pmatrix} \begin{pmatrix} 1 \\ -1 \\ 1 \end{pmatrix}\right] = \left[\begin{pmatrix} -4 \\ 0 \\ 2 \end{pmatrix}\right],$$

that is, the Point with homogeneous coordinates $[-4, 0, 2]$ or (equivalently) $[-2, 0, 1]$.

Similarly, the image of the Point $[1, -2, 2]$ under t_2 is

$$\left[\begin{pmatrix} -8 & -6 & -2 \\ -3 & 4 & 7 \\ 6 & 0 & -4 \end{pmatrix} \begin{pmatrix} 1 \\ -2 \\ 2 \end{pmatrix}\right] = \left[\begin{pmatrix} 0 \\ 3 \\ -2 \end{pmatrix}\right],$$

that is, the Point $[0, 3, -2]$.

Finally, the image of the Point $[-1, 2, -1]$ under t_2 is

$$\left[\begin{pmatrix} -8 & -6 & -2 \\ -3 & 4 & 7 \\ 6 & 0 & -4 \end{pmatrix} \begin{pmatrix} -1 \\ 2 \\ -1 \end{pmatrix}\right] = \left[\begin{pmatrix} -2 \\ 4 \\ -2 \end{pmatrix}\right],$$

that is, the Point with homogeneous coordinates $[-2, 4, -2]$ or (equivalently) $[1, -2, 1]$.

7. We use the strategy preceding the problem.

(a) Let $\mathbf{A}$ be the matrix

$$\begin{pmatrix} -u & -3v & 2w \\ 0 & 2v & 0 \\ 0 & 0 & 4w \end{pmatrix}.$$

We wish to choose u, v, w such that

$$\begin{pmatrix} -u & -3v & 2w \\ 0 & 2v & 0 \\ 0 & 0 & 4w \end{pmatrix}\begin{pmatrix} 1 \\ 1 \\ 1 \end{pmatrix} = \begin{pmatrix} 1 \\ 2 \\ -5 \end{pmatrix},$$

that is,

$$\begin{pmatrix} -u - 3v + 2w \\ 2v \\ 4w \end{pmatrix} = \begin{pmatrix} 1 \\ 2 \\ -5 \end{pmatrix}.$$

It follows that $w = -\frac{5}{4}$ and $v = 1$. Also, $-u - 3v + 2w = 1$, so $u = -\frac{13}{2}$.

Thus

$$\mathbf{A} = \begin{pmatrix} \frac{13}{2} & -3 & -\frac{5}{2} \\ 0 & 2 & 0 \\ 0 & 0 & -5 \end{pmatrix}.$$

A simpler matrix for the projective transformation is the matrix

$$2\mathbf{A} = \begin{pmatrix} 13 & -6 & -5 \\ 0 & 4 & 0 \\ 0 & 0 & -10 \end{pmatrix}.$$

(b) Let $\mathbf{A}$ be the matrix

$$\begin{pmatrix} u & 0 & 0 \\ 0 & 0 & w \\ 0 & v & 0 \end{pmatrix}.$$

We wish to choose u, v, w such that

$$\begin{pmatrix} u & 0 & 0 \\ 0 & 0 & w \\ 0 & v & 0 \end{pmatrix}\begin{pmatrix} 1 \\ 1 \\ 1 \end{pmatrix} = \begin{pmatrix} 3 \\ 4 \\ 5 \end{pmatrix},$$

that is,

$$\begin{pmatrix} u \\ w \\ v \end{pmatrix} = \begin{pmatrix} 3 \\ 4 \\ 5 \end{pmatrix}.$$

It follows that $u = 3$, $v = 5$ and $w = 4$.

Thus

$$\mathbf{A} = \begin{pmatrix} 3 & 0 & 0 \\ 0 & 0 & 4 \\ 0 & 5 & 0 \end{pmatrix}.$$

(c) Let $\mathbf{A}$ be the matrix

$$\begin{pmatrix} 2u & v & 0 \\ u & 0 & 3w \\ 0 & -v & -w \end{pmatrix}.$$

We wish to choose u, v, w such that

$$\begin{pmatrix} 2u & v & 0 \\ u & 0 & 3w \\ 0 & -v & -w \end{pmatrix}\begin{pmatrix} 1 \\ 1 \\ 1 \end{pmatrix} = \begin{pmatrix} 3 \\ -1 \\ 2 \end{pmatrix},$$

that is,

$$\begin{pmatrix} 2u + v \\ u + 3w \\ -v - w \end{pmatrix} = \begin{pmatrix} 3 \\ -1 \\ 2 \end{pmatrix}.$$

It follows that

$$2u + v = 3, \qquad (1)$$

$$u + 3w = -1, \qquad (2)$$

$$-v - w = 2. \qquad (3)$$

Adding equations (1) and (3) in order to eliminate v, we obtain

$$2u - w = 5. \qquad (4)$$

Subtracting equation (4) from twice equation (2) in order to eliminate u, we obtain $7w = -7$ or $w = -1$.

It follows from equation (4) that $u = 2$, and from equation (1) that $v = -1$.

Thus

$$\mathbf{A} = \begin{pmatrix} 4 & -1 & 0 \\ 2 & 0 & -3 \\ 0 & 1 & 1 \end{pmatrix}.$$

8. We use the strategy preceding Example 6.

(a) By Problem 7(a), a matrix associated with the projective transformation t_1 that maps $[1, 0, 0]$, $[0, 1, 0]$, $[0, 0, 1]$, $[1, 1, 1]$ to the Points $[-1, 0, 0]$, $[-3, 2, 0]$, $[2, 0, 4]$, $[1, 2, -5]$, respectively, is

$$\mathbf{A}_1 = \begin{pmatrix} 13 & -6 & -5 \\ 0 & 4 & 0 \\ 0 & 0 & -10 \end{pmatrix}.$$

(b) By Problem 7(c), a matrix associated with the projective transformation t_2 that maps $[1, 0, 0]$, $[0, 1, 0]$, $[0, 0, 1]$, $[1, 1, 1]$ to the Points $[2, 1, 0]$, $[1, 0, -1]$, $[0, 3, -1]$, $[3, -1, 2]$, respectively, is

$$\mathbf{A}_2 = \begin{pmatrix} 4 & -1 & 0 \\ 2 & 0 & -3 \\ 0 & 1 & 1 \end{pmatrix}.$$

(c) A matrix associated with t_1^{-1} is $\mathbf{A}_1^{-1}$, which we can calculate to be

$$\mathbf{A}_1^{-1} = \begin{pmatrix} \frac{1}{13} & \frac{3}{26} & -\frac{1}{26} \\ 0 & \frac{1}{4} & 0 \\ 0 & 0 & -\frac{1}{10} \end{pmatrix}.$$

However, a simpler matrix associated with t_1^{-1} is

$$260\mathbf{A}_1^{-1} = \begin{pmatrix} 20 & 30 & -10 \\ 0 & 65 & 0 \\ 0 & 0 & -26 \end{pmatrix}.$$

Hence a matrix for the projective transformation which maps $[-1, 0, 0]$, $[-3, 2, 0]$, $[2, 0, 4]$, $[1, 2, -5]$ to $[2, 1, 0]$, $[1, 0, -1]$, $[0, 3, -1]$, $[3, -1, 2]$, respectively, is given by

$$\begin{pmatrix} 4 & -1 & 0 \\ 2 & 0 & -3 \\ 0 & 1 & 1 \end{pmatrix} \begin{pmatrix} 20 & 30 & -10 \\ 0 & 65 & 0 \\ 0 & 0 & -26 \end{pmatrix},$$

which equals

$$\begin{pmatrix} 80 & 55 & -40 \\ 40 & 60 & 58 \\ 0 & 65 & -26 \end{pmatrix}.$$

9. We use the strategies in Subsection 3.3.3. First, it follows from the first strategy that a matrix for the projective transformation t_1 that maps $[1, 0, 0]$, $[0, 1, 0]$, $[0, 0, 1]$, $[1, 1, 1]$ to the Points $[1, 0, -3]$, $[1, 1, -2]$, $[3, 3, -5]$, $[6, 4, -13]$, respectively, is

$$\begin{pmatrix} u & v & 3w \\ 0 & v & 3w \\ -3u & -2v & -5w \end{pmatrix},$$

where u, v, w satisfy

$$\begin{pmatrix} u & v & 3w \\ 0 & v & 3w \\ -3u & -2v & -5w \end{pmatrix} \begin{pmatrix} 1 \\ 1 \\ 1 \end{pmatrix} = \begin{pmatrix} 6 \\ 4 \\ -13 \end{pmatrix},$$

that is,

$$\begin{pmatrix} u + v + 3w \\ v + 3w \\ -3u - 2v - 5w \end{pmatrix} = \begin{pmatrix} 6 \\ 4 \\ -13 \end{pmatrix}.$$

By comparing the first two rows of these matrices, it is clear that we must have $u = 2$. It then follows from the last two rows that

$$v + 3w = 4, \quad (5)$$

$$-2v - 5w = -7. \quad (6)$$

If we add twice equation (5) to equation (6), we obtain $w = 1$. Equation (5) then gives $v = 1$. A matrix for the projective transformation t_1 is therefore

$$\mathbf{A}_1 = \begin{pmatrix} 2 & 1 & 3 \\ 0 & 1 & 3 \\ -6 & -2 & -5 \end{pmatrix}.$$

Next, it follows from the first strategy that a matrix for the projective transformation t_2 that maps $[1, 0, 0]$, $[0, 1, 0]$, $[0, 0, 1]$, $[1, 1, 1]$ to the Points $[3, -5, 3]$, $[\frac{1}{2}, -1, 0]$, $[3, -5, 6]$, $[8, -13, 12]$, respectively, is

$$\begin{pmatrix} 3u & \frac{1}{2}v & 3w \\ -5u & -v & -5w \\ 3u & 0 & 6w \end{pmatrix},$$

where u, v, w satisfy

$$\begin{pmatrix} 3u & \frac{1}{2}v & 3w \\ -5u & -v & -5w \\ 3u & 0 & 6w \end{pmatrix} \begin{pmatrix} 1 \\ 1 \\ 1 \end{pmatrix} = \begin{pmatrix} 8 \\ -13 \\ 12 \end{pmatrix},$$

that is,

$$\begin{pmatrix} 3u + \frac{1}{2}v + 3w \\ -5u - v - 5w \\ 3u + 6w \end{pmatrix} = \begin{pmatrix} 8 \\ -13 \\ 12 \end{pmatrix}.$$

Equating corresponding entries in these matrices and multiplying them by 2,1 and $\frac{1}{3}$, respectively, we deduce that

$$6u + v + 6w = 16, \quad (7)$$

$$-5u - v - 5w = -13, \quad (8)$$

$$u + 2w = 4. \quad (9)$$

Adding equations (7) and (8), we obtain

$$u + w = 3. \tag{10}$$

It follows from equations (9) and (10) that $w = 1$ and $u = 2$. Equation (8) then gives $v = -2$.

Hence a matrix for the projective transformation t_2 is

$$\mathbf{A}_2 = \begin{pmatrix} 6 & -1 & 3 \\ -10 & 2 & -5 \\ 6 & 0 & 6 \end{pmatrix}.$$

Finally, we follow the second strategy.
A matrix associated with t_1^{-1} is

$$\mathbf{A}_1^{-1} = \begin{pmatrix} \frac{1}{2} & -\frac{1}{2} & 0 \\ -9 & 4 & -3 \\ 3 & -1 & 1 \end{pmatrix},$$

so a simple matrix associated with t_1^{-1} is

$$2\mathbf{A}_1^{-1} = \begin{pmatrix} 1 & -1 & 0 \\ -18 & 8 & -6 \\ 6 & -2 & 2 \end{pmatrix}.$$

Hence a matrix for the projective transformation that maps $[1, 0, -3]$, $[1, 1, -2]$, $[3, 3, -5]$, $[6, 4, -13]$ to $[3, -5, 3]$, $\left[\frac{1}{2}, -1, 0\right]$, $[3, -5, 6]$, $[8, -13, 12]$, respectively, is given by

$$\begin{pmatrix} 6 & -1 & 3 \\ -10 & 2 & -5 \\ 6 & 0 & 6 \end{pmatrix} \begin{pmatrix} 1 & -1 & 0 \\ -18 & 8 & -6 \\ 6 & -2 & 2 \end{pmatrix},$$

which equals

$$\begin{pmatrix} 42 & -20 & 12 \\ -76 & 36 & -22 \\ 42 & -18 & 12 \end{pmatrix},$$

So we may take as a simple matrix associated with the transformation

$$\begin{pmatrix} 21 & -10 & 6 \\ -38 & 18 & -11 \\ 21 & -9 & 6 \end{pmatrix}.$$

Section 3.4

1. Since the problem is concerned exclusively with projective properties, we shall prove the corresponding projective result. By the Fundamental Theorem of Projective Geometry, we may take A, B, C and U to be the Points $[1, 0, 0]$, $[0, 1, 0]$, $[0, 0, 1]$ and $[1, 1, 1]$, respectively.

First, we find the homogeneous coordinates of the Points A', B' and C'.

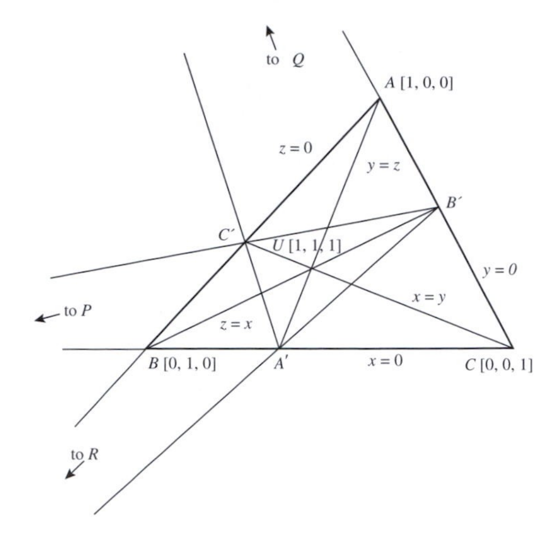

Since the Line AU passes through $[1, 0, 0]$ and $[1, 1, 1]$ it must have equation $y = z$. Also, BC passes through $[0, 1, 0]$ and $[0, 0, 1]$ and therefore has equation $x = 0$. It follows that AU and BC intersect at the Point A' with homogeneous coordinates $[0, y, y]$, that is, at $[0, 1, 1]$.

Since the Line BU passes through $[0, 1, 0]$ and $[1, 1, 1]$ it must have equation $z = x$. Also, AC passes through $[1, 0, 0]$ and $[0, 0, 1]$ and therefore has equation $y = 0$. It follows that BU and AC intersect at the Point B' with homogeneous coordinates $[x, 0, x]$, that is, at $[1, 0, 1]$.

Since the Line CU passes through $[0, 0, 1]$ and $[1, 1, 1]$ it must have equation $x = y$. Also, AB passes through $[1, 0, 0]$ and $[0, 1, 0]$ and therefore has equation $z = 0$. It follows that CU and AB intersect at the Point C' with homogeneous coordinates $[x, x, 0]$, that is, at $[1, 1, 0]$.

Next, we determine the homogeneous coordinates of the Points P, Q and R.

The Line $B'C'$ has equation

$$\begin{vmatrix} x & y & z \\ 1 & 0 & 1 \\ 1 & 1 & 0 \end{vmatrix} = 0,$$

that is, $x = y + z$. Thus the Lines $B'C'$(with equation $x = y + z$) and BC (with equation $x = 0$) meet at the Point P with homogeneous coordinates $[0, y, -y]$, that is, at $P = [0, 1, -1]$.

The Line $A'C'$ has equation

$$\begin{vmatrix} x & y & z \\ 0 & 1 & 1 \\ 1 & 1 & 0 \end{vmatrix} = 0,$$

that is, $y = x + z$. Thus the Lines $A'C'$ (with equation $y = x + z$) and AC (with equation $y = 0$) meet at the Point Q with homogeneous coordinates $[x, 0, -x]$, that is, at $Q = [1, 0, -1]$.

The Line $A'B'$ has equation

$$\begin{vmatrix} x & y & z \\ 0 & 1 & 1 \\ 1 & 0 & 1 \end{vmatrix} = 0,$$

that is, $z = x + y$. Thus the Lines $A'B'$ (with equation $z = x + y$) and AB (with equation $z = 0$) meet at the Point R with homogeneous coordinates $[x, -x, 0]$, that is, at $R = [1, -1, 0]$.

Finally, the Points P, Q and R are collinear if

$$\begin{vmatrix} 0 & 1 & -1 \\ 1 & 0 & -1 \\ 1 & -1 & 0 \end{vmatrix} = 0.$$

Since

$$\begin{vmatrix} 0 & 1 & -1 \\ 1 & 0 & -1 \\ 1 & -1 & 0 \end{vmatrix} = 0\begin{vmatrix} 0 & -1 \\ -1 & 0 \end{vmatrix} - 1\begin{vmatrix} 1 & -1 \\ 1 & 0 \end{vmatrix} + (-1)\begin{vmatrix} 1 & 0 \\ 1 & -1 \end{vmatrix}$$

$$= 0 - 1(0 + 1) - 1(-1 - 0)$$

$$= 0,$$

it follows that the Points P, Q and R are collinear, as required.

2. The following figure illustrates Desargues' Theorem on an embedding plane π for which Q (the Point of intersection of AC and $A'C'$) is an ideal Point. Since P and R are collinear with the ideal Point Q where AC and $A'C'$ meet, it follows that PR is parallel to AC and $A'C'$.

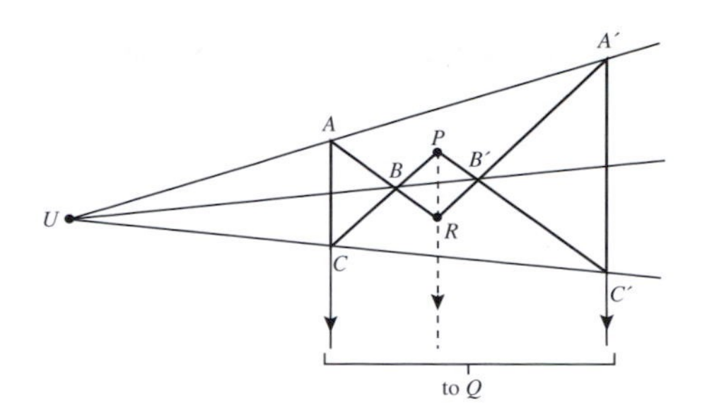

3.

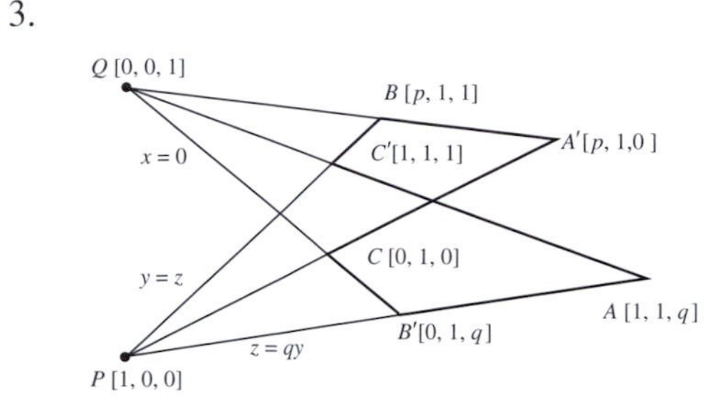

Let P, C, Q, C' be the Points $[1, 0, 0]$, $[0, 1, 0]$, $[0, 0, 1]$, $[1, 1, 1]$, respectively. Then the Line PC' passes through the Points $[1, 0, 0]$ and $[1, 1, 1]$, and therefore has equation $y = z$.

Since B is a Point on PC', it must have homogeneous coordinates of the form $[a, b, b]$, for some real numbers a and b. Now, $b \neq 0$, since if $b = 0$ we must have $B = P$. It follows that we may write the homogeneous coordinates of the Point B in the form $[p, 1, 1]$ (where $p = a/b$).

Similarly, the line QC' has equation $x = y$, so that the homogeneous coordinates of the Point A are $[1, 1, q]$.

We now find the Point B' where PA meets QC. The Line QC has equation $x = 0$. Since the Line PA passes through the Points $P = [1, 0, 0]$ and $A = [1, 1, q]$ it must have equation

$$\begin{vmatrix} x & y & z \\ 1 & 0 & 0 \\ 1 & 1 & q \end{vmatrix},$$

which we may rewrite as $-qy + z = 0$.

It follows that at the Point B' of intersection of the Lines PA and QC we must have $z = qy$ and $x = 0$, so that the Point B' has homogeneous coordinates $[0, 1, q]$.

Similarly, the Point A' is the Point of intersection of the Lines PC and QB. Since $P = [1, 0, 0]$

and $C = [0,1,0]$, the Line PC has equation $z = 0$; and since $Q = [0,0,1]$ and $B = [p,1,1]$, the Line QB has equation $x = py$. Hence the Point A' has homogeneous coordinates $[p,1,0]$.

Since the Line AA' passes through $A = [1,1,q]$ and $A' = [p,1,0]$ it must have equation

$$\begin{vmatrix} x & y & z \\ 1 & 1 & q \\ p & 1 & 0 \end{vmatrix} = 0,$$

which we may rewrite as

$$-qx + pqy + (1-p)z = 0.$$

Since the Line BB' passes through $B = [p,1,1]$ and $B' = [0,1,q]$ it must have equation

$$\begin{vmatrix} x & y & z \\ p & 1 & 1 \\ 0 & 1 & q \end{vmatrix} = 0,$$

which we may rewrite as

$$(q-1)x - pqy + pz = 0.$$

The Line CC' has equation $x = z$, so AA', BB' and CC' will be concurrent if the equations

$$\begin{aligned} -qx + pqy + (1-p)z &= 0, \\ (q-1)x - pqy + pz &= 0, \\ x - z &= 0, \end{aligned}$$

have a non-zero solution. This happens if

$$\begin{vmatrix} -q & pq & 1-p \\ q-1 & -pq & p \\ 1 & 0 & -1 \end{vmatrix} = 0.$$

But the determinant on the left is equal to

$$\begin{aligned} & -qpq - pq(1-q-p) + (1-p)(pq) \\ & = -pq^2 - pq + pq^2 + p^2q + pq - p^2q \\ & = 0, \end{aligned}$$

so AA', BB' and CC' are concurrent.

4. The statement of the collinearity strategy (Subsection 3.2.2) is follows:

Three Points $[a,b,c]$, $[d,e,f]$, $[g,h,k]$

are collinear if and only if $\begin{vmatrix} a & b & c \\ d & e & f \\ g & h & k \end{vmatrix} = 0.$

To make it easier to dualize this result, by breaking the sentence up into shorter portions, we first rephrase it in the following way:

Three Points

$$[a,b,c],\ [d,e,f],\ [g,h,k]$$

are collinear

if and only if $\begin{vmatrix} a & b & c \\ d & e & f \\ g & h & k \end{vmatrix} = 0.$

We first interchange 'Point' and 'Line', and 'collinear' and 'concurrent'; this gives the statement:

Three Lines

$$[a,b,c],\ [d,e,f],\ [g,h,k]$$

are concurrent

if and only if $\begin{vmatrix} a & b & c \\ d & e & f \\ g & h & k \end{vmatrix} = 0.$

Next, we make the changes necessary for this statement to make sense; it becomes:

Three Lines

$$ax + by + cz = 0,\ dx + ey + fz = 0,$$

$$gx + hy + kz = 0$$

are concurrent

if and only if $\begin{vmatrix} a & b & c \\ d & e & f \\ g & h & k \end{vmatrix} = 0.$

This is the result dual to the collinearity strategy.

Section 3.5

1. (a) First, we find real numbers α and β such that the following vector equation holds:

$$(3,5,-5) = \alpha(1,-1,-1) + \beta(1,3,-2).$$

Comparing corresponding coordinates on both sides of this vector equation, we deduce that

$$\begin{aligned} 3 &= \alpha + \beta, \\ 5 &= -\alpha + 3\beta, \\ -5 &= -\alpha - 2\beta. \end{aligned}$$

By adding the first two equations, we obtain $8 = 4\beta$, so $\beta = 2$. The first equation then gives $\alpha = 1$. (As a check, notice that these values of α and β satisfy the third equation too.)

Next we find real numbers γ and δ such that the vector equation

$$(1, -5, 0) = \gamma(1, -1, -1) + \delta(1, 3, -2)$$

holds. Comparing corresponding coordinates on both sides, we deduce that

$$\begin{aligned} 1 &= \gamma + \delta, \\ -5 &= -\gamma + 3\delta, \\ 0 &= -\gamma - 2\delta. \end{aligned}$$

By adding the first two equations, we obtain $-4 = 4\delta$, so $\delta = -1$. The first equation then gives $\gamma = 2$. (As a check, notice that these values of γ and δ satisfy the third equation too.)

It then follows from the definition of cross-ratio that

$$(ABCD) = \frac{\beta}{\alpha} \bigg/ \frac{\delta}{\gamma} = \frac{2}{1} \bigg/ \frac{-1}{2} = -4.$$

(b) First, we find real numbers α and β such that the following vector equation holds:

$$(-3, -5, -8) = \alpha(1, 2, 3) + \beta(2, 2, 4).$$

Comparing corresponding coordinates on both sides of this vector equation, we deduce that

$$\begin{aligned} -3 &= \alpha + 2\beta, \\ -5 &= 2\alpha + 2\beta, \\ -8 &= 3\alpha + 4\beta. \end{aligned}$$

By subtracting the first equation from the second, we obtain $\alpha = -2$. The first equation then gives $2\beta = -1$, or $\beta = -\frac{1}{2}$. (As a check, notice that these values of α and β satisfy the third equation too.)

Next, we find real numbers γ and δ such that the vector equation

$$(3, -3, 0) = \gamma(1, 2, 3) + \delta(2, 2, 4)$$

holds. Comparing corresponding coordinates on both sides, we deduce that

$$\begin{aligned} 3 &= \gamma + 2\delta, \\ -3 &= 2\gamma + 2\delta, \\ 0 &= 3\gamma + 4\delta. \end{aligned}$$

By subtracting the first equation from the second, we obtain $\gamma = -6$. The first equation then gives $2\delta = 9$, or $\delta = \frac{9}{2}$. (As a check, notice that these values of γ and δ satisfy the third equation too.)

It then follows from the definition of cross-ratio that

$$(ABCD) = \frac{\beta}{\alpha} \bigg/ \frac{\delta}{\gamma} = \frac{-\frac{1}{2}}{-2} \bigg/ \frac{\frac{9}{2}}{-6} = -\frac{1}{3}.$$

2. We have

$$\begin{aligned} A &= [1, -1, -1], \quad B = [1, 3, -2], \\ C &= [3, 5, -5], \quad D = [1, -5, 0]. \end{aligned}$$

To determine $(BACD)$, we first find real numbers α and β such that the following vector equation holds:

$$(3, 5, -5) = \alpha(1, 3, -2) + \beta(1, -1, -1).$$

It follows from the result of Problem 1(a) that $\alpha = 2$ and $\beta = 1$ (essentially, the letters α and β in that Problem have been interchanged).

Next, we find real numbers γ and δ such that the vector equation

$$(1, -5, 0) = \gamma(1, 3, -2) + \delta(1, -1, -1)$$

holds. It follows from the result of Problem 1(a) that $\gamma = -1$ and $\delta = 2$ (in this case, the letters γ and δ in that Problem have been interchanged.)

Hence from the definition of cross-ratio,

$$(BACD) = \frac{\beta}{\alpha} \bigg/ \frac{\delta}{\gamma} = \frac{1}{2} \bigg/ \frac{2}{-1} = -\frac{1}{4}.$$

To determine $(ACBD)$, we first find real numbers α and β such that the following vector equation holds:

$$(1, 3, -2) = \alpha(1, -1, -1) + \beta(3, 5, -5).$$

Comparing corresponding coordinates on both sides of this vector equation, we deduce that

$$\begin{aligned}1 &= \alpha + 3\beta,\\ 3 &= -\alpha + 5\beta,\\ -2 &= -\alpha - 5\beta.\end{aligned}$$

By adding the first two equations, we obtain $4 = 8\beta$, so $\beta = \frac{1}{2}$. The first equation then gives $\alpha = -\frac{1}{2}$. (As a check, notice that these values of α and β satisfy the third equation too.)

Next, we find real numbers γ and δ such that the vector equation

$$(1, -5, 0) = \gamma(1, -1, -1) + \delta(3, 5, -5)$$

holds. Comparing corresponding coordinates on both sides, we deduce that

$$\begin{aligned}1 &= \gamma + 3\delta,\\ -5 &= -\gamma + 5\delta,\\ 0 &= -\gamma - 5\delta.\end{aligned}$$

By adding the first two equations, we obtain $-4 = 8\delta$, so $\delta = -\frac{1}{2}$. The first equation then gives $\gamma = \frac{5}{2}$. (As a check, notice that these values of γ and δ satisfy the third equation too.)

It follows from the definition of cross-ratio that

$$(ACBD) = \frac{\beta}{\alpha} \Big/ \frac{\delta}{\gamma} = \frac{\frac{1}{2}}{-\frac{1}{2}} \Big/ \frac{-\frac{1}{2}}{\frac{5}{2}} = 5.$$

3. From the solution to Problem 1(a) we know that $(ABCD) = -4$, so by Theorem 2,

$$\begin{aligned}(ABDC) &= 1/(-4) = -\tfrac{1}{4},\\ (DBCA) &= 1 - (-4) = 5,\\ (ACBD) &= 1 - (-4) = 5.\end{aligned}$$

4.

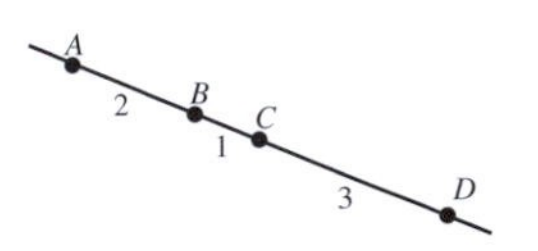

Using the sign convention for ratios of lengths along a line, we obtain

$$\begin{aligned}(ABCD) &= \frac{AC}{CB} \Big/ \frac{AD}{DB}\\ &= \left(-\frac{3}{1}\right) \Big/ \left(-\frac{6}{4}\right)\\ &= 2\end{aligned}$$

and

$$\begin{aligned}(DBCA) &= \frac{DC}{CB} \Big/ \frac{DA}{AB}\\ &= \left(\frac{3}{1}\right) \Big/ \left(-\frac{6}{2}\right)\\ &= -1.\end{aligned}$$

5. In each case we use Theorem 2 to bring the ideal Point to the front of the cross-ratio, and then apply equation (11).

(a) If D is an ideal Point then

$$\begin{aligned}(ABCD) &= 1 - (DBCA) && \text{(swap outer two Points)}\\ &= (DCBA) && \text{(swap middle two Points)}\\ &= \frac{AC}{BC} && \text{(from equation (11)).}\end{aligned}$$

(b) If C is an ideal Point then

$$\begin{aligned}(ABCD) &= 1 - (ACBD) && \text{(swap middle two Points)}\\ &= (DCBA) && \text{(swap outer two Points)}\\ &= 1/(CDBA) && \text{(swap first two Points)}\\ &= 1 \Big/ \left(\frac{AD}{BD}\right) && \text{(from equation (11))}\\ &= \frac{BD}{AD}.\end{aligned}$$

6. Since B is an ideal Point,

$$\begin{aligned}(ABCD) &= \frac{CA}{DA}\\ &= \frac{5}{2}.\end{aligned}$$

7.

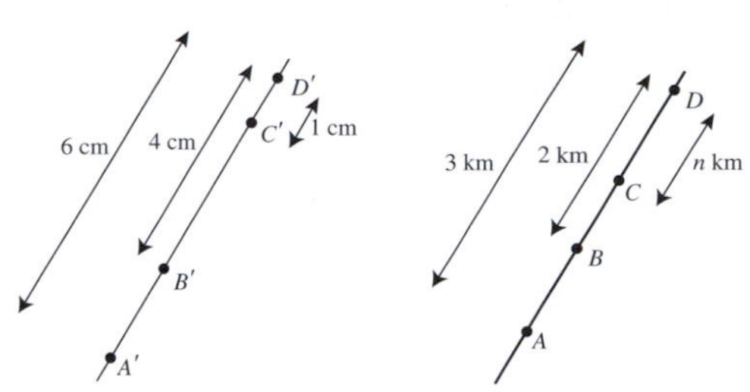

Let A and B denote the signs, C denote the car, D denote the junction, and let A', B', C', D' be their images on the film. Then

$$\begin{aligned}(A'B'C'D') &= \frac{A'C'}{C'B'} \Big/ \frac{A'D'}{D'B'} \\ &= \left(-\frac{5}{3}\right) \Big/ \left(-\frac{6}{4}\right) = \frac{10}{9}.\end{aligned}$$

Now let the car be n km from the junction. Then

$$\begin{aligned}(ABCD) &= \frac{AC}{CB} \Big/ \frac{AD}{DB} \\ &= \left(-\frac{3-n}{2-n}\right) \Big/ \left(-\frac{3}{2}\right) \\ &= \frac{2(3-n)}{3(2-n)}.\end{aligned}$$

Since $(ABCD)$ and $(A'B'C'D')$ must be equal, it follows that

$$\frac{2(3-n)}{3(2-n)} = \frac{10}{9}.$$

Hence

$$54 - 18n = 60 - 30n,$$

and so $12n = 6$, or $n = \frac{1}{2}$. The car is therefore $\frac{1}{2}$ km from the junction.

8.

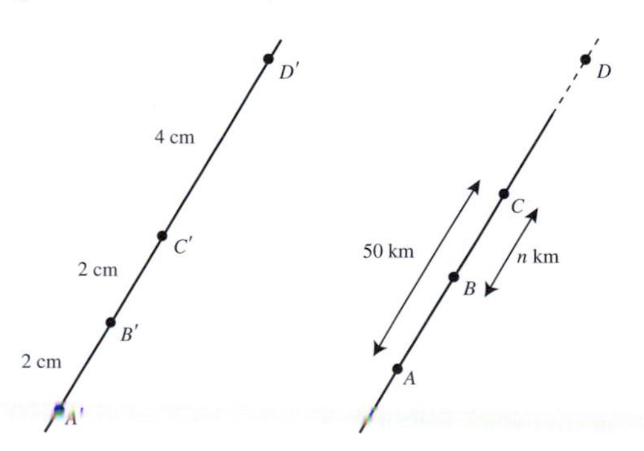

Let A and C denote the stations, B denote the train, and D denote the ideal Point where the railway lines 'meet', and let A', B', C', D' be their images on the film. Then

$$\begin{aligned}(A'B'C'D') &= \frac{A'C'}{C'B'} \Big/ \frac{A'D'}{D'B'} \\ &= \left(-\frac{4}{2}\right) \Big/ \left(-\frac{8}{6}\right) \\ &= \frac{3}{2}.\end{aligned}$$

Now let the train be n km from C. Since D is an ideal Point,

$$\begin{aligned}(ABCD) &= \frac{AC}{BC} \\ &= \frac{50}{n}.\end{aligned}$$

But $(ABCD)$ and $(A'B'C'D')$ must be equal, so

$$\frac{50}{n} = \frac{3}{2}.$$

Hence $n = \frac{100}{3}$. The train is therefore $33\frac{1}{3}$ km from the next station.

Chapter 4

Section 4.1

1.

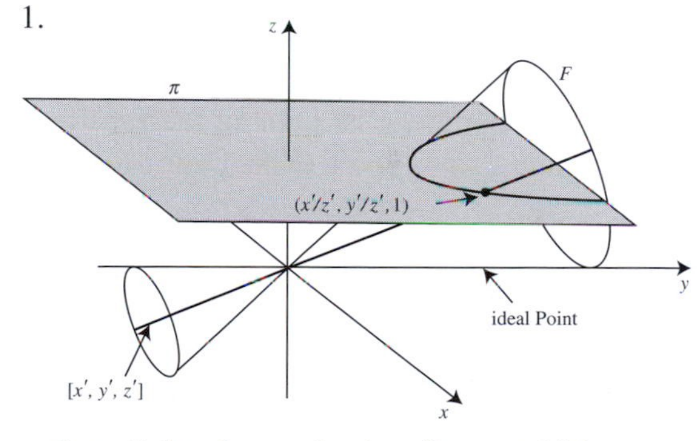

Let F be the projective figure which corresponds to the parabola $\{(x, y, z) : y = x^2, z = 1\}$ in the standard embedding plane π. Any Point $[x', y', z']$ on F must pierce π at a point $(x'/z', y'/z', 1)$ on the parabola, and so

$$\frac{y'}{z'} = \left(\frac{x'}{z'}\right)^2.$$

Since z' is non-zero, we can multiply by $(z')^2$ and drop the dashes to obtain the equivalent equation

$$yz = x^2, \quad z \neq 0.$$

By dropping the constraint that $z \neq 0$, we can include additional ideal Points of the form $[x, y, 0]$; for these points we must have $x^2 = 0$. In fact, there is just one such Point, namely $[0, 1, 0]$.

2. (a) This does not define a projective conic because it includes linear terms in x and y.
   (b) This defines a projective conic because it has the form of equation (2) with $A = B = C = G = 1$ and $F = H = 0$.
   (c) This defines a projective conic because it has the form of equation (2) with $A = B = G = H = 0, C = 1$ and $F = -1$.
   (d) This does not define a projective conic because it includes a constant term.
3. (a) This is true since

$$(1 \cdot 0) + (1 \cdot 0) + (0 \cdot 0) = 0.$$

   (b) This is true since

$$\begin{aligned} &2 \cdot (1)^2 - (2)^2 + (1 \cdot 2) + (1 \cdot 0) + (0)^2 \\ &= 2 - 4 + 2 = 0. \end{aligned}$$

   (c) This is false since

$$3 \cdot 1^2 + 2 \cdot 2^2 - 3^2 = 3 + 8 - 9 = 2 \neq 0.$$

4. First, we rewrite the equation of the hyperbola $2u^2 - 6v^2 + 5v - 1 = 0$ in the form

$$u^2 - 3v^2 + \frac{5}{2}v - \frac{1}{2} = 0,$$

so that we can apply the Eccentricity Formula.

Then, if we apply the Eccentricity Formula with $C = -3$, $G = \frac{5}{2}$ and $H = -\frac{1}{2}$, it follows that the eccentricity e of the hyperbola is given by the formula

$$e^2 = 1 - (-3) = 4,$$

so that $e = 2$.

Section 4.2

1. Here we use Joachimsthal's notation, with $(x_1, y_1) = (1, 0), (x_2, y_2) = (2, 1)$ and

$$s = 2x^2 + 3xy - y^2 + x + 2y + 1.$$

Thus

$$\begin{aligned} s_{11} &= 2x_1^2 + 3x_1y_1 - y_1^2 + x_1 + 2y_1 + 1 \\ &= 2 + 0 - 0 + 1 + 0 + 1 \\ &= 4, \\ s_{22} &= 2x_2^2 + 3x_2y_2 - y_2^2 + x_2 + 2y_2 + 1 \\ &= 8 + 6 - 1 + 2 + 2 + 1 \\ &= 18, \\ s_{12} &= 2x_1x_2 + 3\left(\frac{x_1y_2 + x_2y_1}{2}\right) - y_1y_2 \\ &\quad + \left(\frac{x_1 + x_2}{2}\right) + 2\left(\frac{y_1 + y_2}{2}\right) + 1 \\ &= 4 + 3\left(\frac{1+0}{2}\right) - 0 + \left(\frac{1+2}{2}\right) \\ &\quad + 2\left(\frac{0+1}{2}\right) + 1 \\ &= 4 + \tfrac{3}{2} + \tfrac{3}{2} + 1 + 1 \\ &= 9, \end{aligned}$$

and

$$\begin{aligned} s_1 &= 2x_1x + 3\left(\frac{x_1y + xy_1}{2}\right) - y_1y \\ &\quad + \left(\frac{x_1 + x}{2}\right) + 2\left(\frac{y_1 + y}{2}\right) + 1 \\ &= 2x + 3\left(\frac{y+0}{2}\right) - 0 + \left(\frac{1+x}{2}\right) \\ &\quad + 2\left(\frac{0+y}{2}\right) + 1 \\ &= \tfrac{5}{2}x + \tfrac{5}{2}y + \tfrac{3}{2}. \end{aligned}$$

2. Here we use the results of Problem 1 and Joachimsthal's Section Equation $s_{22}k^2 + 2s_{12}k + s_{11} = 0$.

It follows that the ratio in which the hyperbola divides the segment is $k : 1$, where

$$18k^2 + 18k + 4 = 0;$$

it follows that

$$9k^2 + 9k + 2 = 0,$$

or

$$(3k + 2)(3k + 1) = 0,$$

so that $k = -\frac{2}{3}$ or $k = -\frac{1}{3}$.

Hence the hyperbola divides the line segment at two points, in the ratios $-\frac{2}{3} : 1$ and $-\frac{1}{3} : 1$, that is, in the ratios $-2{:}3$ and $-1{:}3$. The minus signs indicate that the intersection points lie outside the actual segment.

The minus signs indicate that the intersection points lie outside the actual segment.

3. Here we use the formula $s_1 = 0$.

(a) In this case,

$$s = x^2 - xy + 2y - 7,$$

so that at the point $(x_1, y_1) = (-1, 2)$,

$$s_1 = (-1)x - \left(\frac{2x - y}{2}\right) + 2\left(\frac{y + 2}{2}\right) - 7$$
$$= -2x + \tfrac{3}{2}y - 5.$$

The equation of the tangent at $(-1, 2)$ is therefore

$$-2x + \tfrac{3}{2}y - 5 = 0,$$

or

$$4x - 3y + 10 = 0.$$

(b) In this case,

$$s = 3x^2 + 2xy - y^2 + x - 2y - 3,$$

so that at the point $(x_1, y_1) = (1, 1)$,

$$s_1 = 3 \cdot 1 \cdot x + 2\left(\frac{y + x}{2}\right) - 1 \cdot y$$
$$+ \left(\frac{1 + x}{2}\right) - 2\left(\frac{1 + y}{2}\right) - 3$$
$$= \tfrac{9}{2}x - y - \tfrac{7}{2}.$$

It follows that the equation of the tangent at $(1, 1)$ is

$$\tfrac{9}{2}x - y - \tfrac{7}{2} = 0,$$

or

$$9x - 2y - 7 = 0.$$

4. We use the formula $s_1^2 = s \cdot s_{11}$ for pairs of tangents.

Here,

$$s = 4xy + 1,$$

so that at the point $(x_1, y_1) = (2, 1)$,

$$s_{11} = 4 \cdot 2 \cdot 1 + 1 = 9$$

and

$$s_1 = 4\left(\frac{2y + x}{2}\right) + 1$$
$$= 2x + 4y + 1.$$

It follows that the equation of the tangent pair is

$$(2x + 4y + 1)^2 = (4xy + 1) \cdot 9.$$

Multiplying this out, we obtain

$$4x^2 + 16y^2 + 1 + 2(8xy + 2x + 4y) = 36xy + 9,$$

so that

$$x^2 - 5xy + 4y^2 + x + 2y - 2 = 0. \quad (1)$$

Since one of the tangents has equation $y = x - 1$, it follows that $(x - y - 1)$ must be a factor of equation (1). Using this fact, it is straightforward to check that equation (1) may be factorized as

$$(x - y - 1)(x - 4y + 2) = 0.$$

Hence the equation of the other tangent from $(2, 1)$ is

$$x - 4y + 2 = 0.$$

5. Here we use the formula $s_1 = 0$ for the polar.

The equation of the hyperbola is $s = 0$, where

$$s = 2x^2 + xy - 3y^2 + x - 6,$$

so that at the point $(x_1, y_1) = (1, -1)$,

$$s_1 = 2 \cdot 1 \cdot x + \left(\frac{y - x}{2}\right) - 3 \cdot (-1)y$$
$$+ \left(\frac{1 + x}{2}\right) - 6$$
$$= 2x + \tfrac{7}{2}y - \tfrac{11}{2}.$$

It follows that the equation of the polar of $(1, -1)$ is

$$4x + 7y - 11 = 0.$$

6. Let $s = y^2 + z^2 + 2xy - 4yz + zx$, so that the equation of the projective conic is $s = 0$.

(a) At a general Point $[x_1, y_1, z_1]$,

$$s_1 = yy_1 + zz_1 + (x_1y + xy_1)$$
$$- 2(y_1z + yz_1) + \tfrac{1}{2}(z_1x + zx_1).$$

So at the Point $[1, 0, 0]$, which lies on E, we have

$$s_1 = 0 + 0 + (y + 0) - 2(0 + 0) + \tfrac{1}{2}(0 + z)$$
$$= y + \tfrac{1}{2}z.$$

Hence the equation of the tangent to E at $[1, 0, 0]$ is

$$y + \tfrac{1}{2}z = 0,$$

or

$$2y + z = 0.$$

The Point $[0, 1, -2]$ lies on this tangent, since

$$2 \cdot 1 + (-2) = 0.$$

(b) At a general Point $[x_1, y_1, z_1]$,

$$s_1 = yy_1 + zz_1 + (x_1y + xy_1)$$
$$- 2(y_1z + yz_1) + \tfrac{1}{2}(z_1x + zx_1)$$

and

$$s_{11} = y_1^2 + z_1^2 + 2x_1y_1 - 4y_1z_1 + z_1x_1.$$

Thus at the Point $[0, 1, -2]$,

$$s_1 = y - 2z + (0 + x) - 2(z - 2y)$$
$$+ \tfrac{1}{2}(-2x + 0)$$
$$= 5y - 4z$$

and

$$s_{11} = 1 + 4 + 0 - 4 \cdot (-2) + 0 = 13.$$

Now, the pair of tangents from the Point $[0, 1, -2]$ to E is given by the equation $s_1^2 = s \cdot s_{11}$, that is,

$$(5y - 4z)^2 = (y^2 + z^2 + 2xy - 4yz + zx) \cdot 13.$$

Multiplying this out, we obtain

$$25y^2 - 40yz + 16z^2 = 13y^2 + 13z^2 + 26xy$$
$$- 52yz + 13zx,$$

so that

$$12y^2 + 3z^2 - 26xy + 12yz - 13zx = 0. \quad (2)$$

But we know that the Point $[0, 1, -2]$ lies on the tangent in part (a), so $2y + z$ must be a factor of the left-hand side of equation (2). (This fact serves as a useful check on the working so far.) Hence equation (2) can be expressed in the form

$$(2y + z)(6y + 3z - 13x) = 0.$$

It follows that the equations of the two tangents to the projective conic are

$$2y + z = 0 \quad \text{and} \quad 6y + 3z - 13x = 0.$$

(c) The polar of the Point $[0, 1, -2]$ is given by the equation $s_1 = 0$, and we know from the results of part (b) that at this Point

$$s_1 = 5y - 4z.$$

Thus the equation of the polar of $[0, 1, -2]$ with respect to E is

$$5y - 4z = 0.$$

Section 4.3

1. Let the projective conic have equation

$$Ax^2 + Bxy + Cy^2 + Fxz + Gyz + Hz^2 = 0.$$

Since $[1, 0, 0]$ lies on the projective conic, we must have $A = 0$. Similarly, since $[0, 1, 0]$ and $[0, 0, 1]$ lie on the projective conic, we must also have $C = 0$ and $H = 0$. Thus the equation of the projective conic reduces to the form

$$Bxy + Fxz + Gyz = 0.$$

Since $[1, 1, 1]$ and $[-2, 3, 1]$ also lie on the projective conic, we deduce that

$$B + F + G = 0 \quad (1)$$

and

$$-6B - 2F + 3G = 0. \quad (2)$$

Adding 6 times equation (1) to equation (2), we obtain $4F + 9G = 0$, or $F = -\frac{9}{4}G$. Substituting for F into equation (1), we deduce that $B = \frac{5}{4}G$.

It follows that the equation of the projective conic must be of the form

$$\tfrac{5}{4}Gxy - \tfrac{9}{4}Gxz + Gyz = 0,$$

or

$$5xy - 9xz + 4yz = 0.$$

2. Let any such projective conic have equation

$$Ax^2 + Bxy + Cy^2 + Fxz + Gyz + Hz^2 = 0.$$

Then, as in Problem 1, it follows from the fact that $[1,0,0]$, $[0,1,0]$ and $[0,0,1]$ lie on the projective conic, that its equation must reduce to the form

$$Bxy + Fxz + Gyz = 0.$$

Since $[1,2,3]$ also lies on the projective conic, we deduce that

$$2B + 3F + 6G = 0.$$

Any projective conic satisfying this condition contains the given four Points.

There are many such possibilities. For example, taking $B = 3, F = 2$ and $G = -2$ gives the projective conic

$$3xy + 2xz - 2yz = 0;$$

while the choice $B = 6, F = -2, G = -1$ gives the projective conic

$$6xy - 2xz - yz = 0.$$

3. (a) Let $\mathbf{x}' = \mathbf{A}\mathbf{x}$, so that

$$\begin{pmatrix} x' \\ y' \\ z' \end{pmatrix} = \begin{pmatrix} 1 & 2 & 3 \\ 2 & 3 & 4 \\ 3 & 4 & 6 \end{pmatrix} \begin{pmatrix} x \\ y \\ z \end{pmatrix}.$$

Since

$$\begin{pmatrix} 1 & 2 & 3 \\ 2 & 3 & 4 \\ 3 & 4 & 6 \end{pmatrix} \begin{pmatrix} -2 \\ 0 \\ 1 \end{pmatrix} = \begin{pmatrix} 1 \\ 0 \\ 0 \end{pmatrix},$$

$$\begin{pmatrix} 1 & 2 & 3 \\ 2 & 3 & 4 \\ 3 & 4 & 6 \end{pmatrix} \begin{pmatrix} 0 \\ -3 \\ 2 \end{pmatrix} = \begin{pmatrix} 0 \\ -1 \\ 0 \end{pmatrix},$$

and

$$\begin{pmatrix} 1 & 2 & 3 \\ 2 & 3 & 4 \\ 3 & 4 & 6 \end{pmatrix} \begin{pmatrix} 1 \\ -2 \\ 1 \end{pmatrix} = \begin{pmatrix} 0 \\ 0 \\ 1 \end{pmatrix},$$

it follows that the images under t of $[-2,0,1]$, $[0,-3,2]$, $[1,-2,1]$ are $[1,0,0]$, $[0,-1,0] = [0,1,0]$ (since the coordinates are homogeneous coordinates), $[0,0,1]$, respectively.

(b) Since

$$\begin{pmatrix} 1 & 2 & 3 \\ 2 & 3 & 4 \\ 3 & 4 & 6 \end{pmatrix} \begin{pmatrix} -2 & 0 & 1 \\ 0 & 3 & -2 \\ 1 & -2 & 1 \end{pmatrix} = \begin{pmatrix} 1 & 0 & 0 \\ 0 & 1 & 0 \\ 0 & 0 & 1 \end{pmatrix},$$

it follows that $\begin{pmatrix} -2 & 0 & 1 \\ 0 & 3 & -2 \\ 1 & -2 & 1 \end{pmatrix}$ is the inverse of $\begin{pmatrix} 1 & 2 & 3 \\ 2 & 3 & 4 \\ 3 & 4 & 6 \end{pmatrix}$.

(c) Next, $\mathbf{x} = \mathbf{A}^{-1}\mathbf{x}'$ so that

$$\begin{pmatrix} x \\ y \\ z \end{pmatrix} = \begin{pmatrix} -2 & 0 & 1 \\ 0 & 3 & -2 \\ 1 & -2 & 1 \end{pmatrix} \begin{pmatrix} x' \\ y' \\ z' \end{pmatrix};$$

thus,

$$x = -2x' + z',$$
$$y = 3y' - 2z',$$

and

$$z = x' - 2y' + z'.$$

It follows that t maps the given projective conic onto the projective conic with equation

$$\begin{aligned} &17(-2x' + z')^2 + 47(-2x' + z') \\ &\quad \times (3y' - 2z') + 32(3y' - 2z')^2 \\ &\quad + 67(-2x' + z')(x' - 2y' + z') \\ &\quad + 92(3y' - 2z')(x' - 2y' + z') \\ &\quad + 66(x' - 2y' + z')^2 = 0. \end{aligned}$$

After simplifying this equation (with much cancellation), we find that it can be written in the form

$$2x'y' - 3y'z' - z'x' = 0.$$

(d) Finally, let $\mathbf{B} = \begin{pmatrix} -1/3 & 0 & 0 \\ 0 & -1 & 0 \\ 0 & 0 & 1/2 \end{pmatrix}$; then a matrix for a suitable projective transformation is

$$\mathbf{BA} = \begin{pmatrix} -\frac{1}{3} & 0 & 0 \\ 0 & -1 & 0 \\ 0 & 0 & \frac{1}{2} \end{pmatrix} \begin{pmatrix} 1 & 2 & 3 \\ 2 & 3 & 4 \\ 3 & 4 & 6 \end{pmatrix}$$
$$= \begin{pmatrix} -\frac{1}{3} & -\frac{2}{3} & -1 \\ -2 & -3 & -4 \\ \frac{3}{2} & 2 & 3 \end{pmatrix}.$$

Since the projective transformation is unaltered if we multiply this matrix by -6, then another matrix associated with this projective transformation is

$$\begin{pmatrix} 2 & 4 & 6 \\ 12 & 18 & 24 \\ -9 & -12 & -18 \end{pmatrix}.$$

The projective transformation t with which this matrix is associated maps the given projective conic E onto the standard projective conic $xy + yz + zx = 0$.

4. This case corresponds to a representation of Pascal's Theorem in an embedding plane for which R is an ideal Point. Since P and Q are collinear with the ideal Point R where AB' and $A'B$ meet, it follows that PQ is parallel to AB' and $A'B$.

The statement of Pascal's Theorem becomes the following.

Let A, B, C, A', B' and C' be six distinct points on a plane conic, with BC' and $B'C$ intersecting at P, CA' and $C'A$ intersecting at Q, and AB' parallel to $A'B$. Then PQ is parallel to AB' and $A'B$.

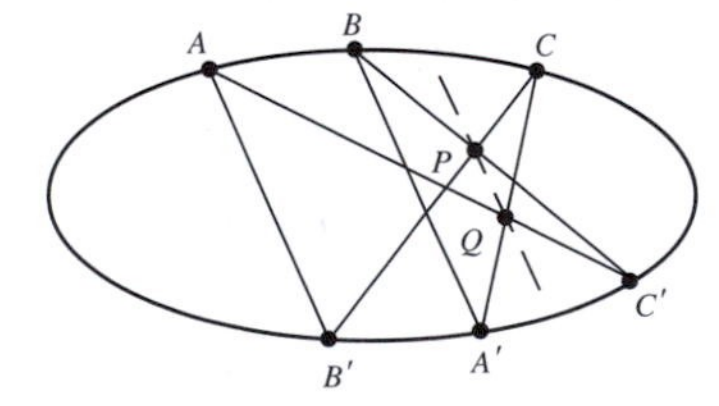

5. (a) If the Line with equation $91x - 60y - 109z = 0$ touches the projective conic with equation $x^2 + y^2 - z^2 = 0$ at some Point $P = [a, b, c]$, say, then its equation must be $ax + by - cz = 0$ (or some multiple of this). Comparing the equations $91x - 60y - 109z = 0$ and $ax + by - cz = 0$, we see that P must have homogeneous coordinates $[91, -60, 109]$. Since

$$\begin{aligned} & 91^2 + 60^2 - 109^2 \\ &= 8281 + 3600 - 11881 \\ &= 0, \end{aligned}$$

it follows that $[91, -60, 109]$ does lie on the projective conic. Thus the given Line is a tangent to the projective conic, and the Point of tangency is $[91, -60, 109]$.

(b) If the Line with equation $4x + 5y + 3z = 0$ touches the projective conic with equation $x^2 + y^2 - z^2 = 0$ at some Point $P = [a, b, c]$, say, then its equation must be $ax + by - cz = 0$ (or some multiple of this). Comparing the equations $4x + 5y + 3z = 0$ and $ax + by - cz = 0$, we see that P must have homogeneous coordinates $[4, 5, -3]$.

However, since

$$\begin{aligned} 4^2 + 5^2 - (-3)^2 &= 16 + 25 - 9 \\ &= 32 \neq 0, \end{aligned}$$

it follows that $[4, 5, -3]$ does not lie on the projective conic. Thus the given Line is not a tangent to the projective conic.

6. (a) As in Problem 1, the Points $[1, 0, 0]$, $[0, 1, 0]$ and $[0, 0, 1]$ lie on E_1. The Points $[2, 2, -1]$ and $[2, -1, 2]$ also lie on E_1 since

$$(2)(2) + (2)(-1) + (-1)(2) = 0$$

and

$$(2)(-1) + (-1)(2) + (2)(2) = 0.$$

(b) Since

$$\begin{pmatrix}1 & -1 & 0\\ 0 & 0 & 2\\ 1 & 1 & 2\end{pmatrix}\begin{pmatrix}1\\0\\0\end{pmatrix}=\begin{pmatrix}1\\0\\1\end{pmatrix},$$

$$\begin{pmatrix}1 & -1 & 0\\ 0 & 0 & 2\\ 1 & 1 & 2\end{pmatrix}\begin{pmatrix}0\\1\\0\end{pmatrix}=\begin{pmatrix}-1\\0\\1\end{pmatrix},$$

$$\begin{pmatrix}1 & -1 & 0\\ 0 & 0 & 2\\ 1 & 1 & 2\end{pmatrix}\begin{pmatrix}0\\0\\1\end{pmatrix}=\begin{pmatrix}0\\2\\2\end{pmatrix},$$

$$\begin{pmatrix}1 & -1 & 0\\ 0 & 0 & 2\\ 1 & 1 & 2\end{pmatrix}\begin{pmatrix}2\\2\\-1\end{pmatrix}=\begin{pmatrix}0\\-2\\2\end{pmatrix},$$

and

$$\begin{pmatrix}1 & -1 & 0\\ 0 & 0 & 2\\ 1 & 1 & 2\end{pmatrix}\begin{pmatrix}2\\-1\\2\end{pmatrix}=\begin{pmatrix}3\\4\\5\end{pmatrix},$$

the images under t_1 of the five Points in part (a) are $[1,0,1]$, $[-1,0,1]$, $[0,2,2]=[0,1,1]$, $[0,-2,2]=[0,-1,1]$ and $[3,4,5]$, respectively.

(c) It is easy to verify that the five Points $[1,0,1]$, $[-1,0,1]$, $[0,1,1]$, $[0,-1,1]$ and $[3,4,5]$ all lie on the projective conic E_2 since their coordinates satisfy its equation $x^2+y^2-z^2=0$.

By the Five Points Theorem (Subsection 4.3.1, Theorem 1) five distinct Points uniquely determine a non-degenerate projective conic. It follows from the results of parts (a) and (b) that the projective transformation t_1 with associated matrix $\mathbf{A}=\begin{pmatrix}1 & -1 & 0\\ 0 & 0 & 2\\ 1 & 1 & 2\end{pmatrix}$ maps E_1 onto E_2.

(d) Since

$$\mathbf{BA}=\begin{pmatrix}-1 & 1 & -1\\ 1 & 1 & -1\\ 0 & -1 & 0\end{pmatrix}\begin{pmatrix}1 & -1 & 0\\ 0 & 0 & 2\\ 1 & 1 & 2\end{pmatrix}$$

$$=\begin{pmatrix}-2 & 0 & 0\\ 0 & -2 & 0\\ 0 & 0 & -2\end{pmatrix}$$

$$=(-2)\begin{pmatrix}1 & 0 & 0\\ 0 & 1 & 0\\ 0 & 0 & 1\end{pmatrix},$$

a matrix associated with the projective transformation $t_2 \circ t_1$ is $-2\mathbf{I}$. Hence another matrix associated with the projective transformation $t_2 \circ t_1$ is the identity matrix $\mathbf{I}$; it follows that $t_2 \circ t_1$ must be the identity projective transformation, and so t_2 is the inverse of t_1.

Hence since $\mathbf{B}$ is a matrix associated with t_2, it is a matrix associated with the projective transformation that maps E_2 onto E_1.

7. We saw in Problem 3, part (d), that a matrix associated with a projective transformation that maps the projective conic E with equation

$17x^2+47xy+32y^2+67xz+92yz+66z^2=0$

onto the projective conic E' with equation $xy+yz+zx=0$ is

$$\mathbf{A}_1=\begin{pmatrix}2 & 4 & 6\\ 12 & 18 & 24\\ -9 & -12 & -18\end{pmatrix},$$

and in Problem 6, part (c), that a matrix associated with a projective transformation that maps E' onto the projective conic E'' with equation $x^2+y^2-z^2=0$ is

$$\mathbf{A}_2=\begin{pmatrix}1 & -1 & 0\\ 0 & 0 & 2\\ 1 & 1 & 2\end{pmatrix}.$$

It follows that a matrix associated with a projective transformation that maps the projective conic E onto the projective conic E'' is

$$\mathbf{A_2A_1}=\begin{pmatrix}1 & -1 & 0\\ 0 & 0 & 2\\ 1 & 1 & 2\end{pmatrix}\begin{pmatrix}2 & 4 & 6\\ 12 & 18 & 24\\ -9 & -12 & -18\end{pmatrix}$$

$$=\begin{pmatrix}-10 & -14 & -18\\ -18 & -24 & -36\\ -4 & -2 & -6\end{pmatrix}.$$

Since the projective transformation is unaltered if we multiply the matrix by the non-zero factor $-\frac{1}{2}$, we conclude that a matrix associated with this projective transformation is

$$-\frac{1}{2}\mathbf{A_2A_1} = \begin{pmatrix} 5 & 7 & 9 \\ 9 & 12 & 18 \\ 2 & 1 & 3 \end{pmatrix}.$$

8. The Point $\left[1, 2\sqrt{2}, 3\right]$ lies on the projective conic with equation $x^2 + y^2 = z^2$; first, we express its homogeneous coordinates in the form $[x, y, 1] = \left[\frac{1-t^2}{1+t^2}, \frac{2t}{1+t^2}, 1\right]$ as described before Theorem 9 in Subsection 4.3.4. Since $\left[1, 2\sqrt{2}, 3\right] = \left[\frac{1}{3}, \frac{2\sqrt{2}}{3}, 1\right]$, we have

$$\frac{1-t^2}{1+t^2} = \frac{1}{3} \quad \text{and} \quad \frac{2t}{1+t^2} = \frac{2\sqrt{2}}{3}.$$

From the first equation we have $3 - 3t^2 = 1 + t^2$, so that $2 = 4t^2$ or $t = \pm(1/\sqrt{2})$. Since, from the second equation above, $2t/(1+t^2)$ is positive, it follows that we must have t positive. Hence, at the Point $P, t = 1/\sqrt{2}$.
Remark It is important *not* to simply assert that

$$1 = 1 - t^2, \quad 2\sqrt{2} = 2t \quad \text{and} \quad 3 = 1 + t^2.$$

(No such t exists!) The coordinates are *homogeneous* coordinates, so that it is their relative ratio that matters not the value of any particular coordinate.

9. 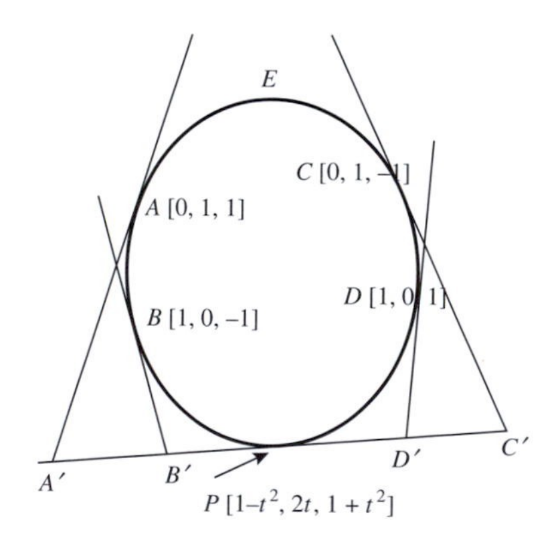

(a) By Joachimsthal's theory, the equation of the tangent at a Point $[x_1, y_1, z_1]$ on E is

$$xx_1 + yy_1 - zz_1 = 0.$$

Hence the equations of the tangents at $A[0, 1, 1]$, $B[1, 0, -1]$, $C[0, 1, -1]$, $D[1, 0, 1]$ and $P[1 - t^2, 2t, 1 + t^2]$ are

$$y - z = 0,$$
$$x + z = 0,$$
$$y + z = 0,$$
$$x - z = 0$$

and

$$(1 - t^2)x + 2ty - (1 + t^2)z = 0,$$

respectively.

(b) At the Point A' we have that $y - z = 0$ and $\left(1 - t^2\right)x + 2ty - \left(1 + t^2\right)z = 0$. It follows that $y = z$ and $\left(1 - t^2\right)x - \left(1 - 2t + t^2\right)z = 0$. So either $t = 1$ (so that $P = [0, 2, 2] = [0, 1, 1] = A$, which we can ignore) or $(1 + t)x - (1 - t)z = 0$. Hence the homogeneous coordinates of A' are $[1 - t, 1 + t, 1 + t]$.

At the Point B' we have that $x + z = 0$ and $\left(1 - t^2\right)x + 2ty - \left(1 + t^2\right)z = 0$. It follows that $x = -z$ and $2ty - 2z = 0$. Hence the homogeneous coordinates of B' are $[t, -1, -t]$.

At the Point C' we have that $y + z = 0$ and $\left(1 - t^2\right)x + 2ty - \left(1 + t^2\right)z = 0$. It follows that $y = -z$ and $\left(1 - t^2\right)x - \left(1 + 2t + t^2\right)z = 0$. So either $t = -1$ (so that $P = [0, -2, 2] = [0, -1, 1] = C$, which we can ignore) or $(1 - t)x - (1 + t)z = 0$. Hence the homogeneous coordinates of C' are $[1 + t, -1 + t, 1 - t]$.

At the Point D' we have that $x - z = 0$ and $\left(1 - t^2\right)x + 2ty - \left(1 + t^2\right)z = 0$; thus, $x = z$ and $2ty - 2t^2z = 0$. So either $t = 0$ (so that $P = [1, 0, 1] = D$, which we can ignore) or $y - tz = 0$. Hence the homogeneous coordinates of D' are $[1, t, 1]$.

(c) We now find the cross-ratio $(A'B'C'D')$, using the definition of cross-ratio in Subsection 3.5.1.

Firstly, we have to find real numbers α and β such that the following vector equation holds:

$$\begin{aligned}&(1+t,-1+t,1-t)\\&\quad=\alpha(1-t,1+t,1+t)+\beta(t,-1,-t).\end{aligned}$$

Comparing corresponding coordinates on both sides, we deduce that we must have

$$1+t=\alpha(1-t)+\beta t, \qquad (3)$$

$$-1+t=\alpha(1+t)-\beta, \qquad (4)$$

and

$$1-t=\alpha(1+t)-\beta t. \qquad (5)$$

Adding equation (3) and equation (5) we get that $2=2\alpha$, so that $\alpha=1$; substituting for α into equation (4) we get that $\beta=1+t+1-t=2$.

Next, we have to find real numbers γ and δ such that the following vector equation holds:

$$(1,t,1)=\gamma(1-t,1+t,1+t)+\delta(t,-1,-t).$$

Comparing corresponding coordinates on both sides, we deduce that we must have

$$1=\gamma(1-t)+\delta t, \qquad (6)$$

$$t=\gamma(1+t)-\delta \qquad (7)$$

and

$$1=\gamma(1+t)-\delta t. \qquad (8)$$

Adding equation (6) and equation (8) we get that $2=2\gamma$, so that $\gamma=1$; substituting for γ into equation (7) we get that $\delta=-t+1+t=1$.

It then follows from the definition of cross-ratio that

$$\begin{aligned}(A'B'C'D')&=\frac{\beta\gamma}{\alpha\delta}\\&=\frac{2\cdot 1}{1\cdot 1}\\&=2.\end{aligned}$$

Section 4.4

1. In each case we use the definition of associated matrix given before the problem.

   (a) Here a suitable matrix is

$$\begin{pmatrix}1 & -\frac{1}{2} & -1\\ -\frac{1}{2} & 3 & \frac{3}{2}\\ -1 & \frac{3}{2} & -\frac{1}{2}\end{pmatrix}.$$

   (b) Here a suitable matrix is

$$\begin{pmatrix}2 & -\frac{1}{2} & -\frac{3}{2}\\ -\frac{1}{2} & -1 & \frac{1}{2}\\ -\frac{3}{2} & \frac{1}{2} & 4\end{pmatrix}.$$

2. We follow the method in the proof of Theorem 2 of Section 4.4.

   (a) A matrix associated with the projective conic with equation $x^2-4xy+2y^2-4yz+3z^2=0$ is

$$\mathbf{A}=\begin{pmatrix}1 & -2 & 0\\ -2 & 2 & -2\\ 0 & -2 & 3\end{pmatrix}.$$

   (b) We start by diagonalizing the matrix $\mathbf{A}$. Its characteristic equation is

$$\begin{aligned}0&=\det(\mathbf{A}-\lambda\mathbf{I})\\&=\begin{vmatrix}1-\lambda & -2 & 0\\ -2 & 2-\lambda & -2\\ 0 & -2 & 3-\lambda\end{vmatrix}\\&=(1-\lambda)\begin{vmatrix}2-\lambda & -2\\ -2 & 3-\lambda\end{vmatrix}\\&\quad+2\begin{vmatrix}-2 & -2\\ 0 & 3-\lambda\end{vmatrix}+0\\&=(1-\lambda)\left(\lambda^2-5\lambda+2\right)+2(-6+2\lambda)\\&=-\lambda^3+6\lambda^2-3\lambda-10\\&=-(\lambda-5)(\lambda-2)(\lambda+1),\end{aligned}$$

   so that the eigenvalues of $\mathbf{A}$ are 5, 2 and -1.

The eigenvector equations of $\mathbf{A}$ are

$$(1-\lambda)x - 2y = 0,$$
$$-2x + (2-\lambda)y - 2z = 0,$$
$$-2y + (3-\lambda)z = 0.$$

When $\lambda = 5$, these equations become

$$-4x - 2y = 0,$$
$$-2x - 3y - 2z = 0,$$
$$-2y - 2z = 0.$$

From the first equation, we have $y = -2x$; and from the third equation, we have $z = -y$. So we may take as a corresponding eigenvector $\begin{pmatrix} 1 \\ -2 \\ 2 \end{pmatrix}$, which we normalise to have unit length as $\begin{pmatrix} 1/3 \\ -2/3 \\ 2/3 \end{pmatrix}$.

When $\lambda = 2$, these equations become

$$-x - 2y = 0,$$
$$-2x - 2z = 0,$$
$$-2y + z = 0.$$

From the first equation, we have $x = -2y$; and from the third equation, we have $z = 2y$. So we may take as a corresponding eigenvector $\begin{pmatrix} -2 \\ 1 \\ 2 \end{pmatrix}$, which we normalise to have unit length as $\begin{pmatrix} -2/3 \\ 1/3 \\ 2/3 \end{pmatrix}$.

When $\lambda = -1$, these equations become

$$2x - 2y = 0,$$
$$-2x + 3y - 2z = 0,$$
$$-2y + 4z = 0.$$

From the first equation, we have $y = x$; and from the third equation, we have $y = 2z$. So we may take as a corresponding eigenvector $\begin{pmatrix} 2 \\ 2 \\ 1 \end{pmatrix}$, which we normalise to have unit length as $\begin{pmatrix} 2/3 \\ 2/3 \\ 1/3 \end{pmatrix}$.

If we then take $\mathbf{P} = \begin{pmatrix} 1/3 & -2/3 & 2/3 \\ -2/3 & 1/3 & 2/3 \\ 2/3 & 2/3 & 1/3 \end{pmatrix}$, we have

$$\mathbf{P}^T\mathbf{AP} = \begin{pmatrix} 5 & 0 & 0 \\ 0 & 2 & 0 \\ 0 & 0 & -1 \end{pmatrix},$$

where $\mathbf{P}$ is an orthogonal matrix.

(c) Thus the transformation of coordinates given by $\mathbf{x} = \mathbf{Px}'$ or $\mathbf{x}' = \mathbf{P}^T\mathbf{x}$ transforms the projective conic with equation $x^2 - 4xy + 2y^2 - 4yz + 3z^2 = 0$ (that is, $\mathbf{x}^T\mathbf{Ax} = 0$) onto the projective conic with equation

$$5(x')^2 + 2(y')^2 - (z')^2 = 0$$

(that is, $(\mathbf{Px}')^T\mathbf{A}(\mathbf{Px}') = 0$ or $(\mathbf{x}')^T(\mathbf{P}^T\mathbf{AP})\mathbf{x}' = 0$).

Next, the transformation of coordinates $\mathbf{x}' \mapsto \mathbf{x}''$ given by $\mathbf{x}'' = \mathbf{Bx}'$, where

$$\mathbf{B} = \begin{pmatrix} \sqrt{5} & 0 & 0 \\ 0 & \sqrt{2} & 0 \\ 0 & 0 & 1 \end{pmatrix},$$

transforms the equation of the projective conic into the equation $(x'')^2 + (y'')^2 - (z'')^2 = 0$.

Thus, after dropping the dashes, it follows that we can map E onto the projective conic with equation $x^2 + y^2 = z^2$

by the projective transformation $[\mathbf{x}] \mapsto [\mathbf{BP}^T\mathbf{x}]$, where

$$\mathbf{BP}^T = \begin{pmatrix} \sqrt{5} & 0 & 0 \\ 0 & \sqrt{2} & 0 \\ 0 & 0 & 1 \end{pmatrix} \times \begin{pmatrix} 1/3 & -2/3 & 2/3 \\ -2/3 & 1/3 & 2/3 \\ 2/3 & 2/3 & 1/3 \end{pmatrix} = \begin{pmatrix} \sqrt{5}/3 & -2\sqrt{5}/3 & 2\sqrt{5}/3 \\ -2\sqrt{2}/3 & \sqrt{2}/3 & 2\sqrt{2}/3 \\ 2/3 & 2/3 & 1/3 \end{pmatrix}.$$

This is a matrix associated with the desired projective transformation.

Section 4.5

1. The statement of the Five Points Theorem (Subsection 4.3.1, Theorem 1) is follows:

   There is a unique non-degenerate projective conic through any given set of five Points, no three of which are collinear.

   To make it easier to dualize this result, by breaking the sentence up into shorter portions, we first rephrase it in the following way:

   Given any five Points,
   no three of which are collinear,
   there is a unique non-degenerate projective conic through the five Points.

   We first interchange 'Point' and 'Line', and 'collinear' and 'concurrent'; this gives the statement:

   Given any five Lines,
   no three of which are concurrent,
   there is a unique non-degenerate projective conic through the five Lines.

   Next, we make the changes necessary for this statement to make sense; it becomes:

   Given any five Lines,
   no three of which are concurrent,
   there is a unique non-degenerate projective conic that is tangential to the five Lines.

   This is the result dual to the Five Points Theorem.

Chapter 5

Section 5.1

1. The centre of inversion is the origin (0, 0) and the radius of the circle of inversion is 1. In each case, the point and its inverse lie on the same half-line from O to infinity.
   (a) Since $4 \times \frac{1}{4} = 1$, and since $(4, 0)$ and $\left(\frac{1}{4}, 0\right)$ both lie on the positive x-axis, it follows that $\left(\frac{1}{4}, 0\right)$ is the inverse of $(4, 0)$.
   (b) Since $1 \times 1 = 1$, it follows that $(0, 1)$ is its own inverse.
   (c) Since $\left|-\frac{1}{3}\right| \times |-3| = 1$, and since $\left(0, -\frac{1}{3}\right)$ and $(0, -3)$ both lie on the negative y-axis, it follows that $(0, -3)$ is the inverse of $\left(0, -\frac{1}{3}\right)$.
   (d) Since $\frac{1}{4} \times 4 = 1$, and since $\left(\frac{1}{4}, 0\right)$ and $(4, 0)$ both lie on the positive x-axis, it follows that $(4, 0)$ is the inverse of $\left(\frac{1}{4}, 0\right)$.
2. We may use the formula for inversion in $\mathscr{C}$ given in Theorem 2.
   (a) The image under inversion in $\mathscr{C}$ of the point (4, 1) is
   $$\left(\frac{4}{4^2+1^2}, \frac{1}{4^2+1^2}\right) = \left(\tfrac{4}{17}, \tfrac{1}{17}\right).$$
   (b) Similarly, the image of $\left(\frac{1}{2}, -\frac{1}{4}\right)$ is
   $$\left(\frac{\frac{1}{2}}{\left(\frac{1}{2}\right)^2 + \left(-\frac{1}{4}\right)^2}, \frac{-\frac{1}{4}}{\left(\frac{1}{2}\right)^2 + \left(-\frac{1}{4}\right)^2}\right) = \left(\frac{\frac{1}{2}}{\frac{5}{16}}, \frac{-\frac{1}{4}}{\frac{5}{16}}\right) = \left(\tfrac{8}{5}, -\tfrac{4}{5}\right).$$
3. Let ℓ be the line with equation $y = 2x$ punctured at the origin. By the strategy above Example 2, the image of ℓ under inversion in $\mathscr{C}$ is the curve whose equation is
   $$\frac{y}{x^2+y^2} = \frac{2x}{x^2+y^2}.$$
   We may rewrite this equation as $y = 2x$. Just as the origin had to be excluded from the line before we could find its image, so the origin

has to be excluded from the image. It follows that the image of ℓ under the inversion is ℓ itself (punctured at the origin in each case).

4. (a) It follows from the strategy above Example 2 that the image under inversion in $\mathscr{C}$ of the line ℓ with equation $x+y=1$ is the curve whose equation is

$$\frac{x}{x^2+y^2}+\frac{y}{x^2+y^2}=1.$$

We may rewrite this equation in the form

$$x+y=x^2+y^2.$$

By completing the square, we may write this as

$$\left(x-\frac{1}{2}\right)^2+\left(y-\frac{1}{2}\right)^2=\frac{1}{2}. \quad (1)$$

This is the equation of a circle that passes through the origin, so the required image is the circle with centre $\left(\frac{1}{2},\frac{1}{2}\right)$ and radius $\frac{1}{\sqrt{2}}$, punctured at the origin.

(b) The points $(1,0)$ and $(0,1)$ lie on ℓ, since their coordinates satisfy the equation of ℓ.

Since the two points lie on $\mathscr{C}$, they are unchanged under inversion with respect to $\mathscr{C}$. Hence they both also lie on the image of ℓ under the inversion.

(c) The image of ℓ is the circle with equation (1), punctured at the origin. From part (b) this image passes through the points with coordinates $(1,0)$ and $(0,1)$. It follows that the image is uniquely determined by the three points $(0,0)$, $(1,0)$ and $(0,1)$, as shown below.

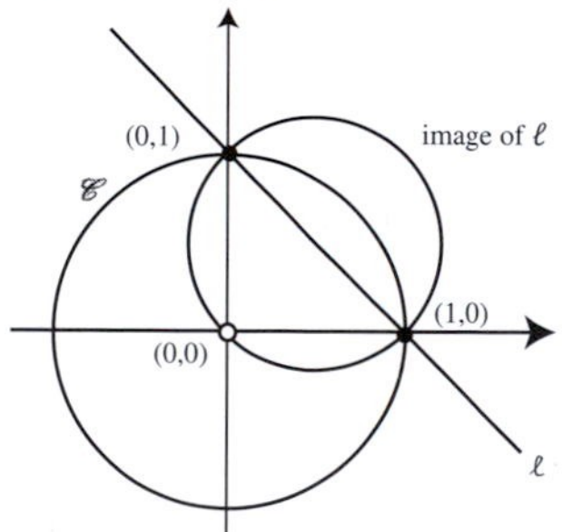

5. The circle with centre $(2,2)$ and radius 1 has equation $(x-2)^2+(y-2)^2=1^2$, which we may rewrite in the form

$$x^2+y^2-4x-4y+7=0.$$

Using the strategy above Example 2, we deduce that the image of this circle under inversion in $\mathscr{C}$ has equation

$$\left(\frac{x}{x^2+y^2}\right)^2+\left(\frac{y}{x^2+y^2}\right)^2-\frac{4x}{x^2+y^2}-\frac{4y}{x^2+y^2}+7=0.$$

We may add together the first pair of terms and the second pair of terms to obtain the equation

$$\frac{1}{x^2+y^2}-\frac{4x+4y}{x^2+y^2}+7=0.$$

We may rearrange this in the form

$$1-4x-4y+7\left(x^2+y^2\right)=0,$$

or

$$x^2+y^2-\frac{4}{7}x-\frac{4}{7}y+\frac{1}{7}=0.$$

By completing the squares, we obtain

$$\left(x-\frac{2}{7}\right)^2+\left(y-\frac{2}{7}\right)^2=\frac{1}{49}.$$

This is the equation of a circle with centre $\left(\frac{2}{7},\frac{2}{7}\right)$ and radius $\frac{1}{7}$.

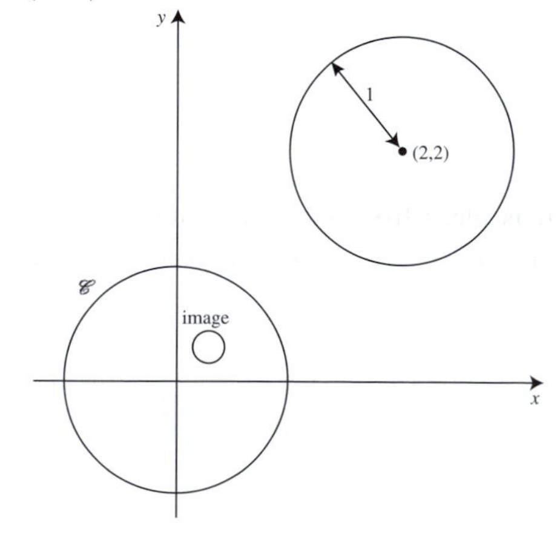

6. The circle with centre $\left(0, -\frac{1}{4}\right)$ and radius $\frac{1}{4}$ has equation $x^2 + \left(y + \frac{1}{4}\right)^2 = \left(\frac{1}{4}\right)^2$, which we may rewrite in the form

$$x^2 + y^2 + \frac{1}{2}y = 0.$$

Using the strategy above Example 2, we deduce that the image of this circle under inversion in $\mathscr{C}$ has equation

$$\left(\frac{x}{x^2+y^2}\right)^2 + \left(\frac{y}{x^2+y^2}\right)^2 + \frac{\frac{1}{2}y}{x^2+y^2} = 0.$$

We may add together the first pair of terms to obtain the equation

$$\frac{1}{x^2+y^2} + \frac{\frac{1}{2}y}{x^2+y^2} = 0,$$

which we may rewrite as

$$y = -2.$$

This is the equation of a horizontal line with y-intercept -2.

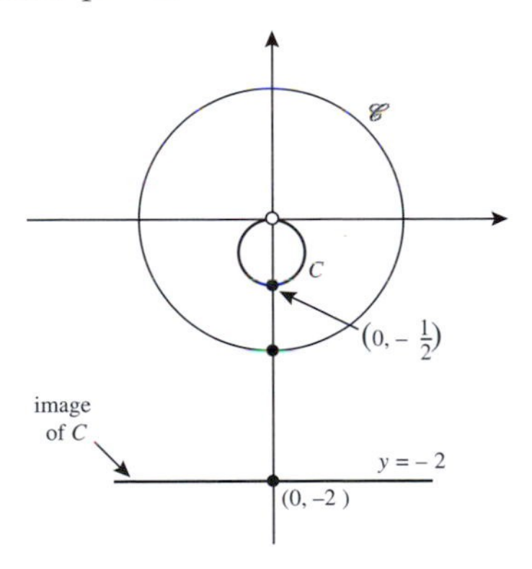

It is clear from the figure that each point of this line is the image of some point on the punctured circle $C - \{O\}$.

7. (a) From the summary box before Example 5, we know that ℓ maps to a circle punctured at the origin. This circle passes through the point $(0, 1)$ since this point is fixed by the inversion. Since ℓ is symmetrical about the y-axis, it follows that the image circle must also be symmetrical about the y-axis. The only circle that fulfils all these criteria is the circle with radius $\frac{1}{2}$ and centre $\left(0, \frac{1}{2}\right)$.

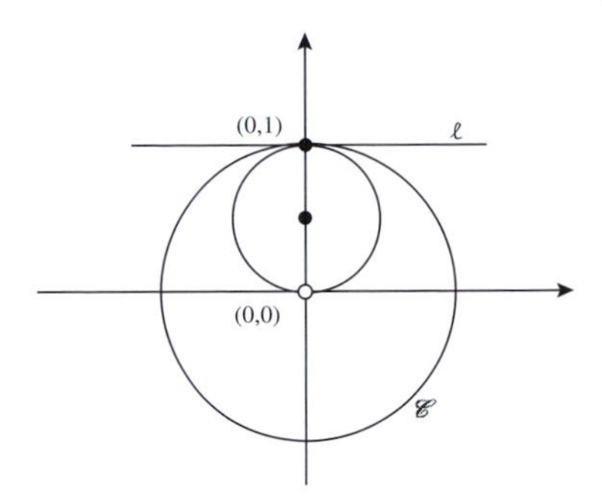

(b) From the summary box before Example 5, since the circle C passes through the origin we know that the image of C is a line that does not pass through the origin. It must be symmetrical about the y-axis (because C is), and it must pass through the point $\left(0, \frac{1}{2}\right)$ (the image of $(0, 2)$). The only line that fulfils all these criteria is the line $y = \frac{1}{2}$.

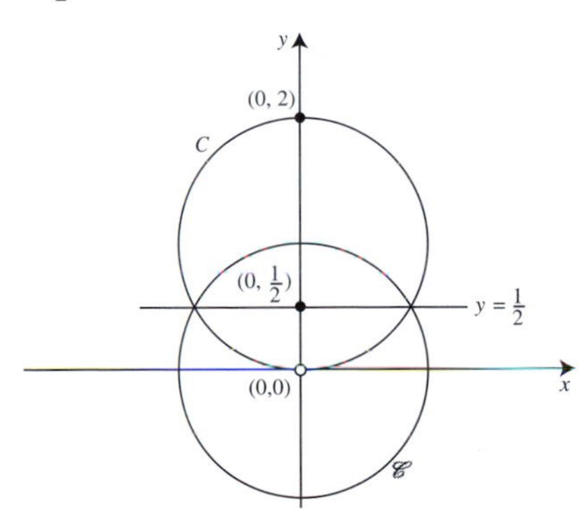

8. We follow the proof of Theorem 5 as far as possible, making the necessary modifications to the line that passes through the centre of inversion O.

Let ℓ_1 be the line that passes through O, and let the second line ℓ_2 meet ℓ_1 at A.

Under the inversion, $\ell_1 - \{O\}$ maps to itself, and ℓ_2 maps to some circle C_2, punctured at O. Then C_2 meets ℓ_1 at O and at the point A', the image of A under the inversion.

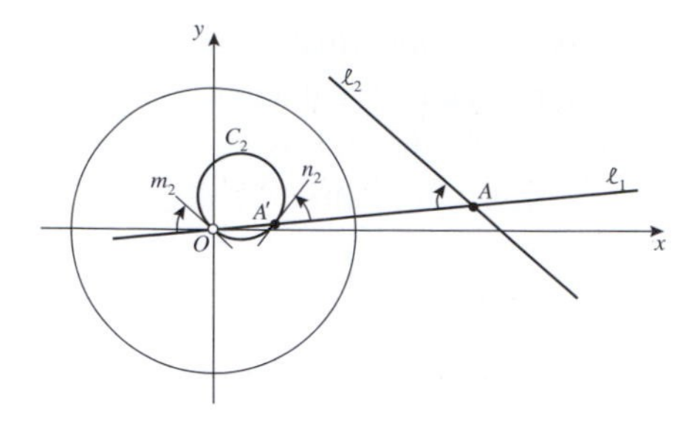

By the Symmetry Lemma, ℓ_2 is parallel to the tangent m_2 to C_2 at O. It follows that the angle from ℓ_1 to ℓ_2 must be equal in magnitude and direction to the angle we have shown from ℓ_1 to m_2 at O.

Next observe that the reflection in the perpendicular bisector of OA' maps ℓ_1 to itself and C_2 to itself, and sends the tangent m_2 at O to the tangent n_2 at A'. Since the reflection preserves the magnitude of an angle but changes its orientation, we conclude that the angle from ℓ_1 to n_2 at A' must be equal in magnitude but opposite in orientation to the angle from ℓ_1 to m_2 at O.

Overall, we have shown that the angle from ℓ_1 to C_2 at A' must be equal in magnitude but opposite in orientation to the angle from ℓ_1 to ℓ_2 at A. This completes the proof.

9. By the conclusion to Example 6, the image under inversion of each punctured circle touching the x-axis is a line parallel to the x-axis (that is, a horizontal line).

   Similarly, the punctured circles that touch the y-axis map to lines parallel to the y-axis.

   It follows that, under the inversion, the circles map to horizontal and vertical lines. Indeed, the only horizontal and vertical lines that are not images of circles are the x- and y-axes.

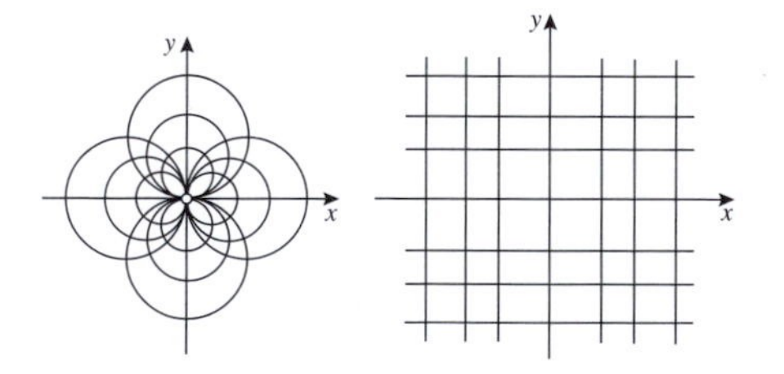

10. Let ℓ be the line through O and A. Under inversion in the unit circle, $\ell - \{0\}$ maps to itself. In particular, the point A on ℓ maps to some point A' on ℓ. Since all the circles pass through A and are punctured at O, their images under the inversion must be straight lines that intersect at A'.

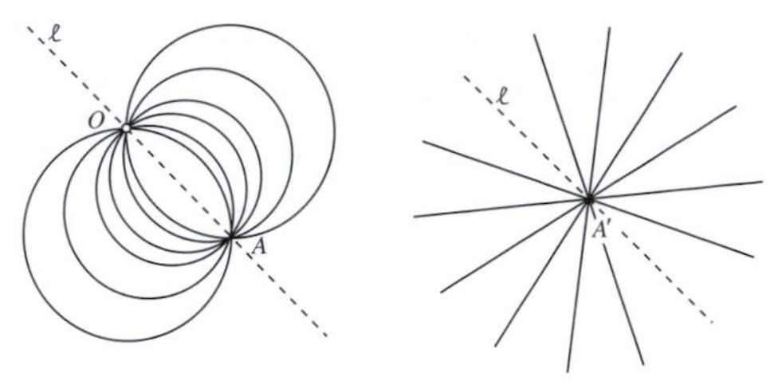

In fact, apart from ℓ, every straight line through A' must be the image of one of the circles in the family. For, as shown below, if a line ℓ_1 is at an angle θ from ℓ, then, by the Angle Theorem, it must be the image of the (punctured) circle C_1 that is at an angle $-\theta$ from ℓ.

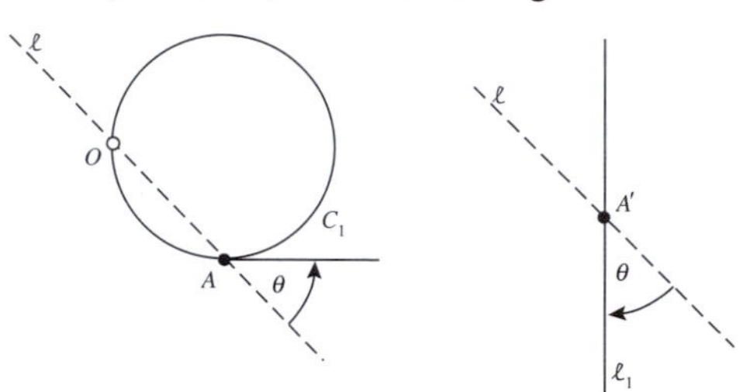

11. Since the circles C_1 and C_2 pass through the origin, it follows from Theorem 4, part (b), that their images under inversion are two lines, ℓ_1 and ℓ_2. Next observe that the common tangent to C_1 and C_2 at the origin maps to itself. Since this tangent and the circles C_1 and C_2 meet only at the origin, it follows that their images do not meet in $\mathbb{C}$. In other words, ℓ_1 and ℓ_2 are both parallel to the tangent.

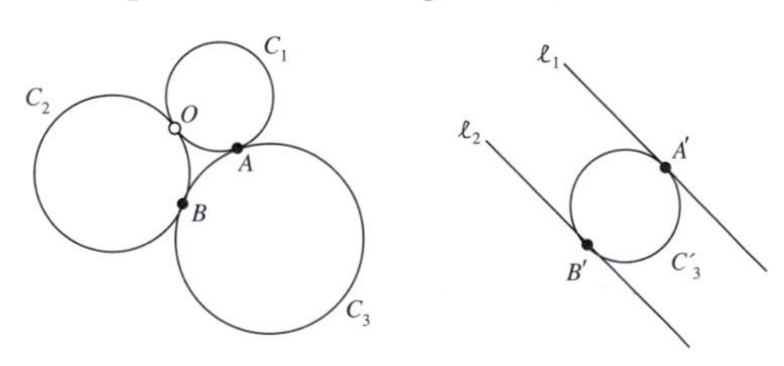

Let A and B have images A' and B', respectively, under the inversion. Then it follows from Theorem 4, part (a), that the image of C_3 under inversion is a circle C_3'. Since A and B lie on C_3, A' and B' must lie on C_3'.

Since angles are preserved (in magnitude) under inversion, the images of tangential curves (which may be considered as curves intersecting at a zero angle) are also tangential curves. Hence C_3' must be tangential to ℓ_1 at A' and to ℓ_2 at B'.

Section 5.2

1. (a) With $z_1 = 2 - 3i$ and $z_2 = -3 + 4i$, we have the following.

(i) $z_1 + z_2 = -1 + i$.

(ii) $z_1 - z_2 = 5 - 7i$.

(iii) $z_1 z_2 = (2-3i)(-3+4i)$.
$= -6 + 8i + 9i + 12$
$= 6 + 17i$.

(iv) Here $\frac{z_1}{z_2} = \frac{2-3i}{-3+4i}$.

Multiplying the numerator and denominator by the complex conjugate of $-3+4i$, we obtain

$$\begin{aligned}\frac{z_1}{z_2} &= \frac{(2-3i)(-3-4i)}{(-3+4i)(-3-4i)}\\ &= \frac{-6-8i+9i-12}{3^2+4^2}\\ &= \tfrac{1}{25}(-18+i).\end{aligned}$$

(v) $\overline{z_1} = 2 + 3i$.

(vi) $\overline{z_2} = -3 - 4i$.

(b) Here

$$|z_1| = \sqrt{2^2 + (-3)^2} = \sqrt{13}$$

and

$$|z_2| = \sqrt{(-3)^2 + 4^2} = \sqrt{25} = 5.$$

2. We can write $z_1 = r(\cos\theta + i\sin\theta)$ where

$$r = |z_1| = \sqrt{1^2 + (-1)^2} = \sqrt{2}$$

and

$$\cos\theta = \tfrac{1}{\sqrt{2}} \quad \text{and} \quad \sin\theta = -\tfrac{1}{\sqrt{2}},$$

so that the principal argument is $\theta = -\pi/4$. Hence the required polar form of $z_1 = 1 - i$ is

$$\sqrt{2}(\cos(-\pi/4) + i\sin(-\pi/4)).$$

Similarly, we can write $z_2 = r(\cos\theta + i\sin\theta)$ where

$$r = |z_2| = \sqrt{\left(-\sqrt{3}\right)^2 + 1^2} = \sqrt{4} = 2$$

and

$$\cos\theta = -\tfrac{\sqrt{3}}{2} \quad \text{and} \quad \sin\theta = \tfrac{1}{2},$$

so that the principal argument is $\theta = 5\pi/6$. Hence the required polar form of $z_2 = -\sqrt{3} + i$ is

$$2(\cos(5\pi/6) + i\sin(5\pi/6)).$$

3. From Problem 2 we know that

$$z_1 = \sqrt{2}\left(\cos\left(-\frac{\pi}{4}\right) + i\sin\left(-\frac{\pi}{4}\right)\right)$$

and

$$z_2 = 2\left(\cos\left(\frac{5\pi}{6}\right) + i\sin\left(\frac{5\pi}{6}\right)\right).$$

(a) Using the strategy of multiplying moduli and adding arguments, we obtain

$$\begin{aligned}z_1 z_2 &= 2\sqrt{2}\left(\cos\left(-\frac{\pi}{4} + \frac{5\pi}{6}\right)\right.\\ &\qquad \left.+ i\sin\left(-\frac{\pi}{4} + \frac{5\pi}{6}\right)\right)\\ &= 2\sqrt{2}\left(\cos\left(\frac{7\pi}{12}\right) + i\sin\left(\frac{7\pi}{12}\right)\right).\end{aligned}$$

(b) Using the strategy of dividing moduli and subtracting arguments, we obtain

$$\begin{aligned}\frac{z_1}{z_2} &= \frac{\sqrt{2}}{2}\left(\cos\left(-\frac{\pi}{4} - \frac{5\pi}{6}\right)\right.\\ &\qquad \left.+ i\sin\left(-\frac{\pi}{4} - \frac{5\pi}{6}\right)\right)\\ &= \frac{1}{\sqrt{2}}\left(\cos\left(-\frac{13\pi}{12}\right) + i\sin\left(-\frac{13\pi}{12}\right)\right)\\ &= \frac{1}{\sqrt{2}}\left(\cos\left(\frac{11\pi}{12}\right) + i\sin\left(\frac{11\pi}{12}\right)\right),\end{aligned}$$

where in the last line we have added 2π to the argument to obtain its principal value.

4. (a) The coefficient of z has modulus $|-i| = 1$. By Theorem 1, it follows that t is an isometry.

   (b) In the formula that defines t, the multiplication by $-i$ corresponds to a *rotation* about the origin through the angle $-\pi/2$, and the addition of $6-4i$ corresponds to a *translation* through the vector $(6, -4)$.

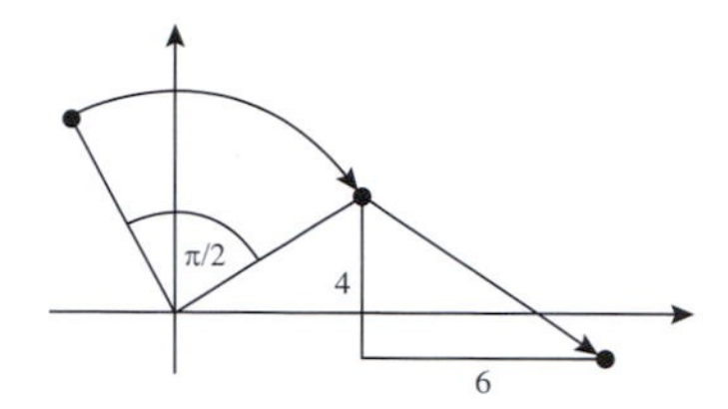

   (c) The *rotation* through $-\pi/2$ corresponds to the composite $r_2 \circ r_1$ where

   r_1 is the reflection in the x-axis,
   r_2 is the reflection in the line $y = -x$.

   (Here $y = -x$ is the line through the origin that makes an angle $-\pi/4$ with the x-axis.)

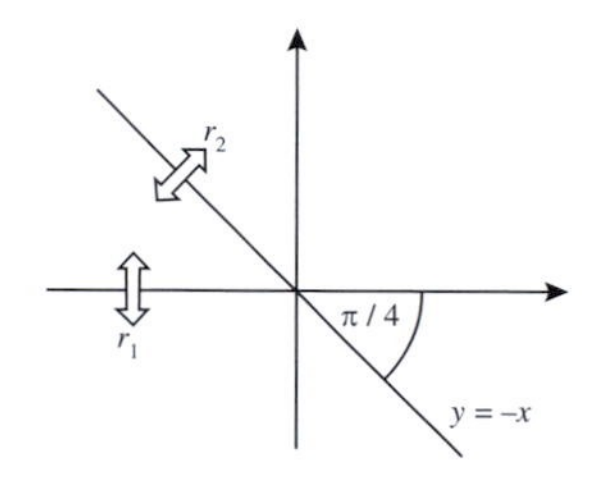

   The *translation* through the vector $(6, -4)$ corresponds to the composite $r_4 \circ r_3$ where

   r_3 is the reflection in the line $6x - 4y = 0$,
   r_4 is the reflection in the line $6x - 4y = 26$.

   (Here $6x - 4y = 0$ is the equation of the line through the origin that is perpendicular to the vector $(6, -4)$, and $6x - 4y = 26$ is the equation of the parallel line that passes through the midpoint $(3, -2)$ of the position vector of $(6, -4)$.)

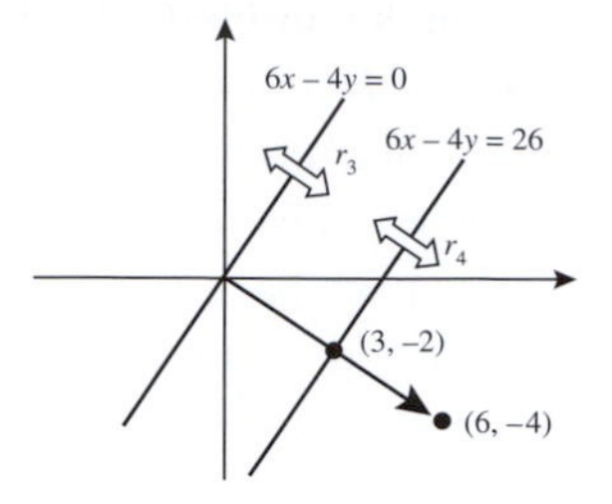

   Overall, $t = r_4 \circ r_3 \circ r_2 \circ r_1$.

5. Here we use the formula for inversion of points in $\mathscr{C}$ given by Theorem 3.

   (a) The image of $-\sqrt{3}+i$ under inversion in $\mathscr{C}$ is

$$\frac{1}{\overline{-\sqrt{3}+i}} = \frac{1}{-\sqrt{3}-i} = \frac{-\sqrt{3}+i}{(-\sqrt{3}-i)(-\sqrt{3}+i)} = \tfrac{1}{4}(-\sqrt{3}+i).$$

   (b) The image of $-3-4i$ under inversion in $\mathscr{C}$ is

$$\frac{1}{\overline{-3-4i}} = \frac{1}{-3+4i} = \frac{-3-4i}{(-3+4i)(-3-4i)} = -\tfrac{1}{25}(3+4i).$$

6. By Theorem 3, with $r = 2$ and $c = 0$, the inversion is given by

$$t(z) = \frac{4}{\bar{z}} \quad (z \in \mathbb{C} - \{O\}).$$

7. (a) Here we use part (a) of the definition of inversion. By Problem 6, the inversion can be written in the form

$$t(z) = \begin{cases} \frac{4}{\bar{z}}, & \text{if } z \in \mathbb{C} - \{O\}, \\ \infty, & \text{if } z = 0, \\ 0, & \text{if } z = \infty. \end{cases}$$

(b) In this case we use part (b) of the definition of inversion. Under reflection in the real axis, a point $z \in \mathbb{C}$ maps to $\bar{z}$. The inversion can therefore be written in the form

$$t(z) = \begin{cases} \bar{z}, & \text{if } z \in \mathbb{C}, \\ \infty, & \text{if } z = \infty. \end{cases}$$

8. Here

$$t_1(z) = \begin{cases} \frac{1}{\bar{z}}, & \text{if } z \in \mathbb{C} - \{O\}, \\ \infty, & \text{if } z = 0, \\ 0, & \text{if } z = \infty, \end{cases}$$

and

$$t_2(z) = \begin{cases} \frac{4}{\bar{z}}, & \text{if } z \in \mathbb{C} - \{O\}, \\ 0, & \text{if } z = \infty, \\ \infty, & \text{if } z = 0. \end{cases}$$

Hence

$$t(\infty) = t_2 \circ t_1(\infty) = t_2(0) = \infty,$$
$$t(0) = t_2 \circ t_1(0) = t_2(\infty) = 0.$$

For the remaining values of $z \in \mathbb{C} - \{O\}$, we have

$$t(z) = t_2 \circ t_1(z) = t_2\left(\frac{1}{\bar{z}}\right) = \frac{4}{\overline{(1/\bar{z})}} = 4z.$$

It follows that $t = t_2 \circ t_1$ is the function

$$t(z) = \begin{cases} 4z, & \text{if } z \in \mathbb{C}, \\ \infty, & \text{if } z = \infty. \end{cases}$$

This fixes the point at infinity, and scales elements of $\mathbb{C}$ by the factor 4.

If the circle of radius 2 were replaced by a circle of radius $\sqrt{k}$, then t would fix the point at infinity and scale elements of $\mathbb{C}$ by the factor k.

9. In addition to mapping ∞ to ∞, the transformation t scales the complex plane by the factor $|-9| = 9$, rotates it through the angle $\text{Arg}(-9) = \pi$, and then translates it through the vector $(6, -10)$.

By the last part of Problem 8, the *scaling* by the factor 9 can be decomposed into the composite $t_2 \circ t_1$ where

t_1 is the inversion in the unit circle $\mathscr{C}$,
t_2 is the inversion in the circle of radius 3 centred at the origin.

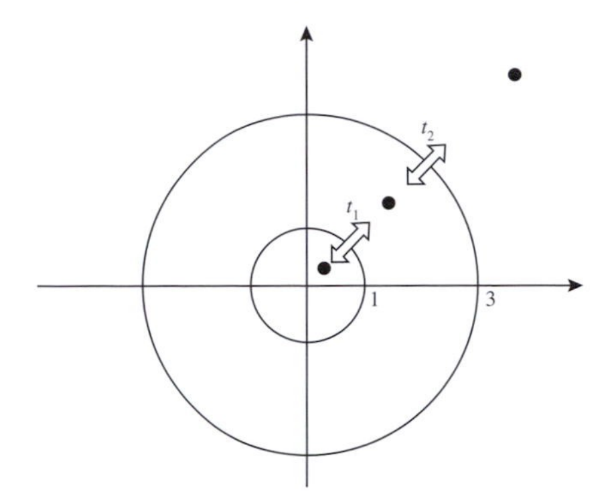

The *rotation* through the angle π can be decomposed into the composite $t_4 \circ t_3$ where

t_3 is the inversion in the extended x-axis,
t_4 is the inversion in the extended y-axis.

(The y-axis makes an angle $\pi/2$ with the x-axis.)

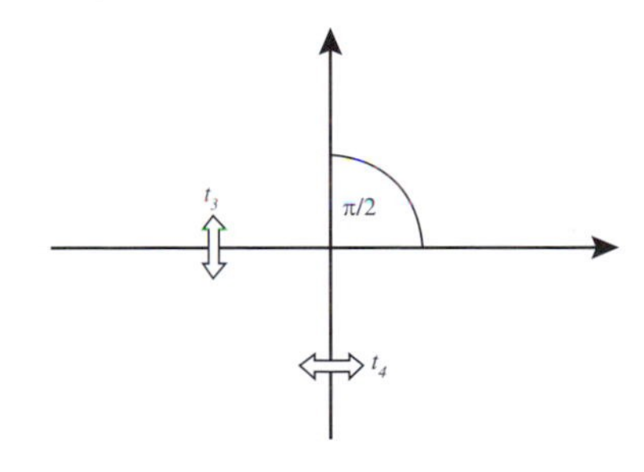

The *translation* through the vector $(6, -10)$ can be decomposed into the composite $t_6 \circ t_5$ where

t_5 is the inversion in the extended line $\ell_5 \cup \{\infty\}$, where ℓ_5 is the line $6x - 10y = 0$,
t_6 is the inversion in the extended line $\ell_6 \cup \{\infty\}$, where ℓ_6 is the line $6x - 10y = 68$.

(Note that ℓ_5 is the line through the origin that is perpendicular to the vector $(6, -10)$, and ℓ_6 is the parallel line that passes through $(3, -5)$.)

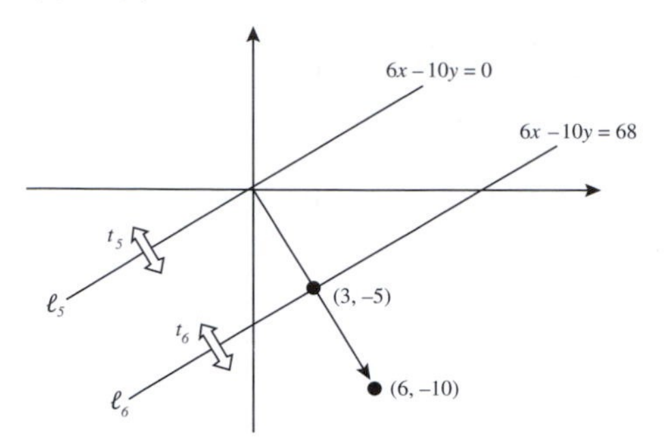

Since t_6, t_5, t_4, t_3 and $t_2 \circ t_1$ all map ∞ to itself, it follows that $t = t_6 \circ t_5 \circ t_4 \circ t_3 \circ t_2 \circ t_1$.

10. We use the following formulas from the proof of Theorem 6 of Subsection 5.2.4:

$$X = \frac{2x}{x^2 + y^2 + 1}, \quad Y = \frac{2y}{x^2 + y^2 + 1},$$

$$Z = \frac{x^2 + y^2 - 1}{x^2 + y^2 + 1}.$$

(a) Substituting for *X, Y* and *Z* from the above formulas, we get that the image under stereographic projection onto $\hat{\mathbb{C}}$ of the circle on $\mathbb{S}$ with equation $X = \frac{1}{2}$ is the set in the plane for which

$$\frac{2x}{x^2 + y^2 + 1} = \frac{1}{2},$$

that is,

$$4x = x^2 + y^2 + 1$$

or

$$(x - 2)^2 + y^2 = 3.$$

Thus, the image under the stereographic projection of the circle on $\mathbb{S}$ is the circle in the plane with centre (2,0) and radius $\sqrt{3}$.

(b) Substituting for *X, Y* and *Z* from the above formulas, we get that the image under stereographic projection onto $\hat{\mathbb{C}}$ of the circle on $\mathbb{S}$ with equation $3X + 2Y + Z = 1$ is the set in the extended plane for which

$$\frac{6x}{x^2 + y^2 + 1} + \frac{4y}{x^2 + y^2 + 1} + \frac{x^2 + y^2 - 1}{x^2 + y^2 + 1} = 1,$$

that is,

$$6x + 4y + x^2 + y^2 - 1 = x^2 + y^2 + 1$$

or

$$3x + 2y - 1 = 0.$$

Thus, the image under the stereographic projection of the circle on $\mathbb{S}$ is the extended line in the extended plane through the points $\left(0, \frac{1}{2}\right)$ and $\left(\frac{1}{3}, 0\right)$. (It is the extended line in $\hat{\mathbb{C}}$ since N maps onto ∞.)

Section 5.3

1. In each case we have to determine whether the formula has the form

$$M(z) = \frac{az + b}{cz + d}, \tag{1}$$

where $a, b, c, d \in \mathbb{C}$ and $ad - bc \neq 0$.

(a) This is a Möbius transformation. It has the form (1), with $a = 0, b = 5, c = 1$ and $d = 0$. Also,

$$ad - bc = -5 \neq 0.$$

By the convention, $M(\infty) = a/c = 0$.

(b) This is a Möbius transformation. It has the form (1), with $a = -1, b = 2i, c = 3$ and $d = -4i$. Also,

$$ad - bc = 4i - 6i = -2i \neq 0.$$

By the convention, $M(\infty) = a/c = -\frac{1}{3}$.

(c) This is not a Möbius transformation. The transformation can be expressed in the form

$$M(z) = \frac{-3z^2 + i}{z + 0},$$

and so clearly cannot be expressed in the required form (1).

(d) This is a Möbius transformation. The transformation can be expressed in the form

$$M(z) = \frac{(z + 2i) + 5}{z + 2i} = \frac{z + (5 + 2i)}{z + 2i}.$$

This has the form (1), with $a = 1, b = 5 + 2i, c = 1$ and $d = 2i$. Also,

$$ad - bc = 2i - (5 + 2i) = -5 \neq 0.$$

By the convention, $M(\infty) = a/c = 1$.

2. (a) All matrices associated with M_1 are non-zero multiples of

$$\mathbf{A} = \begin{pmatrix} 0 & 2i \\ i & 2 \end{pmatrix}.$$

Comparing the zero entries of $\mathbf{A}$ with those of $\mathbf{A}_1$, $\mathbf{A}_2$ and $\mathbf{A}_3$, it is clear that $\mathbf{A}_2$ is the only matrix of the three that could be a multiple of $\mathbf{A}$.

In fact, $\mathbf{A}_2 = -i\mathbf{A}$, so the matrix $\mathbf{A}_2$ is associated with M_1.

(b) All matrices associated with M_2 are non-zero multiples of

$$\mathbf{A} = \begin{pmatrix} 0 & 2i \\ 1 & 0 \end{pmatrix}.$$

Comparing the zero entries of $\mathbf{A}$ with those of $\mathbf{A}_1, \mathbf{A}_2$ and $\mathbf{A}_3$, it is clear that $\mathbf{A}_1$ is the only matrix of the three that could be a multiple of $\mathbf{A}$.

In fact, $\mathbf{A}_1 = \frac{1}{2i}\mathbf{A}$, so the matrix $\mathbf{A}_1$ is associated with M_2.

(c) All matrices associated with M_3 are non-zero multiples of

$$\mathbf{A} = \begin{pmatrix} i & 2 \\ 2 & -i \end{pmatrix}.$$

Since each of the matrices $\mathbf{A}_1, \mathbf{A}_2$ and $\mathbf{A}_3$ has at least one zero entry, whereas $\mathbf{A}$ has no zero entries, it is clear that none of $\mathbf{A}_1, \mathbf{A}_2$ and $\mathbf{A}_3$ is a multiple of $\mathbf{A}$.

It follows that none of the matrices $\mathbf{A}_1, \mathbf{A}_2$ and $\mathbf{A}_3$ is associated with M_3.

3. Matrices associated with the M_1 and M_2 in Example 2 are

$$\mathbf{A}_1 = \begin{pmatrix} i & 1 \\ 2 & -2 \end{pmatrix} \quad \text{and} \quad \mathbf{A}_2 = \begin{pmatrix} 1 & i \\ 2 & -1 \end{pmatrix},$$

respectively. So

$$\mathbf{A}_1\mathbf{A}_2 = \begin{pmatrix} i & 1 \\ 2 & -2 \end{pmatrix}\begin{pmatrix} 1 & i \\ 2 & -1 \end{pmatrix} = \begin{pmatrix} 2+i & -2 \\ -2 & 2+2i \end{pmatrix}.$$

This is a matrix associated with $M_1 \circ M_2$.

4. As in Example 3, matrices associated with the Möbius transformations M_1 and M_2 are

$$\mathbf{A}_1 = \begin{pmatrix} 3 & 1 \\ i & -2 \end{pmatrix} \quad \text{and} \quad \mathbf{A}_2 = \begin{pmatrix} 2i & 3 \\ 1 & -2 \end{pmatrix}.$$

(a) It follows that a matrix associated with $M_2 \circ M_1$ is

$$\mathbf{A}_2\mathbf{A}_1 = \begin{pmatrix} 2i & 3 \\ 1 & -2 \end{pmatrix}\begin{pmatrix} 3 & 1 \\ i & -2 \end{pmatrix} = \begin{pmatrix} 9i & -6+2i \\ 3-2i & 5 \end{pmatrix},$$

so that

$$M_2 \circ M_1(z) = \frac{9iz - 6 + 2i}{(3-2i)z + 5}.$$

(b) A matrix associated with $M_1 \circ M_1$ is

$$\mathbf{A}_1\mathbf{A}_1 = \begin{pmatrix} 3 & 1 \\ i & -2 \end{pmatrix}\begin{pmatrix} 3 & 1 \\ i & -2 \end{pmatrix} = \begin{pmatrix} 9+i & 1 \\ i & 4+i \end{pmatrix},$$

so that

$$M_1 \circ M_1(z) = \frac{(9+i)z + 1}{iz + (4+i)}.$$

5. Matrices associated with the Möbius transformations M_1 and M_2 are

$$\mathbf{A}_1 = \begin{pmatrix} 1 & -i \\ i & 2 \end{pmatrix} \quad \text{and} \quad \mathbf{A}_2 = \begin{pmatrix} 2 & i \\ -i & 1 \end{pmatrix}.$$

A matrix associated with $M_1 \circ M_2$ is therefore

$$\mathbf{A}_1\mathbf{A}_2 = \begin{pmatrix} 1 & -i \\ i & 2 \end{pmatrix}\begin{pmatrix} 2 & i \\ -i & 1 \end{pmatrix} = \begin{pmatrix} 1 & 0 \\ 0 & 1 \end{pmatrix},$$

so that

$$M_1 \circ M_2(z) = z.$$

6. Here we use the formula for the inverse of a Möbius transformation given in Theorem 7.

(a) $M_1^{-1}(z) = \dfrac{-3iz - 2}{-z - 3i} = \dfrac{3iz + 2}{z + 3i}$

(b) $M_2^{-1}(z) = \dfrac{4i}{-z} = -\dfrac{4i}{z}$

(c) $M_3^{-1}(z) = \dfrac{-4iz}{-z + 4i} = \dfrac{4iz}{z - 4i}.$

7. (a) To keep the calculations simple, we pick the points $\infty, 2i$ and 2 on the extended line, as shown below. Now

$$t(\infty) = 1, \quad t(2i) = 0$$

and

$$t(2) = \tfrac{2-2i}{4} = \tfrac{1}{2} - \tfrac{1}{2}i.$$

So the image of the extended line is the generalized circle that passes through the points 1, 0 and $\frac{1}{2} - \frac{1}{2}i$. This is the circle of radius $\frac{1}{2}$ with centre $\frac{1}{2}$.

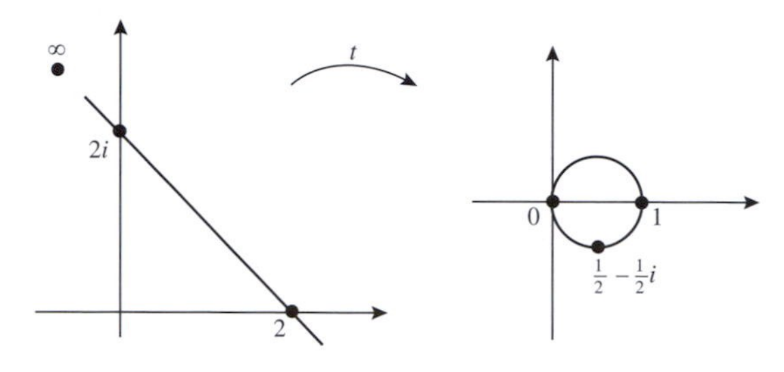

(b) In this case, we pick the points $-2, 0$ and $-1 + i$ on the circle, as shown below. Now

$$t(-2) = \infty, \quad t(0) = -i$$

and

$$t(-1+i) = \frac{-1-i}{1+i} = -1.$$

So the image of the circle is the generalized circle that passes through the points $\infty, -i$ and -1. This is the extended line with slope -1 through the point -1.

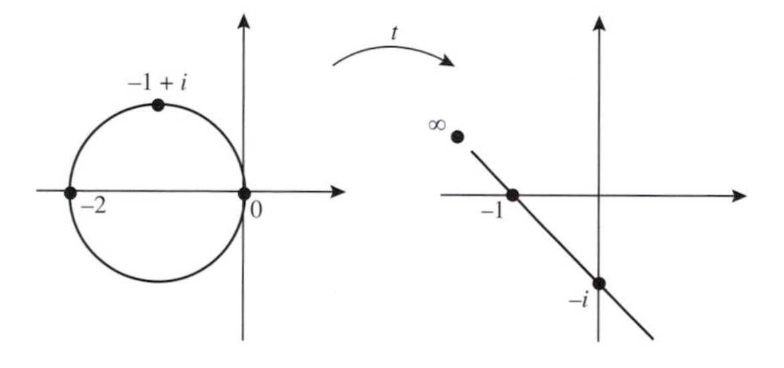

Section 5.4

1. We use the preceding strategy.

(a) The required Möbius transformation is of the form

$$M(z) = K\frac{z-(-1)}{z-0} = K\frac{z+1}{z},$$

for some complex number K.

Since $M(-3) = 1$, we must have

$$1 = K\tfrac{-3+1}{-3},$$

so that $K = \frac{3}{2}$. It follows that the required Möbius transformation is given by

$$M(z) = \frac{3(z+1)}{2z}.$$

(b) The required Möbius transformation is of the form

$$M(z) = K\frac{z-\frac{3}{2}}{z-1},$$

for some complex number K.

Since $M(2) = 1$, we must have

$$1 = K\frac{2-\frac{3}{2}}{2-1} = \tfrac{1}{2}K,$$

so that $K = 2$. It follows that the required Möbius transformation is given by

$$M(z) = 2\frac{z-\frac{3}{2}}{z-1} = \frac{2z-3}{z-1}.$$

(c) The required Möbius transformation is of the form

$$M(z) = \frac{K}{z-2},$$

for some complex number K.

Since $M(-3) = 1$, we must have

$$1 = \frac{K}{-3-2},$$

so that $K = -5$. It follows that the required Möbius transformation is given by

$$M(z) = \frac{-5}{z-2} = \frac{5}{2-z}.$$

(d) The required Möbius transformation is of the form

$$M(z) = K\left(z - \tfrac{3}{2}\right),$$

for some complex number K.

Since $M(2) = 1$, we must have

$$1 = K\left(2 - \tfrac{3}{2}\right) = \tfrac{1}{2}K,$$

so that $K = 2$. It follows that the required Möbius transformation is given by

$$M(z) = 2\left(z - \tfrac{3}{2}\right) = 2z - 3.$$

2. We use the strategy preceding Example 2.

(a) The Möbius transformation M_1 that maps $-1, i, 1$ to $0, 1, \infty$, respectively, is of the form

$$M_1(z) = K\frac{z - (-1)}{z - 1} = K\frac{z+1}{z-1},$$

for some complex number K. Since $M_1(i) = 1$, we must have

$$1 = K\frac{i+1}{i-1},$$

so that

$$K = \frac{i-1}{i+1} = \frac{(i-1)(-i+1)}{(i+1)(-i+1)} = \frac{2i}{2} = i.$$

It follows that

$$M_1(z) = i\frac{z+1}{z-1}.$$

The solution to Problem 1, part (a), shows that the Möbius transformation that maps $-1, -3, 0$ to $0, 1, \infty$, respectively, is given by

$$M_2(z) = \frac{3(z+1)}{2z}.$$

Now by Theorem 7 of Subsection 5.3.6,

$$M_2^{-1}(z) = \frac{3}{2z - 3}.$$

Now matrices associated with M_1 and M_2^{-1} are

$$\begin{pmatrix} i & i \\ 1 & -1 \end{pmatrix} \text{ and } \begin{pmatrix} 0 & 3 \\ 2 & -3 \end{pmatrix},$$

so that a matrix associated with $M_2^{-1} \text{o} M_1$ is

$$\begin{pmatrix} 0 & 3 \\ 2 & -3 \end{pmatrix}\begin{pmatrix} i & i \\ 1 & -1 \end{pmatrix} = \begin{pmatrix} 3 & -3 \\ -3+2i & 3+2i \end{pmatrix}.$$

Hence the required Möbius transformation is given by

$$M(z) = M_2^{-1} \text{ o } M_1(z) = \frac{3z - 3}{(-3+2i)z + (3+2i)}.$$

(b) The Möbius transformation M_1 that maps $3, \infty, -2$ to $0, 1, \infty$, respectively, is of the form

$$M_1(z) = \frac{z-3}{z+2}.$$

The Möbius transformation M_2 that maps $3, \frac{7}{3}, 1$ to $0, 1, \infty$, respectively, is of the form

$$M_2(z) = K\frac{z-3}{z-1},$$

for some complex number K. Since $M_2\left(\frac{7}{3}\right) = 1$, we must have

$$1 = K\frac{\frac{7}{3} - 3}{\frac{7}{3} - 1} = -\frac{1}{2}K,$$

so that $K = -2$. It follows that

$$M_2(z) = -2\frac{z-3}{z-1} = \frac{-2z+6}{z-1}.$$

Using the formula for inverses we have

$$M_2^{-1}(z) = \frac{-z-6}{-z-2} = \frac{z+6}{z+2}.$$

Now matrices associated with M_1 and M_2^{-1} are

$$\begin{pmatrix} 1 & -3 \\ 1 & 2 \end{pmatrix} \text{ and } \begin{pmatrix} 1 & 6 \\ 1 & 2 \end{pmatrix},$$

so that a matrix associated with $M_2^{-1} \text{ o } M_1$ is

$$\begin{pmatrix} 1 & 6 \\ 1 & 2 \end{pmatrix}\begin{pmatrix} 1 & -3 \\ 1 & 2 \end{pmatrix} = \begin{pmatrix} 7 & 9 \\ 3 & 1 \end{pmatrix}.$$

Hence the required Möbius transformation is given by

$$M(z) = M_2^{-1} \text{ o } M_1(z) = \frac{7z+9}{3z+1}.$$

3. We use the preceding strategy.

(a) First we determine the Möbius transformation M that maps $0, -4, -2i$ to $0, 1, \infty$, respectively. Following the usual strategy, we observe that this transformation must be of the form

$$M(z) = K\frac{z-0}{z+2i},$$

for some complex number K. Since $M(-4) = 1$, we must have

$$1 = K\frac{-4}{-4+2i},$$

so that $K = 1 - \frac{1}{2}i$. Thus the transformation M is given by

$$M(z) = \frac{\left(1 - \frac{1}{2}i\right)z}{z+2i}.$$

It follows that

$$M(-1-3i) = \frac{\left(1 - \frac{1}{2}i\right)(-1-3i)}{-1-i}$$
$$= \frac{-\frac{5}{2} - \frac{5}{2}i}{-1-i} = \frac{5}{2}.$$

Since this is a real number, it follows that the four points $0, -4, -2i, -1-3i$ lie on a generalized circle. In fact, they lie on a circle because $M(\infty) = 1 - \frac{1}{2}i$ is not real.

(b) First we determine the Möbius transformation M that maps $-1, -i, i$ to $0, 1, \infty$, respectively.

Following the usual strategy, we observe that this transformation must be of the form

$$M(z) = K\frac{z+1}{z-i},$$

for some complex number K. Since $M(-i) = 1$, we must have

$$1 = K\frac{-i+1}{-2i},$$

so that

$$K = \frac{2i}{i-1}.$$

Thus the transformation M is given by

$$M(z) = \frac{2i(z+1)}{(i-1)(z-i)}.$$

It follows that

$$M(2-i) = \frac{2i(3-i)}{(i-1)(2-2i)}$$
$$= \frac{2i(3-i)}{4i} = \frac{1}{2}(3-i).$$

Since this is not a real number, it follows that the four points $-1, -i, i, 2-i$ do not lie on a generalized circle, and so do not lie on a circle.

Section 5.5

1. Let P be a point (x, y) in the plane whose distance from the point $(0, -1)$ is k times its distance from the point $(0, 2)$. Then, if we use the Euclidean formula for distance between points in the plane, it follows that

$$x^2 + (y+1)^2 = k^2(x^2 + (y-2)^2).$$

For each value of k, this yields an equation for the corresponding Apollonian circle. The point $(1, 1)$ lies on the Apollonian circle for which

$$1^2 + (1+1)^2 = k^2(1^2 + (1-2)^2),$$

that is, $k = \sqrt{\frac{5}{2}}$.

The equation of the Apollonian circle through the point $(1, 1)$ is therefore

$$x^2 + (y+1)^2 = \tfrac{5}{2}\left(x^2 + (y-2)^2\right),$$

which simplifies to

$$-\tfrac{3}{2}x^2 - \tfrac{3}{2}y^2 + 12y - 9 = 0,$$

or

$$x^2 + y^2 - 8y + 6 = 0.$$

2. 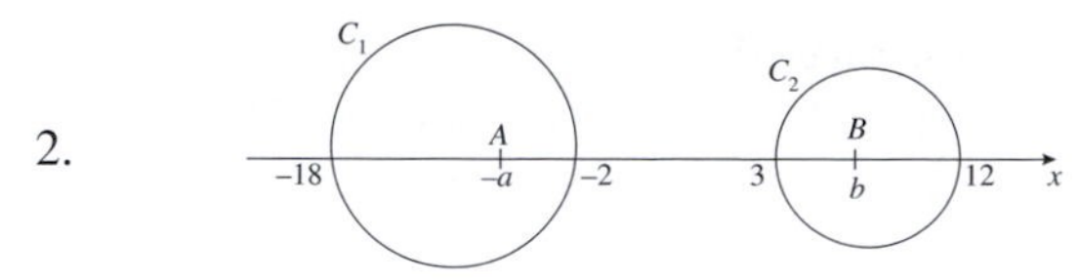

We use the definition of circles in an Apollonian family. Let the family have point circles $A(-a,0)$ and $B(b,0)$; we take these to lie in the open intervals $(-18,-2)$ and $(3,12)$, respectively.

Since the points $(-18,0)$ and $(-2,0)$ lie on the same circle in the family, the ratios of their distances from the points A and B are equal. It follows that

$$\frac{18-a}{b+18} = \frac{a-2}{b+2},$$

so that

$$(18-a)(b+2) = (a-2)(b+18);$$

after multiplying this out and simplifying, we get

$$2ab + 20a - 20b - 72 = 0. \qquad (1)$$

Next, since the points $(3,0)$ and $(12,0)$ lie on the same circle in the family, the ratios of their distances from the points A and B are equal. It follows that

$$\frac{3+a}{b-3} = \frac{12+a}{12-b},$$

so that

$$(3+a)(12-b) = (12+a)(b-3);$$

after multiplying out and simplifying, we get

$$2ab - 15a + 15b - 72 = 0. \qquad (2)$$

We now solve equations (1) and (2) for a and b. If we subtract equation (2) from equation (1), we get

$$35a - 35b = 0;$$

in other words, $b = a$. Substituting for b into equation (1), we get that $2a^2 - 72 = 0$; so that $a = \pm 6$.

Since $-a \in (-18,-2)$, by assumption, it follows that $a = 6$. Finally, since $b = a$, we have that $b = 6$. It follows that the point circles in the Apollonian family are $(-6,0)$ and $(6,0)$.

Then, using the Euclidean formula for distance between points in the plane and the definition of an Apollonian circle, we deduce that the general form of equations of circles in this Apollonian family is

$$(x+6)^2 + y^2 = k^2\left((x-6)^2 + y^2\right).$$

The point $(6,9)$ lies on that circle for which

$$(6+6)^2 + 9^2 = k^2\left((6-6)^2 + 9^2\right),$$

from which we get that $k^2 = 25/9$. The equation of this particular Apollonian circle is therefore

$$(x+6)^2 + y^2 = \tfrac{25}{9}\left((x-6)^2 + y^2\right),$$

which after multiplying out and simplifying becomes

$$4x^2 + 4y^2 - 102x + 144 = 0.$$

Remark
Another method of determining this last equation is as follows. Let P be the point (6,9). Then $PA^2 = (6+6)^2 + 9^2 = 225$, so that $PA = 15$, and $PB^2 = (6-6)^2 + 9^2 = 81$, so that $PB = 9$.

$$BP^2 = (6-6)^2 + 9^2 = 81,$$

so that $BP = 9$.

Then the ratio $PA{:}PB$ equals 15:9, or $\frac{5}{3}$:1. Hence, if we use the Euclidean formula for distance between points in the plane and the definition of an Apollonian circle, we deduce that the equation of the circle in this Apollonian family through the point $(6,9)$ is

$$(x+6)^2 + y^2 = \tfrac{25}{9}\left((x-6)^2 + y^2\right).$$

3.

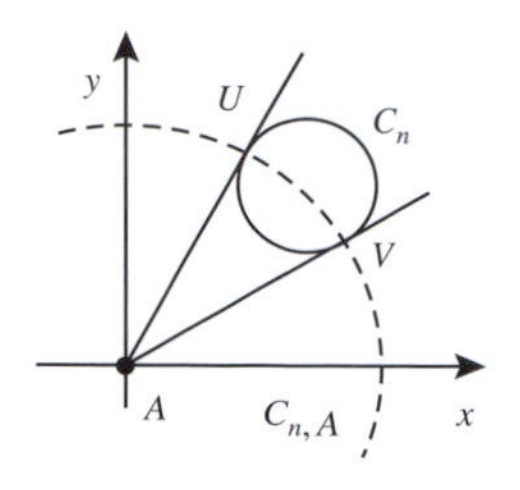

First, we verify that there exists a circle, $C_{n,A}$ say, with centre A that intersects C_n at right angles.

Let the two tangents from A to C_n touch C_n at the points U and V, say. Then the circle $C_{n,A}$ with centre A and radius AU intersects C_n at right angles since AU and AV are tangential to C_n and are radii of $C_{n,A}$.

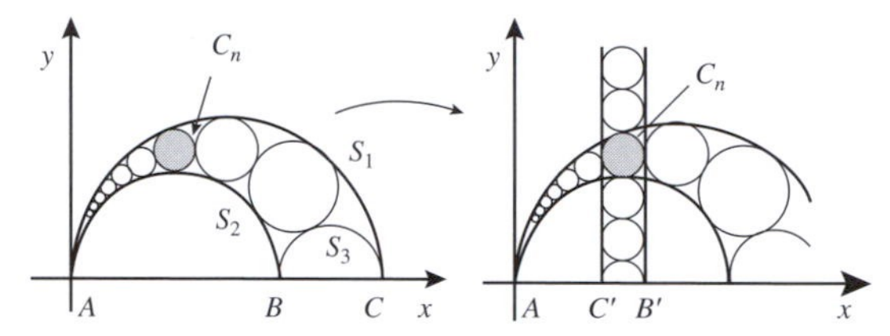

Now invert the figure in Theorem 10 in the circle $C_{n,A}$. This inversion maps C_n to itself, by Corollary 3 of Subsection 5.5.3, since $C_{n,A}$ and C_n intersect at right angles.

Since S_1 and S_2 are circles tangential to C_n, their images under inversion are generalized circles tangential to the image of C_n - which is just C_n itself. Further, since S_1 and S_2 pass through the centre A of the inversion, their images must pass through ∞. Hence the images of S_1 and S_2 must be extended lines tangential to C_n.

Now, the inversion maps the positive x-axis to itself; and, since S_1 and S_2 meet the positive x-axis at right angles, their images must also meet the positive x-axis at right angles. It follows that the images of S_1 and S_2 must be vertical half-lines that are tangential to C_n.

The other circles in the chain also invert to circles that are tangential to these two vertical half-lines, and each successive circle in the image also touches the previous and following circles in the chain.

Similarly, the image of S_3 is the semicircle in the upper half-plane with diameter $C'B'$.

Since all the circles in the image of the chain and the semicircle image of S_3 touch both vertical lines, they must all have the same diameter – namely, the diameter of C_n. It then follows that the height of the centre of C_n above the x-axis

$$\begin{aligned} &= \text{radius of image of } S_3 \\ &\quad + (n-1) \cdot \text{diameter of } C_n + \text{radius of } C_n \\ &= n \cdot \text{diameter of } C_n, \end{aligned}$$

as required.

4.

The points $F\left(\frac{1}{2}b, 0\right)$ and $F'\left(\frac{1}{2}c, 0\right)$ are the centres of the circles S_1 and S_2, respectively.

Let S be any circle in the chain constructed in Theorem 10, and U its centre. Let V be the point of contact of S_1 and S. Since S_1 and S have a common tangent at V and their radii FV and UV are perpendicular to this tangent, the points F, V and U are collinear.

Next, let W be the point of contact of S_2 and S. Since S_2 and S have a common tangent at W and their radii $F'W$ and UW are perpendicular to this tangent, the points F', U and W are collinear.

It follows that

$$\begin{aligned} FU + F'U &= (\text{radius of } S_1 + \text{radius of } S) \\ &\quad + (\text{radius of } S_2 - \text{radius of } S) \\ &= \text{radius of } S_1 + \text{radius of } S_2 \\ &= \tfrac{1}{2}(b+c). \end{aligned}$$

Hence (by Theorem 5 of Subsection 1.1.4) U lies on (the upper half of) an ellipse whose sum of focal distances is $\frac{1}{2}(b+c)$.

From the figure, it is clear that one end of the major axis of this ellipse must be the point $A(0, 0)$. Since (by Theorem 5 of Subsection 1.1.4) the length of the major axis of the ellipse is the sum of the focal distances, it follows that the other end of the major axis must be the point $\left(\frac{1}{2}(b+c), 0\right)$, as required.

Chapter 6

Section 6.1

1.

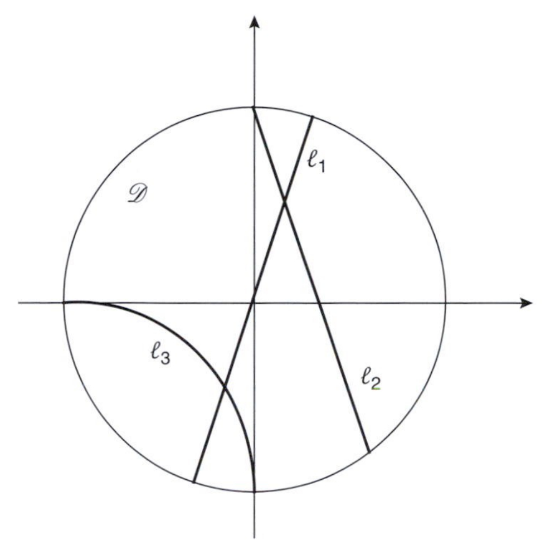

From the figure, ℓ_1 and ℓ_3 are d-lines; ℓ_2 is not. (ℓ_1 and ℓ_3 meet $\mathscr{C}$ at angles; ℓ_2 does not.)

2.

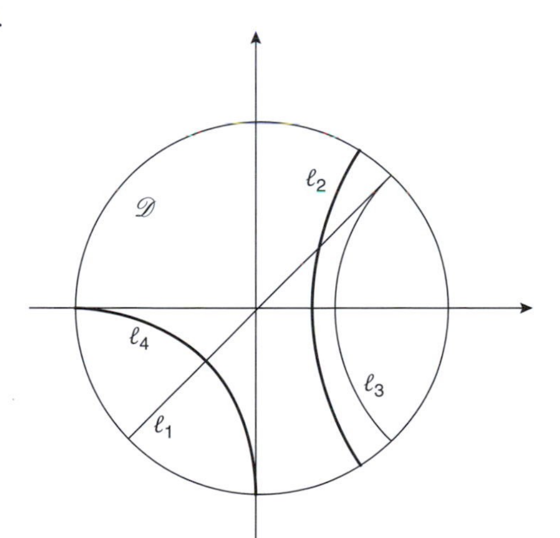

Intersections (in $\mathscr{D}$) : ℓ_1 and ℓ_2, ℓ_1 and ℓ_4.
Parallel: ℓ_1 and ℓ_3.
Ultra-parallel: ℓ_2 and ℓ_3, ℓ_2 and ℓ_4, ℓ_3 and ℓ_4.

3.

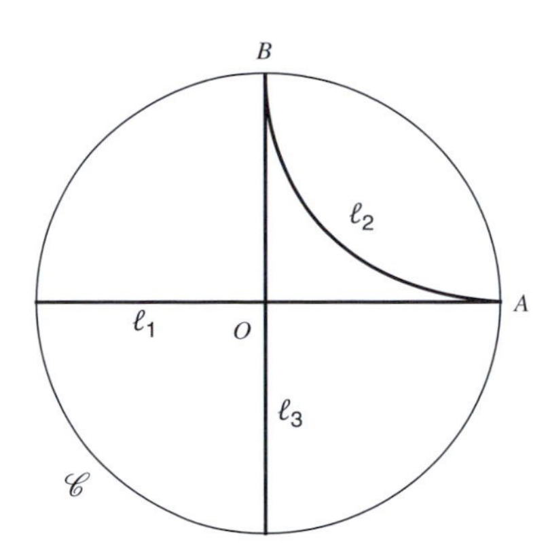

One solution is shown. The d-lines ℓ_1 and ℓ_2 are parallel because they (are parts of generalized circles that) have a common boundary point, A. The d-lines ℓ_2 and ℓ_3 are parallel because they (are parts of generalized circles that) have a common boundary point, B. But ℓ_1 and ℓ_3 meet, at O, and so are not parallel.

4. By the Origin Lemma, there are hyperbolic transformations r_1 and r_2, say, that map A_1 and A_2, respectively, to the origin O. Since $G_{\mathscr{D}}$ is a group, it follows that the composite mapping $r = r_2^{-1} \circ r_1$ is also a hyperbolic transformation; moreover, r maps A_1 to A_2, as required.

5. We use the Origin Lemma to obtain a hyperbolic reflection r that maps the point P to the origin, O, and the d-line ℓ to some d-line ℓ'. Suppose that the boundary points of ℓ are A and B, and the boundary points of ℓ' are A' and B', where $r(A) = A'$ and $r(B) = B'$. Then the diameters from A' and B' are the only two d-lines parallel to ℓ' through O. The reflection r^{-1} sends O to P, and A' and B' to A and B, the boundary points of ℓ, so it sends the diameters from A' and B' to d-lines through P that are parallel to ℓ.

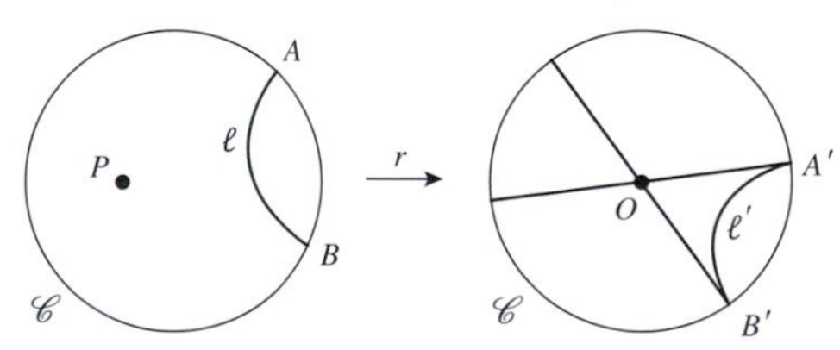

To see that the images under r^{-1} of these diameters are the only such d-lines, suppose that

ℓ'' is a d-line through P and A, and so parallel to ℓ. Then $r(\ell'')$ is a d-line parallel to ℓ' through O and so must be the *unique* diameter of $\mathscr{D}$ through A'. It follows that there is a unique d-line through P and A, as required. (Similarly, there is a unique d-line through P and B.)

Section 6.2

1. Reflection in the d-line obtained from α is given by

$$\rho(z) = \frac{\alpha\bar{z} - 1}{\bar{z} - \bar{\alpha}},$$

so $\rho(z) = 0$ if and only if $\bar{z} = 1/\alpha$, that is, if and only if $z = 1/\bar{\alpha}$.

Let a point $A \in \mathscr{D}$ be represented by the complex number β. Then it follows from the above result that the reflection in the d-line obtained from $z = 1/\bar{\beta}$ maps A to the origin. (As a check, we note that this reflection is given by

$$\rho_1(z) = \frac{(1/\bar{\beta})\bar{z} - 1}{\bar{z} - \overline{(1/\bar{\beta})}}$$

and that

$$\rho_1(\beta) = \frac{(1/\bar{\beta})\bar{\beta} - 1}{\bar{\beta} - \overline{(1/\bar{\beta})}} = 0.)$$

Thus we have proved the Origin Lemma.

2. Since $M(z) = \frac{az+b}{\bar{b}z+\bar{a}}$ is a Möbius transformation, we have $a\bar{a} - b\bar{b} = |a|^2 - |b|^2 \neq 0$. If $M(0) = 0$, then (since a and b cannot both be zero)

$$a \cdot 0 + b = 0,$$

which implies that $b = 0$.

3. Because the direct hyperbolic transformations we seek map 0 to 0, they must be (Euclidean) rotations about 0. So they are of the form

$$M(z) = Kz,$$

where $K = \cos\theta + i\sin\theta$, θ being the anticlockwise angle of rotation.

Now the equation $y = x/\sqrt{3}$ can be rewritten in the form $y = x\tan\frac{\pi}{6}$, and the equation $y = \sqrt{3}x$ can be rewritten as $y = x\tan\frac{\pi}{3}$. So the two rotations we seek must be through angles of $\pi/6$ and $-5\pi/6$. Thus $M(z) = Kz$, where

$$K = \cos\tfrac{\pi}{6} + i\sin\tfrac{\pi}{6} = \tfrac{1}{2}\left(\sqrt{3} + i\right)$$

or

$$K = \cos\left(-\tfrac{5\pi}{6}\right) + i\sin\left(-\tfrac{5\pi}{6}\right) = -\tfrac{1}{2}(\sqrt{3} + i).$$

4. It follows from Theorem 4 that the direct hyperbolic transformations with these properties are as follows.

   (a) $M(z) = K\frac{z-i/4}{1+iz/4} = K\frac{4z-i}{4+iz}$, where $|K| = 1$.

   (b) $M(z) = K\frac{z-(-1/3+2i/3)}{1-(-1/3-2i/3)z} = K\frac{3z+1-2i}{3+(1+2i)z}$, where $|K| = 1$.

5. The general direct hyperbolic transformation M which maps $\frac{3}{4}$ to 0 is

$$M(z) = K\frac{z - \frac{3}{4}}{1 - \frac{3}{4}z} \quad \text{(by Theorem 4)}$$

$$= \frac{Kz - \frac{3}{4}K}{-\frac{3}{4}z + 1}.$$

Hence, by equation (2) in Subsection 6.2.2, the general form of the inverse of M is

$$M^{-1}(z) = \frac{z + \frac{3}{4}K}{\frac{3}{4}z + K},$$

where $|K| = 1$.

6. From Theorem 4, the direct hyperbolic transformation M is necessarily of the form

$$M(z) = K\frac{z - m}{1 - \bar{m}z}, \tag{1}$$

where $|K| = 1$ and $m \in \mathscr{D}$.

Now, it follows from equation (1) that $M(m) = 0$; hence, since M maps $(-1, 1)$ one–one onto itself, the number m must be real and in $(-1, 1)$. Thus

$$M(z) = K\frac{z - m}{1 - mz}, \tag{2}$$

where $|K| = 1$ and $m \in (-1, 1)$.

Next, it follows from equation (2) that $M(0) = -Km$. Since $M(0)$ and m are real, it follows that K is real; so $K = \pm 1$. Thus M must be of the required form

$$M(z) = \pm \frac{z-m}{1-mz}, \quad \text{where } m \in (-1, 1).$$

7. (a) We follow the strategy in Subsection 6.2.2.

First, it follows from Theorem 4 that the general form of the direct hyperbolic transformation M_1 which maps $-\frac{1}{3}i$ to 0 is

$$M_1(z) = K\frac{z+\frac{1}{3}i}{1-\frac{1}{3}iz},$$

where $|K| = 1$; a matrix associated with M_1 is

$$\mathbf{A}_1 = \begin{pmatrix} K & \frac{1}{3}iK \\ -\frac{1}{3}i & 1 \end{pmatrix}.$$

Also, by Theorem 4, the direct hyperbolic transformation

$$M_2(z) = \frac{z-\frac{2}{3}}{1-\frac{2}{3}z}$$

maps $\frac{2}{3}$ to 0; a matrix associated with M_2 is

$$\mathbf{A}_2 = \begin{pmatrix} 1 & -\frac{2}{3} \\ -\frac{2}{3} & 1 \end{pmatrix}.$$

Now the inverse of $\mathbf{A}_2$ is

$$\mathbf{A}_2^{-1} = \frac{9}{5}\begin{pmatrix} 1 & \frac{2}{3} \\ \frac{2}{3} & 1 \end{pmatrix}.$$

Thus a matrix for the required hyperbolic transformation is

$$\begin{aligned}\mathbf{A}_2^{-1}\mathbf{A}_1 &= \frac{9}{5}\begin{pmatrix} 1 & \frac{2}{3} \\ \frac{2}{3} & 1 \end{pmatrix}\begin{pmatrix} K & \frac{1}{3}iK \\ -\frac{1}{3}i & 1 \end{pmatrix} \\ &= \frac{9}{5}\begin{pmatrix} K-\frac{2}{9}i & \frac{1}{3}iK+\frac{2}{3} \\ \frac{2}{3}K-\frac{1}{3}i & \frac{2}{9}iK+1 \end{pmatrix}.\end{aligned}$$

It follows that the general form of the required direct hyperbolic transformation is

$$M(z) = \frac{\left(K-\frac{2}{9}i\right)z+\left(\frac{1}{3}iK+\frac{2}{3}\right)}{\left(\frac{2}{3}K-\frac{1}{3}i\right)z+\left(\frac{2}{9}iK+1\right)}, \quad (1)$$

where $|K| = 1$.

(b) Now $M(i) = 1$ if

$$\begin{aligned}1 &= \frac{\left(K-\frac{2}{9}i\right)i+\left(\frac{1}{3}iK+\frac{2}{3}\right)}{\left(\frac{2}{3}K-\frac{1}{3}i\right)i+\left(\frac{2}{9}iK+1\right)} \\ &= \frac{\frac{4}{3}iK+\frac{8}{9}}{\frac{8}{9}iK+\frac{4}{3}}.\end{aligned}$$

This holds if we choose K such that $Ki = 1$; that is, if $K = 1/i = -i$. It follows from equation (1) that the transformation we require is given by

$$\begin{aligned}M(z) &= \frac{\left(-i-\frac{2}{9}i\right)z+\left(\frac{1}{3}+\frac{2}{3}\right)}{\left(-\frac{2}{3}i-\frac{1}{3}i\right)z+\left(\frac{2}{9}+1\right)} \\ &= \frac{9-11iz}{11-9iz}.\end{aligned}$$

8. We follow the strategy in Subsection 6.2.2.

First, it follows from Theorem 4 that the general form of the direct hyperbolic transformation M_1 which maps $\frac{1}{2}$ to 0 is

$$M_1(z) = K\frac{z-\frac{1}{2}}{1-\frac{1}{2}z},$$

where $|K| = 1$; a matrix associated with M_1 is

$$\mathbf{A}_1 = \begin{pmatrix} K & -\frac{1}{2}K \\ -\frac{1}{2} & 1 \end{pmatrix}.$$

Then, it follows from the above discussion that a particular direct hyperbolic transformation that maps $\frac{1}{2}$ to 0 is

$$M_2(z) = \frac{z-\frac{1}{2}}{1-\frac{1}{2}z},$$

and a matrix associated with M_2 is

$$\mathbf{A}_2 = \begin{pmatrix} 1 & -\frac{1}{2} \\ -\frac{1}{2} & 1 \end{pmatrix}.$$

Now the inverse of $\mathbf{A}_2$ is

$$\mathbf{A}_2^{-1} = \frac{4}{3}\begin{pmatrix} 1 & \frac{1}{2} \\ \frac{1}{2} & 1 \end{pmatrix}.$$

Thus a matrix for the required hyperbolic transformation is

$$\mathbf{A}_2^{-1}\mathbf{A}_1 = \tfrac{4}{3}\begin{pmatrix} 1 & \frac{1}{2} \\ \frac{1}{2} & 1 \end{pmatrix}\begin{pmatrix} K & -\frac{1}{2}K \\ -\frac{1}{2} & 1 \end{pmatrix}$$
$$= \tfrac{4}{3}\begin{pmatrix} K-\frac{1}{4} & -\frac{1}{2}K+\frac{1}{2} \\ \frac{1}{2}K-\frac{1}{2} & -\frac{1}{4}K+1 \end{pmatrix}.$$

It follows that the general form of the required direct hyperbolic transformation is

$$M(z) = \frac{\left(K-\frac{1}{4}\right)z + \left(-\frac{1}{2}K+\frac{1}{2}\right)}{\left(\frac{1}{2}K-\frac{1}{2}\right)z + \left(-\frac{1}{4}K+1\right)}$$
$$= \frac{(4K-1)z + 2(-K+1)}{2(K-1)z + (-K+4)},$$

where $|K| = 1$.

9. (a) A direct hyperbolic transformation mapping $\frac{1}{2}i$ to 0 is of the form

$$M(z) = K\frac{z-\frac{1}{2}i}{1+\frac{1}{2}iz} = K\frac{2z-i}{2+iz},$$

where $|K| = 1$.
Now $M(0) = -\frac{1}{2}Ki$, so

$$M(0) = \tfrac{1}{2} \Leftrightarrow K = i.$$

Hence

$$M(z) = i\frac{2z-i}{2+iz}.$$

(b) A direct hyperbolic transformation mapping $\frac{1}{2}$ to 0 is of the form

$$M(z) = K\frac{z-\frac{1}{2}}{1-\frac{1}{2}z} = K\frac{2z-1}{2-z},$$

where $|K| = 1$.
Now $M\left(\frac{2}{3}\right) = K\frac{\frac{4}{3}-1}{2-\frac{2}{3}} = \frac{1}{4}K$, so

$$M(\tfrac{2}{3}) = \tfrac{1}{2} \Leftrightarrow K = 2.$$

But K must have modulus 1, so no such direct hyperbolic transformation exists.

(c) A direct hyperbolic transformation mapping $\frac{1}{3}(1+i)$ to 0 is of the form

$$M(z) = K\frac{z-\frac{1}{3}(1+i)}{1-\frac{1}{3}(1-i)z} = K\frac{3z-(1+i)}{3-(1-i)z},$$

where $|K| = 1$.
Now

$$M(\tfrac{1}{3}(1-i)) = K\frac{1-i-(1+i)}{3-\frac{1}{3}(1-i)^2}$$
$$= K\frac{-6i}{9+2i},$$

so

$$M(\tfrac{1}{3}(1-i)) = \tfrac{1}{2} \Leftrightarrow K\frac{-6i}{9+2i} = \frac{1}{2}$$
$$\Leftrightarrow K = \frac{9+2i}{-12i}.$$

But $|K|^2 = \frac{81+4}{144} \neq 1$, so no such direct hyperbolic transformation exists.

Section 6.3

1. We use the formula for the hyperbolic distance $d(0,z)$. Thus

$$d\left(0, \tfrac{i}{3}\right) = \tanh^{-1}\left(\tfrac{1}{3}\right)$$
$$\simeq 0.3466$$

and

$$d(0.8, 0.9) = d(0, 0.9) - d(0, 0.8)$$
$$= \tanh^{-1} 0.9 - \tanh^{-1} 0.8$$
$$\simeq 1.4722 - 1.0986$$
$$= 0.3736.$$

2. We use the formula for the hyperbolic distance $d(0,z)$. Thus, if $d(0,a) = 1.6$ and $a > 0$, then

$$a = \tanh 1.6 \simeq 0.9217;$$

and, if $d(0,a) = 3.2$ and $a > 0$, then

$$a = \tanh 3.2 \simeq 0.9967.$$

3. Let m be the hyperbolic midpoint of $[0.5, 0.8]$. Now

$$d(0, 0.5) = \tanh^{-1}\ 0.5 \simeq 0.5493$$

and

$$d(0, 0.8) = \tanh^{-1}\ 0.8 \simeq 1.0986.$$

Since m lies half-way between 0.5 and 0.8, measured in terms of hyperbolic distances, and 0.5, m and 0.8 all lie on the same side of the origin, it follows that

$$\begin{aligned} d(0, m) &= \tfrac{1}{2}(d(0, 0.5) + d(0, 0.8)) \\ &\simeq \tfrac{1}{2}(0.5493 + 1.0986) \\ &= 0.823\,95, \end{aligned}$$

so

$$\begin{aligned} m &\simeq \tanh\ 0.823\,95 \\ &\simeq 0.6772. \end{aligned}$$

Next, let m be the hyperbolic midpoint of $[-0.2, 0.8]$. Now,

$$d(0, -0.2) = \tanh^{-1}\ 0.2 \simeq 0.2027$$

and

$$d(0, 0.8) = \tanh^{-1}\ 0.8 \simeq 1.0986.$$

Clearly m lies between -0.2 and 0.8, on the same side of the origin as does 0.8. It follows that

$$\begin{aligned} d(0, m) &= \tfrac{1}{2}(d(0, 0.8) - d(0, -0.2)) \\ &\simeq \tfrac{1}{2}(1.0986 - 0.2027) \\ &= 0.447\,95, \end{aligned}$$

so

$$\begin{aligned} m &\simeq \tanh\ 0.447\,95 \\ &\simeq 0.4202. \end{aligned}$$

4.

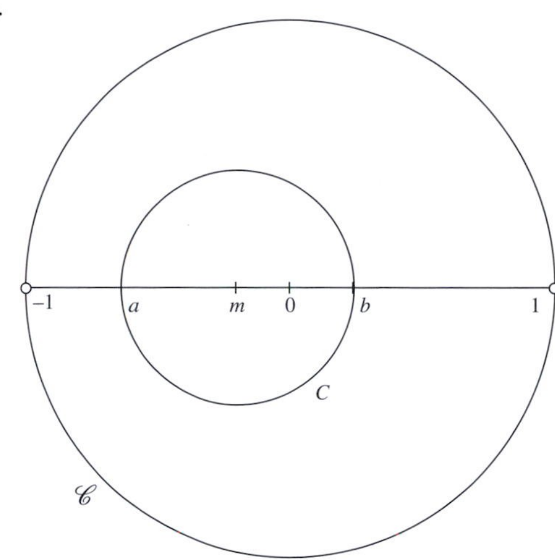

We use the first strategy in Subsection 6.3.3.

1. The line through 0 and the hyperbolic centre, $-\frac{1}{4}$, of C meets C at a and b, where a and b are real numbers with $a < b$. Now

$$d\left(0, -\frac{1}{4}\right) = \tanh^{-1}\left(\frac{1}{4}\right) \simeq 0.2554.$$

Since a lies to the left of the origin and

$$d(0, a) \simeq 0.2554 + 0.5 \simeq 0.7554,$$

it follows that

$$a \simeq -\tanh\ 0.7554 \simeq -0.6384.$$

Next, since the hyperbolic radius of C is greater than the hyperbolic distance of its centre from the origin, we have that $b > 0$. Thus

$$d(0, b) \simeq 0.5 - 0.2554 = 0.2446,$$

so that

$$b \simeq \tanh\ 0.2446 \simeq 0.2398.$$

2. The Euclidean centre of C is the Euclidean midpoint of $[a, b]$, namely the point

$$\tfrac{1}{2}(0.2398 - 0.6384) = -0.1993.$$

3. The Euclidean radius of C is

$$\tfrac{1}{2}|a-b| \simeq \tfrac{1}{2}|-0.6384-0.2398| = 0.4391.$$

5.

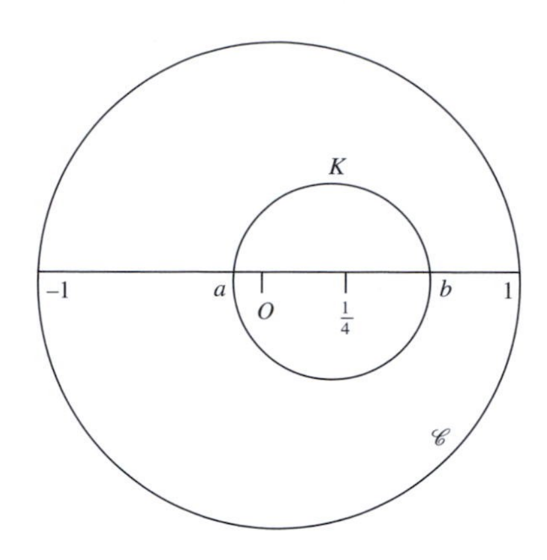

We use the second strategy in Subsection 6.3.3.

The Euclidean radius p is the point $\frac{1}{4}$, and the Euclidean radius is $\frac{1}{2}$, so Op meets K at the points $a = -\frac{1}{4}$ and $b = \frac{3}{4}$. Thus

$$d(0,a) = \tanh^{-1}\left(\left|-\frac{1}{4}\right|\right) = \tanh^{-1}(0.25) \simeq 0.255...$$

and

$$d(0,b) = \tanh^{-1}\left(\left|-\frac{3}{4}\right|\right) = \tanh^{-1}(0.75) \simeq 0.973....$$

Since a and b lie on opposite sides of O, the hyperbolic centre m of K is given by

$$d(0,m) = \frac{1}{2}(d(0,b) - d(0,a)) \simeq 0.359,$$

so that

hyperbolic centre $m \simeq \tanh 0.359 \simeq 0.344$;

hyperbolic radius $= \frac{1}{2}|(d(0,a)+d(0,b)| \simeq 0.614$.

6. (a) Using the Reflection Lemma with $p = 0.8$ and $q = 0.5$, we find that

$$\alpha = \frac{0.3 + 0.8 \cdot 0.5 \cdot 0.3}{0.64 - 0.25} \simeq 1.0769.$$

So the equation of the d-line ℓ which is the (hyperbolic) perpendicular bisector of $[0.5, 0.8]$ is

$$x^2 + y^2 - 2ax + 1 = 0, \qquad (1)$$

where $a \simeq 1.0769$.

(b) We use the fact that angles and hyperbolic lengths are unaltered under hyperbolic transformations; in particular, under reflections in diameters of the unit disc and under rotations of the unit disc about the origin.

The segment $[-0.8, -0.5]$ is obtained by reflecting the segment $[0.5, 0.8]$ in the diameter $(-i, i)$, and so we can obtain its perpendicular bisector by also reflecting the d-line ℓ in the diameter $(-i, i)$.

It then follows from equation (1) that the perpendicular bisector of $[-0.8, -0.5]$ has equation

$$x^2 + y^2 + 2ax + 1 = 0,$$

where $a \simeq 1.0769$.

Next, the segment $[0.5i, 0.8i]$ is obtained from the segment $[0.5, 0.8]$ by rotating the unit disc through an angle $\pi/2$ anticlockwise about the origin; and so we can obtain its perpendicular bisector by rotating the d-line ℓ through an angle of $\pi/2$ anticlockwise about the origin, that is, by applying the mapping $(x, y) \longmapsto (-y, x)$.

It then follows from equation (1) that the perpendicular bisector of $[0.5i, 0.8i]$ has equation

$$x^2 + y^2 - 2by + 1 = 0,$$

where $b \simeq 1.0769$.

7. (a)

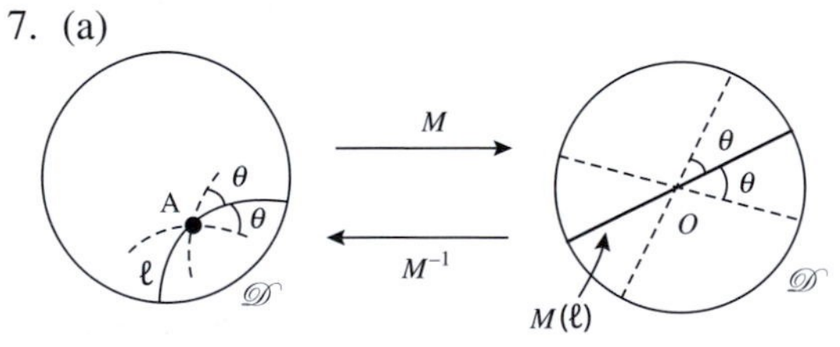

By the Origin Lemma (Subsection 6.1.2, Lemma 2), we may apply a preliminary hyperbolic transformation M to map A to O; this preserves d-lines and angles.

Then the image of the d-line ℓ under M is a diameter of $\mathscr{D}$, and it is obvious that there are exactly two diameters of $\mathscr{D}$ that make an angle θ with the diameter $M(\ell)$. By then applying the hyperbolic transformation M^{-1} (which preserves d-lines and angles), we deduce that there are exactly two d-lines through $A = M^{-1}(O)$ that make an angle θ with ℓ.

(b)

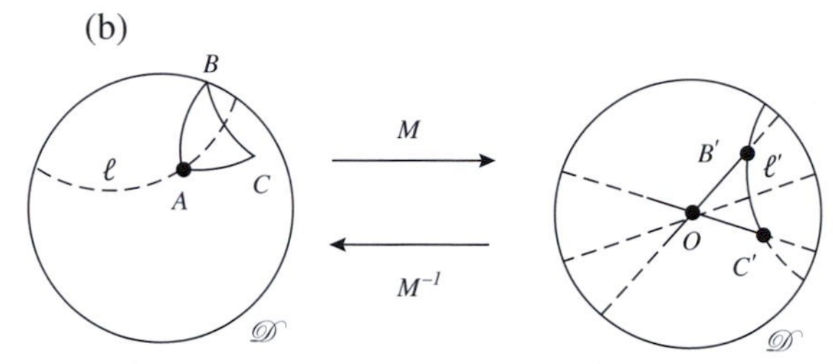

By the Origin Lemma (Subsection 6.1.2, Lemma 2), we may apply a preliminary hyperbolic transformation M to map A to O; this preserves d-lines and angles. Let $B' = M(B)$ and $C' = M(C)$.

The images under M of the d-lines that contain the d-line segments AB and AC are diameters of $\mathscr{D}$, and it is obvious that there is a unique diameter ℓ' of $\mathscr{D}$ that bisects the angle $\angle B'OC'$. By then applying the hyperbolic transformation M^{-1} (which preserves d-lines and angles), we deduce that there is a unique d-line $\ell = M^{-1}(\ell')$ that bisects the angle $\angle BAC$.

Clearly reflection in ℓ' of the diameters of $\mathscr{D}$ that contain OB' and OC' map these two diameters onto each other. By then applying the hyperbolic transformation M^{-1}, we deduce that reflection in ℓ maps the d-lines containing BA and CA onto each other.

Section 6.4

1. Let ΔABC be any d-triangle. Denote the angles of the d-triangle by α, β and γ, respectively, as shown, and let θ denote the external angle of the d-triangle at A. Thus, since the angle between two d-lines is equal to the angle between their tangents (see the definition of the angle between two curves in $\mathscr{D}$ given at the end of Subsection 6.1.1), we have $\alpha + \theta = \pi$.

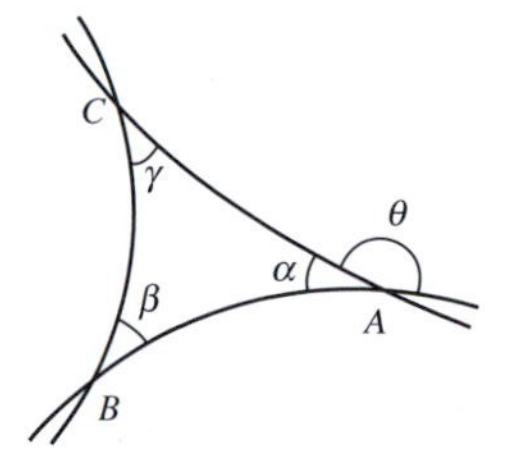

Now, by Theorem 1 in Subsection 6.4.1 for the d-triangle ΔABC, we must have

$$\alpha + \beta + \gamma < \pi,$$

so that

$$\alpha + \beta + \gamma < \alpha + \theta;$$

it follows that $\beta + \gamma < \theta$, as required.

2. From Problem 1,

$$\gamma > \beta' + \gamma'.$$

Hence

$$\alpha + \beta + \gamma > \alpha + \beta + \beta' + \gamma'.$$

3. Let D be a point on the d-line segment BC such that the d-line segment AD is part of the d-line that bisects $\angle BAC$. (This bisector exists, by Problem 7 of Subsection 6.3.4.)

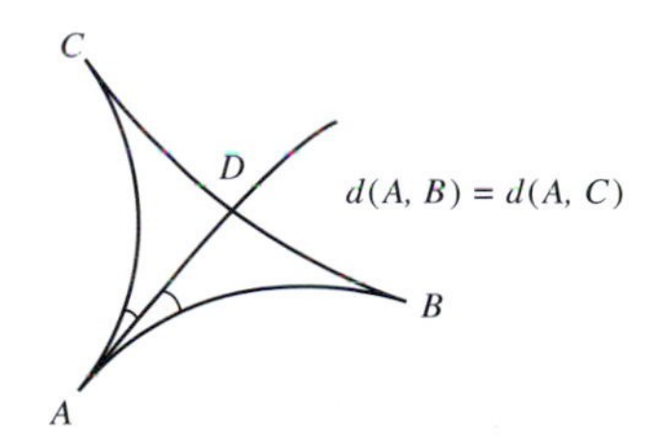

Reflection in this d-line exchanges the d-lines BA and CA, so it maps B to a point B' somewhere on CA. Since reflection preserves length, we have

$$d(A, B) = d(A, B').$$

But

$$d(A, B) = d(A, C),$$

so

$$d(A, B') = d(A, C),$$

and therefore B' and C coincide.

Hence the reflection maps the d-triangle ΔABC onto the d-triangle ΔACB and exchanges the angles at B and C. Since reflection preserves angles, it follows that these angles must be equal, as required.

4. Let $\triangle ABC$ and $\triangle PQR$ be any two doubly asymptotic triangles, with the vertices in $\mathscr{D}$ being at A and P, respectively.

   Suppose, first, that the angles at A and P are equal. Then there is hyperbolic transformation that maps $\triangle ABC$ onto a triangle Δ_1, say, with A going to the origin O; since hyperbolic transformations do not alter the size of angles, the angle of Δ_1 at O equals the original angle A. Since there is no hyperbolic transformation that maps points of $\mathscr{C}$ to points of $\mathscr{D}$, or vice-versa, Δ_1 is also a doubly asymptotic triangle. The triangle $\triangle ABC$ is therefore d-congruent to Δ_1.

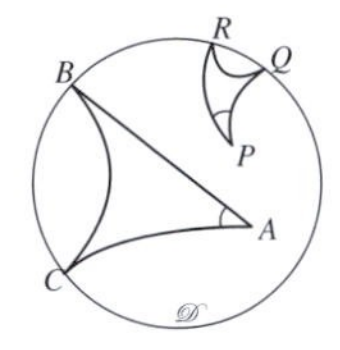

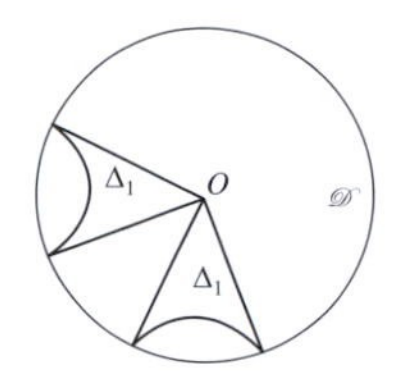

   Similarly, the triangle $\triangle PQR$ is d-congruent to a doubly asymptotic triangle Δ_2 with one vertex at O.

   Then there is a rotation of $\mathscr{D}$ round the origin (or a reflection in a diameter followed by a rotation) that maps the radial sides of Δ_1 and Δ_2 onto each other; and, since there is only one d-line that ends at two given points of $\mathscr{D}$, this mapping must map the two curved sides of Δ_1 and Δ_2 onto each other. Finally, since Δ_1 and Δ_2 are d-congruent, it follows that the triangles $\triangle ABC$ and $\triangle PQR$ must also be d-congruent.

   Suppose, next, that the two doubly asymptotic triangles $\triangle ABC$ and $\triangle PQR$ are d-congruent. By mapping A to O and P to O (as above), and rotating suitably, it follows from the d-congruence that the angles at A and P must be equal.

   Remark
   There is no hyperbolic transformation that maps points of $\mathscr{C}$ to points of $\mathscr{D}$, or vice-versa; hence a d-triangle can only be d-congruent to a doubly asymptotic triangle if it itself is a doubly asymptotic triangle.

5.

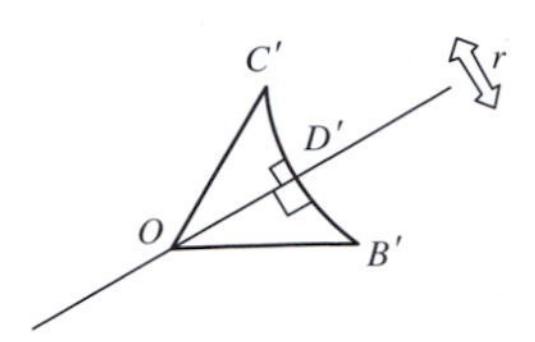

   The two d-triangles $\Delta OB'D'$ and $\Delta OC'D'$ have side OD', which lies along a diameter, in common. Let r be the hyperbolic reflection in that diameter.

   Since the angles $\angle OD'B'$ and $\angle OD'C'$ are equal (both are $\pi/2$), the image under the reflection r of the d-line segment $D'C'$ will lie along the d-line segment $D'B'$. Also, since the angles $\angle B'OD'$ and $\angle C'OD'$ are equal (OD' is the bisector of the angle $\angle B'OC'$), the image under the reflection r of side OC' will lie along side OB'. It follows that B' must reflect onto C'. Hence the d-triangles $\Delta OB'D'$ and $\Delta OC'D'$ are d-congruent.

6.

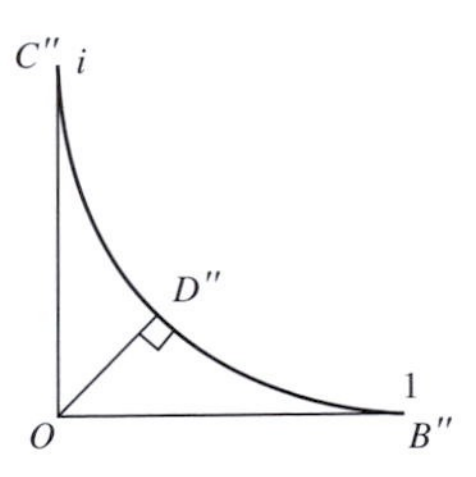

   It follows from the argument in the proof of Theorem 7, with $\theta = \pi/2$, that the required upper bound must be the length of the d-line segment OD''.

   Let m be the d-line that ends at $B'' = 1$ and $C'' = i$. Now m is part of the Euclidean circle with centre C at the point $1+i$ and radius 1. The Euclidean distance OC is $\sqrt{2}$, so the Euclidean distance OD'' is $\sqrt{2} - 1$.

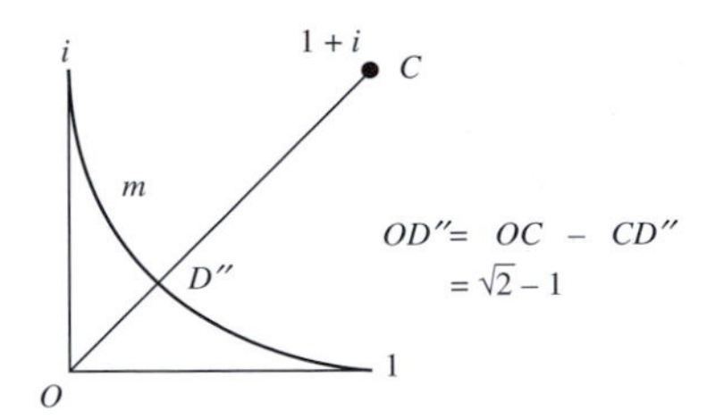

It follows that the hyperbolic length of the altitude OD'' is

$$\tanh^{-1}\left(\sqrt{2}-1\right) \simeq 0.44069.$$

So the hyperbolic length of altitude AD is less than 0.4407.

7.

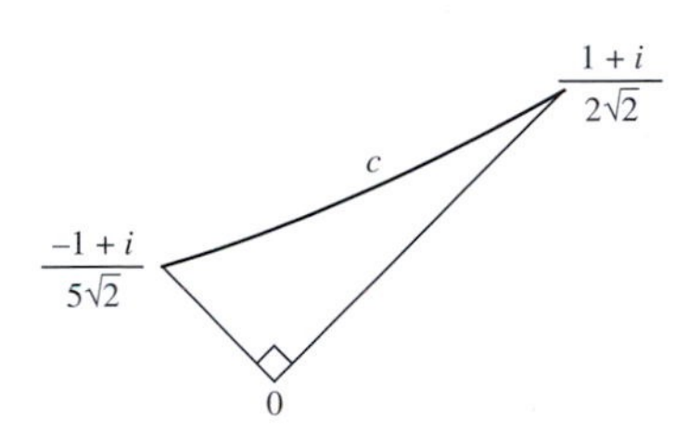

It follows from the Distance Formula for hyperbolic lengths that the hyperbolic lengths of the two sides of the triangle which meet at right angles at the origin are

$$\begin{aligned} d\left(0,(1+i)/2\sqrt{2}\right) &= d\left(0,|1+i|/2\sqrt{2}\right) \\ &= d\left(0,\tfrac{1}{2}\right) \\ &= \tanh^{-1} 0.5 \\ &\simeq 0.5493 \end{aligned}$$

and

$$\begin{aligned} d\left(0,(-1+i)/5\sqrt{2}\right) &= d\left(0,|-1+i|/5\sqrt{2}\right) \\ &= d(0,0.2) \\ &= \tanh^{-1} 0.2 \\ &\simeq 0.2027. \end{aligned}$$

Then, if c is the hyperbolic length of the third side (the 'hypotenuse'), it follows from Pythagoras' Theorem that

$$\begin{aligned} \cosh 2c &\simeq \cosh(2\times 0.5493) \\ &\quad \times \cosh(2\times 0.2027) \\ &= \cosh(1.0986)\times\cosh(0.4054) \\ &\simeq 1.6667\times 1.0833 \\ &\simeq 1.8055. \end{aligned}$$

Hence the required hyperbolic length is

$$\begin{aligned} c &\simeq \tfrac{1}{2}\cosh^{-1}(1.8055) \\ &\simeq 0.5983. \end{aligned}$$

8. (a)

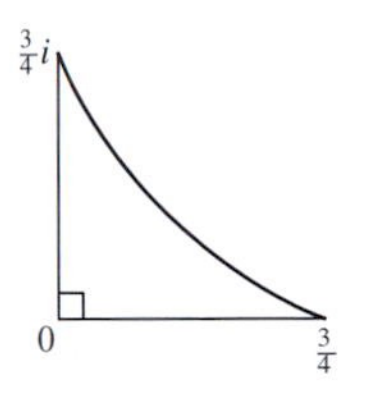

It follows from the Distance Formula for hyperbolic lengths that the hyperbolic lengths of the two sides of the triangle which meet at right angles at the origin are $d\left(0,\tfrac{3}{4}\right)$ and $d\left(0,\tfrac{3}{4}i\right)$. These two are equal to the common value

$$d\left(0,\tfrac{3}{4}\right) = \tanh^{-1}(0.75) \simeq 0.9730.$$

Then, if c is the hyperbolic length of the third side (the 'hypotenuse'), it follows from Pythagoras' Theorem that

$$\begin{aligned} \cosh 2c &\simeq \cosh^2(2\times 0.9730) \\ &= \cosh^2(1.9460) \\ &\simeq 12.755. \end{aligned}$$

Hence the required hyperbolic length is

$$\begin{aligned} c &\simeq \tfrac{1}{2}\cosh^{-1}(12.755) \\ &\simeq 1.6188. \end{aligned}$$

(b)

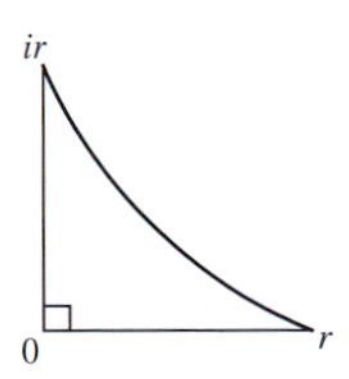

It follows from the Distance Formula that the hyperbolic lengths of the two sides of the triangle which meet at right angles at the origin are equal to the common value

$$d(0,r) = \tanh^{-1} r.$$

Then, if c is the hyperbolic length of the third side (the 'hypotenuse'), it follows from Pythagoras' Theorem that

$$\begin{aligned}\cosh 2c &= \cosh^2 2d\,(0,\ r)\\ &= \cosh^2\left(2\ \tanh^{-1} r\right).\end{aligned}$$

Hence the required hyperbolic length is

$$c = \tfrac{1}{2}\cosh^{-1}\left(\cosh^2\left(2\tanh^{-1} r\right)\right).$$

9.

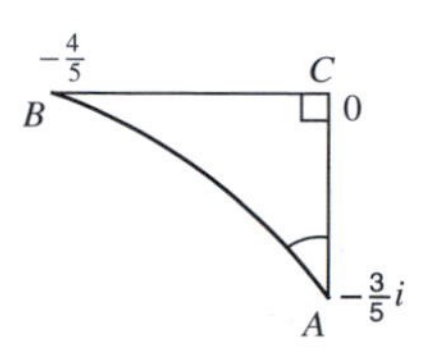

It follows from Lobachevskii's Formula that

$$\begin{aligned}\tan A &= \frac{\tanh\left(2d\left(0,-\frac{4}{5}\right)\right)}{\sinh\left(2d\left(0,-\frac{3}{5}i\right)\right)}\\ &= \frac{\tanh(2\ \tanh^{-1} 0.8)}{\sinh(2\ \tanh^{-1} 0.6)}\\ &\simeq \frac{\tanh(2.1972)}{\sinh(1.3863)}\\ &\simeq \frac{0.9756}{1.875}\\ &\simeq 0.5203,\end{aligned}$$

so that

$$\begin{aligned}A &\simeq \tan^{-1}(0.5203)\\ &\simeq 0.4798 \text{ radians.}\end{aligned}$$

10.

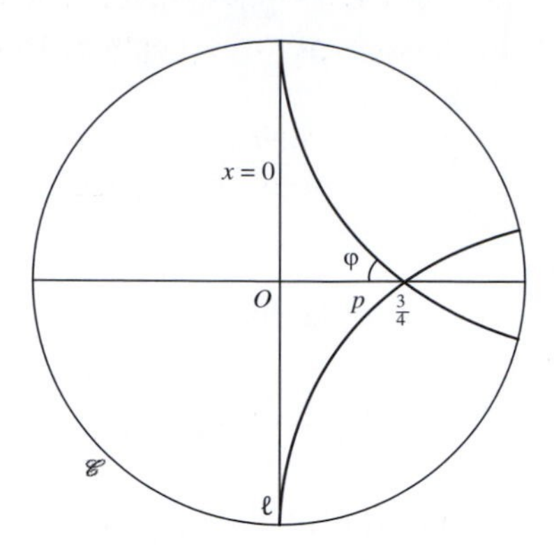

We use the Angle of Parallelism Formula.

The hyperbolic distance of the point $\frac{3}{4}$ from the d-line ℓ with equation $x = 0$ is

$$\begin{aligned}p &= d\left(0,\tfrac{3}{4}\right) = \tanh^{-1} 0.75\\ &\simeq 0.9730.\end{aligned}$$

It follows from the Formula that the required angle φ is such that

$$\begin{aligned}\tan\varphi &= \frac{1}{\sinh\left(2d\left(0,\frac{3}{4}\right)\right)}\\ &\simeq \frac{1}{\sinh(2\times 0.9730)}\\ &= 0.2917,\end{aligned}$$

so that

$$\varphi \simeq 0.2838 \text{ radians.}$$

11.

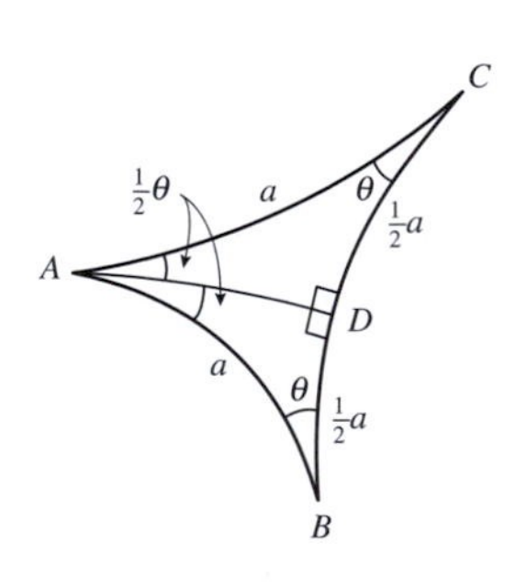

Using the hint, we apply the Sine Formula to the d-triangle ΔDBA, whose angles are

$$\angle ADB = \frac{\pi}{2},\quad \angle ABD = \theta,\quad \angle DAB = \tfrac{1}{2}\theta,$$

and in which side DB has hyperbolic length $\frac{1}{2}a$ and side AB has hyperbolic length a. Thus

$$\sin\left(\tfrac{1}{2}\theta\right) = \frac{\sinh a}{\sinh 2a} = \frac{1}{2\cosh a}.$$

Hence

$$a = \cosh^{-1}\left(\frac{1}{2\sin\frac{1}{2}\theta}\right),$$

which is a function of θ, as required.

Remark

This answer is very plausible when a, the length of the sides, is very small, for then $\sin\left(\frac{1}{2}\theta\right) \simeq \frac{1}{2}$ and so $\theta \simeq \pi/3$, which agrees closely with the value of the angle in a Euclidean equilateral triangle.

It is also very plausible when a is very large, for then $\sin\left(\frac{1}{2}\theta\right)$ is very small and so $\theta \simeq 0$, which is consistent with the fact that a large d-triangle has small angles.

Section 6.5

1.

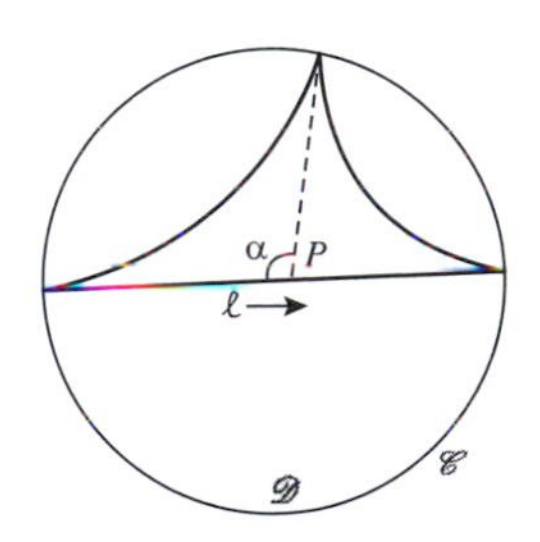

Let the triangle be a trebly asymptotic d-triangle, and let ℓ be one of its sides. Then as a point P slides along the d-line ℓ, the angle α varies from π to zero. In particular, there must be a point on ℓ for which $\alpha = \pi/2$, as required.

2. (a)

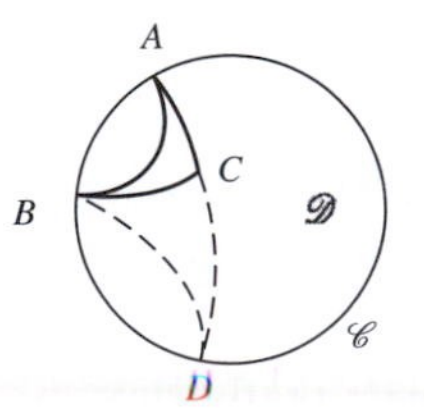

Let ΔABC be a doubly asymptotic d-triangle in $\mathscr{D}$. First, we construct a trebly asymptotic d-triangle ΔABD as follows: extend the segment AC beyond C to meet $\mathscr{C}$ at D. Then draw the trebly asymptotic d-triangle ΔABD.

Now, the d-line containing the points A, C, D cuts off a minor arc of $\mathscr{C}$ that contains B. The d-line joining B and D lies on the opposite side of the segment BC from A.

Hence ΔABC is contained in the trebly asymptotic d-triangle ΔABD.

(b)

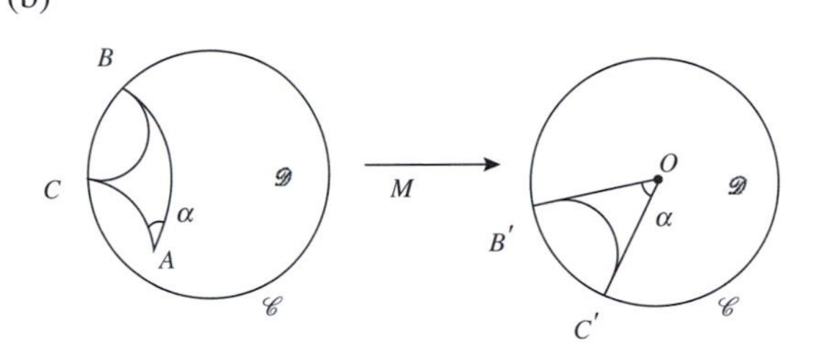

Let ΔABC be a doubly asymptotic d-triangle in $\mathscr{D}$, with A being the vertex of the d-triangle that does not lie on $\mathscr{C}$; and let $\alpha = \angle BAC$.

Then there is a Möbius transformation M that maps $\mathscr{D}$ to itself, with $\mathscr{C}$ mapping to $\mathscr{C}$, that maps A to the origin O. Let M map B, C onto B', C', respectively. Then M maps the doubly asymptotic d-triangle ΔABC onto another doubly asymptotic d-triangle, $\Delta OB'C'$; and, since Möbius transformations preserve angles, we have that $\angle B'OC' = \alpha$.

Any other doubly asymptotic d-triangle whose angle at its vertex in $\mathscr{D}$ is α can also be mapped by a suitable Möbius transformation onto another doubly asymptotic d-triangle that has its angle at the origin also of magnitude α; and then onto the previous doubly asymptotic d-triangle $\Delta OB'C'$ by a further rotation – which is again a Möbius transformation.

So every doubly asymptotic d-triangle whose angle at its vertex in $\mathscr{D}$ is α is d-congruent to the d-triangle $\Delta OB'C'$, and hence has the same area as $\Delta OB'C'$.

Hence the area of a doubly asymptotic d-triangle depends only on the angle at its vertex in $\mathscr{D}$.

Chapter 7

Section 7.1

1. (a) Such a rotation of S^2 exists, by the result of Example 2.

   (b)

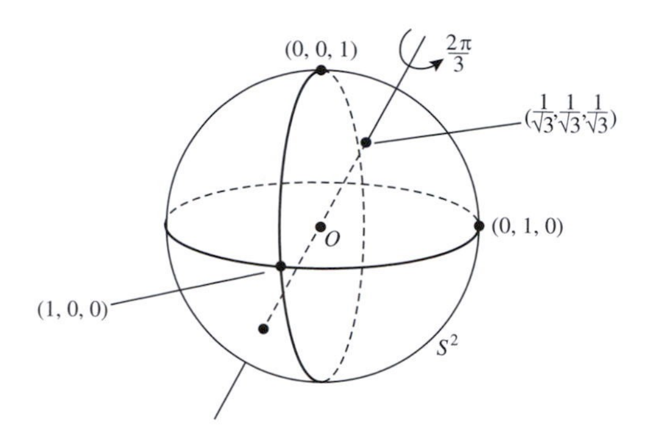

   Rotation about the axis through the points $\left(\frac{1}{\sqrt{3}}, \frac{1}{\sqrt{3}}, \frac{1}{\sqrt{3}}\right)$ and $\left(-\frac{1}{\sqrt{3}}, -\frac{1}{\sqrt{3}}, -\frac{1}{\sqrt{3}}\right)$ through an angle of $\frac{2\pi}{3}$, as shown, maps $(1,0,0)$ to $(0,1,0)$ and $(0,1,0)$ to $(0,0,1)$.

   (c)

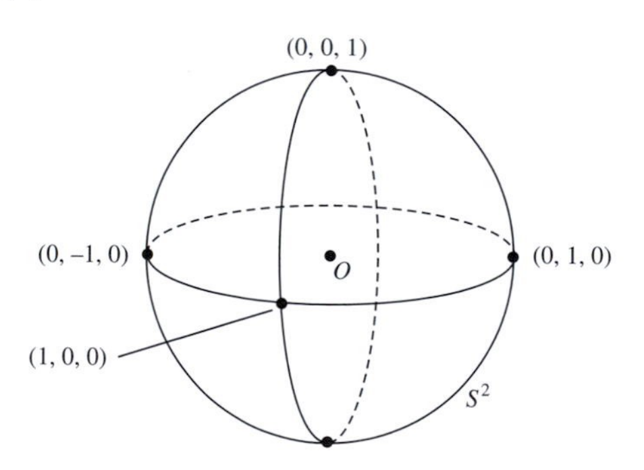

   The line joining the points $(1,0,0)$ and $(0,0,1)$ is a quarter of a circumference in length, whereas the line joining the points $(0,-1,0)$ and $(0,1,0)$ is a half-circumference in length. Hence there is no rotation that maps $(1,0,0) \mapsto (0,-1,0)$ and $(0,0,1) \mapsto (0,1,0)$.

2. First, let $(\cos\phi\sin\theta, \sin\phi\sin\theta, \cos\theta) = \left(0, -\frac{\sqrt{3}}{2}, \frac{1}{2}\right)$; then, we must have

$$\cos\phi\sin\theta = 0, \qquad (1)$$

$$\sin\phi\sin\theta = -\tfrac{\sqrt{3}}{2}, \qquad (2)$$

and

$$\cos\theta = \tfrac{1}{2}. \qquad (3)$$

From equation (3) we have $\theta = \frac{\pi}{3}$. It follows from equation (1) that $\cos\phi = 0$, and from equation (2) that $\sin\phi \cdot \frac{\sqrt{3}}{2} = -\frac{\sqrt{3}}{2}$, so that $\sin\phi = -1$; hence $\phi = \frac{3\pi}{2}$.

Next, let $(\cos\phi\sin\theta, \sin\phi\sin\theta, \cos\theta) = \left(\frac{1}{\sqrt{14}}, -\frac{2}{\sqrt{14}}, \frac{3}{\sqrt{14}}\right)$; then, we must have

$$\cos\phi\sin\theta = \tfrac{1}{\sqrt{14}}, \qquad (4)$$

$$\sin\phi\sin\theta = \tfrac{-2}{\sqrt{14}}, \qquad (5)$$

and

$$\cos\theta = \tfrac{3}{\sqrt{14}}. \qquad (6)$$

From equation (6) we have $\theta = \cos^{-1}\left(\frac{3}{\sqrt{14}}\right) \simeq \cos^{-1}(0.8018) \simeq 0.64$ radians. In particular, we have that $\sin\theta > 0$, so that $\sin\theta = \sqrt{1 - \left(\frac{3}{\sqrt{14}}\right)^2} = \sqrt{\frac{5}{14}}$. It follows from equation (4) that $\cos\phi = \frac{1}{\sqrt{14}} \cdot \frac{\sqrt{14}}{\sqrt{5}} = \frac{1}{\sqrt{5}}$, and from equation (5) that $\sin\phi = \frac{-2}{\sqrt{14}} \cdot \frac{\sqrt{14}}{\sqrt{5}} = \frac{-2}{\sqrt{5}}$; hence $\phi \simeq 5.18$ radians.

3. For the point $\left(\frac{1}{2\sqrt{2}}, \frac{\sqrt{3}}{2\sqrt{2}}, \frac{1}{\sqrt{2}}\right)$ we know that $\theta = \frac{\pi}{4}$ and $\phi = \frac{\pi}{3}$. It follows that the latitude of this point is $\theta' = \frac{\pi}{2} - \frac{\pi}{4} = \frac{\pi}{4}$, the colatitude is $\theta = \frac{\pi}{4}$, and the longitude is $\phi = \frac{\pi}{3}$.
4. For the point 45° W of the Greenwich meridian and 30° S, we have that $\phi = 2\pi - \frac{\pi}{4} = \frac{7\pi}{4}$ and $\theta = \frac{\pi}{2} + \frac{\pi}{6} = \frac{2\pi}{3}$. It follows that the point has spherical polar coordinates

$$\begin{aligned}(\cos\phi\sin\theta, \sin\phi\sin\theta, \cos\theta) &= \left(\cos\tfrac{7\pi}{4}\sin\tfrac{2\pi}{3}, \sin\tfrac{7\pi}{4}\sin\tfrac{2\pi}{3}, \cos\tfrac{2\pi}{3}\right)\\ &= \left(\left(\tfrac{1}{\sqrt{2}}\right)\left(\tfrac{\sqrt{3}}{2}\right), \left(-\tfrac{1}{\sqrt{2}}\right)\left(\tfrac{\sqrt{3}}{2}\right), -\tfrac{1}{2}\right)\\ &= \left(\tfrac{\sqrt{3}}{2\sqrt{2}}, -\tfrac{\sqrt{3}}{2\sqrt{2}}, -\tfrac{1}{2}\right).\end{aligned}$$

5.

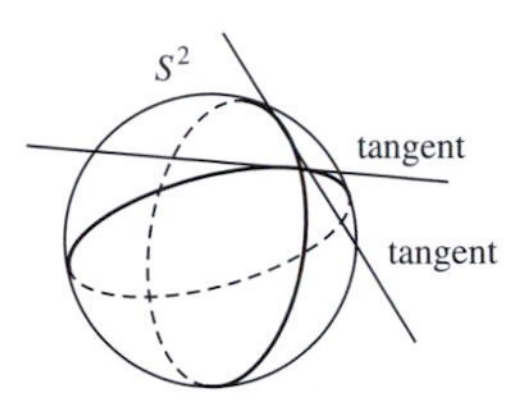

The tangents to the given two great circles form two lines in the tangent plane at the point. The two lines that bisect these angles are both tangent lines to great circles; these great circles bisect the angles between the original great circles.

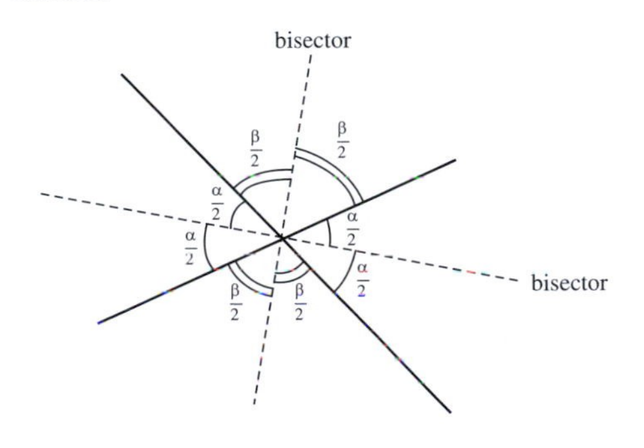

If the angles between the original two tangents are α and β, then clearly $2\alpha + 2\beta = 2\pi$. From the above diagram it is obvious that the angle between the two bisectors is $\frac{1}{2}\alpha + \frac{1}{2}\beta$, that is $\frac{1}{2}\pi$, as required.

Section 7.2

1. The matrix of the transformation $R\left(Y, \frac{\pi}{4}\right)$ is

$$\begin{pmatrix}\cos\frac{\pi}{4} & 0 & \sin\frac{\pi}{4}\\ 0 & 1 & 0\\ -\sin\frac{\pi}{4} & 0 & \cos\frac{\pi}{4}\end{pmatrix} = \begin{pmatrix}\frac{1}{\sqrt{2}} & 0 & \frac{1}{\sqrt{2}}\\ 0 & 1 & 0\\ -\frac{1}{\sqrt{2}} & 0 & \frac{1}{\sqrt{2}}\end{pmatrix}.$$

Now,

$$\begin{pmatrix}\frac{1}{\sqrt{2}} & 0 & \frac{1}{\sqrt{2}}\\ 0 & 1 & 0\\ -\frac{1}{\sqrt{2}} & 0 & \frac{1}{\sqrt{2}}\end{pmatrix}\begin{pmatrix}1\\0\\0\end{pmatrix} = \begin{pmatrix}\frac{1}{\sqrt{2}}\\ 0\\ -\frac{1}{\sqrt{2}}\end{pmatrix},$$

so that the image of $(1,0,0)$ under $R\left(Y, \frac{\pi}{4}\right)$ is $\left(\frac{1}{\sqrt{2}}, 0, -\frac{1}{\sqrt{2}}\right)$.

Similarly, the image of $(0,1,0)$ is $(0,1,0)$ and the image of $(0,0,1)$ is $\left(\frac{1}{\sqrt{2}}, 0, \frac{1}{\sqrt{2}}\right)$.

Finally, since

$$\begin{pmatrix}\frac{1}{\sqrt{2}} & 0 & \frac{1}{\sqrt{2}}\\ 0 & 1 & 0\\ -\frac{1}{\sqrt{2}} & 0 & \frac{1}{\sqrt{2}}\end{pmatrix}\begin{pmatrix}\frac{1}{\sqrt{14}}\\ \frac{-2}{\sqrt{14}}\\ \frac{3}{\sqrt{14}}\end{pmatrix} = \begin{pmatrix}\frac{2}{\sqrt{7}}\\ \frac{-2}{\sqrt{14}}\\ \frac{1}{\sqrt{7}}\end{pmatrix},$$

the image of $\left(\frac{1}{\sqrt{14}}, -\frac{2}{\sqrt{14}}, \frac{3}{\sqrt{14}}\right)$ is $\left(\frac{2}{\sqrt{7}}, -\frac{2}{\sqrt{14}}, \frac{1}{\sqrt{7}}\right)$.

2. Here

$$R(Z,\gamma) = \begin{pmatrix}\cos\gamma & -\sin\gamma & 0\\ \sin\gamma & \cos\gamma & 0\\ 0 & 0 & 1\end{pmatrix}$$

and

$$R(Y,\beta) = \begin{pmatrix}\cos\beta & 0 & \sin\beta\\ 0 & 1 & 0\\ -\sin\beta & 0 & \cos\beta\end{pmatrix}.$$

Hence,

$$\begin{aligned}R(Z,\gamma)\cdot R(Y,\beta) &= \begin{pmatrix}\cos\gamma & -\sin\gamma & 0\\ \sin\gamma & \cos\gamma & 0\\ 0 & 0 & 1\end{pmatrix}\\ &\quad\times\begin{pmatrix}\cos\beta & 0 & \sin\beta\\ 0 & 1 & 0\\ -\sin\beta & 0 & \cos\beta\end{pmatrix}\\ &= \begin{pmatrix}\cos\gamma\cos\beta & -\sin\gamma & \cos\gamma\sin\beta\\ \sin\gamma\cos\beta & \cos\gamma & \sin\gamma\sin\beta\\ -\sin\beta & 0 & \cos\beta\end{pmatrix},\end{aligned}$$

and

$$R(Y,\beta)\cdot R(Z,\gamma)=\begin{pmatrix}\cos\beta & 0 & \sin\beta\\ 0 & 1 & 0\\ -\sin\beta & 0 & \cos\beta\end{pmatrix}$$

$$\times\begin{pmatrix}\cos\gamma & -\sin\gamma & 0\\ \sin\gamma & \cos\gamma & 0\\ 0 & 0 & 1\end{pmatrix}$$

$$=\begin{pmatrix}\cos\beta\cos\gamma & -\cos\beta\sin\gamma & \sin\beta\\ \sin\gamma & \cos\gamma & 0\\ -\sin\beta\cos\gamma & \sin\beta\sin\gamma & \cos\beta\end{pmatrix}.$$

3. The matrix of $R(X,\alpha)$ is $\mathbf{A}=\begin{pmatrix}1 & 0 & 0\\ 0 & \cos\alpha & -\sin\alpha\\ 0 & \sin\alpha & \cos\alpha\end{pmatrix}$. Then

$$\mathbf{A}^T=\begin{pmatrix}1 & 0 & 0\\ 0 & \cos\alpha & \sin\alpha\\ 0 & -\sin\alpha & \cos\alpha\end{pmatrix}$$

$$=\begin{pmatrix}1 & 0 & 0\\ 0 & \cos(-\alpha) & -\sin(-\alpha)\\ 0 & \sin(-\alpha) & \cos(-\alpha)\end{pmatrix},$$

which is the matrix of $R(X,-\alpha)$.

The matrix of $R(Y,\beta)$ is $\mathbf{A}=\begin{pmatrix}\cos\beta & 0 & \sin\beta\\ 0 & 1 & 0\\ -\sin\beta & 0 & \cos\beta\end{pmatrix}$. Then

$$\mathbf{A}^\mathbf{T}=\begin{pmatrix}\cos\beta & 0 & -\sin\beta\\ 0 & 1 & 0\\ \sin\beta & 0 & \cos\beta\end{pmatrix}$$

$$=\begin{pmatrix}\cos(-\beta) & 0 & \sin(-\beta)\\ 0 & 1 & 0\\ -\sin(-\beta) & 0 & \cos(-\beta)\end{pmatrix},$$

which is the matrix of $R(Y,-\beta)$.

The matrix of $R(Z,\gamma)$ is $\mathbf{A}=\begin{pmatrix}\cos\gamma & -\sin\gamma & 0\\ \sin\gamma & \cos\gamma & 0\\ 0 & 0 & 1\end{pmatrix}$. Then

$$\mathbf{A}^T=\begin{pmatrix}\cos\gamma & \sin\gamma & 0\\ -\sin\gamma & \cos\gamma & 0\\ 0 & 0 & 1\end{pmatrix}$$

$$=\begin{pmatrix}\cos(-\gamma) & -\sin(-\gamma) & 0\\ \sin(-\gamma) & \cos(-\gamma) & 0\\ 0 & 0 & 1\end{pmatrix}$$

which is the matrix of $R(Z,-\gamma)$.

4. As you saw in the solution to Problem 3 of Subsection 7.1.2, the point $P\left(\frac{1}{2\sqrt{2}},\frac{\sqrt{3}}{2\sqrt{2}},\frac{1}{\sqrt{2}}\right)$ has spherical polar coordinates of the form $(\cos\phi\sin\theta,\sin\phi\sin\theta,\cos\theta)$, where $\phi=\frac{\pi}{3}$ and $\theta=\frac{\pi}{4}$; then also $\theta'=\frac{\pi}{2}-\theta=\frac{\pi}{4}$. It follows from Theorem 2 that the rotation that maps $A(1,0,0)$ to P is

$$\mathbf{A}=R(Z,\phi)\,R(Y,-\theta')$$

$$=R\left(Z,\frac{\pi}{3}\right)\,R\left(Y,-\frac{\pi}{4}\right)$$

$$=\begin{pmatrix}\cos\frac{\pi}{3} & -\sin\frac{\pi}{3} & 0\\ \sin\frac{\pi}{3} & \cos\frac{\pi}{3} & 0\\ 0 & 0 & 1\end{pmatrix}$$

$$\times\begin{pmatrix}\cos\frac{\pi}{4} & 0 & -\sin\frac{\pi}{4}\\ 0 & 1 & 0\\ \sin\frac{\pi}{4} & 0 & \cos\frac{\pi}{4}\end{pmatrix}$$

$$=\begin{pmatrix}\cos\frac{\pi}{3}\cos\frac{\pi}{4} & -\sin\frac{\pi}{3} & -\cos\frac{\pi}{3}\sin\frac{\pi}{4}\\ \sin\frac{\pi}{3}\cos\frac{\pi}{4} & \cos\frac{\pi}{3} & -\sin\frac{\pi}{3}\sin\frac{\pi}{4}\\ \sin\frac{\pi}{4} & 0 & \cos\frac{\pi}{4}\end{pmatrix}$$

$$=\begin{pmatrix}\frac{1}{2\sqrt{2}} & -\frac{\sqrt{3}}{2} & -\frac{1}{2\sqrt{2}}\\ \frac{\sqrt{3}}{2\sqrt{2}} & \frac{1}{2} & -\frac{\sqrt{3}}{2\sqrt{2}}\\ \frac{1}{\sqrt{2}} & 0 & \frac{1}{\sqrt{2}}\end{pmatrix}.$$

This matrix performs the desired rotation, since

$$\begin{pmatrix}\frac{1}{2\sqrt{2}} & -\frac{\sqrt{3}}{2} & -\frac{1}{2\sqrt{2}}\\ \frac{\sqrt{3}}{2\sqrt{2}} & \frac{1}{2} & -\frac{\sqrt{3}}{2\sqrt{2}}\\ \frac{1}{\sqrt{2}} & 0 & \frac{1}{\sqrt{2}}\end{pmatrix}\begin{pmatrix}1\\0\\0\end{pmatrix}=\begin{pmatrix}\frac{1}{2\sqrt{2}}\\ \frac{\sqrt{3}}{2\sqrt{2}}\\ \frac{1}{\sqrt{2}}\end{pmatrix},$$

as required.

5. As you saw in Problem 4, the matrix

$$\begin{pmatrix} \frac{1}{2\sqrt{2}} & -\frac{\sqrt{3}}{2} & -\frac{1}{2\sqrt{2}} \\ \frac{\sqrt{3}}{2\sqrt{2}} & \frac{1}{2} & -\frac{\sqrt{3}}{2\sqrt{2}} \\ \frac{1}{\sqrt{2}} & 0 & \frac{1}{\sqrt{2}} \end{pmatrix}$$

corresponds to a rotation of S^2 that maps $A(1,0,0)$ to $\left(\frac{1}{2\sqrt{2}}, \frac{\sqrt{3}}{2\sqrt{2}}, \frac{1}{\sqrt{2}}\right)$, so its inverse (which is also its transpose)

$$\begin{pmatrix} \frac{1}{2\sqrt{2}} & \frac{\sqrt{3}}{2\sqrt{2}} & \frac{1}{\sqrt{2}} \\ -\frac{\sqrt{3}}{2} & \frac{1}{2} & 0 \\ -\frac{1}{2\sqrt{2}} & -\frac{\sqrt{3}}{2\sqrt{2}} & \frac{1}{\sqrt{2}} \end{pmatrix}$$

corresponds to a rotation of S^2 that maps $\left(\frac{1}{2\sqrt{2}}, \frac{\sqrt{3}}{2\sqrt{2}}, \frac{1}{\sqrt{2}}\right)$ to $A(1,0,0)$.

Now the matrix of the rotation that maps $A(1,0,0)$ to $N(0,0,1)$ is

$$R\left(Y, -\tfrac{\pi}{2}\right) = \begin{pmatrix} 0 & 0 & -1 \\ 0 & 1 & 0 \\ 1 & 0 & 0 \end{pmatrix};$$

so, by composition, the matrix of the rotation that maps P to N is

$$\begin{pmatrix} 0 & 0 & -1 \\ 0 & 1 & 0 \\ 1 & 0 & 0 \end{pmatrix} \begin{pmatrix} \frac{1}{2\sqrt{2}} & \frac{\sqrt{3}}{2\sqrt{2}} & \frac{1}{\sqrt{2}} \\ -\frac{\sqrt{3}}{2} & \frac{1}{2} & 0 \\ -\frac{1}{2\sqrt{2}} & -\frac{\sqrt{3}}{2\sqrt{2}} & \frac{1}{\sqrt{2}} \end{pmatrix} = \begin{pmatrix} \frac{1}{2\sqrt{2}} & \frac{\sqrt{3}}{2\sqrt{2}} & -\frac{1}{\sqrt{2}} \\ -\frac{\sqrt{3}}{2} & \frac{1}{2} & 0 \\ \frac{1}{2\sqrt{2}} & \frac{\sqrt{3}}{2\sqrt{2}} & \frac{1}{\sqrt{2}} \end{pmatrix}.$$

This matrix performs the desired rotation, since

$$\begin{pmatrix} \frac{1}{2\sqrt{2}} & \frac{\sqrt{3}}{2\sqrt{2}} & -\frac{1}{\sqrt{2}} \\ -\frac{\sqrt{3}}{2} & \frac{1}{2} & 0 \\ \frac{1}{2\sqrt{2}} & \frac{\sqrt{3}}{2\sqrt{2}} & \frac{1}{\sqrt{2}} \end{pmatrix} \begin{pmatrix} \frac{1}{2\sqrt{2}} \\ \frac{\sqrt{3}}{2\sqrt{2}} \\ \frac{1}{\sqrt{2}} \end{pmatrix} = \begin{pmatrix} 0 \\ 0 \\ 1 \end{pmatrix},$$

as required.

6. The spherical polar coordinates of the point $Q\left(\frac{1}{2}, -\frac{1}{2}, -\frac{1}{\sqrt{2}}\right)$ are of the form $(\cos\phi\sin\theta, \sin\phi\sin\theta, \cos\theta)$, where $\cos\theta = -\frac{1}{\sqrt{2}}$; it follows that $\theta = \frac{3\pi}{4}$; hence $\sin\theta = \frac{1}{\sqrt{2}}, \theta' = \frac{\pi}{2} - \theta = \frac{\pi}{2} - \frac{3\pi}{4} = -\frac{\pi}{4}, \cos\phi = \frac{1}{\sqrt{2}}$, and $\sin\phi = -\frac{1}{\sqrt{2}}$, so that $\phi = \frac{7\pi}{4}$.

It follows, from Theorem 2, that the rotation of S^2 that maps $A(1,0,0)$ to Q is $R\left(Z, \frac{7\pi}{4}\right) R\left(Y, \frac{\pi}{4}\right)$. Hence the rotation of S^2 that maps Q to $A(1,0,0)$ is

$$\begin{aligned} &\left(R\left(Z, \tfrac{7\pi}{4}\right) \circ R\left(Y, \tfrac{\pi}{4}\right)\right)^{-1} \\ &\quad = \left(R\left(Y, \tfrac{\pi}{4}\right)\right)^{-1} \circ \left(R\left(Z, \tfrac{7\pi}{4}\right)\right)^{-1} \\ &\quad = R\left(Y, -\tfrac{\pi}{4}\right) \circ R\left(Z, -\tfrac{7\pi}{4}\right); \end{aligned}$$

this has matrix
(matrix of $R(Y, \frac{\pi}{4})$)^T × (matrix of $R(Z, \frac{7\pi}{4})$)^T

$$= \begin{pmatrix} \frac{1}{\sqrt{2}} & 0 & -\frac{1}{\sqrt{2}} \\ 0 & 1 & 0 \\ \frac{1}{\sqrt{2}} & 0 & \frac{1}{\sqrt{2}} \end{pmatrix} \begin{pmatrix} \frac{1}{\sqrt{2}} & -\frac{1}{\sqrt{2}} & 0 \\ \frac{1}{\sqrt{2}} & \frac{1}{\sqrt{2}} & 0 \\ 0 & 0 & 1 \end{pmatrix}$$

$$= \begin{pmatrix} \frac{1}{2} & -\frac{1}{2} & -\frac{1}{\sqrt{2}} \\ \frac{1}{\sqrt{2}} & \frac{1}{\sqrt{2}} & 0 \\ \frac{1}{2} & -\frac{1}{2} & \frac{1}{\sqrt{2}} \end{pmatrix}.$$

Therefore, by composition (using the result of Problem 4), the matrix of the rotation of S^2 that maps Q to P is

$$\begin{pmatrix} \frac{1}{2\sqrt{2}} & -\frac{\sqrt{3}}{2} & -\frac{1}{2\sqrt{2}} \\ \frac{\sqrt{3}}{2\sqrt{2}} & \frac{1}{2} & -\frac{\sqrt{3}}{2\sqrt{2}} \\ \frac{1}{\sqrt{2}} & 0 & \frac{1}{\sqrt{2}} \end{pmatrix} \begin{pmatrix} \frac{1}{2} & -\frac{1}{2} & -\frac{1}{\sqrt{2}} \\ \frac{1}{\sqrt{2}} & \frac{1}{\sqrt{2}} & 0 \\ \frac{1}{2} & -\frac{1}{2} & \frac{1}{\sqrt{2}} \end{pmatrix}$$

$$= \begin{pmatrix} -\frac{\sqrt{3}}{2\sqrt{2}} & -\frac{\sqrt{3}}{2\sqrt{2}} & -\frac{1}{2} \\ \frac{1}{2\sqrt{2}} & \frac{1}{2\sqrt{2}} & -\frac{\sqrt{3}}{2} \\ \frac{1}{\sqrt{2}} & -\frac{1}{\sqrt{2}} & 0 \end{pmatrix}.$$

This matrix performs the desired rotation, since

$$\begin{pmatrix} -\frac{\sqrt{3}}{2\sqrt{2}} & -\frac{\sqrt{3}}{2\sqrt{2}} & -\frac{1}{2} \\ \frac{1}{2\sqrt{2}} & \frac{1}{2\sqrt{2}} & -\frac{\sqrt{3}}{2} \\ \frac{1}{\sqrt{2}} & -\frac{1}{\sqrt{2}} & 0 \end{pmatrix} \begin{pmatrix} \frac{1}{2} \\ -\frac{1}{2} \\ -\frac{1}{\sqrt{2}} \end{pmatrix} = \begin{pmatrix} \frac{1}{2\sqrt{2}} \\ \frac{\sqrt{3}}{2\sqrt{2}} \\ \frac{1}{\sqrt{2}} \end{pmatrix},$$

as required.

7. Here

$$\det \mathbf{A} = \begin{vmatrix} 1 & 0 & 0 \\ 0 & 1 & 0 \\ 0 & 0 & -1 \end{vmatrix} = -1.$$

8. First, since $3^2 + 4^2 + (-5)^2 = 50 = \left(5\sqrt{2}\right)^2$, we write the equation of the plane $\pi : 3x + 4y - 5z = 0$ in the form $ax + by + cz = 0$, where $a = \frac{3}{5\sqrt{2}}, b = \frac{4}{5\sqrt{2}}$ and $c = \frac{-5}{5\sqrt{2}} = -\frac{1}{\sqrt{2}}$. Then, by Theorem 3, the matrix associated with reflection in the plane π is

$$\begin{pmatrix} 1 - \frac{18}{50} & -\frac{24}{50} & \frac{30}{50} \\ -\frac{24}{50} & 1 - \frac{32}{50} & \frac{40}{50} \\ \frac{30}{50} & \frac{40}{50} & 1 - \frac{50}{50} \end{pmatrix} = \begin{pmatrix} \frac{16}{25} & -\frac{12}{25} & \frac{3}{5} \\ -\frac{12}{25} & \frac{9}{25} & \frac{4}{5} \\ \frac{3}{5} & \frac{4}{5} & 0 \end{pmatrix}.$$

9. A reflection of S^2 in a great circle is an isometry, so the composite of an even number of reflections in great circles is also an isometry of S^2. Each reflection in a great circle reverses the orientation of S^2, so the composite of an even number of reflections in great circles does not alter the orientation; hence the composite is a direct isometry.

   It follows from part (b) of Theorem 7 that the composite of an even number of reflections of S^2 in great circles is a rotation of S^2.

Section 7.3

1.

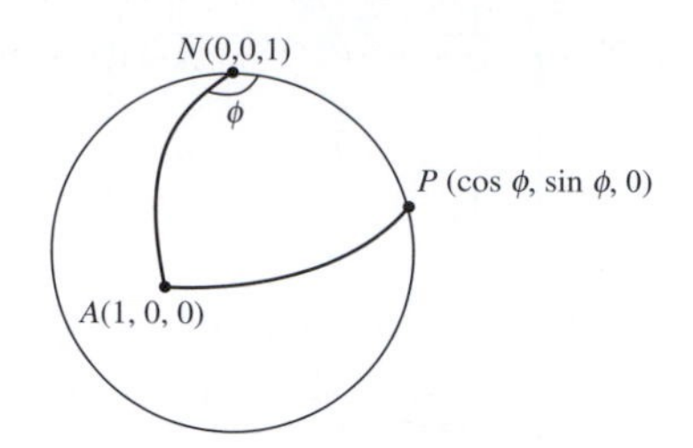

Let A and N be the points $(1, 0, 0)$ and $(0, 0, 1)$, respectively; and let P be the point $(\cos\phi, \sin\phi, 0)$. Then the spherical triangle $\triangle APN$ has angles $\frac{\pi}{2}$, $\frac{\pi}{2}$ and ϕ, so that in particular it has angular excess ϕ.

(a) Let $\phi = 0.01$ radians; this gives the required spherical triangle.

(b) Let $\phi = 3.14$ radians (note that $3.14 < \pi$, so that the triangle $\triangle APN$ does not 'cross back on itself'); this gives the required spherical triangle.

2.

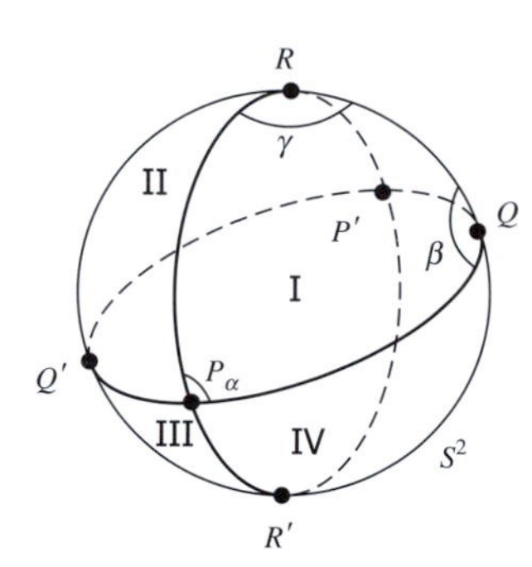

Triangle II Clearly $\angle Q'PR = \pi - \alpha$ and $\angle PRQ' = \pi - \gamma$. Also, by the symmetry of the intersections of the great circles $Q'PQ$ and $QRQ'R'$,

$$\begin{aligned} \angle RQ'P &= \angle R'QP' \text{ (by symmetry)} \\ &= \beta \text{ (vertically opposite).} \end{aligned}$$

Triangle III Clearly $\angle R'PQ' = \angle QPR$ (vertically opposite) $= \alpha$ and $\angle PQ'R' = \pi - \angle PQ'R = \pi - \beta$. Finally, by the symmetry of the intersections of the great circles RPR' and $QRQ'R'$,

$$\begin{aligned} \angle Q'R'P &= \angle QRP' \text{ (by symmetry)} \\ &= \pi - \gamma. \end{aligned}$$

Triangle IV Clearly $\angle QPR' = \pi - \alpha$, $\angle PQR' = \pi - \beta$, and (from the result of Triangle III) $\angle PR'Q = \pi - \angle PR'Q' = \pi - (\pi - \gamma) = \gamma$.

3.

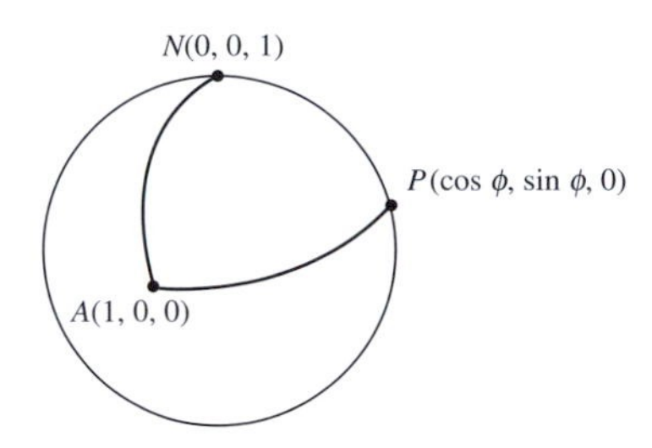

Let A and N be the points $(1, 0, 0)$ and $(0, 0, 1)$, respectively; and let P be the point $(\cos\phi, \sin\phi, 0)$.

Then the spherical triangle ΔAPN has angles $\frac{\pi}{2}$, $\frac{\pi}{2}$ and ϕ, so that in particular it has angular excess ϕ and so (by Theorem 3 of Subsection 7.3.1) has area ϕ.

Let $\phi = \frac{3\pi}{4}$ radians; this gives the required spherical triangle.

4.

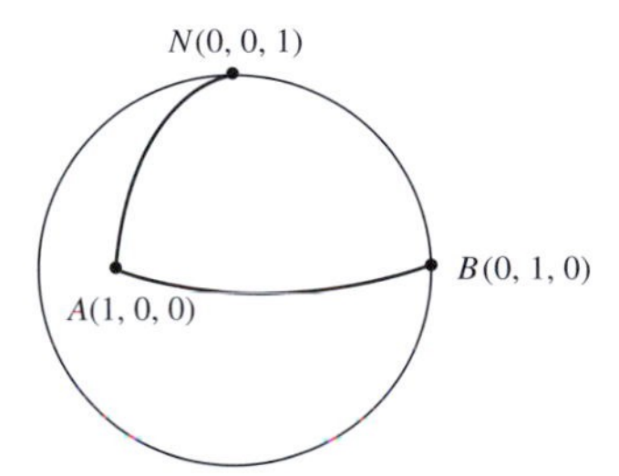

The spherical triangle ΔABN has angles $\frac{\pi}{2}, \frac{\pi}{2}$ and $\frac{\pi}{2}$, so that any dual triangle Δ' has sides each of length $\left(\pi - \frac{\pi}{2}\right) = \frac{\pi}{2}$. The sides of the spherical triangle ΔABN also each have length $\frac{\pi}{2}$.

Similarly, the spherical triangle ΔABN has sides $\frac{\pi}{2}, \frac{\pi}{2}$ and $\frac{\pi}{2}$, so that any dual triangle Δ' has angles each of magnitude $\left(\pi - \frac{\pi}{2}\right) = \frac{\pi}{2}$. The angles of the spherical triangle ΔABN also each have magnitude $\frac{\pi}{2}$.

It follows that ΔABN and any dual triangle Δ' each have angles of the same magnitude and sides of the same magnitude. So clearly ΔABN and the dual triangle Δ' are congruent.

5. We use the result of Theorem 5 of Subsection 7.3.3.

The colatitude of Tokyo is $90° - 36° = 54°$, so that the coordinates of the corresponding point P on S^2 are

$$\begin{aligned}
&(\cos(140°)\sin(54°), \sin(140°)\sin(54°), \\
&\quad \cos(54°)) \\
&\simeq (-0.7660 \cdot 0.8090, 0.6428 \cdot 0.8090, 0.5878) \\
&\simeq (-0.6197, 0.5200, 0.5878).
\end{aligned}$$

The colatitude of Rio de Janeiro is $90° + 23° = 113°$, so that the coordinates of the corresponding point Q on S^2 are

$$\begin{aligned}
&(\cos(-43°)\sin(113°), \sin(-43°)\sin(113°), \\
&\quad \cos(113°)) \\
&\simeq (0.7314 \cdot 0.9205, -0.6820 \cdot 0.9205, \\
&\quad -0.3907) \\
&\simeq (0.6733, -0.6278, -0.3907).
\end{aligned}$$

Hence the distance between Tokyo and Rio de Janeiro is approximately $4000d$ miles, where

$$\begin{aligned}
\cos d &= -0.6197 \cdot 0.6733 - 0.5200 \cdot 0.6278 \\
&\quad - 0.5878 \cdot 0.3907 \\
&\simeq -0.4172 - 0.3265 - 0.2297 \\
&= -0.9734.
\end{aligned}$$

Thus $d \simeq \cos^{-1}(-0.9734) \simeq 2.9104$ radians, and the required distance is

$$4000d \simeq 4000 \cdot 2.9104 \simeq 11\,642 \text{ miles.}$$

Remark

To determine this distance in terms of metric measurements, we replace the figure of 4000 miles above by 6378 km; thus the required distance is

$$\begin{aligned}
6378d &\simeq 6378 \cdot 2.9104 \\
&\simeq 18\,563 \text{ km.}
\end{aligned}$$

6.

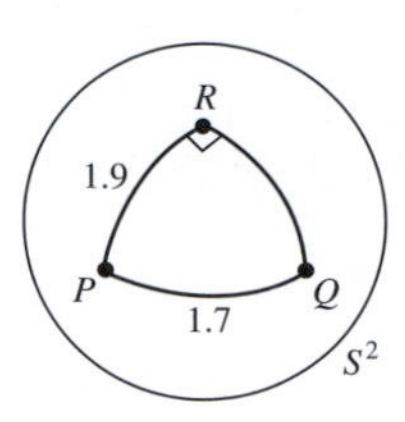

Let d be the length of QR. By applying Pythagoras' Theorem to the triangle ΔPQR, we obtain

$$\cos 1.7 \simeq \cos 1.9 \times \cos d,$$

so that

$$\begin{aligned}\cos d &\simeq \frac{\cos 1.7}{\cos 1.9}\\ &\simeq \frac{-0.1288}{-0.3233}\\ &\simeq 0.3985;\end{aligned}$$

it follows that

$$d \simeq \cos^{-1} 0.3985 \simeq 1.16.$$

7.

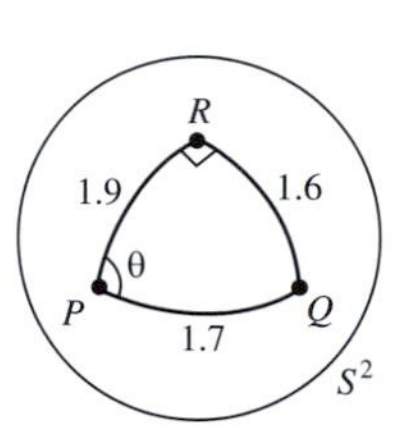

Let θ denote the angle $\angle QPR$. We saw in Problem 6 that $QR \simeq 1.16$. It follows from the second formula in Theorem 7 that

$$\begin{aligned}\sin\theta &\simeq \frac{\sin 1.16}{\sin 1.7}\\ &\simeq \frac{0.9168}{0.9917}\\ &\simeq 0.9245;\end{aligned}$$

it follows that

$$\theta \simeq \sin^{-1} 0.9245 \simeq 1.18 \text{ radians}.$$

8.

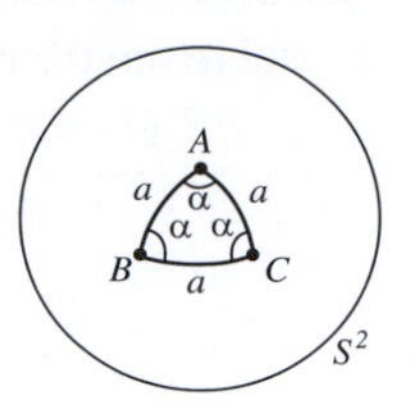

We use here the identities

$$\cos 2x = 1 - 2\sin^2 x$$

and

$$\sin 2x = 2\sin x\cos x.$$

For $x \in \mathbb{R}$, we deduce from the result of part (b) of Example 3 that

$$\begin{aligned}\cos\alpha &= \frac{\tan\left(\frac{1}{2}a\right)}{\tan a}\\ &= \frac{\sin\left(\frac{1}{2}a\right)}{\cos\left(\frac{1}{2}a\right)}\cdot\frac{\cos a}{\sin a}\\ &= \frac{\cos a}{2\cos^2\left(\frac{1}{2}a\right)} = \frac{\cos a}{1+\cos a}.\end{aligned}$$

Section 7.4

1. (a)

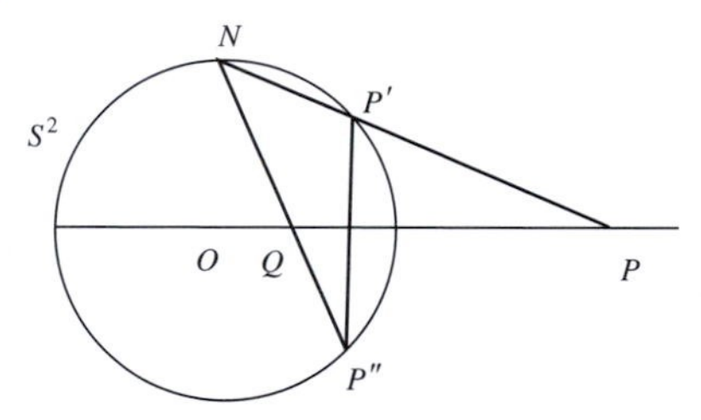

The inverse of the stereographic projection applied to the point $P(x, y)$ gives the coordinates

$$\pi^{-1}(x, y) = \left(\frac{2x}{x^2+y^2+1}, \frac{2y}{x^2+y^2+1}, \frac{x^2+y^2-1}{x^2+y^2+1}\right)$$

for P'.
Then P'' has coordinates

$$\left(\frac{2x}{x^2+y^2+1}, \frac{2y}{x^2+y^2+1}, -\frac{x^2+y^2-1}{x^2+y^2+1}\right);$$

and the formula for stereographic projection gives the following coordinates for Q:

$$\left(\frac{2x}{(x^2+y^2+1)+(x^2+y^2-1)}, \frac{2y}{(x^2+y^2+1)+(x^2+y^2-1)}\right)$$
$$= \left(\frac{x}{x^2+y^2}, \frac{y}{x^2+y^2}\right).$$

(b) The point (x, y) and $\left(\frac{x}{x^2+y^2}, \frac{y}{x^2+y^2}\right)$ are inverse points with respect to the unit circle $x^2+y^2=1$ (see Theorem 2, Subsection 5.1.2).

2.

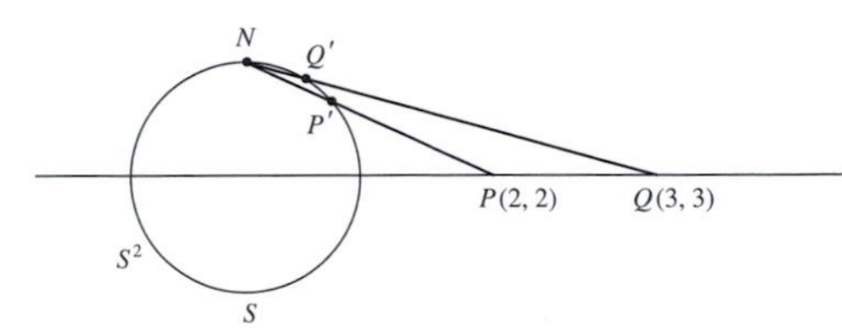

The coordinates of $P(2,2)$ are

$$\left(2\sqrt{2}\cos\tfrac{\pi}{4}, 2\sqrt{2}\sin\tfrac{\pi}{4}\right),$$

and the coordinates of $Q(3,3)$ are

$$\left(3\sqrt{2}\cos\tfrac{\pi}{4}, 3\sqrt{2}\sin\tfrac{\pi}{4}\right).$$

Then it follows from Theorem 1 that the spherical distances SP' and SQ' are

$$2\tan^{-1}\left(2\sqrt{2}\right) \quad \text{and} \quad 2\tan^{-1}\left(3\sqrt{2}\right),$$

respectively. But S, N, P' and Q' all lie in the same vertical plane, so that the spherical distance $P'Q'$ is

(spherical distance SQ') – (spherical distance SP')

$$= 2\tan^{-1}\left(3\sqrt{2}\right) - 2\tan^{-1}\left(2\sqrt{2}\right).$$

It then follows from the identity

$$\tan^{-1}x - \tan^{-1}y = \tan^{-1}\frac{x-y}{1+xy},$$

for $x, y \in \mathbb{R}$, that the required distance can be expressed in the form

$$2\tan^{-1}\left(\frac{3\sqrt{2}-2\sqrt{2}}{1+\left(3\sqrt{2}\right)\left(2\sqrt{2}\right)}\right)$$

$$= 2\tan^{-1}\left(\frac{\sqrt{2}}{13}\right) \simeq 0.2167.$$

3. Let one of the fixed points of the rotation of S^2 be $P'(X,Y,Z)$; then the other must be $Q'(-X,-Y,-Z)$. Under stereographic projection these map to the points $P\left(\frac{X}{1-Z}, \frac{Y}{1-Z}\right)$ and $Q\left(\frac{-X}{1+Z}, \frac{-Y}{1+Z}\right)$ in $\mathbb{R}^2$, respectively; we can express these coordinates as complex numbers $s = \frac{X}{1-Z} + i\frac{Y}{1-Z} = \frac{X+iY}{1-Z}$ and $t = \frac{-X}{1+Z} + i\frac{-Y}{1+Z} = \frac{-X-iY}{1+Z}$.
Then

$$\bar{s}\cdot t = \overline{\left(\frac{X+iY}{1-Z}\right)} \cdot \frac{-X-iY}{1+Z}$$
$$= \frac{X-iY}{1-Z} \cdot \frac{-X-iY}{1+Z}$$
$$= \frac{-X^2-Y^2}{1-Z^2} = -1,$$

since $X^2+Y^2+Z^2=1$. It follows that $t = -1/\bar{s}$.

4. A rotation $R(X,\alpha)$ of S^2 fixes the points $(1,0,0)$ and $(-1,0,0)$, which map under stereographic projection to the points 1 and -1, respectively.

By Theorem 3, a conjugate Möbius transformation must be of the form

$$M(z) = \frac{cz+d}{-\bar{d}z+\bar{c}}$$

(where we use c and d to avoid using the letters a and b for different things).

This mapping fixes 1 and -1 if

$$1 = \frac{c+d}{-\bar{d}+\bar{c}} \quad \text{and} \quad -1 = \frac{-c+d}{\bar{d}+\bar{c}},$$

which we can rewrite in the form

$$-\bar{d}+\bar{c} = c+d \quad \text{and} \quad -\bar{d}-\bar{c} = -c+d.$$

Adding these equations gives $-2\bar{d} = 2d$, so that d is imaginary, and subtracting the equations gives $2\bar{c} = 2c$, so that c is real.

It follows that M has the desired form $M(z) = \frac{az+ib}{ibz+a}$, with a in place of c and ib in place of d.

5. There are no fixed points of the Möbius transformation $z \mapsto z+1$ in $\mathbb{C}$, since the equation $z = z+1$ has no solutions in $\mathbb{C}$. However the Möbius transformation maps ∞ to itself, so ∞ is a fixed point in $\hat{\mathbb{C}}$ of the transformation.

   The fixed points in $\mathbb{C}$ of the Möbius transformation $z \mapsto \frac{-4}{z+4}$ are the solutions of the equation $z = \frac{-4}{z+4}$. We may rewrite this equation in the form $z^2+4z+4=0$, or $(z+2)^2=0$, so that -2 is the only fixed point in $\mathbb{C}$ of the transformation. Since $\infty \mapsto 0, -2$ is also the only fixed point in $\hat{\mathbb{C}}$ of the transformation.

   The fixed points in $\mathbb{C}$ of the Möbius transformation $z \mapsto \frac{-4}{z+5}$ are the solutions of the equation $z = \frac{-4}{z+5}$. We may rewrite this equation in the form $z^2+5z+4=0$, or $(z+4)(z+1)=0$, so that -4 and -1 are the fixed points in $\mathbb{C}$ of the transformation. Since $\infty \mapsto 0, -4$ and -1 are also the only fixed points in $\hat{\mathbb{C}}$ of the transformation.

6. Let C denote the circle in $\mathbb{R}^2$ with equation
$$x^2+y^2+2\alpha x+2\beta y+\gamma=0,$$
which we may rewrite in the more convenient form
$$(x^2+y^2+1)+2\alpha x+2\beta y+(\gamma-1)=0$$
or
$$1+\alpha\frac{2x}{x^2+y^2+1}+\beta\frac{2y}{x^2+y^2+1}+\frac{\gamma-1}{x^2+y^2+1}=0.$$
Now under the inverse mapping $\pi^{-1}:(x,y)\mapsto(X,Y,Z)$ of stereographic projection,
$$X=\frac{2x}{x^2+y^2+1},$$
$$Y=\frac{2y}{x^2+y^2+1},$$
and
$$Z=\frac{x^2+y^2-1}{x^2+y^2+1};$$
in particular, $Z = 1-\frac{2}{x^2+y^2+1}$ so that $\frac{1}{x^2+y^2+1}=\frac{1}{2}(1-Z)$.

   It follows that C corresponds under the inverse mapping π^{-1} of stereographic projection to a plane with equation
$$1+\alpha X+\beta Y+\tfrac{1}{2}(\gamma-1)(1-Z)=0,$$
or
$$2\alpha X+2\beta Y+(1-\gamma)Z+(1+\gamma)=0.$$

Index